Applied

Combinatorics

FRED S. ROBERTS

Rutgers University

Prentice-Hall, Inc., Englewood Cliffs, New Jersey 07632

Library of Congress Cataloging in Publication Data

Roberts, Fred S.
    Applied combinatorics.

    Includes bibliographies and index.
    1. Combinatorial analysis.   I. Title.
QA164.R6 1984   511′.6   83-19052
ISBN 0-13-039313-4

Editorial/production supervision
  and interior design: Paula Martinac
Cover design: Edsal Enterprises
Manufacturing buyer: John Hall

Printed in the United States of America

10 9 8 7 6 5 4 3

ISBN 0-13-039313-4

Prentice-Hall International, Inc., *London*
Prentice-Hall of Australia Pty. Limited, *Sydney*
Editora Prentice-Hall do Brasil, Ltda., *Rio de Janeiro*
Prentice-Hall Canada Inc., *Toronto*
Prentice-Hall of India Private Limited, *New Delhi*
Prentice-Hall of Japan, Inc., *Tokyo*
Prentice-Hall of Southeast Asia Pte. Ltd., *Singapore*
Whitehall Books Limited, *Wellington, New Zealand*

To Helen

Contents

PART II   The Counting Problem   149

4   Generating Functions and Their Applications   149

5   Recurrence Relations   194

13 Optimization Problems for Graphs and Networks 526

Preface

Perhaps the fastest growing area of modern mathematics is combinatorics. A major reason for this rapid growth is its wealth of applications, to computer science, communications, transportation, genetics, experimental design, scheduling, and so on. This book introduces the reader to the tools of combinatorics from an applied point of view.

Much of the growth of combinatorics has gone hand in hand with the development of the computer. Today's high-speed computers make it possible to implement solutions to practical combinatorial problems from a wide variety of fields, solutions that could not be implemented until quite recently. This has resulted in increased emphasis on the development of solutions to combinatorial problems. At the same time, the development of computer science has brought with it numerous challenging combinatorial problems of its own. Thus, it is hard to separate combinatorial mathematics from computing. The reader will see the emphasis on computing here by the frequent use of examples from computer science, the frequent discussion of algorithms, and so on. On the other hand, the general point of view taken in this book is that combinatorics has a wealth of applications to a large number of subjects, and this book has tried to emphasize the variety of these applications rather than just focusing on one.

Many of the mathematical topics presented here are relatively standard topics from the rapidly growing textbook literature of combinatorics. Others are taken from the current research literature, or are chosen because they illustrate interesting applications of the subject. The book is distinguished, I believe, by its wide-ranging treatment of applications. Entire sections are devoted to such applications as switching functions, the use of enzymes to uncover unknown RNA chains, searching and sorting problems of information retrieval, construction of error-correcting codes, counting of chemical compounds, calculation of power in voting situations, and uses of Fibonacci numbers. There are entire sections on applications of recurrences involving convolutions, applications of eulerian chains, applications of Ramsey theory, and so on, which are unique to the literature.

The book is divided into four parts. The first part (Chapters 2 and 3) introduces the basic tools of combinatorics and their applications. It introduces fundamental counting rules, and the tools of graph theory. The remaining three parts are organized around the three basic problems of combinatorics: the counting problem, the existence problem, and the optimization problem. These problems are discussed in Chapter 1. Part II of the book is concerned with more advanced tools for dealing with the counting problem: generating functions, recurrences, inclusion/exclusion, and Polya Theory. Part III deals with the existence problem. It discusses the pigeonhole principle, experimental design, and coding theory. It also begins a series of three chapters on graphs and networks (Chapters 11–13) and begins an introduction to graph algorithms. Part IV deals with combinatorial optimization, illustrating the basic ideas through a continued study of graphs and networks. It begins with a transitional chapter on matching and covering, which starts with the existence problem and ends with the optimization problem. Then Part IV ends with a discussion of optimization problems for graphs and networks. The division of the book into four parts is somewhat arbitrary, and many topics illustrate several different aspects of combinatorics, for instance both existence and optimization questions. However, dividing the book into four parts seemed to be a reasonable way to organize the large amount of material that is modern combinatorics.

This book can be used at a variety of levels. Most of the book is written for a junior-senior audience, in a course populated by math and computer science majors and nonmajors. It could also be appropriate for sophomores with sufficient mathematical maturity. (Topics which can be omitted in elementary treatments are indicated throughout.) On the other hand, at a fast pace, there is more than enough material for a challenging graduate course. In the undergraduate courses for which I have used this material at Rutgers, the majority of the enrollees come from mathematics and computer science, and the rest from such disciplines as business, economics, biology, and psychology. The prerequisites for these courses, and for the book, include familiarity with the language of functions and sets usually attained by taking at least one course in calculus. Infinite sequences and series are used in Chapters 4 and 5 (though much of Chapter 5 uses only the most elementary facts about infinite sequences, and does not require the notion of limit). Other traditional topics of the calculus are not needed. However, the mathematical sophistication attained by taking a course like calculus is a prerequisite. Also required are some tools of linear algebra, specifically familiarity with matrix manipulations. An understanding of mathematical induction is also assumed. (There are those instructors who will want to review mathematical induction in some detail at an early point in their course, or who will want to quickly review the language of sets.) A few optional sections of the book require probability beyond what is developed in the text. Other sections introduce topics in modern algebra, such as groups and finite fields. These sections are self-contained, but they would be too fast-paced for a student without sufficient background.

Many parts of the book put an emphasis on algorithms. This is inevitable, as combinatorics is increasingly connected to the development of precise and efficient procedures for solving complicated problems, and because the development of combinatorics is so closely tied to computer science. My aim is to introduce students to the notion of an algorithm and to introduce them to some important examples of algorithms. For the most part, I have adopted a relatively informal style in presenting algorithms. The style presumes little exposure to the notion of an algorithm and how to describe it. The major goal is to present the basic idea of a procedure, without attempting to present it in its most concise or most computer-oriented form. There are those who will disagree with this method of presenting algorithms. Indeed, one of the prepublication reviewers says that he would either leave algorithms out or do them "right" (for example, in Pascal). My own view is that no combinatorics course is going to replace the learning of algorithms. The computer science student needs a separate course in algorithms, which includes discussion of implementing the data structures for the algorithms presented. However, all students of combinatorics need to be exposed to the idea of algorithm, and to the algorithmic way of thinking, a way of thinking that is so central and basic to the subject. I realize that my compromise on how to present algorithms will not make everyone happy. However, it should be pointed out that for students with a background in computer science, it would make for interesting, indeed important, exercises to translate the informal algorithms of the text into more precise computer algorithms or even computer programs.

Courses in combinatorics or discrete mathematics are typically being offered by departments in the mathematical or computing sciences, at a variety of levels and with a variety of prerequisites. These courses are frequently required of computer science majors or students majoring in mathematics with computer science emphasis. There is a growing feeling in the mathematical community that *all* mathematical sciences majors should be exposed to these kinds of topics. This book is appropriate for a variety of courses at a variety of levels. I have used the material of the book for several courses, in particular a one-semester course entitled Combinatorics and a one-semester course entitled Applied Graph Theory. The combinatorics course, taught to juniors and seniors, covers much of the material of Chapters 1, 2, 3, 4, 5, and 6, Section 8.1, and Chapters 9 and 10, omitting the sections indicated by footnotes in the text. (These are often proofs.) A faster-paced course, which I have used with first year graduate students, puts more emphasis on proofs, includes many of the optional sections, and also covers the material either of Chapter 7 or of Sections 8.2 through 8.3 and Chapter 12. In either an undergraduate or a graduate course, the instructor could also substitute for Chapters 9 and 10 either Chapter 7 or Chapter 11 and parts of Chapters 12 and 13. Including Chapter 11 is especially recommended at institutions that do not have a separate course in graph theory. Similarly, including parts of Chapter 13 is especially recommended for institutions that do not have a course in operations research. At Rutgers, we have separate (both undergraduate and graduate) courses which cover much of the material of Chapters 11 and 13.

Other one-semester or one-quarter courses could be designed from this material, as most of the chapters are relatively independent. (See the discussion below.) The applied graph theory course that I teach is built around Chapters 3 and 11, supplemented with graph-theoretical topics from the rest of the book (Chapters 12 and 13) and elsewhere. (A quick treatment of Sections 2.1 through 2.7, plus perhaps Section 2.18, is needed background.) Chapters 3, 11, 12, and 13 would also be appropriate for a course introducing graph algorithms or a course called Graphs and Networks. The entire book would make a very appropriate one-year introduction to modern combinatorial mathematics and its applications.

Increasingly, the mathematical and computer science community is encouraging students to become exposed to the topics of combinatorics early in their college careers. This book could be used for a one-semester or one-quarter sophomore level course. Such a course would cover much of Chapters 1, 2, and 3, skip Chapter 4, and cover only Sections 5.1 and 5.2 of Chapter 5. It would then cover Chapter 6, Section 8.1, and Chapter 11. Starred sections and most proofs would be omitted. Other topics would be added at the discretion of the instructor.

In organizing any course, the instructor will wish to take note of the relative independence of the topics here. There is no well-accepted order in which to present an introduction to the subject matter of combinatorics, and there is no universal agreement on the topics that make up such an introduction. I have tried to write this book in such a way that the chapters are quite independent and can be covered in various orders.

Chapter 2 is basic to the book. It introduces the basic counting rules that are used throughout. Chapter 3 develops just enough graph theory to introduce the subject. It emphasizes graph-theoretical topics that illustrate the counting rules developed in Chapter 2. The ideas introduced in Chapter 3 are referred to in places throughout the book, and most heavily in Chapters 11, 12, and 13. It is possible to use this book for a one-semester or one-quarter course in combinatorics without covering Chapter 3. However, in my opinion, at least the material on graph coloring (Sections 3.3 and 3.4) should be included. The *major* dependencies beyond Chapter 3 are that Chapter 5 after Section 5.2 depends on Chapter 4; Chapter 6 refers to examples developed in Chapters 3 and 5; Chapters 11, 12, and 13 depend on Chapter 3; and Section 10.5 depends on Chapter 9. Very elementary graph-theoretical ideas are also used in Chapter 8 and ideas from Chapter 12 are used in two places in Chapter 13: Sections 13.3.8 and 13.4.2.

The exercises play a central role in this book. They test routine ideas, introduce new concepts and applications, and attempt to challenge the reader to use the combinatorial techniques developed in the text. It is the nature of combinatorics, indeed the nature of most of mathematics, that it is best mastered by doing many problems. I have tried to include a wide variety of both applied and theoretical exercises, of varying degrees of difficulty, throughout the book. Answers to selected exercises are included at the end of the book.

Finally, it should be emphasized that combinatorics is a rapidly growing subject and one whose techniques are being rapidly developed and whose applications are being rapidly explored. Many of the topics presented here are close to the frontiers of research. It is typical of the subject that it is possible to bring a newcomer to the frontiers very quickly. I have tried to include references to the literature of combinatorics and its applications which will allow the interested reader to delve more deeply into the topics discussed here.

I first started on this book in 1976, when I produced a short set of notes for my undergraduate course in combinatorics at Rutgers. Over the years that this book has changed and grown, I have used it regularly as the text for that course and for the other courses described earlier. It has also been a great benefit to me that others have used this material as the text for their courses and have sent me extensive comments. I would particularly like to thank Midge Cozzens, who used this material at Northeastern, Fred Hoffman, who used it at Florida Atlantic, and Doug West, who used it at Princeton, for their very helpful input.

I would especially like to thank my present and former students who have helped in numerous ways in the preparation of this book, by proofreading, checking exercises, catching numerous mistakes, and making nasty comments. I want to acknowledge the help of Midge Cozzens, Shelly Leibowitz, Bob Opsut, Arundhati Ray-Chaudhuri, Sam Rosenbaum, and Jeff Steif.

I have received comments on this material from many people. I would specifically like to thank the following individuals, who made extremely helpful comments at various stages during the reviewing process as well as at other times: John Cozzens, Paul Duvall, Marty Golumbic, Fred Hoffman, Steve Maurer, Ronald Mullin, Robert Tarjan, Tom Trotter, and Alan Tucker. Although I have received a great deal of help with this material, errors will almost surely remain. I alone am responsible for them.

As this book grew, it was typed and retyped, copied and recopied, cut, pasted together, uncut, glued, and on and on. I had tremendous help with this from Lynn Braun, Carol Brouillard, Mary Anne Jablonski, Kathy King, Annette Roselli, and Dotty West-gate.

I would like to thank Prentice-Hall for permission to use material freely from my 1976 book, *Discrete Mathematical Models, with Applications to Social, Biological, and Environmental Problems*. The following material is reproduced directly from *Discrete Mathematical Models*: Table 2.1, Figures 1.2, 1.3, 1.4, 1.5, 1.6, 1.7, 1.8, 1.9, 1.10, 1.11, 2.2, 2.3, 2.7, 2.8, 2.9, 2.10, 2.11, 2.13, 2.17, 2.18, 2.20, 3.16, 3.17, 3.18, 3.19, 3.20, 3.21, 3.28, 3.29, 3.66, 3.67, 3.68, 3.69, and 3.71. From my 1978 NSF-CBMS Monograph, *Graph Theory and its Applications to Problems of Society*, published by SIAM, I have reproduced Figures 5.4 and 7.1 and modified Sections 8.3, 8.4, 8.5, and 8.6. I would like to thank SIAM for its cooperation.

Finally, I would like to thank my family for its support. Those who have written a book will understand the number of hours it takes away from one's family: cutting short telephone calls to proofread, canceling trips to write, postponing outings to create exercises, stealing away to make just one more improvement. My family has been extremely understanding and helpful. I would like to thank my parents for their love and support. I would like to thank Lily Marcus for her assistance, technical and otherwise. I would like to thank my wife, Helen, who, it seems, is always a "book widow." She has helped me not only by her continued support and guidance, and inspiration, but she has also co-authored one chapter of this book, and introduced me to a wide variety of topics and examples which she developed for her courses and which I have freely scattered throughout this book. Finally, I would like to thank Sarah and David, for being Sarah and David. I do not need the counting techniques of combinatorics to count my blessings.

Fred S. Roberts

NOTATION

Set-theoretic Notation

$\cup$	union	$\varnothing$	empty set
$\cap$	intersection	$\{\cdots\}$	the set ...
$\subseteq$	subset (contained in)	$\{\cdots : \cdots\}$	the set of all ... such that ...
$\subsetneqq$	proper subset		
$\nsubseteq$	is not a subset	A^c	complement of A
$\supseteq$	contains (superset)	$A - B$	$A \cap B^c$
$\in$	member of	$\vert A \vert$	cardinality of A, the number of elements in A
$\notin$	not a member of		

Logical Notation

$\sim$	not
$\Rightarrow$	implies
$\Leftrightarrow$	if and only if (equivalence)
iff	if and only if

Miscellaneous

$\lceil x \rceil$	the least integer greater than or equal to x	$[a, b]$	the closed interval consisting of all real numbers c with $a \leq c \leq b$
$\lfloor x \rfloor$	the greatest integer less than or equal to x	$\approx$	approximately equal to
$f \circ g$	composition of the two functions f and g	$\equiv$	congruent to
		$\mathbf{A}^T$	the transpose of the matrix $\mathbf{A}$
$f(A)$	the image of the set A under the function f; that is, $\{f(a) : a \in A\}$	$\prod$	product
		$\sum$	sum
(a, b)	the open interval consisting of all real numbers c with $a < c < b$	$\int$	integral

1 What is Combinatorics?

1.1 THE THREE PROBLEMS OF COMBINATORICS

Perhaps the fastest-growing area of modern mathematics is combinatorics. Combinatorics is concerned with the study of arrangements, patterns, designs, assignments, schedules, connections, and configurations. In the modern world, people in almost every area of activity find it necessary to solve problems of a combinatorial nature. A computer scientist considers *patterns* of digits and switches to encode complicated statements. A shop supervisor prepares *assignments* of workers to tools or to work areas. An agronomist *assigns* crops to different fields. An electrical engineer considers alternative *configurations* for a circuit. A banker studies alternative *patterns* for electronically transferring funds, and a space scientist studies such patterns for transferring messages to distant satellites. An industrial engineer considers alternative production *schedules* and workplace *configurations* to maximize efficient production. A university scheduling officer *arranges* class meeting times and students' *schedules*. A chemist considers possible *connections* between various atoms and molecules, and *arrangements* of atoms into molecules. A transportation officer *arranges* bus or plane *schedules*. A linguist considers *arrangements* of words in unknown alphabets. A geneticist considers *arrangements* of bases into chains of DNA, RNA, and so on. A statistician considers alternative *designs* for an experiment.

There are three basic problems of combinatorics. They are the *existence problem*, the *counting problem*, and the *optimization problem*. The existence problem deals with the question: Is there at least one arrangement of a particular kind? The counting problem asks: How many arrangements are there? The optimization problem is concerned with choosing, among all possible arrangements, that which is best according to some criterion. We shall illustrate these three problems with a number of examples.

Example 1.1 Design of Experiments

Let us consider an experiment designed to test the effect on human beings of five different drugs. Let the drugs be labeled 1, 2, ..., 5. We could pick out five subjects and give each subject a different drug. Unfortunately, certain subjects might be allergic to a particular drug, or immune to its effects. Thus, we could get very biased results. A more effective use of five subjects would be to give each subject each of the drugs, say on five consecutive days. Table 1.1 shows one possible arrangement of the experiment. What is wrong with this arrangement? For one thing, the day of the week a drug is taken may affect the result. (People with Monday morning hangovers may never respond well to a drug on Monday.) Also, drugs taken earlier might affect the performance of drugs taken later. Thus, giving each subject the drugs in the same order might lead to biased results. One way around these problems is simply to require that no two people get the same drug on the same day. Then the experimental design calls for a 5×5 table, with each entry being one of the integers 1, 2, 3, 4, 5, and with each row having all its entries different and each column having all its entries different. This is a particular kind of pattern. The crucial question for the designer of the drug experiment is this: Does such a design exist? This is the existence problem of combinatorics.

Table 1.1[a] A Design for a Drug Experiment

		Day				
		M	Tu	W	Th	F
	A	1	2	3	4	5
	B	1	2	3	4	5
Subject	*C*	1	2	3	4	5
	D	1	2	3	4	5
	E	1	2	3	4	5

[a]The entry in the row corresponding to a given subject and the column corresponding to a given day shows the drug taken by that subject on that day.

Let us formulate the problem more generally. We define a *Latin square* as an $n \times n$ table that uses the numbers 1, 2, ..., n as entries, and does so in such a way that no number appears more than once in the same row or column. Equivalently, it is required that each number appear exactly once in each row and column. A typical existence problem is the following: Is there a 2×2 Latin square? The answer is yes; Table 1.2 shows such a square. Similarly, one may ask if there is a 3×3 Latin square. Again, the answer is yes; Table 1.3 shows one.

Table 1.2 A 2×2 Latin Square

1	2
2	1

Table 1.3 A 3×3 Latin Square

1	2	3
2	3	1
3	1	2

Table 1.4 A 5 × 5
Latin Square

1	2	3	4	5
2	3	4	5	1
3	4	5	1	2
4	5	1	2	3
5	1	2	3	4

Our specific question asks whether or not there is a 5 × 5 Latin square. Table 1.4 shows that the answer is yes. (Is there an $n \times n$ Latin square for every n? The answer is left to the reader.)

Note that the Latin square is still not a complete solution to the problem that order effects may take place. To avoid any possible order effects, we should ideally have enough subjects so each possible ordering of the 5 drugs can be tested. How many such orderings are there? This is the *counting problem*, the second basic type of problem encountered in combinatorics. It turns out that there are $5! = 60$ such orderings, as will be clear from the methods of Section 2.3. Thus, we would need 60 subjects. If only 5 subjects are available, we could try to avoid order effects by choosing the Latin square we use at random. How many possible 5 × 5 Latin squares are there from which to choose? We shall address this counting problem in Section 5.1.3.

As this very brief discussion suggests, questions of experimental design have been a major stimulus to the development of combinatorics. We shall return to experimental design in detail in Chapter 9.

Example 1.2   Bit Strings and Binary Codes

A *bit* or *binary digit* is a zero or a one. A *bit string* is defined to be a sequence of bits, such as 0001, 1101, or 1010. Bit strings are the crucial carriers of information in modern computers. A bit string can be used to encode detailed instructions, and in turn is translated into a sequence of on-off instructions for switches in the computer. A *binary code* (*binary block code*) for a collection of symbols assigns a different bit string to each of the symbols. Let us consider a binary code for the 26 letters in the alphabet. A typical such code is the Morse code, which in its more traditional form uses dots for zeros and dashes for ones. Some typical letters in Morse code are given as follows:

$$O: \quad 111$$

$$A: \quad 01$$

$$K: \quad 101$$

$$C: \quad 1010$$

If we are restricted to bit strings consisting of either one or two bits, can we encode all 26 letters of the alphabet? The answer is no, for the only possible strings are the following:

$$0, \quad 1, \quad 00, \quad 01, \quad 10, \quad 11.$$

There are only six such strings. Notice that to answer the question posed, we had to *count* the number of possible arrangements. This was an example of a solution to a counting problem. In this case we counted by *enumerating* or listing all possible arrangements. Usually, this will be too tedious or time consuming for us, and we will want to develop shortcuts for counting without enumerating. Let us ask if bit strings of three or fewer bits would do for encoding all 26 letters of the alphabet. The answer is again no. A simple enumeration shows that there are only 14 such strings. (List them.) However, strings of four or fewer bits will suffice. (How many such strings are there?) The Morse code, indeed, uses only strings of four or fewer symbols. Not every possible string is used. (Why?) In Section 2.1 we shall encounter a very similar counting problem in studying the genetic code. DNA chains encode the basic genetic information required to determine long strings of amino acids called proteins. We shall try to explain how long a segment in a DNA chain is required to encode for an amino acid. Codes will arise in other parts of this book as well, not just in the context of genetics or of communication with modern computers. For instance, in Chapter 10 we shall study the error-correcting codes that are used to send and receive messages to and from distant space probes, to fire missiles, and so on.

Example 1.3 The Best Design for a Gas Pipeline

The flow of natural gas through a pipe depends on the diameter of the pipe, its length, the pressures at the end points, the temperature, various properties of the gas, and so on. The problem of designing an offshore gas pipeline system involves, among other things, decisions about what sizes (diameters) of pipe to use at various junctions or links so as to minimize total cost of both construction and operation. A standard approach to this problem has been to use "engineering judgment" to pick reasonable sizes of pipe and then to hope for the best. Any chance of doing better seems, at first glance, to be hopeless. For example, a modest network of 40 links, with 7 possible pipe sizes for each link, would give rise to 7^{40} possible networks, as we shall show in Section 2.1. Now 7^{40}, as we shall see, is a very large number. Our problem is to find the least expensive network out of these 7^{40} possibilities. This is an example of the third kind of combinatorial problem, an *optimization problem*, a problem where we seek to find the optimum (best, maximum, minimum) design or pattern or arrangement.

It should be pointed out that progress in solving combinatorial optimization problems has gone hand in hand with the development of the computer. Today it is possible to solve on a machine problems whose solution would have seemed inconceivable only a few years ago. Thus, the development of the computer has been a major impetus behind the very rapid development of the field of combinatorial optimization. However, there are limitations to what a computing machine can accomplish. We shall see this next.

Now any finite problem can be solved in principle by considering all possibilities. However, how long would this particular problem take to solve by enumerating all possible pipeline networks? To get some idea, note that 7^{40} is approximately 6×10^{33}, that is, 6 followed by 33 zeros. This is a huge number. Indeed, even a computer that could analyze 1 billion different pipeline networks in 1 second (one each nanosecond), would take $1.9 \times 10^{17} = 190,000,000,000,000,000$ years to analyze all 7^{40} possible pipeline networks!*

Much of modern combinatorics is concerned with developing procedures or *algorithms* for solving existence, counting, or optimization problems. From a practical point of view, it is a very important problem in computer science to analyze an algorithm for solving a problem in terms of how long it would take to solve or how much storage capacity would be required to solve it. Before embarking on a computation (such as trying all possibilities) on a machine, we would like to know that the computation can be carried out within a reasonable time or within the available storage capacity of the machine. We return to these points in our discussion of computational complexity in Sections 2.4 and 2.18.

The pipeline problem we have been discussing is a problem that, even with the use of today's high-speed computer tools, does not seem tractable by examining all cases. Any foreseeable improvements in computing speed would make a negligible change in this conclusion. However, a simple procedure gives rise to a method for finding the optimum network in only about $7 \times 40 = 280$ steps, rather than 7^{40} steps. The procedure has been implemented in the Gulf of Mexico at a savings of millions of dollars. See Frank and Frisch [1970], Kleitman [1976], Rothfarb *et al.* [1970], or Zadeh [1973] for references. This is an example of the power of techniques for combinatorial optimization.

Example 1.4   Scheduling Meetings of Legislative Committees

Committees in a state legislature are to be scheduled for a regular meeting once each week. In assigning meeting times, the aide to the Speaker of the legislature must be careful not to schedule simultaneous meetings of two committees that have a member in common. Let us suppose that in a hypothetical situation, there are only three meeting times available: Tuesday, Wednesday, and Thursday mornings. The committees whose meetings must be scheduled are Finance, Environment, Health, Transportation, Education, and Housing. Let us suppose that Table 1.5 summarizes which committees have a common member. A convenient way to represent the information of Table 1.5 is to draw a picture in which the committees are represented by dots or points and two points are joined by an undirected line if and only if the corresponding committees have a common member. The resulting diagram is called a *graph*. Figure 1.1 shows the graph obtained in this way for the data of Table 1.5. Graphs of this kind have a large number of applications, for instance in computer science, operations research, electrical engineering, ecology, policy and decision science, and in the social sciences. We shall discuss graphs and their applications in detail in Chapters 3 and 11 and elsewhere.

Our first question is this. Given the three available meeting times, can we find an assignment of committees to meeting times so that no member has to be at two meetings at once? This is an existence question. In terms of the graph we have

---

*There are 3.15×10^7 seconds per year, so $3.15 \times 10^7 \times 10^9$ or 3.15×10^{16} networks could be analyzed in a year. Then the number of years it takes to check 6×10^{33} networks is

$$\frac{6 \times 10^{33}}{3.15 \times 10^{16}} \approx 1.9 \times 10^{17}.$$

Table 1.5 Common Membership in Committees[a]

	Finance	Environment	Health	Transportation	Education	Housing
Finance	0	0	0	0	0	1
Environment	0	0	1	0	1	0
Health	0	1	0	1	1	1
Transportation	0	0	1	0	0	1
Education	0	1	1	0	0	1
Housing	1	0	1	1	1	0

[a]The i, j entry is 1 if committees i and j have a common member, and 0 otherwise. (The diagonal entries are taken to be 0 by convention.)

drawn, we would like to assign a meeting time to each point so that if two points are joined by a line, they get different meeting times. Can we find such an assignment? The answer in our case, after some analysis, is yes. One assignment that works is this: Let the Housing and Environment committees meet on Tuesday, the Education and Transportation committees on Wednesday, and the Finance and Health committees on Thursday.

Problems analogous to the one we have been discussing arise in scheduling final exams or class meeting times in a university, in scheduling job assignments in a factory, and in many other scheduling problems. We shall return to such problems in Chapter 3, when we look at these questions as questions of graph coloring.

We might ask next: Suppose that each committee chair indicates his or her first choice for a meeting time. What is the assignment of meeting times that satisfies our original requirements (if there is such an assignment) and gives the largest number of committee chairs their first choice? This is an optimization question. Let us again take a hypothetical situation and analyze how we might answer this question. Suppose that Table 1.6 gives the first choice of each committee chair. One approach to the optimization question is simply to try to identify all possible satisfactory assignments of meeting times and for each to count how many committee chairs get their first choice. Before implementing any approach to a combinatorial problem, as we have observed before, we would like to get a feeling for how long the approach will take. How many possibilities will have to be analyzed? This is a counting problem. We shall solve this counting problem by enumeration. It is easy to see from the graph of Figure 1.1 that Housing, Education, and Health must get different times. (Each one has a line joining it to the other two.) Similarly,

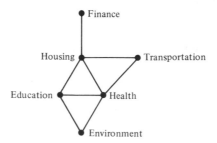

Figure 1.1 The graph obtained from the data of Table 1.5.

Table 1.6 First Choice for Meeting Times

Committee	Finance	Environment	Health	Transportation	Education	Housing
Chair's first choice	Tuesday	Thursday	Thursday	Tuesday	Tuesday	Wednesday

Transportation must get a different time from Housing and Health. (Why?) Hence, since only three meeting times are available, Transportation must meet at the same time as Education. Similarly, Environment must meet at the same time as Housing. Finally, Finance cannot meet at the same time as Housing, and therefore as Envi-

Table 1.7 Possible Assignments of Meeting Times

Assignment number	Tuesday	Wednesday	Thursday	Number of committee chairs getting their first choice
1	Transportation–Education	Environment–Housing	Finance–Health	4
2	Transportation–Education	Finance Health	Environment–Housing	3
3	Environment–Housing	Transportation–Education	Finance–Health	1
4	Environment–Housing	Finance–Health	Transportation–Education	0
5	Finance–Health	Transportation–Education	Environment–Housing	2
6	Finance–Health	Environment–Housing	Transportation–Education	2
7	Transportation–Education–Finance	Environment–Housing	Health	5
8	Transportation–Education–Finance	Health	Environment–Housing	4
9	Environment–Housing	Transportation–Education–Finance	Health	1
10	Environment–Housing	Health	Transportation–Education–Finance	0
11	Health	Transportation–Education–Finance	Environment–Housing	1
12	Health	Environment–Housing	Transportation–Education–Finance	1

ronment, but could meet simultaneously with any of the other committees. Thus, there are only two possible meeting patterns. They are as follows.

Pattern 1. Transportation and Education meet at one time, Environment and Housing at a second time, and Finance and Health meet at the third time.

Pattern 2. Transportation, Education, and Finance meet at one time, Environment and Housing meet at a second time, and Health meets at the third time.

It follows that Table 1.7 gives all possible assignments of meeting times. In all, there are 12 possible assignments. Our counting problem has been solved by enumerating all possibilities. (In Section 3.4.1 we shall do this counting another way.) It should be clear from Example 1.3 that enumeration could not always suffice for solving combinatorial problems. Indeed, if there are more committees and more possible meeting times, the problem we have been discussing gets completely out of hand.

Having succeeded in enumerating in our example, we can easily solve the optimization problem. Table 1.7 shows the number of committee chairs getting their first choices under each assignment. Clearly, assignment number 7 is the best from this point of view. Here, only the chair of the Environment committee does not get his or her first choice. For further reference on assignment of meeting times for state legislative committees, see Bodin and Friedman [1971]. For recent work on other scheduling problems where the schedule is repeated periodically (for example, every week), see, for instance, Baker [1976], Bartholdi et al. [1980], Karp and Orlin [1981], Orlin [1980], or Tucker [1975]. For a recent survey of workforce scheduling algorithms, see Tien and Kamiyama [1982].

This book is organized around the three basic problems of combinatorics that we have been discussing. It has four parts. After an introductory part, the remaining three parts deal with these three problems.

1.2 THE HISTORY AND APPLICATIONS OF COMBINATORICS*

The four examples described in Section 1.1 illustrate some of the problems with which combinatorics is concerned. They were chosen from a variety of fields to illustrate the variety of applications of combinatorics in modern times.

Although combinatorics has achieved its greatest impetus in modern times, it is an old branch of mathematics. According to legend, the Chinese Emperor Yu (in approximately 2200 B.C.) observed a magic square on the back of a divine tortoise. (A *magic square* is a square array of numbers in which the sum of all rows, all columns, and all diagonals is the same. An example of such a square is shown in Table 1.8. The reader might wish to find a different 3×3 magic square.)

*For a more detailed discussion of the history of combinatorics, see David [1962]. For the history of graph theory, see Biggs *et al.* [1976].

Table 1.8 A Magic Square

4	9	2
3	5	7
8	1	6

Permutations or arrangements in order were known in China before 1100 B.C. The binomial expansion [the expansion of $(a + b)^n$] was known to Euclid about 300 B.C. for the case $n = 2$. Applications of the formula for the number of permutations of an n-element set can be found in an anonymous Hebrew work *Sefer Yetzirah*, written between A.D. 200 and 500. The formula itself was known at least 2500 years ago. In A.D. 1100, Rabbi Ibn Ezra knew the formula for the number of combinations of n things taken r at a time, the binomial coefficient. Shortly thereafter, Chinese, Hindu, and Arab works began mentioning binomial coefficients in a primitive way.

In more modern times, the seventeenth-century scholars Pascal and Fermat pursued studies of combinatorial problems in connection with gambling—among other things, they figured out odds. (Pascal's famous triangle was in fact known to Chu Shih-Chieh in China in 1303.) The work of Pascal and Fermat laid the groundwork for probability theory; in the eighteenth century, Laplace defined probability in terms of number of favorable cases. Also in the eighteenth century, Euler invented graph theory in connection with the famous Königsberg bridge problem and Bernoulli published the first book presenting combinatorial methods, *Ars Conjectandi*. In the eighteenth and nineteenth centuries, combinatorial techniques were applied to study puzzles and games, by Hamilton and others. In the nineteenth century, Kirchhoff developed a graph-theoretical approach to electrical networks and Cayley developed techniques of enumeration to study organic chemistry. In modern times, the techniques of combinatorics have come to have far-reaching, significant applications in computer science, transportation, information processing, industrial planning, electrical engineering, experimental design, sampling, coding, genetics, political science, and a variety of other important fields. In this book we shall always keep the applications close at hand, remembering that they are not only a significant benefit derived from the development of the mathematical techniques, but they are also a stimulus to the continuing development of these techniques.

EXERCISES FOR CHAPTER 1

1. Find a 4 × 4 Latin square.
2. Find all possible 3 × 3 Latin squares.
3. Describe how to find an $n \times n$ Latin square.
4. (Liu [1972]) Suppose that we have two types of drugs to test simultaneously, such as headache remedies and fever remedies. In this situation, we might try to design an experiment in which each type of drug is tested using a Latin square design. However, we also want to make sure that, if at all possible, all combinations of headache and fever remedies are tested. For example, Table 1.9 shows two Latin square designs if we have 3 headache remedies and 3 fever remedies. Also shown in Table 1.9 is a third square, which lists as its i, j entry the i, j entries from both of

the first two squares. We demand that each entry of this third square be different. This is not true in Table 1.9.

(a) Find an example with 3 headache and 3 fever drugs where the combined square has the desired property.

(b) Find another example with 4 headache and 4 fever drugs. (In Chapter 9 we shall observe that with 6 headache and 6 fever drugs, this is impossible. The existence problem has a negative solution.) *Note:* If you start with one Latin square design for the headache drugs and cannot find one for the fever drugs so that the combined square has the desired property, you should start with a different design for the headache drugs.

Table 1.9 A Latin square design for testing headache drugs 1, 2, and 3, a Latin square design for testing fever drugs a, b, and c, and a combination of the two[a]

		Day					Day					Day		
		1	2	3			1	2	3			1	2	3
	1	1	2	3		1	a	b	c		1	1, a	2, b	3, c
Subject	2	2	3	1	Subject	2	b	c	a	Subject	2	2, b	3, c	1, a
	3	3	1	2		3	c	a	b		3	3, c	1, a	2, b
		Headache drugs					Fever drugs					Combination		

[a]The third square has as its i, j entry the headache drug and the fever drug shown in the i, j entries of the first two squares, respectively.

5. Show by enumeration that there are 14 bit strings of length at most 3.

6. Use enumeration to find the number of bit strings of length at most 4.

7. Suppose that we want to build a *trinary code* for the 26 letters of the alphabet, using strings in which each symbol is 0, 1, or -1.
   (a) Could we encode all 26 letters using strings of length at most 2? Answer this question by enumeration.
   (b) What about using strings of length exactly 3?

8. The genetic code embodied in the DNA molecule, a code we shall describe in Section 2.1, consists of strings of symbols, each of which is one of the four letters T, C, A, or G. Find by enumeration the number of different code words or strings using these letters and having length 3 or less.

9. Suppose that in designing a gas pipeline network, we have 2 possible pipe sizes, small (*S*) and large (*L*). If there are 4 possible links, enumerate all possible pipeline networks. (A typical one could be abbreviated *LSLL*, where the *i*th letter tells the size of the *i*th pipe.)

10. In Example 1.3, suppose that a computer could analyze as many as 100 billion different pipeline networks in a second, a 100-fold improvement over the speed we assumed in the text. Would this make a significant difference in our conclusions? Why? (Do a computation in giving your answer.)

11. Tables 1.10 and 1.11 give data of overlap in class rosters for several courses in a university.
    (a) Translate Table 1.10 into a graph as in Example 1.4.
    (b) Repeat for Table 1.11.

12. **(a)** Suppose that there are only two possible final examination times for the courses considered in Table 1.10. Is there an assignment of final exam times so that any two classes having a common member get a different exam time? If so, find such an assignment. If not, why not?
    (b) Repeat part (a) for Table 1.10 if there are three possible final exam times.

Table 1.10[a]

	English	Calculus	History	Physics
English	0	1	0	0
Calculus	1	0	1	1
History	0	1	0	1
Physics	0	1	1	0

[a]The i, j entry is 1 if the ith and jth courses have a common member, and 0 otherwise.

Table 1.11[a]

	English	Calculus	History	Physics	Economics
English	0	1	0	0	0
Calculus	1	0	1	1	1
History	0	1	0	1	1
Physics	0	1	1	0	1
Economics	0	1	1	1	0

[a]The i, j entry is 1 if the ith and jth courses have a common member, and 0 otherwise.

(c) Repeat part (a) for Table 1.11 if there are three possible final exam times.

(d) Repeat part (a) for Table 1.11 if there are four possible final exam times.

13. Suppose that there are three possible final exam times, Tuesday, Wednesday, and Thursday mornings. Suppose that each instructor of the courses listed in Table 1.10 requests Tuesday morning as a first choice for final exam time. What assignment (assignments) of exam times, if any exist, gives the largest number of instructors their first choices?

REFERENCES FOR CHAPTER 1

BAKER, K. R., "Workforce Allocation in Cyclical Scheduling Problems," *Oper. Res. Quart., 27* (1976), 155–167.

BARTHOLDI, J. J., III, ORLIN, J. B., and RATLIFF, H. D., "Cyclic Scheduling via Integer Programs with Circular Ones," *Oper. Res., 28* (1980), 1074–1085.

BIGGS, N. L., LLOYD, E. K., and WILSON, R. J., *Graph Theory 1736–1936*, Oxford University Press, London, 1976.

BODIN, L. D., and FRIEDMAN, A. J., "Scheduling of Committees for the New York State Assembly," Tech. Report USE No. 71-9, Urban Science and Engineering, State University of New York, Stony Brook, 1971.

DAVID, F. N., *Games, Gods, and Gambling*, Hafner Press, New York, 1962.

FRANK, H., and FRISCH, I. T., "Network Analysis," *Sci. Amer., 223* (1970), 94–103.

KARP, R. M., and ORLIN, J. B., "Parametric Shortest Path Algorithms with an Application to Cyclic Staffing," *Discrete Appl. Math., 3* (1981), 37–45.

KLEITMAN, D. J., "Comments on the First Two Days' Sessions and a Brief Description of a Gas Pipeline Network Construction Problem," in F. S. Roberts (ed.), *Energy: Mathematics and Models*, SIAM, Philadelphia, 1976, pp. 239–252.

LIU, C. L., *Topics in Combinatorial Mathematics*, Mathematical Association of America, Washington, D.C., 1972.

ORLIN, J. B., "Minimizing the Number of Vehicles to Meet a Fixed Periodic Schedule: An Application of Periodic Posets," Tech. Report OR 102-80, Operations Research Center, Massachusetts Insitute of Technology, Cambridge, Mass., 1980.

ROTHFARB, B., FRANK, H., ROSENBAUM, D. M., STEIGLITZ, K., and KLEITMAN, D. J., "Optimal Design of Offshore Natural-Gas Pipeline Systems," *Oper. Res.*, *18* (1970), 992–1020.

TIEN, J. M., and KAMIYAMA, A., "On Manpower Scheduling Algorithms," *SIAM Rev.*, *24* (1982), 275–287.

TUCKER, A. C., "Coloring a Family of Circular Arcs," *SIAM J. Appl. Math.*, *29* (1975), 493–502.

ZADEH, N., "Construction of Efficient Tree Networks: The Pipeline Problem," *Networks*, *3* (1973), 1–32.

PART I The Basic Tools of Combinatorics

2 Basic Counting Rules

by Helen Marcus-Roberts and Fred S. Roberts

2.1 THE PRODUCT RULE

Some basic counting rules underlie all of combinatorics. We summarize them in this chapter. The reader who is already familiar with these rules may wish to review them rather quickly. This chapter also introduces a variety of applications that may not be as familiar, many of which are returned to in later chapters. In reading this chapter the reader already familiar with counting may wish to concentrate on these applications.

Example 2.1 Bit Strings and Binary Codes Revisited

Let us return to our binary code example (Example 1.2), and ask again how many letters of the alphabet can be encoded if there are exactly two bits. Let us get the answer by drawing a tree diagram. We do that in Figure 2.1. There are 4 possible strings of 2 bits, as we noted before. The reader will observe that there are 2 choices for the first bit, and for each of these choices, there are 2 choices for the second bit, and 4 is 2×2.

Example 2.2 DNA

Deoxyribonucleic acid, DNA, is the basic building block of inheritance. DNA is a chain consisting of bases. Each link or base is one of four possible chemicals: thymine, T; cytosine, C; adenine, A; guanine, G. The sequence of bases encodes certain genetic information. In particular, it determines long chains of amino acids which are known as proteins. There are 20 basic amino acids. A part of a DNA chain will encode one such amino acid. How long does a part of a DNA chain have to be for there to be enough possible sequences to encode 20 different amino acids?

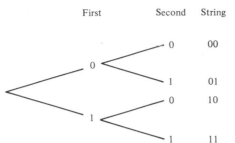

Figure 2.1 A tree diagram for counting the number of bit strings of length 2.

For example, can 2-element DNA chains encode for the 20 different basic amino acids? To answer this, we need to ask: How many 2-element DNA chains are there? The answer to this question is again given by a tree diagram, as shown in Figure 2.2. We see that there are 16 possible 2-element DNA chains. There are 4 choices for the first element, and, for each of these choices, there are 4 choices for the second element; the reader will notice that 16 is 4 × 4. Notice that there are not enough 2-element chains to encode for all 20 different basic amino acids. In fact, a chain of 3 elements does the encoding in practice. A simple counting procedure has shown why at least 3 elements are needed.

The two examples given above illustrate the following basic rule.

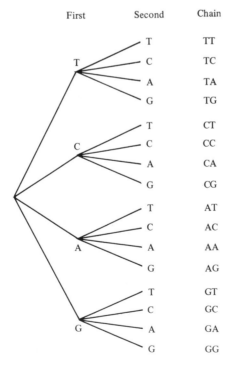

Figure 2.2 A tree diagram for counting the number of 2-element DNA chains.

PRODUCT RULE. If something can happen in n_1 ways, and no matter how the first thing happens, a second thing can happen in n_2 ways, then the two things together can happen in $n_1 \times n_2$ ways. More generally, if something can happen in n_1 ways, and no matter how the first thing happens, a second thing can happen in n_2 ways, and no matter how the first two things happen, a third thing can happen in n_3 ways, and ..., then all the things together can happen in $n_1 \times n_2 \times n_3 \times \cdots$ ways.

Returning to bit strings, we see immediately by the product rule that the number of strings of exactly 3 bits is given by $2 \times 2 \times 2 = 2^3 = 8$. Similarly, in the pipeline problem of Example 1.3, if there are 7 choices of pipe size for each link and 3 links, there are

$$7 \times 7 \times 7 = 7^3 = 343$$

different possible networks. If there are 40 links, there are

$$7 \times 7 \times \cdots \times 7 = 7^{40}$$

different possible networks. Note that, by our observations in Chapter 1, it is impossible to count the number of possible pipeline networks by enumerating them. Some method of counting other than enumeration is needed. The product rule gives such a method. In the early part of this book, we shall be concerned with such simple methods of counting.

Next, suppose that A is a set of a objects and B is a set of b objects. Then the number of ways to pick one object from A and then one object from B is $a \times b$. This statement is a more precise version of the product rule.

To give one final example, the number of 3-element DNA chains is

$$4 \times 4 \times 4 = 4^3 = 64.$$

That is why there are enough different 3-element chains to encode for all 20 different basic amino acids; indeed, several different chains encode for the same amino acid—this is different from the situation in Morse code, where strings of up to 4 bits are required to encode for all 26 letters of the alphabet, but not every possible string is used. In Section 2.10 we will consider Gamow's [1954a, b] suggestion that two 3-element chains encode the same amino acid if and only if they have the same bases, independent of order. We will see that this coding gives rise to exactly 20 distinct amino acids. (Unfortunately, it was later discovered that this is not how things work.)

Continuing with the DNA chains, we see that the number of chains of 4 bases is 4^4, the number with 100 bases is 4^{100}. How long is a full-fledged DNA chain? The answer is given in Table 2.1. Notice that in a chicken, a DNA chain has 5×10^9 bases. Thus, the

Table 2.1 The Number of Possible DNA Chains for Different Organisms[a]

Organism	Number bases per DNA chain	Number possible DNA chains
Chicken	5×10^9	$4^{5 \times 10^9} > 10^{3 \times 10^9}$
Mouse	1.3×10^{10}	$4^{1.3 \times 10^{10}} > 10^{7.8 \times 10^9}$
Guinea pig	1.7×10^{10}	$4^{1.7 \times 10^{10}} > 10^{1.02 \times 10^{10}}$
Human	2.1×10^{10}	$4^{2.1 \times 10^{10}} > 10^{1.26 \times 10^{10}}$

[a]*Source:* Spector [1956].

number of such chains is

$$4^{5 \times 10^9},$$

which is greater than

$$10^{3 \times 10^9}.$$

This number is 1 followed by 3×10^9 zeros or 3 billion zeros! It is a number that is too large to comprehend. Similar results hold for other organisms. By a simple counting of all possibilities, we can understand the tremendous possible variation in genetic makeup. It is not at all surprising, given the number of possible DNA chains, that there is such an amazing variety in nature, and that two individuals are never the same. It should be noted once more that given the tremendous magnitude of the number of possibilities, it would not have been possible to count these possibilities by the simple expedient of enumerating them. It was necessary to develop rules or procedures for counting, which counted the number of possibilities without simply listing them. That is one of the three basic problems in combinatorics: developing procedures for counting without enumerating.

Example 2.3   Telephone Numbers

At one time, a local telephone number was given by a sequence of two letters followed by five numbers. How many different telephone numbers were there? Using the product rule, one is led to the answer:

$$26 \times 26 \times 10 \times 10 \times 10 \times 10 \times 10 = 26^2 \times 10^5.$$

However, this is incorrect, for two letters on the same place on the dial lead to the same telephone numbers. The reader might wish to envision the telephone dial. (A rendering of it is given in Figure 2.3.) There are three letters on all digits, except that 1 and 0 have no letters. Hence, letters A, B, and C were equivalent; so were W, X, and Y; and so on. There were, in effect, only 8 different letters. The number of

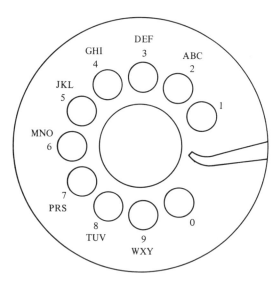

Figure 2.3   A telephone dial.

different telephone numbers was therefore

$$8^2 \times 10^5 = 6.4 \times 10^6.$$

Thus, there were a little over 6 million such numbers. In the 1950s and 1960s, most local numbers were changed to become simply seven-digit numbers, with the restriction that neither of the first two digits could be 0 or 1. The number of telephone numbers was still $8^2 \times 10^5$. Direct distance dialing was accomplished by adding a three-digit area code. The area code cannot begin with a 0 or 1, and it must have 0 or 1 in the middle. Using these restrictions, we compute that the number of possible telephone numbers is

$$8 \times 2 \times 10 \times 8^2 \times 10^5 = 1.024 \times 10^9.$$

This is enough to service over 1 billion customers.

Example 2.4   Switching Functions

Let B_n be the set of all bit strings of length n. A *switching function* (*Boolean function*) *of n variables* is a function that assigns to each bit string of length n a number 0 or 1. For instance, let $n = 2$. Then $B_2 = \{00, 01, 10, 11\}$. Two switching functions S and T defined on B_2 are given in Table 2.2. The problem of making a detailed design of a digital computer usually involves finding a practical circuit implementation of certain functional behavior. A computer device implements a switching function of two, three, or four variables. Now every switching function can be realized in numerous ways by an electrical network of interconnections. Rather than trying to figure out from scratch an efficient design for a given switching function, a computer engineer would like to have a catalog that lists, for every switching function, an efficient network realization. Unfortunately, this seems at first to be an impractical goal. For how many switching functions of n variables are there? There are 2^n elements in the set B_n, by a generalization of Example 2.1. Hence, by the product rule, there are $2 \times 2 \times \cdots \times 2$ different n-variable switching functions, where there are 2^n terms in the product. In sum, there are 2^{2^n} different n-variable switching functions. Even the number of such functions for $n = 4$ is 65,536, and the number grows astronomically fast. Fortunately, by taking advantage of symmetries, we can consider certain switching functions equivalent as far as what they compute is concerned. Then we need not identify the best design for every switching function; we need do it only for enough switching functions so that every other switching function is equivalent to one of the ones for which we have identified the best design. While the first com-

Table 2.2 Two Switching Functions

Bit string x	$S(x)$	$T(x)$
00	1	0
01	0	0
10	0	1
11	1	1

puters were being built, a team of researchers at Harvard painstakingly enumerated all possible switching functions of 4 variables, and determined which were equivalent. They discovered that it was possible to reduce every switching function to one of 222 types (Harvard Computation Laboratory Staff [1951]). In Chapter 7 we will show how to derive results such as this from a powerful theorem due to George Polya. For a more detailed discussion of switching functions, see Deo [1974, Ch. 12], Harrison [1965], Hill and Peterson [1968], Kohavi [1970], Liu [1977], Prather [1976], or Stone [1973].

EXERCISES FOR SECTION 2.1*

1. The population of Westfield, New Jersey, is about 35,000. If each resident has three initials, is it true that there must be at least two individuals with the same initials? Give a justification of your answer.

2. A library has 5000 books, and the librarian wants to encode each using a codeword consisting of 3 letters followed by 3 numbers. Are there enough codewords to encode all 5000 books with different codewords?

3. (a) Continuing with Exercise 7 of Chapter 1, compute the maximum number of strings of length at most 3 in a trinary code.
   (b) Repeat for length at most 4.
   (c) Repeat for length exactly 4, but beginning with a 0 or 1.

4. In our discussion of telephone numbers, suppose that we maintain the restrictions on area code as in Example 2.3. Suppose that we lengthen the local phone number, allowing it to be any eight-digit number with the restriction that none of the first three digits can be 0 or 1. How many local phone numbers are there? How many phone numbers are there including area code?

5. If we want to use bit strings of length at most n to encode not only all 26 letters of the alphabet, but also all 10 decimal digits, what is the smallest number n that works? (What is n for Morse code?)

6. How many $m \times n$ matrices are there each of whose entries is 0 or 1?

7. A committee has to have at least one member. It can contain at most one plumber, at most one piano teacher, at most one truck driver, at most one college professor, at most one firefighter, and at most two doctors. How many possible committees are there if we consider any two plumbers indistinguishable, and the same for other professions, and hence call two committees the same if they have the same number of members of each profession? Justify your answer.

8. How many numbers less than 1 million contain the digit 2?

9. Enumerate all switching functions of 2 variables.

10. If a function assigns 0 or 1 to each switching function of n variables, how many such functions are there?

11. A switching function S is called *self-dual* if the value S of a bit string is unchanged when 0's and 1's are interchanged. For instance, the function S of Table 2.2 is self-dual, but the function T of that table is not. How many self-dual switching functions of n variables are there?

Note to reader: In the exercises in Chapter 2, exercises after each section can be assumed to use techniques of *some* previous (nonoptional) section, not necessarily exactly the techniques just introduced. Also, there are additional exercises at the end of the chapter. Indeed, sometimes an exercise is included which does *not* make use of the techniques of the current section. To understand a new technique, one must understand when it does not apply as well as when it applies.

12. (Stanat and McAllister [1977]) In some computers, an integer (positive or negative) is represented by using bit strings of length p. The last bit in the string represents the sign, and the first $p - 1$ bits are used to encode the integer. What is the largest number of distinct integers that can be represented in this way for a given p? What if 0 must be one of these integers? (The sign of 0 is $+$ or $-$.)

13. (Stanat and McAllister [1977]) Every integer can be represented (nonuniquely) in the form $a \times 2^b$, where a and b are integers. The *floating-point representation* for an integer uses a bit string of length p to represent an integer by using the first m bits to encode a and the remaining $p - m$ bits to encode b, with the latter two encodings performed as described in Exercise 12.
   (a) What is the largest number of distinct integers that can be represented using the floating-point notation for a given p?
   (b) Repeat part (a) if the floating-point representation is carried out in such a way that the leading bit for encoding the number a is 1.
   (c) Repeat part (a) if 0 must be included.

2.2 THE SUM RULE

We turn now to the second fundamental counting rule. Consider the following example.

Example 2.5   Congressional Delegations

There are 100 senators and 435 members of the House of Representatives. A delegation is being selected to see the President. In how many different ways can such a delegation be picked if it consists of one senator *and* one representative? The answer, by the product rule, is

$$100 \times 435 = 43,500.$$

What if the delegation is to consist of one member of the Senate *or* one member of the House? Then there are

$$100 + 435 = 535$$

possible delegates. This computation illustrates the second basic rule of counting, the sum rule.

SUM RULE. If one event can occur in n_1 ways and a second event in n_2 (different) ways, then there are $n_1 + n_2$ ways in which either the first event or the second event can occur (but not both). More generally, if one event can occur in n_1 ways, a second event can occur in n_2 (different) ways, a third event can occur in n_3 (still different) ways, …, then there are

$$n_1 + n_2 + n_3 + \cdots$$

ways in which (exactly) one of the events can occur.

In Example 2.5 we have italicized the words "and" and "or." These key words usually indicate whether the sum rule or the product rule is appropriate. The word "and" suggests the product rule, the word "or" the sum rule.

Example 2.6   International Committees

An international committee consists of 3 Russians, 4 Frenchmen, and 5 Germans. To pick a Russian and a Frenchman to represent the committee, there are $3 \times 4 = 12$ ways, by the product rule. How many ways are there to pick two representatives if they must be from different countries? We can pick either a Russian and a Frenchman, or a Russian and a German, or a Frenchman and a German. There are by previous computation 12 ways of doing the first; there are also 15 ways of doing the second (why?), and 20 ways of doing the third (why?). Hence, by the sum rule, the number of ways of choosing the representatives is

$$12 + 15 + 20 = 47.$$

Example 2.7   Variables in BASIC

A variable name in the programming language BASIC can either be a letter or a letter followed by a decimal digit, that is, one of the numbers 0, 1, ..., 9. By the product rule, there are $26 \times 10 = 260$ names of the latter kind. By the sum rule, there are $26 + 260 = 286$ variable names in all.

In closing this section, let us restate the sum rule this way. Suppose that A and B are disjoint sets and we wish to pick exactly one element, picking it from A or from B. Then the number of ways to pick this element is the number of elements in A plus the number of elements in B.

EXERCISES FOR SECTION 2.2

1. How many bit strings have length 3, 4, or 5?

2. A committee is to be chosen from among 8 scientists, 6 laypersons, and 13 clerics. If the committee is to have two members of different backgrounds, how many such committees are there?

3. How many numbers are there which have five digits, each being a number in $\{1, 2, ..., 9\}$, and either having all digits odd or having all digits even?

4. Each customer of Mobil Credit Corporation is given a 9-digit number for computer identification purposes. If each digit can be any number between 0 and 9, are there enough different account numbers for 10 million credit-card holders? Would there be if the digits were only 0 or 1?

5. How many 5-letter words either start with f or do not have the letter f?

6. In how many ways can we get a sum of 3 or a sum of 4 when two dice are rolled?

7. Suppose that a pipeline network is to have 30 links. For each link, there are 2 choices: The pipe may be any one of 7 sizes and any one of 3 materials. How many different pipeline networks are there?

2.3 PERMUTATIONS

In combinatorics we will frequently talk about n-element sets, sets consisting of n elements. It will be convenient to call these *n-sets*. A *permutation* of an n-set is an arrangement of the elements of the set in order. It is often important to count the number of permutations of an n-set.

Example 2.8   Job Interviews

Three people, Smith, Jones, and Brown, are scheduled for job interviews. In how many different orders can they be interviewed? We can list all possible orders, as follows:

Smith, Jones, Brown

Smith, Brown, Jones

Jones, Smith, Brown

Jones, Brown, Smith

Brown, Smith, Jones

Brown, Jones, Smith

We see that there are 6 possible orders. Alternatively, we can observe that there are 3 choices for the first person being interviewed. For each of these choices, there are 2 remaining choices for the second person. For each of these choices, there is 1 remaining choice for the third person. Hence, by the product rule, the number of possible orders is

$$3 \times 2 \times 1 = 6.$$

Each order is a permutation. We are asking for the number of permutations of a 3-set, the set consisting of Smith, Jones, and Brown.

If there are 5 people to be interviewed, counting the number of possible orders can still be done by enumeration; however, that is rather tedious. It is easier to observe that now there are 5 possibilities for the first person, 4 remaining possibilities for the second person, and so on, resulting in

$$5 \times 4 \times 3 \times 2 \times 1 = 120$$

possible orders in all.

The computations of Example 2.8 generalize to give us the following result: The number of permutations of an n-set is given by

$$n \times (n - 1) \times (n - 2) \times \cdots \times 1 = n!$$

In Example 1.1 we discussed the number of orders in which to take 5 different drugs. This is the same as the number of permutations of a 5-set, so it is $5! = 60$. To see once again why counting by enumeration rapidly becomes impossible, we show in Table 2.3 the values of $n!$ for several values of n. The number $25!$, to give an example, is already so

Table 2.3   Values of $n!$ for n from 0 to 10

n	0	1	2	3	4	5	6	7	8	9	10
$n!$	1	1	2	6	24	120	720	5040	40,320	362,800	3,628,000

large that it is incomprehensible. To see this, note that

$$25! \approx 1.55 \times 10^{25}.$$

A computer checking 1 billion permutations per second would require almost half a billion years to look at 1.55×10^{25} permutations.*

In spite of the result above, there are occasions where it is useful to enumerate all permutations of an n-set. In Section 2.17 we present an algorithm for doing so.

The number $n!$ can be approximated by computing $\sqrt{2\pi n}\,(n/e)^n = s_n$. The approximation $n! \sim s_n$ is called *Stirling's approximation*. To see how good the approximation is, note that it approximates 5! as $s_5 = 118.02$ and 10! as $s_{10} = 3,598,600$. (Compare the real values in Table 2.3.) The ratio of $n!$ to s_n approaches 1 as n approaches ∞. For a proof, see such advanced calculus texts as Buck [1965].

EXERCISES FOR SECTION 2.3

1. List all permutations of
   (a) $\{1, 2, 3\}$;   (b) $\{1, 2, 3, 4\}$.

2. How many permutations of $\{1, 2, 3, 4, 5\}$ begin with 5?

3. How many permutations of $\{1, 2, \ldots, n\}$ begin with 1 and end with n?

4. Compute s_n and compare it to $n!$ if
   (a) $n = 4$;   (b) $n = 6$;   (c) $n = 8$.

5. How many permutations of $\{1, 2, 3, 4\}$ begin with an odd number?

6. (Cohen [1978]) (a) In a six-cylinder engine, the even-numbered cylinders are on the left and the odd-numbered cylinders are on the right. A good firing order is a permutation of the numbers 1 to 6 in which right and left sides are alternated. How many possible good firing orders are there which start with a left cylinder?
   (b) Repeat for a $2n$-cylinder engine.

7. Ten job applicants have been invited for interviews, five having been told to come in the morning and five having been told to come in the afternoon. In how many different orders can the interviews be scheduled?

2.4 COMPLEXITY OF COMPUTATION

We have already observed that not all problems of combinatorics can be solved on the computer, at least not by enumeration. Suppose that a computer program implements an algorithm for solving a combinatorial problem. Before running such a program, we like to know if the program will run in a "reasonable" amount of time and will use no more than "reasonable" (or allowable) amount of storage or memory. The time or storage a program requires depends on the input. To measure how expensive a program is to run, we

---

*To see why, note that there are 3.15×10^7 seconds in a year. Thus,

$$3.15 \times 10^7 \times 10^9 = 3.15 \times 10^{16}$$

permutations can be checked in a year. Hence, the number of years required to check 1.55×10^{25} permutations is

$$\frac{1.55 \times 10^{25}}{3.15 \times 10^{16}} \approx 4.9 \times 10^8.$$

try to calculate a *cost function* or a *complexity function*. This is a function f that measures the cost, in terms of time required or storage required, as a function of the size n of the input problem. For instance, we might ask how many operations are required to multiply two square matrices of n rows and columns each. This number of operations is $f(n)$.

Usually, the cost of running a particular computer program on a particular machine will vary with the skill of the programmer and the characteristics of the machine. Thus there is a big emphasis in modern computer science on comparison of algorithms rather than programs, and on estimation of the complexity $f(n)$ of an algorithm, independent of the particular program or machine used to implement the algorithm. The desire to calculate complexity of algorithms is a major stimulus for the development of techniques of combinatorics.

Example 2.9 The Traveling Salesman Problem

A salesman wishes to visit n different cities, starting and ending his business trip at the first city. He does not care in which order he visits the cities. What he does care about is to minimize the total cost of his trip. Assume that the cost of traveling from city i to city j is c_{ij}. The problem is to find an algorithm for computing the cheapest route, where the cost of a route is the sum of the c_{ij} for links used in the route. This is a typical combinatorial optimization problem. Analogous problems arise when a bank courier is asked to visit all branches of a bank in one day or when a robot in an automated warehouse must visit n different locations to fill an order.*

For the traveling salesman problem, we shall be concerned with the enumeration algorithm: Enumerate all possible routes and calculate the cost of each route. We shall try to compute the complexity $f(n)$ of this algorithm, where n is the size of the input, that is, the number of cities. We shall assume that identifying a route and computing its cost is comparable for each route, and takes 1 unit of time.

Now any route starting and ending at city 1 corresponds to a permutation of the remaining $n - 1$ cities. Hence, there are $(n - 1)!$ such routes, so $f(n) = (n - 1)!$ units of time. We have already shown that this number can be extremely high. When n is 26 and $n - 1$ is 25, we showed that $f(n)$ is so high that it is infeasible to perform this algorithm by computer. We shall return to the traveling salesman problem in Section 11.5.

The traveling salesman problem is an example of a problem that has defied the efforts of researchers to find a "good" algorithm. Indeed, it belongs to a class of problems known as *NP-complete* or *NP-hard problems*, problems for which it is unlikely there will be a good algorithm in a very precise sense of the word *good*. We shall return to this point in Section 2.18, where we briefly define NP-completeness and where we shall define an algorithm to be a *good algorithm* if its complexity function $f(n)$ is bounded by a polynomial in n. Such an algorithm is called a *polynomial algorithm*.

Example 2.10 Scheduling a Computer System†

A computer center has n programs to run. Each program requires certain resources, such as a compiler, a segment of main memory, and certain disk and tape drives. We

---

*For recent work on the latter application, see Elsayed [1981] and Elsayed and Stern [1983].
†This example is due to Stanat and McAllister [1977].

shall refer to the required resources as a *configuration* corresponding to the program. The conversion of the system from the ith configuration to the jth configuration has a cost associated with it, say c_{ij}. For instance, if two programs require a similar configuration, it makes sense to run them consecutively. The computer center would like to minimize the total costs associated with running the n programs. The fixed cost of running each program does not change with different orders of running the programs. The only things that change are the conversion costs c_{ij}. Hence, the center wants to find an order in which to run the programs such that the total conversion costs are minimized. Similar questions arise in many scheduling problems in operations research. We discuss them further in Example 11.3 and Section 11.6.3. As in the traveling salesman problem, the algorithm of enumerating all possible orders of running the programs is infeasible, for it clearly has a computational complexity of $n!$. [Why $n!$ and not $(n-1)!$?] Indeed, from a formal point of view, this problem and the traveling salesman problem are almost equivalent—simply replace cities by configurations. Any algorithm for solving one of these problems is readily translatable into an algorithm for solving the other problem. It is one of the major motivations for using mathematical techniques to solve real problems that we can solve one problem and then immediately have techniques that are applicable to a large number of other problems, which on the surface seem quite different.

Example 2.11 Searching through a File

In determining computational complexity, we do not always know exactly how long a computation will take. For instance, consider the problem of searching through a list of n keys (identification numbers) and finding the key of a particular person in order to access that person's file. Now it is possible that the key in question will be first in the list. However, in the *worst case*, the key will be last on the list. The cost of handling the worst possible case is sometimes used as a measure of computational complexity called the *worst case complexity*. Here $f(n)$ would be proportional to n. On the other hand, another perfectly appropriate measure of computational complexity is the *average* cost of handling a case, the *average case complexity*. Assuming that all cases are equally likely, this is computed by calculating the cost of handling each case, summing up these costs, and dividing by the number of cases. In our example, the average case complexity is proportional to $(n+1)/2$, assuming that all keys are equally likely to be the object of a search, for the sum of the costs of handling the cases is given by $1 + 2 + \cdots + n$. Hence, using a standard formula for this sum, we have

$$f(n) = \frac{1}{n}(1 + 2 + \cdots + n) = \frac{1}{n}\frac{n(n+1)}{2} = \frac{n+1}{2}.$$

In Section 3.6 we will discuss the use of binary search trees for storing files, and argue that the computational complexity of finding a file with a given key can be reduced significantly by using a binary search tree.

EXERCISES FOR SECTION 2.4

1. If a computer could consider 1 billion orders a second, how many years would it take to solve the computer configuration problem of Example 2.10 by enumeration if n is 25?

2. If a computer could consider 100 billion orders a second instead of just 1 billion, how many years would it take to solve the traveling salesman problem by enumeration if $n = 26$? (Does the improvement in computer speed make a serious difference in conclusions based on the footnote on page 22?)

3. Consider the problem of scheduling n people in order for job interviews in n consecutive time slots. Each person indicates which time slot is his or her first choice, and we seek to schedule the interviews so that the number of people receiving their first choice is as large as possible. Suppose that we solve this problem by enumerating all possible schedules, and for each we compute the number of people receiving their first choice. What is the computational complexity of this procedure? (Make an assumption about the number of steps required to compute the number of people receiving their first choice.)

4. Suppose that we have n different products to manufacture in a factory over the period of a month, and there is a certain cost of converting the machinery setup from manufacturing product i to manufacturing product j. We wish to schedule the order of manufacturing the different products in such a way as to minimize conversion costs. Discuss the computational complexity of the most naive algorithm for solving this problem.

5. Solve the traveling salesman problem by enumeration if $n = 4$ and the cost c_{ij} is given in the following matrix:

$$(c_{ij}) = \begin{array}{c} \\ 1 \\ 2 \\ 3 \\ 4 \end{array} \begin{array}{cccc} 1 & 2 & 3 & 4 \\ \left(\begin{array}{cccc} - & 1 & 8 & 11 \\ 16 & - & 3 & 6 \\ 4 & 9 & - & 11 \\ 8 & 3 & 2 & - \end{array} \right) \end{array}.$$

6. Solve the computer system scheduling problem of Example 2.10 if $n = 3$ and the cost of converting from the ith configuration to the jth is given by

$$(c_{ij}) = \begin{array}{c} \\ 1 \\ 2 \\ 3 \end{array} \begin{array}{ccc} 1 & 2 & 3 \\ \left(\begin{array}{ccc} - & 8 & 11 \\ 12 & - & 4 \\ 3 & 6 & - \end{array} \right) \end{array}.$$

7. Suppose that it takes 3×10^{-9} seconds to examine each key in a list. If there are n keys, and we search through them in order until we find the right one, find
   (a) the worst case complexity;
   (b) the average case complexity.

8. Repeat Exercise 7 if it takes 3×10^{-11} seconds to examine each key.

9. (Hopcroft [1981]) Suppose that L is a collection of bit strings of length n. Suppose that A is an algorithm which determines, given a bit string of length n, whether or not it is in L. Suppose that A always takes 2^n seconds to provide an answer. Then A has the same worst case and average case computational complexity, 2^n. Suppose that $\hat{L}$ consists of all bit strings of the form

$$x_1 x_2 \cdots x_n x_1 x_2 \cdots x_n,$$

where $x_1 x_2 \cdots x_n$ is in L. For instance, if $L = \{00, 10\}$, then $\hat{L} = \{0000, 1010\}$. Consider the following algorithm B for determining, given a bit string $y = y_1 y_2 \cdots y_{2n}$ of length $2n$, whether or not it is in $\hat{L}$. First, determine if y is of the form $x_1 x_2 \cdots x_n x_1 x_2 \cdots x_n$. This is easy to check.

Assume for the sake of discussion that it takes essentially 0 seconds to answer this question. If y is not of the proper form, stop and say that y is not in $\hat{L}$. If y is, check if the first n digits of y form a bit string in L.

(a) Compute the worst case complexity of algorithm B.

(b) Compute the average case complexity of algorithm B.

(c) Do your answers suggest that average case complexity might not be a good measure? Why?

2.5 r-PERMUTATIONS

Given an n-set, suppose that we want to pick out r elements and arrange them in order. Such an arrangement is called an *r-permutation of the n-set*. $P(n, r)$ will count the number of r-permutations of an n-set. For example, the number of 3-letter words without repeated letters can be calculated by observing that we want to choose 3 different letters out of 26 and arrange them in order; hence we want $P(26, 3)$. Similarly, if a student has 4 experiments to perform and 10 periods in which to perform them (each experiment taking one period to complete), the number of different schedules he can make for himself is $P(10, 4)$. Note that $P(n, r) = 0$ if $n < r$: There are no r-permutations of an n-set in this case. In what follows, it will usually be understood that $n \geq r$.

To see how to calculate $P(n, r)$, let us note that in the case of the 3-letter words, there are 26 choices for the first letter; for each of these there are 25 remaining choices for the second letter; and for each of these there are 24 remaining choices for the third letter. Hence, by the product rule,

$$P(26, 3) = 26 \times 25 \times 24.$$

In the case of the experiment schedules, we have 10 choices for the first experiment, 9 for the second, 8 for the third, and 7 for the fourth, giving us

$$P(10, 4) = 10 \times 9 \times 8 \times 7.$$

By the same reasoning, if $n \geq r,$*

$$P(n, r) = n \times (n - 1) \times (n - 2) \times \cdots \times (n - r + 1).$$

If $n > r$, this can be simplified as follows:

$$P(n, r) = \frac{[n \times (n - 1) \times \cdots \times (n - r + 1)] \times [(n - r) \times (n - r - 1) \times \cdots \times 1)]}{(n - r) \times (n - r - 1) \times \cdots \times 1}.$$

Hence, we obtain the result

$$P(n, r) = \frac{n!}{(n - r)!}. \qquad (2.1)$$

We have derived (2.1) under the assumption $n > r$. It clearly holds for $n = r$ as well. (Why?)

*This formula even holds if $n < r$. Why?

EXERCISES FOR SECTION 2.5

1. Find
   (a) $P(3, 2)$; (b) $P(5, 3)$; (c) $P(8, 5)$; (d) $P(1, 3)$.
2. Let $A = \{0, 1, 2, 3, 4, 5, 6\}$.
   (a) Find the number of sequences of length 3 using elements of A.
   (b) Repeat part (a) if no element of A is to be used twice.
   (c) Repeat part (a) if the first element of the sequence is 4.
   (d) Repeat part (a) if the first element of the sequence is 4 and no element of A is used twice.
3. Let $A = \{a, b, c, d, e, f, g\}$.
   (a) Find the number of sequences of length 4 using elements of A.
   (b) Repeat part (a) if no letter is repeated.
   (c) Repeat part (a) if the first letter in the sequence is b.
   (d) Repeat part (a) if the first letter is b and the last is d and no letters are repeated.
4. In how many different orders can five cards be drawn (one at a time without replacement) from a deck of 52 cards?
5. If a campus telephone extension has four digits, how many different extensions are there with no repeated digits
   (a) if the first digit cannot be 0?
   (b) if the first digit cannot be 0 and the second cannot be 1?

2.6 SUBSETS

Example 2.12   The Pizza Problem

A pizza shop advertises that it offers over 500 varieties of pizza. The local consumer protection bureau is suspicious. At the pizza shop, it is possible to have on a pizza a choice of any combination of the following ingredients:

> pepperoni, mushrooms, peppers, sardines, sausage,
> anchovies, salami, onions, bacon.

Is the pizza shop telling the truth in its advertisements? We shall be able to answer this question with some simple applications of the product rule.

To answer the question raised in Example 2.12, let us consider the set $\{a, b, c\}$. Let us ask how many subsets there are of this set. The answer can be obtained by enumeration, and we find there are 8 such subsets. They are the following:

$$\varnothing, \{a\}, \{b\}, \{c\}, \{a, b\}, \{a, c\}, \{b, c\}, \{a, b, c\}.$$

The answer can also be obtained using the product rule. We think of building up a subset in steps. First, we think of either including element a or not. There are 2 choices. Then we either include element b or not. There are again 2 choices. Finally, we either include element c or not. There are again 2 choices. The total number of ways of building up the subset is, by the product rule,

$$2 \times 2 \times 2 = 2^3 = 8.$$

Similarly, the number of subsets of a 4-set is

$$2 \times 2 \times 2 \times 2 = 2^4 = 16,$$

and the number of subsets of an n-set is

$$\underbrace{2 \times 2 \times \cdots \times 2}_{n \text{ times}} = 2^n.$$

Do these considerations help with the pizza problem? We can think of a particular pizza as a subset of the set of ingredients. Alternatively, we can think, for each ingredient, of either including it or not. Either way, we see that there are $2^9 = 512$ possible pizzas. Thus, the pizza shop has not advertised falsely.

EXERCISES FOR SECTION 2.6

1. Enumerate the 16 subsets of $\{a, b, c, d\}$.
2. A magazine subscription service deals with 35 magazines. A subscriber may order any number of them. The subscription service is trying to computerize its billing procedure and wishes to assign a different computer key (identification number) to two different people unless they subscribe to exactly the same magazines. How much storage is required, that is, how many different code numbers?
3. If A is a set of 10 elements, how many nonempty subsets does A have?
4. If A is a set of 8 elements, how many subsets of more than one element does A have?
5. A *value function* on a set A assigns 0 or 1 to each subset of A.
   (a) If A has 3 elements, how many different value functions are there on A?
   (b) What if A has n elements?
6. In a simple game (see Section 2.16), every subset of players is identified as either winning or losing.
   (a) If there is no restriction on this identification, how many distinct simple games are there with 3 players?
   (b) With n players?

2.7 COMBINATIONS

An *r-combination of an n-set* is a selection of r elements from the set. Order does not count. Thus, an r-combination is an r-element subset. $C(n, r)$ will denote the number of r-combinations of an n-set. For example, the number of ways to choose a committee of 3 from a set of 4 people is given by $C(4, 3)$. If the 4 people are Smith, Jones, Brown, and White, the possible committees are

$$\{\text{Smith, Jones, Brown}\}$$

$$\{\text{Smith, Jones, White}\}$$

$$\{\text{Smith, Brown, White}\}$$

$$\{\text{Jones, Brown, White}\}.$$

Hence, $C(4, 3) = 4$. We shall prove some simple theorems about $C(n, r)$. Note that $C(n, r)$ is 0 if $n < r$: There are no r-combinations of an n-set in this case. Henceforth, $n \geq r$ will usually be understood. It is assumed in all the theorems in this section.

Theorem 2.1. $P(n, r) = C(n, r) \times P(r, r)$.

Proof. An ordered arrangement of r objects out of n can be obtained by first choosing r objects [this can be done in $C(n, r)$ ways] and then ordering them [this can be done in $P(r, r) = r!$ ways]. The theorem follows by the product rule. Q.E.D.

Corollary 2.1.1. $$C(n, r) = \frac{n!}{r!(n-r)!}. \tag{2.2}$$

Proof. $$C(n, r) = \frac{P(n, r)}{P(r, r)} = \frac{n!/(n-r)!}{r!} = \frac{n!}{r!(n-r)!}.$$ Q.E.D.

Corollary 2.2.2. $$C(n, r) = C(n, n-r).$$

Proof. $$\frac{n!}{r!(n-r)!} = \frac{n!}{(n-r)!r!} = \frac{n!}{(n-r)![n-(n-r)]!}.$$ Q.E.D.

Note: The number

$$\frac{n!}{r!(n-r)!}$$

is often denoted by

$$\binom{n}{r}$$

and called a *binomial coefficient*. This is because, as we shall see below, this number arises in the binomial expansion. Corollary 2.2.2 states the result that

$$\binom{n}{r} = \binom{n}{n-r}.$$

In what follows we use $C(n, r)$ and $\binom{n}{r}$ interchangeably.

Theorem 2.2. $C(n, r) = C(n-1, r-1) + C(n-1, r)$.

Proof. Mark one of the n objects with a $*$. The r objects can be selected either to include the object $*$ or not to include it. There are $C(n-1, r-1)$ ways to do the former, since this is equivalent to choosing $r-1$ objects out of the $n-1$ non-$*$ objects. There are $C(n-1, r)$ ways to do the latter, since this is equivalent to choosing r objects out of the $n-1$ non-$*$ objects. Hence, the sum rule yields the theorem. Q.E.D.

Note: This proof can be described as a "combinatorial" proof. This theorem can also be proved by algebraic manipulation, using the formula (2.2). Here is such an "algebraic" proof.

Second Proof of Theorem 2.2.

$$C(n-1, r-1) + C(n-1, r) = \frac{(n-1)!}{(r-1)![(n-1)-(r-1)]!} + \frac{(n-1)!}{r![(n-1)-r]!}$$

$$= \frac{(n-1)!}{(r-1)!(n-r)!} + \frac{(n-1)!}{r!(n-r-1)!}$$

$$= \frac{r(n-1)!}{r!(n-r)!} + \frac{(n-r)(n-1)!}{r!(n-r)!}$$

$$= \frac{r(n-1)! + (n-r)(n-1)!}{r!(n-r)!}$$

$$= \frac{(n-1)![r+n-r]}{r!(n-r)!}$$

$$= \frac{n!}{r!(n-r)!}$$

$$= C(n, r). \qquad\qquad \text{Q.E.D.}$$

Let us give some quick applications of our new formulas and our basic rules so far.

1. In the pizza problem (Example 2.12), the number of pizzas with exactly 3 different ingredients is

$$C(9, 3) = \frac{9!}{3!6!} = 84.$$

2. The number of pizzas with at most 3 different ingredients is, by the sum rule,

$$C(9, 0) + C(9, 1) + C(9, 2) + C(9, 3).$$

3. The number of ways to choose 2 courses to take out of 6 is

$$C(6, 2) = \frac{6!}{2!4!} = 15.$$

4. In basketball, the number of ways of choosing 5 starters (independent of position) from a 7-member team is

$$C(7, 5) = \frac{7!}{5!2!} = 21.$$

5. The number of 5-member committees from a group of 9 people is

$$C(9, 5) = 126.$$

6. The number of 7-member committees from the U.S. Senate is

$$C(100, 7).$$

7. The number of delegations to the President consisting of 2 senators and 2 repre-

sentatives is

$$C(100, 2) \times C(435, 2).$$

8. The number of 9-digit bit strings with 5 1's and 4 0's is $C(9, 5)$. To see why, think of having 9 unknown digits and choosing 5 of them to be 1's.

EXERCISES FOR SECTION 2.7

1. How many ways are there to choose 3 books to buy from a group of 5?

2. How many ways can 7 award-winners be chosen from a group of 50 nominees?

3. Compute
   (a) $C(6, 3)$;   **(b)** $C(7, 4)$;   **(c)** $C(5, 1)$;   **(d)** $C(2, 4)$.

4. Find $C(n, 1)$.

5. Compute $C(5, 2)$ and check your answer by enumeration.

6. Compute $C(6, 2)$ and check your answer by enumeration.

7. Check by computation that
   (a) $C(7, 2) = C(7, 5)$;
   (b) $C(6, 4) = C(6, 2)$.

8. **(a)** In how many ways can 8 blood samples be divided into 2 groups to be sent to different laboratories for testing if there are 4 samples in each group? (Assume that the laboratories are distinguishable.)
   (b) In how many ways can the 8 samples be divided into 2 groups if there is at least 1 item in each group? (Assume that the laboratories are distinguishable.)

9. There are 6 candidates for promotion and an office manager has a budget for at most 3 promotions. In how many ways can the office manager choose the candidates to be promoted?

10. **(a)** In how many ways can 10 food items be divided into 2 groups to be sent to different laboratories for purity testing if there are 5 items in each group?
    (b) In how many ways can the 10 items be divided into 2 groups if there is at least 1 item in each group?

11. How many 8-letter words with no repeated letters can be constructed using the 26 letters of the alphabet if each word contains 3, 4, or 5 vowels?

12. How many odd numbers between 1000 and 9999 have distinct digits?

13. A committee is to be chosen from a set of 7 women and 4 men. How many ways are there to form the committee if
    (a) the committee has 5 people, 3 women and 2 men?
    (b) the committee can be any size (except empty), but it must have equal numbers of women and men?
    (c) the committee has 4 people and 1 of them must be Mr. Smith?
    (d) the committee has 4 people, 2 of each sex, and Mr. and Mrs. Smith cannot both be on the committee?

14. **(a)** A computer center has 9 different programs to run. Four of them use the language ALGOL and 5 use the language BASIC. The ALGOL programs are considered indistinguishable and so are the BASIC programs. Find the number of possible orders for running the programs if
    (i)   there are no restrictions;
    (ii) the ALGOL programs must be run consecutively;
    (iii) the ALGOL programs must be run consecutively and the BASIC programs must be run consecutively;
    (iv) the languages must alternate.

(b) Suppose that the cost of switching from an ALGOL configuration to a BASIC configuration is 10 units, the cost of switching from a BASIC configuration to an ALGOL configuration is 5 units, and there is no cost to switch from ALGOL to ALGOL or BASIC to BASIC. What is the most efficient (least cost) ordering in which to run the programs?

(c) Repeat part (a) if the ALGOL programs are all distinguishable from each other and so are the BASIC programs.

15. A certain company has 30 female employees, including 3 in the management ranks, and 150 male employees, including 12 in the management ranks. A committee consisting of 3 women and 3 men is to be chosen. How many ways are there to choose it if
   (a) there is at least 1 person of management rank of each sex?
   (b) there is at least 1 person of management rank?

16. Show that

$$\binom{n}{m}\binom{m}{k} = \binom{n}{k}\binom{n-k}{m-k}.$$

17. Prove the following identity (using a combinatorial proof if possible). The identity is called Vandermonde's identity.

$$\binom{n+m}{r} = \binom{n}{0}\binom{m}{r} + \binom{n}{1}\binom{m}{r-1} + \binom{n}{2}\binom{m}{r-2} + \cdots + \binom{n}{r}\binom{m}{0}.$$

2.8 PASCAL'S TRIANGLE

A convenient method of calculating the numbers $C(n, r)$ is to use the array shown in Figure 2.4. The number $C(n, r)$ appears in the nth row, rth diagonal. Each element in a given position is obtained by summing the two elements in the row above it which are just to the left and just to the right. For example, $C(5, 2)$ is given by summing up the numbers 4 and 6, which are circled in Figure 2.4. The array of Figure 2.4 is called Pascal's triangle, after the famous French philosopher and mathematician Blaise Pascal. Pascal was one of the inventors of probability theory and discovered many interesting combinatorial techniques.

Why does Pascal's triangle work? The answer is that it depends on the relation

$$C(n, r) = C(n - 1, r - 1) + C(n - 1, r). \tag{2.3}$$

This is exactly the relation that was proved in Theorem 2.2. The relation (2.3) is an

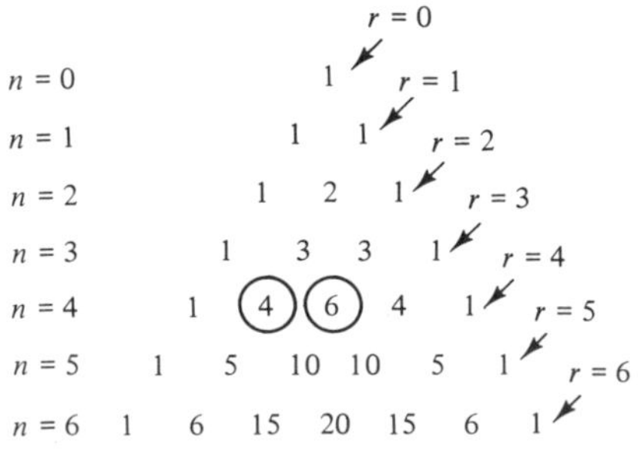

Figure 2.4 Pascal's triangle. The circled numbers are added to give $C(5, 2)$.

example of a recurrence relation. We shall see many such relations later, especially in Chapter 5. Obtaining such relations allows one to reduce calculations of complicated numbers to earlier steps, and therefore allows the computation of these numbers in stages.

EXERCISES FOR SECTION 2.8

1. Extend Figure 2.4 by adding one more row.
2. Compute $C(5, 3)$, $C(4, 2)$, and $C(4, 3)$ and verify that formula (2.3) holds.
3. Repeat Exercise 2 for $C(7, 5)$, $C(6, 4)$, and $C(6, 5)$.
4. How would you find the sum $\binom{n}{0} + \binom{n}{1} + \binom{n}{2} + \cdots + \binom{n}{n}$ from Pascal's triangle? Do so for $n = 2, 3$, and 4. Guess at the answer in general.
5. Show that

$$\binom{n}{0} + \binom{n+1}{1} + \cdots + \binom{n+r}{r} = \binom{n+r+1}{r}.$$

6. Following Cohen [1978], define $\left\langle {}^{n}_{r} \right\rangle$ to be $\binom{n+r-1}{r}$. Show that

$$\left\langle {n \atop r} \right\rangle = \left\langle {n \atop r-1} \right\rangle + \left\langle {n-1 \atop r} \right\rangle.$$

7. If $\left\langle {}^{n}_{r} \right\rangle$ is defined as in Exercise 6, show that

$$\left\langle {n \atop r} \right\rangle = \frac{n}{r} \left\langle {n+1 \atop r-1} \right\rangle = \frac{n+r-1}{r} \left\langle {n \atop r-1} \right\rangle.$$

8. Note that the entries in any row of Pascal's triangle increase for awhile and then decrease. A sequence of numbers $a_0, a_1, a_2, \ldots, a_n$ is called *unimodal* if for some integer t, $a_0 \le a_1 \le \cdots \le a_t$, and $a_t \ge a_{t+1} \ge \cdots \ge a_n$.
   (a) Show that if $a_0, a_1, a_2, \ldots, a_n$ is unimodal, t is not necessarily unique.
   (b) Show that if $n > 0$, the sequence $\binom{n}{0}, \binom{n}{1}, \binom{n}{2}, \ldots, \binom{n}{n}$ is unimodal.
   (c) Show that the largest entry in the sequence in part (b) is $\binom{n}{\lfloor n/2 \rfloor}$, where $\lfloor x \rfloor$ is the greatest integer less than or equal to x.

2.9 PROBABILITY

The history of combinatorics is closely intertwined with the history of the theory of probability. The theory of probability was developed to deal with uncertain events, events that might or might not occur. In particular, this theory was developed by Pascal, Fermat, Laplace, and others in connection with the outcomes of certain gambles. In his *Théorie Analytique des Probabilités*, first published in 1812, Laplace defined probability as follows: The probability of an event is the number of possible outcomes whose occurrence signals the event divided by the total number of possible outcomes. For instance, suppose that we consider choosing a 2-digit bit string at random. There are 4 such strings, 00, 01, 10, and 11. What is the probability that the string chosen has a 0? The answer is $\frac{3}{4}$, because 3 of the possible outcomes signal the event in question, that is, have a 0, and there are 4 possible outcomes in all. This definition of Laplace's is appropriate only if all the possible outcomes are equally likely, as we shall quickly observe.

Let us make things a little more precise. We shall try to formalize the notion of probability by thinking of an experiment that produces one of a number of possible outcomes. The set of possible outcomes is called the *sample space*. An *event* corresponds to a subset of the set of outcomes, that is, of the sample space; it corresponds to those outcomes that signal that the event has taken place. Laplace's definition says that if E is an event in the sample space S, then

$$\text{probability of } E = \frac{n(E)}{n(S)},$$

where $n(E)$ is the number of outcomes in E and $n(S)$ is the number of outcomes in S. Note that it follows that the probability of E is a number between 0 and 1.

Let us apply this definition to a gambling situation. We toss a die—this is the experiment. We wish to compute the probability that the outcome will be an even number. The sample space is the set of possible outcomes, $\{1, 2, 3, 4, 5, 6\}$. The event in question is the set of all outcomes which are even, that is, the set $\{2, 4, 6\}$. Then we have

$$\text{probability of even} = \frac{n\{2, 4, 6\}}{n\{1, 2, 3, 4, 5, 6\}} = \frac{3}{6} = \frac{1}{2}.$$

Notice that this result would not hold unless all the outcomes in the sample space were equally likely. If we have a weighted die that always comes up 1, then the probability of getting an even number is not $\frac{1}{2}$.

Let us consider a family with two children. What is the probability that the family will have at least one boy? There are three possibilities for such a family: It can have two boys, two girls, or a boy and a girl. Let us take the set of these three possibilities as our sample space. The first and third outcomes make up the event "having at least one boy," and hence, by Laplace's definition,

$$\text{probability of having at least one boy} = \frac{2}{3}.$$

Is this really correct? It is not. If we look at families with two children, more than $\frac{2}{3}$ of them have at least one boy. That is because there are four ways to build up a family of two children: we can have first a boy and then another boy, first a girl and then another girl, first a boy and then a girl, or first a girl and then a boy. Thus, there are more ways to have a boy and a girl than there are ways to have two boys, and the outcomes in our sample space were not equally likely. However, the outcomes BB, GG, BG, and GB, to use an obvious abbreviation, are equally likely, so we can take them as our sample space. Now the event "having at least one boy" has 3 outcomes in it out of 4, and we have

$$\text{probability of having at least one boy} = \frac{3}{4}.$$

Even this conclusion is not quite accurate, because it is slightly more likely to have a boy than a girl; so the four events we have chosen are not exactly equally likely. In particular, BB is more likely than GG. However, the conclusion we have reached is a good working approximation.

We shall limit computations of probability in this book to situations where the outcomes in the sample space are equally likely. Note that our definition of probability applies only to the case where the sample space is finite. In the infinite case, the Laplace definition obviously has to be modified. For a discussion of the not-equally-likely case and the infinite case, the reader is referred to almost any textbook on probability theory, for instance Feller [1968] or Parzen 1960].

Let us continue by giving several more applications of our definition. Suppose that a family is known to have 4 children. What is the probability that half of them are boys? The answer is not $\frac{1}{2}$. To obtain the answer we observe that the sample space is all sequences of B's and G's of length 4; a typical such sequence is $BGGB$. How many such sequences have exactly 2 B's? There are 4 positions, and 2 of these must be chosen for B's. Hence, there are $C(4, 2)$ such sequences. How many sequences are there in all? By the product rule, there are 2^4. Hence,

$$\text{probability that half are boys} = \frac{C(4, 2)}{2^4} = \frac{6}{16} = \frac{3}{8}.$$

The reader might wish to write out all 16 possible outcomes and note the 6 that signal the event having exactly 2 boys.

Next, suppose that a fair coin is tossed 5 times. What is the probability that there will be at least 2 heads? The sample space consists of all possible sequences of heads and tails of length 5, that is, it consists of sequences such as HHHTH, to use an obvious abbreviation. How many such sequences have at least 2 heads? The answer is that $C(5, 2)$ sequences have exactly 2 heads, $C(5, 3)$ have exactly 3 heads, and so on. Thus, the number of sequences having at least 2 heads is given by

$$C(5, 2) + C(5, 3) + C(5, 4) + C(5, 5) = 26.$$

The total number of possible sequences is $2^5 = 32$. Hence,

$$\text{probability of having at least two heads} = \frac{26}{32} = \frac{13}{16}.$$

Example 2.13   Reliability of Systems

Imagine that a system has n components, each of which can work or fail to work. Let x_i be 1 if the ith component works and 0 if it fails. Let the bit string $x_1 x_2 \ldots x_n$ describe the system. Thus, the bit string 0011 describes a system with four components, with the first two failing and the third and fourth working. Since many systems have built-in redundancy, the system as a whole can work even if some components fail. Let $F(x_1 x_2 \ldots x_n)$ be 1 if the system described by $x_1 x_2 \ldots x_n$ works and 0 if it fails. Then F is a function from bit strings of length n to $\{0, 1\}$, that is, an n-variable switching function (Example 2.4). For instance, suppose that we have a highly redundant system with three identical components, and the system works if and only if at least two components work. Then F is given by Table 2.4. We shall study other specific examples of functions F in Section 3.2.4 and Exercise 15, Section 13.3. Suppose that components in a system are equally likely to work or not to work.* Then any two bit strings are equally likely to be the bit string $x_1 x_2 \ldots x_n$ describing the system. Now we may ask: What is the probability that the system works, that is, what is the probability that $F(x_1 x_2 \ldots x_n) = 1$? This is a measure of

---

*In a more general analysis, we would first estimate the probability p_i that the ith component works.

Table 2.4. The Switching Function F which is 1 if and only if Two or Three Components of a System Work

$x_1 x_2 x_3$	$F(x_1 x_2 x_3)$
111	1
110	1
101	1
100	0
011	1
010	0
001	0
000	0

the *reliability* of the system. In our example, 4 of the 8 bit strings, 111, 110, 101, and 011, signal the event that $F(x_1 x_2 x_3) = 1$. Since all bit strings are equally likely, the probability that the system works is $\frac{4}{8} = \frac{1}{2}$. For more on this approach to system reliability, see Karp and Luby [1983] and Barlow and Proschan [1975].

To close this section, we observe that some common statements about probabilities of events correspond to operations on the associated subsets. Thus, we have

probability that E does not occur is probability of E^c,

probability that E or F occurs is probability of $E \cup F$,

probability that E and F occur is probability of $E \cap F$.

It is also easy to see from the definition of probability that

$$\text{probability of } E^c = 1 - \text{probability of } E, \tag{2.4}$$

if E and F are disjoint,

$$\text{probability of } E \cup F = \text{probability of } E + \text{probability of } F, \tag{2.5}$$

and, in general,

$$\text{probability of } E \cup F = \text{probability of } E + \text{probability of } F \\ - \text{probability of } E \cap F. \tag{2.6}$$

To illustrate these observations, let us consider the die-tossing experiment. Then the probability of not getting a 3 is 1 minus the probability of getting a 3, that is, it is $1 - \frac{1}{6} = \frac{5}{6}$. What is the probability of getting a 3 or an even number? Since $E = \{3\}$ and $F = \{2, 4, 6\}$ are disjoint, (2.5) implies it is probability of E plus probability of $F = \frac{1}{6} + \frac{3}{6} = \frac{2}{3}$. Finally, what is the probability of getting a number larger than 4 or an even number? The event in question is the set $\{2, 4, 5, 6\}$, which has probability $\frac{4}{6} = \frac{2}{3}$. Note that this is not the same as the probability of a number larger than 4 plus the probability of an even number $= \frac{2}{6} + \frac{3}{6} = \frac{5}{6}$. This is because $E = \{5, 6\}$ and $F = \{2, 4, 6\}$ are not disjoint, for $E \cap F = \{6\}$. Applying (2.6), we have probability of $E \cup F = \frac{2}{6} + \frac{3}{6} - \frac{1}{6} = \frac{2}{3}$, which agrees with our first computation.

EXERCISES FOR SECTION 2.9

1. Are the outcomes in the following experiments equally likely?
   (a) A citizen of the United States is chosen at random and his or her state of residence is recorded.
   (b) Two red balls and three black balls are placed in a box, and one ball is chosen at random and its color is noted.
   (c) Two fair dice are tossed and the sum of the numbers appearing is recorded.
   (d) A bit string of length 3 is chosen at random and the sum of its digits is observed.

2. Calculate the probability that when a die is tossed, the outcome will be
   (a) an odd number;
   (b) a number less than or equal to 2;
   (c) a number divisible by 3.

3. Calculate the probability that a family of 3 children has
   (a) exactly 2 boys;
   (b) at least 2 boys;
   (c) at least 1 boy and at least 1 girl.

4. Calculate the probability that in 4 tosses of a fair coin, there are at most 3 heads.

5. Calculate the probability that if a DNA chain of length 5 is chosen at random, it will have at least four A's.

6. If a card is drawn at random from a deck of 52, what is the probability that it is an ace?

7. Suppose that a card is drawn at random from a deck of 52, the card is replaced, and then another card is drawn at random. What is the probability of getting two aces?

8. If a bit string of length 4 is chosen at random, what is the probability of having at least three 1's?

9. What is the probability that a bit string of length 3, chosen at random, does not have two consecutive 0's?

10. Suppose that a system has four independent components, each of which is equally likely to work or not to work. Suppose that the system works if and only if at least three components work. What is the probability that the system works?

11. Repeat Exercise 10 if the system works if and only if the fourth component works and at least two of the other components work.

12. A shop can operate only if at least one sorter is present and at least one packager. There are three sorters and two packagers, and each worker is equally likely to show up for work on a given day or to stay home. Assuming that each worker decides independently whether or not to come to work, what is the probability that the shop can operate?

13. Suppose that we have 10 different pairs of shoes. From the 20 shoes, 4 are chosen at random. What is the probability of getting at least one pair?

14. Use rules (2.4)–(2.6) to calculate the probability of getting, in six tosses of a fair coin,
   (a) two heads or three heads;
   (b) two heads or two tails;
   (c) two heads or a head on the first toss;
   (d) an even number of heads or at least nine heads;
   (e) an even number of heads and a head on the first toss.

15. Use the definition of probability to verify rules
   (a) (2.4);   **(b)** (2.5);   **(c)** (2.6).

2.10 SAMPLING WITH REPLACEMENT

A numbers lottery is sometimes held by choosing a 5-digit number, no zeros allowed. How many such numbers are there? By the product rule, the answer is 9^5. The lottery often takes place in the following way. We put balls having the digits 1 through 9 in a container and pick one out for the first digit. We then replace the ball chosen and pick again for the second digit. And so on. We say that we are *sampling with replacement*. We are choosing a 5-permutation out of a 9-set, but with replacement. Equivalently, we are allowing repetition. Let $P^R(m, r)$ be the number of r-permutations of an m-set, replacement or repetition allowed. Then the product rule gives us

$$P^R(m, r) = m^r. \tag{2.7}$$

The number $P(m, r)$ counts the number of r-permutations of an m-set if we are sampling without replacement or repetition.

   We can make a similar distinction in the case of r-combinations. Let $C^R(m, r)$ be the number of r-combinations of an m-set if we sample with replacement or repetition. For instance, the 4-combinations of a 2-set $\{a, b\}$ if replacement is allowed are given by

$$\{a, a, a, a\}, \quad \{a, a, a, b\}, \quad \{a, a, b, b\}, \quad \{a, b, b, b\} \quad \{b, b, b, b\}.$$

Thus, $C^R(2, 4) = 5$. We now state a formula for $C^R(m, r)$.

   Theorem 2.3. $C^R(m, r) = C(m + r - 1, r)$.

We shall prove Theorem 2.3 at the end of this section. Here, let us illustrate it with some examples.

Example 2.14   The Candy Store

Suppose that there are three kinds of fillings for candies at a candy store: cherry (c), orange (o), and vanilla (v). As a sales gimmick, the store allows a customer to design a candy box by choosing a dozen candies, each with one filling. How many different candy boxes are there? We can think of having a 3-set, $\{c, o, v\}$, and picking a 12-combination from it, with replacement. Thus, the number of candy boxes is

$$C^R(3, 12) = C(3 + 12 - 1, 12) = C(14, 12).$$

Example 2.15   DNA Chains: Gamow's Encoding

In Section 2.1 we studied DNA chains, strings on the alphabet $\{T, C, A, G\}$, and the minimum length required for such a chain to encode for an amino acid. We noted that there are 20 different amino acids, and showed in Section 2.1 that there are only 16 different strings of length 2, so a string of length at least 3 is required. But there are $4^3 = 64$ different strings of length 3. Gamow [1954a, b] suggested that two 3-element DNA chains encode the same amino acid if and only if they have the same bases independent of order. For example, GGA and GAG and AGG would all

Table 2.5 Choosing a Sample of r Elements from a Set of m Elements

Order counts?	Repetitions allowed?	The sample is called	Number of ways to choose the sample	Reference
No	No	r-combination	$C(m, r) = \dfrac{m!}{r!(m-r)!}$	Corollary 2.1.1
Yes	No	r-permutation	$P(m, r) = \dfrac{m!}{(m-r)!}$	Eq. (2.1)
No	Yes	r-combination with replacement	$C^R(m, r) = C(m + r - 1, r)$	Theorem 2.3
Yes	Yes	r-permutation with replacement	$P^R(m, r) = m^r$	Eq. (2.7)

encode the same amino acid. If Gamow's suggestion were right, how many amino acids could be encoded using only DNA chains of length 3? This is an application of Theorem 2.3. We have $m = 4$ objects and we wish to choose $r = 3$ objects, with replacement. This can be done in

$$C(m + r - 1, r) = C(6, 3) = 20$$

different ways. Thus, it would be possible to encode exactly the 20 different amino acids using a 3-element DNA chain in which only the number of bases of each kind mattered. Unfortunately, it was later discovered that this is not the way the coding works. See Golomb [1962] for a discussion.

Our discussion of sampling with and without replacement is summarized in Table 2.5.

Proof of Theorem 2.3. * Suppose that the m-set has elements $a_1, a_2, \ldots, a_m$. Then any sample of r of these objects can be described by listing how many a_1 are in it, how many a_2, and so on. For instance, if $r = 7$ and $m = 5$, typical samples are $a_1 a_1 a_2 a_3 a_4 a_4 a_5$ and $a_1 a_1 a_1 a_2 a_4 a_5 a_5$. We can also represent these samples by putting a vertical line after the last a_i, for $i = 1, 2, \ldots, m - 1$. Thus, these two samples would be written as $a_1 a_1 | a_2 | a_3 | a_4 a_4 | a_5$ and $a_1 a_1 a_1 | a_2 \| a_4 | a_5 a_5$, where in the second case we have two consecutive vertical lines since there is no a_3. Now if we use this notation to describe a sample of r objects, we can omit the subscripts. For instance, $aa | aa \,\|\, aaa$ represents $a_1 a_1 | a_2 a_2 \,\|\, a_5 a_5 a_5$. Then the number of samples of r objects is just the number of different arrangements of r letters a and $m - 1$ vertical lines. Such an arrangement has $m + r - 1$ elements in it, and we determine it by choosing r positions for the a's. Hence, there are $C(m + r - 1, r)$ such arrangements. Q.E.D.

EXERCISES FOR SECTION 2.10

1. If replacement is allowed, find all
   (a) 4-permutations of a 2-set;

*The proof can be omitted.

    (b) 2-permutations of a 3-set;

    (c) 4-combinations of a 2-set;

    (d) 2-combinations of a 3-set.

2. Check your answers in Exercise 1 by using Equation (2.7) or Theorem 2.3.

3. If replacement is allowed, compute the number of

    (a) 8-permutations of a 3-set;

    (b) 6-combinations of a 4-set.

4. In how many ways can we choose 8 bottles of soda if there are 4 brands available?

5. In how many different ways can we choose 12 cans of soup if there are 5 different varieties available?

6. Suppose that a codeword of length 8 consists of letters A, B, C, D, or E and cannot start with A. How many such codewords are there?

7. How many DNA chains of length 6 have at least one of each base T, C, A, and G? Answer this question under the following assumptions:

    (a) Only the number of bases of a given kind matter.

    (b) Only the number of bases of a given kind and which base comes first matter.

2.11 OCCUPANCY PROBLEMS*

2.11.1 The Types of Occupancy Problems

In the history of combinatorics and probability theory, problems of placing *balls* into *cells* or *urns* have played an important role. Such problems are called *occupancy problems*. Occupancy problems have numerous applications. In classifying types of accidents according to the day of the week in which they occur, the balls are the types of accidents and the cells are the days of the week. In cosmic-ray experiments, the balls are the particles reaching a Geiger counter and the cells are the counters. In computer science, the possible distributions of keypunching errors on k cards are obtained by studying the cards as cells and the balls as errors. A similar application occurs in book publishing, with the distribution of misprints on k pages. In the study of irradiation in biology, the light particles hitting the retina correspond to balls, the cells of the retina to the cells. In coupon collecting, the balls correspond to particular coupons, the cells to the types of coupons. We shall return in various places to these applications. See Feller [1968, pp. 10–11] for other applications.

    In occupancy problems, it makes a big difference whether or not we regard two balls as distinguishable and whether or not we regard two cells as distinguishable. For instance, suppose that we have two distinguishable balls, *a*, and *b*, and three distinguishable cells, 1, 2, and 3. Then the possible distributions of balls to cells are shown in Table 2.6. There are nine distinct distributions. However, suppose that we have two indistinguishable balls. We can label them both *a*. Then the possible distributions to three distinguishable cells are shown in Table 2.7. There are just six of them. Similarly, if the cells are not distinguishable but the balls are, distributions 1, 2, and 3 of Table 2.6 are considered the same: two balls in one cell, none in the others. Similarly, distributions 4–9 are considered the same: two cells with one ball, one cell with no balls. There are then just two distinct distri-

---

*For a quick reading of this section, it suffices just to read Section 2.11.1.

Table 2.6. The Distributions of Two Distinguishable Balls to Three Distinguishable Cells

		Distribution								
		1	2	3	4	5	6	7	8	9
Cell	1	ab			a	a		b	b	
	2		ab		b		a	a		b
	3			ab		b	b		a	a

butions. Finally, if neither the balls nor the cells are distinguishable, then distributions 1–3 of Table 2.7 are considered the same and distributions 4–6 are as well, so there are two distinct distributions.

It is also common to distinguish between occupancy problems where the cells are allowed to be empty and those where they are not. For instance, if we have two distinguishable balls and two distinguishable cells, then the possible distributions are given by Table 2.8. There are four of them. However, if no cell can be empty, there are only two, distributions 3 and 4 of Table 2.8.

Table 2.7. The Distributions of Two Indistinguishable Balls to Three Distinguishable Cells

		Distribution					
		1	2	3	4	5	6
Cell	1	aa			a	a	
	2		aa		a		a
	3			aa		a	a

The possible cases of occupancy problems are summarized in Table 2.9. The notation and terminology in the fourth column which has not yet been defined will be defined below. We shall now discuss the different cases.

Case 1a is covered by the product rule: There are k choices of cell for each ball. If $k = 3$ and $n = 2$, we get $k^n = 9$, which is the number of distributions shown in Table 2.6. Case 1b will be discussed in Section 2.11.3.

Table 2.8 The Distributions of Two Distinguishable Balls to Two Distinguishable Cells

		Distribution			
		1	2	3	4
Cell	1	ab		a	b
	2		ab	b	a

Table 2.9 Classification of Occupancy Problems

	Distinguished balls?	Distinguished cells?	Can cells be empty?	Number of ways to place n balls into k cells
Case 1				
1a	Yes	Yes	Yes	k^n
1b	Yes	Yes	No	$k!S(n, k)$
Case 2				
2a	No	Yes	Yes	$C(k + n - 1, n)$
2b	No	Yes	No	$C(n - 1, k - 1)$
Case 3				
3a	Yes	No	Yes	$S(n, 1) + S(n, 2) + \cdots + S(n, k)$
3b	Yes	No	No	$S(n, k)$
Case 4				
4a	No	No	Yes	Number of partitions of n into k or fewer parts
4b	No	No	No	Number of partitions of n into exactly k parts

2.11.2 Case 2: Indistinguishable Balls and Distinguishable Cells*

Case 2a follows from Theorem 2.3, for we have the following result.

Theorem 2.4. The number of ways to distribute n indistinguishable balls into k distinguishable cells is $C(k + n - 1, n)$.

Proof. Suppose that the cells are labeled $C_1, C_2, \ldots, C_k$. A distribution of balls into cells can be summarized by listing for each ball the cell into which it goes. Then, a distribution corresponds to a collection of n cells with repetition allowed. For instance, in Table 2.7, distribution 1 corresponds to the collection $\{C_1, C_1\}$ and distribution 5 to the collection $\{C_1, C_3\}$. If there are four balls, the collection $\{C_1, C_2, C_3, C_3\}$ corresponds to the distribution that puts one ball into cell C_1, one ball into cell C_2, and two balls into cell C_3. Because a distribution corresponds to a collection $\{C_{i_1}, C_{i_2}, \ldots, C_{i_n}\}$, the number of ways to distribute the balls into cells is the same as the number of n-combinations of the k-set $\{C_1, C_2, \ldots, C_k\}$ in which repetition is allowed. This is given by Theorem 2.3 to be $C(k + n - 1, n)$. Q.E.D.

Theorem 2.4 is illustrated by Table 2.7. We have $k = 3$, $n = 2$, and $C(k + n - 1, n) = C(4, 2) = 6$.

The result in case 2b now follows from the result in case 2a. Given n indistinguishable balls and k distinguishable cells, we first place one ball in each cell. There is one way to do this. It leaves $n - k$ indistinguishable balls. We wish to place these into k distinguishable cells, with no restriction as to cells being nonempty. By Theorem 2.4 this can be done in

$$C(k + (n - k) - 1, n - k) = C(n - 1, k - 1)$$

---

*The rest of Section 2.11 can be omitted.

ways. We now use the product rule to derive the result for case 2b of Table 2.9. Note that $C(n-1, k-1)$ is 0 if $n < k$. There is no way to assign n balls to k cells with at least one ball in each cell.

2.11.3 Case 3: Distinguishable Balls and Indistinguishable Cells

Let us turn next to case 3b. Let $S(n, k)$ be defined to be the number of ways to distribute n distinguishable balls into k indistinguishable cells with no cell empty. The number $S(n, k)$ is called a *Stirling number of the second kind*. In Section 4.5.3 we show that

$$S(n, k) = \frac{1}{k!} \sum_{i=0}^{k} (-1)^i \binom{k}{i} (k - i)^n. \tag{2.8}$$

To illustrate this result, let us consider the case $n = 2$, $k = 2$. Then

$$S(n, k) = S(2, 2) = \frac{1}{2}[2^2 - 2 \cdot 1^2 + 0] = 1.$$

There is only one distribution of two distinguishable balls a and b to two indistinguishable cells such that each cell has at least one ball: one ball in each cell.

The result in case 3a now follows from the result in case 3b by the sum rule. For to distribute n distinguishable balls into k indistinguishable cells with no cell empty, either one cell is not empty or two cells are not empty or … . The result in case 1b now follows also, since putting n distinguishable balls into k distinguishable cells with no cells empty can be accomplished by putting n distinguishable balls into k indistinguishable cells with no cells empty [which can be done in $S(n, k)$ ways] and then labeling the cells [which can be done in $k!$ ways]. For instance, if $k = n = 2$, then by our previous computation, $S(2, 2) = 1$. Thus, the number of ways to put two distinguishable balls into two distinguishable cells with no cells empty is $2!S(2, 2) = 2$. This is the observation we made earlier from Table 2.8.

2.11.4 Case 4: Indistinguishable Balls and Indistinguishable Cells

To handle cases 4a and 4b, we define a *partition* of a positive integer n to be a collection of positive integers that sum to n. For instance, the integer 5 has the partitions

$$\{1, 1, 1, 1, 1\}, \quad \{1, 1, 1, 2\}, \quad \{1, 2, 2\}, \quad \{1, 1, 3\}, \quad \{2, 3\}, \quad \{1, 4\}, \quad \{5\}$$

Note that $\{3, 2\}$ is considered the same as $\{2, 3\}$. We are interested only in what integers are in the collection, not in their order. The number of ways to distribute n indistinguishable balls into k indistinguishable cells is clearly the same as the number of ways to partition the integer n into at most k parts. This gives us the result in case 4a of Table 2.9. For instance, if $n = 5$ and $k = 3$, there are five possible partitions, all but the first two listed above. If $n = 2$ and $k = 3$, there are two possible partitions, $\{1, 1\}$ and $\{2\}$. This corresponds in Table 2.7 to the two distinct distributions: Two cells with one ball in each or one cell with two balls in it. The result in case 4b of Table 2.9 follows similarly: The number of ways to distribute n indistinguishable balls into k indistinguishable cells with no cell empty is clearly the same as the number of ways to partition the integer n into exactly k parts. To illustrate this, if $n = 2$ and $k = 3$, there is no way.

We will explore partitions of integers briefly in the exercises and return to them in the exercises of Sections 4.3 and 4.4, where we approach them using the method of generating functions. For a detailed discussion of partitions, see most number theory books, for instance, Hardy and Wright [1960]. See also Berge [1971] or Riordan [1958].

2.11.5 Examples

We now give a number of examples, applying the results of Table 2.9. The reader should notice that whether or not balls or cells are distinguishable is often a matter of judgment, depending on the interpretation and in what we are interested.

Example 2.16   Hospital Deliveries

Suppose that 80 babies are born in the month of September in a hospital and we record the day each baby is born. In how many ways can this event occur? The babies are the balls and the days the cells. If we do not distinguish between 2 babies but do distinguish between days, we are in case 2, $n = 80$, $k = 30$, and the answer is given by $C(109, 80)$. The answer is given by $C(79, 29)$ if we count only the number of ways this can happen with each day having at least 1 baby. If we do not care about what day a particular number of babies is born but only about the number of days in which 2 babies are born, the number in which 3 are born, and so on, we are in case 4 and we need to consider partitions of the integer 80 into 30 or fewer parts.

Example 2.17   Keypunching Errors

In monitoring the work of a keypuncher, suppose that we keep a record of errors. The keypuncher punches 100 cards and makes 30 errors. In how many ways could this happen? The errors are the balls and the cards are the cells. It seems reasonable to disregard the distinction between errors and concentrate on whether more errors occur later in the job (because of weariness or boredom). Then cards are distinguished. Hence, we are in case 2, and the answer is given by $C(129, 30)$.

Example 2.18   Sex Distribution

Suppose that we record the sex of the first 1000 people to get a degree in computer science at a new school. The people correspond to the balls and the two sexes (M and F) are the cells. We certainly distinguish cells. If we distinguish individuals, that is, if we distinguish between individual 1 being male and individual 2 being male, then we are in case 1. However, if we are interested only in the number of people of each sex, we are in case 2. In the former case, the number of possible distributions is 2^{1000}. In the latter case, the number of possible distributions is given by $C(1001, 1000) = 1001$.

Example 2.19   Elevators

An elevator with 8 passengers stops at 6 different floors. The passengers correspond to the balls, the floors to the cells. If we are interested only in the passengers who get off together, we can consider the balls distinguishable and the cells indistinguishable. Then we are in case 3. The number of possible distributions is given by

$$S(8, 1) + S(8, 2) + \cdots + S(8, 6).$$

Example 2.20   Statistical Mechanics

In statistical mechanics, suppose that we have a system of t particles. Suppose that there are p different states or levels, for example energy levels, in which each of the particles can be. The state of the system is described by giving the distribution of particles to levels. In all, if the particles are distinguishable, there are p^t possible distributions. For instance, if we have 4 particles and 3 levels, there are $3^4 = 81$ different arrangements. One of these has particle 1 at level 1, particle 2 at level 3, particle 3 at level 2, and particle 4 at level 3. Another has particle 1 at level 2, particle 2 at level 1, and particles 3 and 4 at level 3. If we consider any distribution of particles to levels to be equally likely, then the probability of any given arrangement is $1/p^t$. In this case we say that the particles obey the *Maxwell–Boltzmann statistics*. Unfortunately, apparently no known physical particles exhibit these Maxwell–Boltzmann statistics; the p^t different arrangements are not equally likely. It turns out that for many different particles, in particular photons and nuclei, a relatively simple change of assumption gives rise to an empirically accurate model. Namely, suppose that we consider the particles as indistinguishable. Then we are in case 2: Two arrangements of particles to levels are considered the same if the same number of particles is assigned to the same level. Thus, the two arrangements described above are considered the same, as they each assign one particle to level 1, one to level 2, and two to level 3. By Theorem 2.4, the number of distinguishable ways to arrange t particles into p levels is now given by $C(p + t - 1, t)$. If we consider any distribution of particles to levels to be equally likely, then the probability of any one arrangement is

$$\frac{1}{C(p + t - 1, t)}$$

In this case, we say the particles satisfy the *Bose–Einstein statistics*. A third model in statistical mechanics arises if we consider the particles indistinguishable but add the assumption that there can be no more than one particle at a given level. Then we get the *Fermi–Dirac statistics* (see Exercise 19). See Feller [1968] or Parzen [1960] for a more detailed discussion of all the cases we have described.

EXERCISES FOR SECTION 2.11

Note to the reader: When it is unclear whether balls or cells are distinguishable, you should state your interpretation, give a reason for it, and then proceed.

1. Write down all the distributions of
    (a) 3 distinguishable balls a, b, c into 2 distinguishable cells, 1, 2;
    (b) 4 distinguishable balls a, b, c, d into 2 distinguishable cells 1, 2;
    (c) 2 distinguishable balls a, b into 4 distinguishable cells 1, 2, 3, 4;
    (d) 3 indistinguishable balls a, a, a into 2 distinguishable cells 1, 2;

(e) 4 indistinguishable balls a, a, a, a into 2 distinguishable cells 1, 2;

(f) 2 indistinguishable balls a, a into 4 distinguishable cells 1, 2, 3, 4.

2. In Exercise 1, which of the distributions are distinct if the cells are indistinguishable?

3. Use the results of Table 2.9 to compute the number of distributions in each case in Exercise 1 and check the result by comparing the distributions you have written down.

4. Repeat Exercise 3 if the cells are indistinguishable.

5. Use the results of Table 2.9 to compute the number of distributions with no empty cell in each case in Exercise 1. Check the result by comparing the distributions you have written down.

6. Repeat Exercise 5 if the cells are indistinguishable.

7. Find all partitions of
(a) 4; (b) 6; (c) 7.

8. Find all partitions of
(a) 8 into four or fewer parts;
(b) 10 into three or fewer parts.

9. Compute
(a) $S(n, 0)$; (b) $S(n, 1)$; (c) $S(n, 2)$; (d) $S(n, n - 1)$; (e) $S(n, n)$.

10. In checking the work of a proofreader, we look for 5 kinds of misprints in a textbook. In how many ways can we find 12 misprints?

11. In Exercise 10, suppose that we do not distinguish the types of misprints but we do keep a record of the page on which a misprint occurred. In how many different ways can we find 25 misprints in 75 pages?

12. If bullets are fired at 5 targets, how many ways are there for 10 bullets to hit? (You do not have to assume that each bullet hits a target.)

13. A Geiger counter records the impact of 6 different kinds of radioactive particles over a period of time. How many ways are there to obtain a count of 30?

14. Find the number of ways to distribute 10 prizes to 7 children so that each child gets at least 1 prize.

15. Find the number of ways to pair off 10 police officers into partners for a patrol.

16. Find the number of ways to assign 6 jobs to 4 workers so that each job gets a worker and each worker gets at least 1 job.

17. Find the number of ways to partition a set of 25 elements into exactly 4 subsets.

18. In Example 2.20, suppose that there are 8 photons and 4 energy levels, with 2 photons at each energy level. What is the probability of this occurrence under the assumption that the particles are indistinguishable (the Bose–Einstein case)?

19. Show that in Example 2.20, if particles are indistinguishable but no two particles can be at the same level, then there are $C(p, t)$ possible arrangements of t particles into p levels. (Assume that $t \leq p$.)

20. (a) Show by a combinatorial argument that

$$S(n, k) = kS(n - 1, k) + S(n - 1, k - 1).$$

(b) Use the result in part (a) to describe how to compute Stirling numbers of the second kind by a method similar to Pascal's triangle.

(c) Apply your result in part (b) to compute $S(6, 3)$.

21. Show by a combinatorial argument that

$$S(n + 1, k) = C(n, 0)S(0, k - 1) + C(n, 1)S(1, k - 1) + \cdots + C(n, n)S(n, k - 1).$$

22. (a) If order counts in a partition, then $\{3, 2\}$ is different from $\{2, 3\}$. Find the number of partitions of 5 if order counts.

(b) Find the number of partitions of 5 into exactly 2 parts where order counts.

(c) Show that the number of partitions of n into exactly k parts where order counts is given by $C(n-1, k-1)$.

2.12 MULTINOMIAL COEFFICIENTS

2.12.1 Occupancy Problems with a Specified Distribution

In this section we consider the occupancy problem of distributing n distinguishable balls into k distinguishable cells. In particular, we consider the situation where we distribute n_1 balls into the first cell, n_2 into the second cell, ..., n_k into the kth cell. Let

$$C(n; n_1, n_2, \ldots, n_k)$$

be the number of ways this can be done. This section is devoted to the study of the number $C(n; n_1, \ldots, n_k)$, which is sometimes also written as

$$\binom{n}{n_1 n_2 \cdots n_k}$$

and called the *multinomial coefficient*.

Example 2.21   Campus Housing

> The university housing office is having a problem. It has 11 students to squeeze into 2 rooms, 5 in the first and 6 in the second. In how many ways can this be done? The answer is $C(11; 5, 6)$. Now there are $C(11, 5)$ choices for the first room; for each of these there are $C(6, 6)$ choices for the second room. Hence, by the product rule, the number of ways to assign rooms is
>
> $$C(11; 5, 6) = C(11, 5) \times C(6, 6) = \frac{11!}{5!6!} \times \frac{6!}{6!0!} = \frac{11!}{5!6!},$$
>
> since $0! = 1$. Of course, $C(6, 6) = 1$, so the answer is equivalent to $C(11, 5)$. The reason for this is that once the 5 students for the first room have been chosen, there is only one way to choose the remaining 6.
>
> Note that if room assignments for 11 students are made at random, there are 2^{11} possible assignments: For each student, there are 2 choices of room. Hence, the probability of having 5 students in the first room and 6 in the second is given by
>
> $$\frac{C(11; 5, 6)}{2^{11}}.$$
>
> In general, suppose that $\Pr(n; n_1, n_2, \ldots, n_k)$ denotes the probability that n balls are distributed at random into k cells, with n_i balls in cell i, $i = 1, 2, \ldots, k$. Then
>
> $$\Pr(n; n_1, n_2, \ldots, n_k) = \frac{C(n; n_1, n_2, \ldots, n_k)}{k^n}.$$

(Why?)

Continuing with our example, suppose that suddenly a third room becomes available. The housing office now wishes to put 3 people into the first room, 6 into the second, and 2 into the third. In how many ways can this be done? Of the 11 students, 3 must be chosen for the first room; of the remaining 8 students, 6 must be chosen for the second room; finally, the remaining 2 must be put into the third room. The total number of ways of making the assignments is

$$C(11; 3, 6, 2) = C(11, 3) \times C(8, 6) \times C(2, 2) = \frac{11!}{3!8!} \times \frac{8!}{6!2!} \times \frac{2!}{2!0!}$$

$$= \frac{11!}{3!6!2!0!}$$

$$= \frac{11!}{3!6!2!}.$$

Let us derive a formula for $C(n; n_1, n_2, \ldots, n_k)$. By reasoning analogous to that used in Example 2.21,

$$C(n; n_1, n_2, \ldots, n_k) = C(n, n_1) \times C(n - n_1, n_2) \times C(n - n_1 - n_2, n_3) \times \cdots$$

$$\times C(n - n_1 - n_2 - \cdots - n_{k-1}, n_k)$$

$$= \frac{n!}{n_1!(n - n_1)!} \times \frac{(n - n_1)!}{n_2!(n - n_1 - n_2)!} \times \frac{(n - n_1 - n_2)!}{n_3!(n - n_1 - n_2 - n_3)!} \times \cdots$$

$$\times \frac{(n - n_1 - n_2 - \cdots - n_{k-1})!}{n_k!(n - n_1 - n_2 - \cdots - n_k)!}$$

$$= \frac{n!}{n_1!n_2!\cdots n_k!(n - n_1 - n_2 - \cdots - n_k)!}.$$

Since $n_1 + n_2 + \cdots + n_k = n$, and since $0! = 1$, we have the following result.

Theorem 2.5. $C(n; n_1, n_2, \ldots, n_k) = \dfrac{n!}{n_1!n_2!\cdots n_k!}.$ (2.9)

We now give several applications of this theorem.

1. The number of 3-digit bit strings consisting of two 1's and one 0 is $C(3; 2, 1)$: Out of three places, we choose two for the digit 1 and one for the digit 0. Hence, the number of such strings is given by

$$C(3; 2, 1) = \frac{3!}{2!1!} = 3.$$

The three such strings are 110, 101, and 011.

2. The number of 5-digit numbers consisting of two 2's, two 3's, and one 1 is

$$C(5; 2, 2, 1) = \frac{5!}{2!2!1!} = 30.$$

3. Notice that $C(n; n_1, n_2) = C(n, n_1)$. Why should this be true?

4. A football season consists of 11 games. The number of ways the season can end in 7 wins, 2 losses, and 2 ties is

$$C(11; 7, 2, 2) = \frac{11!}{7!2!2!}.$$

5. RNA is a messenger molecule whose links are defined from DNA. An RNA chain has at each link one of four bases. The possible bases are the same as those in DNA (Example 2.2), except that the base uracil (U) replaces the base thymine (T). How many possible RNA chains of length 6 are there consisting of 3 cytosines (C) and 3 adenines (A)? To answer this question, we think of 6 positions in the chain, and of dividing these positions into two sets, 3 into the C set and 3 into the A set. The number of ways this can be done is given by

$$C(6; 3, 3) = 20.$$

6. There are 4^6 possible RNA chains of length 6. The probability of obtaining one with 3 C's and 3 A's if the RNA chain were produced at random (that is, all possibilities equally likely) would be

$$\Pr(6; 3, 3) = \frac{C(6; 3, 3)}{4^6} = \frac{20}{4096} = \frac{5}{1024} \approx .005.$$

7. The number of 10-link RNA chains consisting of 3 A's, 2 C's, 2 U's, and 3 G's is

$$C(10; 3, 2, 2, 3) = 25,200.$$

8. The number of RNA chains as described in (7) which end in AAG is

$$C(7; 1, 2, 2, 2) = 630,$$

since there are now only the first 7 positions to be filled, and two of the A's and one of the G's is already used up. Notice how knowing the end of a chain can reduce dramatically the number of possible chains. In the next section we shall see how, by judicious use of various enzymes which decompose RNA chains, we might further limit the number of possible chains until, by a certain amount of detective work, we can uncover the original RNA chain without actually observing it.

2.12.2 Permutations with Classes of Indistinguishable Objects

Applications 1, 2, and 5–8 suggest the following general notion: Suppose that there are n objects, n_1 of type 1, n_2 of type 2, ..., n_k of type k, with $n_1 + n_2 + \cdots + n_k = n$. Suppose that objects of the same type are *indistinguishable*. The number of *distinguishable permutations* of these objects is denoted $P(n; n_1, n_2, \ldots, n_k)$. We use the word *distinguishable* here because we assume objects of the same type are indistinguishable. For instance, suppose that $n = 3$ and there are two type 1 objects, a and a, and one type 2 object, b. Then there are $3! = 6$ permutations of the three objects, but several of these are indistinguishable. For example, *baa* in which the first of the two a's comes second is indistinguishable from *baa* in which the second of the two a's comes second. There are only three distinguishable permutations, *baa*, *aba*, and *aab*.

Theorem 2.6. $P(n; n_1, n_2, \ldots, n_k) = C(n; n_1, n_2, \ldots, n_k)$.

Proof. We have n positions or places to fill in the permutation, and we assign n_1 of these to type 1 objects, n_2 to type 2 objects, and so on. Q.E.D.

We return to permutations with classes of indistinguishable objects in Section 2.14.

EXERCISES FOR SECTION 2.12

1. Compute
   (a) $C(7; 2, 2, 2, 1)$; (b) $C(8; 3, 3, 2)$; (c) $C(6; 1, 1, 1, 2, 1)$;
   (d) Pr $(6; 2, 2, 2)$; (e) Pr $(10; 2, 2, 1, 2, 3)$; (f) $P(9; 6, 1, 2)$; (g) $P(7; 4, 1, 2)$.

2. Find $C(n; 1, 1, 1, \ldots, 1)$.

3. Find
   (a) $P(n; 1, n - 1)$; (b) Pr $(n; 1, n - 1)$.

4. A code is being written using the four symbols $+$, $-$, $\times$, and $\#$.
   (a) How many 8-digit codewords are there which use exactly 2 of each symbol?
   (b) If an 8-digit codeword is chosen at random, what is the probability it will use exactly 2 of each symbol?

5. A code is being written using three symbols, a, b, and c.
   (a) How many 6-digit codewords can be written using exactly 3 a's, 1 b, and 2 c's?
   (b) If a 6-digit codeword is chosen at random, what is the probability it will use exactly 3 a's, 1 b, and 2 c's?

6. A code is being written using the four letters a, b, c, and d.
   (a) How many 12-digit codewords are there which use exactly 3 of each letter?
   (b) If a 12-digit codeword is chosen at random, what is the probability it will use exactly 3 of each letter?

7. How many RNA chains have the same makeup of bases as the chain

   CGCCAUCCGAC?

8. How many different "words" can be formed using all the letters of the word *renegotiate*?

9. How many ways are there to form a sequence of 10 letters from 4 a's, 4 b's, 4 c's, and 4 d's if each letter must appear at least twice?

10. How many distinguishable permutations are there of the symbols $a, a, a, a, a, a, b, c, d, e$, and f if no two a's are adjacent?

11. Repeat Exercise 18 of Section 2.11 under the assumption that the particles are distinguishable (the Maxwell–Boltzmann case).

12. (a) Suppose that we distinguish 4 different light particles hitting the retina. In how many ways could these be distributed among three cells, with two hitting the first cell and one hitting each of the other cells?
    (b) If we know there are 4 different light particles distributed among the three cells, what is the probability that they will be distributed as in part (a)?

13. Suppose that 30 radioactive particles hit a Geiger counter with 50 counters. In how many different ways can this happen with all but the 30th particle hitting the first counter?

14. Suppose that 6 individuals have accidents.
    (a) How many different ways are there for 2 of them to have an accident on Monday, 2 on Wednesday, and 2 on Saturday?

(b) Given the 6 accidents, what is the probability that the accidents will be distributed as in part (a)?

(c) How many ways are there for the accidents to be distributed into 3 days, 2 per day?

15. Suppose that we have 4 elements. How many distinguishable ways are there to assign these to 4 distinguishable sets, 1 to each set, if the elements are
(a) a, b, c, d;   **(b)** a, a, a, b;   **(c)** a, a, b, b;   **(d)** a, a, b, c.

2.13 COMPLETE DIGEST BY ENZYMES*

As an aside, let us consider the problem of discovering what a given RNA chain looks like without actually observing the chain itself (RNA chains were introduced in Section 2.12). Some enzymes break up an RNA chain into fragments after each G link. Others break up the chain after each C or U link. For example, suppose that we have the chain

$$CCGGUCCGAAAG.$$

Applying the G enzyme breaks the chain into the following fragments:

G fragments: CCG, G, UCCG, AAAG.

We then know that these are the fragments, but we do not know in what order they appear. How many possible chains have these four fragments? The answer is that $4! = 24$ chains do: There is one chain corresponding to each of the different permutations of the fragments. One such chain (different from the original) is the chain

$$UCCGGCCGAAAG.$$

Suppose that we next apply the U, C enzyme, the enzyme that breaks up a chain after each C link or U link. We obtain the following fragments:

U, C fragments: C, C, GGU, C, C, GAAAG.

Again, we know that these are the fragments, but we do not know in what order they appear. How many chains are there with these fragments? One is tempted to say that there are 6! chains, but that is not right. For example, if the fragments were

$$C, C, C, C, C, C,$$

there would not be 6! chains with these fragments, but only one, the chain

$$CCCCCC.$$

The point is that some of the fragments are indistinguishable. To count the number of distinguishable chains with the given fragments, we note that there are six fragments. Four of these are C fragments, one is GGU, and one is GAAAG. Thus, by Theorem 2.6, the number of possible chains with these as fragments is

$$P(6; 4, 1, 1) = C(6; 4, 1, 1) = \frac{6!}{4!1!1!} = 30.$$

---

*This section can be omitted without loss of continuity. The material here is not needed again until Section 11.4.4. However, this section includes a detailed discussion of an applied topic, and I always include it in my courses.

Actually, this computation is still a little off. Notice that the fragment GAAAG among the U, C fragments could not have appeared except as the terminal fragment because it does not end in U or C. Hence, we know that the chain ends

GAAAG.

There are five remaining U, C fragments: C, C, C, C, and GGU. The number of chains (beginning segments of chains) with these as fragments is

$$C(5; 4, 1) = 5.$$

The possible chains are obtained by adding GAAAG to one of these 5 beginning chains. The possibilities are

CCCCGGUGAAAG

CCCGGUCGAAAG

CCGGUCCGAAAG

CGGUCCCGAAAG

GGUCCCCGAAAG.

We have not yet combined our knowledge of both G and U, C fragments. Can we learn anything about the original chain by using our knowledge of both? Which of the 5 chains that we have listed has the proper G fragments? The first does not, for it would have a G fragment CCCCG, which does not appear among the fragments when the G enzyme is applied. A similar analysis shows that only the third chain,

CCGGUCCGAAAG,

has the proper set of G fragments. Hence, we have recovered the initial chain from among those that have the given U, C fragments.

This is an example of recovery of an RNA chain given a *complete enzyme digest*, that is, a split up after every G link and another after every U or C link. It is remarkable that we have been able to limit the large number of possible chains for any one set of fragments to only one possible chain by considering both sets of fragments. This result is more remarkable still if we consider trying to guess the chain knowing just its bases but not their order. Then we have

$$C(12; 4, 4, 3, 1) = 138,600 \text{ possible chains!}$$

Let us give another example. Suppose we are told that an RNA chain gives rise to the following fragments after complete digest by the G enzyme and the U, C enzyme:

G fragments: UG, ACG, AC

U, C fragments: U, GAC, GAC.

Can we discover the original chain? To begin with, we ask again whether or not the U, C fragments tell us which part of the chain must come last. The answer is that, in this case, they do not. However, the G fragments do: AC could only have arisen as a G fragment if it came last. Hence, the two remaining G fragments can be arranged in any order, and the possible chains with the given G fragments are

UGACGAC

ACGUGAC.

Now the latter chain would give rise to AC as one of the U, C fragments. Hence, the former must be the correct chain.

It is not always possible to completely recover the original RNA chain knowing the G fragments and U, C fragments. Sometimes the complete digest by these two enzymes is ambiguous in the sense that there are two RNA chains with the same set of G fragments and the same set of U, C fragments. We ask the reader to show this as an exercise (Exercise 8).

The reader who is interested in more details about complete digests by enzymes should read Mosimann [1968] or Hutchinson [1969] or Mosimann *et al.* [1966]. We return to this problem in Section 11.4.4.

EXERCISES FOR SECTION 2.13

1. An RNA chain has the following fragments after being subjected to complete digest by G and U, C enzymes:

    G fragments: CUG, CAAG, G, UC

    U, C fragments: C, C, U, AAGC, GGU.

    (a) How many RNA chains are there with these G fragments?
    (b) How many RNA chains are there with these U, C fragments?
    (c) Find *all* RNA chains that have these G and U, C fragments.

2. In Exercise 1, find the number of RNA chains with the same bases as those of the chains with the given G fragments.

3. Repeat Exercise 1 for the following G and U, C fragments:

    G fragments: G, UCG, G, G, UU

    U, C fragments: GGGU, U, GU, C.

4. In Exercise 3, find the number of RNA chains with the same bases as those of the chains with the given G fragments.

5. Repeat Exercise 1 for the following G and U, C fragments:

    G fragments: G, G, CC, CUG, G

    U, C fragments: GGGC, U, C, GC.

6. In Exercise 5, find the number of RNA chains with the same bases as those of the chains with the given G fragments.

7. A bit string is broken up after every 1 and after every 0. The resulting pieces (not necessarily in proper order) are as follows:

    break up after 1:   0, 001, 01, 01

    break up after 0:   0, 10, 0, 10, 10.

(a) How many bit strings are there which have these pieces after breakup following each 1?

(b) After each 0?

(c) Find all bit strings having both of these sets of pieces.

8. Find an RNA chain which is ambiguous in the sense that there is another chain with the same G fragments and the same U, C fragments. (Can you find one with six or fewer links?)

9. What is the shortest possible RNA chain that is ambiguous in the sense of Exercise 8?

10. Can a bit string be ambiguous if it is broken up as in Exercise 7? Why?

2.14 PERMUTATIONS WITH CLASSES OF INDISTINGUISHABLE OBJECTS REVISITED

In Sections 2.12 and 2.13 we encountered the problem of counting the number of permutations of a set of objects in which some of the objects were indistinguishable. In this section we develop an alternative procedure for counting in this situation.

Example 2.22   Sick Trees and Well Trees

A forester has observed that of 10 trees in a row, 4 are sick and 6 are well. However, all 4 sick trees come first. Thus, the forester's observation can be abbreviated as

$$SSSSWWWWWW.$$

Is this observation a coincidence, or does it suggest that the sickness is contagious? To answer this question, let us assume that the sickness is not contagious, that is, that a neighboring tree of a sick one is no more likely to get the sickness than a nonneighboring tree. Thus, let us assume that the sickness occurs at random, and each tree has the same probability of being sick, independent of what happens to the other trees. It follows that all possible orderings of 4 sick and 6 well trees are equally likely.* How many such orderings are there? The answer, to use the notation of Section 2.12, is

$$P(10;\ 6,\ 4) = C(10;\ 6,\ 4) = \frac{10!}{4!6!} = 210.$$

To derive this directly, note that there are 10 positions, and we wish to assign 4 of these to S and 6 to W. Thus, the number of such orderings is $C(10;\ 6,\ 4) = 210$. If all such orderings are equally likely, the probability of seeing the specific arrangement

$$SSSSWWWWWW$$

is 1 out of 210. Of course, this is the probability of seeing any one given arrangement. What is more interesting to calculate is the probability of seeing 4 sick trees together out of 10 trees. In how many arrangements of 10 trees, 4 sick and 6 well, do the 4 sick ones occur together? To answer this, let us consider the 4 sick trees as one

---

*It does not follow that all orderings of 10 sick and well trees are equally likely. For instance, even if sickness occurs at random, if sickness is very unlikely, then the sequence $WWWWWWWWWW$ is much more likely than the sequence $SSSSSSSSSS$.

unit $S*$. Then we wish to consider the number of orders of 1 $S*$ and 6 W's. There are

$$C(7; 1, 6) = \frac{7!}{1!6!} = 7$$

such orders. These correspond to the orders

$SSSSWWWWWW$	(which is $S*WWWWWW$)
$WSSSSWWWWW$	(which is $WS*WWWWW$)
$WWSSSSWWWW$	(which is $WWS*WWWW$)
$WWWSSSSWWW$	(which is $WWWS*WWW$)
$WWWWSSSSWW$	(which is $WWWWS*WW$)
$WWWWWSSSSW$	(which is $WWWWWS*W$)
$WWWWWWSSSS$	(which is $WWWWWWS*$).

The probability of seeing 4 sick trees and 6 well trees with all the sick trees together, given that there are 4 sick trees and 6 well ones, is therefore 7 out of 210, or $\frac{1}{30}$. This is quite small. Hence, since seeing all the sick trees together is unlikely, we expect that perhaps this is not a random occurrence, and the sickness is probably contagious.

Before leaving this example, it is convenient to repeat the calculation of the number of ways of ordering 4 sick trees and 6 well ones. Suppose we label the trees so that they are distinguishable:

$$S_a, S_b, S_c, S_d, W_a, W_b, W_c, W_d, W_e, W_f.$$

There are 10! permutations of these 10 labels. For each such permutation, we can reorder the 4 S's arbitrarily; there are 4! such reorderings. Each reordering gives rise to an ordering which is considered the same as far as we are concerned. Similarly, we can reorder the 6 W's in 6! ways. Thus, groups of 4! × 6! orderings are ths same, and each permutation corresponds to 4! × 6! similar ones. The number of indistinguishable permutations is

$$\frac{10!}{4!6!}.$$

The reasoning in Example 2.22 generalizes to give us another proof of the result (Theorem 2.6) that $P(n; n_1, n_2, \ldots, n_k) = C(n; n_1, n_2, \ldots, n_k)$.

EXERCISES FOR SECTION 2.14

1. Suppose a researcher observes that of 12 petri dishes, 5 have growths and 7 do not. The 5 dishes with growths are next to each other. Assuming that 5 of the 12 petri dishes have growths and that all orderings of these dishes are equally likely, what is the probability that the 5 dishes with growths will all be next to each other?

2. Suppose a realtor observes that of 12 houses on a block, 4 are red and 8 are blue. The 4 red houses are next to each other. Assuming that 4 of the 12 houses are red and that all orderings of these 12 houses are equally likely, what is the probability that the 4 red houses are next to each other?

3. Suppose a forester observes that some trees are sick (S), some well (W), and some questionable (Q). Assuming that of 30 trees, 10 are sick or questionable, what is the probability that these 10 appear consecutively? Assume that all sequences of sick, well, and questionable with 10 sick or questionable are equally likely.

4. Suppose that of 11 houses lined up in one block, 6 have a person ill with the flu.
   (a) In how many ways can the presence or absence of flu occur so that these 6 houses are next to each other?
   (b) In how many ways can this occur so that none of these 6 houses are next to each other?
   (c) In how many ways can we schedule an order of visits that go to two of these houses in which there is no flu?
   (d) In how many ways can we schedule an order of visits that go to two of these houses if at most one house that we visit can have the flu?

5. How many distinct ways are there to seat 10 people around a circular table? (Clarify what "distinct" means here.)

6. If an RNA chain of length 4 is chosen at random, what is the probability that it has
   (a) (at least) three consecutive C's?
   (b) (at least) two consecutive C's?
   (c) a consecutive AG?
   (d) a consecutive AUC?

7. How many bit strings of length 21 have every 1 followed by 0 and have seventeen 0's and four 1's?

8. How many RNA chains of length 18 have four A's, three U's, five C's, and six G's and have every C followed by G?

9. There are 20 individuals whose records are stored in order in a file. We want to choose 4 of these at random in performing a survey, making sure not to choose two consecutive individuals. In how many ways can this be done? (*Hint:* Either we choose the last individual or we do not.)

2.15 THE BINOMIAL EXPANSION

As an application of the ideas considered in this chapter, let us develop a formula for $(a + b)^n$ which will be very useful in the following.

Theorem 2.7 (Binomial Expansion).   For $n \geq 0$,

$$(a + b)^n = \sum_{k=0}^{n} C(n, k)a^k b^{n-k} = \sum_{k=0}^{n} \binom{n}{k} a^k b^{n-k}.$$

Proof. * Note that

$$(a + b)^n = \underbrace{(a + b)(a + b) \cdots (a + b)}_{n \text{ times}}.$$

*For an alternative proof, see Exercise 10.

In multiplying out, we pick one term from each factor $(a + b)$. Hence, we only obtain terms of the form $a^k b^{n-k}$. To find the coefficient of $a^k b^{n-k}$, note that to obtain $a^k b^{n-k}$, we need to choose k of the terms from which to choose a. This can be done in $\binom{n}{k}$ ways. Q.E.D.

In particular, we have

$$(a + b)^2 = a^2 + 2ab + b^2$$

$$(a + b)^3 = a^3 + 3a^2 b + 3ab^2 + b^3$$

$$(a + b)^4 = a^4 + 4a^3 b + 6a^2 b^2 + 4ab^3 + b^4.$$

The reader might wish to compare the coefficients here with those numbers appearing in the Pascal triangle.

It is not hard to generalize the binomial expansion of Theorem 2.7 to an expansion of

$$(a + b + c)^n$$

and more generally of

$$(a_1 + a_2 + \cdots + a_k)^n.$$

We leave the generalization to the reader (Exercises 5 and 6).

Let us give a few applications of the binomial expansion here. The coefficient of x^{20} in the expansion of $(1 + x)^{30}$ is obtained by taking $a = 1$ and $b = x$ in Theorem 2.7. We are seeking the coefficient of $1^{10} x^{20}$, that is, $C(30, 10)$.

Theorem 2.8.   For $n \geq 0$,

$$\binom{n}{0} + \binom{n}{1} + \cdots + \binom{n}{n} = 2^n.$$

Proof. Note that

$$2^n = (1 + 1)^n.$$

Hence, by the binomial expansion,

$$2^n = \sum_{k=0}^{n} \binom{n}{k} 1^k 1^{n-k}. \qquad \text{Q.E.D.}$$

Another way of looking at Theorem 2.8 is the following. The number 2^n counts the number of subsets of an n-set. Also, the left-hand side of Theorem 2.8 counts the number of 0-element subsets of an n-set plus the number of 1-element subsets plus ... plus the number of n-element subsets. Each subset is counted once and only once in this way.

Theorem 2.9.   For $n > 0$,

$$\binom{n}{0} - \binom{n}{1} + \binom{n}{2} - \cdots + (-1)^k \binom{n}{k} + \cdots + (-1)^n \binom{n}{n} = 0.$$

Proof.

$$0 = (1 - 1)^n = (-1 + 1)^n = \sum_{k=0}^{n} \binom{n}{k} (-1)^k (1)^{n-k}. \qquad \text{Q.E.D.}$$

Corollary 2.9.1. For $n > 0$,

$$\binom{n}{0} + \binom{n}{2} + \cdots = \binom{n}{1} + \binom{n}{3} + \cdots .$$

The interpretation of this corollary is that the number of ways to select an even number of objects from n equals the number of ways to select an odd number.

EXERCISES FOR SECTION 2.15

1. Write out
   (a) $(x + y)^5$; **(b)** $(a + 2b)^3$; **(c)** $(2u + 3v)^4$.
2. Find the coefficient of x^{11} in the expansion of
   (a) $(1 + x)^{15}$; **(b)** $(2 + x)^{13}$; **(c)** $(2x + 3y)^{11}$.
3. What is the coefficient of x^{10} in the expansion of $(1 + x)^{12}(1 + x)^4$?
4. What is the coefficient of x^9 in the expansion of $(1 + x)^{10}(1 + x)^6$?
5. Find a formula for $(a + b + c)^n$.
6. Find a formula for $(a_1 + a_2 + \cdots + a_k)^n$.
7. What is the coefficient of a^3bc in the expansion of $(a + b + c)^5$?
8. What is the coefficient of xy^2z^2w in the expansion of $(x + y + z + 2w)^6$?
9. What is the coefficient of $a^3b^2cd^4$ in the expansion of $(a + 5b + 2c + 2d)^{10}$?
10. Prove Theorem 2.7 by induction on n.
11. Find $\binom{8}{0} + \binom{8}{2} + \binom{8}{4} + \binom{8}{6} + \binom{8}{8}$.
12. A bit string has *even parity* if it has an even number of 1's. How many bit strings of length n have even parity?
13. Find $\sum_{k=0}^{n} 2^k \binom{n}{k}$.
14. Find $\sum_{k=0}^{n} 4^k \binom{n}{k}$.
15. Find $\sum_{k=0}^{n} x^k \binom{n}{k}$.
16. Find $\sum_{k=1}^{n} k \binom{n}{k}$. [*Hint:* Differentiate the expansion of $(x + 1)^n$ and set $x = 1$.]
17. Find $\sum_{k=2}^{n} k(k - 1) \binom{n}{k}$.
18. Show that $\sum_{k=1}^{n} k \binom{n}{k} 2^{k-1} 2^{n-k} = n(4^{n-1})$.
19. Show that $\sum_{k=1}^{n} k \binom{n}{k} 2^{n-k} = n(3^{n-1})$.

2.16 POWER IN SIMPLE GAMES*

2.16.1 Examples of Simple Games

In this section we apply some of the counting rules previously described to the analysis of multiperson games. Now in modern applied mathematics, a game has come to mean more than just Monopoly, chess, or poker. It is any situation where a group of players is

*This section can be omitted without loss of continuity. The formal prerequisites for this section are Sections 2.1–2.7 and 2.9.

competing for different rewards or payoffs. In this sense, politics is a game, the economic marketplace is a game, the international bargaining arena is a game, and so on. We shall take this broad view of games here.

Let us think of a game as having a set I of n players. We shall be interested in possible cooperation among the players and, accordingly, we shall study *coalitions* of players, which correspond to subsets of the set I. We shall concentrate on *simple games*, games in which each coalition is either winning or losing. We can define a simple game by giving a value function v which assigns the number 0 or 1 to each coalition S, with $v(S)$ equal to 0 if S is a losing coalition and 1 if S is a winning coalition. It is usually assumed in game theory that a subset of a losing coalition cannot be winning, and we shall make that assumption. It is also usually assumed that for all S, either S or $I - S$ is losing. We shall assume that.

Very important examples of simple games are the weighted majority games. In a *weighted majority game*, there are n players, player i has v_i votes, and a coalition is winning if and only if it has at least q votes. We denote this game by

$$[q; v_1, v_2, \ldots, v_n].$$

The assumption that either S or $I - S$ loses places restrictions on the allowable q, v_1, v_2, $\ldots$, v_n. For instance, [3; 4, 4] does not satisfy this requirement. Weighted majority games arise in corporations, where the players are the stockholders and a stockholder has one vote for each share owned. Most legislatures are weighted majority games of the form $[q; 1, 1, \ldots, 1]$, where each player has one vote. However, some legislatures give a legislator a number of votes corresponding to the population of his or her district. For example, in 1964 the Nassau County New York Board of Supervisors was the weighted majority game [59; 31, 31, 21, 28, 2, 2] (Banzhaf [1965]).

Perhaps the most elementary weighted majority game is the game [2; 1, 1, 1]. In this game there are three players, each having one vote, and a simple majority of the players forms a winning coalition. Thus, the winning coalitions are the sets

$$\{1, 2\}, \quad \{1, 3\}, \quad \{2, 3\}, \quad \{1, 2, 3\}.$$

Another example of a simple game is the U.N. Security Council. Here there are 15 players: 5 permanent members (China, France, the Soviet Union, the United Kingdom, the United States) and 10 nonpermanent members. A coalition is winning if and only if it has all 5 permanent members (since they have veto power) and at least 4 of the 10 nonpermanent members. It is interesting to note that the Security Council can be looked at as a weighted majority game. For consider the game

$$[39; 7, 7, 7, 7, 7, 1, 1, 1, 1, 1, 1, 1, 1, 1, 1],$$

where the first five players correspond to the permanent members. The winning coalitions in this game are exactly the same as those in the Security Council, as is easy to check. Hence, even though weighted votes are not explicitly assigned in the Security Council, the game can be considered a weighted majority game. (The reader might wish to think about how to obtain numbers, such as 39, 7, and 1, which translate a simple game into a weighted majority game.)

A similar situation arises for the Australian government. In making national decisions, 6 states and the federal government play a role. In effect, a measure passes if and only if it has the support of at least 5 states or at least 2 states and the federal government. As is easy to see, this simple game corresponds (in the sense of having the same winning coalitions) to the game [5; 1, 1, 1, 1, 1, 1, 3], where the seventh player is the federal government.

Not every simple game is a weighted majority game. A bicameral legislature is an example (see Exercise 5).

2.16.2 The Shapley–Shubik Power Index

We shall be concerned with measuring the *power* of a player in a simple game, his or her ability to maneuver into a winning coalition. Note first that power is not necessarily proportional to the number of votes a player has. For example, compare the two games [2; 1, 1, 1] and [51; 49, 48, 3]. In each game, there are 3 players, and any coalition of 2 or more players wins. Thus, in both games, player 3 is in the same winning coalitions, and hence has essentially the same power. These two games might be interpreted as a legislature with 3 parties. In the first legislature, there are 3 equal parties and 2 out of 3 must go along for a measure to pass. In the second legislature, there are 2 large parties and a small party. However, assuming that party members vote as a bloc, it is still necessary to get 2 out of 3 parties to go along to pass a measure, so in effect the third small party has as much power as it does in the first legislature.

There have been a number of alternative approaches to the measurement of power in simple games. We shall refer to the Banzhaf and Coleman power indices in Section 4.7. The former has come to be used in the courts, in "one-person, one-vote" cases. Here we shall concentrate on the Shapley–Shubik power index, introduced in its original form by Shapley [1953] and in the form we present by Shapley and Shubik [1954]. (In its more general original form, it is called the *Shapley value*.) For a survey of the literature of the Shapley–Shubik index, see Shapley [1981].

Let us think of building up a coalition by adding one player at a time until we reach a winning coalition. The player whose addition throws the coalition over from losing to winning is called pivotal. More formally, let us consider any permutation of the players and call a player i *pivotal* for that permutation if the set of players preceding i is losing, but the set of players up to and including player i is winning. For example, in the game [2; 1, 1, 1], player 2 is pivotal in the permutation 1, 2, 3 and in the permutation 3, 2, 1. The *Shapley–Shubik power index* p_i for player i in a simple game is defined as follows:

$$p_i = \frac{\text{number of permutations of the players in which } i \text{ is pivotal}}{\text{number of permutations of the players}}.$$

If we think of one permutation of the players being chosen at random, the Shapley–Shubik power index for player i is the probability that player i is pivotal. In the game [2; 1, 1, 1], for example, there are three players and hence 3! permutations. Each player is pivotal in two of these. For example, as we have noted, player 2 is pivotal in 1, 2, 3 and 3, 2, 1. Hence, each player has power $2/3! = 1/3$. In the game [51; 49, 48, 3], player 1 is pivotal in the permutations 2, 1, 3 and 3, 1, 2. For in the first he or she brings in enough votes to change player 2's 48 votes into 97 and in the second he or she brings in enough votes to change player 3's 3 votes to 52. Thus,

$$p_1 = \frac{2}{3!} = \frac{1}{3}.$$

Similarly, player 2 is pivotal in permutations 1, 2, 3 and 3, 2, 1, and player 3 in permutations 1, 3, 2 and 2, 3, 1. Hence,

$$p_2 = p_3 = \frac{1}{3}.$$

Thus, as anticipated, the small third party has the same power as the two larger parties.

In the game [51; 40, 30, 15, 15], the possible permutations are shown in Table 2.10, and the pivotal player in each is circled. Player 1 is pivotal 12 times, so his or her power is $12/4! = \frac{1}{2}$. Players 2, 3, and 4 are each pivotal 4 times, so they each have power $4/4! = \frac{1}{6}$.

Let us compute the Shapley–Shubik index for the Australian government, the game [5; 1, 1, 1, 1, 1, 1, 3]. In this and the following examples, the enumeration of all permutations is not the most practical method for computing the Shapley–Shubik index. We proceed by another method. The federal government (player 7) is pivotal in a given permutation if and only if it is third, fourth, or fifth. By symmetry, we observe that the federal government is picked in the ith position in a permutation of the 7 players in exactly 1 out of every 7 permutations. Hence, it is picked third, fourth, or fifth in exactly 3 out of every 7 permutations. Thus, the probability that the federal government is pivotal is $\frac{3}{7}$, that is,

$$p_7 = \frac{3}{7} = \frac{9}{21}.$$

We can also see this by observing that the number of permutations of the seven players in which the federal government is third is 6!, for we have to order the remaining players. Similarly, the number of permutations in which the federal government is fourth (fifth) is also 6!. Thus,

$$p_7 = \frac{3(6!)}{7!} = \frac{3}{7}.$$

Now it is easy to see that

$$p_1 + p_2 + \cdots + p_7 = 1.$$

Table 2.10 All permutations of the players in the game [51; 40, 30, 15, 15], with pivotal player circled

1 ② 3 4	2 ① 3 4	3 ① 2 4	4 ① 2 3
1 ② 4 3	2 ① 4 3	3 ① 4 2	4 ① 3 2
1 ③ 2 4	2 3 ① 4	3 2 ① 4	4 2 ① 3
1 ③ 4 2	2 3 ④ 1	3 2 ④ 1	4 2 ③ 1
1 ④ 2 3	2 4 ① 3	3 4 ① 2	4 3 ① 2
1 ④ 3 2	2 4 ③ 1	3 4 ② 1	4 3 ② 1

It is always the case that

$$\sum_i p_i = 1.$$

(Why?) Hence,

$$p_1 + p_2 + p_3 + p_4 + p_5 + p_6 = 1 - \frac{3}{7} = \frac{4}{7}.$$

Since by symmetry

$$p_1 = p_2 = \cdots = p_6,$$

each of these numbers is equal to

$$\frac{4}{7} \div 6 = \frac{2}{21}.$$

Thus, although the federal government has only 3 times the number of votes of a state, it has $4\frac{1}{2}$ times the power.

2.16.3 The U.N. Security Council

Let us turn next to the Shapley–Shubik index for the U.N. Security Council. Let us fix a nonpermanent player i. Player i is pivotal in exactly those permutations in which all permanent players precede i and exactly three nonpermanent players precede i. How many such permutations are there? To find such a permutation, we first choose the three nonpermanent players who precede i; for each such choice, we order all eight players who precede i (the five permanent and three nonpermanent ones); for each choice and ordering, we order the remaining six nonpermanent players who follow i. The number of ways to make the first choice is $C(9, 3)$, the number of ways to order the preceding players is $8!$, and the number of ways to order the following players is $6!$. Thus, by the product rule, the number of permutations in which i is pivotal is given by

$$C(9, 3) \times 8! \times 6! = \frac{9!}{3!6!} \times 8! \times 6! = \frac{9!8!}{3!}.$$

Thus, since the total number of permutations of the 15 players is $15!$,

$$p_i = \frac{9!8!}{3!} \div 15! \approx .001865.$$

It follows that the sum of the powers of the nonpermanent players is 10 times this number, that is, it is $.01865$. Thus, since all the powers add to 1, the sum of the powers of the permanent players is $.98135$. There are five permanent players, each of whom, by symmetry, has equal power. It follows that each has power

$$p_i = \frac{.98135}{5} = .1963.$$

Hence, permanent members have more than 100 times the power of nonpermanent members. (The idea of calculating power in the Security Council and other legislative bodies in this manner was first introduced in Shapley and Shubik [1954]).)

2.16.4 Bicameral Legislatures

To give a more complicated example, suppose that we have a bicameral legislature with n_1 members in the first house and n_2 members in the second house.* Suppose that a measure can pass only if it has a majority in each house of the legislature, and suppose for simplicity that n_1 and n_2 are both odd. Let I be the union of the sets of members of both houses, and let π be any permutation of I. A player i in the jth house is pivotal in π if he is the $[(n_j + 1)/2]$th player of his house in π and a majority of players in the other house precede him. However, for every permutation π in which a player in the first house is pivotal, the reverse permutation makes a player in the second house pivotal. (Why?) Moreover, every permutation has some player as pivotal. Thus, some player in house number 1 is pivotal in exactly $\frac{1}{2}$ of all permutations. Since all players in house number 1 are treated equally, any one of these players is pivotal in exactly $1/(2n_1)$ of the permutations. Similarly, any player of house number 2 will be pivotal in exactly $1/(2n_2)$ of the permutations. Thus, each player of house number 1 has power $1/(2n_1)$ and each player of house number 2 has power $1/(2n_2)$. In the U.S. House and Senate, $n_1 = 435$ and $n_2 = 101$, including the Vice-President, who votes in case of a tie. According to our calculation, each representative has power $1/870 \approx .0011$ and each senator (including the Vice-President) has power $1/202 \approx .005$. Thus, a senator has about five times as much power as a representative.

Next, let us add an executive (a governor, the President) who can veto the vote in the two houses, but let us assume that there is no possibility of overriding the veto. Now there are $n_1 + n_2 + 1$ players and a coalition is winning if and only if it contains the executive and a majority from each house. Assuming that n_1 and n_2 are large, Shapley and Shubik [1954] argue that the executive will be pivotal in approximately one-half of the permutations. (This argument is a bit complicated and we omit it.) The two houses divide the remaining power almost equally. Finally, if the possibility of overriding the veto with a two-thirds majority of both houses is added, a similar discussion implies that the executive has power approximately one-sixth, and the two houses divide the remaining power almost equally. The reader is referred to Shapley and Shubik's paper for details.

Similar calculations can be made for the relative power various states wield in the electoral college. Mann and Shapley [1964a, b] calculated this using the distribution of electoral votes as of 1961. New York had 43 out of the total of 538 electoral votes, and had a power of .0841. This compared to a power of .0054 for states like Alaska, which had three electoral votes. According to the Shapley–Shubik index, the power of New York exceeded its percentage of the vote, whereas that of Alaska lagged behind its percentage.

Similar results for the distribution of electoral votes as of 1972 were obtained by Boyce and Cross [unpublished observations, 1973]. In the 1972 situation, New York had a total of 41 electoral votes (the total was still 538) and a power of .0797, whereas Alaska still had three electoral votes and a power of .0054. For a more comprehensive discussion of power in electoral games, see Brams et al. [1983], Lucas [1983], Shapley [1981], and Straffin [1980].

*This example is also due to Shapley and Shubik [1954].

2.16.5 Characteristic Functions

We have concentrated in this section on simple games, games that can be defined by giving each coalition S a value $v(S)$ equal to 0 or 1. If the value function or *characteristic function* $v(S)$ can take on arbitrary real numbers as values, the game is called a game in *characteristic function form*. Such games have in recent years found a wide variety of applications, to decisionmaking about water and air pollution; to the determination of airport landing fees; to assessment of costs in running a library; to telephone billing rates; to transportation; and so on. For a summary of applications, see Brams *et al.* [1979] or Lucas [1981a].

For more on game theory in general, see, for example, Jones [1980], Owen [1968], Lucas [1981b], or Roberts [1976].

EXERCISES FOR SECTION 2.16

1. For each of the following weighted majority games, describe all winning coalitions.
   (a) [51; 50, 30, 20];
   (b) [51; 60, 30, 10];
   (c) The Board of Supervisors in Nassau County, New York in 1964: [59; 31, 31, 21, 28, 2, 2];
   (d) [201; 100, 100, 100, 100, 1];
   (e) [151; 100, 100, 100, 1].

2. For the following weighted majority games, identify all *minimal* winning coalitions, that is, winning coalitions with the property that a removal of any player results in a losing coalition.
   (a) [7; 3, 3, 5, 6, 1];
   (b) [15; 14, 1, 1, 1, 1, 5];
   (c) All games of Exercise 1.

3. Calculate the Shapley–Shubik power index for each player in the following weighted majority games.
   (a) [51; 49, 47, 4];
   (b) [201; 100, 100, 100, 100, 1];
   (c) [151; 100, 100, 100, 1];
   (d) [12; 5, 5, 2, 1] (*Hint:* Is player 4 ever pivotal?);
   (e) [51; 26, 26, 26, 22].

4. Calculate the Shapley–Shubik power index for the following games.
   (a) [16; 9, 9, 7, 3, 1, 1]. (This game arose in the Nassau County, New York, Board of Supervisors in 1958; see Banzhaf [1965].)
   (b) [59; 31, 31, 21, 28, 2, 2]. (This game arose in the Nassau County, New York, Board of Supervisors in 1964; again see Banzhaf [1965].)

5. Consider a conference committee consisting of three senators, x, y, and z, and three members of the House of Representatives, a, b, and c. A measure passes this committee if and only if it receives the support of at least two senators and at least two representatives.
   (a) Identify the winning coalitions of this simple game.
   (b) Show that this game is not a weighted majority game. That is, we cannot find votes $v(x)$, $v(y)$, $v(z)$, $v(a)$, $v(b)$, and $v(c)$ and a quota q such that a measure passes if and only if the sum of the votes in favor of it is at least q. (*Note:* A similar argument shows that, in general, a bicameral legislature cannot be thought of as a weighted majority game.)

6. Which of the following defines a weighted majority game in the sense that there is a weighted majority game with the same winning coalitions? Give a proof of your answer.

(a) Three players, and a coalition wins if and only if player 1 is in it.

(b) Four players, a, b, x, y; a coalition wins if and only if at least a or b and at least x or y is in it.

(c) Four players and a coalition wins if and only if at least three players are in it.

7. Suppose that a state has 3 counties. The number of representatives of each county in the state legislature is given as follows: county A has 6, county B has 7, and county C has 2. If all representatives of a county vote alike, and a two-thirds majority of votes is needed to win, find the power of each county using the Shapley–Shubik index.

8. Calculate the Shapley–Shubik power index for the conference committee (Exercise 5).

9. (Lucas [1983]) In the original Security Council, there were five permanent members and only six nonpermanent members. The winning coalitions consisted of all five permanent members plus at least two nonpermanent members. Formulate this as a weighted majority game and calculate the Shapley–Shubik power index.

10. (Lucas [1983]) It has been suggested that Japan be added as a sixth permanent member of the Security Council. If this were the case, assume that there would still be 10 nonpermanent members and winning coalitions would consist of all six permanent members plus at least four nonpermanent members. Formulate this as a weighted majority game and calculate the Shapley–Shubik index.

11. Compute the Shapley–Shubik power index of a player with 1 vote in the game in which 6 players have 11 votes each, 12 players have 1 vote each, and 71 votes are needed to win.

12. If we do not require that every subset of a losing coalition is a losing coalition or that for all S, either S or $I - S$ is losing, then how many different simple games are there on a set of n players?

13. In a simple game, if p_i is the Shapley–Shubik power index for player i and $\sum_{i \in S} p_i$ is greater than $\frac{1}{2}$, is S necessarily a winning coalition? Why?

14. Suppose that $v(S)$ gives 1 if coalition S is winning and 0 if S is losing. Show that

$$p_i = \sum \{\gamma(s)[v(S) - v(S - \{i\})]\colon \quad S \text{ such that } i \in S\},$$

where

$$s = |S| \quad \text{and} \quad \gamma(s) = \frac{(s - 1)\,!\,(n - s)!}{n!}.$$

15. Apply the formula in Exercise 14 to compute the Shapley–Shubik index for each of the weighted majority games in Exercise 3.

2.17 AN ALGORITHM FOR GENERATING PERMUTATIONS*

In Examples 2.9 and 2.10 we discussed algorithms that would proceed by examining every possible permutation of a set. We did not comment there on the problem of determining in what order to examine the permutations, because we were making the point that such algorithms are not usually very efficient. However, there are occasions when such algorithms are useful. In connection with them, we need a procedure to generate all permutations of a set. In this section we describe such a procedure.

A natural order in which to examine permutations is the *lexicographic order*. To describe this order, suppose that $\pi = \pi_1 \pi_2 \pi_3$ and $\sigma = \sigma_1 \sigma_2 \sigma_3$ are two permutations of the set $\{1, 2, 3\}$. We say that π *precedes* σ if $\pi_1 < \sigma_1$ or if $\pi_1 = \sigma_1$ and $\pi_2 < \sigma_2$. (Note that if $\pi_1 = \sigma_1$ and $\pi_2 = \sigma_2$, then $\pi_3 = \sigma_3$.) For instance, $\pi = 123$ precedes $\sigma = 231$ since $1 < 2$, and $\pi = 123$ precedes $\sigma = 132$ because $1 = 1$ and $2 < 3$. More generally, suppose that

*This section can be omitted.

$\pi = \pi_1 \pi_2 \cdots \pi_n$ and $\sigma = \sigma_1 \sigma_2 \cdots \sigma_n$ are two permutations of the set $\{1, 2, \ldots, n\}$. Then π precedes σ if $\pi_1 < \sigma_1$ or if $\pi_1 = \sigma_1$ and $\pi_2 < \sigma_2$ or if $\pi_1 = \sigma_1$ and $\pi_2 = \sigma_2$ and $\pi_3 < \sigma_3$ or ... or if $\pi_1 = \sigma_1$ and $\pi_2 = \sigma_2$ and ... and $\pi_k = \sigma_k$ and $\pi_{k+1} < \sigma_{k+1}$ or if .... Thus, $\pi = 42135$ precedes $\sigma = 42153$ because $4 = 4$, $2 = 2$, $1 = 1$, and $3 < 5$. In this lexicographic order, we order as we do words in a dictionary, considering first the first "letter," then in case of ties the second "letter," and so on. The following lists all permutations of $\{1, 2, 3\}$ in lexicographic order:

$$123, \quad 132, \quad 213, \quad 231, \quad 312, \quad 321.$$

We shall describe an algorithm for listing all permutations in lexicographic order. The key step is to determine, given a permutation $\pi = \pi_1 \pi_2 \cdots \pi_n$, what permutation comes next. The last permutation in the lexicographic order is n, $n - 1$, $n - 2$, ..., 2, 1. This has no next permutation in the order. Any other permutation π has $\pi_i < \pi_{i+1}$ for some i. If $\pi_{n-1} < \pi_n$, the next permutation in the order is obtained by interchanging π_{n-1} and π_n. For instance, if $\pi = 43512$, then the next permutation is 43521. Now suppose that $\pi_{n-1} > \pi_n$. If $\pi_{n-2} < \pi_{n-1}$, then we rearrange the last three entries of π to obtain the next permutation in the order. Specifically, we consider π_{n-1} and π_n and find the smallest of these which is larger than π_{n-2}. We put this in the $(n-2)$nd position. We then order the remaining two of the last three digits in increasing order. For instance, suppose that $\pi = 15243$. Then $\pi_{n-1} > \pi_n$ but $\pi_{n-2} < \pi_{n-1}$. Both π_{n-1} and π_n are larger than π_{n-2}, and 3 is the smaller of π_{n-1} and π_n. Thus, we put 3 in the third position, and put 2 and 4 in increasing order, obtaining the permutation 15324. If π is 15342, we switch 4 into the third position, not 2, since $2 < 3$, and obtain 15423.

In general, if $\pi \neq n$, $n - 1$, $n - 2$, ..., 2, 1, there must be a rightmost i so that $\pi_i < \pi_{i+1}$. Then the elements from π_i on must be rearranged to find the next permutation in the order. This is accomplished by examining all π_j for $j > i$ and finding the smallest such π_j that is larger than π_i. Then π_i and π_j are interchanged. Having made the interchange, the numbers following π_j after the interchange are placed in increasing order. They are now in decreasing order, so simply reversing them will suffice. For instance, suppose that $\pi = 412653$. Then $\pi_i = 2$ and $\pi_j = 3$. Interchanging π_i and π_j gives us 413652. Then reversing gives us 413256, which is the next permutation in the lexicographic order.

The steps of the algorithm are summarized as follows.

Algorithm 2.1. Generating All Permutations of $\{1, 2, \ldots, n\}$

Input: n

Output: A list of all $n!$ permutations of $\{1, 2, \ldots, n\}$, in lexicographic order.

Step 1. Set $\pi = 12 \cdots n$ and output π.

Step 2. If $\pi_i > \pi_{i+1}$ for all i, stop. (The list is complete.)

Step 3. Find the largest i so that $\pi_i < \pi_{i+1}$.

Step 4. Find the smallest π_j so that $i < j$ and $\pi_i < \pi_j$.

Step 5. Interchange π_i and π_j.

Step 6. Reverse the numbers following π_j in the new order, let π denote the resulting permutation, output π, and return to step 2.

Note that Algorithm 2.1 can be modified so that as a permutation is generated, it is examined for one purpose or another. For details of a computer implementation of Algorithm 2.1, see for example Reingold et al. [1977]. See this same reference for similar algorithms for generating combinations. An early but comprehensive paper on generating permutations and combinations is Lehmer [1964].

EXERCISES FOR SECTION 2.17

1. For each of the following pairs of permutations, determine which comes first in the lexicographic order.
   (a) 2341 and 3412;
   (b) 3124 and 3241;
   (c) 231564 and 235164;
   (d) 7613254 and 7613542.

2. For each of the following permutations, find the permutation immediately following it in the lexicographic order.
   (a) 123456;
   (b) 3456712;
   (c) 124563;
   (d) 87612543;
   (e) 3456721;
   (f) 5437621.

3. List all permutations of $\{1, 2, 3, 4\}$ in lexicographic order.

4. List all permutations of $\{u, v, x, y\}$ in lexicographic order.

5. Let $f_n(\pi)$ be i if π is the ith permutation in the lexicographic order of all permutations of the set $\{1, 2, \ldots, n\}$. Compute
   (a) $f_2(12)$; (b) $f_3(213)$; (c) $f_4(1243)$; (d) $f_5(54321)$.

6. Suppose that $f_n(\pi)$ is defined as in Exercise 5 and that permutation π' is obtained from permutation $\pi = \pi_1 \pi_2 \cdots \pi_n$ by deleting π_1 and reducing by 1 all elements π_j such that $\pi_j > \pi_1$. Show that $f_n(\pi) = (\pi_1 - 1)(n - 1)! + f_{n-1}(\pi')$.

2.18 GOOD ALGORITHMS*

2.18.1 Asymptotic Analysis

We have already observed in Section 2.4 that some algorithms for solving combinatorial problems are not very good. In this section we try to make precise what we mean by a good algorithm. As we pointed out in Section 2.4, the cost of running a particular computer program on a particular machine will vary with the skill of the programmer and the characteristics of the machine. Thus, in modern computer science, the emphasis is

---

*This section should be omitted in elementary treatments.

on analyzing algorithms for solving problems rather than on analyzing particular computer programs, and that will be our emphasis here.

In analyzing how good an algorithm is, we try to estimate a complexity function $f(n)$, to use the terminology of Section 2.4. If n is relatively small, then $f(n)$ is usually relatively small, too. Most any algorithm will suffice for a small problem. We shall be mainly interested in comparing complexity functions $f(n)$ for n relatively large.

The crucial concept in the analysis of algorithms is the following. Suppose that F is an algorithm with complexity function $f(n)$ and that $g(n)$ is any function of n. We write that F or f is $O(g)$, and say that F or f is "big oh of g," if there is an integer r and a positive constant k so that for all $n \geq r, f(n) \leq kg(n)$. [If f is $O(g)$, we sometimes say that g *asymptotically dominates f*.] If f is $O(g)$, then for problems of input size at least r, an algorithm with complexity function f will never be more than k times as costly as an algorithm with complexity function g. To give some examples, $100n$ is $O(n^2)$ because for $n \geq 100$, $100n \leq n^2$. Also, $n + 1/n$ is $O(n)$, because for $n \geq 1$, $n + 1/n \leq 2n$. An algorithm that is $O(n)$ is called *linear*, an algorithm that is $O(n^2)$ is called *quadratic*, and an algorithm that is $O(g)$ for g a polynomial is called *polynomial*. Other important classes of algorithms in computer science are algorithms that are $O(\log n)$, $O(n \log n)$, $O(c^n)$ for $c > 1$, and $O(n!)$. We discuss these below or in the exercises.

An algorithm whose complexity function is c^n, $c > 1$, is called *exponential*. Note that every exponential algorithm is $O(c^n)$, but not every $O(c^n)$ algorithm is exponential. For an algorithm whose complexity function is n is $O(c^n)$ for any $c > 1$. This is because $n \leq c^n$ for n sufficiently large.

A generally accepted principle is that an algorithm is *good* if it is polynomial. This idea is originally due to Edmonds [1965]. See Garey and Johnson [1979], Lawler [1976], or Reingold *et al.* [1977] for a good discussion of it. We shall try to give a very quick justification here.*

Since we are interested in $f(n)$ and $g(n)$ only for n relatively large, we introduce the constant r in defining the concept "f is $O(g)$." But where does the constant k come from?

Consider algorithms F and G whose complexity functions are, respectively, $f(n) = 20n$ and $g(n) = 40n$. Now clearly algorithm F is preferable, because $f(n) \leq g(n)$ for all n. However, if we could just improve a particular computer program for implementing algorithm G so that it would run in $\frac{1}{2}$ the time, or if we could implement G on a faster machine so that it would run in $\frac{1}{2}$ the time, then $f(n)$ and $g(n)$ would be the same. Since the constant $\frac{1}{2}$ is independent of n, it is not farfetched to think of improvements by this constant factor to be a function of the implementation rather than of the algorithm. In this sense, since $f(n)/g(n)$ equals a constant, that is, since $f(n) = kg(n)$, the functions $f(n)$ and $g(n)$ are considered the same for all practical purposes.

Now, to say that f is $O(g)$ means that $f(n)$ is at most $kg(n)$ (for n relatively large). But since $kg(n)$ and $g(n)$ are considered the same for all practical purposes, $f(n) \leq kg(n)$ says that $f(n) \leq g(n)$ for all practical purposes. Thus, to say that f is $O(g)$ says that an algorithm of complexity g is no more efficient than an algorithm of complexity f.

Before justifying the criterion of polynomial boundedness, we summarize some basic results in the following theorem.

---

*The reader who only wants to understand the definition may skip the rest of this subsection.

Theorem 2.10.

(a) If c is a positive constant, then f is $O(cf)$ and cf is $O(f)$.

(b) n is $O(n^2)$, n^2 is $O(n^3)$, ..., n^{p-1} is $O(n^p)$, ... Moreover, n^p is not $O(n^{p-1})$.

(c) If $f(n) = a_q n^q + a_{q-1} n^{q-1} + \cdots + a_0$ is a polynomial of degree q, with $a_q > 0$, and if $a_i \geq 0$, all i, then f is $O(n^q)$.

(d) If $c > 1$ and $p \geq 0$, then n^p is $O(c^n)$. Moreover, c^n is not $O(n^p)$.

Part (a) of Theorem 2.10 shows that, just as we have assumed, algorithms of complexity f and cf are considered equally efficient. Part (b) asserts that an $O(n^p)$ algorithm is more efficient the smaller the value of p. Part (c) asserts that the degree of the polynomial tells the relative complexity of a polynomial algorithm. Part (d) asserts that polynomial algorithms are always more efficient than exponential algorithms. This is why polynomial algorithms are treated as good, whereas exponential ones are not. The results of Theorem 2.10 are vividly demonstrated in Table 2.11, which shows how rapidly different complexity functions grow. Notice how much faster the exponential complexity function 2^n grows in comparison to the other complexity functions.

Table 2.11 Growths of Different Complexity Functions

Input size n		Complexity function $f(n)$		
	n	n^2	$10n^2$	2^n
5	5	25	250	32
10	10	10^2	10^3	$1024 \approx 1.02 \times 10^2$
20	20	400	4,000	$1,048,576 \approx 1.05 \times 10^6$
30	30	900	9,000	1.07×10^9
50	50	2,500	25,000	1.13×10^{15}
$100 = 10^2$	10^2	10^4	10^5	1.27×10^{30}
$1,000 = 10^3$	10^3	10^6	10^7	$>10^{300}$
$10,000 = 10^4$	10^4	10^8	10^9	$>10^{3000}$

Proof of Theorem 2.10.

(a) Clearly, cf is $O(f)$. Take $k = c$. Next, f is $O(cf)$, because $f(n) \leq (1/c)cf(n)$ for all n.

(b) Since $n^p \geq n^{p-1}$ for $n \geq 1$, n^{p-1} is $O(n^p)$. Now, n^p is not $O(n^{p-1})$. For $n^p \leq cn^{p-1}$ only for $n \leq c$.

(c) Note that since $a_i \geq 0$, $a_i n^i \leq a_i n^q$, all i, all $n \geq 1$. Hence, it follows that $f(n) \leq (a_0 + a_1 + \cdots + a_q)n^q$, all $n \geq 1$.

(d) This is a standard result from calculus or advanced calculus. It can be derived by noting that $n^p/c^n \to 0$ as $n \to \infty$. This result is obtained by applying l'Hospital's rule p times. Since $n^p/c^n \to 0$, $n^p/c^n \leq k$ for n sufficiently large. A similar analysis shows that $c^n/n^p \to \infty$ as $n \to \infty$, so c^n could not be $\leq kn^p$ for $n \geq r$.   Q.E.D.

Before closing this subsection, we should note again that our results depend on the crucial "equivalence" between algorithms of complexities f and cf, and on the idea that the size n of the input is relatively large. In practice, an algorithm of complexity $100n$ is definitely worse than an algorithm of complexity n. Moreover, it is also definitely worse, for small values of n, than an algorithm of complexity 2^n. Thus, an $O(n)$ algorithm, in practice, can be worse than an $O(2^n)$ algorithm. The results of this section, and the emphasis on polynomial algorithms, must be interpreted with care.

2.18.2 NP-Complete Problems

In studying algorithms, it is convenient to distinguish between deterministic procedures and nondeterministic ones. An algorithm may be thought of as passing from state to state. A *deterministic algorithm* may move to only one new state at a time, while a *nondeterministic algorithm* may move to several new states at once. That is, a nondeterministic algorithm may explore several possibilities simultaneously. In this book we concentrate exclusively on deterministic algorithms, and indeed, when we use the term *algorithm*, we shall mean deterministic. The class of problems for which there is a deterministic algorithm whose complexity is polynomial is called P. The class of problems for which there is a nondeterministic algorithm whose complexity is polynomial is called NP. Clearly, every problem in P is also in NP. To this date, no one has discovered a problem in NP that can be shown not to be in P. However, there are many problems known to be in NP which may or may not be in P. Many of these problems are extremely common and seemingly difficult problems, for which it would be very important to find a polynomial (deterministic) algorithm. Cook [1971] discovered the remarkable fact that there were some problems L, known as NP-*hard* problems, with the following property: If L can be solved by a deterministic polynomial algorithm, then so can every problem in NP. The traveling salesman problem discussed in Example 2.9 is such an NP-hard problem. Indeed, it is an NP-*complete* problem, an NP-hard problem which belongs to the class NP. Karp [1972] showed that there were a great many NP-complete problems. Now many people doubt that every problem for which there is a nondeterministic polynomial algorithm also will have a deterministic polynomial algorithm. Hence, they doubt whether it will ever be possible to find deterministic polynomial algorithms for such NP-hard (NP-complete) problems as the traveling salesman problem. Thus, NP-hard (NP-complete) problems are hard in a very real sense. See Garey and Johnson [1979] for a comprehensive discussion of NP-completeness. See also Reingold et al. [1977].

Since real-world problems have to be solved, we cannot simply stop seeking a solution when we find that a problem is NP-complete or NP-hard. We make compromises, for instance by dealing with special cases of the problem which might not be NP-hard. Alternatively, we seek good algorithms which approximate the solution to the problem with which we are dealing. An increasingly important activity in present-day combinatorics is to find good algorithms which come close to the (optimal) solution to a problem, for instance the traveling salesman problem.

EXERCISES FOR SECTION 2.18

1. In each of the following cases determine if f is $O(g)$ and justify your answer *from the definition*.
   (a) $f = 3^n$, $g = 4^n$;
   (b) $f = n + 2/n$, $g = n^3$;
   (c) $f = 10n$, $g = n^2$;

(d) $f = 3^{2n}, g = 3^{5n} - 100$;
(e) $f = n^2 + n, g = \frac{1}{2} n^3$.

2. Prove that if f is $O(h)$ and g is $O(h)$, then $f + g$ is $O(h)$.

3. Prove that if f is $O(g)$ and g is $O(h)$, then f is $O(h)$.

4. Prove that if f is $O(g)$ and g is $O(f)$, then for all h, h is $O(f)$ if and only if h is $O(g)$.

5. **(a)** Show that $\log_2 n$ is $O(n)$.
   (b) Show that n is not $O(\log_2 n)$.

6. Suppose that $c > 1$.
   (a) Show that c^n is $O(n!)$.
   (b) Show that $n!$ is not $O(c^n)$.

7. **(a)** Is it true that n is $O(n \log_2 n)$? Why?
   (b) Is it true that $n \log_2 n$ is $O(n)$? Why?

8. For each of the following functions f and g, determine if f is $O(g)$ and justify your answer. You may use the definition, a theorem from the text, or the result of a previous exercise.
   (a) $f = 3n, g = n^2$;
   (b) $f = n^2, g = 3n$;
   (c) $f = 5n^3 + 2n, g = n^3$;
   (d) $f = n, g = 2^n$;
   (e) $f = n^3, g = n^5$;
   (f) $f = 7n^7, g = 3^n$;
   (g) $f = 10 \log_2 n, g = n$;
   (h) $f = n^2 + 2^n, g = 4^n$;
   (i) $f = \log_2 n + 25, g = n^3$;
   (j) $f = n^2, g = 2^n + n$;
   (k) $f = 8^n, g = 3n^{100} + 25n^{99} + 100$;
   (l) $f = 3n^4 + 2n^2 + 1, g = 2^n$;
   (m) $f = n + n \log_2 n, g = 2^n$;
   (n) $f = n^3 + 3^n, g = n!$.

9. Explain what functions are $O(1)$, that is, $O(g)$ for $g \equiv 1$.

10. Let $f(n) = \sum_{i=1}^{n} i$. Show that f is $O(n^2)$.

11. Which of the following functions are "big oh" of the others?

$$4n \log_2 n + \log_2 n, \qquad \frac{n}{2} + \frac{3}{n^2}, \qquad \log_2 \log_2 n.$$

12. Repeat Exercise 11 for $(\log_2 n)^2$, $5n^4$, and $4n^2 \log_2 n + 2^n$.

ADDITIONAL EXERCISES FOR CHAPTER 2

1. There are 1000 applicants for admission to a college that plans to admit 300. How many possible ways are there for the college to choose the 300 applicants admitted.?

2. An octapeptide is a chain of 8 amino acids, each of which is one of 20 naturally occurring amino acids. How many octapeptides are there?

3. In an RNA chain of 15 bases, there are 4 A's, 6 U's, 4 G's, and 1 C. If the chain begins with AU and ends with UG, how many such chains are there?

4. How many functions are there each of which assigns a number 0 or a number 1 to each $m \times n$ matrix of 0's and 1's?

5. How many switching functions of 4 variables either assign 1 to all bit strings that start with a 1 or assign 0 to all bit strings that start with a 1?

6. In scheduling chest X-rays, eight people have been told to come between 9 and 10 A.M., and nine others between 10 and 11 A.M.. In how many different orders can we perform the X-rays?

7. If a campus mail "ZIP code" has 5 digits with no repetitions, how many different ZIP codes are possible?

8. In an RNA chain of 18 bases, there are 4 A's, 5 U's, 5 G's, and 4 C's. If the chain begins either AU or GC, how many such chains are there?

9. How many distinguishable permutations are there of the letters in the word *optimization*?

10. A chain of 20 amino acids has 5 histidines, 6 arginines, 4 glycines, 1 aspargine, 3 lysines, and 1 glutamic acid. How many such chains are there?

11. A system of 8 components works if at least 3 of the first 4 components work and at least 3 of the second 4 components work. In how many ways can the system work?

12. Of 15 computer programs to be run in a day, 5 of them are short, 4 are long, and 6 are of intermediate length. If the 15 programs are all distinguishable, in how many different orders can they be run so that
   (a) all the short programs are run at the beginning?
   (b) all the programs of the same length class are run consecutively?

13. A family with 8 children has 2 children with black hair, 3 with brown hair, 1 with red hair, and 2 with blond hair. How many different birth orders can give rise to such a family?

14. An ice cream parlor offers 29 different flavors. How many different triple cones are possible if each scoop on the cone has to be a different flavor?

15. A man has 6 different pairs of shoes. In how many ways can he choose a right shoe and a left shoe that do not match?

16. Suppose that of 11 houses on a block, 6 have termites.
   (a) In how many ways can the presence or absence of termites occur so that the houses with termites are all next to each other?
   (b) In how many ways can the presence or absence of termites occur so that none of the houses with termites are next to each other?

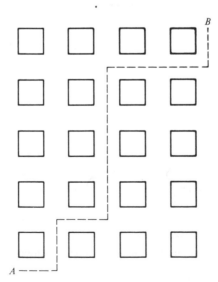

Figure 2.5 A grid of city streets. The dashed line gives a typical route.

17. How many ways are there to distribute 8 patients to 5 doctors?

18. How many ways are there to run an experiment in which each of 5 subjects is given some pills to take if there are 20 pills and each subject must take at least 1?

19. How many ways are there to pick 20 people to call in a 7-day period if there is a list of 100 people to choose from, if the order in which people are called does not count, and if repeat calls are acceptable?

20. (Polya) Consider a city with a grid of streets as shown in Figure 2.5. In an experiment, each subject starts at corner A and is told to proceed to corner B, which is four blocks east and five blocks north of A. He or she is given a route that takes exactly nine blocks to walk. The experimenter wants to use 100 subjects and give each one a different route to walk. Is it possible to perform this experiment? Why?

21. (Liu [1968]) We print one 5-digit number on a slip of paper. We include numbers beginning with 0's, for example, 00158. Since the digits 0, 1, and 8 look the same upside down, and since 6 and 9 are interchanged when a slip of paper is turned upside down, 5-digit numbers such as 61891 and 16819 can share the same slip of paper. If we want to include all possible 5-digit numbers, but allow this kind of sharing, how many different slips of paper do we need?

REFERENCES FOR CHAPTER 2

BANZHAF, J. F., III, "Weighted Voting Doesn't Work: A Mathematical Analysis," *Rutgers Law Rev.*, *19* (1965), 317–343.

BARLOW, R. E., and PROSCHAN, F., *Statistical Theory of Reliability and Life Testing: Probability Models*, Holt, Rinehart and Winston, New York, 1975.

BERGE, C., *Principles of Combinatorics*, Academic Press, New York, 1971.

BOYCE, D. M., and CROSS, M. J., "An Algorithm for the Shapley–Shubik Voting Power Index for Weighted Voting," unpublished Bell Telephone Laboratories manuscript, 1973.

BRAMS, S. J., LUCAS, W. F., and STRAFFIN, P. D. (eds.), *Political and Related Models*, Vol. 2 of *Modules in Applied Mathematics*, Springer-Verlag, New York, 1983.

BRAMS, S. J., SCHOTTER, A., and SCHWÖDIANER, G. (eds.), *Applied Game Theory*, IHS-Studies No. 1, Physica-Verlag, Würzburg, 1979.

BUCK, R. C., *Advanced Calculus*, 2d ed., McGraw-Hill, New York, 1965.

COHEN, D. I. A., *Basic Techniques of Combinatorial Theory*, Wiley, New York, 1978.

COOK, S. A., "The Complexity of Theorem-Proving Procedures," *Proceedings of the Third ACM Symposium on Theory of Computing*, Association for Computing Machinery, New York, 1971, pp. 151–158.

DEO, N., *Graph Theory with Applications to Engineering and Computer Science*, Prentice-Hall, Englewood Cliffs, N.J., 1974.

EDMONDS, J., "Paths, Trees, and Flowers," *Canad. J. Math.*, *17* (1965), 449–467.

ELSAYED, E. A., "Algorithms for Optimal Material Handling in Automatic Warehousing Systems," *Int. J. Prod. Res.*, *19* (1981), 525–535.

ELSAYED, E. A., and STERN, R. G., "Computerized Algorithms for Order Processing in Automated Warehousing Systems," *Int. J. Prod. Res.*, *21* (1983), 579–586.

FELLER, W., *An Introduction to Probability Theory and Its Applications*, 3d ed., Wiley, New York, 1968.

GAMOW, G., "Possible Mathematical Relation between Deoxyribonucleic Acid and Proteins," *K. Dan. Vidensk. Selsk. Biol. Medd.*, *22* (1954a), 1–13.

GAMOW, G., "Possible Relations between Deoxyribonucleic Acid and Protein Structures," *Nature*, *173* (1954b), 318.

GAREY, M. R., and JOHNSON, D. S., *Computers and Intractability: A Guide to the Theory of NP-Completeness*, W. H. Freeman, San Francisco, 1979.

GOLOMB, S., "Efficient Coding for the Deoxyribonucleic Channel," in *Mathematical Problems in the Biological Sciences*, Proceedings of Symposia in Applied Mathematics, Vol. 14, American Mathematical Society, Providence, R.I., 1962, pp. 87–100.

HARDY, G. H., and WRIGHT, E. M., *An Introduction to the Theory of Numbers*, 4th ed., Oxford University Press, London, 1960.

HARRISON, M. A., *Introduction to Switching and Automata Theory*, McGraw-Hill, New York, 1965.

Harvard Computation Laboratory Staff, *Synthesis of Electronic Computing and Control Circuits*, Harvard University Press, Cambridge, Mass., 1951.

HILL, F. J., and PETERSON, G. R., *Switching Theory and Logical Design*, Wiley, New York, 1968.

HOPCROFT, J. E., "Recent Developments in Random Algorithms," paper presented at SIAM National Meeting, Troy, N.Y., June 1981.

HUTCHINSON, G., "Evaluation of Polymer Sequence Fragment Data Using Graph Theory," *Bull. Math. Biophys.*, *31* (1969), 541–562.

JONES, A. J., *Game Theory: Mathematical Models of Conflict*, Wiley, New York, 1980.

KARP, R. M., "Reducibility among Combinatorial Problems," in R. E. Miller and J. W. Thatcher (eds.), *Complexity of Computer Computations*, Plenum Press, New York, 1972, pp. 85–103.

KARP, R. M., and LUBY, M. G., "A New Monte-Carlo Method for Estimating the Failure Probability of an n-Component System," Report No. UCB/CSD 83/117, Computer Science Division, University of California, Berkeley, 1983. (Submitted to *IEEE Trans. on Reliability*.)

KOHAVI, Z., *Switching and Finite Automata Theory*, McGraw-Hill, New York, 1970.

LAWLER, E. L., *Combinatorial Optimization: Networks and Matroids*, Holt, Rinehart and Winston, New York, 1976.

LEHMER, D. H., "The Machine Tools of Combinatorics," in E. F. Beckenbach (ed.), *Applied Combinatorial Mathematics*, Wiley, New York, 1964, pp. 5–31.

LIU, C. L., *Introduction to Combinatorial Mathematics*, McGraw-Hill, New York, 1968.

LIU, C. L., *Elements of Discrete Mathematics*, McGraw-Hill, New York, 1977.

LUCAS, W. F., "Applications of Cooperative Games to Equitable Allocation," in W. F. Lucas (ed.), *Game Theory and Its Applications*, Proceedings of Symposia in Applied Mathematics, Vol. 24, American Mathematical Society, Providence, R.I., 1981a.

LUCAS, W. F. (ed.), *Game Theory and Its Applications*, Proceedings of Symposia in Applied Mathematics, Vol. 24, American Mathematical Society, Providence, R. I., 1981b.

LUCAS, W. F., "Measuring Power in Weighted Voting Systems," in S. J. Brams, W. F. Lucas, and P. D. Straffin (eds.), *Political and Related Models*, Vol. 2 of *Modules in Applied Mathematics*, Springer-Verlag, New York, 1983, pp. 183–238.

MANN, I., and SHAPLEY, L. S., "Values of Large Games IV: Evaluating the Electoral College by Monte Carlo Techniques," Rand Corporation Memorandum RM-2651, September 1960; reproduced in M. Shubik (ed.), *Game Theory and Related Approaches to Social Behavior*, Wiley, New York, 1964a.

MANN, I., and SHAPLEY, L. S., "Values of Large Games VI: Evaluating the Electoral College Exactly," Rand Corporation Memorandum RM-3158-PR, May 1962; reproduced in part in M. Shubik (ed.), *Game Theory and Related Approaches to Social Behavior*, Wiley, New York, 1964b.

MOSIMANN, J. E., *Elementary Probability for the Biological Sciences*, Appleton-Century-Crofts, New York, 1968.

MOSIMANN, J. E., SHAPIRO, M. B., MERRIL, C. R., BRADLEY, D. F., and VINTON, J. E., "Reconstruction of Protein and Nucleic Acid Sequences IV: The Algebra of Free Monoids and the Fragmentation Stratagem," *Bull. Math. Biophys.*, *28* (1966), 235–260.

OWEN, G., *Game Theory*, Saunders, Philadelphia, 1968 (2nd ed., Academic Press, 1982).

PARZEN, E., *Modern Probability Theory and Its Applications*, Wiley, New York, 1960.

PRATHER, R. E., *Discrete Mathematical Structures for Computer Science*, Houghton Mifflin, Boston, 1976.

REINGOLD, E. M., NIEVERGELT, J., and DEO, N., *Combinatorial Algorithms: Theory and Practice*, Prentice-Hall, Englewood Cliffs, N.J., 1977.

RIORDAN, J., *An Introduction to Combinatorial Analysis*, Wiley, New York, 1958.

ROBERTS, F. S., *Discrete Mathematical Models, with Applications to Social, Biological, and Environmental Problems*, Prentice-Hall, Englewood Cliffs, N.J., 1976.

SHAPLEY, L. S., "A Value for *n*-Person Games," in H. W. Kuhn and A. W. Tucker (eds.), *Contributions to the Theory of Games*, Vol. 2, Annals of Mathematics Studies No. 28, Princeton University Press, Princeton, N.J., 1953, pp. 307–317.

SHAPLEY, L. S., "Measurement of Power in Political Systems," in W. F. Lucas (ed.), *Game Theory and Its Applications*, Proceedings of Symposia in Applied Mathematics, Vol. 24, American Mathematical Society, Providence, R.I., 1981, pp. 69–81.

SHAPLEY, L. S., and Shubik, M., "A Method for Evaluating the Distribution of Power in a Committee System," *Amer. Polit. Sci. Rev.*, *48* (1954), 787–792.

SPECTOR, W. S. (ed.), *Handbook of Biological Data*, Saunders, Philadelphia, 1956.

STANAT, D. F., and McALLISTER, D. F., *Discrete Mathematics in Computer Science*, Prentice-Hall, Englewood Cliffs, N.J., 1977.

STONE, H. S., *Discrete Mathematical Structures and Their Applications*, Science Research Associates, Chicago, 1973.

STRAFFIN, P. D., Jr., *Topics in the Theory of Voting*, Birkhäuser-Boston, Cambridge, Mass., 1980.

3 Introduction to Graph Theory

3.1 FUNDAMENTAL CONCEPTS*

3.1.1 Some Examples

In Example 1.4 we introduced informally the notion of a graph. In this chapter we study graphs and directed analogues of graphs, called digraphs, and their numerous applications. Graphs are a fundamental tool in solving problems of combinatorics. In turn, many of the counting techniques of combinatorics are especially useful in solving problems of graph theory. The theory of graphs is an old subject that has been undergoing a tremendous growth in interest in recent years. From the beginning, the subject has been closely tied to applications. It was invented by Euler [1736] in the process of settling the famous Königsberg bridge problem, which we discuss in Section 11.3.1. Graph theory was later applied by Kirchhoff [1847] to the study of electrical networks, by Cayley [1857, 1874] to the study of organic chemistry, by Hamilton to the study of puzzles, and by

---

*The topics in graph theory introduced in this chapter were chosen for three reasons. First, they illustrate quickly the nature and variety of applications of the subject. Second, they can be used to illustrate the counting techniques introduced in Chapter 2. Third, they will be used to illustrate the counting and existence results in Chapters 4–8. We return to graph theory more completely in Chapter 11, which begins a sequence of three chapters on graphs and networks and begins an introduction to the algorithmic aspects of graph theory.

In my undergraduate combinatorics course at Rutgers, I do not cover much graph theory, as there is a separate undergraduate graph theory course. Accordingly, I go through this chapter very rapidly. I cover Sections 3.1.1 and 3.1.2; all of Section 3.2 (but in about 30 minutes, with little emphasis on the exercises); 3.3.1, 3.3.3; 3.4.1, 3.4.2; 3.5.1, 3.5.2, 3.5.4., 3.5.5 (only the proof of Theorem 3.11); and 3.5.6 (without sketching the proof of Cayley's Theorem). Other sections can be added to expand on the material covered. In a graph theory course or in a combinatorics course with more emphasis on graphs or on computing or on theory, more material from this chapter should be included.

many mathematicians and nonmathematicians to the study of maps and map coloring. In the twentieth century, graph theory has been increasingly used in electrical engineering, computer science, chemistry, political science, ecology, genetics, transportation, information processing, and a variety of other fields.

To illustrate applications of graph theory, and to motivate the formal definitions of graph and digraph that we introduce, let us consider several examples.

Example 3.1   Transportation Networks

Graphs and digraphs arise in many transportation problems. For example, consider any set of locations in a given area, between which it is desired to transport goods, people, cars, and so on. The locations could be cities, warehouses, street corners, airfields, and the like. Represent the locations as points, as shown in the example of Figure 3.1, and draw an arrow or directed line (or curve) from location x to location y if it is possible to move the goods, people, and so on, directly from x to y. The situation where all the links are two-way can be more simply represented by drawing a single undirected line between two locations which are directly linked, rather than drawing two arrows for each pair of locations (again see Figure 3.1). Interesting questions about transportation networks are how to design them to move traffic efficiently, how to make sure that they are not vulnerable to disruption, and so on.

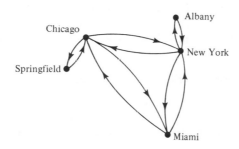

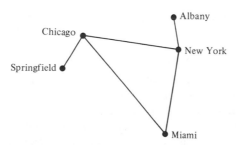

Figure 3.1   Direct air links.

Example 3.2   Communication Networks

Graphs are also used in the study of communications. Consider a committee, a corporate body, or any similar organization in which communication takes place. Let each member of the organization be represented by a point, as in Figure 3.2, and draw a line with an arrow from member x to member y if x can communicate

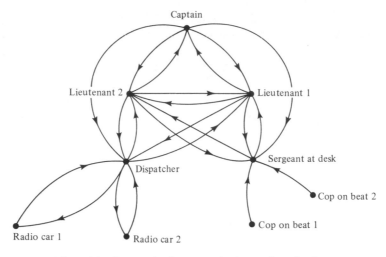

Captain

Lieutenant 2

Lieutenant 1

Dispatcher

Sergeant at desk

Cop on beat 2

Cop on beat 1

Radio car 1

Radio car 2

Figure 3.2 Communication network of part of a police force.

directly with *y*. For example, in the police force of Figure 3.2, the captain can communicate directly with the dispatcher, who in turn can reach the captain via either of the lieutenants.* Typical questions asked about such a "communication network" are similar to questions about transportation networks: How can the network be designed efficiently, how easy is it to disrupt communications, and so on?

Example 3.3   Physical Networks: Electrical, Telephone, and Pipeline

Graphs often correspond to physical systems. For example, an electrical network can be thought of as a graph, with points as the electrical components, and two points joined by an undirected line if and only if there is a wire connecting them. Similarly, in telephone networks, we can think of the switching centers and the individual telephones as points, with two points joined by a line if there is a direct telephone line between them. Oil or gas pipelines of the kind studied in Example 1.3 can also be translated into graphs in this way. We usually seek the most efficient or least expensive design of a network meeting certain interconnection requirements, or seek a network that is least vulnerable to disruption.

Example 3.4   Analysis of Computer Programs

Graphs have extensive applications in the analysis of computer programs. One approach is to subdivide a large program into subprograms, as an aid to understanding it, or to document it, or to detect structural errors. The subprograms considered are *program blocks*, or sequences of computer instructions with the property that whenever any instruction in the sequence is executed, all instructions in the sequence are executed. Let each program block be represented by a point, much as we represented cities in Figure 3.1. If it is possible to transfer control from the last instruction in program block *x* to the first instruction in program block *y*, draw a

*See Kemeny and Snell [1962, Ch. 8] for a more detailed discussion of a similar communication network of a police force.

line with an arrow from x to y. The resulting diagram is called a *program digraph*. (A *flowchart* is a special case where each program block has one instruction.) Certain program blocks are designated as starting and stopping blocks. To detect errors in a program, we might use a compiler to ask if there are points (program blocks) from which it is never possible to reach a stopping point by following arrows. Or we might use the compiler to ask if there are points that can never be reached from a starting point. (If so, these correspond to subroutines which are never called.) The program digraph can also be used to estimate running time for the program. Graphs have many other uses in computer science, for instance in the design and analysis of computers and digital systems, in data structures, in the design of fault-tolerant systems, and in the fault diagnosis of digital systems.

Example 3.5   Competition among Species

Graphs also arise in the study of ecosystems. Consider a number of species that make up an ecosystem. Represent the species (or groups of species) as points, as in Figure 3.3, and draw an undirected line between species x and species y if and only if x and y compete. The resulting diagram is called a *competition graph* (or *niche overlap graph*). Questions one can ask about competition graphs include questions about their structural properties (for instance, how "connected" are they and what are their connected "pieces"); and about the "density" of lines (ratio of the number of lines present to the number of lines possible).

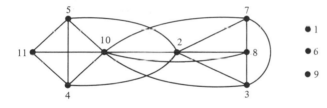

Figure 3.3   A competition graph for species in Malaysian rain forest. (From data of Harrison [1962], as adapted by Cohen [1978]. Graph from Roberts [1978].)

Key:

1. Canopy: leaves, fruit, flowers
2. Canopy animals: birds, fruit bats, and other mammals
3. Upper air mammals: birds and bats, insectivorous
4. Insects
5. Large ground animals: large mammals and birds
6. Trunk, fruit, flowers
7. Middle-zone scansorial animals: mammals in both canopy and ground zones
8. Middle-zone flying animals: birds and insectivorous bats
9. Ground: roots, fallen fruit, leaves, and trunks
10. Small ground animals: birds and small mammals
11. Fungi

Example 3.6   Tournaments

To give yet another application of graphs, consider a round-robin tournament* in tennis, where each player must play every other player exactly once, and no ties are allowed. One can represent the players as points and draw an arrow from player x to player y if x "beats" y, as in Figure 3.4. Similar tournaments arise in a surprisingly large number of places in the social, biological, and environmental sciences. Psychologists perform a pair comparison preference experiment on a set of alternatives by asking a subject, for each pair of alternatives, to state which he or

*This is not the more common elimination tournament.

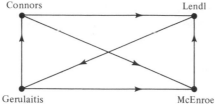

Connors   Lendl

Gerulaitis   McEnroe   **Figure 3.4** A tournament.

she prefers. This defines a tournament, if we think of the alternatives as corresponding to the players and "prefers" as corresponding to "beats." Tournaments also arise in biology. In the farmyard, for every pair of chickens, it has been found that exactly one "dominates" the other. This "pecking order" among chickens again defines a tournament. In studying tournaments, a basic problem is to decide on the "winner" and to rank the "players." Graph theory will help with this problem, too.

Example 3.7   Information Retrieval*

In an information retrieval system on a computer, each document being indexed is labeled with a number of *index terms* or *descriptors*. Let the index terms be drawn as points and join two points with a line if the corresponding index terms are closely related, as in Figure 3.5. The diagram that results is called a *similarity graph*. It can be used to produce a classification of documents and to help in information retrieval: One provides some index terms and the information retrieval system produces a list of related terms and the corresponding documents.

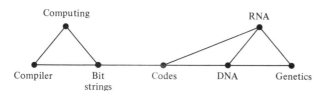

Computing   RNA

Compiler   Bit   Codes   DNA   Genetics
          strings   **Figure 3.5**   Part of a similarity graph.

3.1.2 Definition of Digraph and Graph

These examples all give rise to directed or undirected graphs. To be precise, let us define a *directed graph* or *digraph D* as a pair (V, A), where V is a set and A is a set of ordered pairs of elements of V. V will be called the set of *vertices* and A the set of *arcs*. (Some authors use the terms *node*, *point*, and so on, in place of *vertex*, and the terms *arrow*, *directed line*, *directed edge*, or *directed link* in place of *arc*.) If more than one digraph is being considered, we will use the notation $V(D)$ and $A(D)$ for the vertex set and the arc set of D, respectively. Usually, digraphs are represented by simple diagrams such as those of Figure 3.6. Here, the vertices are represented by points and there is a directed line (or curve or arrow) heading from u to v if and only if (u, v) is in A. For example, in the digraph D_1 of Figure 3.6, V is $\{u, v, w, x\}$ and A is the set

$$\{(u, v), (u, w), (v, w), (w, x), (x, u)\}.$$

*This example is due to Deo [1974].

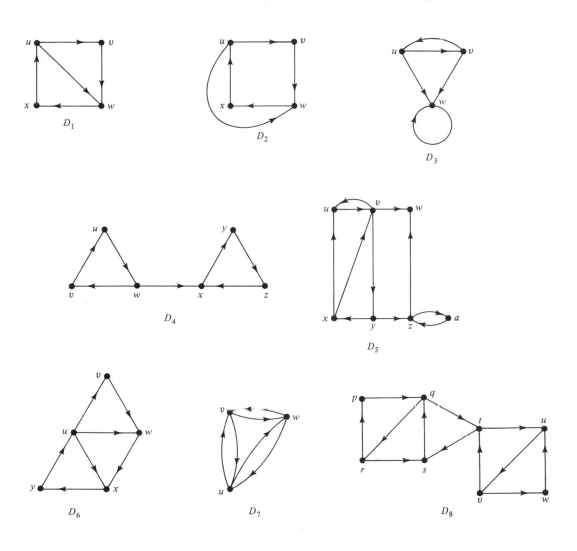

Figure 3.6 Digraphs.

If there is an arc from vertex u to vertex v, we shall say that u is *adjacent* to v. Thus, in Figure 3.6, in digraph D_1, v is adjacent to w, w is adjacent to x, and so on.

The reader should notice that the particular placement of the vertices in a diagram of a digraph is unimportant. The distances between the vertices have no significance, the nature of the lines joining them is unimportant, and so on. Moreover, whether or not two arcs cross is also unimportant; the crossing point is not necessarily a vertex of the digraph. All the information in a diagram of a digraph is included in the observation of whether or not a given pair of vertices is joined by a directed line or arc and in which direction the arc goes. Thus, digraphs D_1 and D_2 of Figure 3.6 are the same digraph, only drawn differently. In the next subsection, we shall say that D_1 and D_2 are isomorphic.

In a digraph, it is perfectly possible to have arcs in both directions, from u to v and from v to u, as shown in digraph D_3 of Figure 3.6, for example. It is possible to have an arc from a vertex to itself, as is also shown in digraph D_3. Such an arc is called a *loop*. It is not possible, however, to have more than one arc from u to v. Often in the theory and applications of digraphs, such multiple arcs are useful—this is true in the study of chemical bonding, for example—and then one studies *multigraphs* or better, *multidigraphs*, rather than digraphs.

Very often, there is an arc from u to v whenever there is an arc from v to u. In this case we say that the digraph (V, A) is a *graph*. Figure 3.7 shows several graphs. In the diagram of a graph, it is convenient to disregard the arrows and to replace a pair of arcs between vertices u and v by a single nondirected line joining u and v. (In the case of a directed loop, it is replaced by an undirected one.) We shall call such a line an *edge* of the graph and think of it as an unordered pair of vertices $\{u, v\}$. (The vertices u and v do not have to be distinct.) If there is an edge $\{u, v\}$ in the graph, we call u and v *neighbors*. The graph diagrams obtained from those of Figure 3.7 in this way are shown in Figure 3.8. Thus, a graph G may be defined as a pair (V, E), where V is a set of vertices and E is a set of unordered pairs of elements from V, the edges. If more than one graph is being considered, we will use the notation $V(G)$ and $E(G)$ for the vertex set and the edge set of G, respectively.

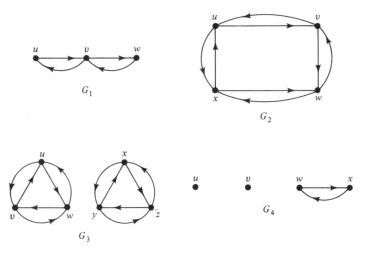

Figure 3.7 Graphs.

At this point, let us make explicit several assumptions about our digraphs and graphs. Many graph theorists make explicit the following assumption: There are no multiple arcs or edges, that is, no more than one arc or edge from vertex u to vertex v. For us, this assumption is contained in our definition of a digraph or graph. We shall assume, at least at first, that digraphs and graphs have no loops. (Almost everything we say for loopless digraphs and graphs will be true for digraphs and graphs with loops.) We shall also limit ourselves to digraphs or graphs with finite vertex sets. Let us summarize these assumptions as follows:

Assumptions: Unless otherwise specified, all digraphs and graphs referred to in this book have finite vertex sets, have no loops, and are not allowed to have multiple arcs or edges.

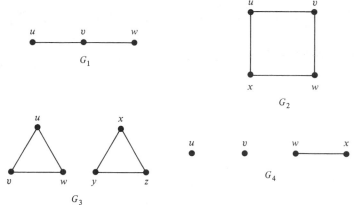

Figure 3.8 The graphs of Figure 3.7 with arcs replaced by edges.

3.1.3 Labeled Digraphs and the Isomorphism Problem*

A *labeled digraph* or *graph* of n vertices is a digraph or graph which has the integers 1, 2, ..., n assigned, one to each vertex. Two labeled digraphs or graphs can be, for all practical purposes, the same. For instance, Figure 3.9 shows an unlabeled graph G and three labelings of the vertices of G. The first two labelings are considered the same in the following sense: Their edge sets are the same. However, the first and third labelings are different, because, for instance, the first has an edge $\{3, 4\}$ while the third does not.

Figure 3.9 A graph G and three labelings of its vertices.

As a simple exercise in counting, let us ask how many distinct labeled graphs there are which have n vertices, for $n \geq 2$. The answer is most easily obtained if we observe that a labeled graph with n vertices can have at most $C(n, 2)$ edges, for $C(n, 2)$ is the number of unordered pairs of vertices from the n vertices. Let us suppose that the graph has e edges. Then we must choose e edges out of these $C(n, 2)$ possible edges. Hence, we see that the number $L(n, e)$ of labeled graphs with n vertices and e edges is given by

$$L(n, e) = C[C(n, 2), e] = \left(\binom{\binom{n}{2}}{e} \right). \tag{3.1}$$

Thus, by the sum rule, the number $L(n)$ of labeled graphs of n vertices is given by

$$L(n) = \sum_{e=0}^{C(n,\ 2)} L(n, e). \tag{3.2}$$

*This subsection is optional. It is placed here as a good application of the counting techniques of Chapter 2 and the concepts are used occasionally. The material can be returned to when it is needed.

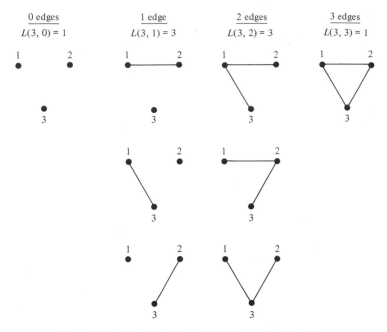

0 edges	1 edge	2 edges	3 edges
$L(3, 0) = 1$	$L(3, 1) = 3$	$L(3, 2) = 3$	$L(3, 3) = 1$

Figure 3.10 The eight different labeled graphs of 3 vertices.

For instance, if $n = 3$, then $L(3, 0) = 1$, $L(3, 1) = 3$, $L(3, 2) = 3$, $L(3, 3) = 1$, and $L(3) = 8$. Figure 3.10 shows these eight labeled graphs of 3 vertices.

Note that Equation (3.2) implies that the number of distinct labeled graphs grows very fast as n grows. To see that, note that if $r = C(n, 2)$, then using Theorem 2.8,

$$L(n) = \sum_{e=0}^{r} C(r, e)$$

$$= 2^r,$$

so

$$L(n) = 2^{n(n-1)/2}. \tag{3.3}$$

The number given by (3.3) grows exponentially fast as n grows. There are just too many graphs to answer most graph-theoretical questions by enumerating graphs.

Two labeled digraphs are considered the same if their arc sets are the same. How many different labeled digraphs are there with n vertices? By the product rule, there are $n(n - 1)$ possible arcs. The number $M(n, a)$ of labeled digraphs with n vertices and a arcs is given by

$$M(n, a) = C[n(n - 1), a]. \tag{3.4}$$

The number $M(n)$ of labeled digraphs with n vertices is thus given by

$$M(n) = \sum_{a=0}^{n(n-1)} C[n(n - 1), a]. \tag{3.5}$$

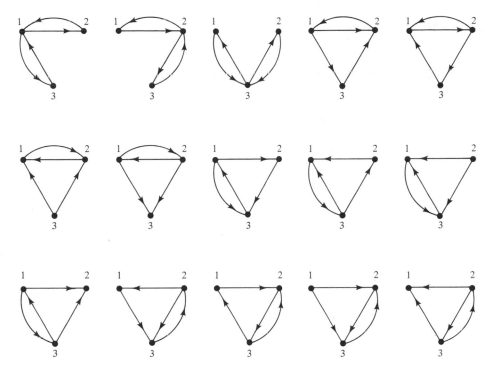

Figure 3.11 The 15 labeled digraphs of 3 vertices and 4 arcs.

For instance, if $n = 3$, then $M(3, 0) = 1$, $M(3, 1) = 6$, $M(3, 2) = 15$, $M(3, 3) = 20$, $M(3, 4) = 15$, $M(3, 5) = 6$, $M(3, 6) = 1$, and $M(3) = 64$. Figure 3.11 shows the 15 labeled digraphs with 3 vertices and 4 arcs.

Two unlabeled graphs (digraphs) G and H, each having n vertices, are considered the same if the vertices of both can be labeled with the integers $1, 2, \ldots, n$ so that the edge sets (arc sets) consist of the same unordered (ordered) pairs, that is, if the two graphs (digraphs) can each be given a labeling that shows them to be the same as labeled graphs (digraphs). If this can be done, we say that G and H are *isomorphic*. For instance, the graphs G and H of Figure 3.12 are isomorphic, as is shown by labeling vertex u as 1, v as 2, w as 3, x as 4, a as 1, b as 2, c as 3, and d as 4. The digraphs D_1 and D_2 of Figure 3.6 are also isomorphic. Labeling u as 1, v as 2, w as 3 and x as 4 in both digraphs demonstrates this.

Although it is easy to tell whether or not two labeled graphs or digraphs are the same, the problem of determining whether or not two unlabeled graphs or digraphs are

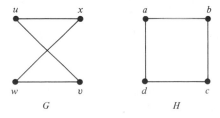

Figure 3.12 Graphs G and H are isomorphic.

the same, that is, isomorphic, is a very difficult one indeed. This is called the *isomorphism problem*, and it is one of the most important problems in graph theory. The most naive algorithm for determining if two graphs G and H of n vertices are isomorphic would fix a labeling of G using the integers $1, 2, \ldots, n$, and then try out all possible labelings of H using these integers. Thus, this algorithm has computational complexity $f(n) = n!$, and we have already seen in Sections 2.3 and 2.4 that even for moderate n, such as $n = 25$, considering this many cases is infeasible. Although better algorithms are known, the best algorithms known for solving the isomorphism problem have computational complexity which is exponential in the size of the problem; that is, they require an infeasibly large number of steps to compute if the number of vertices gets moderately large. See Reingold *et al.* [1977, Sec. 8.5] and Deo [1974, Sec. 11-7] for some discussion.*

EXERCISES FOR SECTION 3.1

1. In the digraph of Figure 3.1, identify
   (a) the set of vertices;
   (b) the set of arcs.
2. Repeat Exercise 1 for the digraph of Figure 3.4.
3. Repeat Exercise 1 for the digraph D_4 of Figure 3.6.
4. In each of the graphs of Figure 3.8, identify
   (a) the set of vertices;
   (b) the set of edges.
5. In digraph D_5 of Figure 3.6, find a vertex adjacent to vertex y.
6. In the graph of Figure 3.1, find all neighbors of the vertex New York.
7. Draw a transportation network with the cities New York, Paris, Vienna, Washington, D.C., and Algiers as vertices, and an edge joining two cities if it is possible to travel between them by road.
8. Draw a communication network for a team fighting a forest fire.
9. Draw a digraph representing the following football tournament. The teams are Princeton, Harvard, and Yale. Princeton beats Harvard, Harvard beats Yale, and Yale beats Princeton.
10. Figure 3.13 shows a graph and three labelings of its vertices.
    (a) Are the first two labelings the same? Why?
    (b) Are the first and the third? Why?

Figure 3.13   A graph and three labelings of its vertices.

11. Figure 3.14 shows a digraph and three labelings of its vertices.
    (a) Are the first two labelings the same? Why?
    (b) Are the first and the third? Why?

*Luks [1980] has shown that there is a polynomial algorithm for solving the isomorphism problem for graphs where the number of neighbors of a vertex is bounded.

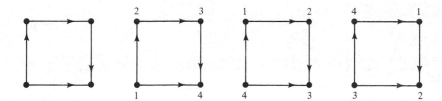

Figure 3.14   A digraph and three labelings of its vertices.

12. Find the number of labeled graphs with 4 vertices and 2 edges by using Equation (3.1). Check by drawing all such graphs.
13. How many different labeled graphs are there with 4 vertices and an even number of edges?
14. Find the number of labeled digraphs with 4 vertices and 2 arcs by using Equation (3.4). Check by drawing all such digraphs.
15. Are the graphs of Figure 3.15 isomorphic? Why?

Figure 3.15   Graphs for Exercise 15, Section 3.1.

16. Are the graphs of Figure 3.16 isomorphic? Why?

Figure 3.16   Graphs for Exercises 16 and 20, Section 3.1.

17. Are the digraphs of Figure 3.17 isomorphic? Why?

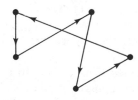

Figure 3.17   Digraphs for Exercise 17, Section 3.1.

18. Are the digraphs of Figure 3.18 isomorphic? Why?

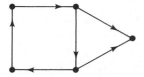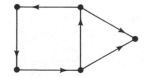

Figure 3.18   Digraphs for Exercise 18, Section 3.1.

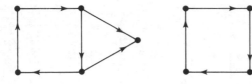

Figure 3.19 Digraphs for Exercise 19, Section 3.1.

19. Are the digraphs of Figure 3.19 isomorphic? Why?
20. An *orientation* of a graph arises by replacing each edge $\{x, y\}$ by one of the arcs (x, y) or (y, x). For instance, the digraph of Figure 3.14 is an orientation of graph H of Figure 3.12. For each of the graphs of Figure 3.16, find all nonisomorphic orientations.
21. Suppose that G and H are two graphs with the same number of vertices and the same number of edges. Suppose that α_k is the number of vertices in G with exactly k neighbors, and β_k is the number of vertices in H with exactly k neighbors. Suppose that $\alpha_k = \beta_k$ for all k. Are G and H necessarily isomorphic? Why?
22. Repeat Exercise 21 if $\alpha_2 = \beta_2 = |V(G)| = |V(H)|$ and $\alpha_k = \beta_k = 0$ for $k \neq 2$.

3.2 CONNECTEDNESS

3.2.1 Reaching in Digraphs

In a communication network, a natural question to ask is: Can one person initiate a message to another person? In a transportation network, an analogous question is: Can a car move from location u to location v? In a program digraph, we are interested in determining if from every vertex it is possible to follow arcs and ultimately hit a stopping vertex. All of these questions have in common the following idea of reachability in a digraph $D = (V, A)$: Can we reach vertex v by starting at vertex u and following arrows?

To make this concept precise, let us introduce some definitions. A *path* in D is a sequence

$$u_1, a_1, u_2, a_2, \ldots, u_t, a_t, u_{t+1}, \tag{3.6}$$

where $t \geq 0$, each u_i is in V, that is, is a vertex, and each a_i is in A, that is, is an arc, and a_i is the arc (u_i, u_{i+1}). That is, arc a_i goes from u_i to u_{i+1}. Since t might be 0, u_1 alone is a path, a path from u_1 to u_1. The path (3.6) is called a *simple path* if we never use the same vertex more than once.* For example, in digraph D_5 of Figure 3.6, $u, (u, v), v (v, w), w$ is a simple path and $u, (u, v), v, (v, y), y, (y, x), x, (x, v), v, (v, w), w$ is a path that is not a simple path since it uses vertex v twice. Naming the arcs is superfluous when referring to a path, so we simply speak of (3.6) as the path $u_1, u_2, \ldots, u_t, u_{t+1}$.

A path (3.6) is called *closed* if $u_{t+1} = u_1$. In a closed path, we end at the starting point. If the path (3.6) is closed and the vertices $u_1, u_2, \ldots, u_t$ are distinct, then (3.6) is called a *cycle* (a simple closed path†). (The reader should note that if the vertices u_i, $i \leq t$, are distinct, the arcs a_i must also be distinct.)

---

*One of the difficulties in learning graph theory is the large number of terms that have to be mastered early. To help the reader overcome this difficulty, we have included the terms *path*, *simple path*, and so on, in succinct form in Table 3.1.

†A simple closed path is, strictly speaking, not a simple path.

Table 3.1 Reaching and Joining

Digraph D	Graph G
$u_1, a_1, u_2, a_2, \ldots, u_t, a_t, u_{t+1}$	$u_1, e_1, u_2, e_2, \ldots, u_t, e_t, u_{t+1}$
Reaching	Joining
Path: $\quad a_i$ is (u_i, u_{i+1})	*Chain:* $\quad e_i$ is $\{u_i, u_{i+1}\}$
Simple path: $\quad$ Path and $\quad u_i$ distinct	*Simple chain:* $\quad$ Chain and $\quad u_i$ distinct
Closed path: $\quad$ Path and $u_{t+1} = u_1$	*Closed chain:* $\quad$ Chain and $u_{t+1} = u_1$
Cycle (simple closed path): $\quad$ Path $\quad u_{t+1} = u_1$ $\quad u_i$ distinct, $i \le t$ $\quad (a_i$ distinct)[a]	*Circuit (simple closed chain):* $\quad$ Chain $\quad u_{t+1} = u_1$ $\quad u_i$ distinct, $i \le t$ $\quad e_i$ distinct

[a]This follows from u_i distinct, $i \le t$.

To give some examples, the path u, v, w, x, u in digraph D_1 of Figure 3.6 is a cycle, as is the path u, v, w, x, y, u in digraph D_6. But the closed path u, v, y, x, v, y, x, u of D_5 is not a cycle, since there are repeated vertices. In general, in counting or listing cycles of a digraph, we shall not distinguish two cycles that use the same vertices and arcs in the same order, but start at a different vertex. Thus, in digraph D_6 of Figure 3.6, the cycle w, x, y, u, v, w is considered the same as the cycle u, v, w, x, y, u. The *length* of a path, simple path, cycle, and so on is the number of *arcs* in it. Thus, the path (3.6) has length t. In digraph D_5 of Figure 3.6, u, v, y, x, v is a path of length 4, u, v, y, z is a simple path of length 3, and u, v, y, x, u is a cycle of length 4. We say that v is *reachable* from u if there is a path from u to v. Thus, in D_5, z is reachable from u. However, u is not reachable from z.

A digraph $D = (V, A)$ is called *complete symmetric* if for every $u \ne v$ in V, the ordered pair (u, v) is an arc of D. For instance, the digraph D_7 of Figure 3.6 is complete symmetric. A complete symmetric digraph on n vertices has $n(n-1)$ arcs. Let us ask how many simple paths it has of a given length k, if $k \le n$, the number of vertices. The answer is that to find such a path, we choose any $k + 1$ vertices, and then order them. Thus, we have $P(n, k + 1)$ such paths.

3.2.2 Joining in Graphs

Suppose that $G = (V, E)$ is a graph. Terminology analogous to that for digraphs can be introduced. A *chain* in G is a sequence

$$u_1, e_1, u_2, e_2, \ldots, u_t, e_t, u_{t+1}, \tag{3.7}$$

where $t \ge 0$, each u_i is a vertex, and each e_i is the edge $\{u_i, u_{i+1}\}$. A chain is called *simple* if all the u_i are distinct and *closed* if $u_{t+1} = u_1$. A closed chain (3.7) in which $u_1, u_2, \ldots, u_t$

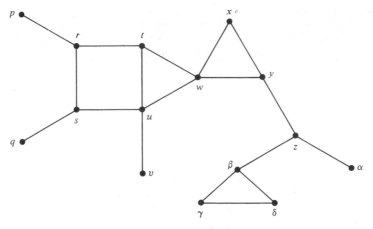

Figure 3.20 A graph.

are distinct and $e_1, e_2, \ldots, e_t$ are distinct is called a *circuit* (a simple closed chain).* The *length* of a chain, circuit, and so on of form (3.7) is the number of edges in it. We say that u and v are *joined* if there is a chain from u to v.

To give some examples, in the graph of Figure 3.20, r, $\{r, t\}$, t, $\{t, w\}$, w, $\{w, u\}$, u, $\{u, t\}$, t, $\{t, r\}$, r, $\{r, s\}$, s, $\{s, u\}$, u, $\{u, t\}$, t, $\{t, w\}$, w is a chain. This chain can be written without reference to the edges as r, t, w, u, t, r, s, u, t, w. A simple chain is given by r, t, u, w, x. A circuit is given by r, t, u, s, r. Finally, p, $\{p, r\}$, r, $\{r, p\}$, p is not considered a circuit, since $e_1 = e_t$. (The edges are unordered, so $\{p, r\} = \{r, p\}$.) Note that without the restriction that the edges be distinct, this would be a circuit. In the analogous case of digraphs, we did not have to assume arcs distinct for a cycle, since that follows from vertices distinct.

The *complete graph* on n vertices, denoted K_n, is defined to be the graph in which every pair of vertices is joined by an edge. In K_n, there are $C(n, 2)$ edges. Thus, in a competition graph of n vertices and e edges, the "density" of edges is $e/C(n, 2)$. The number of simple chains of length k in K_n is given by $P(n, k + 1)$.

3.2.3 Strongly Connected Digraphs and Connected Graphs

One reason graph theory is so useful is that its geometric point of view allows us to define various structural concepts. One of these concepts is connectedness. A digraph is said to be *strongly connected* if for every pair of vertices u and v, v is reachable from u and u is reachable from v. Thus, digraph D_6 of Figure 3.6 is strongly connected, but digraph D_5 is not. If a communication network is strongly connected, then every person can *initiate* a communication to every other person. If a transportation network is not strongly connected, then there are two locations u and v so that one cannot go from the first to the second, or vice versa. In Section 11.2 we shall study how to obtain strongly connected transportation networks. A program digraph is never strongly connected, for there are no arcs leading out of a stopping vertex. A tournament can be strongly connected (see Figure 3.4); however, it does not have to be. In a strongly connected tournament, it is hard to

---

*We shall see below why the restriction that the edges be distinct is added. Also, it should be noted that a simple closed chain is, strictly speaking, not simple.

rank the players, since we get situations where u_1 beats u_2, who beats u_3, ..., who beats u_t, who beats u_1.

We say that a graph is *connected* if between every pair of vertices u and v there is a chain. This notion of connectedness coincides with the one used in topology: The graph has one "piece." In Figure 3.8, graphs G_1 and G_2 are connected while G_3 and G_4 are not. Physical networks (electrical, telephone, pipeline) are usually connected. Indeed, we try to build them so that an outage at one edge does not result in a disconnected graph.

Algorithms to test whether or not a graph is connected or strongly connected have been designed in a variety of ways. The fastest are very good. They have computational complexity that is linear in the number of vertices n plus the number of edges e. In the language of Section 2.18, they take on the order of $n + e$ steps. [They are $O(n + e)$.] Since a graph has at most $\binom{n}{2}$ edges, we have

$$e \leq \binom{n}{2} = \frac{n(n-1)}{2} \leq n^2.$$

Thus, $n + e \leq n + n^2$ and so these algorithms take a number of steps of the order of $n + n^2$, which is a polynomial in n. [In the notation of Section 2.18, they are $O(n^2)$. We say they are quadratic in n.] We discuss some of these algorithms in Section 11.1. See also Aho *et al.* [1974], Even [1979], Golumbic [1980], or Reingold *et al.* [1977].

3.2.4 Subgraphs

In what follows it will sometimes be useful to look at parts of a graph. Formally, suppose that $G = (V, E)$ is a graph. A *subgraph* $H = (W, F)$ is a graph such that W is a subset of V and F is a set of unordered pairs of vertices of W which is a subset of E. Thus, to define a subgraph of G, we choose from G some vertices and some edges joining the chosen vertices. For instance, in Figure 3.21, graphs H and H' are both subgraphs of graph G. In H', the edge set consists of all edges of G joining vertices of $W = \{a, c, d\}$. We say that H' is the *subgraph generated* or *induced* by the vertices of W.

Similar concepts apply to digraphs. If $D = (V, A)$ is a digraph, then a *subgraph* $E = (W, B)$ of D is a digraph with W as a subset of V and B a set of ordered pairs of vertices of W which is a subset of A. E is a *generated subgraph* if B is all arcs of D that join vertices in W. For instance, in Figure 3.22, digraph E is a subgraph of digraph D and digraph E' is the subgraph generated by the vertices a, c, and d.

As a simple application of these ideas, let us ask how many subgraphs of k vertices there are if we start with the complete symmetric digraph on n vertices. To find such a subgraph, we first choose the k vertices; this can be done in $C(n, k)$ ways. These vertices are joined by $k(k-1)$ arcs in D. We may choose any subset of this set of arcs for the

G

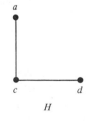

H

H'

Figure 3.21 H is a subgraph of G and H' is a generated subgraph.

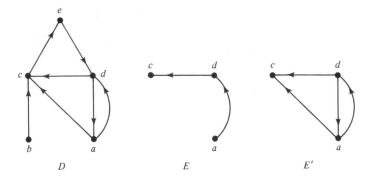

Figure 3.22 E is a subgraph of D and E' is the subgraph generated by vertices a, c, and d.

subgraph; that is, we may choose arcs for the subgraph in $2^{k(k-1)}$ ways. Thus, by the product rule, there are

$$C(n, k)2^{k(k-1)}$$

subgraphs of k vertices.

Example 3.8 Reliability of Systems Revisited

In Example 2.13 we studied systems consisting of components that might or might not work, and introduced rules for determining, given which components are working, whether or not the system works. In studying reliability of such systems, we commonly represent a system by a graph G and let each edge correspond to a component. Then in many situations it makes sense to say that the system works if and only if the subgraph H consisting of all vertices of G and the working edges of G is connected. For instance, the vertices of G might be locations and the edges might be road connections; then the system works if and only if it is possible to get from any location x to any location y by a system of "working" roads. Consider, for example, the graph G of Figure 3.23(a). There are three components, labeled x_1, x_2, and x_3. Clearly, the system works if and only if at least two of these components work. Similarly, if G is as in Figure 3.23(b), then the system works if and only if component x_1 works and at least two of the remaining three components work.

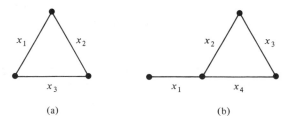

(a)   (b)   **Figure 3.23** Two systems.

3.2.5 Connected Components

Suppose that $G = (V, E)$ is a graph. A *connected component* or a *component* of G is a connected, generated subgraph H of G which is maximal in the sense that no larger connected generated subgraph K of G contains all the vertices of H. For example, in the

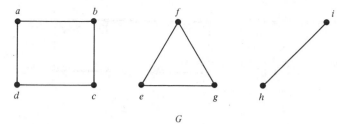

Figure 3.24 There are three components, the subgraphs generated by vertices a, b, c, and d, by vertices e, f, and g, and by vertices h and i.

G

graph G of Figure 3.24, the subgraph generated by the vertices a, b, and c is connected, but it is not a component since the subgraph generated by the vertices a, b, c, and d is a connected generated subgraph containing all the vertices of the first subgraph. This second subgraph is a component. There are three components in all, the other two being the subgraphs generated by vertices e, f, and g and by h and i. These components correspond to the "pieces" of the graph. In the information retrieval situation of Example 3.7, to give one simple application, components produce a natural classification of documents. In the competition graph of Figure 3.3, there are four components (three consisting of one vertex). Real-world competition graphs tend to have at least two components. Concepts for digraphs analogous to connected components are studied in the exercises.

EXERCISES FOR SECTION 3.2

1. For the digraph D_8 of Figure 3.6:
   (a) Find a path that is not a simple path.
   (b) Find a closed path.
   (c) Find a simple path of length 4.
   (d) Determine if u, (u, v), v, (v, w), w, (w, u), u is a cycle.
   (e) Find a cycle of length 3 containing vertex t.

2. For the graph of Figure 3.20:
   (a) Find a closed chain that is not a circuit.
   (b) Find the longest circuit.
   (c) Find a chain different from the one in the text which is not simple.
   (d) Find a closed chain of length 6.

3. Give an example of a digraph and a path in that digraph which is not a simple path but has no repeated arcs.

4. Give an example of a graph in which the shortest circuit has length 5 and the longest circuit has length 8.

5. For each digraph of Figure 3.6, determine if it is strongly connected.

6. Which of the graphs of Figure 3.25 are connected?

7. For each digraph of Figure 3.26:
   (a) Find a subgraph that is not a generated subgraph.
   (b) Find the subgraph generated by vertices 5, 8, and 9.
   (c) Find a strongly connected generated subgraph.

8. For the graph of Figure 3.27:
   (a) Find a subgraph that is not a generated subgraph.
   (b) Find a generated subgraph that is connected but not a connected component.
   (c) Find all connected components.

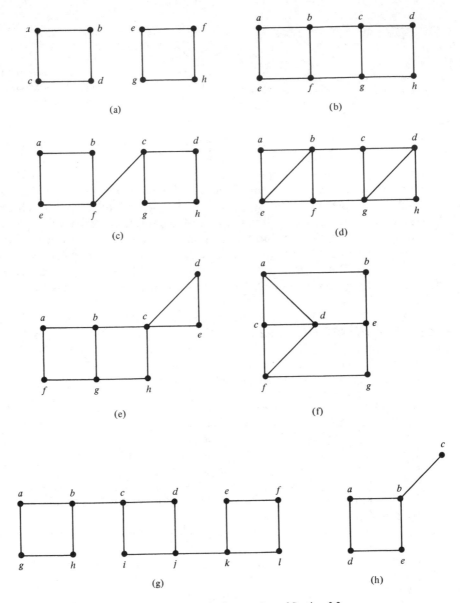

Figure 3.25 Graphs for exercises of Section 3.2.

9. A digraph is *unilaterally connected* if for every pair of vertices u and v, either v is reachable from u or u is reachable from v, but not necessarily both.
   (a) Give an example of a digraph that is unilaterally connected but not strongly connected.
   (b) For each digraph of Figure 3.6, determine if it is unilaterally connected.
10. A digraph is *weakly connected* if, when all directions on arcs are disregarded, the resulting graph (multigraph) is connected.
   (a) Give an example of a digraph that is weakly connected but not unilaterally connected.

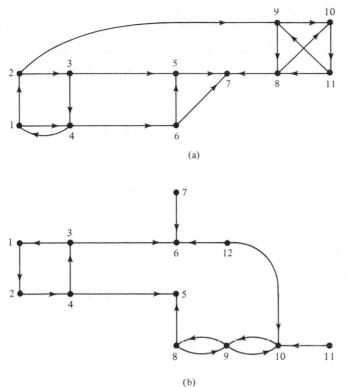

(a)

(b)

Figure 3.26 Digraphs for exercises of Section 3.2.

(b) Give an example of a digraph that is not weakly connected.

(c) For each digraph of Figure 3.6, determine if it is weakly connected.

11. Prove that if v is reachable from u in digraph D, then there is a simple path from u to v in D.

12. Suppose that a system defined by a graph G works if and only if the vertices of G and the working edges form a connected subgraph of G. Under what circumstances does each of the systems given in Figure 3.28 work?

13. In a digraph D, a *strong component* is a strongly connected, generated subgraph which is maximal in the sense that it is not contained in any larger strongly connected, generated subgraph. For example, in digraph D_5 of Figure 3.6, the subgraph generated by vertices x, y, v is strongly connected, but not a strong component since the subgraph generated by x, y, v, u is also strongly connected. The latter is a strong component. So is the subgraph generated by the single vertex w and the subgraph generated by the vertices z and a. There are no other strong components. (For applications of strong components to communication networks, to energy

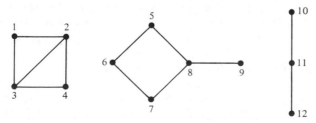

Figure 3.27 Graph for exercises of Section 3.2.

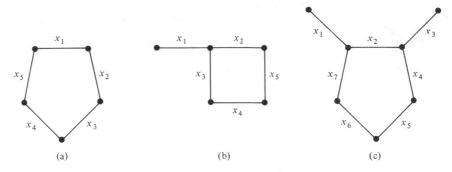

Figure 3.28 Systems for Exercise 12, Section 3.2.

demand, and to Markov models of probabilistic phenomena, see Roberts [1976].) Find all strong components of each digraph of Figure 3.26.

14. Find all strong components for the police force of Figure 3.2.

15. In a digraph D, show that:
    (a) every vertex is in some strong component;
    (b) every vertex is in at most one strong component.

16. Show that a graph is connected if and only if it has a chain going through all the vertices.

17. Prove that a digraph is strongly connected if and only if it has a closed path going through all the vertices.

18. Prove that in a unilateral digraph D, in any set of vertices, there is a vertex that can reach (using arcs of D) all others in the set.

19. Show from the result of Exercise 18 that a digraph is unilaterally connected if and only if it has a path going through all the vertices.

20. (a) Give an example of a strongly connected digraph that has no cycle through all the vertices.
    (b) Does every unilaterally connected digraph have a simple path through all the vertices?

21. A *weak component* of a digraph is a maximal weakly connected generated subgraph. For each digraph of Figure 3.26, find all weak components.

22. A *unilateral component* of a digraph is a maximal unilaterally connected generated subgraph.
    (a) Find a unilateral component with five vertices in the digraph (b) of Figure 3.26.
    (b) Is every vertex of a digraph in at least one unilateral component?
    (c) Can it be in more than one?

23. A digraph is *unipathic* if whenever v is reachable from u, then there is exactly one simple path from u to v.
    (a) Is the digraph D_4 of Figure 3.6 unipathic?
    (b) What about the digraph of Figure 3.14?

24. For a digraph that is strongly connected and has n vertices, what is the least number of arcs? What is the most? (Observe that a digraph which is strongly connected with the least number of arcs is very vulnerable to disruption. How many links is it necessary to sever in order to disrupt communications?)

25. (Harary *et al.* [1965]) Refer to the definition of unipathic in Exercise 23. Can two cycles of a unipathic digraph have a common arc? (Give a proof or counterexample.)

26. (Harary *et al.* [1965]) If D is strongly connected and has at least two vertices, does every vertex have to be on a cycle? (Give a proof or counterexample.)

27. Suppose that a digraph D is not weakly connected.

(a) If D has four vertices, what is the maximum number of arcs?

(b) What if D has n vertices?

28. Do Exercise 27 for digraphs that are unilaterally connected but not strongly connected.

29. Do Exercise 27 for digraphs that are weakly connected but not unilaterally connected.

30. (Harary *et al.* [1965]) If D is a digraph, define the *complementary digraph* D^c as follows: $V(D^c) = V(D) = V$ and an ordered pair (u, v) from $V \times V$ (with $u \neq v$) is in $A(D^c)$ if and only if it is not in $A(D)$. For example, if D is the digraph of Figure 3.29, then D^c is the digraph shown. Give examples of digraphs D that are weakly connected, not unilaterally connected, and such that:

(a) D^c is strongly connected.

(b) D^c is unilaterally connected but not strongly connected.

(c) D^c is weakly connected but not unilaterally connected.

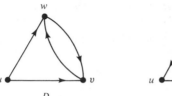 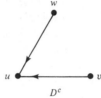

Figure 3.29 A digraph D and its complementary digraph D^c.

31. Find the number of distinct cycles of length k in the complete symmetric digraph of n vertices if two cycles are considered the same if one can be obtained from the other by changing the starting vertex.

3.3 GRAPH COLORING AND ITS APPLICATIONS

3.3.1 Some Applications

In Example 1.4 we considered the problem of scheduling meetings of committees in a state legislature, and translated that into a problem concerning graphs. In this section we formulate the graph problem as a problem of coloring a graph. We shall remark on a number of applications of graph coloring. In the next section, we shall apply the counting tools of Chapter 2 to count the number of graph colorings.

Example 3.9   Committee Scheduling Revisited

In the committee scheduling problem, we draw a graph G as follows. The vertices of G are all the committees that need to be assigned regular meeting times. Two committees are joined by an edge if and only if they have a member in common. Now we would like to assign a meeting time to each committee in such a way that if two committees have a common member, that is, if the corresponding vertices are joined by an edge in G, then the committees get different meeting times. Instead of assigning meeting times, let us think of assigning a color (corresponding to a meeting time) to each vertex of G, in such a way that if two vertices are joined by an edge, they get a different color. If such an assignment can be carried out using at most k colors, we call it a *k-coloring* of G and say G is *k-colorable*. The smallest

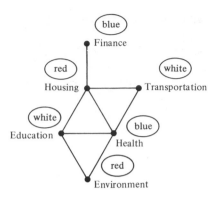

Figure 3.30 A 3-coloring for the graph G of Figure 1.1. The color assigned to a vertex is circled.

number k such that G is k-colorable is called the *chromatic number* of G and is denoted $\chi(G)$. To illustrate these ideas, let us return to the graph of Figure 1.1 and call that graph G. A 3-coloring of G is shown in Figure 3.30. The three colors used are red, white, and blue. Note that this graph also has a 4-coloring. Indeed, by our definition, this figure shows a 4-coloring—we do not require that each color be used. A 4-coloring that uses four colors would be obtained from this coloring by changing the color of the Finance vertex to green. We can always increase the number of colors used (up to the number of vertices). Thus, the emphasis is on finding the smallest number of colors we can use, that is, $\chi(G)$. Here, $\chi(G)$ obviously equals 3. For the three vertices Education, Housing, and Health must all get different colors. In this section we describe applications of graph coloring. As we remarked in Example 1.4, other applications with the same flavor as scheduling committee meetings involve scheduling final exam and class meeting times in a university, scheduling job assignments in a factory, and many other such problems.

Example 3.10 Index Registers and Optimizing Compilers (Tucker [1980])

In an optimizing compiler, it is more efficient to temporarily store the values of frequently used variables in index registers in the central processor, rather than in the regular memory, when computing in loops in a program. We wish to know how many index registers are required for storage in connection with a given loop. We let the variables that arise in the loop be vertices of a graph G, and draw an edge in G between two variables if at some step in the loop, they will both have to be stored. Then we wish to assign an index register to each variable in such a way that if two variables are joined by an edge in G, they must be assigned to different registers. The number of registers required is then given by the chromatic number of G.

Example 3.11 Channel Assignments

Television transmitters in a region are to be assigned a channel over which to operate. If two transmitters are within 100 miles of each other, they must get different channels. The problem of assigning channels can be looked at as a graph coloring problem. Let the vertices of a graph G be the transmitters and join two transmitters by an edge if and only if they are within 100 miles of each other. Assign a color (channel) to each vertex so that if two vertices are joined by an edge, they get different colors. How many channels are needed for a given region? This is the chromatic number of G. (For more information on applications of graph coloring to television or radio-frequency assignments, see, for example, Cozzens and Roberts [1982], Hale [1980], Opsut and Roberts [1981], or Pennotti [1976].)

Example 3.12 Routing Garbage Trucks

Let us next consider a routing problem posed by the Department of Sanitation of the City of New York (see Beltrami and Bodin [1973] and Tucker [1973]).* It should be clear that techniques like those to be discussed can be applied to other routing problems, for example milk routes, air routes, and so on. A garbage truck can visit a number of sites on a given day. A *tour* of such a truck is a schedule (an ordering) of sites it visits on a given day, subject to the restriction that the tour can be completed in one working day. We would like to find a set of tours with the following properties:

1. Each site i is visited a specified number k_i times in a week.
2. The tours can be partitioned among the six days of the week (Sunday is a holiday) in such a way that (a) no site is visited twice on one day† and (b) no day is assigned more tours than there are trucks.
3. The total time involved for all trucks is minimal.

In one method proposed for solving this problem, one starts with any given set of tours and successively improves the set as far as total time is concerned. (In the present state of the art, the method comes close to a minimal set, but does not always reach one.) At each step, the given improved collection of tours must be tested to see if it can be partitioned in such a way as to satisfy condition (2a), that is, partitioned among the six days of the week in such a way that no site is visited twice on one day. Thus, we need an efficient test for partitionability which can be applied over and over. Formulation of such a test reduces to a problem in graph coloring, and that problem will be the one on which we concentrate. (The reader is referred to Beltrami and Bodin [1973] and to Tucker [1973] for a description of the treatment of the total problem.)

To test if a given collection of tours can be partitioned so as to satisfy condition (2a), let us define a graph G, the *tour graph*, as follows. The vertices of G are the tours in the collection, and two distinct tours are joined by an edge if and only if they service some common site. Then the given collection of tours can be partitioned into six days of the week in such a way that condition (2a) is satisfied if and only if the collection of vertices $V(G)$ can be partitioned into six classes with the property that no edge of G joins vertices in the same class. It is convenient to speak of this question in terms of colors. Each class in the partition is assigned one of six colors and we ask for an assignment of colors to vertices such that no two vertices of the same color are joined by an edge.‡ The question about tours can now be rephrased as follows: Is the tour graph 6-colorable?

*Other applications of graph theory to sanitation will be discussed in Section 11.4.3.

†Requirement (a) is included to guarantee that garbage pickup is spread out enough to make sure that there is no accumulation.

‡This idea is due to Tucker [1973].

Example 3.13 Map Coloring and Planarity

The problem of coloring maps is an old and important problem which has been one of the prime stimulants for the development of graph theory. To explain the map coloring problem, let us consider the map of Figure 3.31. It is desired to color the countries on the map in such a way that if two countries share a common boundary, they get a different color. Of course, each country can be colored in a different color. However, for many years, cartographers have been interested in coloring maps with a small number of colors if possible. We can start coloring the countries in the map of Figure 3.31 by coloring country 1 red (see Figure 3.32). Then country 2, which shares a boundary with country 1, must get a different color, say blue. Country 3 shares a boundary with each of the other countries colored so far, so it must get still a different color, say green. Country 4 shares a boundary with all of the first three countries, so it must get still a fourth color, say yellow. Country 5 shares a boundary with countries 1, 2, and 4, but not with 3. Thus, it is possible to color country 5 green. Finally, country 6 cannot be blue or green. In Figure 3.32, we have colored it red. Notice that the map has been colored with four colors. No one has ever found a map for which more than four colors are needed, provided that "map" and "boundary" are defined precisely so as to eliminate such things as countries having two pieces, countries whose common boundary is a single point, and so on. For more than 100 years, it was conjectured that every map could be colored in four or fewer colors. However, despite the work of some of the best mathematical minds in the world, this *four-color conjecture* was neither proved nor disproved, and the four-color problem remained unsolved. Finally, in 1977, the four-color conjecture was proved (see Appel and Haken [1977], Appel *et al.* [1977]). The original proof of the four-color theorem involved the use of high-speed computers to check certain difficult cases, and involved some 1200 hours of computer time. One of the major steps in handling the map-coloring problem and k-colorings of maps was to translate the map-coloring problem into an equivalent but somewhat more tractable problem. Let the nation's capital of each country be represented by a point. Join two of these capitals by a (dashed) line if the corresponding countries share a common boundary. This gives rise to the lines of Figure 3.33. In Figure 3.34 the diagram is redrawn with just the capitals and the lines joining them left. This diagram defines a graph.

Instead of coloring a whole country, we can think of just coloring its capital. In terms of a graph such as that in Figure 3.34, the requirement is that if two

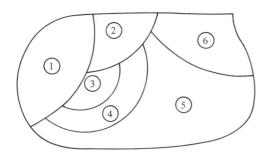

Figure 3.31   A map.

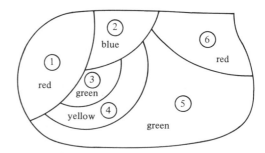

Figure 3.32   A coloring of the map in Figure 3.31.

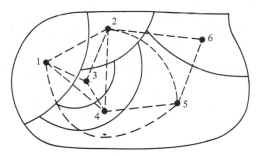

Figure 3.33 A dashed line joins two capitals if and only if their corresponding countries share a common boundary.

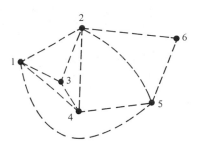

Figure 3.34 The graph of linked capitals from the map of Figure 3.33.

capitals or vertices are joined by an edge, they must get different colors. Thus, a map is colorable in k colors if and only if the corresponding graph is k-colorable.

The graph of Figure 3.34 has the property that no two edges cross except at vertices of the graph. A graph which has this property, or which has an equivalent (isomorphic) redrawing with this property, is called *planar*. Every map gives rise to a planar graph, and, conversely every planar graph comes from a map. Thus, the four-color theorem can be stated as the following theorem in graph theory: Every planar graph is 4-colorable. The first graph of Figure 3.35 is planar, even though its drawing has edges crossing. The second graph of Figure 3.35, which is equivalent (isomorphic) to the first graph, is drawn without edges crossing. The first graph in Figure 3.35 is the complete graph K_4. Thus, K_4 is planar. The graph of Figure 3.36(a), which is K_5, is not planar. No matter how you locate five points in the plane, it is impossible to connect them all with lines without two of these lines crossing. The reader is encouraged to try this. The graph of Figure 3.36(b) is another example of a graph that is not planar. This graph is called the *water–light–gas graph*. We think of three houses and three utilities, and try to join each house to each utility. It is impossible to do this without some lines crossing. Again, the reader is encouraged to try this. The problem of determining if a graph is planar has a variety of applications. For instance, in electrical engineering, the planar graphs correspond exactly to the possible printed circuits. In Section 11.6.4, we will show the use of planar graphs in a problem of facilities design. Kuratowski [1930] showed

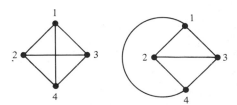

Figure 3.35 The first graph is planar, as is demonstrated by the second graph and the isomorphism shown by the vertex labelings.

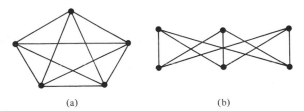

(a)   (b)

Figure 3.36 Two nonplanar graphs, K_5 and the water–light–gas graph.

that in some sense K_5 and the water–light–gas graph are the only nonplanar graphs. In Section 3.3.2 we present a formal statement of Kuratowski's Theorem.

Graph coloring and its generalizations have numerous applications in addition to those described here, for instance to time sharing in computer science, phasing traffic lights in transportation science, and various scheduling and maintenance problems in operations research. See Opsut and Roberts [1981] for descriptions of some of these problems.

3.3.2* Kuratowski's Theorem

To make precise the sense in which K_5 and the water–light–gas graph are the only nonplanar graphs, let us say that graph G' is obtained from graph G by *subdivision* if we obtain G' by adding vertices on one edge of G. In Figure 3.37, graph G_i' is always obtained from graph G_i by subdivision. Two graphs G and G' are called *homeomorphic* if both can be obtained from the same graph H by a sequence of subdivisions of edges. For example, any two simple chains are homeomorphic. Figure 3.38 shows two graphs G and G' obtained from a graph H by a sequence of subdivisions. Thus, G and G' are homeomorphic (and, incidentally, homeomorphic to H).

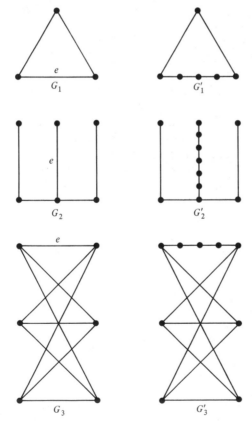

Figure 3.37 Graph G_i' is obtained from graph G_i by subdivision of edge e.

*This subsection can be omitted.

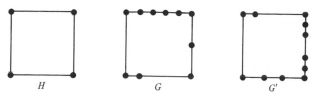

Figure 3.38 *G* and *G'* are homeomorphic because they are each obtained from *H* by subdivision.

Theorem 3.1 (Kuratowski [1930]). A graph is planar if and only if it has no subgraph* homeomorphic to K_5 or the water–light–gas graph.

For a proof of this theorem, we refer the reader to Harary [1969] or Bondy and Murty [1976]. According to Kuratowski's Theorem, the graph *G* of Figure 3.39 is not planar because it is homeomorphic to K_5 and the graph *G'* is not planar because it has a subgraph *H* homeomorphic to the water–light–gas graph.

Before closing this subsection, we note that Kuratowski's Theorem does not give a good algorithm for testing a graph for planarity. However, there are such algorithms. For a good discussion of them, see Even [1979].

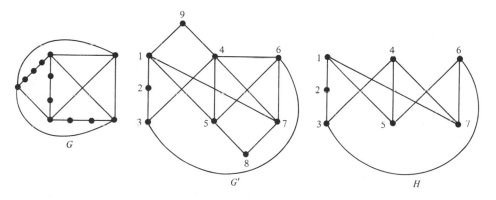

Figure 3.39 *G* is homeomorphic to K_5 and *G'* has the subgraph *H* which is homeomorphic to the water–light–gas graph.

3.3.3 Calculating the Chromatic Number

Let us study the colorability of various graphs. The graph K_4 is obviously colorable in four colors but not in three or fewer. Thus, $\chi(K_4) = 4$. The graph K_5 is colorable in five colors, but not in four or fewer (why?). Thus, $\chi(K_5) = 5$. (This is not a counterexample to the four-color theorem, since we have pointed out that K_5 is not a planar graph and so could not arise from a map.) The water–light–gas graph of Figure 3.36(b) is colorable in two colors: Color the top three vertices red and the bottom three blue. Since clearly two colors are needed, the chromatic number of the water–light–gas graph is 2.

Let us return briefly to the tour graph problem of Example 3.12. In general, to apply the procedure for finding a minimal set of tours, one has to have an algorithm, which can

*Not necessarily a generated subgraph.

be applied quickly over and over again, for deciding whether a given graph is k-colorable. Unfortunately, there is not always a "good," that is, polynomial, algorithm for solving this problem. Indeed, in general, it is not known whether there is a "good" algorithm (in the sense of Section 2.4) for deciding if a given graph is k-colorable. This problem is NP-complete in the sense of Section 2.18, so is difficult in a precise sense. The garbage truck routing problem has thus reduced to a difficult mathematical question. However, formulation in precise mathematical terms has made it clear why this is a hard problem, and it has also given us many tools to use in solving it, at least in special cases. Let us remark that in a real-world situation, it is not sufficient to say that a problem is unsolvable or hard. Imagine the $500,000 consultant walking into the mayor's office and reporting that after careful study, he or she has concluded that the problem of routing garbage trucks is hard! Garbage trucks must be routed. So what can you do in such a situation? The answer is, you develop partial solutions, you develop solutions that are applicable only to certain special situations, you modify the problem, or in some cases, you even "lie." You lie by using results that are not necessarily true but seem to work. One such result is the Strong Perfect Graph Conjecture or the Strong Berge Conjecture, which goes back to Claude Berge [1961, 1962]. (To understand the following reasoning, it is not important to know what this conjecture says. See Golumbic [1980] for a detailed treatment of the conjecture or Roberts [1976, 1978] or Tucker [1973] for a treatment of the conjecture and its applications to garbage trucks and routing.) There is a great deal of evidence in favor of the Strong Berge Conjecture, but no one has found a proof. However, as Tucker [1973] points out, if the conjecture is true, there is an efficient algorithm for determining if a given graph is colorable in a given number of colors, at least in the context of the tour graph problem, where the tour graph is changed each time only locally and not globally. Thus, Tucker argues, it pays to "lie" and to use the Strong Berge Conjecture in routing garbage trucks. What could go wrong? The worst thing that could happen, says Tucker, is the following. One applies the conjecture to garbage truck routing and finds a routing that is supposedly assignable to the 6 days of the week, but which in fact cannot be so assigned. In this worst case, think of the boon to mathematics: We would have found a counterexample to the Strong Berge Conjecture!

3.3.4 2-Colorable Graphs

Let us note next that there is one value of k for which it is easy to determine if G is k-colorable. This is the case $k = 2$.* A graph is 2-colorable if and only if the vertices can be partitioned into two classes so that all edges in the graph join vertices in two different classes. (Why?) A graph with this kind of a partition is called *bipartite*. The depth-first search procedure to be described in Section 11.1 gives a polynomial algorithm for testing if a graph is bipartite (see Reingold *et al.* [1977, pp. 399–400]).

There is also a useful characterization of 2-colorable graphs, which we shall state next. Let Z_p be the graph that consists of just a single circuit of p vertices. Figure 3.40 shows Z_3, Z_4, Z_5, and Z_6. It is easy enough to show that Z_4 and Z_6 are 2-colorable. A 2-coloring for each is shown. Clearly Z_3 is not 2-colorable. Z_5 is also not 2-colorable.

---

*For a discussion of other cases where there are good algorithms for determining if G is k-colorable, see Garey and Johnson [1979] and Golumbic [1980].

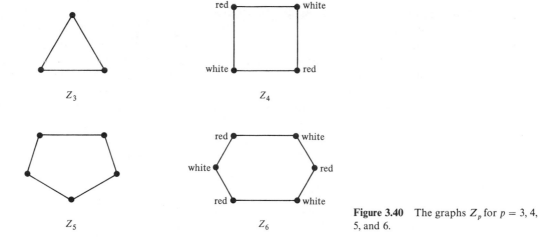

Figure 3.40 The graphs Z_p for $p = 3, 4, 5,$ and 6.

This takes a little proving, and we leave the proof to the reader. In general, it is easy to see that Z_p is 2-colorable if and only if p is even.

Now suppose that we start with Z_5, possibly add some edges, and then add new vertices and edges joining these vertices or joining them to vertices of Z_5. We might get graphs such as those in Figure 3.41. Now none of these graphs is 2-colorable. For a 2-coloring of the whole graph would automatically give a 2-coloring of Z_5. This is a general principle: A k-coloring of any graph G is a k-coloring of all subgraphs of G. Thus, any graph containing Z_5 as a subgraph is not 2-colorable. The same is true for Z_3, Z_7, Z_9, and so on. If G has any circuit of odd length, the circuit defines a subgraph of the form Z_p, for p odd; thus G could not be 2-colorable. The converse of this statement is also true, and we formulate the result as a theorem.

Theorem 3.2 (König [1936]). A graph is 2-colorable if and only if it has no circuits of odd length.

To prove the converse part of Theorem 3.2, we start with a graph G with no circuits of odd length and we present an algorithm for finding a 2-coloring of G. We may suppose that G is connected. (We can color each connected component separately.) Pick an

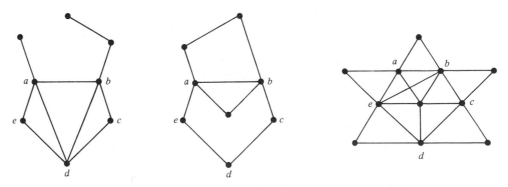

Figure 3.41 Graphs containing Z_5 as a subgraph. The vertices of Z_5 are labeled $a, b, c, d,$ and e.

arbitrary vertex x. Color x blue. Color all neighbors of x red. For each of these neighbors, color its uncolored neighbors blue. Continue in this way until all vertices are colored. The algorithm is illustrated by the graph of Figure 3.42, which is connected and has no odd length circuits. Here, x is chosen to be a. The 2-coloring is shown.

To implement this algorithm formally, vertices which have been colored but whose neighbors have not yet been colored are saved in an ordered list called a *queue*. At each stage of the algorithm, we find the first vertex y in the queue and remove it from the queue. We find its uncolored neighbors and color them the opposite color of y. We then add these neighbors to the end of the queue. We continue until all vertices have been colored. The algorithm is formally stated as Algorithm 3.1. Here, Q is the queue.

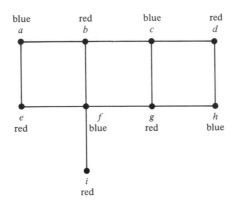

Figure 3.42 A connected graph without odd length circuits. A 2-coloring obtained by first coloring vertex a is shown.

Algorithm 3.1. Two Color

> *Input:* A graph $G = (V, E)$ which is connected and has no odd length circuits.
> *Output:* A coloring of the vertices of G using the two colors blue and red.

Step 1. Initially, all vertices of V are uncolored and Q is empty.

Step 2. Pick x in V, color x blue, and put x in Q.

Step 3. Let y be the first vertex in Q. Remove y from Q.

Step 4. Find all uncolored neighbors of y. Color each in the color opposite of that used on y. Add them to the end of Q in arbitrary order.

Step 5. If all vertices are colored, stop. Otherwise, return to Step 3.

To illustrate Algorithm 3.1, consider the graph of Figure 3.42. The steps are summarized in Table 3.2. Pick x to be vertex a, color a blue, and put a into the queue. Find the uncolored neighbors of a, namely b and e. Color them red (the opposite of a's color). Remove a from the queue Q and add b and e, say in the order b,e. Pick the first vertex in Q; here it is b. Remove it from Q. Find its uncolored neighbors; they are c and f. Color these the opposite color of that used for b, namely, blue. Add them to the end of Q in arbitrary order. If c is added first, then Q now is e,c,f. Remove the first vertex in Q, namely e. It has no uncolored neighbors. Q is now c,f. Go to the first vertex in Q, namely c, and remove it. Find its uncolored neighbors, d and g, and color them the opposite of c,

Table 3.2. Applying Algorithm 3.1 to the Graph of Figure 3.42

Vertex currently considered	Vertices colored	New queue Q
	a (blue)	a
a	b,e (red)	b,e
b	c,f (blue)	e,c,f
e	none	c,f
c	d,g (red)	f,d,g
f	i (red)	d,g,i
d	h (blue)	g,i,h

namely, red. Add d and g to the end of Q, say d first. Q is now f,d,g. Next remove f from the head of Q, color its uncolored neighbor i red, and add i to the end of Q. Q is now d,g,i. Finally, remove d from the head of Q, color its uncolored neighbor h blue, and add h to the end of Q. Stop since all vertices have now been colored.

The procedure we have used to visit all the vertices is called *breadth-first search.* It is a very efficient computer procedure which has many applications in graph theory. We shall return to breadth-first search, and the related procedure called depth-first search, in Section 11.1, when we discuss algorithms for testing a graph for connectedness. Algorithm 3.1 is a "good" algorithm in the sense of Sections 2.4 and 2.18. It is not hard to show that its complexity is of the order $n + e$, where n is the number of vertices of the graph and e the number of edges. Since a graph has at most $\binom{n}{2}$ edges, we reason as in Section 3.2.3 to conclude that

$$e \le \binom{n}{2} = \frac{n(n-1)}{2} \le n^2.$$

Thus, Algorithm 3.1 takes at most a number of steps of the order of $n + n^2$, which is a polynomial in n. [In the notation of Section 2.18, the algorithm is $O(n^2)$.]

To show that Algorithm 3.1 works, we have to show that every vertex eventually gets colored and that we attain a graph coloring this way. Connectedness of the graph G guarantees that every vertex eventually gets colored. (We omit a formal proof of this fact.) To show that we get a graph coloring, suppose u and v are neighbors in G. Could they get the same color? The easiest way to see that they could not is to define the distance $d(a, b)$ between two vertices a and b in a connected graph to be the length of the shortest chain between them. Then one can show that vertex z gets colored red if $d(x, z)$ is odd and blue if $d(x, z)$ is even.* (The proof is left as Exercise 25.) Now if two neighbors u and v are both colored red, then there is a shortest chain C_1 from x to u of odd length and a shortest chain C_2 from x to v of odd length. It follows that C_1 plus edge $\{u, v\}$ plus C_2 (followed backwards) forms a closed chain from x to x of odd length. But if G has an odd length closed chain, it must have an odd length circuit (Exercise 26). Thus, we have a contradiction. We reach a similar contradiction if u and v are both colored blue.

*The shortest proof of Theorem 3.2 is to simply *define* the coloring this way.

1. Consider the following four tours of garbage trucks on the West Side of New York City. Tour 1 visits sites from 21st Street to 30th, tour 2 visits sites from 28th to 40th Streets, tour 3 visits sites from 35th to 50th Streets, and tour 4 visits sites from 80th to 110th Streets. Draw the corresponding tour graph.

2. In Exercise 1, can the tours each be scheduled on Monday or Tuesday in such a way that no site is visited twice on the same day?

3. For each graph of Figure 3.43, determine if it is 3-colorable.

4. For each graph of Figure 3.43, determine its chromatic number $\chi(G)$.

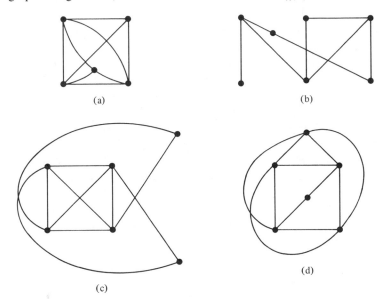

Figure 3.43 Graphs for exercises of Section 3.3.

5. A local zoo wants to take visitors on animal feeding tours, and has hit on the following tours. Tour 1 visits the lions, elephants, and ostriches; tour 2 the monkeys, birds, and deer; tour 3 the elephants, zebras, and giraffes; tour 4 the birds, reptiles, and bears; and tour 5 the kangaroos, monkeys, and seals. If animals should not get fed more than once a day, can these tours be scheduled using only Monday, Wednesday, and Friday?

6. The following tours of garbage trucks in New York City are being considered (behind the mayor's back). Tour 1 picks up garbage at the Empire State Building, then Madison Square Garden, and then Pier 42 on the Hudson River. Tour 2 visits Greenwich Village, Pier 42, the Empire State, and the Metropolitan Opera House. Tour 3 visits Shea Stadium, the Bronx Zoo, and the Brooklyn Botanic Garden. Tour 4 goes to the Statue of Liberty and Pier 42; tour 5 to the Statue of Liberty, the New York Stock Exchange, and the Empire State; tour 6 to Shea Stadium, Yankee Stadium, and the Bronx Zoo; and tour 7 to the New York Stock Exchange, Columbia University, and the Bronx Zoo. Assuming that sanitation workers refuse to work more than three days a week, can these tours be partitioned so that no site is visited more than once on a given day?

7. The following committees need to have meetings scheduled. Are three meeting times sufficient to schedule the committees so that no member has to be at two meetings simultaneously? Why?

$$A = \{\text{Smith, Jones, Brown, Green}\}$$

$$B = \{\text{Jones, Wagner, Chase}\}$$

$$C = \{\text{Harris, Oliver}\}$$

$$D = \{\text{Harris, Jones, Mason}\}$$

$$E = \{\text{Oliver, Cummings, Larson}\}.$$

8. In assigning frequencies to mobile radio telephones, a zone gets a frequency to be used by all vehicles in the zone. Two zones that interfere (because of proximity or meteorological reasons) must get different frequencies. How many different frequencies are required if there are 6 zones, a, b, c, d, e, and f, and zone a interferes with zone b only; b interferes with a, c, and d; c with b, d, and e; d with b, c, and e; e with c, d, and f; and f with e only?

9. In assigning work areas to workers, we want to be sure that if two such workers will interfere with each other, they will get different work areas. How many work areas are required if there are six workers, a, b, c, d, e, and f, and worker a interferes with workers b, e, and f; worker b interferes with workers a, c, and f; worker c with b, d, and f; worker d with c, e, and f; e with a, d, and f; and f with all other workers?

10. In a given loop of a program, six variables arise. Variable A must be stored in steps 1 through 4, variable B in steps 3 through 6, variable C in steps 4 through 7, variable D in steps 6 through 9, variable E in steps 8 and 9, and variable F in steps 9 and 10. How many index registers are required for storage?

11. Find the graphs corresponding to the maps of Figure 3.44. Note that a single common point does not qualify as a common boundary.

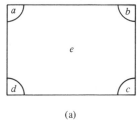 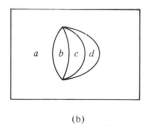 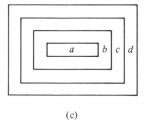

(a) (b) (c)

Figure 3.44 Maps.

12. Translate the map of Figure 3.45 into a graph G and calculate $\chi(G)$.

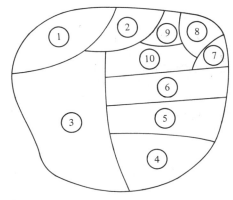

Figure 3.45 Map for exercises of Section 3.3.

13. Can any graph containing K_5 as a subgraph be planar? Why?

14. Suppose that there are four houses and four utilities, and each house is joined by an edge to each utility. Is the resulting graph planar? Why?

15. Could K_6 come from a map? Why?

16. Can K_3 be obtained by subdivision from K_2?

17. Are K_2 and K_3 homeomorphic?

18. Which of the graphs of Figure 3.43 are planar?

19. Let the *degree* of a vertex u in a graph G be defined to be the number of neighbors of u. Let $\Delta(G)$ be the maximum degree of a vertex in G. Show that $\chi(G) \leq 1 + \Delta(G)$.

20. Give an example of a graph G such that $\chi(G) < 1 + \Delta(G)$, where $\Delta(G)$ is defined in Exercise 19.

21. A *clique* in a graph G is a subgraph that is a complete graph. The *clique number* $\omega(G)$ is the largest p such that G has a clique of p vertices. State a relationship between χ and ω.

22. A graph G in which $\chi(G) = \omega(G)$ is called *weakly γ-perfect*. Give an example of a graph that is weakly γ-perfect and a graph that is not weakly γ-perfect.

23. Show that every 2-colorable graph is weakly γ-perfect.

24. Illustrate Algorithm 3.1 on graphs (c) and (g) of Figure 3.25.

25. Show that under Algorithm 3.1, vertex z gets colored red if $d(x, z)$ is odd and blue if $d(x, z)$ is even.

26. Show that if a graph has an odd length closed chain, then it has an odd length circuit.

27. In the situation of Exercise 8, we can imagine each zone getting m different radio frequencies, so that if two zones interfere, none of their frequencies match. This suggests the idea of a *multicoloring* or *m-tuple coloring* of a graph G, an assignment of a set $S(x)$ of m colors to each vertex x of G, with the requirement that

$$\{x, y\} \in E(G) \Rightarrow S(x) \cap S(y) = \varnothing.$$

Figure 3.46 shows a 2-tuple coloring of the graph Z_4. Here, $S(a) = \{\text{red, white}\}$, to give one example. This 2-tuple coloring uses six colors in all. Find a 2-tuple coloring using fewer colors.

{ red, white } a b { blue, green }

{ blue, purple } d c { red, orange }   **Figure 3.46** A 2-tuple coloring of Z_4.

28. Let $\chi_m(G)$ be the smallest k so that G has an m-tuple coloring using k colors in all. Find
   (a) $\chi_2(Z_4)$; (b) $\chi_2(Z_3)$; (c) $\chi_2(K_4)$; (d) $\chi_3(Z_4)$; (e) $\chi_3(K_4)$.

29. If G is 2-colorable and has at least one edge, show that $\chi_m(G) = 2m$.

30. A graph G is *k-edge-colorable* if you can color the edges with k colors so that two edges with common vertices get a different color. Let $\chi'(G)$, the *edge chromatic number*, be the smallest k so that G is k-edge-colorable. State a relation between $\chi'(G)$ and the number $\Delta(G)$ defined in Exercise 19. (For applications of edge coloring, see Fiorini and Wilson [1977].)

31. A set of vertices in a graph G is *independent* if there are no edges between any two vertices in the set. The *vertex independence number* $\alpha(G)$ is the size of the largest independent set of vertices. If G has n vertices, show that

$$\frac{n}{\alpha(G)} \leq \chi(G) \leq n - \alpha(G) + 1.$$

32. A graph G is called *k-critical* if $\chi(G) = k$ but $\chi(G - u) < k$ for each vertex $u \in V(G)$.
   (a) Find all 2-critical graphs.
   (b) Give an example of a 3-critical graph.
   (c) Can you identify *all* 3-critical graphs?

33. If $G = (V, E)$ is a graph, its *complement* G^c is the graph with vertex set V and an edge between $x \neq y$ in V if and only if $\{x, y\} \notin E$.
   (a) Comment on the relationship between the clique number $\omega(G)$ and the vertex independence number $\alpha(G^c)$.
   (b) Let $\theta(G)$ be the smallest number of cliques covering all vertices of G. Show that $\chi(G) = \theta(G^c)$.
   (c) Say that G is *weakly α-perfect* if $\theta(G) = \alpha(G)$. Give an example of a graph that is weakly α-perfect, and an example of a graph that is not weakly α-perfect.

34. G is said to be *γ-perfect (α-perfect)* if every generated subgraph of G is weakly γ-perfect (weakly α-perfect). Give examples of graphs that are
   (a) γ-perfect;
   (b) α-perfect;
   (c) weakly γ-perfect, but not γ-perfect;
   (d) weakly α-perfect, but not α-perfect.

35. Lovász [1972a, b] shows that a graph G is γ-perfect if and only if it is α-perfect. Hence, a graph that is γ-perfect (or α-perfect) is called *perfect*. For more on perfect graphs and their many applications, see Golumbic [1980].
   (a) Show that it is not true that G is weakly γ-perfect if and only if G is weakly α-perfect.
   (b) Show that G is γ-perfect if and only if G^c is γ-perfect. (You may use Lovász's result.)

36. (Tutte [1954], Kelly and Kelly [1954], Zykov [1949]) Show that for any integer k, there is a graph G such that $\omega(G) = 2$ and $\chi(G) = k$.

3.4 CHROMATIC POLYNOMIALS*

3.4.1 Definitions and Examples

Suppose that G is a graph and $P(G, x)$ counts the number of ways to color G in at most x colors. The related idea of counting the number of ways to color a map (see Exercise 3) was introduced by Birkhoff [1912] in an attack on the four color conjecture—we shall discuss this below. The numbers $P(G, x)$ were introduced by Birkhoff and Lewis [1946]. Note that $P(G, x)$ is 0 if it is not possible to color G using x colors. Now $\chi(G)$ is the smallest positive integer x such that $P(G, x) \neq 0$. One of the primary reasons for studying the numbers $P(G, x)$ is to learn something about $\chi(G)$. In this section we shall study these numbers $P(G, x)$ in some detail, making heavy use of the counting techniques of Chapter 2.

$$q \quad\quad\quad v$$

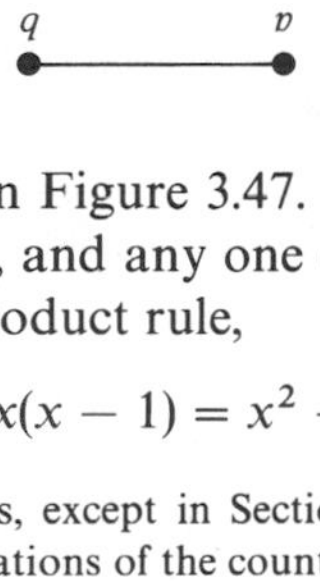

Figure 3.47 The graph K_2.

Consider first the graph K_2 shown in Figure 3.47. If x colors are available, then any one of them can be used to color vertex a, and any one of the remaining $x - 1$ colors can be used to color vertex b. Hence, by the product rule,

$$P(K_2, x) = x(x - 1) = x^2 - x.$$

---

*This section is not needed for what follows, except in Section 6.1.6. However, the reader is strongly encouraged to include it, for it provides many applications of the counting techniques of Chapter 2.

Table 3.3 Colorings of graph K_2 of Figure 3.47 with colors red (R), green (G), blue (B), and yellow (Y)

a	R	R	R	G	G	G	B	B	B	Y	Y	Y
b	G	B	Y	R	B	Y	R	G	Y	R	G	B

In particular,

$$P(K_2, 4) = 16 - 4 = 12.$$

The 12 ways of coloring K_2 in at most 4 colors are shown in Table 3.3.

Consider next the graph K_4 of Figure 3.48. If there are x colors available, then there are x choices of color for vertex a. For each of these choices, there are $x - 1$ choices for vertex b; for each of these there are $x - 2$ choices for vertex c, since c has edges to both a and b; for each of these there are $x - 3$ choices for vertex d, since d has edges to each of a, b, and c. Hence,

$$P(K_4, x) = x(x - 1)(x - 2)(x - 3) = x^4 - 6x^3 + 11x^2 - 6x.$$

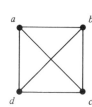

Figure 3.48 The graph K_4.

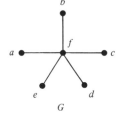

Figure 3.49 A graph G.

To color the graph G of Figure 3.49 in x or fewer colors, there are x choices for vertex f. Then there are $x - 1$ choices left for vertex a, and for each of these also $x - 1$ choices for vertex b (since b may get the same color as a), and for each of these $x - 1$ choices for vertex c, and for each of these $x - 1$ choices for vertex d, and for each of these $x - 1$ choices for vertex e. Hence,

$$P(G, x) = x(x - 1)^5 = x^6 - 5x^5 + 10x^4 - 10x^3 + 5x^2 - x.$$

Let us turn now to the graph Z_4 of Figure 3.50. Here a and b must get different colors, but a and c could get the same color. Similarly, b and d could get the same color. And so on. If there are x colors available, there are x choices of color for a. Then b and d can either get the same color or a different one. It is convenient to think of two cases.

Figure 3.50 The graph Z_4.

Case 1. b and d get the same color.

Case 2. b and d get different colors.

In case 1, c can get any of the colors not used for b and d. Hence, there are x choices for a, for each of these $x - 1$ choices for the common color for b and d, and for each of these $x - 1$ choices for the color for c. Hence, the number of colorings with x or fewer colors in which b and d get the same color is

$$x(x - 1)^2.$$

In case 2, there are x choices for a, $x - 1$ for b, $x - 2$ for d (since it must get a different color than b), and then $x - 2$ choices for c (since it cannot receive either of the colors used on b and d, but it can receive a's color). Hence, the number of colorings in which b and d get different colors is

$$x(x - 1)(x - 2)^2.$$

Since either case 1 or case 2 holds, the sum rule gives us

$$P(Z_4, x) = x(x - 1)^2 + x(x - 1)(x - 2)^2 = x^4 - 4x^3 + 6x^2 - 3x. \tag{3.8}$$

The reader will notice that in each of the examples we have given, $P(G, x)$ is a polynomial in x. This will always be the case. Hence, it makes sense to call $P(G, x)$ the *chromatic polynomial.*

Theorem 3.3.* $P(G, x)$ is always a polynomial.

We prove this theorem below. Recall that $\chi(G)$ is the smallest positive integer x such that $P(G, x) \neq 0$, that is, such that x is not a root of the polynomial $P(G, x)$. Thus, the chromatic number can be estimated by finding roots of a polynomial. Birkhoff's approach to the four-color problem was based on the idea of trying to characterize what polynomials were chromatic polynomials, in particular of maps (or planar graphs), and then seeing if 4 was a root of any of these polynomials. To this day, the problem of characterizing the chromatic polynomials is not yet solved. We return to this problem below.

Example 3.9 (Committee Scheduling) Continued

Let us count the number of colorings of the graph G of Figure 3.30 using three or fewer colors. We start by computing $P(G, x)$. If there are x choices for the color of Education, there are $x - 1$ for the color of Housing, and then $x - 2$ for the color of Health. This leaves $x - 2$ choices for the color of Transportation, and then $x - 2$ choices for the color of Environment, and then $x - 1$ choices for the color of Finance. Hence,

$$P(G, x) = x(x - 1)^2(x - 2)^3$$

and

$$P(G, 3) = 12.$$

---

*This theorem was discovered for maps (equivalently for graphs arising from maps) by Birkhoff [1912]. For all graphs, it is due to Birkhoff and Lewis [1946].

This result agrees with our conclusion in our discussion of Example 1.4. There we described in Table 1.7 the 12 possible colorings in three or fewer colors.

Next, we state two simple but fundamental results about chromatic polynomials. One of these is about the graph I_n with n vertices and no edges, the *empty graph*.

Theorem 3.4.
(a) If G is K_n, then

$$P(G, x) = x(x - 1)(x - 2) \cdots (x - n + 1). \tag{3.9}$$

(b) If G is I_n, then $P(G, x) = x^n$.

Proof. (a) There are x choices for the color of the first vertex, $x - 1$ choices for the color of the second vertex, and so on.
    (b) Obvious.   Q.E.D.

The expression on the right-hand side of (3.9) will occur so frequently that it is convenient to give it a name. We call it $x^{(n)}$.

3.4.2 Reduction Theorems

A very common technique in combinatorics is to reduce large computations to a set of smaller ones. We shall employ this technique often in this book, and in particular in Chapter 5 when we study recurrence relations and reduce calculation of x_n to values of x_k for k smaller than n. This turns out to be a very useful technique for the computation of chromatic polynomials. In this subsection we develop and apply a number of reduction theorems that can be used to reduce the computation of any chromatic polynomial to that of a chromatic polynomial for a graph with fewer edges, and eventually down to the computation of chromatic polynomials of complete graphs or empty graphs.
    We are now ready to state the first theorem.

Theorem 3.5 (The Two-Pieces Theorem).   Let the vertex set of G be partitioned into disjoint sets W_1 and W_2 and let G_1 and G_2 be the subgraphs generated by W_1 and W_2, respectively. Suppose that in G, no edge joins a vertex of W_1 to a vertex of W_2. Then

$$P(G, x) = P(G_1, x)P(G_2, x).$$

Proof. If there are x colors available, there are $P(G_1, x)$ colorings of G_1; for each of these there are $P(G_2, x)$ colorings of G_2. This is because a coloring of G_1 does not affect a coloring of G_2, since there are no edges joining the two pieces. The theorem follows by the product rule.   Q.E.D.

To illustrate the theorem, we note that if G is the graph shown in Figure 3.51, then

$$P(G, x) = [x^{(3)}]^2 = [x(x - 1)(x - 2)]^2.$$

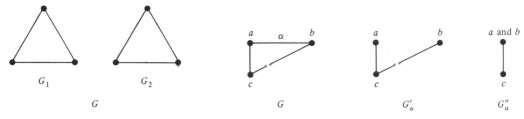

Figure 3.51 A graph with two pieces.

Figure 3.52 The graphs G'_α and G''_α.

To state our next reduction theorem, the crucial one, suppose that α is an edge of the graph G, joining vertices a and b. We define two new graphs from G. The graph G'_α is obtained by deleting the edge α but retaining the vertices a and b. The graph G''_α is obtained by identifying the two vertices a and b. In this case, the new combined vertex is joined to all those vertices to which either a or b were joined. (If both a and b were joined to a vertex c, only one of the edges from the combined vertex is included.) Figures 3.52 and 3.53 illustrate the two new graphs.

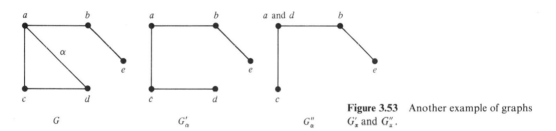

Figure 3.53 Another example of graphs G'_α and G''_α.

Theorem 3.6 (Fundamental Reduction Theorem).*

$$P(G, x) = P(G'_\alpha, x) - P(G''_\alpha, x). \tag{3.10}$$

Proof. Suppose that we use up to x colors to color G'_α. Then either a and b receive different colors or a and b receive the same color. The number of ways of coloring G'_α so that a and b receive different colors is simply the same as the number of ways of coloring G, that is, $P(G, x)$. The number of ways of coloring G'_α so that a and b get the same color is the same as the number of ways of coloring G''_α, namely $P(G''_\alpha, x)$. For we know that in G''_α, a and b would get the same color, and moreover the joint vertex a and b is forced to get a different color from that given to vertex c if and only if one of a and b, and hence both, is forced to get a different color from that given to vertex c. The result now follows by the sum rule:

$$P(G'_\alpha, x) = P(G, x) + P(G''_\alpha, x). \qquad \text{Q.E.D.}$$

To illustrate the theorem, let us consider the graph G of Figure 3.54 and use the Fundamental Reduction Theorem to calculate its chromatic polynomial. We choose the edge between vertices 1 and 3 to use as α, and so obtain G'_α and G''_α as shown in the figure.

*This theorem is due to Birkhoff and Lewis [1946].

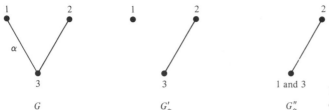

Figure 3.54 An application of the Fundamental Reduction Theorem.

Now the graph G''_α of Figure 3.54 is the complete graph K_2, and hence by Theorem 3.4, we know that

$$P(G''_\alpha, x) = x(x - 1). \tag{3.11}$$

The graph G'_α has two pieces, K_1 and K_2. By the Two-Pieces Theorem,

$$P(G'_\alpha, x) = P(K_1, x)P(K_2, x). \tag{3.12}$$

By Theorem 3.4, the first expression on the right-hand side of (3.12) is x and the second expression is $x(x - 1)$. Hence,

$$P(G'_\alpha, x) = x \cdot x(x - 1) = x^2(x - 1). \tag{3.13}$$

Substituting (3.11) and (3.13) into (3.10), we obtain

$$P(G, x) = x^2(x - 1) - x(x - 1)$$
$$= x(x - 1)[x - 1]$$
$$= x(x - 1)^2.$$

This expression could of course have been derived directly. However, it is a good illustration of the use of the Fundamental Reduction Theorem. Incidentally, applying the Fundamental Reduction Theorem again to $G''_\alpha = K_2$, we have

$$P(G''_\alpha, x) = P(K_2, x) = P(I_2, x) - P(I_1, x), \tag{3.14}$$

as is easy to verify. Also, since $K_1 = I_1$, (3.12) and (3.14) give us

$$P(G'_\alpha, x) = P(I_1, x)[P(I_2, x) - P(I_1, x)]. \tag{3.15}$$

Finally, plugging (3.14) and (3.15) into (3.10) gives us

$$P(G, x) = P(I_1, x)[P(I_2, x) - P(I_1, x)] - [P(I_2, x) - P(I_1, x)]. \tag{3.16}$$

We have reduced $P(G, x)$ to an expression (3.16) that requires only the knowledge of the polynomials $P(I_k, x)$ for different values of k.

As a second example, let us use the Fundamental Reduction Theorem to calculate $P(K_3, x)$. Note from Figure 3.55 that if $H = K_3$, then H'_α is the graph G of Figure 3.54 and H''_α is K_2. Thus,

$$P(K_3, x) = P(G, x) - P(K_2, x)$$
$$= x(x - 1)^2 - x(x - 1),$$

where the first expression arises from our previous computation, and the second from the

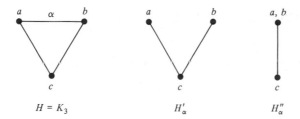

$H = K_3$ H'_α H''_α

Figure 3.55 A second application of the Fundamental Reduction Theorem.

formula for $P(K_2, x)$. Simplifying, we obtain

$$P(K_3, x) = x(x - 1)[(x - 1) - 1]$$
$$= x(x - 1)(x - 2)$$
$$= x^{(3)},$$

which agrees with Theorem 3.4.

The reader should note that each application of the Fundamental Reduction Theorem reduces the number of edges in each graph that is left. Hence, by repeated uses of the Fundamental Reduction Theorem, we must eventually end up with graphs with no edges, namely graphs of the form I_k. We illustrated this point with our first example. In any case, this shows that Theorem 3.3 must indeed hold, that is, that $P(G, x)$ is always a polynomial in x. For $P(I_k, x) = x^k$. Hence, we eventually reduce $P(G, x)$ to an expression that is a sum, difference, or product of terms each of which is of the form x^k, some k. (The proof may be formalized by arguing in terms of induction on the number of edges in the graph.)

Figure 3.56 gives one final illustration of the Fundamental Reduction Theorem. In that figure, we simply draw a graph to stand for the chromatic polynomial of the graph. The edge α is indicated at each step.

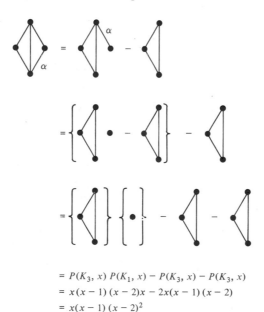

$= P(K_3, x)\, P(K_1, x) - P(K_3, x) - P(K_3, x)$
$= x(x - 1)\,(x - 2)x - 2x(x - 1)\,(x - 2)$
$= x(x - 1)\,(x - 2)^2$

Figure 3.56 Another application of the Fundamental Reduction Theorem.

3.4.3 Properties of Chromatic Polynomials*

We have already pointed out that one of Birkhoff's hopes in introducing chromatic polynomials was to be able to tell what polynomials were chromatic, and then to study the roots of those polynomials which were chromatic. In this section we study the properties of the chromatic polynomials. We shall discover that we can learn a great deal about a graph by knowing its chromatic polynomial. Exercises 16–20 outline the proofs of the theorems stated here, and present further properties of chromatic polynomials.

The first theorem summarizes some elementary properties of chromatic polynomials. These can be checked for all the examples of Sections 3.4.1 and 3.4.2.

Theorem 3.7. Suppose that G is a graph of n vertices and

$$P(G, x) = a_p x^p + a_{p-1} x^{p-1} + \cdots + a_1 x + a_0 .$$

Then:

(a) The degree of $P(G, x)$ is n, that is, $p = n$.
(b) The coefficient of x^n is 1, that is, $a_n = 1$.
(c) The constant term is 0, that is, $a_0 = 0$.
(d) Either $P(G, x) = x^n$ or the sum of the coefficients in $P(G, x)$ is 0.

Theorem 3.8 (Whitney [1932]). $P(G, x)$ is the sum of consecutive powers of x and the coefficients of these powers alternate in sign. That is, for some I,

$$P(G, x) = x^n - \alpha_{n-1} x^{n-1} + \alpha_{n-2} x^{n-2} - + \cdots \pm \alpha_0 , \qquad (3.17)$$

with $\alpha_i > 0$ for $i \geq I$ and $\alpha_i = 0$ for $i < I$.

Theorem 3.9. In $P(G, x)$, the absolute value of the coefficient of x^{n-1} is the number of edges of G.

Unfortunately, the properties of chromatic polynomials that we have listed in Theorems 3.7 and 3.8 do not characterize chromatic polynomials. There are polynomials $P(x)$ that satisfy all these conditions, but which are not chromatic polynomials of any graph. For instance, consider

$$P(x) = x^4 - 4x^3 + 3x^2.$$

Note that the coefficient of x^n is 1, the constant term is 0, the sum of the coefficients is 0, and the coefficients alternate in sign until from the coefficient of x^1 on, they are 0. However, $P(x)$ is not a chromatic polynomial of any graph. If it were, the number of vertices of the graph would have to be 4, by part (a) of Theorem 3.7. The number of edges would also have to be 4, by Theorem 3.9. The only graphs with four vertices and four edges do not have this polynomial as their chromatic polynomial, as is easy to check.

For further results on chromatic polynomials, see, for example, Liu [1972] or Read [1968].

*This subsection can be omitted.

EXERCISES FOR SECTION 3.4

1. Find the chromatic polynomial of each graph in Figure 3.57.

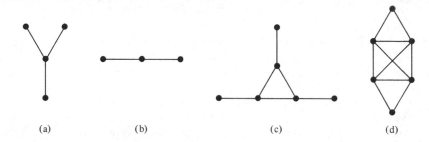

(a)   (b)   (c)   (d)

Figure 3.57 Graphs for exercises of Section 3.4.

2. For each graph in Figure 3.57, find the number of ways to color the graph in at most three colors.

3. The chromatic polynomial $P(M, x)$ of a map M is the number of ways to color M in x or fewer colors. Find $P(M, x)$ for the map of Figure 3.58.

Figure 3.58 Map for exercises of Section 3.4.

4. Let L_n be the graph consisting of a simple chain of n vertices. Find a formula for $P(L_n, x)$.

5. For each of the graphs of Figure 3.59, find the chromatic polynomial using reduction theorems. (You may reduce to graphs with previously known chromatic polynomials.)

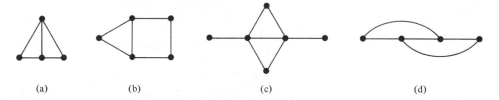

(a)   (b)   (c)   (d)

Figure 3.59 Graphs for exercises of Section 3.4.

6. If G is the graph of part (a) of Figure 3.59, express $P(G, x)$ in terms of polynomials $P(I_k, x)$ for various k.

7. If L_n is as defined in Exercise 4, what is the relation among $P(Z_n, x)$, $P(Z_{n-1}, x)$, and $P(L_n, x)$?

8. Use reduction theorems to find the chromatic polynomial of the map of Figure 3.60 (see Exercise 3). You may use the result of Exercise 4.

9. Let $N(G, x)$ be the number of ways of coloring G in exactly x colors. Find $N(G, x)$ for each of the following graphs G and the given values of x.
(a) Z_5, $x = 4$; (b) K_5, $x = 6$; (c) L_5, $x = 3$.

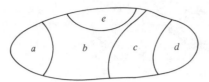

Figure 3.60 Map for exercises of Section 3.4.

10. Find an expression for $P(G, x)$ in terms of the numbers $N(G, r)$ for $r \le x$.

11. If we have a coloring of some vertices of G, we call this a *subcoloring* of G. A coloring of all the vertices of G that agrees with the subcoloring on the vertices that subcoloring colors is called an *extension* of the subcoloring. Figure 3.61 shows a graph G and three subcolorings of G. If there is just one more color available, say blue, then the first subcoloring can be extended to G in just one way, namely by coloring vertex a blue and vertex b red. However, the second subcoloring can be extended to a subcoloring of G in two ways, by coloring a blue and b red, or by coloring a red and b blue. How many extensions are there of the third subcoloring shown in Figure 3.61?

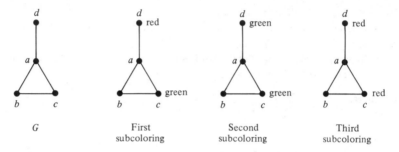

Figure 3.61 A graph G and three subcolorings of G.

12. Consider the graph G of Figure 3.62 and the subcoloring of the vertices a, b, and c shown in that figure. How many extensions are there of this subcoloring to all of G if the colors green, red, blue, and brown are available?

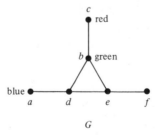

G **Figure 3.62** A graph and a subcoloring.

13. Consider the graphs of Figure 3.63 and the subcolorings shown in that figure. Find the number

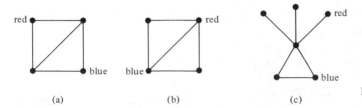

(a)   (b)   (c)

Figure 3.63 Graphs and subcolorings.

of extensions of each subcoloring to a coloring of the whole graph in three or fewer colors, if red, blue, and green are the three colors available.

14. Repeat Exercise 13, finding the number of extensions using at most x colors, $x \geq 3$.

15. Show that the following could not be chromatic polynomials.
    (a) $P(x) = x^8 - 1$;
    (b) $P(x) = x^5 - x^3 + 2x$;
    (c) $P(x) = 2x^3 - 3x^2$;
    (d) $P(x) = x^3 + x^2 + x$;
    (e) $P(x) = x^3 - x^2 + x$;
    (f) $P(x) = x^4 - 3x^3 + 3x^2$;
    (g) $P(x) = x^9 + x^8 - x^7 - x^6$.

16. Prove parts (c) and (d) of Theorem 3.7.

17. Prove parts (a) and (b) of Theorem 3.7 together, by induction on the number e of edges and by use of the Fundamental Reduction Theorem.

18. Prove Theorem 3.8 by induction on the number e of edges and by use of the Fundamental Reduction Theorem.

19. Prove Theorem 3.9 from Theorem 3.8, by induction on the number e of edges.

20. (a) If G has k connected components, show that the smallest i such that x^i has a nonzero coefficient in $P(G, x)$ is at least k.
    (b) Is this smallest i necessarily equal to k? Why?

21. Suppose that W_n is the wheel of $n + 1$ vertices, that is, the graph obtained from Z_n by adding one vertex and joining it to all vertices of Z_n. W_4 and W_5 are shown in Figure 3.64. Find $P(W_n, x)$. You may leave your answer in terms of $P(Z_n, x)$.

W_4 W_5 **Figure 3.64** The wheels W_4 and W_5.

22. If Z_n is the circuit of length n:
    (a) Show that for $n \geq 3$, $(-1)^n \{P(Z_n, x) - (x - 1)^n\}$ is constant, independent of n.
    (b) Solve for $P(Z_n, x)$ by evaluating the constant in part (a).

23. Suppose that H is a clique (Exercise 21, Section 3.3) of G and that we have two different subcolorings of H in at most x colors. Show that the number of extensions to a coloring of G in at most x colors is the same for each coloring.

24. The following is another reduction theorem. Suppose that H and K are generated subgraphs of G, with $V(G) = V(H) \cup V(K)$ and $E(G) = E(H) \cup E(K)$, and that $V(H) \cap V(K)$ is a clique of G of p vertices. Then

$$P(G, x) = \frac{P(H, x)P(K, x)}{x^{(p)}}.$$

    (a) Illustrate this result on the graph G of Figure 3.62, if H is the subgraph generated by $\{a, d, e, f\}$ and K the subgraph generated by $\{c, b, d, e\}$. (Disregard the subcoloring.)
    (b) Make use of the result of Exercise 23 to prove the theorem.

25. If the chromatic polynomial $P(K_n, x)$ is expanded out, the coefficient of x^k is denoted $s(n, k)$ and called a *Stirling number of the first kind*. Exercises 25–27 will explore these numbers. Find
  (a) $s(n, 0)$;   **(b)** $s(n, n)$;   **(c)** $s(n, 1)$;   **(d)** $s(n, n - 1)$.

26. Show that
$$s(n, k) = (n - 1) \cdot s(n - 1, k) + s(n - 1, k - 1).$$

27. Use the result in Exercise 26 to describe how to compute Stirling numbers of the first kind by a method similar to Pascal's triangle and apply your ideas to compute $s(6, 3)$.

3.5 TREES

3.5.1 Definition of a Tree and Examples

In this section and the next we consider one of the most useful concepts in graph theory, that of a tree. A *tree* is a graph T that is connected and has no circuits. Figure 3.65 shows some trees.

Example 3.14   Telephone Trees

Many companies and other organizations have prearranged telephone chains to notify their employees in case of an emergency, such as a snowstorm that will shut down the company. In such a telephone chain, a person in charge makes a decision (for example, to close because of snow) and calls several designated people, who each call several designated people, who each call several designated people, and so on. We let the people in the company be vertices of a graph and include an edge from a to b if a calls b (we include an undirected edge even though calling is not symmetric). The resulting graph is a tree, such as that shown in Figure 3.66.

Example 3.15   Sorting Mail*

Mail intended for delivery in the United States carries on it, if the sender follows Postal Service instructions, a ZIP code consisting of a certain number of decimal digits. Mail arriving at a post office is first sorted into 10 piles by the most significant digit. Then each pile is divided into 10 piles by the next most significant digit. And so on. The sorting procedure can be summarized by a tree, part of which is shown in Figure 3.67. To give a simpler example, suppose that we sort mail within a large organization by giving a mail code not unlike a ZIP code, but consisting of only three digits, each being 0 or 1. Then the sort tree is shown in Figure 3.68.

In this section and the next we consider a variety of other applications of trees, in particular emphasizing applications to organic chemistry and to searching and sorting problems in computer science.

3.5.2 Properties of Trees

Perhaps the most famous property of trees is obtained by noting the relationship between the number of vertices and the number of edges of a tree.

*This example is based on Deo [1974].

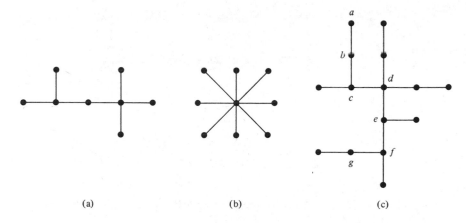

(a)   (b)   (c)

Figure 3.65 Some trees.

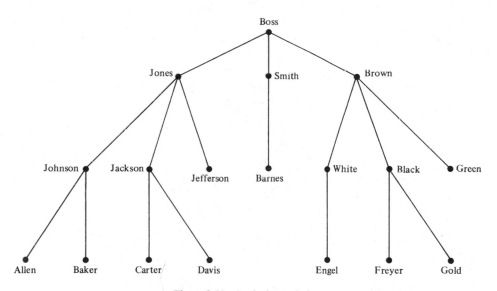

Figure 3.66 A telephone chain.

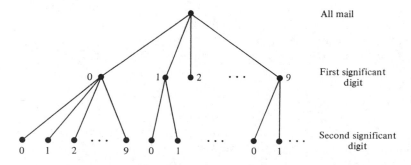

Figure 3.67 Part of the sort tree for sorting mail by ZIP code.

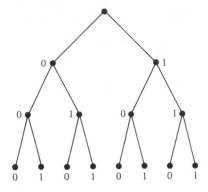

Figure 3.68 The sort tree for mail coded by three binary digits.

Theorem 3.10. If T is a tree of n vertices and e edges, then $n = e + 1$.

Theorem 3.10 is illustrated by any of the trees we have drawn. It will be proved in Section 3.5.3.

Note that the property $n = e + 1$ does not characterize trees. There are graphs that have $n = e + 1$ and are not trees (Exercise 5). However, we have the following result, which will be proved in Section 3.5.5.

Theorem 3.11. Suppose that G is a graph of n vertices and e edges. Then G is a tree if and only if G is connected and $n = e + 1$.

We next note an interesting consequence of Theorem 3.10. Let the *degree* of vertex u of graph G, deg (u), count the number of neighbors of u. Note that if we sum up the degrees of all vertices of G, we count each edge twice, once for each vertex on it. Thus, we have the following theorem.

Theorem 3.12. If G is any graph of e edges, we have $\sum \deg (u) = 2e$.

We can now derive the following result. This result will be used in counting the number of trees.

Theorem 3.13. If T is a tree with more than one vertex, there are at least two vertices of degree 1.

Proof. Since T is connected, every vertex must have degree ≥ 1. (Why?) Now by Theorems 3.10 and 3.12, $\sum \deg (u) = 2e = 2n - 2$. If $n - 1$ vertices had degree ≥ 2, the sum of the degrees would have to be at least $2(n - 1) + 1 = 2n - 1$, which is greater than $2n - 2$. Thus, no more than $n - 2$ vertices can have degree ≥ 2. Q.E.D.

3.5.3 Proof of Theorem 3.10*

Theorem 3.14. In a tree T, if x and y are any two vertices, then there is one and only one simple chain joining x and y.

*This subsection can be omitted.

Proof. We know there is a chain between x and y, since T is connected. A shortest chain between x and y must be a simple chain. For if $x, u, \ldots, v, \ldots, v, \ldots, w, y$ is a shortest chain between x and y with a repeated vertex v, we can skip the part of the chain between repetitions of v, thus obtaining a shorter chain from x to y. Thus, there is a simple chain joining x and y. Suppose next that $C_1 = x, x_1, x_2, \ldots, x_k, y$ is a shortest simple chain joining x and y. We show that there can be no other simple chain joining x and y. Suppose C_2 is such a chain. Let x_{p+1} be the first vertex of C_1 on which C_1 and C_2 differ and let x_q be the next vertex of C_1 following x_p on which C_1 and C_2 agree. Then we obtain a circuit by following C_1 from x_p to x_q and C_2 back from x_q to x_p.   Q.E.D.

To illustrate this theorem, we note that in tree (c) of Figure 3.65, the unique simple chain joining vertices a and g is given by a, b, c, d, e, f, g.

Proof of Theorem 3.10. The proof is by induction on n. If $n = 1$, the result is trivial. The only graph with one vertex has no edges. Now suppose that the result is true for all trees of fewer than n vertices and that tree T has n vertices. Pick any edge $\{u, v\}$ in T. By Theorem 3.14, u, v is the only simple chain between u and v. If we take edge $\{u, v\}$ away from G (but leave vertices u and v), we get a new graph H. Now in H, there can be no chain between u and v. For if there is, it is easy to find a simple chain between u and v. But this is also a simple chain in G, and G has only one simple chain between u and v.

Now since H has no chain between u and v, H is disconnected. It is not difficult to show (Exercise 21) that H has exactly two connected components. Call these H_1 and H_2. Since each of these is connected and can have no circuits (why?), each is a tree. Moreover, each has fewer vertices than G. By inductive hypothesis, if n_i and e_i are the number of vertices and edges of H_i, we have $n_1 = e_1 + 1$ and $n_2 = e_2 + 1$. Now $n = n_1 + n_2$ and $e = e_1 + e_2 + 1$ (add edge $\{u, v\}$). We conclude that

$$n = n_1 + n_2 = (e_1 + 1) + (e_2 + 1) = (e_1 + e_2 + 1) + 1 = e + 1. \qquad \text{Q.E.D.}$$

3.5.4 Spanning Trees*

Suppose that $G = (V, E)$ is a graph and $H = (W, F)$ is a subgraph. We say H is a *spanning subgraph* if $W = V$. A spanning subgraph that is a tree is called a *spanning tree*. For example, in the graph G of Figure 3.69, H and K as shown in the figure are spanning subgraphs because they have the same vertices as G. K is a spanning tree. Spanning trees have a wide variety of applications in combinatorics, which we shall investigate shortly. Analysis of an electrical network reduces to finding all spanning trees of the corresponding graph (see Deo [1974]). Spanning trees can also be used, through the program digraph (see Example 3.4), to estimate program running time (Deo [1974, p. 442]). They arise in connection with seriation problems in political science and archaeology (Roberts [1979], Wilkinson [1971]). They also form the basis for a large number of algorithms in network flows and the solution of minimal cost problems in operations research (Chapter 13). In a recent paper, Graham and Hell [1982] mention applications of spanning trees to design of computer and communication networks, power networks, leased-line telephone networks, wiring connections, links in a transportation network, piping in a flow network,

---

*This subsection can be omitted until just before Section 11.1 or Section 13.1.

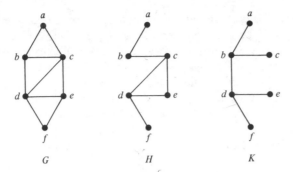

Figure 3.69 H and K are spanning subgraphs of G, K a spanning tree.

network reliability, picture processing, automatic speech recognition, clustering and classification problems, and so on. Their paper also gives many references to the literature of spanning trees and their applications.

We now give several applications of spanning trees in more detail.

Example 3.16 Highway Construction

Suppose that a number of towns grow up in a remote region where there are no good highways. It is desired to build enough highways so that it is possible to travel from any of the towns to any other of the towns by highway either directly or by going through another town. Let the towns be vertices of a graph G and join each pair of towns by an edge. We wish to choose a subset F of the edges of G representing highways to be built, so that if we use the vertices of G and the edges of F, we have a connected graph. Thus, we seek a connected, spanning subgraph of G.* If we wish to minimize costs, then certainly the highways we choose to build correspond to a connected spanning subgraph H of G with the property that removal of any edge leads to a spanning subgraph which is no longer connected. But in any connected spanning subgraph H of G, if there is a circuit, removal of any edge of the circuit cannot disconnect H. (Why?) Thus, H is connected and has no circuits; that is, H is a tree, a spanning tree. If, in addition, each edge of G has a weight or number on it, representing the cost of building the corresponding highway, then we wish to find a spanning tree F which is *minimum* in the sense that the sum of the weights on its edges is no larger than that on any other spanning tree. We shall study algorithms for finding minimum spanning trees in Section 13.1. We shall sometimes also be interested in finding maximum spanning trees. The procedure for finding them is analogous to the procedure for minimum spanning trees. It should be noted that a similar application of minimum spanning trees arises if the links are gas pipelines, electrical wire connections, railroads, sewer lines, and so on.

Example 3.17 Telephone Lines

In a remote region, isolated villages are linked by road, but there is not yet any telephone service. We wish to put in telephone lines so that every pair of villages is linked by a (not necessarily direct) telephone connection. It is cheapest to put the lines in along existing roads. Along which stretches of road should we put telephone

---

*This approach assumes that we do not allow several highways to link up at points other than the towns in question. This is implicit in the assumption that we want to go either directly between two towns or link them up through another town.

lines so as to make sure that every pair of villages is linked and the total number of miles of telephone line (which is probably proportional to the total cost of installation) is minimized? Again we seek a minimum spanning tree. (Why?)

Example 3.18   Measuring the Homogeneity of Bimetallic Objects (Shier [1982], Filliben *et al.* [1983], Goldman [1981])

Minimum spanning trees have recently been finding applications at the U.S. National Bureau of Standards and elsewhere, in determining to what extent a bimetallic object is homogeneous in composition. Given such an object, build a graph by taking as vertices a set of sample points in the material. Measure the composition at each sample point and join physically adjacent sample points by an edge. Place on the edge $\{x, y\}$ a weight equal to the physical distance between sample points x and y multiplied by a homogeneity factor between 0 and 1. This factor is 0 if the composition of the two points is exactly alike and 1 if it is dramatically opposite, and otherwise is a number in between. In the resulting graph, find a minimum spanning tree. The sum of the weights in this tree is 0 if all the sample points have exactly the same composition. A high value says that the material is quite inhomogeneous. This statistic, the sum of the weights of the edges in a minimum spanning tree, is a very promising statistic in measuring homogeneity. According to Goldman [1981], it will probably come into standard use over an (appropriately extended) period of time.

We now present one main result about spanning trees.

Theorem 3.15.   A graph G is connected if and only if it has a spanning tree.

Proof. Suppose that G is connected. Then there is a connected spanning subgraph H of G with a minimum number of edges. Now H can have no circuits. For if C is a circuit of H, removal of any edge of C (without removing the corresponding vertices) leaves a spanning subgraph of G which is still connected and has one less edge than H. This is impossible by choice of H. Thus, H has no circuits. Also, by choice, H is connected. Thus, H is a spanning tree.

Conversely, if G has a spanning tree, it is clearly connected.   Q.E.D.

The proof of Theorem 3.15 can be reworded as follows. Suppose that we start with a connected graph G. If G has no circuits, it is already a tree. If it has a circuit, remove any edge of the circuit and a connected graph remains. If there is still a circuit, remove any edge and a connected graph remains. And so on.

Note that Theorem 3.15 gives us a method for determining if a graph G is connected: simply test if G has a spanning tree. Algorithms for doing this are discussed in Section 13.1.

3.5.5. Proof of Theorem 3.11 and a Related Result*

We now present a proof of Theorem 3.11 and then give a related theorem.

*This subsection can be omitted.

Proof of Theorem 3.11. We have already shown, in Theorem 3.10, one direction of this equivalence. To prove the other direction, suppose that G is connected and $n = e + 1$. By Theorem 3.15, G has a spanning tree T. Then T and G have the same number of vertices. By Theorem 3.10, T has $n - 1$ edges. Since G also has $n - 1$ edges, and since all edges of T are edges of G, $T = G$.   Q.E.D.

We also have another result, similar to Theorem 3.11, which will be needed in Chapter 13.

Theorem 3.16. Suppose that G is a graph with n vertices and e edges. Then G is a tree if and only if G has no circuits and $n = e + 1$.

Proof. Again it remains to prove one direction of this theorem. Suppose that G has no circuits and $n = e + 1$. If G is not connected, let $K_1, K_2, \ldots, K_p$ be its connected components, $p > 1$. Let K_i have n_i vertices. Then K_i is connected and has no circuits, so K_i is a tree. Thus, by Theorem 3.10, K_i has $n_i - 1$ edges. We conclude that the number of edges of G is given by

$$(n_1 - 1) + (n_2 - 1) + \cdots + (n_p - 1) = \sum n_i - p = n - p.$$

But since $p > 1$, $n - p < n - 1$, so $n \neq e + 1$. This is a contradiction.   Q.E.D.

3.5.6 Chemical Bonds and the Number of Trees

In 1857, Arthur Cayley discovered trees while he was trying to enumerate the isomers of the saturated hydrocarbons, the chemical compounds of the form C_kH_{2k+2}. This work was the forerunner of a large amount of graph-theoretical work in chemistry and biochemistry. We present Cayley's approach here. For good references on graph theory and chemistry, see Balaban [1976] and Rouvray and Balaban [1979]. We give other applications of combinatorics to chemistry in Section 5.4 and in Chapter 7.

The isomers C_kH_{2k+2} can be represented by representing a carbon atom with a letter C and a hydrogen atom with a letter H, and linking two atoms if they are bonded in the given compound. For example, methane and ethane are shown in Figure 3.70. We can replace these diagrams with graphs by replacing each letter with a vertex, as shown in Figure 3.71. We call these graphs *bond graphs*. Note that given a bond graph of a saturated hydrocarbon, the vertices can be relabeled with letters C and H in an unambiguous fashion: A vertex is labeled C if it is bonded to four other vertices (carbon has chemical valence 4), and H if it is bonded to one other vertex (hydrogen has valence 1).

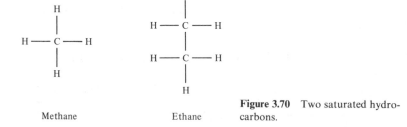

Methane   Ethane

Figure 3.70 Two saturated hydrocarbons.

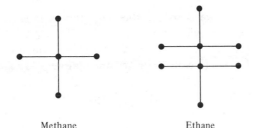

Methane   Ethane

Figure 3.71 Bond graphs for the saturated hydrocarbons of Figure 3.70.

Every vertex of one of our bond graphs is either bonded to one or to four other vertices. The bond graphs of some other saturated hydrocarbons are shown in Figure 3.72.

The graphs of Figures 3.71 and 3.72 are all trees. We shall show that this is no accident, that is, that the bond graph of every saturated hydrocarbon is a tree.

Recall that the degree of a vertex in a graph is the number of its neighbors, and by Theorem 3.12, the sum of the degrees of the vertices is twice the number of edges. Now in the bond graph for $C_k H_{2k+2}$, there are

$$k + 2k + 2 = 3k + 2$$

vertices. Moreover, each carbon vertex has degree 4 and each hydrogen vertex has degree 1. Thus, the sum of the degrees is

$$4k + 1(2k + 2) = 6k + 2.$$

The number of edges is half this number, or $3k + 1$. Thus, the number of edges is exactly one less than the number of vertices. Since the bond graph for a chemical compound $C_k H_{2k+2}$ must be connected, it follows by Theorem 3.11 that it is a tree.

Now Cayley abstracted the problem of enumerating all possible saturated hydrocarbons to the problem of enumerating all trees in which every vertex has degree 1 or 4. He found it easier to begin by enumerating all trees. In the process, he discovered abstractly the bond graphs of some saturated hydrocarbons which were previously unknown, and predicted their existence. They were later discovered.

It clearly makes a big difference in counting the number of distinct graphs of a

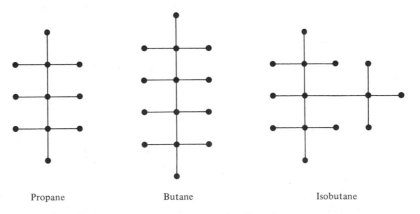

Propane   Butane   Isobutane

Figure 3.72 More bond graphs.

certain type whether or not we consider these graphs as labeled (see the discussion in Section 3.1.3). In particular, Cayley discovered that for $n \geq 2$, there were n^{n-2} distinct labeled trees with n vertices. We shall present one proof of this result here. For a survey of other proofs, see Moon [1967] (see also Harary and Palmer [1973]).

Theorem 3.17 (Cayley [1857]). If $n \geq 2$, there are n^{n-2} distinct labeled trees of n vertices.

Let $N(d_1, d_2, \ldots, d_n)$ count the number of labeled trees with n vertices with the vertex labeled i having degree $d_i + 1$. We first note the following result, which is proved in Exercise 31.

Theorem 3.18. If $n \geq 2$ and all d_i are nonnegative integers, then

$$N(d_1, d_2, \ldots, d_n) = \begin{cases} 0 & \text{if } \sum_{i=1}^{n} d_i \neq n - 2 \\ C(n-2; d_1, d_2, \ldots, d_n) & \text{if } \sum_{i=1}^{n} d_i = n - 2. \end{cases} \tag{3.18}$$

In this theorem,

$$C(n-2; d_1, d_2, \ldots, d_n) = \frac{(n-2)!}{d_1! d_2! \ldots d_n!}$$

is the multinomial coefficient studied in Section 2.12.

To illustrate Theorem 3.18, note that

$$N(1, 3, 0, 0, 0, 0) = C(4; 1, 3, 0, 0, 0, 0) = \frac{4!}{1!3!0!0!0!0!} = 4.$$

There are four labeled trees of six vertices with vertex 1 having degree 2, vertex 2 having degree 4, and the remaining vertices having degree 1. Figure 3.73 shows the four trees.

Cayley's Theorem now follows as a corollary of Theorem 3.18. For the number $T(n)$ of labeled trees of n vertices is given by

$$T(n) = \sum \left\{ N(d_1, d_2, \ldots, d_n): \ d_i \geq 0, \ \sum_{i=1}^{n} d_i = n - 2 \right\},$$

which is the same as

$$T(n) = \sum \left\{ C(n-2; d_1, d_2, \ldots, d_n): \ d_i \geq 0, \ \sum_{i=1}^{n} d_i = n - 2 \right\}.$$

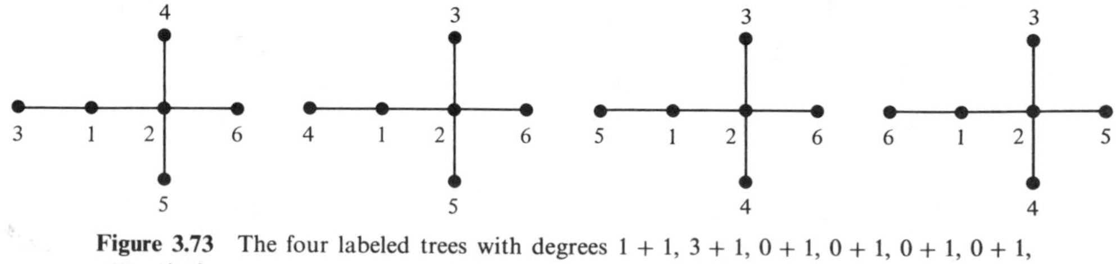

Figure 3.73 The four labeled trees with degrees $1 + 1$, $3 + 1$, $0 + 1$, $0 + 1$, $0 + 1$, $0 + 1$, respectively.

Now it is easy to see that the binomial expansion (Theorem 2.7) generalizes to the following *multinomial expansion*: If p is a positive integer, then

$$(a_1 + a_2 + \cdots + a_k)^p = \sum \left\{ C(p; d_1, d_2, \ldots, d_k) a_1^{d_1} a_2^{d_2} \cdots a_k^{d_k} : \quad d_i \geq 0, \quad \sum_{i=1}^{k} d_i = p \right\}.$$

Thus, taking $k = n$ and $p = n - 2$ and $a_1 = a_2 = \cdots = a_k = 1$, we have

$$T(n) = (1 + 1 + \cdots + 1)^{n-2} = n^{n-2}.$$

This is Cayley's Theorem.

EXERCISES FOR SECTION 3.5

1. Find all nonisomorphic trees of four vertices.
2. Find all nonisomorphic trees of five vertices.
3. Find
   (a) the number of vertices in a tree of 10 edges;
   (b) the number of edges in a tree of 10 vertices.
4. Check Theorem 3.12 for every graph of Figure 3.69.
5. Give an example of a graph G with $n = e + 1$ but such that G is not a tree.
6. For each graph of Figure 3.25, either find a spanning tree or argue that none exists.
7. A *forest* is a graph each of whose connected components is a tree. If a forest has n vertices and k components, how many edges does it have?
8. Show that in a graph G with n vertices and e edges, there is a vertex of degree at least $2e/n$.
9. A simpler proof of uniqueness in Theorem 3.14 would be as follows. Suppose C_1 and C_2 are two distinct simple chains joining x and y. Then C_1 followed by C_2 is a closed chain. But if a graph has a closed chain, it must have a circuit. Show that this latter statement is false.
10. In a connected graph with 15 edges, what is the maximum possible number of vertices?
11. In a connected graph with 25 vertices, what is the minimum possible number of edges?
12. What is the maximum number of vertices in a graph with 15 edges and three components?
13. Prove the converse of Theorem 3.14, that is, that if G is any graph and any two vertices are joined by exactly one simple chain, then G is a tree.
14. Prove that if two nonadjacent vertices of a tree are joined by an edge, the resulting graph will have a circuit.
15. Prove that if any edge is deleted from a tree, the resulting graph will be disconnected.
16. If we have a (connected) electrical network with e elements (edges) and n nodes (vertices), what is the minimum number of elements we have to remove to eliminate all circuits in the network?
17. If G is a tree of n vertices, show that its chromatic polynomial is given by

$$P(G, x) = x(x - 1)^{n-1}.$$

18. *Use the result of Exercise 17* to determine the chromatic number of a tree.
19. Find the chromatic number of a tree by showing that every tree is bipartite.
20. Show that the converse of the result in Exercise 17 is true, that is, that if

$$P(G, x) = x(x - 1)^{n-1},$$

then G is a tree.

21. Suppose that G is a tree, $\{u, v\}$ is an edge of G, and H is obtained from G by deleting edge $\{u, v\}$, but not vertices u and v. Show that H has exactly two connected components. (You may not assume any of the theorems of this section except possibly Theorem 3.14.)

22. (Peschon and Ross [1982]) In an electrical distribution system, certain locations are joined by connecting electrical lines. A system of switches is used to open or close these lines. The collection of open lines has to have two properties: (1) every location has to be on an open line, and (2) there can be no circuits of open lines, for a short on one open line in an open circuit would shut down all lines in the circuit. Discuss the mathematical problem of finding which switches to open.

23. Suppose that a chemical compound C_kH_m has a bond graph which is connected and has no circuits. Show that m must be $2k + 2$.

24. Find the number of spanning trees of K_n.

25. Check Cayley's Theorem by finding all labeled trees of
    (a) three vertices;
    (b) four vertices.

26. Is there a tree of seven vertices
    (a) with each vertex having degree 1?
    (b) with two vertices having degree 1 and five vertices having degree 2?
    (c) with five vertices having degree 1 and two vertices having degree 2?
    (d) with vertices having degrees 2, 2, 2, 3, 1, 1, 1?

27. Is there a tree of five vertices with two vertices of degree 3?

28. In each of the following cases, find the number of labeled trees satisfying the given degree conditions by our formula and draw the trees in question.
    (a) Vertices 1, 2, and 3 have degree 2, and vertices 4 and 5 have degree 1.
    (b) Vertex 1 has degree 2, vertex 2 has degree 3, and vertices 3, 4, and 5 have degree 1.
    (c) Vertex 1 has degree 3, vertices 2 and 3 have degree 2, and vertices 4, 5, and 6 have degree 1.

29. Find the number of labeled trees of
    (a) six vertices, four having degree 2;
    (b) eight vertices, six having degree 2;
    (c) five vertices, exactly three of them having degree 1;
    (d) six vertices, exactly three of them having degree 1.

30. Prove that $N(d_1, d_2, \ldots, d_n) = 0$ if $\sum_{i=1}^{n} d_i \neq n - 2$. [*Hint*: Count edges.]

31. This exercise sketches a proof of Theorem 3.18. Define $M(d_1, d_2, \ldots, d_n)$ by the right-hand side of Equation (3.18). It suffices to prove that if $n \geq 2$ and all d_i are nonnegative and $\sum_{i=1}^{n} d_i = n - 2$, then

$$N(d_1, d_2, \ldots, d_n) = M(d_1, d_2, \ldots, d_n). \tag{3.19}$$

    (a) Under the given assumptions, verify (3.19) for $n = 2$.
    (b) Under the given assumptions, show that $d_i = 0$, some i.
    (c) Suppose i in part (b) is n. Show that

$$N(d_1, d_2, \ldots, d_{n-1}, 0) = N(d_1 - 1, d_2, d_3, \ldots, d_{n-1}) + N(d_1, d_2 - 1, d_3, \ldots, d_{n-1})$$
$$+ \cdots + N(d_1, d_2, \ldots, d_{n-2}, d_{n-1} - 1), \tag{3.20}$$

    where a term $N(d_1, d_2, \ldots, d_{k-1}, d_k - 1, d_{k+1}, \ldots, d_{n-1})$ appears in the right-hand side of (3.20) if and only if $d_k > 0$.
    (d) Show that M also satisfies (3.20).
    (e) Verify (3.19) by induction on n. (In the language of Chapter 5, the argument essentially amounts to showing that if M and N satisfy the same recurrence and the same initial condition, then $M = N$.)

3.6 APPLICATIONS OF TREES TO SEARCHING AND SORTING PROBLEMS*

3.6.1 Search Trees

Using trees to search through a table or a file is one of the most important operations in computer science. In Example 2.11 we discussed the problem of searching through a file to find the key (identification number) of a particular individual, and pointed out that there were more efficient ways to search than to simply go through the list of keys from beginning to end. In this section we show how to do this using search trees. Then we show how trees can be used for another important problem in computer science: sorting a collection into its natural order, given a list of its members.

Before defining search trees, let us note that each of the examples of trees in Figures 3.66–3.68 is drawn in the following way. Each vertex has a *level*, 0, 1, 2, ..., k. There is exactly one vertex, the *root*, which is at level 0. All adjacent vertices differ by exactly one level and each vertex at level $i + 1$ is adjacent to exactly one vertex at level i. Such a tree is called a *rooted tree*. The number k is called the *height* of the tree. In the example of Figure 3.66, the level 0 vertex is Boss, the level 1 vertices are Jones, Smith, and Brown; the level 2 vertices are Johnson, Jackson, and so on; and the level 3 vertices are Allen, Baker, and so on. The height is 3. In a rooted tree, all vertices adjacent to vertex u and at a level below u's are called the *sons* of u. For instance, in Figure 3.66, the sons of Brown are White, Black, and Green. All vertices that are joined to u by a chain of vertices at levels below u's in the tree are called *descendants* of u. Thus, in our example, the descendants of Brown are White, Black, Green, Engel, Freyer, and Gold.

A rooted tree is called *m-ary* if every vertex has either no sons or m sons. A 2-ary rooted tree is called a *binary tree*. Figure 3.74 shows a binary tree. In a binary tree, we shall assume that we separate the two sons of each vertex into a *left son* and a *right son*.

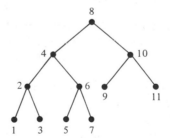

Figure 3.74 A binary search tree.

Suppose that we have a file of n names. Let T be any (rooted) binary tree. Label each vertex u of T with a key (a real number) in such a way that if u has a son, then its left son and all descendants of its left son get keys which are lower numbers than the key of u, and its right son and all descendants of its right son get keys which are higher numbers than the key of u. A binary tree with such a labeling will be called a *binary search tree*. (Such a labeling can be found for every binary tree. See below.) Figure 3.74 shows a binary search tree, where the keys are the integers 1, 2, ..., 11. Now given a particular key k, to locate it, we search through the binary search tree, starting with the root. At any

*This section can be omitted.

stage, we look at the key of the vertex x being examined. At the start, x is the root. We ask if k is equal to, less than, or greater than the key of x. If equal, we have found the desired file. If less, then we look next at the left son of x; if greater, we look next at the right son of x. We continue the same procedure. For instance, to find the key 7 using the binary search tree of Figure 3.74, we start with key 8, that of the root. Then we go to key 4 (since 7 is less than 8), then to key 6 (since 7 is higher than 4), then to key 7 (since 7 is higher than 6). Notice that our search took four steps rather than the seven steps it would have taken to go through the list 1, 2, ..., 11 in order. See Exercise 15 for another application of binary search trees.

Suppose that we can find a binary search tree. What is the computational complexity of a file search in the worst case? It is the number of vertices in the longest chain descending from the root to a vertex with no sons; that is, it is one more than the height of the binary search tree. Obviously, the computational complexity is minimized if we find a binary search tree of minimum height. Now any binary tree can be made into a binary search tree. For a proof, and an algorithm, see, for example, Reingold *et al.* [1977]. Hence, we are left with the following question. Given n, what is the smallest h such that there is a binary tree on n vertices with height h? We shall answer this question through the following theorem, which is proved in Section 3.6.2.

Theorem 3.19. The minimum height of a binary tree on n vertices is $\lceil \log_2 (n+1) \rceil - 1$, where $\lceil a \rceil$ is the least integer greater than or equal to a.

The binary tree of Figure 3.74 is a binary tree on 11 vertices which has the minimum height, 3, since

$$\lceil \log_2 (n + 1) \rceil = \lceil \log_2 12 \rceil = \lceil 3.58 \rceil = 4.$$

In sum, Theorem 3.19 gives a logarithmic bound

$$\lceil \log_2 (n + 1) \rceil - 1 + 1 = \lceil \log_2 (n + 1) \rceil$$

on the computational complexity of file search using binary search trees. This bound can of course be attained by finding a binary search tree of minimum height, which can always be done (see the proof of Theorem 3.19). To summarize:

Corollary 3.19.1. The computational complexity of file search using binary search trees is $\lceil \log_2 (n + 1) \rceil$.

This logarithmic complexity is in general a much better complexity than the complexity n we obtained in Example 2.11 for file search by looking at the entries in a list in order. For $\lceil \log_2 (n + 1) \rceil$ becomes much less than n as n increases.

3.6.2 Proof of Theorem 3.19*

To prove Theorem 3.19, we first prove the following.

Theorem 3.20. If T is a binary tree of n vertices and height h, then $n \leq 2^{h+1} - 1$.

*This subsection can be omitted.

Proof. There is one vertex at level 0, there are at most $2 = 2^1$ vertices at level 1, there are at most $2^2 = 4$ vertices at level 2, there are at most $2^3 = 8$ vertices at level 3, ..., and there are at most 2^h vertices at level h. Hence,

$$n \leq 1 + 2^1 + 2^2 + 2^3 + \cdots + 2^h. \tag{3.21}$$

Now we use the general formula

$$1 + x + x^2 + \cdots + x^h = \frac{1 - x^{h+1}}{1 - x}, \tag{3.22}$$

$x \neq 1$, which will be a very useful tool throughout this book. Substituting $x = 2$ into (3.22) and using (3.21), we obtain

$$n \leq \frac{1 - 2^{h+1}}{1 - 2} = 2^{h+1} - 1. \qquad \text{Q.E.D.}$$

Proof of Theorem 3.19. We have

$$2^{h+1} \geq n + 1,$$

$$h + 1 \geq \log_2 (n + 1),$$

$$h \geq \log_2 (n + 1) - 1.$$

Since h is an integer,

$$h \geq \lceil \log_2 (n + 1) \rceil - 1.$$

Thus, every binary tree on n vertices has height at least $\lceil \log_2 (n + 1) \rceil - 1$. It is straightforward to show that there is always a binary tree of n vertices whose height is exactly $p = \lceil \log_2 (n + 1) \rceil - 1$. Indeed, any binary tree in which the only vertices with no sons are at level p or $p - 1$ will suffice to demonstrate this. (Such a binary tree is called *balanced*. Figure 3.74 is an example of a balanced binary tree.) Q.E.D.

3.6.3 Sorting

A basic problem in computer science is the problem of placing a set of items in their natural order, usually according to some numerical value. This problem we shall call the *sorting problem*. We shall study the problem of sorting by making comparisons of pairs of items.

Any algorithm for sorting by comparisons can be represented by a binary tree called a *decision tree*. Figure 3.75 shows a decision tree for a computer program which would sort the three numbers a, b, and c. In each case, a vertex of the tree corresponds to a test question or an output. At a test question vertex, control moves to the left son if the question is answered "yes" and to the right son if "no." Output vertices are shown by squares, test vertices by circles. The complexity of the algorithm represented by the decision tree T of Figure 3.75 is the number of steps (comparisons) required to reach a decision in the worst case. Since outputs correspond to vertices with no sons, the complexity is obtained by finding one less than the number of vertices in the longest chain from the root to a vertex of T with no sons, that is by finding the height of the binary tree T. In our example, the height is 3.

In a rooted tree, let us call vertices with no sons *leaves*. Note that to sort a set of p

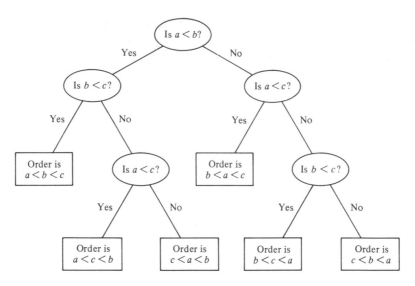

Figure 3.75 A decision tree T for sorting the set $\{a, b, c\}$.

(distinct) items, there are $p!$ possible orders, so a decision tree for the sorting will have at least $p!$ leaves. (We say "at least" because several chains may lead from a root to the same order.) Now we have the following theorem, whose proof is left as an exercise (Exercise 24).

Theorem 3.21. A binary tree of height h has at most 2^h leaves.

This theorem is easily illustrated by the binary trees of Figures 3.68 or 3.74.

Theorem 3.22. Any algorithm for sorting $p \geq 4$ items by pairwise comparisons requires in the worst case at least $cp \log_2 p$ comparisons, where c is a positive constant.

Proof. We have already observed that any decision tree T which sorts p items must have at least $p!$ leaves. Thus, the number of comparisons in the worst case, which is the height of the tree T, has to be at least $\log_2 p!$. (For $2^{\log_2 p!} = p!$). Now for $p \geq 1$,

$$p! \geq p(p - 1)(p - 2) \cdots \left(\left\lceil \frac{p}{2} \right\rceil \right) \geq \left(\frac{p}{2} \right)^{p/2}.$$

Thus, for $p \geq 4$,

$$\log_2 p! \geq \frac{p}{2} \log_2 \left(\frac{p}{2} \right) \geq \frac{p}{4} \log_2 p = \frac{1}{4} p \log_2 p. \qquad \text{Q.E.D.}$$

There are a variety of sorting algorithms which actually achieve the bound in Theorem 3.22, that is, can be carried out in a constant times $p \log_2 p$ steps. Among the better known ones is heap sort. In the text and exercises, we shall discuss two well-known sorting algorithms, bubble sort and quik sort, which do not achieve the bound. For careful descriptions of all three of these algorithms, see, for example, Aho *et al.* [1974],

Baase [1978], Knuth [1973], or Rcingold *et al.* [1977]. Note that $cp \log_2 p \leq cp^2$, so an algorithm that takes $cp \log_2 p$ steps is certainly a polynomially bounded algorithm.

In the algorithm known as *bubble sort*, we begin with an ordered set of p (distinct) items. We wish to put them in their proper (increasing) order. We successively compare the ith item to the $(i + 1)$st item in the list, interchanging them if the ith item is larger than the $(i + 1)$st. The algorithm is called bubble sort because the larger items rise to the top,

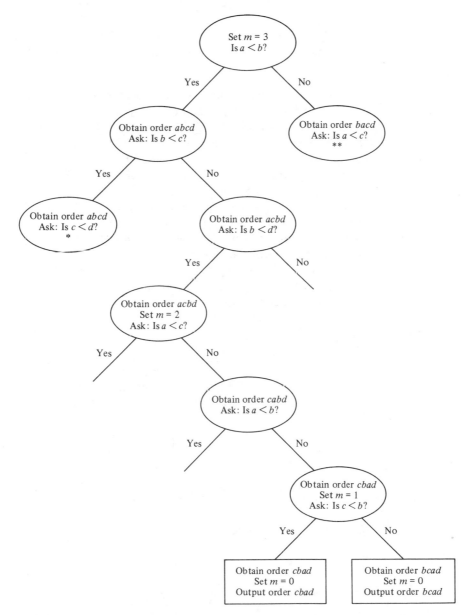

Figure 3.76 Part of the decision tree for bubble sort on an ordered set of four items, *abcd*. (The labels ∗ and ∗∗ are for Exercise 19.)

Sec. 3.6   Applications of Trees to Searching and Sorting Problems   **137** much like bubbles in a glass of champagne. Here is a more formal statement of the algorithm.

Algorithm 3.2. Bubble Sort

Input: An ordered list $a_1, a_2, \ldots, a_p$ of p items.
Output: A listing of $\{a_1, a_2, \ldots, a_p\}$ in increasing order.

Step 1. Set $m = p - 1$.

Step 2. For $i = 1, 2, \ldots, m$, if $a_i > a_{i+1}$, interchange a_i and a_{i+1}.

Step 3. Decrease m by 1. If m is now 0, stop and output the order. If not, return to step 2.

To illustrate bubble sort, suppose that we start with the order 5 1 6 4 2 3. We first set $m = p - 1 = 5$. We compare 5 to 1, and interchange them, getting 1 5 6 4 2 3. We then compare 5 to 6, leaving this order as is. We compare 6 to 4, and interchange them, getting 1 5 4 6 2 3. Next, we - compare 6 to 2, and interchange them, getting 1 5 4 2 6 3. Finally, we compare 6 to 3, interchange them, and get 1 5 4 2 3 6. We decrease m to 4 and repeat the process, getting successively 1 5 4 2 3 6, 1 4 5 2 3 6, 1 4 2 5 3 6, and 1 4 2 3 5 6. Note that we do not have to compare 5 to 6, since m is now 4. We decrease m to 3 and repeat the process, getting 1 4 2 3 5 6, 1 2 4 3 5 6, and 1 2 3 4 5 6. Then m is set equal to 2 and we get 1 2 3 4 5 6 and 1 2 3 4 5 6. Note that no more interchanges are needed. Next, we set $m = 1$. No interchanges are needed. Finally, we set $m = 0$, and we output the order 1 2 3 4 5 6.

Part of the decision tree for bubble sort on an order a, b, c, d is shown in Figure 3.76.

Note that bubble sort requires $p(p - 1)/2$ comparisons. For at the mth iteration or repetition of the procedure, m comparisons are required, and m takes on the values $p - 1$, $p - 2, \ldots, 1$. Thus, using the standard formula for the sum of an arithmetic progression, we see that a total of

$$(p - 1) + (p - 2) + \cdots + 1 = \frac{p(p - 1)}{2}$$

steps are needed. In the language of Section 2.18, the algorithm bubble sort is not as efficient as an algorithm that requires $cp \log_2 p$ steps. For bubble sort is an $O(p^2)$ algorithm and p^2 is not $O(p \log_2 p)$.

EXERCISES FOR SECTION 3.6

1. In the rooted tree of Figure 3.77, find the level of each vertex.

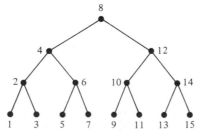

Figure 3.77 Rooted tree for Exercises of Section 3.6.

2. In each rooted tree of Figure 3.78, find the level of each vertex.

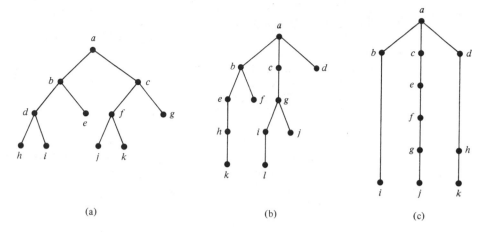

Figure 3.78 Rooted trees for Exercises of Section 3.6.

3. In each rooted tree of Figure 3.78, find the height of the tree.

4. In each rooted tree of Figure 3.78, find all descendants of the vertex b.

5. Is the tree of Figure 3.77 a binary search tree? If so, describe how to find the key 7.

6. Find a balanced binary tree with n vertices where
   (a) $n = 5$;   **(b)** $n = 8$;   **(c)** $n = 12$;   **(d)** $n = 15$.

7. Find a binary tree of
   (a) height 3 and 8 leaves;
   (b) height 3 and fewer than 8 leaves.

8. Find a binary tree of 11 vertices with
   (a) as large a height as possible;
   (b) as small a height as possible.

9. Find a binary search tree with n vertices where n is
   (a) 10;   **(b)** 14;   **(c)** 18;   **(d)** 20.

10. Find the minimum height of a binary tree of n vertices where n is
   (a) 74;   **(b)** 512;   **(c)** 4095.

11. A binary tree can be used to encode bit strings of n bits as follows. At any vertex, its left son is labeled 0 and its right son is labeled 1. A bit string then corresponds to a simple chain from the root to a vertex with no sons. Draw such a tree for $n = 3$ and identify the simple chain corresponding to the string 011.

12. Of all ternary (3-ary) trees on n vertices, what is the least possible height?

13. Suppose we have a *fully balanced binary search tree*, that is, a balanced binary search tree in which every vertex with no sons is at the same level. Assume that a file is equally likely to be at any of the n vertices of T. What is the computational complexity of file search using T if we measure complexity using the average number of steps to find a file, rather than the largest number of steps to find one.

14. In an m-ary rooted tree, if a vertex is chosen at random, show that the probability it has a son is about $1/m$.

15. (Tucker [1980])   In a compiler, a control word is stored as a number. Suppose that the possible control words are GET, DO, ADD, FILL, STORE, REPLACE, and WAIT, and these are represented by the numbers 1, 2, 3, 4, 5, 6, and 7 (in binary notation), respectively. Given an unknown control word X, we wish to test it against the possible control words on this list until we find which word it is. One approach is to compare X's number in order to the numbers 1, 2, ..., 7 corresponding to the control words. Another approach is to build a binary search tree. Describe how the latter approach would work and build such a tree.

16. The *binary search algorithm* searches through an ordered file to see if a key x is in the file. The entries of the file are n numbers, $x_1 < x_2 < \cdots < x_n$. The algorithm compares x to the middle entry x_i in the file, the $\lfloor n/2 \rfloor$th entry. If $x = x_i$, the search is done. If $x < x_i$, then $x_i, x_{i+1}, \ldots$ are eliminated from consideration, and the algorithm searches through the file $x_1, x_2, \ldots, x_{i-1}$, starting with the middle entry. If $x > x_i$, then $x_1, x_2, \ldots, x_i$ are eliminated from consideration, and the algorithm searches through the file $x_{i+1}, x_{i+2}, \ldots, x_n$, starting with the middle entry. The procedure is repeated iteratively. Table 3.4 shows two examples. (See Knuth [1973] for details.) Compute the computational complexity of binary search. (*Hint:* Show that a binary search tree is being used.)

Table 3.4 Two Binary Searches[a]

Searching for 180:
71 97 164 180 285 <u>436</u> 513 519 522 622 663 687	Compare 180 to 436
71 97 <u>164</u> 180 285	Compare 180 to 164
<u>180</u> 285	Compare 180 to 180; 180 has been found

Searching for 515:
71 97 164 180 285 <u>436</u> 513 519 522 622 663 687	Compare 515 to 436
513 519 <u>522</u> 622 663 687	Compare 515 to 522
<u>513</u> 519	Compare 515 to 513
<u>519</u>	Compare 515 to 519; conclude that 515 is not in the file

[a] In each case, the first list contains the whole file, the underlined number is the middle entry, and the next line shows the remaining file to be searched.

17. Apply bubble sort to the order 4 3 1 2 and illustrate the steps required to put this in proper order.

18. Apply bubble sort to the following orders.
(a) 5 1 6 3 2 4;   (b) 3 4 6 1 5 2.

19. In the partial decision tree for bubble sort (Figure 3.76), fill in the part beginning at the vertex labeled
(a) *;   (b) **.

20. Compare the worst case complexity of the bubble sort algorithm to the bound in Theorem 3.22 for p equal to (a) 7;   (b) 15.

21. In the sorting algorithm known as *quik sort*, we start with an ordered list of p items and try to find the proper order. We select the first item from the list and divide the remaining items into two groups, those less than the selected item and those greater than it. We then put the item chosen in between the two groups and repeat the process on each group. We eventually get down to groups of one element, and we stop. For instance, given the order 3 1 5 6 4 8 2 7, we select item 3. Then the items before 3 are listed as 1 2, those after 3 as 5 6 4 8 7. We now apply the algorithm to sort these two groups. For instance, choosing 5 from the second group

gives us the two subgroups 4 and 6 8 7. We now order these. And so on. Apply quik sort to the following ordered lists.

(a) 5 1 7 6 3 2 4;   (b) 9 4 1 2 5 8 3 7 6.

22. How many steps (comparisons) does the algorithm quik sort (Exercise 21) require in the worst case if we start with a list 1 2 3 $\cdots$ p and $p = 5$?

23. Repeat Exercise 22 for arbitrary p.

24. Prove Theorem 3.21 by induction on h.

25. Suppose that D is a digraph. A *level assignment* is an assignment of a level L_i to each vertex i so that if (i, j) is an arc, then $L_i < L_j$. Show that a digraph has a level assignment if and only if it has no cycles.

26. A *graded level assignment* for a digraph D is a level assignment (Exercise 25) such that if there is an arc (i, j) in D, then $L_j = L_i + 1$. Show that if $D = (V, A)$ has a graded level assignment, then it is *equipathic*, that is, for all u, v in V, all simple paths from u to v have the same length.

27. Suppose that T is a rooted tree. Orient T by directing an edge $\{u, v\}$ from lower level to higher level. The resulting digraph is called a *directed rooted tree*.
    (a) Show that every directed rooted tree has exactly one vertex from which every other vertex is reachable by a path.
    (b) Can a directed rooted tree be unilaterally connected?
    (c) Show that every directed rooted tree has a level assignment.
    (d) Suppose that D is a directed rooted tree. Show that D has a graded level assignment if and only if D is equipathic.

28. Find the number of rooted, labeled trees of
    (a) five vertices, two having degree 2;
    (b) five vertices in which the root has degree 2;
    (c) four vertices in which the root has degree 1.

3.7 REPRESENTING A GRAPH IN THE COMPUTER

Many problems in graph theory cannot be solved except on the computer. Indeed, the development of high-speed computing machines has significantly aided the solution of graph-theoretical problems—for instance, the four-color problem—and it has also aided the applications of graph theory to other disciplines. At the same time, the development of computer science has provided graph theorists with a large number of important and challenging problems to solve. In this section we discuss various ways to represent a digraph or graph as input to a computer program.

Note that a diagram of a digraph or graph such as the diagrams we have used is not very practical for large digraphs or graphs and also not very amenable for input into a computer. Rather, some alternative means of entering digraphs and graphs are required. The best way to input a digraph or graph depends on the properties of the digraph or graph and the use that will be made of it. The efficiency of an algorithm depends on the choice of method for entering a digraph or graph and the memory storage required depends on this as well. Here, we shall mention a number of different ways to input a digraph or graph.

Let us concentrate on entering digraphs, recalling that graphs can be thought of as special cases. One of the most common approaches to entering a digraph D into a computer is to give its *adjacency matrix* $\mathbf{A} = \mathbf{A}(D)$. This matrix is obtained by labeling the

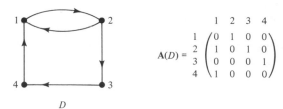

$$A(D) = \begin{array}{c} \\ 1 \\ 2 \\ 3 \\ 4 \end{array} \begin{array}{cccc} 1 & 2 & 3 & 4 \\ \left(\begin{array}{cccc} 0 & 1 & 0 & 0 \\ 1 & 0 & 1 & 0 \\ 0 & 0 & 0 & 1 \\ 1 & 0 & 0 & 0 \end{array}\right) \end{array}$$

Figure 3.79 A digraph with its adjacency matrix $A(D)$.

vertices of D as $1, 2, \ldots, n$, and letting the i, j entry of **A** be 1 if there is an arc from i to j in D, and 0 if there is no such arc. Figure 3.79 shows a digraph and its associated adjacency matrix.* Note that the adjacency matrix can be entered as a 2-dimensional array or as a bit string with n^2 bits, by listing one row after another. Thus, n^2 bits of storage are required, $n(n-1)$ if the diagonal elements are assumed to be 0. The storage requirements are less for a graph, thought of as a symmetric digraph, since its adjacency matrix is symmetric. Specifically, if there are no loops, we require

$$(n-1) + (n-2) + \cdots + 1 = \frac{n(n-1)}{2}$$

bits of storage. For we only need to encode the entries above the diagonal. If a digraph is very sparse, that is, has few arcs, the adjacency matrix is not a very useful representation.

Another matrix that is useful for entering graphs, although not digraphs, is called the incidence matrix. In general, suppose that S is a set and $\mathcal{F}$ is a family of subsets of S. Define the *point-set incidence matrix* **M** as follows. Label the elements of S as $1, 2, \ldots, n$ and the sets in $\mathcal{F}$ as $S_1, S_2, \ldots, S_m$. Then **M** is an $n \times m$ matrix and its i, j entry is 1 if element i is in set S_j and 0 if element i is not in set S_j. For example, if $S = \{1, 2, 3, 4\}$ and

$$\mathcal{F} = \{\{1, 2\}, \{2, 3\}, \{1, 4\}\},$$

then **M** is given by

$$\mathbf{M} = \begin{array}{c} \\ 1 \\ 2 \\ 3 \\ 4 \end{array} \begin{array}{ccc} \{1,2\} & \{2,3\} & \{1,4\} \\ \left[\begin{array}{ccc} 1 & 0 & 1 \\ 1 & 1 & 0 \\ 0 & 1 & 0 \\ 0 & 0 & 1 \end{array}\right] \end{array} \tag{3.23}$$

If G is a graph, its *incidence matrix* is the point-set incidence matrix for the set $S = V(G)$ and the set $\mathcal{F} = E(G)$. The matrix **M** of (3.23) is the incidence matrix for the graph of Figure 3.80. An incidence matrix requires $n \times e$ bits of storage, where e is the number of edges. This tends to be larger than $\frac{1}{2} n(n-1)$, the storage required for the adjacency matrix, and it is always at least $n(n-1) = n^2 - n$ for connected graphs. (Why?) Incidence matrices are widely used in inputting electrical networks and switching networks. It is possible to define an incidence matrix for digraphs as well (see Exercise 16).

Still another approach to inputting a digraph D is simply to give an *arc list*, a list of its arcs. How much storage is required here? Suppose that each vertex label is encoded by a bit string of length t. Then if D has a arcs, since each arc is encoded as a pair of

---

*Of course, the specific matrix obtained depends on the way we list the vertices. But by an abuse of language, we call any of these matrices *the* adjacency matrix.

Figure 3.80 The graph whose incidence matrix is given by (3.23).

codewords, $2at$ bits of storage are required. Note that t must be large enough so that 2^t, the number of bit strings of length t, is at least n, the number of vertices. Thus, $t \geq \log_2 n$. In fact,

$$t \geq \lceil \log_2 n \rceil,$$

where $\lceil x \rceil$ is the least integer greater than or equal to x. Hence, we require at least $2a\lceil \log_2 n \rceil$ bits of storage, since each arc is stored as an ordered pair of vertices. If there are not many arcs, that is, if the adjacency matrix is sparse, this can be a smaller number than n^2 and so require less storage than the adjacency matrix.

There are two variants on the arc list. The first is to input two linear orders or arrays, each having a codewords. If these arrays are $(h_1, h_2, \ldots, h_a)$ and $(k_1, k_2, \ldots, k_a)$, and h_i encodes for the vertex x and k_i for the vertex y, then (x, y) is the ith arc on the arc list. The storage requirements are the same as for an arc list.

Another variant on the arc list is to give n *adjacency lists*, one for each vertex x of D, listing the vertices y such that (x, y) is an arc. The n lists are called an *adjacency structure*. An adjacency structure requires $(n + a)t$ bits of storage, where t is as for the arc list. This is because the adjacency list for a vertex x must be encoded by encoding x and then encoding all the y's so that (x, y) is adjacent to x.

As we have already pointed out, the best way to input a digraph or a graph into a computer will depend on the algorithm we use. To give a simple example, suppose that we ask the question: Is (u_i, u_j) an arc? This question can be answered in one step from an adjacency matrix (look at the i, j entry). To answer it from an adjacency structure requires $\deg(u_i)$ steps in the worst case where u_j is the last vertex to which there is an arc from u_i. On the other hand, let us consider an algorithm that requires marking all vertices to which there is an arc from u_i. This requires n steps from an adjacency matrix (look at all entries in the ith row), but only $\deg(u_i)$ steps from an adjacency structure.

EXERCISES FOR SECTION 3.7

1. For each digraph of Figure 3.6, find
   (a) its adjacency matrix;
   (b) an arc list;
   (c) two linear arrays that can be used to input the arcs;
   (d) an adjacency structure.
2. For each graph of Figure 3.25, find an incidence matrix.
3. Draw the digraph whose adjacency matrix is given by

   (a) $\quad A = \begin{pmatrix} 0 & 1 & 0 & 1 & 0 \\ 1 & 0 & 0 & 1 & 1 \\ 1 & 1 & 0 & 0 & 0 \\ 1 & 0 & 0 & 0 & 0 \\ 0 & 1 & 0 & 0 & 0 \end{pmatrix};$

$$
\textbf{(b)} \quad \mathbf{A} = \begin{pmatrix} 0 & 1 & 0 & 0 & 0 & 0 \\ 0 & 0 & 1 & 0 & 0 & 0 \\ 0 & 0 & 0 & 1 & 0 & 0 \\ 0 & 0 & 0 & 0 & 1 & 0 \\ 0 & 0 & 0 & 0 & 0 & 1 \\ 1 & 0 & 0 & 0 & 0 & 0 \end{pmatrix}.
$$

4. Find the digraph whose vertices are a, b, c, d, e and whose arc list is

$$\{(a, b), (b, d), (e, d), (c, a), (a, c)\}.$$

5. Find the digraph whose vertices are a, b, c, d, e and whose arcs are encoded by the two linear arrays (a, c, d, e, e), (b, d, c, d, a).

6. Find the digraph whose adjacency structure is given by Table 3.5.

Table 3.5 An Adjacency Structure for a Digraph

Vertex x	1	2	3	4	5
Vertices adjacent to x	2, 3	3, 4	4, 5	5, 6	6, 1

7. Calculate the adjacency matrix for each of the graphs of Figure 3.8. (The adjacency matrix is the adjacency matrix of the corresponding digraph.)

8. Suppose that it requires one step to scan an entry in a list or an array. Suppose that a digraph is stored as an adjacency matrix.
   (a) How many steps are required to mark all vertices x adjacent to a particular vertex y, that is, such that (x, y) is an arc?
   (b) How many steps are required to mark or count all arcs?

9. Repeat Exercise 8 if the digraph is stored as an arc list.

10. Repeat Exercise 8 if the digraph is stored as an adjacency structure.

11. If D is a digraph of n vertices, its *reachability matrix* is an $n \times n$ matrix $\mathbf{R}$ whose i, j entry r_{ij} is 1 if vertex j is reachable from vertex i by a path, and 0 otherwise. For each digraph of Figure 3.6, find its reachability matrix. (Note that i is always reachable from i.)

12. Show that D is strongly connected if and only if its reachability matrix (Exercise 11) is $\mathbf{J}$, the matrix of all 1's.

13. If D is a digraph with adjacency matrix $\mathbf{A}$, show by induction on t that the i, j entry of $\mathbf{A}^t$ gives the number of paths of length t in D which lead from i to j.

14. If G is a graph with adjacency matrix $\mathbf{A}$ (Exercise 7), what is the interpretation in graph (as opposed to digraph) language of the i, j entry of $\mathbf{A}^t$?

15. For digraphs D_1, D_2, and D_7 of Figure 3.6, use the result of Exercise 13 to find the number of paths of length 3 from u to v. Identify the paths.

16. If D is a digraph, its *incidence matrix* has rows corresponding to vertices and columns to arcs, with i, j entry equal to 1 if j is the arc (i, k) for some k, -1 if j is the arc (k, i) for some k, and 0 otherwise.
   (a) Find the incidence matrix for each digraph of Figure 3.6.
   (b) How many bits of storage are required for the incidence matrix?
   (c) If $\mathbf{M} = (m_{ij})$ is the incidence matrix of digraph D, what is the significance of the matrix $\mathbf{N} = (n_{ij})$, where $n_{ij} = \sum_k m_{ik} m_{jk}$?

(d) If D is strongly connected, can we ever get away with fewer bits of storage than are needed for the adjacency matrix?

17. If $M = (m_{ij})$ is any matrix of nonnegative entries, let $B(M)$ be the matrix whose i, j entry is 1 if $m_{ij} > 0$ and 0 if $m_{ij} = 0$. Show that if D is a digraph of n vertices with reachability matrix R and adjacency matrix A, and I is the identity matrix, then
    (a) $R = B[I + A + A^2 + \cdots + A^{n-1}]$; **(b)** $R = B[(I + A)^{n-1}]$.

18. Check the results of Exercise 17 on the digraphs D_1 and D_7 of Figure 3.6.

19. **(a)** Show that D is unilaterally connected (Exercise 9, Section 3.2) if and only if $B(R + R^T) = J$, where B is defined in Exercise 17, R in Exercise 11, and J in Exercise 12, and where R^T is the transpose of R.
    (b) Show that D is weakly connected (Exercise 10, Section 3.2) if and only if we have $B[(I + A + A^T)^{n-1}] = J$.

20. Suppose that D is a digraph with reachability matrix $R = (r_{ij})$ and that R^2 is a matrix (s_{ij}). Show that:
    (a) The strong component (Exercise 13, Section 3.2) containing vertex i is given by the entries of 1 in the ith row of $T = (t_{ij})$, where $t_{ij} = r_{ij} \times r_{ij}^{(T)}$ and $r_{ij}^{(T)}$ is the i, j entry of the transpose of R.
    (b) The number of vertices in the strong component containing i is s_{ii}.

21. For each digraph of Figure 3.6, use the results of Exercise 20 to find the strong components.

22. If R is the reachability matrix of a digraph and $c(i)$ is the ith column sum of R, what is the interpretation of $c(i)$?

23. If G is a graph, how would you define directly its reachability matrix $R(G)$?

24. If G is a graph, $R = R(G)$ is its reachability matrix (Exercise 22), and T is as defined in Exercise 20:
    (a) Show that $T = R$.
    (b) What is the interpretation of the 1, 1 entry of R^2?

25. Suppose that R is a matrix of 0's and 1's with 1's down the diagonal (and perhaps elsewhere). Is R necessarily the reachability matrix of some digraph? (Give a proof or counterexample.)

26. (Harary [1969]) Suppose that B is the incidence matrix of a graph G and B^T is the transpose of B. What is the significance of the i, j entry of the matrix $B^T B$?

27. (Harary [1969]) Let G be a graph. The *circuit matrix* C of G is the point-set incidence matrix with S the set of edges of G and $\mathscr{F}$ the family of circuits of G. Let B be the incidence matrix of G. Show that every entry of BC is $\equiv 0 \pmod 2$.

28. Can two graphs have the same incidence matrix and be nonisomorphic? Why?

29. Can two graphs have the same circuit matrix and be nonisomorphic? Why? What if every edge is on a circuit?

REFERENCES FOR CHAPTER 3

AHO, A. V., HOPCROFT, J. E., and ULLMAN, J. D., *The Design and Analysis of Computer Algorithms*, Addison-Wesley, Reading, Mass., 1974.

APPEL, K., and HAKEN, W., "Every Planar Map Is Four Colorable. Part I: Discharging," *Ill. J. Math.*, *21* (1977), 429–490.

APPEL, K., HAKEN, W., and KOCH, J., "Every Planar Map Is Four Colorable. Part II: Reducibility," *Ill. J. Math.*, *21* (1977), 491–567.

BAASE, S., *Computer Algorithms: Introduction to Design and Analysis*, Addison-Wesley, Reading, Mass., 1978.

BALABAN, A. T. (ed.), *Chemical Applications of Graph Theory*, Academic Press, New York, 1976.

BEHZAD, M., CHARTRAND, G., and LESNIAK-FOSTER, L., *Graphs and Digraphs*, Wadsworth, Belmont, Calif., 1979.

BELTRAMI, E., and BODIN, L., "Networks and Vehicle Routing for Municipal Waste Collection," *Networks, 4* (1973), 65–94.

BERGE, C., "Färbung von Graphen, deren sämtliche bzw. deren ungerade Kreise starr sind," *Wiss. Z. Martin-Luther-Univ. Halle-Wittenberg, Math.-Naturwiss. Reihe, 10* (1961), 114.

BERGE, C., "Sur une conjecture relative au problème des codes optimaux," *Commun. 13ème Assemblée Générale de l'URSI* (International Scientific Radio Union), Tokyo, 1962.

BERGE, C., *Graphs and Hypergraphs*, American Elsevier, New York, 1973.

BIRKHOFF, G. D., "A Determinant Formula for the Number of Ways of Coloring a Map," *Ann. Math., 14* (1912), 42–46.

BIRKHOFF, G. D., and LEWIS, D. C., "Chromatic Polynomials," *Trans. Amer. Math. Soc., 60* (1946), 355–451.

BONDY, J. A., and MURTY, U. S. R., *Graph Theory with Applications*, Elsevier, New York/MacMillan, London, 1976.

CAYLEY, A., "On the Theory of the Analytical Forms Called Trees," *Philos. Mag., 13* (1857), 172–176. [Also *Math. Papers*, Cambridge, *3* (1891), 242–246.]

CAYLEY, A., "On the Mathematical Theory of Isomers," *Philos. Mag., 67* (1874), 444–446. [Also *Math. Papers*, Cambridge, *9* (1895), 202–204.]

CAYLEY, A., "A Theorem on Trees," *Quart. J. Math., 23* (1889), 376–378. [Also *Math. Papers*, Cambridge, *13* (1897), 26–28.]

COHEN, J. E., *Food Webs and Niche Space*, Princeton University Press, Princeton, N.J., 1978.

COZZENS, M. B., and ROBERTS, F. S., "*T*-Colorings of Graphs and the Channel Assignment Problem," *Congressus Numerantium, 35* (1982), 191–208.

DEO, N., *Graph Theory with Applications to Engineering and Computer Science*, Prentice-Hall, Englewood Cliffs, N.J., 1974.

EULER, L., "Solutio Problematis ad Geometriam Situs Pertinentis," *Comment. Acad. Sci. I. Petropolitanae, 8* (1736), 128–140. [Reprinted in *Opera Omnia*, Series I-7 (1766), 1–10.]

EVEN, S., *Graph Algorithms*, Computer Science Press, Potomac, Md., 1979.

FILLIBEN, J. J., KAFADAR, K., and SHIER, D. R., "Testing for Homogeneity of Two-Dimensional Surfaces," *Math. Modelling, 4* (1983), 167–189.

FIORINI, S., AND WILSON, R. J., *Edge Colorings of Graphs*, Pitman, London, 1977.

GAREY, M. R., and JOHNSON, D. S., *Computers and Intractability: A Guide to the Theory of NP-Completeness*, W. H. Freeman, San Francisco, 1979.

GOLDMAN, A. J., "Discrete Mathematics in Government," lecture presented at SIAM Symposium on Applications of Discrete Mathematics, Troy, N.Y., June 1981.

GOLUMBIC, M. C., *Algorithmic Graph Theory and Perfect Graphs*, Academic Press, New York, 1980.

GRAHAM, R. L., and HELL, P., "On the History of the Minimum Spanning Tree Problem," mimeographed, Bell Laboratories, Murray Hill, N.J., 1982. (To appear in *Annals of the History of Computing*.)

HALE, W. K., "Frequency Assignment: Theory and Applications," *Proc. IEEE, 68* (1980), 1497–1514.

HARARY, F., *Graph Theory*, Addison-Wesley, Reading, Mass., 1969.

HARARY, F., NORMAN, R. Z., and CARTWRIGHT, D., *Structural Models: An Introduction to the Theory of Directed Graphs*, Wiley, New York, 1965.

HARARY, F., and PALMER, E. M., *Graphical Enumeration*, Academic Press, New York, 1973.

HARRISON, J. L., "The Distribution of Feeding Habits among Animals in a Tropical Rain Forest," *J. Anim. Ecol., 31* (1962), 53–63.

KELLY, J. B., and KELLY, L. M., "Paths and Circuits in Critical Graphs," *Amer. J. Math., 76* (1954), 786–792.

KEMENY, J. G., and SNELL, J. L., *Mathematical Models in the Social Sciences*, Blaisdell, New York, 1962. (Reprinted by MIT Press, Cambridge, Mass., 1972.)

KIRCHHOFF, G., "Über die Auflösung der Gleichungen, auf welche man bei der Untersuchung der linearen Verteilung galvanischer Ströme geführt wird," *Ann. Phys. Chem., 72* (1847), 497–508.

KNUTH, D. E., *The Art of Computer Programming*, Vol. 3: *Sorting and Searching*, Addison-Wesley, Reading, Mass., 1973.

KÖNIG, D., *Theorie des endlichen und unendlichen Graphen*, Akademische Verlagsgesellschaft, Leipzig, 1936. (Reprinted by Chelsea, New York, 1950.)

Kuratowski, K., "Sur le problème des courbes gauches en topologie," *Fund. Math., 15* (1930), 271–283.

LIU, C. L., *Topics in Combinatorial Mathematics*, Mathematical Association of America, Washington, D. C., 1972.

LOVÁSZ, L., "Normal Hypergraphs and the Perfect Graph Conjecture," *Discrete Math., 2* (1972a), 253–267.

LOVÁSZ, L., "A Characterization of Perfect Graphs," *J. Comb. Theory B, 13* (1972b), 95–98.

LUKS, E. M., "Isomorphism of Bounded Valence Can Be Tested in Polynomial Time," *Proceedings of the Twenty-First Annual Symposium on Foundations of Computer Science*, 1980, pp. 42–49.

MOON, J. W., "Various Proofs of Cayley's Formula for Counting Trees," in F. Harary (ed.), *A Seminar on Graph Theory*, Holt, Rinehart and Winston, New York, 1967, pp. 70–78.

OPSUT, R. J., and ROBERTS, F. S., "On the Fleet Maintenance, Mobile Radio Frequency, Task Assignment, and Traffic Phasing Problems," in G. Chartrand, Y. Alavi, D. L. Goldsmith, L. Lesniak-Foster, and D. R. Lick (eds.), *The Theory and Applications of Graphs*, Wiley, New York, 1981, pp. 479–492.

PENNOTTI, R. J., "Channel Assignment in Mobile Telecommunication Systems," Ph.D. thesis, Polytechnic Institute of New York, 1976.

PESCHON, J., and ROSS, D., "New Methods for Evaluating Distribution, Automation, and Control (DAC) Systems Benefits," *SIAM J. Algebraic Discrete Methods, 3* (1982), 439–452.

READ, R. C., "An Introduction to Chromatic Polynomials," *J. Comb. Theory, 4* (1968), 52–71.

REINGOLD, E. M., NIEVERGELT, J., and DEO, N., *Combinatorial Algorithms: Theory and Practice*, Prentice-Hall, Englewood Cliffs, N.J., 1977.

ROBERTS, F. S., *Discrete Mathematical Models, with Applications to Social, Biological, and Environmental Problems*, Prentice-Hall, Englewood Cliffs, N.J., 1976.

ROBERTS, F. S., *Graph Theory and Its Applications to Problems of Society*, NSF–CBMS Monograph No. 29, SIAM, Philadelphia, 1978.

ROBERTS, F. S., "Indifference and Seriation," *Ann. N.Y. Acad. Sci., 328* (1979), 173–182.

ROUVRAY, D. H., and BALABAN, A. T., "Chemical Applications of Graph Theory," in R. J. Wilson and L. W. Beinecke (eds.), *Applications of Graph Theory*, Academic Press, London, 1979, pp. 177–221.

SHIER, D., "Testing for Homogeneity using Minimum Spanning Trees," *UMAP J., 3* (1982), 273–283.

TUCKER, A. C., "Perfect Graphs and an Application to Optimizing Municipal Services," *SIAM Rev.*, *15* (1973), 585–590.

TUCKER, A., *Applied Combinatorics*, Wiley, New York, 1980.

TUTTE, W. T. (alias B. DESCARTES), "Solution to Advanced Problem No. 4526," *Amer. Math. Monthly*, *61* (1954), 352.

WHITNEY, H., "The Coloring of Graphs," *Ann. Math.*, *33* (1932), 688–718.

WILKINSON, E. M., "Archaeological Seriation and the Traveling Salesman Problem," in F. R. Hodson *et al.* (eds.), *Mathematics in the Archaeological and Historical Sciences*, Edinburgh University Press, Edinburgh, 1971.

ZYKOV, A. A., "On Some Properties of Linear Complexes (Russian)," *Math. Sb. N.S.*, *24* (1949), 163–188. (English transl.: *Amer. Math. Soc. Transl. No. 79*, 1952.)

PART II The Counting Problem

4 Generating Functions and their Applications*

4.1 DEFINITION

Much of combinatorics is devoted to developing tools for counting. We have seen that it is often important to count the number of arrangements or patterns, but in practice it is impossible to list all of these arrangements. Hence, we need tools to help us in counting. In the next four chapters we present a number of tools that are useful in counting. One of the most powerful tools that we shall present is the notion of the generating function. This chapter is devoted to generating functions.

Often in combinatorics, we seek to count a quantity a_k that depends on an input or a parameter, say k. This is true, for instance, if a_k is the number of steps required to perform a computation if the input has size k. We can formalize the dependence on k by speaking of a sequence of unknown values, $a_0, a_1, a_2, \ldots, a_k, \ldots$. We seek to determine the kth term in this sequence. Generating functions provide a simple way to "encode" a sequence such as $a_0, a_1, a_2, \ldots, a_k, \ldots$, which can be readily "decoded" to find the terms of the sequence. The trick will be to see how to compute the encoding or generating function for the sequence without having the sequence. Then we can decode to find a_k. The method will enable us to determine the unknown quantity a_k in an indirect, but highly effective, manner.

The method of generating functions that we shall present is an old one, which has its roots in the work of DeMoivre around 1720, was developed by Euler in 1748 in connection with partition problems, and was extensively treated in the late eighteenth century and early nineteenth century by Laplace, primarily in connection with probability theory.

---

*In an elementary treatment, this chapter should be omitted. Chapters 4 and 5 are the only chapters which really make use of the calculus prerequisites.

In spite of its long history, the method continues to have widespread application, as we shall see. For a more complete treatment of generating functions, see MacMahon [1960] or Riordan [1958] (see also Riordan [1964]).

4.1.1 Power Series

In this chapter we use a fundamental idea from calculus, the notion of power series. The results about power series we shall need are summarized in this subsection. The reader who wants more details, including proofs of these results, can consult most calculus books.

A *power series* is an infinite series of the form $\sum_{k=0}^{\infty} a_k x^k$. Such an infinite series always converges for $x = 0$. Either it does not converge for any other value of x, or there is a positive number R (possibly infinite) so that it converges for all x with $|x| < R$. In the latter case, the largest such R is called the *radius of convergence*. In the former case, we say that 0 is the radius of convergence. A power series $\sum_{k=0}^{\infty} a_k x^k$ can be thought of as a function $f(x)$ of x, which is defined for those values of x for which the infinite sum converges, and is computed by calculating that infinite sum. In most of this chapter we shall not be concerned with matters of convergence. We shall simply assume that x has been chosen so that $\sum_{k=0}^{\infty} a_k x^k$ converges.*

Power series arise in calculus in the following way. There are many functions $f(x)$ for which

$$f(x) = \sum_{k=0}^{\infty} \frac{1}{k!} f^{(k)}(0) x^k = f(0) + f'(0)x + \frac{1}{2!} f''(0)x^2 + \frac{1}{3!} f'''(0)x^3 + \cdots \qquad (4.1)$$

The power series on the right-hand side of (4.1) for all x in some interval containing 0, is called the *Maclaurin expansion* for f, or the *Taylor series expansion* for f about $x = 0$.

Some of the most famous and useful Maclaurin expansions are the following:

$$\frac{1}{1-x} = \sum_{k=0}^{\infty} x^k = 1 + x + x^2 + x^3 + \cdots, \qquad \text{for } |x| < 1, \quad (4.2)$$

$$e^x = \sum_{k=0}^{\infty} \frac{1}{k!} x^k = 1 + x + \frac{1}{2!} x^2 + \frac{1}{3!} x^3 + \cdots, \qquad \text{for } |x| < \infty, \quad (4.3)$$

$$\sin x = \sum_{k=0}^{\infty} \frac{(-1)^k}{(2k+1)!} x^{2k+1} = x - \frac{1}{3!} x^3 + \frac{1}{5!} x^5 - \cdots, \qquad \text{for } |x| < \infty, \quad (4.4)$$

and

$$\ln(1+x) = \sum_{k=1}^{\infty} \frac{(-1)^{k+1}}{k} x^k = x - \frac{1}{2} x^2 + \frac{1}{3} x^3 - \frac{1}{4} x^4 + \cdots, \qquad \text{for } |x| < 1. \quad (4.5)$$

To show, for instance, that (4.3) is a special case of (4.1), it suffices to observe that if $f(x) = e^x$, then $f^{(k)}(x) = e^x$ for all k, and $f^{(k)}(0) = 1$. Readers should check for themselves that Equations (4.2), (4.4), and (4.5) are also special cases of (4.1).

---

*This assumption can be made more precise by thinking of $\sum_{k=0}^{\infty} a_k x^k$ as simply a formal expression, a *formal power series*, rather than as a function, and by performing appropriate manipulations on these formal expressions. For details of this approach, see Niven [1969].

One of the reasons power series are so useful is that they can easily be added, multiplied, divided, composed, differentiated, or integrated. We remind the reader of these properties of power series by formulating several general principles.

Principle 1. Suppose that $f(x) = \sum_{k=0}^{\infty} a_k x^k$ and $g(x) = \sum_{k=0}^{\infty} b_k x^k$. Then $f(x) + g(x)$, $f(x)g(x)$, and $f(x)/g(x)$ can be computed by, respectively, adding term by term, multiplying out, or using long division. [This is true for division only if $g(x)$ is not zero for the values of x in question.] Specifically,

$$f(x) + g(x) = \sum_{k=0}^{\infty} (a_k + b_k)x^k$$

$$= (a_0 + b_0) + (a_1 + b_1)x + (a_2 + b_2)x^2 + \cdots,$$

$$f(x)g(x) = a_0 \sum_{k=0}^{\infty} b_k x^k + a_1 x \sum_{k=0}^{\infty} b_k x^k + a_2 x^2 \sum_{k=0}^{\infty} b_k x^k + \cdots$$

$$= a_0(b_0 + b_1 x + b_2 x^2 + \cdots) + a_1 x(b_0 + b_1 x + b_2 x^2 + \cdots)$$

$$+ a_2 x^2(b_0 + b_1 x + b_2 x^2 + \cdots) + \cdots,$$

$$\frac{f(x)}{g(x)} = \frac{a_0}{\displaystyle\sum_{k=0}^{\infty} b_k x^k} + \frac{a_1 x}{\displaystyle\sum_{k=0}^{\infty} b_k x^k} + \frac{a_2 x^2}{\displaystyle\sum_{k=0}^{\infty} b_k x^k} + \cdots.$$

If the power series for $f(x)$ and $g(x)$ both converge for $|x| < R$, then so do $f(x) + g(x)$ and $f(x)g(x)$. If $g(0) \neq 0$, $f(x)/g(x)$ converges in some interval about 0.

For instance, using (4.2) and (4.3), we have

$$\frac{1}{1-x} + e^x = (1 + x + x^2 + x^3 + \cdots) + \left(1 + x + \frac{1}{2!}x^2 + \frac{1}{3!}x^3 + \cdots\right)$$

$$= (1 + 1) + (1 + 1)x + \left(1 + \frac{1}{2!}\right)x^2 + \left(1 + \frac{1}{3!}\right)x^3 + \cdots$$

$$= \sum_{k=0}^{\infty} \left(1 + \frac{1}{k!}\right)x^k.$$

Also,

$$\frac{1}{1-x} e^x = (1 + x + x^2 + x^3 + \cdots)\left(1 + x + \frac{1}{2!}x^2 + \frac{1}{3!}x^3 + \cdots\right)$$

$$= 1\left(1 + x + \frac{1}{2!}x^2 + \frac{1}{3!}x^3 + \cdots\right) + x\left(1 + x + \frac{1}{2!}x^2 + \frac{1}{3!}x^3 + \cdots\right)$$

$$+ x^2\left(1 + x + \frac{1}{2!}x^2 + \frac{1}{3!}x^3 + \cdots\right) + \cdots$$

$$= 1 + 2x + \tfrac{5}{2}x^2 + \tfrac{5}{3}x^3 + \cdots$$

Power series are also easy to compute under composition of functions.

Principle 2.   If $f(x) = g(u(x))$ and if we know that $g(u) = \sum_{k=0}^{\infty} a_k u^k$, we have $f(x) = \sum_{k=0}^{\infty} a_k [u(x)]^k.$*

Thus, setting $u = x^4$ in Equation (4.5) gives us

$$\ln (1 + x^4) = \sum_{k=1}^{\infty} \frac{(-1)^{k+1}}{k} x^{4k}.$$

[Principle 2 generalizes to the situation where we have a power series for $u(x)$.]

Principle 3.   If a power series $f(x) = \sum_{k=0}^{\infty} a_k x^k$ converges for all $|x| < R$ with $R > 0$, then the derivative and antiderivative of $f(x)$ can be computed by differentiating and integrating term by term. Namely,

$$\frac{df}{dx}(x) = \sum_{k=0}^{\infty} \frac{d}{dx}(a_k x^k) = \sum_{k=0}^{\infty} k a_k x^{k-1} \tag{4.6}$$

and

$$\int_0^x f(t)\, dt = \sum_{k=0}^{\infty} \int_0^x a_k t^k\, dt = \sum_{k=0}^{\infty} \frac{1}{k+1} a_k x^{k+1}. \tag{4.7}$$

The power series in (4.6) and (4.7) also converge for $|x| < R$.

For instance, since

$$\frac{1}{(1-x)^2} = \frac{d}{dx}\left[\frac{1}{1-x}\right],$$

we see from (4.2) and (4.6) that

$$\frac{1}{(1-x)^2} = \sum_{k=0}^{\infty} kx^{k-1} = 1 + 2x + 3x^2 + 4x^3 + \cdots.$$

4.1.2 Generating Functions

Suppose that we are interested in computing the kth term in a sequence (a_k) of numbers. We shall use the convention that (a_k) refers to the sequence and a_k written without parentheses, to the kth entry. The (*ordinary*) *generating function* for the sequence (a_k) is defined to be

$$G(x) = \sum_k a_k x^k = a_0 x^0 + a_1 x^1 + a_2 x^2 + \cdots. \tag{4.8}$$

The sum is finite if the sequence is finite, infinite if the sequence is infinite. In the latter case, we will think of x as having been chosen so that the sum in (4.8) converges.

*If the power series for $g(u)$ converges for $|u| < S$ and $|u(x)| < S$ whenever $|x| < R$, then the power series for $f(x)$ converges for all $|x| < R$.

Example 4.1

Suppose that $a_k = \binom{n}{k}$, $k = 0, 1, \ldots, n$. Then the ordinary generating function for the sequence (a_k) is

$$G(x) = \binom{n}{0} + \binom{n}{1}x + \binom{n}{2}x^2 + \cdots + \binom{n}{n}x^n.$$

By the binomial expansion (Theorem 2.7),

$$G(x) = (1 + x)^n.$$

The advantage of what we have done is that we have expressed $G(x)$ in a simple (encoded) form. Knowing this simple form for $G(x)$, it is now possible to derive a_k by simply remembering this closed form for $G(x)$ and decoding, that is, expanding out, and searching for the coefficient of x^k. Even more useful is the fact that as we have observed before, often we will be able to find $G(x)$ without knowing a_k, and then be able to solve for a_k by expanding out.

Example 4.2

Suppose that $a_k = 1$, for $k = 0, 1, 2, \ldots$. Then

$$G(x) = 1 + x + x^2 + \cdots .$$

By Equation (4.2),

$$G(x) = \frac{1}{1 - x},$$

provided that $|x| < 1$. Again the reader will note the closed form for $G(x)$.

Example 4.3

Often we will know the generating function but not the sequence. We will try to "recover" the sequence from the generating function. For example, suppose that

$$G(x) = \sum_{k=0}^{\infty} a_k x^k$$

and we know

$$G(x) = \frac{x^2}{1 - x}.$$

What is a_k? Using Equation (4.2), we have for $|x| < 1$,

$$G(x) = x^2 \left[\frac{1}{1 - x} \right]$$
$$= x^2(1 + x + x^2 + \cdots)$$
$$= x^2 + x^3 + x^4 + \cdots .$$

Hence,

$$(a_k) = (0, 0, 1, 1, 1, \ldots).$$

In this chapter and Chapter 5 we will study a variety of techniques for expanding out $G(x)$ to obtain the desired sequence (a_k).

Example 4.4

Suppose that $a_k = 1/k!$, $k = 0, 1, 2, \ldots$. Then

$$G(x) = \frac{1}{0!} + \frac{1}{1!} x + \frac{1}{2!} x^2 + \frac{1}{3!} x^3 + \cdots .$$

By Equation (4.3), $G(x) = e^x$ for all values of x.

Example 4.5

Suppose that $G(x) = x \sin (x^2)$ is the ordinary generating function for the sequence (a_k). To find a_k, use Equation (4.4), substitute x^2 for x, and multiply by x, to find that

$$G(x) = x \left[x^2 - \frac{1}{3!} (x^2)^3 + \frac{1}{5!} (x^2)^5 - \cdots \right]$$

$$= x^3 - \frac{1}{3!} x^7 + \frac{1}{5!} x^{11} - \cdots$$

Thus, we see that a_k is the kth term of the sequence

$$0, 0, 0, 1, 0, 0, 0, -\frac{1}{3!}, 0, 0, 0, \frac{1}{5!}, 0, \cdots .$$

Example 4.6

Suppose that $G(x) = \cos x$ is the ordinary generating function for the sequence (a_k). Since $G(x)$ has derivatives of all orders, we can expand out in a Maclaurin series, by calculating $f^{(k)}(0)$ for all k, and we see that

$$G(x) = \sum_{k=0}^{\infty} \frac{(-1)^k}{(2k)!} x^{2k} = 1 - \frac{1}{2!} x^2 + \frac{1}{4!} x^4 - \cdots . \tag{4.9}$$

The verification of this is left as an exercise. An alternative approach is to observe that $G(x) = d(\sin x)/dx$, and so to use Equation (4.4). Then we see that

$$G(x) = \frac{d}{dx} (\sin x)$$

$$= \frac{d}{dx} \left[x - \frac{1}{3!} x^3 + \frac{1}{5!} x^5 - \cdots \right]$$

$$= \frac{d}{dx} [x] + \frac{d}{dx} \left[-\frac{1}{3!} x^3 \right] + \frac{d}{dx} \left[\frac{1}{5!} x^5 \right] + \cdots$$

$$= 1 - \frac{1}{2!} x^2 + \frac{1}{4!} x^4 - \cdots ,$$

which agrees with Equation (4.9).

Example 4.7

If $(a_k) = (1, 1, 1, 0, 1, 1, \ldots)$, then the ordinary generating function is given by

$$G(x) = 1 + x + x^2 + x^4 + x^5 + \cdots$$
$$= (1 + x + x^2 + x^3 + x^4 + x^5 + \cdots) - x^3$$
$$= \frac{1}{1-x} - x^3.$$

Example 4.8

If $(a_k) = (1/2!, 1/3!, 1/4!, \ldots)$, then the ordinary generating function is given by

$$G(x) = \frac{1}{2!} + \frac{1}{3!}x + \frac{1}{4!}x^2 + \cdots$$
$$= \frac{1}{x^2}\left(\frac{1}{2!}x^2 + \frac{1}{3!}x^3 + \frac{1}{4!}x^4 + \cdots\right)$$
$$= \frac{1}{x^2}[e^x - 1 - x].$$

Example 4.9   The Number of Labeled Graphs

In Section 3.1.3 we counted the number $L(n, e)$ of labeled graphs with n vertices and e edges, $n \geq 2$, $e \leq C(n, 2)$. If n is fixed and we let $a_k = L(n, k)$, $k = 0, 1, \ldots, C(n, 2)$, then let us consider the generating function

$$G_n(x) = \sum_{k=0}^{C(n,\,2)} a_k x^k.$$

Note that in Section 3.1.3 we computed $L(n, e) = C[C(n, 2), e]$. Hence, if $r = C(n, 2)$,

$$G_n(x) = \sum_{k=0}^{r} C(r, k)x^k.$$

By the binomial expansion (Theorem 2.7), we have

$$G_n(x) = (1 + x)^r = (1 + x)^{C(n,\,2)}, \tag{4.10}$$

which is a simple way to summarize our knowledge of the numbers $L(n, e)$. In particular, from (4.10) we can derive a formula for the number $L(n)$ of labeled graphs of n vertices. For

$$L(n) = \sum_{k=0}^{r} C(r, k),$$

which is $G_n(1)$. Thus, taking $x = 1$ in (4.10) gives us

$$L(n) = 2^{C(n,\,2)},$$

which is the result we derived in Section 3.1.3.

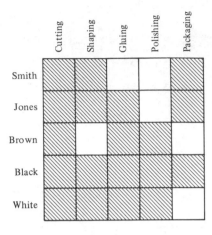

Figure 4.1 A board corresponding to a job assignment problem.

Example 4.10   Rook Polynomials

Suppose that B is any $n \times m$ board such as those in Figures 4.1 and 4.2, with certain squares forbidden and others acceptable, the acceptable ones being darkened. Let $r_k(B)$ be the number of ways to choose k acceptable (darkened) squares, no two of which lie in the same row and no two of which lie in the same column. We can think of B as part of a chessboard. A *rook* is a piece that can travel either horizontally or vertically on the board. Thus, one rook is said to be able to *take* another if the two are in the same row or the same column. We wish to place k rooks on B in acceptable squares in such a way that no rook can take another. Thus, $r_k(B)$ counts the number of ways k nontaking rooks can be placed in acceptable squares of B.

The 5×5 board in Figure 4.1 arises from a job assignment problem. The rows correspond to workers, the columns to jobs, and the i, j position is darkened if worker i is suitable for job j. We wish to determine the number of ways in which each worker can be assigned to one job, no more than one worker per job, so that a worker only gets a job to which he or she is suited. It is easy to see that this is equivalent to the problem of computing $r_5(B)$.

The 5×7 board in Figure 4.2 arises from a problem of storing computer programs. The i, j position is darkened if storage location j has sufficient storage

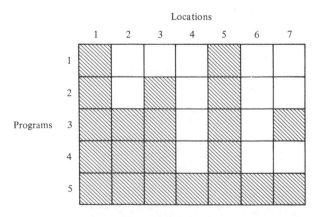

Figure 4.2 A board corresponding to a computer storage problem.

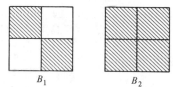

B_1 B_2 **Figure 4.3** Two boards.

capacity for program i. We wish to assign each program to a storage location with sufficient storage capacity, at most one program per location. The number of ways this can be done is again given by $r_5(B)$. We shall compute $r_5(B)$ in these two examples in the text and exercises of Section 6.1.

The expression

$$R(x, B) = r_0(B) + r_1(B)x + r_2(B)x^2 + \cdots$$

is called the *rook polynomial* for the board B. The rook polynomial is indeed a polynomial in x, since $r_k(B) = 0$ for k larger than the number of acceptable squares. The rook polynomial is just the ordinary generating function for the sequence

$$(r_0(B), r_1(B), r_2(B), \ldots).$$

As with generating functions in general, we shall find methods for computing the rook polynomial without explicitly computing the coefficients $r_k(B)$, and then we shall be able to compute these coefficients from the polynomial.

To give some examples, consider the two boards B_1 and B_2 of Fig. 4.3. In board B_1, there is one way to place no rooks (this will be the case for any board), two ways to place one rook (use either darkened square), one way to place two rooks (use both darkened squares), and no way to place more than two rooks. Thus,

$$R(x, B_1) = 1 + 2x + x^2.$$

In board B_2, there is again one way to place no rooks, four ways to place one rook (use any darkened square), two ways to place two rooks (use the diagonal squares or the non-diagonal squares), and no way to place more than two rooks. Thus,

$$R(x, B_2) = 1 + 4x + 2x^2.$$

EXERCISES FOR SECTION 4.1

1. For each of the following functions, find its Maclaurin expansion by computing the derivatives $f^{(k)}(0)$.
   (a) $\cos x$ (b) e^{2x} (c) $\sin (3x)$
   (d) $x + e^x$ (e) xe^x (f) $\ln (1 + 5x)$.

2. For each of the following functions, use known Maclaurin expansions to find the Maclaurin expansion, by adding, composing, differentiating, and so on.

   (a) $x^2 + \dfrac{1}{1 - x}$ (b) $x^3 \dfrac{1}{1 - x}$ (c) $\sin (x^3)$

   (d) $\sin (x^2 + x + 1)$ (e) $\dfrac{1}{(1 - x)^3}$ (f) $7e^x + e^{8x}$

   (g) $\ln (1 + 3x)$ (h) $x^5 \sin (x^2)$ (i) $\dfrac{1}{1 - 2x} e^x$.

3. For the following sequences, find the ordinary generating function and simplify if possible.
   (a) $(1, 1, 1, 0, 0, \ldots)$ **(b)** $(1, 0, 2, 3, 4, 0, 0, \ldots)$
   (c) $(3, 3, 3, \ldots)$ **(d)** $(1, 0, 1, 1, \ldots)$
   (e) $(0, 0, 0, 1, 1, 1, 1, \ldots)$ **(f)** $(0, 0, 4, 4, 4, \ldots)$

   (g) $(1, 1, 1, 2, 1, 1, \ldots)$ **(h)** $(a_k) = \left(\dfrac{2}{k!}\right)$

   (i) $(a_k) = \left(\dfrac{2^k}{k!}\right)$ **(j)** $\left(0, 0, \dfrac{1}{2!}, \dfrac{1}{3!}, \dfrac{1}{4!}, \ldots\right)$

   (k) $\left(1, 1, \dfrac{1}{2!}, \dfrac{1}{3!}, \dfrac{1}{4!}, \ldots\right)$ **(l)** $\left(\dfrac{1}{3!}, \dfrac{1}{4!}, \dfrac{1}{5!}, \ldots\right)$

   (m) $\left(1, -1, \dfrac{1}{2!}, -\dfrac{1}{3!}, \dfrac{1}{4!}, -\dfrac{1}{5!}, \ldots\right)$ **(n)** $(1, 0, 1, 0, 1, 0, \ldots)$

   (o) $(0, 1, 0, 1, 0, 1, \ldots)$ **(p)** $\left(2, 0, -\dfrac{2}{3!}, 0, \dfrac{2}{5!}, \ldots\right)$

   (q) $\left(3, -\dfrac{3}{2}, \dfrac{3}{3}, -\dfrac{3}{4}, \ldots\right)$.

4. Find the sequence whose ordinary generating function is given as follows:
   (a) $(x + 5)^2$ **(b)** $(1 + x)^4$ **(c)** $\dfrac{x^5}{1 - x}$

   (d) $\dfrac{1}{1 - 3x}$ **(e)** $\dfrac{1}{1 + 8x}$ **(f)** e^{4x}

   (g) $1 + \dfrac{1}{1 - x}$ **(h)** $1 + e^x$ **(i)** xe^x

   (j) $x^3 + x^4 + e^{2x}$ **(k)** $\dfrac{1}{1 - x^2}$ **(l)** $2x + e^{-x}$

   (m) e^{-2x} **(n)** $\sin(3x)$ **(o)** $x^2 \ln(1 + 2x) + e^x$

   (p) $\dfrac{1}{1 + x^2}$.

5. Suppose that the ordinary generating function for the sequence (a_k) is given as follows. In each case, find a_3.
   (a) $(x + 2)^3$ **(b)** $\dfrac{4}{1 - x}$ **(c)** e^{7x}.

6. In each case of Exercise 5, find a_4.

7. Suppose that T_n is the number of rooted (unlabeled) trees of n vertices. The ordinary generating function $T(x) = \sum T_n x^n$ is given by

$$T(x) = x + x^2 + 2x^3 + 4x^4 + 9x^5 + 20x^6 + \cdots$$

(Riordan [1958] computes T_n for $n \le 26$. More values of T_n are now known.) Verify the coefficients of x, x^2, x^3, x^4, x^5, and x^6.

8. Let $M(n, a)$ be the number of labeled digraphs with n vertices and a arcs and let $M(n)$ be the number of labeled digraphs with n vertices (see Section 3.1.3). If n is fixed, let $b_k = M(n, k)$ and let

$$D_n(x) = \sum_{k=0}^{n(n-1)} b_k x^k.$$

   (a) Find a simple expression for $D_n(x)$.
   (b) Use this expression to derive a formula for $M(n)$.

9. Suppose that $c_k = R(n, k)$ is the number of labeled graphs with a certain property P and having n vertices and k edges, and $R(n)$ is the number of labeled graphs with property P and n vertices. Suppose that

$$G_n(x) = \sum_{k=0}^{\infty} c_k x^k$$

is the ordinary generating function and we know that $G_n(x) = (1 + x + x^2)^n$. Find $R(n)$.

10. Suppose that $d_k = S(n, k)$ is the number of labeled digraphs with a certain property Q and having n vertices and k arcs, and $S(n)$ is the number of labeled digraphs with property Q and n vertices and at least two arcs. Let the ordinary generating function be given by

$$H_n(x) = \sum_{k=0}^{\infty} d_k x^k.$$

Suppose we know that $H_n(x) = (1 + x^2)^{n+5}$. Find $S(n)$.

11. Compute the rook polynomial for each of the boards of Figure 4.4.

(a)

(b)

(c)

(d)

Figure 4.4 Boards for Exercise 11, Section 4.1.

12. Compute the rook polynomial for the $n \times n$ chessboard with all squares darkened if n is
    (a) 3; (b) 4; (c) 6; (d) 8.

13. A *Latin rectangle* is an $r \times s$ array with entries 1, 2, ..., n, so that no two entries in any row or column are the same. A Latin square (Example 1.1) is a Latin rectangle with $r = s = n$. One way to build a Latin square is to build it up one row at a time, adding rows successively to Latin rectangles. In how many ways can we add a third row to the Latin rectangle of Figure 4.5? Set this up as a rook polynomial problem by observing what symbols can still be included in the jth column. You do not have to solve this problem.

1	2	3	4	5
2	3	1	5	4

Figure 4.5 A 2 × 5 Latin rectangle.

14. *Use rook polynomials* to count the number of permutations of $\{1, 2, 3, 4\}$ in which 1 is not in the second position, 2 is not in the fourth position, and 3 is not in the first or fourth position.

15. Show that if board B' is obtained from board B by deleting rows or columns with no darkened squares, then $r_k(B) = r_k(B')$.

4.2 OPERATING ON GENERATING FUNCTIONS

A sequence defines a unique generating function and a generating function defines a unique sequence; we will be able to pass back and forth between sequences and generating functions. It will be useful to compile a "library" of basic generating functions and their corresponding sequences. Our list can start with the generating functions $1/(1 - x)$, e^x, $\sin x$, $\ln (1 + x)$, whose corresponding sequences can be derived from Equations (4.2)–(4.5). By operating on these generating functions as in Section 4.1.1, namely by adding, multiplying, dividing, composing, differentiating, or integrating, we can add to our basic list. In this section we do so.

In this section we observe how the various operations on generating functions relate to operations on the corresponding sequences. We start with some simple examples. Suppose that (a_k) is a sequence with ordinary generating function $A(x) = \sum_{k=0}^{\infty} a_k x^k$. Then multiplying $A(x)$ by x corresponds to shifting the sequence one place to the right and starting with 0. For $xA(x) = \sum_{k=0}^{\infty} a_k x^{k+1}$ is the ordinary generating function for the sequence $(0, a_0, a_1, a_2, \ldots)$. Similarly, multiplying $A(x)$ by $1/x$ and subtracting a_0/x corresponds to shifting the sequence one place to the left and deleting the first term. For

$$\frac{1}{x} A(x) - \frac{a_0}{x} = \sum_{k=0}^{\infty} a_k x^{k-1} - \frac{a_0}{x} = \sum_{k=1}^{\infty} a_k x^{k-1} = \sum_{k=0}^{\infty} a_{k+1} x^k.$$

To illustrate these two results, note that since $A(x) = e^x$ is the ordinary generating function for the sequence

$$\left(1, 1, \frac{1}{2!}, \frac{1}{3!}, \ldots\right),$$

xe^x is the ordinary generating function for the sequence

$$\left(0, 1, 1, \frac{1}{2!}, \frac{1}{3!}, \ldots\right)$$

and $(1/x)e^x - 1/x$ is the ordinary generating function for the sequence

$$\left(1, \frac{1}{2!}, \frac{1}{3!}, \ldots\right).$$

Similarly, by Equation (4.6), differentiating $A(x)$ with respect to x corresponds to multiplying the kth term of (a_k) by k and shifting by one place to the left. Thus, we saw in Section 4.1.1 that since

$$\frac{1}{(1 - x)^2} = \frac{d}{dx}\left[\frac{1}{1 - x}\right],$$

$1/(1 - x)^2$ is the ordinary generating function for the sequence $(1, 2, 3, \ldots)$.

Suppose that (a_k) and (b_k) are sequences with ordinary generating functions $A(x) = \sum_{k=0}^{\infty} a_k x^k$ and $B(x) = \sum_{k=0}^{\infty} b_k x^k$, respectively. Since two power series can be added term by term, we see that $C(x) = A(x) + B(x)$ is the ordinary generating function for the sequence (c_k) whose kth term is $c_k = a_k + b_k$. This sequence (c_k) is called the *sum* of (a_k) and (b_k) and is denoted $(a_k) + (b_k)$. Thus,

$$\frac{1}{1 - x} + e^x$$

is the ordinary generating function for

$$(1, 1, 1, \ldots) + \left(1, 1, \frac{1}{2!}, \frac{1}{3!}, \ldots\right) = \left(2, 2, 1 + \frac{1}{2!}, 1 + \frac{1}{3!}, \ldots\right).$$

From the point of view of combinatorics, the most interesting case arises from multiplying two generating functions. Suppose that

$$C(x) = A(x)B(x), \tag{4.11}$$

where $A(x)$, $B(x)$, and $C(x)$ are the ordinary generating functions for the sequences (a_k), (b_k), and (c_k), respectively. Does it follow that $c_k = a_k b_k$ for all k? Let $A(x) = 1 + x$ and $B(x) = 1 + x$. Then $C(x) = A(x)B(x)$ is given by $1 + 2x + x^2$. Now $(c_k) = (1, 2, 1, 0, 0, \ldots)$ and $(a_k) = (b_k) = (1, 1, 0, 0, \ldots)$, so $c_1 \neq a_1 b_1$. Thus, $c_k = a_k b_k$ does not follow from (4.11). What we can observe is that if we multiply $A(x)$ by $B(x)$, we obtain a term x^k by combining a term $a_j x^j$ from $A(x)$ with a term $b_{k-j} x^{k-j}$ from $B(x)$. Thus, for all k,

$$c_k = a_0 b_k + a_1 b_{k-1} + \cdots + a_{k-1} b_1 + a_k b_0. \tag{4.12}$$

This is easy to check in the case where both $A(x)$ and $B(x)$ are $1 + x$. Note that if $k = 0$, (4.12) says that $c_0 = a_0 b_0$. Note also that (4.12) for all k implies (4.11). If (4.12) holds for all k, we say the sequence (c_k) is the *convolution* of the two sequences (a_k) and (b_k), and we write $(c_k) = (a_k) * (b_k)$. Our results are summarized as follows.

Theorem 4.1. Suppose that $A(x)$, $B(x)$, and $C(x)$ are, respectively, the ordinary generating functions for the sequences (a_k), (b_k), and (c_k). Then

(a) $C(x) = A(x) + B(x)$ if and only if $(c_k) = (a_k) + (b_k)$.
(b) $C(x) = A(x)B(x)$ if and only if $(c_k) = (a_k) * (b_k)$.

Example 4.11

Suppose that $b_k = 1$, all k. Then (4.12) becomes

$$c_k = a_0 + a_1 + \cdots + a_k.$$

The generating function $B(x)$ is given by $B(x) = 1/(1 - x) = (1 - x)^{-1}$. Hence, by Theorem 4.1,

$$C(x) = A(x)(1 - x)^{-1}.$$

This is the generating function for the sum of the first k terms of a series. For instance, suppose that (a_k) is the sequence $(0, 1, 1, 0, 0, \ldots)$. Then $A(x) = x + x^2$ and

$$C(x) = (x + x^2)[1 + x + x^2 + \cdots]$$
$$= x + 2x^2 + 2x^3 + 2x^4 + \cdots.$$

We conclude that

$$(x + x^2)(1 - x)^{-1}$$

is the ordinary generating function for the sequence (c_k) given by $(0, 1, 2, 2, \ldots)$. This can be checked by noting that

$$a_0 = 0, \qquad a_0 + a_1 = 1, \qquad a_0 + a_1 + a_2 = 2,$$
$$a_0 + a_1 + a_2 + a_3 = 2, \cdots$$

Example 4.12

If $A(x)$ is the generating function for the sequence (a_k), then $A^2(x)$ is the generating function for the sequence (c_k) where

$$c_k = a_0 a_k + a_1 a_{k-1} + \cdots + a_{k-1} a_1 + a_k a_0.$$

This result will also be useful in the enumeration of chemical isomers by counting trees in Section 5.4. In particular, if $a_k = 1$, all k, then $A(x) = (1-x)^{-1}$. It follows that

$$C(x) = A^2(x) = (1-x)^{-2}$$

is the generating function for (c_k) where $c_k = k + 1$. We have obtained this result before, by differentiating $(1-x)^{-1}$.

Example 4.13

Suppose that

$$G(x) = \frac{1 + x + x^2 + x^3}{1 - x}$$

is the ordinary generating function for a sequence (a_k). Can we find a_k? We can write

$$G(x) = (1 + x + x^2 + x^3)(1 - x)^{-1}.$$

Now $1 + x + x^2 + x^3$ is the ordinary generating function for the sequence

$$(b_k) = (1, 1, 1, 1, 0, 0, \ldots)$$

and $(1-x)^{-1}$ is the ordinary generating function for the sequence

$$(c_k) = (1, 1, 1, \ldots).$$

Thus, $G(x)$ is the ordinary generating function for the convolution of these two sequences, that is,

$$a_k = b_0 c_k + b_1 c_{k-1} + \cdots + b_k c_0 = b_0 + b_1 + \cdots + b_k.$$

It is easy to show from this that

$$(a_k) = (1, 2, 3, 4, 4, \ldots).$$

Example 4.14   A Reduction for Rook Polynomials

In computing rook polynomials, it is frequently useful to reduce a complicated computation to a number of smaller ones, a trick we have previously encountered in connection with chromatic polynomials in Section 3.4. Exercise 15 of Section 4.1 shows one such reduction. Here we present another. Suppose I is a set of darkened squares in a board B and B_I is the board obtained from B by lightening the darkened squares not in I. Suppose the darkened squares of B are partitioned into two sets I and J so that no square in I lies in the same row or column as any square of J. In this case, we say that B_I and B_J *decompose* B. Figure 4.6 illustrates this situation.

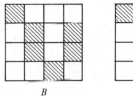
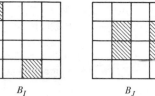

B B_I B_J **Figure 4.6** B_I and B_J decompose B.

If B_I and B_J decompose B, then since all acceptable squares fall in B_I or B_J, and no rook of I can take a rook of J, or vice versa, to place k nontaking rooks on B, we place p nontaking rooks on B_I and then $k - p$ nontaking rooks on B_J, for some p. Thus,

$$r_k(B) = r_0(B_I)r_k(B_J) + r_1(B_I)r_{k-1}(B_J) + \cdots$$
$$+ r_p(B_I)r_{k-p}(B_J) + \cdots + r_k(B_I)r_0(B_J).$$

This implies that the sequence $(r_k(B))$ is simply the convolution of the two sequences $(r_k(B_I))$ and $(r_k(B_J))$. Hence,

$$R(x, B) = R(x, B_I)R(x, B_J).$$

EXERCISES FOR SECTION 4.2

1. In each of the following, the function is the ordinary generating function for a sequence (a_k). Find this sequence.

    (a) $x \sin x$ **(b)** $\dfrac{1}{x} \ln (1 + x)$ **(c)** $x^3 \sin x$

    (d) $\dfrac{1}{x^3} \ln (1 + x)$ **(e)** $\dfrac{2}{1 - x} + 4x^2 + 8x + 1$ **(f)** $\dfrac{1}{1 - 3x} + \dfrac{4x}{1 - x}$

    (g) $\dfrac{1}{1 - x^2} + 3x^3 + 12$ **(h)** $\dfrac{1}{2} [e^x + e^{-x}]$.

2. For each of the following functions $A(x)$, suppose that $B(x) = xA'(x)$. Find the sequence for which $B(x)$ is the ordinary generating function.

    (a) $\dfrac{1}{1 - x}$ **(b)** e^{2x} **(c)** $\dfrac{1}{1 - 2x}$ **(d)** $\ln (1 + x)$.

3. In each of the following, find a formula for the convolution of the two sequences.
    (a) $(1, 1, 1, \ldots)$ and $(1, 1, 1, \ldots)$
    (b) $(1, 1, 1, \ldots)$ and $(0, 1, 2, 3, \ldots)$
    (c) $(1, 1, 1, 0, 0, \ldots)$ and $(0, 1, 2, 3, \ldots)$
    (d) $(1, 2, 3, 0, 0, \ldots)$ and $(1, 2, 3, 0, 0, \ldots)$
    (e) $(0, 0, 0, 1, 0, 0, \ldots)$ and $(8, 9, 10, 11, \ldots)$.

4. In each of the following, the function is the ordinary generating function for a sequence (a_k). Find this sequence.

    (a) $\left(\dfrac{2}{1 - x}\right)\left(\dfrac{3}{1 - x}\right)$ **(b)** $\dfrac{1}{1 - x} e^{2x}$ **(c)** $\dfrac{x^5 + x^6}{1 - x}$

    (d) $\dfrac{x^2 - 3x}{1 - x} + 7$ **(e)** $(1 + x)^q$, where q is a positive integer   **(f)** $xe^{3x} + (1 + x)^2$.

5. If $G(x) = [1/(1 - x)]^2$ is the ordinary generating function for the sequence (a_k), find a_4.

6. If $A(x) = (1 + 10x^2)(1 + 2x + 3x^2 + 4x^3 + \cdots)$ is the ordinary generating function for the sequence (a_k), find a_{11}.

7. Suppose that $A(x)$ is the ordinary generating function for the sequence $(1, 2, 4, 8, 16, \ldots)$ and $B(x)$ is the ordinary generating function for the sequence (b_k). Find (b_k) if $B(x)$ equals

    (a) $A(x) + x$ **(b)** $A(x) + \dfrac{1}{1 - x}$ **(c)** $2A(x)$.

8. In each of the following cases, suppose that $B(x)$ is the ordinary generating function for (b_k) and $A(x)$ is the ordinary generating function for (a_k). Find an expression for $B(x)$ in terms of $A(x)$.

    (a) $b_k = \begin{cases} a_k & \text{if } k \neq 2 \\ 7 & \text{if } k = 2 \end{cases}$

    (b) $b_k = \begin{cases} a_k & \text{if } k \neq 0, 2 \\ 1 & \text{if } k = 0 \\ 4 & \text{if } k = 2. \end{cases}$

9. Suppose that

$$a_k = \begin{cases} b_k & \text{if } k \neq 0, 2 \\ 1 & \text{if } k = 0 \\ 4 & \text{if } k = 2. \end{cases}$$

Find an expression for $A(x)$, the ordinary generating function for the sequence (a_k), in terms of $B(x)$, the ordinary generating function for the sequence (b_k), if $b_0 = 0$ and $b_2 = 2$.

10. Find a simple expression for the ordinary generating function of the following sequences (a_k).
    (a) $a_k = k + 2$ **(b)** $a_k = 7k$ **(c)** $a_k = 5k + 4$.

11. Make use of derivatives to find the ordinary generating function for the following sequences (b_k).

    (a) $b_k = k^2$ **(b)** $b_k = k(k + 1)$ **(c)** $b_k = (k + 1)\dfrac{1}{k!}$.

12. Suppose that $A(x)$ is the ordinary generating function for the sequence (a_k) and $b_k = a_{k+1}$. Find the ordinary generating function for the sequence (b_k).

13. Suppose that

$$a_k = \begin{cases} \displaystyle\sum_{i=0}^{k-2} b_i b_{k-2-i} & \text{if } k \geq 2 \\ 0 & \text{if } k = 0 \text{ or } k = 1. \end{cases}$$

Suppose that $A(x)$ is the ordinary generating function for (a_k) and $B(x)$ is the ordinary generating function for (b_k). Find an expression for $A(x)$ in terms of $B(x)$.

14. Suppose that $A(x)$ is the ordinary generating function for the sequence (a_k) and the sequence (b_k) is defined by taking

$$b_k = \begin{cases} 0 & \text{if } k < i \\ a_{k-i} & \text{if } k \geq i. \end{cases}$$

Find the ordinary generating function for the sequence (b_k) in terms of $A(x)$.

15. Find an ordinary generating function for the sequence whose kth term is

$$a_k = \frac{1}{0!} + \frac{1}{1!} + \frac{1}{2!} + \cdots + \frac{1}{k!}.$$

16. Use the reduction result of Example 4.14 to compute the rook polynomial of the board (d) of Figure 4.4.
17. Use the result of Exercise 15, Section 4.1, to find the rook polynomials of B_I and B_J of Figure 4.6. Then use the reduction result of Example 4.14 to compute the rook polynomial of the board B of Figure 4.6.

4.3 APPLICATIONS TO COUNTING

4.3.1 Sampling Problems

Generating functions will help us in counting. To illustrate how this will work, we first consider sampling problems where the objects being sampled are of different types and objects of the same type are indistinguishable. In the language of Section 2.10, we consider sampling without replacement. For instance, suppose that there are three objects, a, b, and c. Each one can be chosen or not, How many selections are possible? Let a_k be the number of ways to select k objects. Let $G(x)$ be the generating function $\sum a_k x^k$. Now it is easy to see that $a_k = \binom{3}{k}$, and hence

$$G(x) = \binom{3}{0}x^0 + \binom{3}{1}x^1 + \binom{3}{2}x^2 + \binom{3}{3}x^3. \tag{4.13}$$

Let us calculate $G(x)$ another way. We can either pick no a's or one a, and no b's or one b, and no c's or one c. Let us consider the schematic product

$$[(ax)^0 + (ax)^1][(bx)^0 + (bx)^1][(cx)^0 + (cx)^1], \tag{4.14}$$

where addition corresponds to the words "or" italicized in the preceding sentence and multiplication to the words "and" (recall the sum rule and the product rule). The expression (4.14) becomes

$$(1 + ax)(1 + bx)(1 + cx),$$

which equals

$$1 + (a + b + c)x + (ab + ac + bc)x^2 + abcx^3. \tag{4.15}$$

Notice that the coefficient of x lists the ways to get one object: It is a, or b, or c. The coefficient of x^2 lists the ways to get two objects: It is a and b, or a and c, or b and c. The coefficient of x^3 lists the ways to get three objects, and the coefficient of x^0 (namely 1) lists the number of ways to get no objects. If we set $a = b = c = 1$, the coefficient of x^k will count the number of ways to get k objects, that is, a_k. Hence, setting $a = b = c = 1$ in (4.15) gives rise to the generating function

$$G(x) = 1 + 3x + 3x^2 + x^3,$$

which is what we calculated in (4.13).

The same technique works in problems where it is not immediately clear what the coefficients in $G(x)$ are. Then we can calculate $G(x)$ by means of this technique and calculate the appropriate coefficients from $G(x)$.

Example 4.15

Suppose that we have three types of objects, a's, b's, and c's. Suppose that we can pick either 0, 1, or 2 a's, then 0 or 1 b, and finally 0 or 1 c. How many ways are there to pick k objects? The answer is not $\binom{4}{k}$. For example, 2 a's and 1 b is not considered the same as 1 a, 1 b, and 1 c. However, picking the first a and also b is considered the same as picking the second a and also b: The a's are indistinguishable. We want the number of distinguishable ways to pick k objects. Suppose that b_k is the desired number of ways. We shall try to calculate the ordinary generating function $G(x) = \sum b_k x^k$. The right expression to consider here is

$$[(ax)^0 + (ax)^1 + (ax)^2][(bx)^0 + (bx)^1][(cx)^0 + (cx)^1], \qquad (4.16)$$

since we can pick either 0, or 1, or 2 a's, and 0 or 1 b, and 0 or 1 c. The expression (4.16) reduces to

$$(1 + ax + a^2x^2)(1 + bx)(1 + cx),$$

which equals

$$1 + (a + b + c)x + (ab + bc + ac + a^2)x^2 + (abc + a^2b + a^2c)x^3 + a^2bcx^4. \quad (4.17)$$

As in our previous example, the coefficient of x^3 gives the ways of obtaining three objects: a, b, and c; or 2 a's and b; or 2 a's and c. The same thing holds for the other coefficients. Again, taking $a = b = c = 1$ in (4.17) gives the generating function

$$G(x) = 1 + 3x + 4x^2 + 3x^3 + x^4.$$

The coefficient of x^k is b_k. For example, $b_2 = 4$. (The reader should check why this is so.)

In general, suppose that we have p types of objects, with n_1 indistinguishable objects of type 1, n_2 of type 2, $\ldots$, n_p of type p. Let c_k be the number of distinguishable ways of picking k objects if we can pick any number of objects of each type. The ordinary generating function is given by $G(x) = \sum c_k x^k$. To calculate this, we consider the product

$$[(a_1x)^0 + (a_1x)^1 + \cdots + (a_1x)^{n_1}][(a_2 x)^0 + (a_2 x)^1 + \cdots + (a_2 x)^{n_2}] \cdots$$
$$[(a_p x)^0 + (a_p x)^1 + \cdots + (a_p x)^{n_p}].$$

Setting $a_1 = a_2 = \cdots = a_p = 1$, we obtain

$$G(x) = (1 + x + x^2 + \cdots + x^{n_1})(1 + x + x^2 + \cdots + x^{n_2}) \cdots (1 + x + x^2 + \cdots + x^{n_p}).$$

The number c_k is given by the coefficient of x^k in $G(x)$. Thus, we have the following theorem.

Theorem 4.2. Suppose that we have p types of objects, with n_i indistinguishable objects of type i, $i = 1, 2, \ldots, p$. The number of distinguishable ways of picking k objects if we can pick any number of objects of each type is given by the coefficient of x^k in the ordinary generating function

$$(1 + x + x^2 + \cdots + x^{n_1})(1 + x + x^2 + \cdots + x^{n_2}) \cdots (1 + x + x^2 + \cdots + x^{n_p}).$$

Example 4.16 Indistinguishable Men and Indistinguishable Women

Suppose that we have m (indistinguishable) men and n (indistinguishable) women. If we can choose any number of men and any number of women, Theorem 4.2 implies that the number of ways we can choose k people is given by the coefficient of x^k in

$$G(x) = (1 + x + \cdots + x^m)(1 + x + \cdots + x^n). \qquad (4.18)$$

Note: This coefficient is not

$$\binom{m + n}{k},$$

because, for example, having 3 men and $k - 3$ women is the same no matter which men and women you pick. Now $1 + x + \cdots + x^m$ is the generating function for the sequence

$$(a_k) = (1, 1, \ldots, 1, 0, 0, \ldots)$$

and $1 + x + \cdots + x^n$ is the generating function for the sequence

$$(b_k) = (1, 1, \ldots, 1, 0, 0, \ldots),$$

where a_k is 1 for $k = 0, 1, \ldots, m$ and b_k is 1 for $k = 0, 1, \ldots, n$. It follows from (4.18) that $G(x)$ is the generating function for the convolution of the two sequences (a_k) and (b_k). We leave it to the reader (Exercise 10) to compute this convolution in general. For example, if $m = 8$ and $n = 7$, then the number of ways we can choose $k = 9$ people is given by

$$a_0 b_9 + a_1 b_8 + \cdots + a_9 b_0 = 0 + 0 + 1 + \cdots + 1 + 1 + 0 = 7.$$

As a check on the answer, we note that the seven ways are the following: 2 men and 7 women, 3 men and 6 women, ..., 8 men and 1 woman.

Example 4.17 A Sampling Survey

In doing a sampling survey, suppose that we have divided the possible men to be interviewed into various categories, such as teachers, doctors, lawyers, and so on, and similarly for the women. Suppose that in our group we have two men from each category and one woman from each category, and suppose there are q categories. How many distinguishable ways are there of picking a sample of k people? We now want to distinguish people of the same sex if and only if they belong to different categories. The generating function for the number of ways to choose k people is given by

$$G(x) = \underbrace{(1 + x + x^2)(1 + x + x^2) \cdots (1 + x + x^2)}_{q \text{ terms}}\underbrace{(1 + x)(1 + x) \cdots (1 + x)}_{q \text{ terms}}$$

$$= (1 + x + x^2)^q(1 + x)^q.$$

The number of ways to select k people is the coefficient of x^k. For instance, if $q = 2$, then

$$G(x) = x^6 + 4x^5 + 8x^4 + 10x^3 + 8x^2 + 4x + 1.$$

For example, there are 10 ways to pick 3 people. The reader might wish to identify those 10 ways. In general, if there are m categories of men and n categories of women, and there are p_i men in category i, $i = 1, 2, \ldots, m$, and q_j women in category j, $j = 1, 2, \ldots, n$, the reader might wish to express the ordinary generating function for the number of ways to select k people.

Example 4.18 Another Survey

Suppose that our survey team is considering three households. The first household has two people living in it (let us call them "a's"). The second household has one person living in it (let us call the person a "b"). The third household has one person living in it (let us call the person a "c"). In how many ways can k people be selected if we either choose none of the members of a given household or all of the members of the household? Let us again think of the product and sum rule: We can either pick 0 or 2 a's, and either 0 or 1 b, and either 0 or 1 c. Hence, we consider the expression

$$[(ax)^0 + (ax)^2][(bx)^0 + (bx)^1][(cx)^0 + (cx)^1].$$

This becomes

$$(1 + a^2x^2)(1 + bx)(1 + cx) = 1 + (b + c)x + (bc + a^2)x^2$$
$$+ (a^2b + a^2c)x^3 + a^2bcx^4.$$

The ways of choosing two people are given by the coefficient of x^2: We can choose two a's or we can choose b and c. Setting $a = b = c = 1$, we obtain the generating function

$$G(x) = 1 + 2x + 2x^2 + 2x^3 + x^4.$$

The number of ways of choosing k people is given by the coefficient of x^k.

Example 4.19

Suppose that we have p different kinds of objects, each in (for all practical purposes) infinite supply. How many ways are there of picking a sample of k objects? The answer is given by the coefficient of x^k in the ordinary generating function

$$G(x) = \underbrace{(1 + x + x^2 + \cdots)(1 + x + x^2 + \cdots) \cdots (1 + x + x^2 + \cdots)}_{p \text{ terms}}$$

$$= (1 + x + x^2 + \cdots)^p. \tag{4.19}$$

(An alternative approach, leading ultimately to the same result, is to apply Theorem 4.2 with each $n_i = k$.) We shall want to develop ways of finding the coefficient of x^k given an expression for $G(x)$ such as (4.19). To do so, we shall introduce the Binomial Theorem in the next section.

Example 4.20 Keypunching Errors Revisited

In Example 2.17 we considered the problem of checking the work of a keypuncher. Suppose that we check two kinds of errors: omission of a digit (0 or 1) and reversal of a digit (0 to 1 or 1 to 0). In how many ways can we find 30 errors? To answer this question, we can assume that two errors of the same type are indistinguishable. Moreover, for all practical purposes, each type of error is in infinite supply. Thus, we seek the coefficient of x^{30} in the ordinary generating function

$$G(x) = (1 + x + x^2 + \cdots)(1 + x + x^2 + \cdots).$$

This is a special case of (4.19).

Suppose next that we do not distinguish the two types of errors, but we do keep a record of whether an error occurred in the first card punched, the second card punched, and so on. In how many different ways can we find 30 errors if there are 100 cards? This is the question we addressed in Example 2.17. To answer this question, we note that there are 100 types of errors, one type for each card. We can either assume that the number of possible errors per card is, for all practical purposes infinite, or that it is bounded by 30, or that it is bounded by the number of digits per card. In the former case, we consider the ordinary generating function

$$G(x) = (1 + x + x^2 + \cdots)^{100}$$

and look for the coefficient of x^{30}. In the latter two cases, we simply end the terms in $G(x)$ at an appropriate power. The coefficient of x^{30} will, of course, be the same regardless, provided that at least 30 digits appear per card.

Example 4.21 Partitions of Integers

Recall from Section 2.11.4 that a *partition* of a positive integer k is a collection of positive integers which sum to k. For instance, the integer 4 has the partitions $\{1, 1, 1, 1\}$, $\{1, 1, 2\}$, $\{2, 2\}$, $\{1, 3\}$, and $\{4\}$. Suppose that $p(k)$ is the number of partitions of the integer k. Thus, $p(4) = 5$. Exercises 11–15 investigate partitions of integers, using the techniques of this section. The idea of doing so goes back to Euler in 1748. For a detailed discussion of partitions, see most number theory books, for instance Hardy and Wright [1960]. See also Berge [1971] or Cohen [1978] or Riordan [1958].

4.3.2 A Comment on Occupancy Problems

In Section 2.11 we considered occupancy problems, problems of distributing balls to cells. The second part of Example 4.20 involves an occupancy problem: We have 30 balls (errors) and 100 cells (cards). This is a special case of the occupancy problem where we have k indistinguishable balls and we wish to distribute them among p distinguishable cells. If we put no restriction on the number of balls in each cell, then it is easy to generalize the reasoning in Example 4.20 and show that the number of ways to distribute the k balls into the p cells is given by the coefficient of x^k in the ordinary generating function (4.19). By way of contrast, if we allow no more than n_i balls in the ith cell, we see easily that the number of ways is given by the coefficient of x^k in the ordinary generating function of Theorem 4.2.

EXERCISES FOR SECTION 4.3

1. In each of the following, set up the appropriate generating function. Do not calculate an answer. But indicate what you are looking for, for example the coefficient of x^{10}.

   (a) A survey team wants to select at most 3 male students from Harvard, at most 3 female students from Wellesley, at most 2 male students from Princeton, and at most 2 female students from Vassar. In how many ways can 5 students be chosen to interview if only Harvard and Princeton males and Wellesley and Vassar females can be chosen, and 2 students of the same sex from the same school are indistinguishable?

   (b) In how many ways can 5 letters be picked from the letters a, b, c, d if b, c, and d can be picked at most once and a, if picked, must be picked 4 times?

   (c) An athletic director wants to pick at least 3 small college teams as football opponents for a particular season, at least 3 teams from medium-size colleges, and at least 2 large-college teams. He has limited the choice to 7 small college teams, 6 from medium-size colleges, and 4 from large colleges. In how many ways can he pick 11 opponents? (The schedule will be decided upon later.) (*Note:* Do not distinguish 2 small colleges from each other, or 2 medium-size colleges, or 2 large colleges.)

   (d) In how many ways can 8 binary digits be picked if each must be picked an even number of times?

   (e) How many ways are there to choose 10 voters from a group of 5 Republicans, 5 Democrats, and 7 Independents, if we want at least 3 Independents and any two voters of the same political persuasion are indistinguishable?

   (f) A Geiger counter records the impact of five different kinds of radioactive particles over a period of 5 minutes. How many ways are there to obtain a count of 20? Set up a generating function for the answer.

   (g) In checking the work of a proofreader, we look for 4 types of proofreading errors. In how many ways can we find 40 errors?

   (h) In part (g), suppose that we do not distinguish the types of errors, but we do keep a record of the page on which an error occurred. In how many different ways can we find 40 errors in 100 pages?

   (i) How many ways are there to distribute 15 indistinguishable balls into 10 distinguishable cells?

   (j) Repeat part (i) if no cell can be empty.

   (k) A survey team divides the possible people to interview into 5 groups depending on age, and independently into 4 groups depending on geographic location. In how many ways can 8 people be chosen to interview, if 2 people are distinguished only if they belong to different age groups, live in different geographic locations, or are of opposite sex?

   (l) Find the number of solutions to the equation

   $$x_1 + x_2 + x_3 = 12$$

   in which each x_i is a nonnegative integer and $x_i \leq 6$.

2. Suppose that we wish to build up an RNA chain which has length 4 and uses U, C, and A each at most once and G arbitrarily often. How many ways are there to choose the bases (not order them)? Answer this question by setting up a generating function and computing an appropriate coefficient.

3. A small company wants to buy five vehicles, including at most two pickup trucks, at most two station wagons, at most two passenger cars, and at least one van but at most two vans. How many ways are there to buy five vehicles if any two vehicles of the same type, for example, any two vans, are indistinguishable?

4. Suppose that there are p kinds of objects, with n_i indistinguishable objects of the ith kind.

Suppose that we can pick all or none of each kind. Set up a generating function for computing the number of ways to choose k objects.

5. Consider the voting situation of Exercise 7, Section 2.16.
   (a) If all representatives of a county vote alike, set up a generating function for calculating the number of ways to get k votes. (Getting a vote from a representative of county A is considered different from getting a vote from a representative of county B, and so on.)
   (b) Repeat part (a) if representatives of a county do not necessarily vote alike.

6. Consider the following four basic groups of foods: meats, vegetables, starches, and desserts. A dietician wants to choose a daily menu in a cafeteria by choosing 10 items (the limit of the serving space), with at least one item from each category.
   (a) How many ways are there of choosing the basic menu if items in the same group are treated as indistinguishable and it is assumed that there are for all practical purposes an arbitrarily large number of different foods in each group? Answer this by setting up a generating function. Do not do the computation.
   (b) How would the problem be treated if items in a group were treated as distinguishable, and there are, say, 20 items in each group?

7. Suppose that there are p kinds of objects, each in infinite supply. Let a_k be the number of distinguishable ways of choosing k objects if only an even number (including 0) of each kind of object can be taken. Set up a generating function for a_k.

8. In a presidential primary involving all of the New England states on the same day, a presidential candidate would receive a number of electoral votes from each state proportional to the number of voters who voted for the candidate in that state. For example, a candidate who won 50 percent of the vote in Maine and no votes in the other states would receive two electoral votes (or two state votes) in the convention, since Maine has four electoral votes. Votes would be translated into integers by rounding—there are no fractional votes. Set up a generating function for computing the number of ways a candidate could receive 20 votes. (The electoral votes of the states in the region as of the 1980 census are as follows: Maine, 4; New Hampshire, 4; Rhode Island, 4; Vermont, 3; Connecticut, 8; Massachusetts, 13.)

9. In Example 4.17 we showed that if $q = 2$, there are 10 ways to pick 3 people. What are they?

10. Find a general formula for the convolution of the two sequences (a_k) and (b_k) found in Example 4.16.

11. The next five exercises investigate partitions of integers, as defined in Example 4.21. Let $p(k)$ be the number of partitions of integer k and let

$$G(x) = \frac{1}{(1 - x)(1 - x^2)(1 - x^3)(1 - x^4) \cdots}.$$

Show that $G(x)$ is the ordinary generating function for the sequence $(p(k))$.

12. Let $p^*(k)$ be the number of ways to partition the integer k into distinct integers. Thus, $p^*(7) = 5$, using the partitions $\{7\}, \{1, 6\}, \{2, 5\}, \{3, 4\},$ and $\{1, 2, 4\}$. Find an ordinary generating function for $(p^*(k))$.

13. Let $p_o(k)$ be the number of ways to partition integer k into not necessarily distinct odd integers. Find a generating function for $(p_o(k))$.

14. Let $p^*(k)$ and $p_o(k)$ be defined as in Exercises 12 and 13, respectively. Show that $p^*(k) = p_o(k)$.

15. (a) Show that for $|x| < 1$,

$$(1 - x)(1 + x)(1 + x^2)(1 + x^4)(1 + x^8) \cdots (1 + x^{2^k}) \cdots = 1.$$

   (b) Deduce that for $|x| < 1$,

$$1 + x + x^2 + x^3 + \cdots = (1 + x)(1 + x^2)(1 + x^4)(1 + x^8) \cdots (1 + x^{2^k}) \cdots.$$

(c) Conclude that any integer can be written uniquely in binary form, that is, as a sum $a_0 2^0 + a_1 2^1 + a_2 2^2 + \cdots$, where each a_i is 0 or 1. (This conclusion is a crucial one underlying the binary arithmetic that pervades computing machines.)

4.4. THE BINOMIAL THEOREM

In order to expand out the generating function of Equation (4.19), it will be useful to find the Maclaurin series for the function $f(x) = (1 + x)^u$, where u is an arbitrary real number, positive or negative, and not necessarily an integer. We have

$$f'(x) = u(1 + x)^{u-1}$$

$$f''(x) = u(u - 1)(1 + x)^{u-2}$$

$$f^{(r)}(x) = u(u - 1) \cdots (u - r + 1)(1 + x)^{u-r}.$$

Thus, by Equation (4.1), we have the following theorem.

Theorem 4.3 (Binomial Theorem).

$$(1 + x)^u = 1 + ux + \frac{u(u - 1)}{2!} x^2 + \cdots + \frac{u(u - 1) \cdots (u - r + 1)}{r!} x^r + \cdots. \qquad (4.20)$$

One can prove that the expansion (4.20) holds for $|x| < 1$. This expansion can be written succinctly by introducing the *generalized binomial coefficient*

$$\binom{u}{r} = \begin{cases} \dfrac{u(u - 1) \cdots (u - r + 1)}{r!} & \text{if } r > 0 \\ 1 & \text{if } r = 0, \end{cases}$$

which is defined for any real number u and nonnegative integer r. Then (4.20) can be rewritten as

$$(1 + x)^u = \sum_{r=0}^{\infty} \binom{u}{r} x^r. \qquad (4.21)$$

If u is a positive integer n, then $\binom{u}{r}$ is 0 for $r > u$, and (4.21) reduces to the binomial expansion (Theorem 2.7).

Example 4.22

Before returning to generating functions, let us give one quick application of the binomial theorem to the computation of square roots. We have

$$(1 + x)^{1/2} = 1 + \tfrac{1}{2} x + \frac{\tfrac{1}{2}(\tfrac{1}{2} - 1)}{2!} x^2 + \frac{\tfrac{1}{2}(\tfrac{1}{2} - 1)(\tfrac{1}{2} - 2)}{3!} x^3 + \cdots, \qquad (4.22)$$

if $|x| < 1$. Let us use this result to compute $\sqrt{30}$. Note that $|29| \geq 1$, so we cannot use (4.22) directly. However,

$$\sqrt{30} = \sqrt{25 + 5} = 5\sqrt{1 + .2},$$

so we can apply (4.22) with $x = .2$. This gives us

$$\sqrt{30} = 5[1 + \tfrac{1}{2}(.2) - \tfrac{1}{8}(.2)^2 + \tfrac{1}{16}(.2)^3 - \cdots] \approx 5.4775.$$

Returning to Example 4.19, let us apply the Binomial Theorem to find the coefficient of x^k in the expansion of

$$G(x) = (1 + x + x^2 + \cdots)^p.$$

Note that provided that $|x| < 1$,

$$G(x) = \left(\frac{1}{1 - x}\right)^p,$$

using the identity (4.2). $G(x)$ can be rewritten as

$$G(x) = (1 - x)^{-p}.$$

We can now apply the binomial theorem with $-x$ in place of x and with $u = -p$. Then we have

$$G(x) = \sum_{r=0}^{\infty} \binom{-p}{r}(-x)^r.$$

For $k > 0$, the coefficient of x^k is

$$\binom{-p}{k}(-1)^k = \frac{(-p)(-p - 1) \cdots (-p - k + 1)}{k!}(-1)^k,$$

which equals

$$\frac{p(p + 1) \cdots (p + k - 1)}{k!} = \frac{(p + k - 1)(p + k - 2) \cdots p}{k!}$$

$$= \frac{(p + k - 1)!}{k!(p - 1)!}$$

$$= \binom{p + k - 1}{k}.$$

Since $\binom{p + 0 - 1}{0} = 1$, $\binom{p + k - 1}{k}$ gives the coefficient of x^0 also. Hence, we have proved the following theorem.

Theorem 4.4. If there are p types of objects, the number of distinguishable ways to choose k objects if we are allowed unlimited repetition of each type is given by

$$\binom{p + k - 1}{k}.$$

Corollary 4.4.1. Suppose that p is a fixed positive integer. Then the ordinary generating function for the sequence (c_k) where

$$c_k = \binom{p + k - 1}{k}$$

is given by

$$C(x) = \left(\frac{1}{1 - x}\right)^p = (1 - x)^{-p}.$$

Proof. This is a corollary of the proof.

Note that the case $p = 2$ is covered in Example 4.12. In this case,

$$c_k = \binom{2 + k - 1}{k} = \binom{k + 1}{k} = k + 1.$$

Note that in terms of occupancy problems (Section 4.3.2), Theorem 4.4 says that the number of ways to place k indistinguishable balls into p distinguishable cells, with no restriction on the number of balls in a cell, is given by

$$\binom{p + k - 1}{k}.$$

We have already seen this result in Theorem 2.4.

Example 4.23 The Pizza Problem Revisited

In a restaurant that serves pizza, suppose there are nine types of ingredients (see Example 2.12). If a pizza can have at most one kind of ingredient, how many ways are there to sell 100 pizzas? It is reasonable to assume that each ingredient is, for all practical purposes, in infinite supply. Now we have $p = 10$ types of ingredients, including the ingredient "nothing but cheese." Then by Theorem 4.4, the number of distinguishable ways to pick $k = 100$ pizzas is given by $\binom{109}{100}$.

Example 4.24 Rating Job Candidates

Three distinguishable experts rate a job candidate on a scale of 1 to 6. In how many ways can the total of the ratings add up to 12? To answer this question, let us think of each person choosing either 1, 2, 3, 4, 5, or 6 points to award. Hence, the generating function to consider is

$$G(x) = (x + x^2 + \cdots + x^6)^3.$$

Note that we start with x rather than 1 (or x^0) because there must be at least one point chosen. We take the third power because there are three people. We want the coefficient of x^{12}. How can this be found? The answer uses the identity

$$1 + x + x^2 + \cdots + x^s = \frac{1 - x^{s+1}}{1 - x}. \tag{4.23}$$

Then we note that

$$G(x) = [x(1 + x + x^2 + \cdots + x^5)]^3$$

$$= x^3 \left[\frac{1 - x^6}{1 - x} \right]^3$$

$$= x^3(1 - x^6)^3(1 - x)^{-3}. \tag{4.24}$$

We already know that $C(x) = (1 - x)^{-p}$ is the generating function for the sequence (c_k) where

$$c_k = \binom{p + k - 1}{k}.$$

Here

$$c_k = \binom{3 + k - 1}{k}.$$

The expression $B(x) = x^3(1 - x^6)^3$ may be expanded out using the binomial expansion, giving us

$$B(x) = x^3[1 - 3x^6 + 3x^{12} - x^{18}]$$

$$= x^3 - 3x^9 + 3x^{15} - x^{21}.$$

Hence, $B(x)$ is generating function for the sequence

$$(b_k) = (0, 0, 0, 1, 0, 0, 0, 0, 0, -3, 0, 0, 0, 0, 0, 3, 0, 0, 0, 0, 0, -1, 0, 0, \ldots).$$

It follows from (4.24) that $G(x)$ is the generating function for the convolution (a_k) of the sequences (b_k) and (c_k). We wish to find the coefficient a_{12} of x^{12}. This is obtained as

$$a_{12} = b_0 c_{12} + b_1 c_{11} + b_2 c_{10} + \cdots + b_{12} c_0$$

$$= b_3 c_9 + b_9 c_3$$

$$= 1 \cdot \binom{3 + 9 - 1}{9} - 3 \cdot \binom{3 + 3 - 1}{3}$$

$$= \binom{11}{9} - 3 \binom{5}{3}$$

$$= 25.$$

EXERCISES FOR SECTION 4.4

1. Use the binomial theorem to find the coefficient of x^3 in the expansion of
   (a) $\sqrt[3]{1 + x}$ (b) $(1 + x)^{-2}$ (c) $(1 - x)^{-5}$ (d) $(1 + 4x)^{1/2}$.
2. Find the coefficient of x^7 in the expansion of
   (a) $(1 - x)^{-6}x^4$ (b) $(1 - x)^{-4}x^{11}$ (c) $(1 + x)^{1/2}x^3$.

3. If $(1 + x)^{1/3}$ is the ordinary generating function for the sequence (a_k), find a_2.

4. Do the calculation to solve Exercise 6(a) of Section 4.3.

5. Use Theorem 4.4 to compute the number of ways to pick four letters if a and b are the only letters available. Check your answer by writing out all the ways.

6. How many ways are there to choose 12 cans of beer if there are 5 different brands available?

7. How many ways are there to choose 18 candy bars from a machine that has 8 different varieties?

8. A fruit fly is classified as either dominant, hybrid, or recessive for eye color. Ten fruit flies are to be chosen for an experiment. In how many different ways can the genotypes (classifications) dominant, hybrid, and recessive be chosen if you are interested only in the number of dominants, number of hybrids, and number of recessives?

9. If five different colors are available, in how many different ways can 10 flower boxes be painted provided that each box is painted in one solid color and order does not count, that is, we are concerned only with the number of boxes receiving each color?

10. Suppose there are six different kinds of fruit available, each in (theoretically) infinite supply. How many different fruit baskets of 10 pieces of fruit are there?

11. A person drinks one can of beer an evening, choosing one of five different brands. How many different ways are there in which to drink beer over a period of a week if as many cans of a given brand are available as necessary, any two such cans are interchangeable, and we do not distinguish between drinking brand x on Monday and drinking brand x on Tuesday?

12. In studying motor vehicle safety, we classify accidents by the day of the week in which they occur. In how many different ways can we classify 10 accidents?

13. Suppose that there are p different kinds of objects, each in infinite supply. Let a_k be the number of distinguishable ways to pick k of the objects, if we must pick at least one of each kind.
    (a) Set up a generating function for a_k.
    (b) The sequence (a_k) is the convolution of the sequence (b_k) whose generating function is x^p and a sequence (c_k). Find c_k.
    (c) Find a_k for all k.

14. In Exercise 7 of Section 4.3, solve for a_k. (*Hint:* Set $y = x^2$ in the generating function.)

15. Suppose that $B(x)$ is the ordinary generating function for the sequence (b_k). Let

$$Sb_k = b_0 + b_1 + \cdots + b_k$$

and

$$S^2(b_k) = S(Sb_k) = \sum_{j=0}^{k} (b_0 + b_1 + \cdots + b_j).$$

In general, let $a_k = S^p(b_k) = S(S^{p-1}(b_k))$. Then we can show that

$$a_k = b_k + pb_{k-1} + \cdots + \binom{p+j-1}{j} b_{k-j} + \cdots + \binom{p+k-1}{k} b_0.$$

(a) Verify this for $p = 2$.
(b) If $A(x)$ is the ordinary generating function for (a_k), find an expression for $A(x)$ in terms of $B(x)$.

16. Let p_n^r be the number of partitions of the integer n into *exactly* r parts where order counts. For example, there are 10 partitions of 6 into exactly 4 parts where order counts, namely

$$\{3, 1, 1, 1\}, \quad \{1, 3, 1, 1\}, \quad \{1, 1, 3, 1\}, \quad \{1, 1, 1, 3\}, \quad \{2, 2, 1, 1\},$$

$$\{2, 1, 2, 1\}, \quad \{2, 1, 1, 2\}, \quad \{1, 2, 2, 1\}, \quad \{1, 2, 1, 2\}, \quad \{1, 1, 2, 2\}.$$

(a) Set up an ordinary generating function for p_n^r.

(b) Solve for p_n^r.

17. A polynomial in the three variables u, v, w is called *homogeneous* if the total degree of each term $\alpha u^i v^j w^k$ is the same, that is, if $i + j + k$ is constant. For instance,

$$3v^4 + 2uv^2w + 4vw^3$$

is homogeneous with each term having total degree 4. What is the largest number of terms possible in a polynomial of three variables which is homogeneous of total degree n?

18. **(a)** Show that p_n^r as defined in Exercise 16 is the maximum number of terms in a homogeneous polynomial in r variables and having total degree n in which each term has each variable with degree at least 1.

(b) Use this result and the result of Exercise 16 to answer the question in Exercise 17.

19. Three people each roll a die once. In how many ways can the score add up to 11?

4.5 EXPONENTIAL GENERATING FUNCTIONS AND GENERATING FUNCTIONS FOR PERMUTATIONS

4.5.1 Definition of Exponential Generating Function

So far we have used ordinary generating functions to count the number of combinations of objects—we use the word *combination* because order does not count. Let us now try to do something similar if order does count, and we are counting permutations. Recall that $P(n, k)$ is the number of k-permutations of an n-set. The ordinary generating function for $P(n, k)$ with n fixed is given by

$$G(x) = P(n, 0)x^0 + P(n, 1)x^1 + P(n, 2)x^2 + \cdots + P(n, n)x^n$$

$$= \sum_{k=0}^{n} \frac{n!}{(n-k)!} x^k.$$

Unfortunately, there is no good way to simplify this expression. Had we been dealing with combinations, and the number of ways $C(n, k)$ of choosing k elements out of an n-set, we would have been able to simplify, for we would have had the expression

$$C(n, 0)x^0 + C(n, 1)x^1 + C(n, 2)x^2 + \cdots + C(n, n)x^n, \tag{4.25}$$

which by the binomial expansion simplifies to $(1 + x)^n$. By Theorem 2.1,

$$P(n, r) = C(n, r)P(r, r) = C(n, r)r!$$

Hence, the equivalence of (4.25) to $(1 + x)^n$ can be rewritten as

$$P(n, 0)\frac{x^0}{0!} + P(n, 1)\frac{x^1}{1!} + P(n, 2)\frac{x^2}{2!} + \cdots + P(n, n)\frac{x^n}{n!} = (1 + x)^n. \tag{4.26}$$

The number $P(n, k)$ is the coefficient of $x^k/k!$ in the expansion of $(1 + x)^n$.

This suggests the following idea. If (a_k) is any sequence, the *exponential generating function* for the sequence is the function

$$H(x) = a_0 \frac{x^0}{0!} + a_1 \frac{x^1}{1!} + a_2 \frac{x^2}{2!} + \cdots + a_k \frac{x^k}{k!} + \cdots$$

$$= \sum_k a_k \frac{x^k}{k!}.$$

As with the ordinary generating function, we think of x as being chosen so that the sum converges.*

To give an example, if $a_k = 1$, for $k = 0, 1, \ldots$, then, using Equation (4.3), the exponential generating function is

$$H(x) = 1 \cdot \frac{x^0}{0!} + 1 \cdot \frac{x^1}{1!} + 1 \cdot \frac{x^2}{2!} + \cdots$$

$$= 1 + \frac{x^1}{1!} + \frac{x^2}{2!} + \cdots$$

$$= e^x.$$

To give another example, if $a_k = P(n, k)$, we have shown in (4.26) that the exponential generating function is $(1 + x)^n$. To give still one more example, suppose that α is any real number and (a_k) is the sequence $(1, \alpha, \alpha^2, \alpha^3, \ldots)$. Then the exponential generating function for (a_k) is

$$H(x) = \sum_{k=0}^{\infty} \alpha^k \frac{x^k}{k!}$$

$$= \sum_{k=0}^{\infty} \frac{(\alpha x)^k}{k!}$$

$$= e^{\alpha x}.$$

Just as with ordinary generating functions, we will want to go back and forth between sequences and exponential generating functions.

Example 4.25 Eulerian Graphs

A graph without isolated vertices (vertices of degree 0) will be called *eulerian* if it is connected and every vertex has even degree. Eulerian graphs will be very important in a variety of applications in Chapter 11. Harary and Palmer [1973] and Read [1962] show that if u_n is the number of labeled, eulerian graphs of n vertices, then the exponential generating function $U(x)$ for the sequence (u_n) is given by

$$U(x) = x + \frac{x^3}{3!} + \frac{3x^4}{4!} + \frac{38x^5}{5!} + \cdots.$$

Thus, there is one labeled eulerian graph of three vertices and there are three of four vertices. These are shown in Figure 4.7.

---

*As mentioned in the footnote on page 150, we can make this precise using the notion of formal power series.

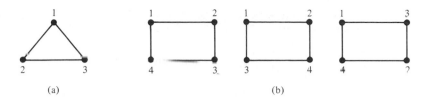

Figure 4.7 The labeled eulerian graphs of (a) three vertices and (b) four vertices.

4.5.2 Applications to Counting Permutations

Example 4.26

A code can use three different letters, a, b, or c. A sequence of five or fewer letters gives a codeword. The codeword can use at most one b, at most one c, and up to three a's. How many possible codewords are there of length k, with $k \leq 5$? Note that order counts in a codeword, so we are interested in counting permutations rather than combinations. However, taking a hint from our previous experience, let us begin by counting combinations, the number of ways of getting k letters if it is possible to pick at most one b, at most one c, and at most three a's. The ordinary generating function is calculated by taking

$$(1 + ax + a^2x^2 + a^3x^3)(1 + bx)(1 + cx),$$

which equals

$$1 + (a + b + c)x + (bc + a^2 + ab + ac)x^2 + (a^3 + abc + a^2b + a^2c)x^3$$
$$+ (a^2bc + a^3b + a^3c)x^4 + a^3bcx^5.$$

The coefficient of x^k gives the ways of obtaining k letters. For example, three letters can be obtained as follows: 3 a's, a and b and c, 2 a's and b, or 2 a's and c. If we make a choice of a and b and c, there are 3! corresponding permutations:

$$abc, \quad acb, \quad bac, \quad bca, \quad cab, \quad cba.$$

For the 3 a's choice, there is only one corresponding permutation: aaa. For the 2 a's and b choice, there are 3 permutations:

$$aab, \quad aba, \quad baa.$$

From our general formula of Theorem 2.6 we see why this is true: The number of distinguishable permutations of 3 objects with 2 of one type and 1 of another is given by

$$\frac{3!}{2!1!}.$$

In general, if we have n_1 a's, n_2 b's, and n_3 c's, the number of corresponding permutations is

$$\frac{n!}{n_1!n_2!n_3!}.$$

In particular, in our schematic, the proper information for the ways to obtain codewords if three letters are chosen is given by

$$\frac{3!}{3!}\,a^3 + \frac{3!}{1!1!1!}\,abc + \frac{3!}{2!1!}\,a^2 b + \frac{3!}{2!1!}\,a^2 c. \tag{4.27}$$

Setting $a = b = c = 1$ would yield the proper count of number of such codewords of three letters. We can obtain (4.27) and the other appropriate coefficients by the trick of using

$$\frac{(ax)^p}{p!} = \frac{a^p}{p!}\,x^p$$

instead of $a^p x^p$ to derive our schematic generating function. In our example, we have

$$\left(1 + \frac{a}{1!}\,x + \frac{a^2}{2!}\,x^2 + \frac{a^3}{3!}\,x^3\right)\left(1 + \frac{b}{1!}\,x\right)\left(1 + \frac{c}{1!}\,x\right),$$

which equals

$$1 + \left(\frac{a}{1!} + \frac{b}{1!} + \frac{c}{1!}\right)x + \left(\frac{bc}{1!1!} + \frac{a^2}{2!} + \frac{ab}{1!1!} + \frac{ac}{1!1!}\right)x^2$$

$$+ \left(\frac{a^3}{3!} + \frac{abc}{1!1!1!} + \frac{a^2 b}{2!1!} + \frac{a^2 c}{2!1!}\right)x^3 \tag{4.28}$$

$$+ \left(\frac{a^2 bc}{2!1!1!} + \frac{a^3 b}{3!1!} + \frac{a^3 c}{3!1!}\right)x^4 + \frac{a^3 bc}{3!1!1!}\,x^5.$$

This is still not a satisfactory schematic—compare the coefficients of x^3 to the expression in (4.27). However, the schematic works if we consider this as an exponential generating function, and choose the coefficient of $x^k/k!$. For the expression (4.28) is equal to

$$1 + 1!\left(\frac{a}{1!} + \frac{b}{1!} + \frac{c}{1!}\right)\frac{x}{1!} + 2!\left(\frac{bc}{1!1!} + \frac{a^2}{2!} + \frac{ab}{1!1!} + \frac{ac}{1!1!}\right)\frac{x^2}{2!}$$

$$+ 3!\left(\frac{a^3}{3!} + \frac{abc}{1!1!1!} + \frac{a^2 b}{2!1!} + \frac{a^2 c}{2!1!}\right)\frac{x^3}{3!} + \cdots.$$

Setting $a = b = c = 1$ and taking the coefficient of $x^k/k!$ gives the appropriate number of codewords (permutations). For example, the number of length 3 is

$$3!\left(\frac{1}{3!} + 1 + \frac{1}{2!} + \frac{1}{2!}\right) = 13.$$

The corresponding codewords are the six we have listed with a, b, and c, the three with 2 a's and 1 b, the three with 2 a's and 1 c, and the one with 3 a's.

The analysis in Example 4.26 generalizes as follows.

Theorem 4.5. Suppose that we have p types of objects, with n_i indistinguishable objects of type i, $i = 1, 2, \ldots, p$. The number of distinguishable permutations of length k

with up to n_i objects of type i is the coefficient of $x^k/k!$ in the exponential generating function

$$\left(1 + x + \frac{x^2}{2!} + \cdots + \frac{x^{n_1}}{n_1!}\right)\left(1 + x + \frac{x^2}{2!} + \cdots + \frac{x^{n_2}}{n_2!}\right) \cdots \left(1 + x + \frac{x^2}{2!} + \cdots + \frac{x^{n_p}}{n_p!}\right).$$

Example 4.27   RNA Chains

To give an application of this result, let us consider the number of 2-link RNA chains if we have available up to 3 A's, up to 3 G's, up to 2 C's, and up to 1 U. Since order counts, we seek an exponential generating function. This is given by

$$\left(1 + x + \frac{x^2}{2!} + \frac{x^3}{3!}\right)^2 \left(1 + x + \frac{x^2}{2!}\right)(1 + x),$$

which turns out to equal

$$1 + 4x + \tfrac{15}{2}x^2 + \tfrac{53}{6}x^3 + \cdots.$$

Here, the coefficient of x^2 is $\tfrac{15}{2}$, so the coefficient of $x^2/2!$ is $2! \, (\tfrac{15}{2}) = 15$. Thus, there are 15 such chains. They are AA, AG, AC, AU, GA, GG, GC, GU, CA, CG, CC, CU, UA, UG, and UC, that is, all but UU. Similarly, the number of 3-link RNA chains made up from these available bases is the coefficient of $x^3/3!$, or $3! \, (\tfrac{53}{6}) = 53$. The reader can readily check this result.

Example 4.28   RNA Chains Continued

To continue with the previous example, suppose that we wish to find the number of RNA chains of length k if we assume an arbitrarily large supply of each base. The exponential generating function is given by

$$\left(1 + x + \frac{x^2}{2!} + \cdots\right)^4 = (e^x)^4$$

$$= e^{4x}$$

$$= \sum \frac{(4x)^k}{k!}$$

$$= \sum 4^k \frac{x^k}{k!}.$$

Thus, the number in question is given by 4^k. This agrees with what we already concluded in Chapter 2, by a simple use of the product rule.

Let us make one modification here, namely, to count the number of RNA chains of length k if the number of U links is even. The exponential generating function is given by

$$H(x) = \left(1 + \frac{x^2}{2!} + \frac{x^4}{4!} + \cdots\right)\left(1 + x + \frac{x^2}{2!} + \cdots\right)^3.$$

Now the second term in $H(x)$ is given by $(e^x)^3 = e^{3x}$. It is also not hard to show that the first term is given by

$$\tfrac{1}{2}(e^x + e^{-x}).$$

Thus,

$$H(x) = \tfrac{1}{2}(e^x + e^{-x})(e^{3x})$$

$$= \tfrac{1}{2}(e^{4x} + e^{2x})$$

$$= \frac{1}{2}\left[\sum 4^k \frac{x^k}{k!} + \sum 2^k \frac{x^k}{k!}\right]$$

$$= \sum \left(\frac{4^k + 2^k}{2}\right)\frac{x^k}{k!}.$$

We conclude that the number of RNA chains in question is

$$\frac{4^k + 2^k}{2}.$$

To check this, note for example that if $k = 2$, this number is 10. The 10 chains are UU, GG, GA, GC, AG, AA, AC, CG, CA, and CC.

4.5.3 Distributions of Distinguishable Balls into Indistinguishable Cells*

Recall from Section 2.11.3 that the Stirling number of the second kind, $S(n, k)$, is defined to be the number of distributions of n distinguishable balls into k indistinguishable cells with no cell empty. Here we shall show that

$$S(n, k) = \frac{1}{k!} \sum_{i=0}^{k}(-1)^i\binom{k}{i}(k - i)^n. \qquad (4.29)$$

Let us first consider the problem of finding the number $T(n, k)$ of ways to put n distinguishable balls into k distinguishable cells labeled $1, 2, \ldots, k$, with no cell empty. Note that

$$T(n, k) = k!S(n, k), \qquad (4.30)$$

since we obtain a distribution of n distinguishable balls into k distinguishable cells with no cell empty by finding a distribution of n distinguishable balls into k indistinguishable cells with no cell empty and then labeling (ordering) the cells. Next we compute $T(n, k)$. Suppose that ball i goes into cell $C(i)$. We can encode the distribution of balls into distinguishable cells by giving a sequence $C(1)C(2) \cdots C(n)$. This is an n-permutation from the k-set $\{1, 2, \ldots, k\}$ with each label j in the k-set used at least once. Thus, $T(n, k)$ is the number of such permutations, and for fixed k, the exponential generating function for $T(n, k)$ is therefore given by

$$H(x) = \left(x + \frac{x^2}{2!} + \frac{x^3}{3!} + \cdots\right)^k = (e^x - 1)^k.$$

*This subsection can be omitted.

$T(n, k)$ is given by the coefficient of $x^n/n!$ in the expansion of $H(x)$. By the binomial expansion (Theorem 2.7),

$$H(x) = \sum_{i=0}^{k} \binom{k}{i}(-1)^i e^{(k-i)x}.$$

Substituting $(k - i)x$ for x in the power series (4.3) for e^x, we obtain

$$H(x) = \sum_{i=0}^{k} \binom{k}{i}(-1)^i \sum_{n=0}^{\infty} \frac{1}{n!} (k - i)^n x^n$$

$$= \sum_{n=0}^{\infty} \frac{x^n}{n!} \sum_{i=0}^{k} (-1)^i \binom{k}{i}(k - i)^n.$$

Finding the coefficient of $x^n/n!$ in the expansion of $H(x)$, we have

$$T(n, k) = \sum_{i=0}^{k} (-1)^i \binom{k}{i}(k - i)^n. \tag{4.31}$$

Now equations (4.30) and (4.31) give us Equation (4.29).

EXERCISES FOR SECTION 4.5

1. For each of the following sequences (a_k), find a simple expression for the exponential generating function.

   (a) $(2, 2, 2, \ldots)$ (b) $a_k = 5^k$ (c) $(1, 0, 1, 1, \ldots)$

   (d) $(0, 0, 1, 1, \ldots)$ (e) $(0, 1, 0, 1, 0, 1, \ldots)$ (f) $(2, 1, 2, 1, 2, 1, \ldots)$

2. For each of the following functions, find a sequence for which the function is the exponential generating function.

   (a) $3 + 3x + 3x^2 + 3x^3 + \cdots$ (b) $\dfrac{1}{1 - x}$ (c) $x^2 + 3x$ (d) e^{6x}

   (e) $2e^x$ (f) $e^x + e^{4x}$ (g) $(1 + x^2)^n$ (h) $\dfrac{1}{1 - 4x}$.

3. A graph is said to be *even* if every vertex has even degree. If e_k is the number of labeled even graphs of k vertices, Harary and Palmer [1973] show that the exponential generating function $E(x)$ for the sequence (e_k) is given by

$$E(x) = x + \frac{x^2}{2!} + \frac{2x^3}{3!} + \frac{8x^4}{4!} + \cdots.$$

   Verify the coefficients of $x^3/3!$ and $x^4/4!$. (Note that e_k can be derived from Exercise 11, Section 11.3, and the results of Section 3.1.3.)

4. In Example 4.27, check by enumeration that there are 53 3-link RNA chains made up from the available bases.

5. Find the number of 3-link RNA chains if the available bases are 2 A's, 2 G's, 3C's, and 1 U. Check your answer by enumeration.

6. In each of the following, set up the appropriate generating function, but do not calculate an answer. Indicate what you are looking for, for example, the coefficient of x^8.

   (a) How many codewords of three letters can be built from the letters a, b, c, and d if b, c, and d can only be picked once?

(b) A codeword consists of at least one of each of the digits 0, 1, 2, and 3, and has length 5. How many such codewords are there?

(c) How many 10-digit numbers consist of at most three 0's, at most three 1's, and at most four 2's?

(d) In how many ways can $3n$ letters be selected from $2n$ A's, $2n$ B's, and $2n$ C's?

(e) If n is a fixed even number, find the number of n-digit words generated from the alphabet $\{0, 1, 2, 3\}$ in each of which the number of 0's and the number of 1's is even.

(f) In how many ways can a total of 100 be obtained if 50 dice are rolled?

(g) Eight municipal bonds are each to be rated as a good risk, a bad risk, or hopeless. In how many different ways can the ratings be assigned?

(h) Three pigeons, two doves, and five sparrows are flying around a telephone wire. If four of them land, in how many different orders could they sit on the telephone wire?

(i) Suppose that with a type A coin, you get 1 point if the coin turns up heads, and 2 points if it turns up tails. With a type B coin, you get 2 points for a head and 3 points for a tail. In how many ways can you get 10 points if you toss 3 type A coins and 4 type B coins?

(j) In how many ways can 200 identical terminals be divided among four computer rooms so that each room will have 20 or 40 or 60 or 80 or 100 terminals?

(k) One way for a ship to communicate with another visually is to hang a sequence of colored flags from a flagpole. The meaning of a signal depends on the order of the flags from top to bottom. If there are available 5 red flags, 3 green ones, 3 yellow ones, and 1 blue one, how many different signals are possible if 10 flags are to be used?

(l) In part (k), how many different signals are possible if at least 10 flags are to be used?

7. Suppose that there are p different kinds of objects, each in infinite supply. Let a_k be the number of permutations of k objects chosen from these objects. Find a_k explicitly by using exponential generating functions.

8. In how many ways can 60 identical terminals be divided among two computer rooms so that each room will have 20 or 40 terminals?

9. If order counts, find an exponential generating function for the number of partitions of integer k (Example 4.21 and Exercise 16, Section 4.4).

10. Find a simple form for the exponential generating function if we have p types of objects, each in infinite supply, and we wish to choose k objects, at least one of each kind, with order counting.

11. Find a simple form for the exponential generating function if we have p types of objects, each in infinite supply, and we wish to choose k objects, with an even number (including 0) of each kind, and order counts.

12. Find the number of codewords of length k from an alphabet $\{a, b, c, d, e\}$ if b occurs an even number of times.

13. Find the number of codewords of length 3 from an alphabet $\{1, 2, 3, 4, 5, 6\}$ if 1 and 3 occur an even number of times.

14. Compute $S(4, 2)$ and $T(4, 2)$ from Equations (4.29) and (4.31), respectively, and check your answers by listing all the appropriate distributions.

15. Exercises 15–18 investigate combinations of exponential generating functions. Suppose that $A(x)$ and $B(x)$ are the exponential generating functions for the sequences (a_k) and (b_k), respectively. Find an expression for the kth term c_k of the sequence (c_k) whose exponential generating function is $C(x) = A(x) + B(x)$.

16. Repeat Exercise 15 for $C(x) = A(x)B(x)$.

17. Find a_3 if the exponential generating function for (a_k) is

(a) $e^x(1 + x)^8$ **(b)** $\dfrac{e^{3x}}{1 - x}$.

18. Suppose that $a_{n+1} = (n + 1)b_n$, with $a_0 = b_0 = 1$. If $A(x)$ is the exponential generating function for the sequence (a_n) and $B(x)$ is the exponential generating function for the sequence (b_n), derive a relation between $A(x)$ and $B(x)$.

4.6. PROBABILITY GENERATING FUNCTIONS*

The simple idea of generating function has interesting uses in probability. In fact, the first complete treatment of generating functions was by Laplace in his *Théorie Analytique des Probabilités*, Paris, 1812, and much of the motivation for the development of generating functions came from probability. Suppose that after an experiment is performed, it is known that one and only one of a (finite or countably infinite) set of possible events will occur. Let p_k be the probability that the kth event occurs, $k = 0, 1, 2, \ldots$. (Of course, this notation does not work if there is a continuum of possible events.) The ordinary generating function

$$G(x) = \sum p_k x^k \tag{4.32}$$

is called the *probability generating function*. [Note that (4.32) converges at least for $|x| \leq 1$, since $p_0 + p_1 + \cdots + p_k + \cdots = 1$.] We shall see that probability generating functions are extremely useful in evaluating experiments, in particular in analyzing roughly what we "expect" the outcomes to be.

Example 4.29   Coin Tossing

Suppose that the experiment is tossing a fair coin. Then the events are heads (H) and tails (T), with p_0, probability of H, equal to $1/2$, and p_1, probability of T, equal to $\frac{1}{2}$. Hence, the probability generating function is

$$G(x) = \tfrac{1}{2} + \tfrac{1}{2}x.$$

Example 4.30   Bernoulli Trials

In Bernoulli trials there are n independent repeated trials of an experiment with each trial leading to a success with probability p and a failure with probability $q = 1 - p$. The experiment could be a test to see if a product is defective or nondefective, a test for the presence or absence of a disease, or a decision about whether to accept or reject a candidate for a job. If S stands for success and F for failure, a typical outcome in $n = 5$ trials is a sequence like $SSFSF$ or $SSFFF$. The probability that in n trials there will be k successes is given by

$$b(k, n, p) = C(n, k)p^k q^{n-k},$$

as is shown in any standard book on probability theory (such as Feller [1968]), or on finite mathematics (such as Goodman and Ratti [1971] or Kemeny *et al.* [1974]).

---

*This section can be omitted without loss of continuity. Although it is essentially self-contained, the reader with some prior exposure to probability theory, at least at the level of a "finite math" book such as Goodman and Ratti [1971] or Kemeny *et al.* [1974], will get more out of this.

The probability generating function for the number of successes in n trials is given by

$$G(x) = \sum_{k=0}^{n} b(k, n, p)x^k$$

$$= \sum_{k=0}^{n} C(n, k)p^k q^{n-k} x^k.$$

By the binomial expansion (Theorem 2.7), we have

$$G(x) = (px + q)^n.$$

Let us note some simple results about probability generating functions.

Theorem 4.6. If G is a probability generating function, then

$$G(1) = 1.$$

Proof. Since the outcomes are mutually exclusive and exhaustive by assumption, we have

$$p_0 + p_1 + \cdots + p_k + \cdots = 1. \qquad \text{Q.E.D.}$$

Corollary 4.6.1. $\qquad \sum_{k=0}^{n} C(n, k)p^k q^{n-k} = 1.$

Proof. In Bernoulli trials (Example 4.30), set $G(1) = 1$.

Note: Corollary 4.6.1 may also be proved directly from the binomial expansion, noting that

$$(p + q)^n = 1^n.$$

Suppose that in an experiment, if the kth event occurs, we get k dollars (or k units of some reward). Then the expression $E = \sum kp_k$ is called the *expected value* or the *expectation*. It is what we expect to "win" *on the average* if the experiment is repeated many times, and we expect 0 dollars a fraction p_0 of the time, 1 dollar a fraction p_1 of the time, and so on. For a more detailed discussion of expected value, see any elementary book on probability theory or on finite mathematics. Note that the expected value is defined only if the sum $\sum kp_k$ converges. If the sum does converge, we say the expected value *exists*. We can have the same expected value in an experiment that always gives 1 dollar and in an experiment that gives 0 dollars with probability $\frac{1}{2}$ and 2 dollars with probability $\frac{1}{2}$. However, there is more variation in outcomes in the second experiment. Probability theorists have introduced the concept of variance to measure this variation. Specifically, the *variance V* is defined to be

$$V = \sum_k k^2 p_k - \left(\sum_k kp_k\right)^2. \tag{4.33}$$

See a probability book such as Feller [1968] for a careful explanation of this concept.

Variance is defined only if the sums in (4.33) converge. In case they do converge, we say the variance *exists*. In the first experiment mentioned above,

$$V = [1^2(1)] - [1(1)]^2 = 0.$$

In the second experiment mentioned above,

$$V = [0^2(\tfrac{1}{2}) + 2^2(\tfrac{1}{2})] - [0(\tfrac{1}{2}) + 2(\tfrac{1}{2})]^2 = 1.$$

Hence, the variance is higher in the second experiment. We shall see how the probability generating function allows us to calculate expected value and variance.

Differentiating (4.32) with respect to x leads to the equation

$$G'(x) = \sum kp_k x^{k-1}.$$

Hence, if $G'(x)$ converges for $x = 1$, that is, if $\sum kp_k$ converges, then

$$G'(1) = \sum kp_k. \tag{4.34}$$

If the kth event gives value k dollars or units, then the expression on the right-hand side of (4.34) is the expected value.

Theorem 4.7. Suppose that $G(x)$ is the probability generating function and the kth event gives value k. If the expected value exists, then $G'(1)$ is the expected value.

Let us apply Theorem 4.7 to the case of Bernoulli trials. We have

$$G(x) = (px + q)^n$$
$$G'(x) = n(px + q)^{n-1}p$$
$$G'(1) = np(p + q)^{n-1}$$
$$= np(1)^{n-1}$$
$$= np.$$

Thus, the expected number of successes in n trials is np. The reader who recalls the "standard" derivation of this fact should be pleased at how simple this derivation is. To illustrate the result, we note that in $n = 100$ tosses of a fair coin, the probability of a head (success) is $p = .5$ and the expected number of heads is $np = 50$.

The next theorem is concerned with variance. Its proof is left to the reader (Exercise 5).

Theorem 4.8. Suppose that $G(x)$ is the probability generating function and the kth event has value k. If the variance V exists, then V is given by $V = G''(1) + G'(1) - [G'(1)]^2$.

Applying Theorem 4.8 to Bernoulli trials, we have

$$G''(x) = n(n - 1)p^2(px + q)^{n-2}.$$

Also,

$$G'(1) = np$$
$$G''(1) = n(n - 1)p^2.$$

Hence,

$$V = G''(1) + G'(1) - [G'(1)]^2$$
$$= n(n-1)p^2 + np - n^2p^2$$
$$= np - np^2$$
$$= np(1-p)$$
$$= npq.$$

This gives npq for the variance, a well-known formula.

EXERCISES FOR SECTION 4.6

1. In each of the following situations, find a simple expression for the probability generating function, and use this to compute expected value and variance.
   (a) $p_0 = p_1 = p_2 = \frac{1}{3}, p_k = 0$ otherwise.
   (b) $p_5 = \frac{2}{3}, p_7 = \frac{1}{3}, p_k = 0$ otherwise.

2. For fixed positive number λ, the *Poisson distribution* with parameter λ has

   $$p_k = \frac{e^{-\lambda}\lambda^k}{k!}, \qquad k = 0, 1, 2, \ldots .$$

   (a) Find a simple expression for the probability generating function.
   (b) Use the methods of generating functions to find the expected value and the variance.

3. In Bernoulli trials suppose that we compute the probability that the first success occurs on trial k. This is given by $p_k = 0$, $k = 0$ (assuming we start with trial 1), and $p_k = (1-p)^{k-1}p$, $k > 0$. The probabilities p_k define the *geometric distribution*. Repeat Exercise 2 for this distribution.

4. Fix a positive integer m. In Bernoulli trials, the probability that the mth success takes place on trial $k + m$ is given by

   $$p_k = \binom{k+m-1}{k}q^k p^m.$$

   The probabilities p_k define the *negative binomial distribution*.
   (a) Show that the probability generating function $G(x)$ for the negative binomial distribution p_k is given by

   $$G(x) = \frac{p^m}{(1-qx)^m}.$$

   (b) Compute expected value and variance.

5. Prove Theorem 4.8.

4.7. THE COLEMAN AND BANZHAF POWER INDICES*

In Section 2.16 we introduced the notion of simple game and the Shapley–Shubik power index. Here, we shall define two alternative power indices and discuss how to use generating functions to calculate them. We defined the *value* $v(S)$ of a coalition S to be 1 if S is

*This section can be omitted without loss of continuity.

winning and 0 if S is losing. Coleman [1971] defines the power of player i as

$$P_i^C = \frac{\sum_S [v(S) - v(S - \{i\})]}{\sum_S v(S)}. \tag{4.35}$$

In calculating this measure, the sums are taken over all coalitions S. The term

$$v(S) - v(S - \{i\})$$

is 1 if removal of i changes S from winning to losing, and it is 0 otherwise. (It cannot be -1, since we assumed a winning coalition can never be contained in a losing one.) Thus, P_i^C is the number of winning coalitions from which removal of i leads to a losing coalition divided by the number of winning coalitions, or the proportion of winning coalitions in which i's defection is critical. This index avoids the seemingly extraneous notion of order that underlies the computation of the Shapley–Shubik index.

It is interesting to note that the Shapley–Shubik index p_i^S can be calculated by a formula similar to (4.35). For Shapley [1953] proved that

$$p_i^S = \sum_S \{\gamma(s)[v(S) - v(S - \{i\})]\}: \quad S \text{ such that } i \in S\}, \tag{4.36}$$

where

$$s = |S| \quad \text{and} \quad \gamma(s) = \frac{(s - 1)!(n - s)!}{n!}.$$

(See Exercise 14, Section 2.16.)

A variant of the Coleman power index is the Banzhaf index (Banzhaf [1965]), defined as

$$P_i^B = \frac{\sum_S [v(S) - v(S - \{i\})]}{\sum_{j=1}^n \sum_S [v(S) - v(S - \{j\})]}. \tag{4.37}$$

This index has the same numerator as Coleman's while the denominator sums the numerators for all players j. Thus, P_i^B is the number of critical defections of player i divided by the total number of critical defections of all players, or player i's proportion of all critical defections.*

To give an example, let us consider the game $[51; 49, 48, 3]$. Here, the winning coalitions are $\{1, 2, 3\}$, $\{1, 2\}$, $\{1, 3\}$, $\{2, 3\}$. Player 1's defection is critical to $\{1, 2\}$, $\{1, 3\}$, so we have the Coleman index

$$P_1^C = \tfrac{2}{4} = \tfrac{1}{2}.$$

---

*For a unifying framework for the Shapley–Shubik, Banzhaf, and Coleman indices, see Straffin [1980]. For a survey of the literature of the Shapley–Shubik index, see Shapley [1981]. For one on the Banzhaf and Coleman indices, see Dubey and Shapley [1979]. For applications of all three indices, see Brams, *et al.* [1983] and Lucas [1983].

Similarly, each player's defection is critical to two coalitions, so

$$P_2^C = \tfrac{2}{4} = \tfrac{1}{2}$$

$$P_3^C = \tfrac{2}{4} = \tfrac{1}{2}.$$

Note that in the Coleman index, the powers P_i^C may not add up to 1. It is the relative values that count. The Banzhaf index is given by

$$P_1^B = \tfrac{2}{6} = \tfrac{1}{3}$$

$$P_2^B = \tfrac{2}{6} = \tfrac{1}{3}$$

$$P_3^B = \tfrac{2}{6} = \tfrac{1}{3}.$$

These two indices agree with the Shapley–Shubik index in saying that all three players have equal power. It is not hard to give examples where these indices may differ from that of Shapley–Shubik (see Exercise 2). (When a number of ways to measure something have been introduced, and they can differ, how do we choose among them? One approach is to lay down conditions or axioms that a reasonable measure should satisfy. We can then test different measures to see if they satisfy the axioms. One set of axioms, which is satisfied only by the Shapley–Shubik index, is due to Shapley [1953]; see Owen [1968], Roberts [1976] or Shapley [1981]. Another set of axioms, which is satisfied only by the Banzhaf index, is due to Dubey and Shapley [1979]. The axiomatic approach is probably the most reasonable procedure to use in convincing legislators or judges to use one measure over another. For legislators can then decide whether they like certain general conditions, rather than argue about a procedure. Incidentally, it is the Banzhaf index which has found use in the courts, in one-person, one-vote cases.)

Generating functions can be used to calculate the numerator of P_i^C and P_i^B in case we have a weighted majority game $[q; v_1, v_2, \ldots, v_n]$. (Exercise 3 asks the reader to describe how to find the denominator of the former. The denominator of the latter is trivial to compute if all the numerators are known.) Suppose that player i has v_i votes. His defection will be critical if it comes from a coalition with q votes, or $q + 1$ votes, or $\ldots$, or $q + v_i - 1$ votes. His defection in these cases will lead to a coalition with $q - v_i$ votes, or $q - v_i + 1$ votes, or $\ldots$, or $q - 1$ votes. Suppose that $a_k^{(i)}$ is the number of coalitions with exactly k votes and not containing player i. Then the number of coalitions in which player i's defection is critical is given by

$$a_{q-v_i}^{(i)} + a_{q-v_i+1}^{(i)} + \cdots + a_{q-1}^{(i)} = \sum_{k=q-v_i}^{q-1} a_k^{(i)}. \tag{4.38}$$

This expression can be substituted for

$$\sum_S [v(S) - v(S - \{i\})]$$

in the computation of the Coleman or Banzhaf indices, provided that we can calculate the numbers $a_k^{(i)}$. Brams and Affuso [1976] point out that the numbers $a_k^{(i)}$ can be found using ordinary generating functions. To form a coalition, player j contributes either 0 votes or v_j votes. Hence, the ordinary generating function for the $a_k^{(i)}$ is given by

$$G^{(i)}(x) = (1 + x^{v_1})(1 + x^{v_2}) \cdots (1 + x^{v_{i-1}})(1 + x^{v_{i+1}}) \cdots (1 + x^{v_n})$$

$$= \prod_{j \neq i}(1 + x^{v_j}).$$

The number $a_k^{(i)}$ is given by the coefficient of x^k.

Let us consider the weighted majority game $[4; 1, 2, 4]$ as an example. We have

$$G^{(1)}(x) = (1 + x^2)(1 + x^4) = 1 + x^2 + x^4 + x^6$$

$$G^{(2)}(x) = (1 + x)(1 + x^4) = 1 + x + x^4 + x^5$$

$$G^{(3)}(x) = (1 + x)(1 + x^2) = 1 + x + x^2 + x^3.$$

Thus, for example, $a_4^{(1)}$ is the coefficient of x^4 in $G^{(1)}(x)$, i.e., it is 1. There is one coalition not containing player 1 which has exactly four votes, namely the coalition consisting of the third player alone. Using (4.38), we obtain

$$\sum_S [v(S) - v(S - \{1\})] = a_{4-1}^{(1)} = a_3^{(1)} = 0$$

$$\sum_S [v(S) - v(S - \{2\})] = a_{4-2}^{(2)} + a_{4-2+1}^{(2)} = a_2^{(2)} + a_3^{(2)} = 0$$

$$\sum_S [v(S) - v(S - \{3\})] = a_{4-4}^{(3)} + a_{4-4+1}^{(3)} + a_{4-4+2}^{(3)} + a_{4-4+3}^{(3)}$$

$$= a_0^{(3)} + a_1^{(3)} + a_2^{(3)} + a_3^{(3)}$$

$$= 4.$$

This immediately gives us

$$P_1^B = \frac{0}{4} = 0$$

$$P_2^B = \frac{0}{4} = 0$$

$$P_3^B = \frac{4}{4} = 1.$$

According to the Banzhaf index, player 3 has all the power. This makes sense: No coalition can be winning without him. The Coleman index and Shapley–Shubik index give rise to the same values. Computation is left to the reader.

EXERCISES FOR SECTION 4.7

1. Calculate the Banzhaf and Coleman power indices for each of the following games, using generating functions to calculate the numerators. Check your answer using the definitions of these indices.
   (a) $[51; 51, 48, 1]$ (b) $[51; 49, 47, 4]$ (c) $[51; 40, 30, 20, 10]$
   (d) $[20; 1, 10, 10, 10]$ (e) $[102; 80, 40, 80, 20]$
   (f) The Australian government, $[5; 1, 1, 1, 1, 1, 1, 3]$.

2. Give an example of a game where:
   (a) the Banzhaf and Coleman power indices differ;
   (b) the Banzhaf and Shapley–Shubik power indices differ;

(c) the Coleman and Shapley–Shubik power indices differ;

(d) all three of these indices differ.

3. Describe how to find $\sum_S v(S)$ by generating functions.

4. Use the formula of Equation (4.36) to calculate the Shapley–Shubik power index of each of the weighted majority games in Exercises 1(a)–(e).

5. (a) Explain how you could use generating functions to compute the Shapley–Shubik power index.

   (b) Apply your results to the games in Exercise 1.

REFERENCES FOR CHAPTER 4

BANZHAF, J. F., III, "Weighted Voting Doesn't Work: A Mathematical Analysis," *Rutgers Law Rev.*, *19* (1965), 317–343.

BERGE, C., *Principles of Combinatorics*, Academic Press, New York, 1971.

BRAMS, S. J., and AFFUSO, P. J., "Power and Size: A Near Paradox," *Theory and Decision, 1* (1976), 68–94.

BRAMS, S. J., LUCAS, W. F., and STRAFFIN, P. D. (eds.), *Political and Related Models*, Modules in Applied Mathematics, Vol. 2, Springer-Verlag, New York, 1983.

COHEN, D. I. A., *Basic Techniques of Combinatorial Theory*, Wiley, New York, 1978.

COLEMAN, J. S., "Control of Collectivities and the Power of a Collectivity to Act," in B. Lieberman (ed.), *Social Choice*, Gordon and Breach, New York, 1971, pp. 269–300.

DUBEY, P., and SHAPLEY, L. S., "Mathematical Properties of the Banzhaf Power Index," *Math. Oper. Res., 4* (1979), 99–131.

FELLER, W., *An Introduction to Probability Theory and Its Applications*, 3d ed., Wiley, New York, 1968.

GOODMAN, A. W., and RATTI, J. S., *Finite Mathematics with Applications*, Macmillan, New York, 1971.

HARARY, F., and PALMER, E. M., *Graphical Enumeration*, Academic Press, New York, 1973.

HARDY, G. H., and WRIGHT, E. M., *An Introduction to the Theory of Numbers*, 4th ed., Oxford University Press, London, 1960.

KEMENY, J. G., SNELL, J. L., and THOMPSON, G. L., *Introduction to Finite Mathematics*, Prentice-Hall, Englewood Cliffs, N.J., 1974.

LUCAS, W. F., "Measuring Power in Weighted Voting Systems," in S. J. Brams, W. F. Lucas, and P. D. Straffin (eds.), *Political and Related Models*, Modules in Applied Mathematics, Vol. 2, Springer-Verlag, New York, 1983, pp. 183–238.

MACMAHON, P., *Combinatory Analysis*, Vols. 1 and 2, 1915. (Reprinted in one volume by Chelsea, New York, 1960.)

NIVEN, I., "Formal Power Series," *Amer. Math. Monthly*, 76 (1969), 871–889.

OWEN, G., *Game Theory*, Saunders, Philadelphia, 1968 (2nd ed., Academic Press, 1982).

READ, R. C., "Euler Graphs on Labelled Nodes," *Canad. J. Math.,* 14 (1962), 482–486.

RIORDAN, J., *An Introduction to Combinatorial Analysis*, Wiley, New York, 1958.

RIORDAN, J., "Generating Functions," in E. F. Beckenbach (ed.), *Applied Combinatorial Mathematics*, Wiley, New York, 1964, pp. 67–95.

ROBERTS, F. S., *Discrete Mathematical Models, with Applications to Social, Biological, and Environmental Problems*, Prentice-Hall, Englewood Cliffs, N.J., 1976.

SHAPLEY, L. S., "A Value for n-Person Games," in H. W. Kuhn and A. W. Tucker (eds.), *Contributions to the Theory of Games*, Vol. 2, Annals of Mathematics Studies No. 28, Princeton University Press, Princeton, N.J., 1953, pp. 307–317.

SHAPLEY, L. S., "Measurement of Power in Political Systems," in W. F. Lucas (ed.), *Game Theory and Its Applications*, Proceedings of Symposia in Applied Mathematics, Vol. 24, American Mathematical Society, Providence, R.I., 1981, pp. 69–81.

STRAFFIN, P. D., JR., *Topics in the Theory of Voting*, Birkhäuser-Boston, Cambridge, Mass., 1980.

5 Recurrence Relations*

5.1 SOME EXAMPLES

At the beginning of Section 4.1, we saw that we frequently want to count a quantity a_k that depends on an input or a parameter, say k. We then studied the sequence of unknown values, $a_0, a_1, a_2, \ldots, a_k, \ldots$. We shall see how to reduce computation of the kth or the $(k + 1)$st member of such a sequence to earlier members of the sequence. In this way we can reduce a bigger problem to a smaller one, or to one solved earlier. In Section 3.4 and in Example 4.14, we did much the same thing by giving reduction theorems, which reduced a complicated computation to simpler ones or ones made earlier. Having seen how to reduce computation of later terms of a sequence to earlier terms, we shall discuss several methods for finding general formulas for the kth term of an unknown sequence. The ideas and methods we present will have a wide variety of applications.

5.1.1 Some Simple Recurrences

Example 5.1 The Grains of Wheat

According to Gamow [1954], the following is the story of King Shirham of India. The King wanted to reward his Grand Vizier, Sissa Ben Dahir, for inventing the game of chess. The Vizier made a modest request: Give me one grain of wheat for the first square on a chessboard, two grains for the second square, four for the third

---

*If Chapter 4 has been omitted, then Sections 5.3 and 5.4 should be omitted. Chapters 4 and 5 are the only chapters which really assume calculus. The only calculus used in Chapter 5 except in Sections 5.3 and 5.4 is elementary knowledge about infinite sequences; even here, the concept of limit is used in only a few applications, and these can be omitted.

square, eight for the fourth square, and so on until all the squares are covered. The King was delighted at the modesty of his Vizier's request, and granted it immediately. Did the King do a very wise thing? To answer this question, let s_k be the number of grains of wheat required for the first k squares, and t_k be the number of grains for the kth square. We have

$$t_{k+1} = 2t_k. \tag{5.1}$$

Equation (5.1) is an example of a *recurrence relation*, a formula reducing later values of a sequence of numbers to earlier ones. Let us see how we can use the recurrence formula to get a general expression for t_k. We know that $t_1 = 1$. This is given to us, and is called an *initial condition*. We know that

$$t_2 = 2t_1,$$

$$t_3 = 2t_2 = 2^2 t_1,$$

$$t_4 = 2t_3 = 2^2 t_2 = 2^3 t_1,$$

and in general

$$t_k = 2t_{k-1} = \cdots = 2^{k-1} t_1.$$

Using the initial condition, we have

$$t_k = 2^{k-1}, \tag{5.2}$$

all k. We have solved the recurrence (5.1) by *iteration* or repeated use of the recurrence. Note that a recurrence like (5.1) will in general have many *solutions*, that is, sequences which satisfy it. However, once sufficiently many *initial conditions* are specified, there will be a unique solution. Here the sequence 1, 2, 4, 8, ... is the unique solution given the initial condition. However, if the initial condition is disregarded, any multiple of this sequence is a solution, as, for instance, 3, 6, 12, 24, ... or 5, 10, 20, 40, ...

We are really interested in s_k. We have

$$s_{k+1} = s_k + t_{k+1}, \tag{5.3}$$

another form of recurrence formula which relates later values of s to earlier values of s and to values of t already calculated. We can reduce (5.3) to a recurrence for s_k alone by using (5.2). This gives us

$$s_{k+1} = s_k + 2^k. \tag{5.4}$$

Let us again use iteration to solve the recurrence relation (5.4) for s_k for all k. We have

$$s_2 = s_1 + 2,$$

$$s_3 = s_2 + 2^2 = s_1 + 2 + 2^2,$$

and in general

$$s_k = s_{k-1} + 2^{k-1} = \cdots = s_1 + 2 + 2^2 + \cdots + 2^{k-1}.$$

Since we have the initial condition $s_1 = 1$, we obtain

$$s_k = 1 + 2 + 2^2 + \cdots + 2^{k-1}.$$

This expression can be simplified if we use the following well-known identity which we have already encountered in Chapter 4:

$$1 + x + x^2 + \cdots + x^p = \frac{1 - x^{p+1}}{1 - x}.$$

Using this identity with $x = 2$ and $p = k - 1$, we have

$$s_k = \frac{1 - 2^k}{1 - 2} = 2^k - 1.$$

Now there are 64 squares on a chessboard. Hence, the number of grains of wheat the Vizier asked for is given by $2^{64} - 1$, which is

$$18,446,744,073,709,551,615,$$

a very large number indeed!

Example 5.2   Computational Complexity

One major use of recurrences in computer science is in the computation of the complexity $f(n)$ of an algorithm (see Section 2.4). Often, computation of the complexity $f(n + 1)$ with input $n + 1$ is reduced to knowledge of the complexities $f(n)$, $f(n - 1)$, and so on. As a trivial example, let us consider the following algorithm for summing the first n entries of a sequence or array A.

Algorithm 5.1.   Summing the First n Entries of a Sequence or Array

Input: A sequence A and a number n.
Output: The sum $A(1) + A(2) + \cdots + A(n)$.

Step 1. Set $i = 1$.
Step 2. Set $T = A(1)$.
Step 3. If $i = n$, stop and output T. Otherwise, set $i = i + 1$ and go to step 4.
Step 4. Set $T = T + A(i)$ and return to step 3.

If $f(n)$ is the number of additions performed in summing the first n entries of A, we have the recurrence

$$f(n) = f(n - 1) + 1. \tag{5.5}$$

Also, we have the initial condition $f(1) = 0$. Thus, by iteration, we have

$$f(n) = f(n - 1) + 1 = f(n - 2) + 1 + 1 = \cdots = f(1) + n - 1 = n - 1.$$

Example 5.3   Simple and Compound Interest

Suppose that a sum of money S_0 is deposited in a bank at *interest rate* r per interest period (say per year), that is, at $100r\%$. If the interest is *simple*, after every interest period a fraction r of the initial deposit S_0 is credited to the account. If S_k is the amount on deposit after k periods, then we have the recurrence

$$S_{k+1} = S_k + rS_0. \tag{5.6}$$

By iteration, we find that

$$S_k = S_{k-1} + rS_0 = S_{k-2} + rS_0 + rS_0 = \cdots = S_0 + krS_0,$$

so

$$S_k = S_0(1 + kr).$$

If interest is compounded each period, then we receive as interest after each period a fraction r of the amount on deposit at the beginning of the period; that is, we have the recurrence

$$S_{k+1} = S_k + rS_k,$$

or

$$S_{k+1} = (1 + r)S_k. \tag{5.7}$$

We find by iteration that

$$S_k = (1 + r)^k S_0.$$

Example 5.4 Legitimate Codewords

Codewords from the alphabet $\{0, 1, 2, 3\}$ are to be recognized as *legitimate* if and only if they have an even number of 0's. How many legitimate codewords of length k are there? Let a_k be the answer. We derive a recurrence for a_k. (Note that a_k could be computed using the method of generating functions of Chapter 4.) Observe that $4^k - a_k$ is the number of illegitimate k-digit codewords, that is, the k-digit words with an odd number of 0's. Consider a legitimate $(k + 1)$-digit codeword. It starts with 1, 2, or 3, or it starts with 0. In the former case, the last k digits form a legitimate codeword of length k, and in the latter case they form an illegitimate codeword of length k. Thus, by the product and sum rules of Chapter 2,

$$a_{k+1} = 3a_k + 1(4^k - a_k),$$

that is,

$$a_{k+1} = 2a_k + 4^k. \tag{5.8}$$

We have the initial condition $a_1 = 3$. One way to solve the recurrence (5.8) is by the method of iteration. This is analogous to the solution of recurrence (5.4) and is left to the reader. An alternative method will be described in Section 5.3. For now, we compute some values of a_k. Note that since $a_1 = 3$, the recurrence gives us

$$a_2 = 2a_1 + 4^1 = 2(3) + 4 = 10$$

and

$$a_3 = 2a_2 + 4^2 = 2(10) + 16 = 36.$$

The reader might wish to check these numbers by writing out the legitimate codewords of lengths 2 and 3. Note how early values of a_k are used to derive later values.

We do not need an explicit solution to use a recurrence to calculate unknown numbers.

Example 5.5 Duration of Messages

Imagine that we transmit messages over a channel using only two signals, a and b (say dots and dashes). A codeword is any sequence from the alphabet $\{a, b\}$. Now suppose that signal a takes 1 unit of time to transmit and signal b takes 2 units of time to transmit. Let N_t be the number of possible codewords which can be transmitted in exactly t units of time. What is N_t? To answer this question, consider a codeword transmittable in t units of time. It either begins in a or in b. If it begins in a, the remainder is any codeword that can be transmitted in $t - 1$ units of time. If it begins in b, the remainder is any codeword that can be transmitted in $t - 2$ units of time. Thus, by the sum rule, for $t \geq 2$,

$$N_t = N_{t-1} + N_{t-2}. \tag{5.9}$$

This is our first example of a recurrence where a given value depends on more than one previous value. For this recurrence, since the tth term depends on two previous values, we need two initial values, N_1 and N_2. Clearly, $N_1 = 1$ and $N_2 = 2$, the latter since aa and b are the two sequences that can be transmitted in 2 units of time. We shall solve the recurrence (5.9) in Section 5.2 after we develop some general tools for solving recurrences. Shannon [1956] defines the *capacity* C of the transmission channel as

$$C = \lim_{t \to \infty} \frac{\log_2 N_t}{t}.$$

This is a measure of the capacity of the channel to transmit information. We return to this in Sections 5.2.2 and 8.3.2.

Example 5.6 Regions in the Plane

A line separates the plane into two regions (see Figure 5.1). Two intersecting lines separate the plane into four regions (again see Figure 5.1). Suppose that we have n lines in "general position"; that is, no two are parallel and no three lines intersect in the same point. Into how many regions do these lines divide the plane? To answer this question, let $f(n)$ be the appropriate number of regions. We have already seen that $f(1) = 2$ and $f(2) = 4$. Figure 5.1 also shows that $f(3) = 7$. To determine $f(n)$, we shall derive a recurrence relation.

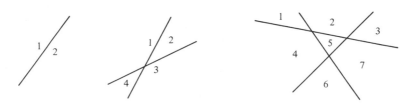

Figure 5.1 Lines dividing the plane into regions.

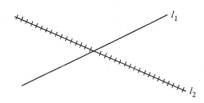

Figure 5.2 Line l_1 divides line l_2 into two segments.

Consider a line l_1 as shown in Figure 5.2. Draw a second line l_2. Line l_1 divides line l_2 into two segments, and each segment divides an existing region into two regions. Hence,

$$f(2) = f(1) + 2.$$

Similarly, if we add a third line l_3, this line is divided by l_1 and l_2 into three segments, with each segment splitting an existing region into two parts (see Figure 5.3). Hence

$$f(3) = f(2) + 3.$$

In general, suppose that we add a line l_{n+1} to already existing lines $l_1, l_2, \ldots, l_n$. The existing lines split l_{n+1} into $n + 1$ segments, each of which splits an existing region into two parts (Figure 5.4). Hence, we have

$$f(n + 1) = f(n) + (n + 1). \tag{5.10}$$

Equation (5.10) gives a recurrence relation which we shall use to solve for $f(n)$. The initial condition is the value of $f(1)$, which is 2. To solve the recurrence relation (5.10), we note that

$$f(2) = f(1) + 2,$$
$$f(3) = f(2) + 3 = f(1) + 2 + 3,$$
$$f(4) = f(3) + 4 = f(2) + 3 + 4 = f(1) + 2 + 3 + 4,$$

and in general

$$f(n) = f(n - 1) + n = \cdots = f(1) + 2 + 3 + \cdots + n.$$

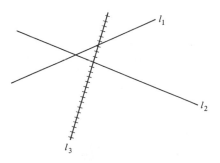

Figure 5.3 Line l_3 is divided by lines l_1 and l_2 into three segments.

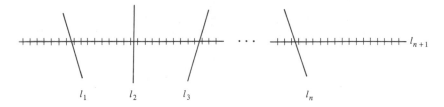

Figure 5.4 Line l_{n+1} is split into $n + 1$ segments.

Since $f(1) = 2$, we have

$$f(n) = 2 + 2 + 3 + 4 + \cdots + n$$
$$= 1 + (1 + 2 + 3 + 4 + \cdots + n)$$
$$= 1 + \frac{n(n + 1)}{2},$$

using the standard formula for sum of an arithmetic progression. For example, we have

$$f(4) = 1 + \frac{4 \cdot 5}{2} = 11.$$

5.1.2 Fibonacci Numbers and Their Applications

In the year 1202, Leonardo of Pisa, better known as Fibonacci, posed the following problem in his book *Liber Abaci*. Suppose that we study the prolific breeding of rabbits. We start with one pair of adult rabbits (of opposite sex). Assume that each pair of adult rabbits produce one pair of young (of opposite sex) each month. A newborn pair of rabbits become adults in two months, at which time they also produce their first pair of young. Assume that rabbits never die. Let F_k count the number of rabbit pairs present at the beginning of the kth month. Table 5.1 lists for each of several values of k the number of adult pairs, the number of one-month-old young pairs, the number of newborn young pairs, and the total number of rabbit pairs. For example, at the beginning of the second

Table 5.1 Rabbit breeding

Month k	Number of adult pairs at beginning of month k	Number of pairs one month old at beginning of month k	Number of newborn pairs at beginning of month k	Total number of pairs at beginning of month $k = F_k$
1	1	0	0	1
2	1	0	1	2
3	1	1	1	3
4	2	1	2	5
5	3	2	3	8
6	5	3	5	13

month, there is one newborn rabbit pair. At the beginning of the third month, the newborns from the previous month are one month old, and there is again a newborn pair. At the beginning of the fourth month, the one-month-olds have become adults and given birth to newborns, so there are now two adult pairs and two newborn pairs. The one newborn pair from month 3 has become a one-month-old pair. And so on.

Let us derive a recurrence relation for F_k. Note that the number of rabbit pairs at the beginning of the kth month is given by the number of rabbit pairs at the beginning of the $(k-1)$st month plus the number of newborn pairs at the beginning of the kth month. But the latter number is the same as the number of adult pairs at the beginning of the kth month, which is the same as the number of all rabbit pairs at the beginning of the $(k-2)$nd month. (It takes exactly 2 months to become an adult.) Hence, we have for $k \geq 3$,

$$F_k = F_{k-1} + F_{k-2}. \tag{5.11}$$

Note that if we define F_0 to be 1, then (5.11) holds for $k \geq 2$. Observe the similarity of recurrences (5.11) and (5.9). We return to this point in Section 5.2.2. Let us compute several values of F_k using the recurrence (5.11). We already know that

$$F_0 = F_1 = 1.$$

Hence,

$$F_2 = F_1 + F_0 = 2,$$
$$F_3 = F_2 + F_1 = 3$$
$$F_4 = F_3 + F_2 = 5,$$
$$F_5 = F_4 + F_3 = 8,$$
$$F_6 = F_5 + F_4 = 13,$$
$$F_7 = F_6 + F_5 = 21,$$
$$F_8 = F_7 + F_6 = 34.$$

In Section 5.2.2 we shall use the recurrence (5.11) to obtain an explicit formula for the number F_k. The sequence of numbers $F_0, F_1, F_2, \ldots$ is called the *Fibonacci sequence* and the numbers F_k the *Fibonacci numbers*. These numbers have remarkable properties and arise in a great variety of places. We shall describe some of their properties and applications here.

The *growth rate* of the sequence (F_k) is defined to be

$$G_k = \frac{F_k}{F_{k-1}}.$$

Then we have

$$G_1 = 1, \quad G_2 = \tfrac{2}{1} = 2, \quad G_3 = \tfrac{3}{2} = 1.5, \quad G_4 = \tfrac{5}{3} = 1.67, \quad G_5 = \tfrac{8}{5} = 1.60,$$
$$G_6 = \tfrac{13}{8} = 1.625, \quad G_7 = \tfrac{21}{13} = 1.615, \quad G_8 = \tfrac{34}{21} = 1.619, \quad \ldots$$

The numbers G_k seem to be converging to a limit between 1.60 and 1.62. In fact, this limit turns out to be exactly

$$\tau = \tfrac{1}{2}(1 + \sqrt{5}) = 1.618034 \ldots$$

The number τ will arise in the development of a general formula for F_k. [This number is called the *golden ratio* or the *divine proportion*. It is the number with the property that if one divides the line AB at C so that $\tau = AB/AC$, then

$$\frac{AB}{AC} = \frac{AC}{CB}.$$

The rectangle with sides in the ratio $\tau : 1$ is called the *golden rectangle*, and is considered by many to be the most aesthetic rectangle. Indeed, a fifteenth-century artist Piero della Francesca wrote a whole book (*De Divina Proportione*) about the applications of τ and the golden rectangle in art, in particular in the work of Leonardo da Vinci.]

Fibonacci numbers have important applications in numerical analysis, in particular in the search for the maximum value of a function $f(x)$ in an interval (a, b). A *Fibonacci search* for the maximum value, performed on a computer, makes use of the Fibonacci numbers to determine where to evaluate the function in getting better and better estimates of the location of the maximum value. When f is concave, this is known to be the best possible search procedure in the sense of minimizing the number of function evaluations for finding the maximum to a desired degree of accuracy. See Kiefer [1953], Adby and Dempster [1974], or Hollingdale [1978].

It is intriguing that Fibonacci numbers appear very frequently in nature. The field of botany that studies the arrangements of leaves around stems, the scales on cones, and so on, is called *phyllotaxis*. Usually, leaves appearing on a given stem or branch point out in different directions. The second leaf is rotated from the first by a certain angle, the third leaf from the second by the same angle, and so on until some leaf points in the same direction as the first. For example, if the angle of rotation is 30°, then the twelfth leaf is the first one pointing in the same direction as the first, since $12 \times 30° = 360°$. If the angle is 144°, then the fifth leaf is the first one pointing in the same direction as the first, for $5 \times 144° = 720°$. Two complete 360° returns have been made before a leaf faces in the same direction as the first. In general, let n count the number of leaves before returning to the same direction as the first, and let m count the number of complete 360° turns which have been made before this leaf is encountered. Table 5.2 shows the values of n and m for various plants. It is a remarkable empirical fact of biology that both n and m take as values the Fibonacci numbers. There is no good theoretical explanation for this fact.

Table 5.2 Values of n and m for Various Plants

Plant	Angle of rotation	n	m
Elm	180°	2	1
Alder, birch	120°	3	1
Rose	144°	5	2
Cabbage	135°	8	3

Source: Schips [1922], Batschelet [1971].

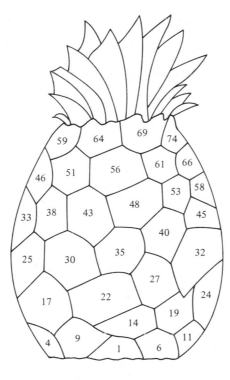

Figure 5.5 Fibonacci numbers and pineapples. (From Coxeter [1961]. Reprinted with permission of John Wiley & Sons, Inc.)

Coxeter [1961] points out that the Fibonacci numbers also arise in the study of scales on a fir cone, whorls on a pineapple, and so on. These whorls (cells) are arranged in fairly visible diagonal rows. The whorls can be assigned integers in such a way that each diagonal row of whorls forms an arithmetic sequence with common difference (difference between successive numbers) a Fibonacci number. This is shown in Figure 5.5. Note, for example, the diagonal

$$9, \quad 22, \quad 35, \quad 48, \quad 61, \quad 74,$$

which has common difference the Fibonacci number 13. Similarly, the diagonal

$$11, \quad 19, \quad 27, \quad 35, \quad 43, \quad 51, \quad 59$$

has common difference 8. Similar remarks hold for the florets of sunflowers, the scales of fir cones, and so on (see Adler [1977]). Again, no explanation for why Fibonacci numbers arise in this aspect of phyllotaxis is known.

5.1.3 Derangements

Example 5.7   The Hatcheck Problem

Imagine that n gentlemen attend a party and check their hats. The checker has a little too much to drink, and returns the hats at random. What is the probability that no gentleman receives his own hat? How does this probability depend on the number of gentlemen? We shall answer this question by studying the notion of a derangement.

Let n objects be labeled $1, 2, \ldots, n$. An arrangement or permutation in which object i is not placed in the ith place for any i is called a *derangement*. For example, if n is 3, then 231 is a derangement, but 213 is not since 3 is in the third place. Let D_n be the number of derangements of n objects.

Derangements arise in a card game (rencontres) where a deck of cards is laid out in a row on the table and a second deck is laid out randomly, one card on top of each of the cards of the first deck. The number of matching cards determines a score. In 1708, the Frenchman P. R. Montmort posed the problem of calculating the probability that no matches will take place, and called it "le problème des rencontres," rencontres meaning "match" in French. This is of course the problem of calculating D_{52}. The problem of computing the number of matching cards will be taken up in Section 6.2. There we will also discuss applications to the analysis of guessing abilities in psychic experiments. The first deck of cards corresponds to an unknown order of things selected or sampled, and the second to the order predicted by a psychic. In testing claims of psychic powers, one would like to compute the probability of getting m matches right. The probability of getting at least one match right is one minus the probability of getting no matches right, that is, one minus the probability of getting a derangement.

Clearly, $D_1 = 0$: There is no arrangement of one element in which the element does not appear in its proper place. $D_2 = 1$: The only derangement is 21. We shall derive a recurrence relation for D_n. Suppose that there are $n + 1$ elements, $1, 2, \ldots, n + 1$. A derangement of these $n + 1$ elements involves a choice of first element and then an ordering of the remaining n. The first element can be any of n different elements: $2, 3, \ldots, n + 1$. Suppose that k is put first. Then either 1 appears in the kth spot or it does not. If 1 appears in the kth spot, then we have left the elements

$$2, 3, \ldots, k - 1, k + 1, \ldots, n + 1,$$

and we wish to order them so none appears in its proper location [see Figure 5.6(a)]. There are D_{n-1} ways to do this, since there are $n - 1$ elements. Suppose next that 1 does not appear in the kth spot. We can think of first putting 1 into the kth spot [as shown in Figure 5.6(b)] and then deranging the elements in the second through $(n + 1)$st spots. There are D_n such derangements. In sum, we have n choices for the element k which appears in the first spot. For each of these, we either choose an arrangement with 1 in the kth spot—which we can do in D_{n-1} ways—or we choose an arrangement with 1 not in the kth spot—which we can do in D_n ways. It follows by the product and sum rules that

$$D_{n+1} = n(D_{n-1} + D_n). \tag{5.12}$$

Equation (5.12) is a recurrence relation which makes sense for $n \geq 2$. If we add the initial conditions $D_1 = 0$, $D_2 = 1$, it turns out that

$$D_n = n!\left(1 - \frac{1}{1!} + \frac{1}{2!} - \frac{1}{3!} + \cdots + (-1)^n \frac{1}{n!} \right) \tag{5.13}$$

or (for $n \geq 2$),

$$D_n = n!\left(\frac{1}{2!} - \frac{1}{3!} + \cdots + (-1)^n \frac{1}{n!} \right). \tag{5.14}$$

We shall see in Section 5.3.2 how to derive these formulas.

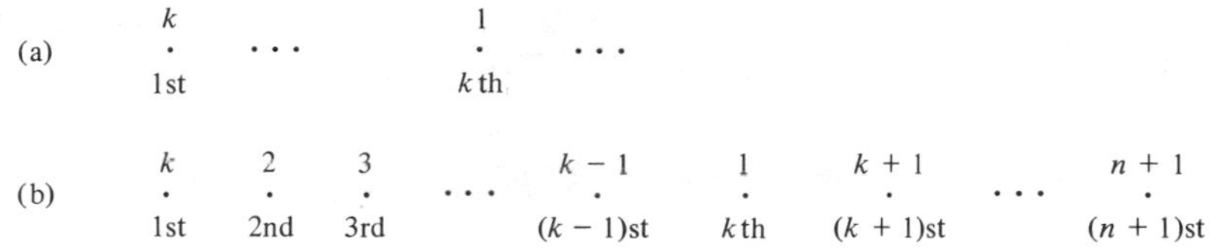

(a)

(b)

Figure 5.6 Derangements with k in the first spot and 1 in the kth spot and other elements (a) in arbitrary order (b) in the proper spots.

Let us now apply these formulas to the hatcheck problem of Example 5.7. Let p_n be the probability that no gentleman gets his hat back if there are n gentlemen. Then (for $n \geq 2$) we have

$$p_n = \frac{\text{number of arrangements with no one receiving his hat}}{\text{number of arrangements}}$$

$$= \frac{D_n}{n!}$$

$$= \left(\frac{1}{2!} - \frac{1}{3!} + \cdots + (-1)^n \frac{1}{n!}\right).$$

Table 5.3 shows values of p_n for several n. Note that p_n seems to be converging rapidly. In fact, we can calculate exactly what p_n converges to. Recall that

$$e^x = 1 + x + \frac{1}{2!} x^2 + \frac{1}{3!} x^3 + \cdots$$

[see (4.3)]. Hence,

$$e^{-1} = \frac{1}{2!} - \frac{1}{3!} + \cdots,$$

so p_n converges to e^{-1}. The convergence is so rapid that p_7 and p_8 already differ only in the fifth decimal place. The probability that no gentleman receives his hat back becomes essentially independent of the number of gentlemen very rapidly.

Example 5.8 Latin Rectangles

In Chapter 1 we talked about Latin squares and their applications to experimental design, and in Exercise 13, Section 4.1, we introduced Latin rectangles and the idea of building up Latin squares one row at a time. Let $L(r, n)$ be the number of $r \times n$ Latin rectangles with entries 1, 2, ..., n. (Recall that such a rectangle is an $r \times n$

Table 5.3 Values of p_n

n	2	3	4	5	6	7	8
p_n	.500000	.333333	.375000	.366667	.368056	.367858	.367883

array with entries 1, 2, ..., n so that no two entries in any row or column are the same.) Let $K(r, n)$ be the number of $r \times n$ Latin rectangles with entries 1, 2, ..., n and first row in the *standard position* 1 2 3 $\cdots$ n. Then

$$L(r, n) = n!K(r, n), \tag{5.15}$$

for one may obtain any Latin rectangle by finding one with a standard first row and then permuting the first row and performing the same permutation on the elements in the remaining rows.

We would like to calculate $L(r, n)$ or $K(r, n)$ for every r and n. By virtue of (5.15), these problems are equivalent. In Example 1.1, we asked for $L(5, 5)$. $K(2, n)$ is easy to calculate. It is simply D_n, the number of derangements of n elements. For we obtain the second row of a Latin rectangle with the first row in the standard position by deranging the elements of the first row. For most values of r and n, it is not known what the numbers $K(r, n)$ and $L(r, n)$ are.* Some counting problems are so hard that they have not been solved even though many attempts have been made. We shall encounter a number of equally difficult problems later.

5.1.4 Recurrences Involving More Than One Sequence

Generalizing Example 5.4, let us define a codeword from the alphabet $\{0, 1, 2, 3\}$ to be *legitimate* if and only if it has an even number of 0's *and* an even number of 3's. Let a_k be the number of legitimate codewords of length k. How do we find a_k? To answer this question, it turns out to be useful, in a manner analogous to the situation of Example 5.4, to consider other possibilities for a word of length k. Let b_k be the number of k-digit words from the alphabet $\{0, 1, 2, 3\}$ with an even number of 0's and an odd number of 3's, c_k the number with an odd number of 0's and an even number of 3's, and d_k the number with an odd number of 0's and an odd number of 3's. Note that

$$d_k = 4^k - a_k - b_k - c_k. \tag{5.16}$$

Observe that we get a legitimate codeword of length $k + 1$ by preceding a legitimate codeword of length k by a 1 or a 2, by preceding a word of length k with an even number of 0's and an odd number of 3's with a 3, or by preceding a word of length k with an odd number of 0's and an even number of 3's with a 0. Hence,

$$a_{k+1} = 2a_k + b_k + c_k. \tag{5.17}$$

Similarly,

$$b_{k+1} = a_k + 2b_k + d_k, \tag{5.18}$$

or, using (5.16),

$$b_{k+1} = b_k - c_k + 4^k. \tag{5.19}$$

---

*See Ryser [1963] for some results and references on $K(r, n)$ and $L(r, n)$. In particular, $L(5, 5)$ is 161,280.

Finally,

$$c_{k+1} = a_k + 2c_k + d_k, \tag{5.20}$$

or

$$c_{k+1} = c_k - b_k + 4^k. \tag{5.21}$$

Equations (5.17), (5.19), and (5.21) can be used together to compute any desired value a_k. We start with the initial conditions $a_1 = 2$, $b_1 = 1$, $c_1 = 1$. From (5.17), (5.19), (5.21), we compute

$$a_2 = 2 \cdot 2 + 1 + 1 = 6,$$
$$b_2 = 1 - 1 + 4^1 = 4,$$
$$c_2 = 1 - 1 + 4^1 = 4.$$

These results are easy to check by listing the corresponding sequences. For instance, the six sequences from $\{0, 1, 2, 3\}$ of length 2 and having an even number of 0's and of 3's are 00, 33, 11, 12, 21, 22. Similarly, one obtains

$$a_3 = 2 \cdot 6 + 4 + 4 = 20,$$
$$b_3 = 4 - 4 + 4^2 = 16,$$
$$c_3 = 4 - 4 + 4^2 = 16.$$

Notice that we have not found a single recurrence relation. However, we have found three relations which may be used simultaneously to compute the desired numbers.

EXERCISES FOR SECTION 5.1

1. Suppose that $a_n = 3a_{n-1} + 2$, $n \geq 1$, and $a_1 = 5$. Derive a_5 and a_6.
2. In Example 5.3 with $r = .1$, $S_0 = \$1000$, and simple interest, use the recurrence successively to compute S_1, S_2, S_3, S_4, S_5, and S_6, and check your answer by using the equation $S_k = S_0(1 + kr)$.
3. Repeat Exercise 2 for compound interest, using the equation $S_k = (1 + r)^k S_0$ to check.
4. In Example 5.4:
   (a) Derive a_4.   (b) Derive a_5.
   (c) Verify that $a_2 = 10$ by listing all legitimate codewords of length 2.
   (d) Repeat part (c) for $a_3 = 36$.
5. In Example 5.6, verify that $f(4) = 11$ by drawing four lines in the plane.
6. Find a solution to the recurrence (5.5) under the initial condition $f(1) = 11$.
7. Find two different solutions to the recurrence (5.6).
8. Find a solution to the recurrence (5.12) different from the sequence defined by (5.13).
9. Find all derangements of $\{1, 2, 3, 4\}$.
10. In the example of Section 5.1.4, use the recurrences (5.17), (5.19), and (5.21) to compute a_4, b_4, and c_4.

11. On the first day, n jobs are to be assigned to n workers. On the second day, the jobs are again to be assigned, but no worker is to get the same job he or she had on the first day. In how many ways can the jobs be assigned for the two days?

12. The labels from seven different cans of soup were unfortunately removed and a new clerk at the supermarket replaced them at random. In how many ways could he do this so that
(a) at least one can gets properly labeled?
(b) all the cans get properly labeled?

13. A lab director can run two different kinds of experiments, the expensive one (E) costing \$4000 and the inexpensive one (I) costing \$2000. If, for example, he has a budget of \$6000, he can perform experiments in the following sequences: *III, IE,* or *EI*. Let $F(n)$ be the number of sequences of experiments he can run spending exactly \$$n$. Thus, $F(\$6000) = 3$.
(a) Find a recurrence for $F(n)$.
(b) Suppose that there are p different kinds of experiments, $E_1, E_2, \ldots, E_p$, with E_i costing d_i dollars to run. Find a recurrence for $F(n)$ in this case.

14. Suppose that we have 4¢ stamps, 6¢ stamps, and 10¢ stamps, each in unlimited supply. Let $f(n)$ be the number of ways of obtaining n cents of postage if the order in which we put on stamps counts. For example, $f(4) = 1$ and $f(8) = 1$ (two 4¢ stamps), while $f(10) = 3$ (one 10¢ stamp, or a 4¢ stamp followed by a 6¢ stamp, or a 6¢ stamp followed by a 4¢ stamp).
(a) If $n > 10$, derive a recurrence for $f(n)$.
(b) *Use the recurrence* of part (a) to find the number of ways of obtaining 14¢ of postage.
(c) Check your answer to part (b) by writing down all the ways.

15. An industrial plant has two sections. Suppose that in one week, nine workers are assigned to nine different jobs in the first section and another nine workers are assigned to nine different jobs in the second section. In the next week, the supervisor would like to reassign jobs so that no worker gets back his or her old job. (No two jobs in the plant are considered the same.)
(a) In how many ways can this be done if each worker stays in the same section of the plant?
(b) In how many ways can this be done if each worker is shifted to a section of the plant different from the one in which he or she previously worked?

16. In predicting future discoveries of a scarce resource such as oil, one (probably incorrect) assumption is to say that the amount discovered next year will be the average of the amount discovered this year and last year. Suppose that a_n is the amount discovered in year n.
(a) Find a recurrence for a_n.
(b) Solve the recurrence if $a_0 = a_1 = 1$.

17. Find the number of derangements of $\{1, 2, 3, 4, 5, 6, 7, 8, 9, 10\}$ in which the first five elements are mapped into
(a) 1, 2, 3, 4, 5 in some order;
(b) 6, 7, 8, 9, 10 in some order.

18. Suppose that $f(p)$ is the number of comparisons required to sort p items using the algorithm bubble sort (Section 3.6.3). Find a recurrence for $f(p)$ and solve.

19. A codeword from the alphabet $\{0, 1, 2\}$ is considered legitimate if and only if no two 0's appear consecutively. Find a recurrence for the number b_n of legitimate codewords of length n.

20. A codeword from the alphabet $\{0, 1, 2\}$ is considered legitimate if and only if there is an even number of 0's and an odd number of 1's. Find simultaneous recurrences from which it is possible to compute the number of legitimate codewords of length n.

21. (a) How many permutations of the integers 1, 2, ..., 11 put each even integer into its natural position and no odd integer into its natural position?
(b) How many permutations of the integers 1, 2, ..., 11 have exactly four numbers in their natural position?

22. Determine a recurrence for $f(n)$ if $f(n)$ is defined as follows. In a singles tennis tournament, $2n$

players are paired off in n matches, and $f(n)$ is the number of different ways in which this pairing can be done.

23. Let $S_{n,t}$ denote the number of ways to partition an n-element set into t nonempty, unordered, subsets. Then $S_{n,t}$ satisfies

$$S_{n,t} = tS_{n-1,t} + S_{n-1,t-1}, \tag{5.22}$$

for $t = 1, 2, \ldots, n - 1$. Equation (5.22) is an example of a recurrence involving two indices. We could use it to solve for any $S_{n,t}$. For instance, suppose that we start with the observation that $S_{n,1} = 1$, all n, and $S_{n,n} = 1$, all n. Then we can compute $S_{n,t}$ for all remaining n and $t \leq n$. For $S_{3,2}$ can be computed from $S_{2,2}$ and $S_{2,1}$; $S_{4,2}$ from $S_{3,2}$ and $S_{3,1}$; $S_{4,3}$ from $S_{3,3}$ and $S_{3,2}$; and so on.
   (a) Compute $S_{5,3}$ by using (5.22).
   (b) Compute $S_{6,3}$.
   (c) Show that (5.22) holds.

24. In Exercise 20 of the Additional Exercises for Chapter 2, consider a grid with m north-south streets and n east-west streets. Let $s(m, n)$ be the number of different routes from point A to point B, if A and B are located as in Figure 2.5. Find a recurrence for $s(m + 1, n + 1)$.

25. Determine a recurrence relation for $f(n)$, the number of regions into which the plane is divided by n circles each pair of which intersect in exactly two points and no three of which meet in a single point.

26. If F_n is the nth Fibonacci number, find a simple expression for

$$F_1 + F_2 + \cdots + F_n$$

which involves F_p for only one p.

27. (a) Suppose that $f(n + 1) = f(n)f(n - 1)$, all $n \geq 1$, and $f(0) = f(1) = 2$. Find $f(n)$.
   (b) Repeat part (a) if $f(0) = f(1) = 1$.

28. (Liu [1968]) A *pattern* in a bit string consists of a number of consecutive digits, for example 011. A pattern is said to *occur* at the kth digit of a bit string if when scanning the string from left to right, the full pattern appears after the kth digit has been scanned. Once a pattern occurs, that is, is observed, scanning begins again. For example, in the bit string 110101010101, the pattern 010 occurs at the fifth and ninth digits, but not at the seventh digit. Let b_n denote the number of n-digit bit strings with the pattern 010 occurring at the nth digit. Find a recurrence for b_n. (*Hint:* Consider the number of bit strings of length n ending in 010, and divide these into those where the 010 pattern occurs at the nth digit and those where it does not.)

29. Suppose B is an $n \times n$ board and $r_n(B)$ is the coefficient of x^n in the rook polynomial $R(x, B)$. Use recurrence relations to compute $r_n(B)$ if
   (a) B has all squares darkened;
   (b) B has only the main diagonal lightened.

30. In the example of Section 5.1.4, find a single recurrence for a_k in terms of earlier values of a_p only.

31. Suppose that C_k is the number of connected, labeled graphs of k vertices. Harary and Palmer [1973] derive the recurrence

$$C_k = 2^{\binom{k}{2}} - \frac{1}{k} \sum_{p=1}^{k-1} \binom{k}{p} C_p p 2^{\binom{k-p}{2}}.$$

Using the fact that $C_1 = 1$, compute C_2, C_3, and C_4, and check your answers by drawing the graphs.

5.2 THE METHOD OF CHARACTERISTIC ROOTS

5.2.1 The Case of Distinct Roots

So far we have derived a number of interesting recurrence relations. Several of these we were able to solve by iterating back to the original values or initial conditions. Indeed, we could do something similar even with some of the more difficult recurrences we have encountered. There are no general methods for solving all recurrences. However, there are methods that work for a very broad class of recurrences. In this section we investigate one such method, the method of characteristic roots. In Section 5.3 we shall show how to make use of the notion of generating function, developed in Chapter 4, for solving recurrences. The methods for solving recurrences were developed originally in the theory of difference equations. Careful treatments of difference equations can be found in Goldberg [1958] or Levy and Lessman [1961].

Consider the recurrence

$$a_n = c_1 a_{n-1} + c_2 a_{n-2} + \cdots + c_p a_{n-p}, \qquad n \geq p, \tag{5.23}$$

where $c_1, c_2, \ldots, c_p$ are constants and $c_p \neq 0$. Such a recurrence is called *linear* because all terms a_k occur to the first power and it is called *homogeneous* because there is no term on the right-hand side which does not involve some a_k, $n - p \leq k \leq n - 1$. Since the coefficients c_i are constants, the recurrence (5.23) is called a *linear homogeneous recurrence relation with constant coefficients*. The recurrences (5.1), (5.7), (5.9), and (5.11) are examples of such recurrences. The recurrences (5.4), (5.5), (5.6), (5.8), and (5.10) are not homogeneous and the recurrence (5.12) does not have constant coefficients. All of the recurrences we have encountered so far are linear.

We shall present a technique for solving linear homogeneous recurrence relations with constant coefficients; it is very similar to that used to solve linear differential equations with constant coefficients, as the reader who is familiar with the latter technique will see.

A recurrence (5.23) has a unique solution once we specify the values of the first p terms, $a_0, a_1, \ldots, a_{p-1}$; these values form the *initial conditions*. From $a_0, a_1, \ldots, a_{p-1}$, we can use the recurrence to find a_p. Then from $a_1, a_2, \ldots, a_p$, we can use the recurrence to find a_{p+1}, and so on.

In general, a recurrence (5.23) has many solutions, if the initial conditions are disregarded. Some of these solutions will be sequences of the form

$$\alpha^0, \alpha^1, \alpha^2, \ldots, \alpha^n, \ldots, \tag{5.24}$$

where α is a number. We begin by finding values of α for which (5.24) is a solution to the recurrence (5.23). In (5.23), let us substitute x^k for a_k and solve for x. Making this substitution, we get the equation

$$x^n - c_1 x^{n-1} - c_2 x^{n-2} - \cdots - c_p x^{n-p} = 0. \tag{5.25}$$

Dividing both sides of (5.25) by x^{n-p}, we obtain

$$x^p - c_1 x^{p-1} - c_2 x^{p-2} - \cdots - c_p = 0. \tag{5.26}$$

Equation (5.26) is called the *characteristic equation* of the recurrence (5.23). It is a polynomial in x of power p, so has p roots $\alpha_1, \alpha_2, \ldots, \alpha_p$. Some of these may be repeated roots and some may be complex numbers. These roots are called the *characteristic roots* of the equation (5.23). For instance, consider the recurrence

$$a_n = 5a_{n-1} - 6a_{n-2}, \tag{5.27}$$

with initial conditions $a_0 = 1$, $a_1 = 1$. Then $p = 2$, $c_1 = 5$, $c_2 = -6$, and the characteristic equation is given by

$$x^2 - 5x + 6 = 0.$$

This has roots $x = 2$ and $x = 3$, so $\alpha_1 = 2$ and $\alpha_2 = 3$ are the characteristic roots.

If α is a characteristic root of the recurrence (5.23), and if we take $a_n = \alpha^n$, it follows that the sequence (a_n) satisfies the recurrence. Thus, corresponding to each characteristic root, we have a solution to the recurrence. In (5.27), $a_n = 2^n$ and $a_n = 3^n$ give solutions. However, neither satisfies both initial conditions $a_0 = 1$, $a_1 = 1$.

The next important observation to be made is that if the sequences (a'_n) and (a''_n) both satisfy the recurrence (5.23) and if λ_1 and λ_2 are constants, then the sequence (a'''_n), where $a'''_n = \lambda_1 a'_n + \lambda_2 a''_n$, also is a solution to (5.23). In other words, *a weighted sum of solutions is a solution.* To see this, note that

$$a'_n = c_1 a'_{n-1} + c_2 a'_{n-2} + \cdots + c_p a'_{n-p} \tag{5.28}$$

and

$$a''_n = c_1 a''_{n-1} + c_2 a''_{n-2} + \cdots + c_p a''_{n-p}. \tag{5.29}$$

Multiplying (5.28) by λ_1 and (5.29) by λ_2 and adding gives us

$$
\begin{aligned}
a'''_n &= \lambda_1 a'_n + \lambda_2 a''_n \\
&= \lambda_1(c_1 a'_{n-1} + c_2 a'_{n-2} + \cdots + c_p a'_{n-p}) + \lambda_2(c_1 a''_{n-1} + c_2 a''_{n-2} + \cdots + c_p a''_{n-p}) \\
&= c_1(\lambda_1 a'_{n-1} + \lambda_2 a''_{n-1}) + c_2(\lambda_1 a'_{n-2} + \lambda_2 a''_{n-2}) + \cdots + c_p(\lambda_1 a'_{n-p} + \lambda_2 a''_{n-p}) \\
&= c_1 a'''_{n-1} + c_2 a'''_{n-2} + \cdots + c_p a'''_{n-p},
\end{aligned}
$$

so a'''_n does satisfy (5.23). In our example (5.27), if we define $a_n = 3 \cdot 2^n + 8 \cdot 3^n$, it follows that a_n satisfies (5.27).

In general, suppose that $\alpha_1, \alpha_2, \ldots, \alpha_p$ are the characteristic roots of recurrence (5.23). Then our reasoning shows that if $\lambda_1, \lambda_2, \ldots, \lambda_p$ are constants, and if

$$a_n = \lambda_1 \alpha_1^n + \lambda_2 \alpha_2^n + \cdots + \lambda_p \alpha_p^n,$$

then a_n satisfies (5.23). It turns out that every solution of (5.23) can be expressed in this form, *provided the roots $\alpha_1, \alpha_2, \ldots, \alpha_p$ are distinct.* For a proof of this fact, see the end of this subsection.

Theorem 5.1. Suppose that a linear homogeneous recurrence (5.23) with constant coefficients has characteristic roots $\alpha_1, \alpha_2, \ldots, \alpha_p$. Then if $\lambda_1, \lambda_2, \ldots, \lambda_p$ are constants, every expression of the form

$$a_n = \lambda_1 \alpha_1^n + \lambda_2 \alpha_2^n + \cdots + \lambda_p \alpha_p^n \tag{5.30}$$

is a solution to the recurrence. Moreover, if the characteristic roots are distinct, then every solution to the recurrence has the form (5.30) for some constants $\lambda_1, \lambda_2, \ldots, \lambda_p$.

We call the expression in (5.30) the *general solution* of the recurrence (5.23).

It follows from Theorem 5.1 that to find the unique solution of a recurrence (5.23) subject to initial conditions $a_0, a_1, \ldots, a_{p-1}$, if the characteristic roots are distinct, we simply need to find values for the constants $\lambda_1, \lambda_2, \ldots, \lambda_p$ in the general solution so that the initial conditions are satisfied. Let us see how to find these λ_i. In (5.27), every solution has the form

$$a_n = \lambda_1 2^n + \lambda_2 3^n.$$

Now we have

$$a_0 = \lambda_1 2^0 + \lambda_2 3^0, \qquad a_1 = \lambda_1 2^1 + \lambda_2 3^1,$$

so from $a_0 = 1$, $a_1 = 1$, we get the system of equations

$$\lambda_1 + \lambda_2 = 1$$
$$2\lambda_1 + 3\lambda_2 = 1.$$

This system has the unique solution $\lambda_1 = 2$, $\lambda_2 = -1$. Hence, since $\lambda_1 \neq \lambda_2$, the unique solution to (5.27) with the given initial conditions is $a_n = 2 \cdot 2^n - 3^n$.

The general procedure works just as in this example. If we define a_n by (5.30), then we use the initial values of $a_0, a_1, \ldots, a_{p-1}$ to set up a system of p simultaneous equations in the p unknowns $\lambda_1, \lambda_2, \ldots, \lambda_p$. One can show that if $\alpha_1, \alpha_2, \ldots, \alpha_p$ are distinct, this system always has a unique solution. The proof of this is the essence of the rest of the proof of Theorem 5.1, which we shall now present.

We close this subsection by sketching a proof of the statement in Theorem 5.1 that if the characteristic roots $\alpha_1, \alpha_2, \ldots, \alpha_p$ are distinct, then every solution of a recurrence (5.23) has the form (5.30) for some constants $\lambda_1, \lambda_2, \ldots, \lambda_p$.* Suppose that

$$b_n = \lambda_1 \alpha_1^n + \lambda_2 \alpha_2^n + \cdots + \lambda_p \alpha_p^n$$

is a solution to (5.23). Using the initial conditions $b_0 = a_0$, $b_1 = a_1$, $\ldots$, $b_{p-1} = a_{p-1}$, we find that

$$\left.\begin{aligned}
\lambda_1 + \lambda_2 + \cdots + \lambda_p &= a_0 \\
\lambda_1 \alpha_1 + \lambda_2 \alpha_2 + \cdots + \lambda_p \alpha_p &= a_1 \\
\lambda_1 \alpha_1^2 + \lambda_2 \alpha_2^2 + \cdots + \lambda_p \alpha_p^2 &= a_2 \\
&\vdots \\
\lambda_1 \alpha_1^{p-1} + \lambda_2 \alpha_2^{p-1} + \cdots + \lambda_p \alpha_p^{p-1} &= a_{p-1}.
\end{aligned}\right\} \qquad (5.31)$$

Equations (5.31) are a system of p linear equations in the p unknowns $\lambda_1, \lambda_2, \ldots, \lambda_p$. Consider now the matrix of coefficients of the system (5.31):

$$\begin{bmatrix}
1 & 1 & & 1 \\
\alpha_1 & \alpha_2 & & \alpha_p \\
\alpha_1^2 & \alpha_2^2 & \cdots & \alpha_p^2 \\
\vdots & \vdots & & \vdots \\
\alpha_1^{p-1} & \alpha_2^{p-1} & & \alpha_p^{p-1}
\end{bmatrix}.$$

*The proof can be omitted.

The determinant of this matrix is the famous *Vandermonde determinant*. One can show that the Vandermonde determinant is given by the product

$$\prod_{1 \le i < j \le p} (\alpha_j - \alpha_i),$$

the product of all terms $\alpha_j - \alpha_i$ with $1 \le i < j \le p$. Since $\alpha_1, \alpha_2, \ldots, \alpha_p$ are distinct, the determinant is not zero. Thus, there is a unique solution $\lambda_1, \lambda_2, \ldots, \lambda_p$ of the system (5.31). Hence, we see that a recurrence (5.23) with initial conditions $a_0, a_1, \ldots, a_{p-1}$ has a solution of the form (5.30). Now the recurrence with these initial conditions has just one solution, so this must be it. That completes the proof.

5.2.2 Computation of the kth Fibonacci Number

Let us illustrate the method with another example, the recurrence (5.11) for the Fibonacci numbers, which we repeat here:

$$F_k = F_{k-1} + F_{k-2}. \tag{5.11}$$

Here $p = 2$ and $c_1 = c_2 = 1$. The characteristic equation is given by $x^2 - x - 1 = 0$. By the quadratic formula, the roots of this equation, the characteristic roots, are given by $\alpha_1 = (1 + \sqrt{5})/2$ and $\alpha_2 = (1 - \sqrt{5})/2$. Because $\alpha_1 \ne \alpha_2$, the general solution is

$$\lambda_1 \left(\frac{1 + \sqrt{5}}{2} \right)^k + \lambda_2 \left(\frac{1 - \sqrt{5}}{2} \right)^k.$$

The initial conditions $F_0 = F_1 = 1$ give us the two equations

$$\lambda_1 + \lambda_2 = 1$$

$$\lambda_1 \left(\frac{1 + \sqrt{5}}{2} \right) + \lambda_2 \left(\frac{1 - \sqrt{5}}{2} \right) = 1.$$

Solving these simultaneous equations for λ_1 and λ_2 gives us

$$\lambda_1 = \frac{1}{\sqrt{5}} \left(\frac{1 + \sqrt{5}}{2} \right), \qquad \lambda_2 = -\frac{1}{\sqrt{5}} \left(\frac{1 - \sqrt{5}}{2} \right).$$

Hence, under the given initial conditions, the solution to (5.11), that is, the kth Fibonacci number, is given by

$$F_k = \frac{1}{\sqrt{5}} \left(\frac{1 + \sqrt{5}}{2} \right) \left(\frac{1 + \sqrt{5}}{2} \right)^k - \frac{1}{\sqrt{5}} \left(\frac{1 - \sqrt{5}}{2} \right) \left(\frac{1 - \sqrt{5}}{2} \right)^k,$$

or

$$F_k = \frac{\left(\dfrac{1 + \sqrt{5}}{2} \right)^{k+1} - \left(\dfrac{1 - \sqrt{5}}{2} \right)^{k+1}}{\sqrt{5}}, \tag{5.32}$$

or

$$F_k = \frac{\tau^{k+1} - (1 - \tau)^{k+1}}{\sqrt{5}},$$

where τ is the golden ratio of Section 5.1.2. In Exercise 8 of Section 5.3, this result is derived using generating functions.

Example 5.5 (Duration of Messages) Revisited

We note next that the recurrences (5.9) and (5.11) are the same. Moreover, the initial conditions are the same. For $F_1 = 1$ and $F_2 = 2$, while $N_1 = 1$ and $N_2 = 2$. Also, for the same reason that we took F_0 to be 0, namely, to maintain the recurrence even for $k = 2$, we take N_0 to be 0. Now as we observed earlier, if we are given a recurrence (5.23) and initial conditions $a_0, a_1, \ldots, a_{p-1}$, the solution is determined uniquely. Hence, it follows that N_t must equal F_t for all $t \ge 0$, so we may use (5.32) to compute N_k. It is not too hard to show from this result that the Shannon capacity defined in Example 5.5 is given by

$$C = \log_2 \left(\frac{1 + \sqrt{5}}{2} \right).$$

5.2.3 The Case of Multiple Roots

Consider the recurrence

$$a_n = 6a_{n-1} - 9a_{n-2}, \tag{5.33}$$

with $a_0 = 1$, $a_1 = 2$. Its characteristic equation is $x^2 - 6x + 9 = 0$, or $(x - 3)^2 = 0$. The two characteristic roots are 3 and 3, that is, 3 is a multiple root. Hence, the second part of Theorem 5.1 does not apply. Whereas it is still true that 3^n is a solution of (5.33), and it is also true that $\lambda_1 3^n + \lambda_2 3^n$ is always a solution, it is not true that every solution of (5.33) takes the form $\lambda_1 3^n + \lambda_2 3^n$. In particular, there is no such solution satisfying our given initial conditions. For these conditions give us the equations

$$\lambda_1 + \lambda_2 = 1$$
$$3\lambda_1 + 3\lambda_2 = 2.$$

There are no λ_1, λ_2 satisfying these two equations.

Suppose that α is a characteristic root of multiplicity u; that is, it appears as a root of the characteristic equation exactly u times. Then it turns out that not only does $a_n = \alpha^n$ satisfy the recurrence (5.23), but so do $a_n = n\alpha^n$, $a_n = n^2\alpha^n$, ..., and $a_n = n^{u-1}\alpha^n$ (see Exercises 26, 27). In our example, 3 is a characteristic root of multiplicity $u = 2$, and both $a_n = 3^n$ and $a_n = n3^n$ are solutions of (5.23). Moreover, since a weighted sum of solutions is a solution and since both $a'_n = 3^n$ and $a''_n = n3^n$ are solutions, so is $a'''_n = \lambda_1 3^n + \lambda_2 n3^n$. Using this expression a'''_n and the initial conditions $a_0 = 1$, $a_1 = 2$, we get the equations

$$\lambda_1 = 1$$
$$3\lambda_1 + 3\lambda_2 = 2.$$

These have the unique solution $\lambda_1 = 1$, $\lambda_2 = -\frac{1}{3}$. Hence, $a_n = 3^n - \frac{1}{3} \cdot n \cdot 3^n$ is a solution to the recurrence (5.33) with the initial conditions $a_0 = 1$, $a_1 = 2$. It follows that this must be the unique solution.

This procedure generalizes as follows. Suppose that a recurrence (5.23) has characteristic roots $\alpha_1, \alpha_2, \ldots, \alpha_q$, with α_i having multiplicity u_i. Then

$$\alpha_1^n, n\alpha_1^n, n^2\alpha_1^n, \ldots, n^{u_1-1}\alpha_1^n, \alpha_2^n, n\alpha_2^n, n^2\alpha_2^n, \ldots, n^{u_2-1}\alpha_2^n, \ldots, \alpha_q^n, n\alpha_q^n, n^2\alpha_q^n, \ldots, n^{u_q-1}\alpha_q^n$$

must all be solutions of the recurrence. Let us call these the *basic solutions*. There are p of these basic solutions in all. Let us denote them $b_1, b_2, \ldots, b_p$. Since a weighted sum of solutions is a solution, for any constants $\lambda_1, \lambda_2, \ldots, \lambda_p$,

$$a_n = \lambda_1 b_1 + \lambda_2 b_2 + \cdots + \lambda_p b_p$$

is also a solution of the recurrence. By a method analogous to that used to prove Theorem 5.1, one can show that every solution has this form for some constants $\lambda_1, \lambda_2, \ldots, \lambda_p$.

Theorem 5.2. Suppose that a linear homogeneous recurrence (5.23) with constant coefficients has basic solutions $b_1, b_2, \ldots, b_p$. Then the general solution is given by

$$a_n = \lambda_1 b_1 + \lambda_2 b_2 + \cdots + \lambda_p b_p, \tag{5.34}$$

for some constants $\lambda_1, \lambda_2, \ldots, \lambda_p$.

The unique solution satisfying initial conditions $a_0, a_1, \ldots, a_{p-1}$ can be computed by setting $n = 0, 1, \ldots, p-1$ in (5.34) and getting p simultaneous equations in the p unknowns $\lambda_1, \lambda_2, \ldots, \lambda_p$.

To illustrate, consider the recurrence

$$a_n = 7a_{n-1} - 16a_{n-2} + 12a_{n-3}, \tag{5.35}$$

$a_0 = 1, a_1 = 2, a_2 = 0$. Then the characteristic equation is

$$x^3 - 7x^2 + 16x - 12 = 0,$$

which factors as $(x-2)(x-2)(x-3) = 0$. The characteristic roots are therefore $\alpha_1 = 2$, with multiplicity $u_1 = 2$, and $\alpha_2 = 3$, with multiplicity $u_2 = 1$. Thus, the general solution to (5.35) has the form

$$\lambda_1 \alpha_1^n + \lambda_2 n\alpha_1^n + \lambda_3 \alpha_2^n = \lambda_1 2^n + \lambda_2 n2^n + \lambda_3 3^n.$$

Setting $n = 0, 1, 2$, we get

$$a_0 = \lambda_1 + \lambda_3 = 1$$
$$a_1 = 2\lambda_1 + 2\lambda_2 + 3\lambda_3 = 2$$
$$a_2 = 4\lambda_1 + 8\lambda_2 + 9\lambda_3 = 0.$$

This system has the unique solution $\lambda_1 = 5, \lambda_2 = 2, \lambda_3 = -4$. Hence, the unique solution to (5.35) with the given initial conditions is

$$5 \cdot 2^n + 2 \cdot n2^n - 4 \cdot 3^n.$$

EXERCISES FOR SECTION 5.2

1. Which of the following recurrences are linear?
   (a) $a_n = 3a_{n-1} + 2a_{n-2} + 5$ (b) $b_n = 8b_{n-1} + 16b_{n-2} + 18b_{n-3} + 24b_{n-4}$
   (c) $c_n = 6c_{n-2} + 84c_{n-5}$ (d) $d_n = 16d_{n-1} - 8d_{n-2}$
   (e) $e_n = 84e_{n-1} + 12e_{n-2}^2$ (f) $f_n = nf_{n-1} + f_{n-2}$

(g) $g_n = n^2 g_{n-2}$ **(h)** $h_n = 2h_{n-3} + 11$
(i) $i_n = 3i_{n-1} + 2^n i_{n-2}$.

2. Which of the recurrences in Exercise 1 are homogeneous?

3. Which of the recurrences in Exercise 1 have constant coefficients?

4. Find the characteristic equation of each of the following recurrences.
   (a) $a_n = 2a_{n-1} - a_{n-2}$ **(b)** $b_k = 10b_{k-1} - 16b_{k-2}$
   (c) $c_n = 3c_{n-1} + 12c_{n-2} - 18c_{n-3}$ **(d)** $d_n = 8d_{n-4} + 16d_{n-5}$
   (e) $e_k = e_{k-2}$ **(f)** $f_{n+1} = -f_n + 2f_{n-1}$
   (g) $g_n = 15g_{n-1} + 12g_{n-2} + 11g_{n-3} - 33g_{n-8}$ **(h)** $h_n = 4h_{n-2}$
   (i) $i_n = 6i_{n-1} - 11i_{n-2} + 6i_{n-3}$ **(j)** $j_n = 2j_{n-1} + j_{n-2} - 2j_{n-3}$.

5. In Exercise 4, find the characteristic roots of the recurrences of parts (a), (b), (e), (f), (h), (i), and (j). [*Hint:* 1 is a root in parts (i) and (j).]

6. **(a)** Show that the recurrence $a_n = 3a_{n-1}$ can have many solutions.
   (b) Show that this recurrence has a unique solution if we know that $a_0 = 10$.

7. Solve the following recurrences using the method of characteristic roots.
   (a) $a_n = 8a_{n-1}, a_0 = 5$
   (b) $t_{k+1} = 2t_k, t_1 = 1$ (this is Example 5.1).

8. Consider the recurrence $a_n = 15a_{n-1} - 44a_{n-2}$. Show that each of the following sequences is a solution.
   (a) (4^n) **(b)** $(3 \cdot 11^n)$ **(c)** $(4^n - 11^n)$ **(d)** (4^{n+1}).

9. In Exercise 8, which of the following sequences is a solution?
   (a) (-4^n) **(b)** $(4^n + 1)$ **(c)** $(3 \cdot 4^n + 12 \cdot 11^n)$ **(d)** $(n4^n)$ **(e)** $(4^n 11^n)$ **(f)** $((-4)^n)$.

10. Consider the recurrence $b_k = 8b_{k-1} - 15b_{k-2}$. Which of the following sequences is a solution?
    (a) (3^k) **(b)** $(5^k - 3^k)$ **(c)** $(3^k + 12)$ **(d)** $(3^k - 5^k)$ **(e)** $(3^k + 5^{k+1})$.

11. Use the method of characteristic roots to solve the following recurrences in Exercise 4 under the following initial conditions.
    (a) That of part (a) if $a_0 = 1, a_1 = 2$
    (b) That of part (b) if $b_0 = 0, b_1 = 1$
    (c) That of part (e) if $e_0 = 1, e_1 = -1$
    (d) That of part (f) if $f_0 = f_1 = 1$
    (e) That of part (h) if $h_0 = 10, h_1 = 2$
    (f) That of part (i) if $i_0 = 0, i_1 = 1, i_2 = 2$
    (g) That of part (j) if $j_0 = 1, j_1 = 1, j_2 = 0$.

12. Suppose that in Example 5.5, a requires 2 units of time to transmit and b requires 3 units of time. Solve for N_t.

13. Solve for a_n in the oil discovery problem (Exercise 16, Section 5.1) if $a_0 = 0, a_1 = 1$.

14. Consider the recurrence $a_n = -a_{n-2}$.
    (a) Show that i and $-i$ are the characteristic roots. (*Note:* $i = \sqrt{-1}$.)
    (b) Is the sequence (i^n) a solution?
    (c) What about the sequence $(2i^n + (-i)^n)$?
    (d) Find the unique solution if $a_0 = 0, a_1 = 1$.

15. Consider the recurrence $a_n = -4a_{n-2}$.
    (a) Find a general solution.
    (b) Find the unique solution if $a_0 = 0, a_1 = 10$.

16. Consider the recurrence $F_n = 4F_{n-1} - 4F_{n-2}$. Show that there is a multiple characteristic root α and that for the initial conditions $F_0 = 1, F_1 = 3$, there are no constants λ_1 and λ_2 so that $F_n = \lambda_1 \alpha^n + \lambda_2 \alpha^n$ for all n.

17. Suppose that (a'_n) and (a''_n) are two solutions of a recurrence (5.23), and that $a'''_n = a'_n - a''_n$. Is (a'''_n) necessarily a solution to (5.23)? Why?

18. Suppose that (a'_n), (a''_n), and (a'''_n) are three solutions to a recurrence (5.23), and that we have $b_n = \lambda_1 a'_n + \lambda_2 a''_n + \lambda_3 a'''_n$. Is (b_n) necessarily a solution to (5.23)? Why?

19. Consider the recurrence

$$a_n = 6a_{n-1} - 12a_{n-2} + 8a_{n-3}.$$

Show that each of the following sequences is a solution.
   (a) (2^n) (b) $(n2^n)$ (c) $(n^2 2^n)$
   (d) $(3 \cdot 2^n)$ (e) $(2^n + n2^n)$ (f) $(4 \cdot 2^n + 8 \cdot n2^n - n^2 2^n)$.

20. Consider the recurrence

$$a_n = 3a_{n-1} - 3a_{n-2} + a_{n-3}.$$

Show that each of the following sequences is a solution:

$$(1, 1, 1, \ldots), (0, 1, 2, 3, \ldots), (0, 1, 4, 9, \ldots).$$

21. Consider the recurrence

$$b_n = 9b_{n-1} - 24b_{n-2} + 20b_{n-3}.$$

Show that each of the sequences (2^n), $(n2^n)$, and (5^n) is a solution.

22. Consider the recurrence

$$c_k = 13c_{k-1} - 60c_{k-2} + 112c_{k-3} - 64c_{k-4}.$$

Show that each of the following sequencies is a solution.
   (a) $(2, 2, \ldots)$,   (b) $(3 \cdot 4^k)$ (c) $(4^k + k4^k + k^2 4^k)$
   (d) $(4^k + 1)$ (e) $(k4^k - 11)$.

23. Find the unique solution to
   (a) the recurrence of Exercise 19 if $a_0 = 0$, $a_1 = 1$, $a_2 = 1$;
   (b) the recurrence of Exercise 20 if $a_0 = 1$, $a_1 = 1$, $a_2 = 2$;
   (c) the recurrence of Exercise 21 if $b_0 = 1$, $b_1 = 2$, $b_2 = 0$;
   (d) the recurrence of Exercise 22 if $c_0 = c_1 = 0$, $c_2 = 10$, $c_3 = 0$.

24. Solve the following recurrence relations under the given initial conditions.
   (a) $a_n = 10a_{n-1} - 25a_{n-2}$, $a_0 = 1$, $a_1 = 2$.
   (b) $b_k = 14b_{k-1} - 49b_{k-2}$, $b_0 = 0$, $b_1 = 10$.
   (c) $c_n = 9c_{n-1} - 15c_{n-2} + 7c_{n-3}$, $c_0 = 0$, $c_1 = 1$, $c_2 = 2$.
       (*Hint*: $x = 1$ is a characteristic root.)
   (d) $d_n = 13d_{n-1} - 40d_{n-2} + 36d_{n-3}$, $d_0 = 1$, $d_1 = 1$, $d_2 = 0$.
       (*Hint*: $x = 2$ is a characteristic root.)
   (e) $e_k = 10e_{k-1} - 37e_{k-2} + 60e_{k-3} - 36e_{k-4}$, $e_0 = e_1 = e_2 = 0$, $e_3 = 5$.
       (*Hint*: $x = 2$ and $x = 3$ are characteristic roots.)

25. Solve the following recurrence under the given initial condition.

$$a_n = -2a_{n-2} - a_{n-4}, \quad a_0 = 0, \quad a_1 = 1, \quad a_2 = 2, \quad a_3 = 3.$$

(*Hint*: $x = i$ and $x = -i$ are characteristic roots.)

26. Suppose that α is a characteristic root of the recurrence (5.23) and α has multiplicity 2. Show that (α^n) and $(n\alpha^n)$ are solutions to (5.23). [*Hint*: If $C(x) = 0$ is the characteristic equation, then $C(x) = (x - \alpha)^2 D(x)$ for some polynomial $D(x)$. If $C_n(x) = x^{n-p} C(x)$, then show that α is a root of the derivative $C'_n(x)$. Substituting α for x in the equation $xC'_n(x) = 0$ shows that $(n\alpha^n)$ is a solution to (5.23).]

27. (a) Suppose that α is a characteristic root of the recurrence (5.23) and α has multiplicity 3. Show that (α^n), $(n\alpha^n)$, and $(n^2\alpha^n)$ are solutions to (5.23). [*Hint:* Generalize the argument in Exercise 26 by noting that $C(x) = (x - \alpha)^3 D(x)$. Consider $C_n(x) = x^{n-p} C(x)$, $A_n(x) = xC'_n(x)$, and $B_n(x) = xA'_n(x)$. Show that $(n\alpha^n)$ is a solution by considering $A_n(x) = 0$, and $(n^2\alpha^n)$ is a solution by considering $B_n(x) = 0$.]

(b) Generalize to the case where α is a characteristic root of multiplicity u.

5.3 SOLVING RECURRENCES USING GENERATING FUNCTIONS

5.3.1 The Method

Another method for solving recurrences uses the notion of generating function developed in Chapter 4. Suppose that $G(x)$ is the ordinary generating function for the sequence (a_k), that is, the function

$$G(x) = \sum_{k=0}^{\infty} a_k x^k.$$

We shall try to find (a_k) by finding its generating function. In particular, if we have a recurrence for a_k, the trick will be to multiply both sides of the recurrence by x^k and then take the sum, giving us an expression which can be used to derive $G(x)$.

Example 5.1 (The Grains of Wheat) Revisited

Let us illustrate the method with the recurrence relation of Example 5.1,

$$t_{k+1} = 2t_k. \tag{5.1}$$

The initial condition was $t_1 = 1$. In this case, t_0 is not defined. However, it will usually be convenient to think of our sequences as beginning with the zeroth term. Hence, we will try to define the early terms from the recurrence if they are not known or given. In particular, by (5.1), it is consistent to take

$$t_0 = \tfrac{1}{2} t_1 = \tfrac{1}{2}.$$

The ordinary generating function is

$$G(x) = \sum_{k=0}^{\infty} t_k x^k.$$

To derive $G(x)$, we start by multiplying both sides of the recurrence (5.1) by x^k, obtaining

$$t_{k+1} x^k = 2t_k x^k.$$

Then, we take sums*:

$$\sum_{k=0}^{\infty} t_{k+1} x^k = 2 \sum_{k=0}^{\infty} t_k x^k. \tag{5.36}$$

---

*The sums may be taken over all values of k for which the recurrence applies. In some cases, it will be better or more appropriate to take the sum from $k = 1$ or from $k = 2$, and so on.

The right-hand side of (5.36) is $2G(x)$. What is the left-hand side? We shall try to reduce that to an expression involving $G(x)$. Note that

$$\sum_{k=0}^{\infty} t_{k+1} x^k = t_1 + t_2 x + t_3 x^2 + \cdots$$

$$= \frac{1}{x} [t_1 x + t_2 x^2 + t_3 x^3 + \cdots]$$

$$= \frac{1}{x} [t_0 + t_1 x + t_2 x^2 + t_3 x^3 + \cdots - t_0]$$

$$= \frac{1}{x} [G(x) - t_0].$$

Hence,

$$\sum_{k=0}^{\infty} t_{k+1} x^k = \frac{G(x) - t_0}{x}. \tag{5.37}$$

Equations (5.36) and (5.37) now give us the equation

$$\frac{G(x) - t_0}{x} = 2G(x).$$

This equation, a *functional equation* for $G(x)$, can be solved for $G(x)$. A little bit of algebraic manipulation gives us

$$G(x) = \frac{t_0}{1 - 2x}.$$

Since we have computed $t_0 = \frac{1}{2}$, we have

$$G(x) = \frac{1}{2} (1 - 2x)^{-1}.$$

Knowing $G(x)$, we can compute the desired value of t_k from it. The number t_k is given by the coefficient of x^k if we expand out $G(x)$. How can we expand $G(x)$? There are two methods. The easiest is to use the identity

$$1 + y + y^2 + \cdots + y^n + \cdots = \frac{1}{1 - y}, \tag{5.38}$$

$|y| < 1$ (see Equation (4.2)). Doing so gives us the result

$$G(x) = \frac{1}{2} [1 + (2x) + (2x)^2 + \cdots + (2x)^n + \cdots]$$

or

$$G(x) = \frac{1}{2} + x + 2x^2 + \cdots + 2^{n-1} x^n + \cdots.$$

In other words,

$$t_k = 2^{k-1},$$

which agrees with our earlier computation. An alternative way of expanding $G(x)$ is to use the Binomial Theorem (Theorem 4.3). We leave it to the reader to try this.

Example 5.4 (Legitimate Codewords) Revisited

Let us now illustrate the method with the following recurrence of Example 5.4:

$$a_{k+1} = 2a_k + 4^k. \tag{5.8}$$

We use the ordinary generating function

$$G(x) = \sum_{k=0}^{\infty} a_k x^k.$$

We know that $a_1 = 3$. From the recurrence, we can derive a_0 even though a_0 is not defined. We obtain

$$a_1 = 2a_0 + 4^0,$$
$$3 = 2a_0 + 1,$$
$$a_0 = 1.$$

We now multiply both sides of the recurrence (5.8) by x^k and sum, obtaining

$$\sum_{k=0}^{\infty} a_{k+1} x^k = 2 \sum_{k=0}^{\infty} a_k x^k + \sum_{k=0}^{\infty} 4^k x^k. \tag{5.39}$$

The left-hand side of (5.39) is given by

$$\frac{1}{x} \sum_{k=0}^{\infty} a_{k+1} x^{k+1} = \frac{1}{x} [G(x) - a_0] = \frac{1}{x} [G(x) - 1].$$

Hence, we obtain

$$\frac{1}{x} G(x) - \frac{1}{x} = 2G(x) + \sum_{k=0}^{\infty} (4x)^k. \tag{5.40}$$

From the identity (5.38), we can rewrite this as

$$\frac{1}{x} G(x) - \frac{1}{x} = 2G(x) + \frac{1}{1 - 4x}.$$

From this functional equation, it is simply a matter of algebraic manipulation to solve for $G(x)$. We obtain

$$G(x) - 1 = 2xG(x) + \frac{x}{1 - 4x},$$

$$G(x)(1 - 2x) = 1 + \frac{x}{1 - 4x},$$

$$G(x) = \frac{1}{1 - 2x} \left(1 + \frac{x}{1 - 4x} \right),$$

$$G(x) = \frac{1}{1 - 2x} + \frac{x}{(1 - 2x)(1 - 4x)}. \tag{5.41}$$

This gives us the generating function for (a_k). How do we find a_k? It is easy enough to expand out the first term on the right-hand side of (5.41). The second term we expand by the method of partial fractions.* Namely, the second term on the right-hand side can be expressed as

$$\frac{a}{1 - 2x} + \frac{b}{1 - 4x},$$

for appropriate a and b. We compute that

$$a = -\tfrac{1}{2}, \qquad b = \tfrac{1}{2}.$$

Thus,

$$G(x) = \frac{1}{1 - 2x} + \frac{-\tfrac{1}{2}}{1 - 2x} + \frac{\tfrac{1}{2}}{1 - 4x},$$

$$G(x) = \frac{\tfrac{1}{2}}{1 - 2x} + \frac{\tfrac{1}{2}}{1 - 4x}. \tag{5.42}$$

To expand (5.42), we again use the identity (5.38), obtaining

$$G(x) = \frac{1}{2} \sum_{k=0}^{\infty} (2x)^k + \frac{1}{2} \sum_{k=0}^{\infty} (4x)^k.$$

Thus, the coefficient of x^k is given by

$$a_k = \tfrac{1}{2}(2)^k + \tfrac{1}{2}(4)^k.$$

In particular, we can check our computation in Section 5.1. We have

$$a_2 = \tfrac{1}{2}(2)^2 + \tfrac{1}{2}(4)^2 = 10,$$

$$a_3 = \tfrac{1}{2}(2)^3 + \tfrac{1}{2}(4)^3 = 36.$$

The reader might wish to check our results in still another way, namely by computing an exponential generating function for a_k directly by the methods of Section 4.5.

Example 5.9 Rook Polynomials†

In Examples 4.10 and 4.14, we introduced rook polynomials and stated a result that would reduce computation of a rook polynomial of a board B to computation of the rook polynomials of smaller boards. Here we state another such result. Suppose that s is any darkened square of the board B. Let B_s be obtained from B by forbidding s (lightening s) and let B'_s be obtained from B by forbidding all squares in the same row or column as s. Figure 5.7 shows a board B, a square s, and the boards B_s and B'_s.

Note that to place $k \geq 1$ rooks on B, we either use square s or we do not. If we do not use square s, then we have to place k rooks on the squares of B_s. If we use

*See most calculus texts for a discussion of this method.

†This example can be omitted if the reader has skipped Chapter 4.

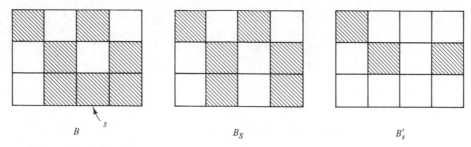

Figure 5.7 A board B, a square s (the $(3, 3)$ square), and the boards B_s and B'_s obtained from board B and square s.

square s, we have $k - 1$ rooks still to place, and we may use any darkened square of B except those in the same row or column as s, that is, we may use any darkened square of B'_s. Thus, by the sum rule of Chapter 2,

$$r_k(B) = r_k(B_s) + r_{k-1}(B'_s),$$ (5.43)

for $k \geq 1$. If we multiply both sides of (5.43) by x^k and sum over all $k \geq 1$, we find that

$$\sum_{k=1}^{\infty} r_k(B)x^k = \sum_{k=1}^{\infty} r_k(B_s)x^k + \sum_{k=1}^{\infty} r_{k-1}(B'_s)x^k.$$ (5.44)

The term on the left-hand side of (5.44) is just

$$R(x, B) - r_0(B) = R(x, B) - 1,$$

since $r_0(B) = 1$ for all boards B. The first term on the right-hand side is

$$R(x, B_s) - r_0(B_s) = R(x, B_s) - 1.$$

The second term on the right-hand side is equal to

$$x \sum_{k=1}^{\infty} r_{k-1}(B'_s)x^{k-1} = x \sum_{k=0}^{\infty} r_k(B'_s)x^k = xR(x, B'_s).$$

Thus, (5.44) gives us

$$R(x, B) - 1 = R(x, B_s) - 1 + xR(x, B'_s),$$

or

$$R(x, B) = R(x, B_s) + xR(x, B'_s).$$ (5.45)

Application of this result to the board B of Figure 5.7 is left as an exercise (Exercise 27).

The method used in the preceding three examples can be applied to solve a variety of recurrences. It will always work on any recurrence like (5.23) which is linear and homogeneous with constant coefficients.* The result will give a generating function $G(x)$

---

*The rest of this subsection can be omitted on first reading.

of the form

$$\frac{p(x)}{q(x)},$$

where $p(x)$ is a polynomial of degree less than p and $q(x)$ is a polynomial of degree p and constant term equal to 1. [The polynomials $p(x)$ and $q(x)$ can be expressed in terms of the coefficients $c_1, c_2, \ldots, c_p$ and initial conditions $a_0, a_1, \ldots, a_{p-1}$ of the recurrence (5.23). See Brualdi [1977] or Exercise 25 for details.] If all the roots of $q(x)$ are real numbers, one can then use the method of partial fractions to express $p(x)/q(x)$ as a sum of terms of the form

$$\frac{\alpha}{(1 - \beta x)^t},$$

where t is a positive integer and α and β are real numbers. In turn, the terms

$$\frac{\alpha}{(1 - \beta x)^t}$$

can be expanded out using the Binomial Theorem, giving us

$$\frac{\alpha}{(1 - \beta x)^t} = \alpha \sum_{k=0}^{\infty} \binom{t + k - 1}{k} \beta^k x^k. \tag{5.46}$$

This also follows directly by using βx in place of x in Corollary 4.4.1.

If $q(x)$ has complex roots, then the method of partial fractions can be used to express $p(x)/q(x)$ as a sum of terms of the form

$$\frac{a}{(x - b)^t} \qquad \text{or} \qquad \frac{ax + b}{(x^2 + cx + d)^t},$$

where t is a positive integer and a, b, c, and d are real numbers. The former terms can be changed into terms of the form

$$\frac{\alpha}{(1 - \beta x)^t}.$$

The latter terms can be manipulated by completing the square in the denominator and then using the expansion for

$$\frac{1}{(1 + y^2)^t} = \frac{1}{[1 - (-y^2)]^t}.$$

We omit the details.

5.3.2 Derangements

Let us next use the techniques of this section to derive the formula for the number of derangements D_n of n elements. We have the recurrence

$$D_{n+1} = n(D_{n-1} + D_n), \tag{5.12}$$

$n \geq 2$. We know that $D_2 = 1$ and $D_1 = 0$. Hence, using the recurrence (5.12), we derive

$D_0 = 1$. With $D_0 = 1$, (5.12) holds for $n \geq 1$. The recurrence (5.12) is inconvenient because it expresses D_{n+1} in terms of both D_n and D_{n-1}. Some algebraic manipulation reduces (5.12) to the recurrence

$$D_{n+1} = (n+1)D_n + (-1)^{n+1}, \tag{5.47}$$

$n \geq 0$. For a detailed verification of this fact, see the end of this subsection.

Let us try to calculate the ordinary generating function

$$G(x) = \sum_{n=0}^{\infty} D_n x^n.$$

We multiply (5.47) by x^n and sum, obtaining

$$\sum_{n=0}^{\infty} D_{n+1} x^n = \sum_{n=0}^{\infty} (n+1)D_n x^n + \sum_{n=0}^{\infty} (-1)^{n+1} x^n. \tag{5.48}$$

The left-hand side of (5.48) is

$$\frac{1}{x}[G(x) - D_0] = \frac{1}{x} G(x) - \frac{1}{x}.$$

The second term on the right-hand side is

$$-\sum_{n=0}^{\infty} (-1)^n x^n = -\sum_{n=0}^{\infty} (-x)^n = -\frac{1}{1+x},$$

using the identity (5.38). Finally, the first term on the right-hand side can be rewritten as

$$\sum_{n=0}^{\infty} nD_n x^n + \sum_{n=0}^{\infty} D_n x^n = x\sum_{n=0}^{\infty} nD_n x^{n-1} + \sum_{n=0}^{\infty} D_n x^n = xG'(x) + G(x).$$

Thus, (5.48) becomes

$$\frac{1}{x} G(x) - \frac{1}{x} = xG'(x) + G(x) - \frac{1}{1+x},$$

or

$$G'(x) + \left(\frac{1}{x} - \frac{1}{x^2}\right)G(x) = \frac{1}{x+x^2} - \frac{1}{x^2}. \tag{5.49}$$

Equation (5.49) is a linear first-order differential equation. Unfortunately, it is not easy to solve.

It turns out that the recurrence (5.47) is fairly easy to solve if we use instead of the ordinary generating function the exponential generating function

$$H(x) = \sum_{n=0}^{\infty} D_n \frac{x^n}{n!}.$$

To find $H(x)$, we multiply (5.47) by $x^{n+1}/(n+1)!$ and sum, obtaining

$$\sum_{n=0}^{\infty} D_{n+1} \frac{x^{n+1}}{(n+1)!} = \sum_{n=0}^{\infty} (n+1)D_n \frac{x^{n+1}}{(n+1)!} + \sum_{n=0}^{\infty} (-1)^{n+1} \frac{x^{n+1}}{(n+1)!}. \tag{5.50}$$

The left-hand side of (5.50) is

$$H(x) - D_0 = H(x) - 1.$$

The first term on the right-hand side is

$$\sum_{n=0}^{\infty} D_n \frac{x^{n+1}}{n!} = x \sum_{n=0}^{\infty} D_n \frac{x^n}{n!} = xH(x).$$

The second term on the right-hand side is

$$-\frac{x}{1!} + \frac{x^2}{2!} - \frac{x^3}{3!} + \cdots,$$

which is $e^{-x} - 1$ [see (4.3)]. Equation (5.50) now becomes

$$H(x) - 1 = xH(x) + e^{-x} - 1.$$

Hence,

$$H(x) = \frac{e^{-x}}{1 - x}.$$

We may expand this out to obtain D_n, which is the coefficient of $x_n/n!$. Writing $H(x)$ as $e^{-x}(1 - x)^{-1}$, we have

$$H(x) = \left[1 - x + \frac{x^2}{2!} - \frac{x^3}{3!} + \cdots\right][1 + x + x^2 + \cdots]. \tag{5.51}$$

It is easy to see directly that

$$H(x) = \sum_{n=0}^{\infty} x^n \left[1 - \frac{1}{1!} + \frac{1}{2!} - \frac{1}{3!} + \cdots + (-1)^n \frac{1}{n!}\right],$$

so that the coefficient of $x^n/n!$ becomes

$$D_n = n!\left[1 - \frac{1}{1!} + \frac{1}{2!} - \frac{1}{3!} + \cdots + (-1)^n \frac{1}{n!}\right]. \tag{5.52}$$

Equation (5.52) agrees with our earlier formula (5.13).

Another way to derive D_n from $H(x)$ is to observe that $H(x)$ is the ordinary generating function of the sequence (c_n) which is the convolution of the sequences $((-1)^n/n!)$ and $(1, 1, 1, \ldots)$. Hence, $H(x)$ is the exponential generating function of $(n!c_n)$. Still a third way to derive D_n from $H(x)$ is explained in Exercise 20.

We close this subsection by deriving the recurrence (5.47). Note that by (5.12),

$$D_{n+1} - (n + 1)D_n = D_{n+1} - nD_n - D_n$$

$$= nD_{n-1} - D_n$$

$$= -[D_n - nD_{n-1}].$$

Thus, we conclude that for all $j \geq 1$ and $k \geq 1$,

$$(-1)^j[D_j - jD_{j-1}] = (-1)^k[D_k - kD_{k-1}].$$

Now

$$(-1)^2[D_2 - 2D_1] = 1[1 - 0] = 1.$$

Thus, we see that for $n \geq 0$,

$$(-1)^{n+1}[D_{n+1} - (n + 1)D_n] = (-1)^2[D_2 - 2D_1] = 1,$$

from which (5.47) follows for $n \geq 0$.

5.3.3 Simultaneous Equations for Generating Functions

In Section 5.1.4 we considered a situation where instead of one sequence, we had to use three sequences to find a satisfactory system of recurrences (5.17), (5.19), and (5.21). The method of generating functions can be applied to solve a system of recurrences. To illustrate, let us first choose a_0, b_0, and c_0 so that (5.17), (5.19), and (5.21) hold. Using $a_1 = 2$, $b_1 = 1$, $c_1 = 1$, we find from (5.17), (5.19) and (5.21) that

$$2 = 2a_0 + b_0 + c_0$$

$$1 = b_0 - c_0 + 1$$

$$1 = c_0 - b_0 + 1.$$

One solution to this system is to take $a_0 = 1$, $b_0 = c_0 = 0$. With these values, we can assume that (5.17), (5.19), and (5.21) hold for $k \geq 0$.

We now multiply both sides of each of our equations by x^k and sum from $k = 0$ to ∞. We get

$$\sum_{k=0}^{\infty} a_{k+1}x^k = 2\sum_{k=0}^{\infty} a_k x^k + \sum_{k=0}^{\infty} b_k x^k + \sum_{k=0}^{\infty} c_k x^k,$$

$$\sum_{k=0}^{\infty} b_{k+1}x^k = \sum_{k=0}^{\infty} b_k x^k - \sum_{k=0}^{\infty} c_k x^k + \sum_{k=0}^{\infty} 4^k x^k,$$

$$\sum_{k=0}^{\infty} c_{k+1}x^k = \sum_{k=0}^{\infty} c_k x^k - \sum_{k=0}^{\infty} b_k x^k + \sum_{k=0}^{\infty} 4^k x^k.$$

If

$$A(x) = \sum_{k=0}^{\infty} a_k x^k, \qquad B(x) = \sum_{k=0}^{\infty} b_k x^k, \qquad \text{and} \qquad C(x) = \sum_{k=0}^{\infty} c_k x^k$$

are the ordinary generating functions for the sequences (a_k), (b_k), and (c_k), respectively, we find that

$$\frac{1}{x}[A(x) - a_0] = 2A(x) + B(x) + C(x),$$

$$\frac{1}{x}[B(x) - b_0] = B(x) - C(x) + \frac{1}{1 - 4x},$$

$$\frac{1}{x}[C(x) - c_0] = C(x) - B(x) + \frac{1}{1 - 4x}.$$

Using $a_0 = 1$, $b_0 = c_0 = 0$, we see from these three equations that

$$A(x) = \frac{1}{1 - 2x} [xB(x) + xC(x) + 1], \tag{5.53}$$

$$B(x) = \frac{1}{1 - x} \left[-xC(x) + \frac{x}{1 - 4x} \right], \tag{5.54}$$

$$C(x) = \frac{1}{1 - x} \left[-xB(x) + \frac{x}{1 - 4x} \right]. \tag{5.55}$$

It is easy to see from (5.54) and (5.55) that

$$B(x) = C(x) = \frac{x}{1 - 4x}. \tag{5.56}$$

It then follows from (5.53) and (5.56) that

$$A(x) = \frac{2x^2 - 4x + 1}{(1 - 2x)(1 - 4x)}. \tag{5.57}$$

By using (5.38), we see that (5.56) implies that

$$B(x) = C(x) = \sum_{k=0}^{\infty} 4^k x^{k+1}.$$

Thus, $b_k = c_k = 4^{k-1}$ for $k > 0$, $b_k - c_k = 0$ for $k = 0$. The right-hand side of (5.57) can be expanded out using the method of partial fractions, and we obtain

$$A(x) = \frac{1 - 3x}{1 - 4x} + \frac{x}{1 - 2x}.$$

This can be rewritten as

$$A(x) = 1 + \frac{x}{1 - 4x} + \frac{x}{1 - 2x}$$

$$= 1 + \sum_{k=0}^{\infty} 4^k x^{k+1} + \sum_{k=0}^{\infty} 2^k x^{k+1}.$$

Thus, $a_k = 4^{k-1} + 2^{k-1}$ for $k > 0$, and $a_0 = 1$. The results can be readily checked. In particular, we have $a_2 = 4 + 2 = 6$, which agrees with the result obtained in Section 5.1.4.

EXERCISES FOR SECTION 5.3

Note to the reader: In each of these exercises, if the denominator of the generating function turns out to have complex roots, it is acceptable to give the generating function as the answer.

   1. Use generating functions to solve the following recurrences.
      (a) (5.5) in Example 5.2 under the initial condition $f(1) = 0$.
      (b) (5.6) in Example 5.3.
      (c) (5.7) in Example 5.3.

2. Use generating functions to solve the following recurrences under the given initial conditions.
   (a) $a_{k+1} = a_k + 2$, $a_0 = 1$
   (b) $a_{k+1} = 3a_k + 1$, $a_1 = 1$
   (c) $a_{k+2} = 2a_{k+1} - a_k$, $a_0 = 0$, $a_1 = 1$.
3. Use generating functions to solve each of the recurrences in Exercise 11, Section 5.2.
4. In each of the following cases, suppose that $G(x)$ is the ordinary generating function for a sequence (a_k). Find a_k.

   (a) $G(x) = \dfrac{1}{(1 - 2x)(1 - 4x)}$

   (b) $G(x) = \dfrac{2x + 1}{(1 - 3x)(1 - 4x)}$

   (c) $G(x) = \dfrac{x^2}{(1 - 4x)(1 - 5x)(1 - 6x)}$

   (d) $G(x) = \dfrac{1}{8x^2 - 6x + 1}$

   (e) $G(x) = \dfrac{x}{x^2 - 3x + 2}$

   (f) $G(x) = \dfrac{1}{6x^3 - 5x^2 + x}$.

5. Use the results of Section 5.3.3 to verify the values we obtained in Section 5.1.4 for
   (a) a_3; (b) b_3; (c) c_3.
6. Consider the oil discovery problem (Exercise 16, Section 5.1). Suppose that $A(x) = \sum_{n=0}^{\infty} a_n x^n$ is the ordinary generating function for (a_n).
   (a) If $a_0 = 0$ and $a_1 = 1$, find $A(x)$.
   (b) In general, find $A(x)$ in terms of a_0 and a_1.
   (c) If $a_0 = a_1$, then use your answer to part (b) to show that a_n is constant. (This is obvious from the recurrence.)
7. In Exercise 19 of Section 5.1, find b_n.
8. This exercise asks the reader to derive the formula for the kth Fibonacci number by the method of generating functions.
   (a) It is useful to define F_{-1}. Use the recurrence (5.11) and the values of F_1 and F_0 to derive an appropriate value for F_{-1}.
   (b) If $G(x) = \sum_{k=0}^{\infty} F_k x^k$ is the ordinary generating function for F_k, derive a functional equation for $G(x)$ by multiplying the recurrence (5.11) by x^k and summing from $k = 0$ to ∞.
   (c) Show that

$$G(x) = \frac{1}{(1 - x - x^2)}.$$

   (d) Find the roots of $1 - x - x^2$ and use them to write $1 - x - x^2$ as $(1 - \lambda x)(1 - \mu x)$.
   (e) Use partial fractions to write

$$\frac{1}{(1 - \lambda x)(1 - \mu x)}$$

   as

$$\frac{A}{1 - \lambda x} + \frac{B}{1 - \mu x}.$$

   (f) From the result in part (e), derive a formula for F_k.

9. Suppose that G_n satisfies the equation

$$G_{n+1} = G_n + G_{n-1},$$

$n \geq 1$. Suppose that $G_0 = 3$ and $G_1 = 4$. Use generating functions to find a formula for G_n.

10. Use generating functions to solve the recurrence (5.10) in Example 5.6.

11. Solve the following recurrence:

$$a_{k+1} = a_k + k + 7, \quad k \geq 0, \quad a_0 = 0.$$

12. Suppose that $a_{n+1} = (n + 1)b_n$, for $n \geq 0$, and $a_0 = 0$. Find a relation involving $A(x)$, $B'(x)$, and $B(x)$, if $A(x)$ and $B(x)$ are the ordinary generating functions for (a_n) and (b_n), respectively.

13. Suppose that

$$y_{k+1} = Ay_k + B,$$

for $k \geq 0$, where A and B are real numbers, $A \neq 1$. Find a formula for y_k in terms of y_0 using the method of generating functions.

14. Suppose that

$$y_{k+2} - 2y_{k+1} + y_k = 2^k,$$

$k \geq 0$, and that $y_0 = 2$, $y_1 = 1$. Find y_k using the method of generating functions.

15. Suppose that Y_t is national income at time t. Following Samuelson [1939], Goldberg [1958] derives the recurrence relation

$$Y_t = \alpha(1 + \beta)Y_{t-1} - \alpha\beta Y_{t-2} + 1,$$

$t \geq 2$, for α and β positive constants. Assuming that $Y_0 = 2$, $Y_1 = 3$, $\alpha = \frac{1}{2}$, and $\beta = 1$, find a generating function for the sequence (Y_t).

16. Repeat Exercise 15 for $\alpha = 2$ and $\beta = 4$.

17. (Goldberg [1958]) In his work on inventory cycles, Metzler [1941] studies the total income i_t produced in the tth time period by an entrepreneur who is producing goods for sales and for inventory. Metzler derives the recurrence

$$i_{t+2} - 2\beta i_{t+1} + \beta i_t = v_0,$$

$t \geq 0$, where β is a constant such that $0 < \beta < 1$ and v_0 is a positive constant. Assuming that $i_0 = i_1 = 0$, find a generating function for the sequence (i_t).

18. If

$$C_{n+1} = 2nC_n + 2C_n + 2,$$

$n \geq 0$, and $C_0 = 1$, find C_n.

19. Solve the recurrence derived in Exercise 28, Section 5.1.

20. Derive D_n from Equation (5.51) for $H(x)$ by observing that $H(x)$ is the product of the exponential generating functions for the sequences (a_k) and (b_k), where $a_k = (-1)^k$ and $b_k = k!$. Use your results from Exercise 16, Section 4.5.

21. Derive a formula for D_n as follows.
    (a) Let

$$C_n = \frac{D_n}{n!} - \frac{D_{n-1}}{(n-1)!}.$$

Find a recurrence for C_{n+1} in terms of C_n.
    (b) Solve the recurrence for C_n by iteration.
    (c) Use the formula for C_n to solve for D_n.

22. Solve the recurrences of Exercise 20, Section 5.1, by the method of Section 5.3.3.

23. Solve simultaneously the recurrences

$$a_{n+1} = a_n + b_n + c_n, \, n \geq 1,$$

$$b_{n+1} = 4^n - c_n, \, n \geq 1,$$

$$c_{n+1} = 4^n - b_n, \, n \geq 1,$$

subject to the initial conditions $a_1 = b_1 = c_1 = 1$.

24. (Anderson [1974]) Suppose that (a_n) satisfies

$$na_n = 2(a_{n-1} + a_{n-2}),$$

$n \geq 2$, and $a_0 = e$, $a_1 = 2e$. Let $A(x)$ be the ordinary generating function for (a_n).
(a) Show that $A'(x) = 2(1 + x)A(x)$.
(b) Find $A(x)$. [*Hint*: Recall the equation $f'(x) = f(x)$.]

25. Consider a linear homogeneous recurrence relation (5.23) with constant coefficients. This exercise explores the relationship between the solution using characteristic roots and the solution using generating functions.
(a) Show that the ordinary generating function $G(x)$ for the sequence (a_n) is given by $G(x) = p(x)/q(x)$, where

$$q(x) = 1 - c_1 x - c_2 x^2 - \cdots - c_p x^p$$

and

$$p(x) = a_0 + (a_1 - c_1 a_0)x + (a_2 - c_1 a_1 - c_2 a_0)x^2$$
$$+ \cdots + (a_{p-1} - c_1 a_{p-2} - \cdots - c_{p-1} a_0)x^{p-1}.$$

(b) Show that if $\alpha_1, \alpha_2, \ldots, \alpha_p$ are the characteristic roots, then

$$q(x) = (1 - \alpha_1 x)(1 - \alpha_2 x) \cdots (1 - \alpha_p x)$$

and the roots of $q(x)$ are $1/\alpha_1, 1/\alpha_2, \ldots, 1/\alpha_p$.
(c) Illustrate these results by using the method of generating functions to solve the recurrence (5.27), and compare to the results in Section 5.2.1.
(d) Illustrate these results by using the method of characteristic roots to solve the recurrence (5.1), and compare to the results in Section 5.3.1.

26. Compute $R(x, B_J)$ for board B_J of Figure 4.6 by using Equation (5.45).

27. Compute the rook polynomial of board B in Figure 5.7 by using the results of Exercise 15 of Section 4.1, Example 4.14, and Equation (5.45).

5.4 SOME RECURRENCES INVOLVING CONVOLUTIONS*

5.4.1 The Number of Simple, Ordered, Rooted Trees

In Section 3.5.6 we noted that Cayley reduced the problem of counting the saturated hydrocarbons to the problem of counting trees. Here, we shall discuss a related problem, the problem of counting the number of *simple, ordered, rooted trees*, or *SOR trees* for

---

*The four subsections of this section are relatively independent and can, in principle, be read in any order. From a purely pedagogical viewpoint, if there is not enough time for all four subsections, then one of Sections 5.4.1, 5.4.2, 5.4.3 should be read—5.4.1 would be best—and then Section 5.4.4 should be read.

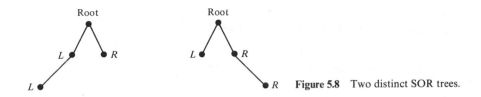

Figure 5.8 Two distinct SOR trees.

short. These are (unlabeled) rooted trees† which are simple in the sense that each vertex has zero, one, or two sons. Also, they are ordered so that the sons of each vertex are labeled left (L) or right (R). We shall distinguish two SOR trees if they are not isomorphic, or if they have different roots, or if they have the same root and are isomorphic, but there is a disagreement on left or right sons. For instance, the two SOR trees of Figure 5.8 are considered different even though they are isomorphic and have the same root.

We shall let u_n be the number of distinct SOR trees of n vertices. Then Figure 5.9 shows that $u_1 = 1$, $u_2 = 2$, and $u_3 = 5$. It is convenient to count the trees with no vertices as an SOR tree. Thus, we have $u_0 = 1$.

Suppose that T is an SOR tree of $n + 1$ vertices, $n \geq 0$. Then the root has at most two sons. If vertices a and b are the left and right sons of the root in an SOR tree, then a and b themselves form the roots of SOR trees T_L and T_R, respectively. (If a or b does not exist, the corresponding SOR tree is the tree with no vertices.) In particular, if T_R has r vertices, then T_L has $n - r$ vertices. Thus, we have the following recurrence:

$$u_{n+1} = u_0 u_n + u_1 u_{n-1} + u_2 u_{n-2} + \cdots + u_n u_0, \tag{5.58}$$

$n \geq 0$. Equation (5.58) gives us a way of computing u_{n+1} knowing all previous values u_i, $i \leq n$.

Note that the right-hand side of (5.58) comes from a convolution. In particular, if the sequence (v_n) is defined to be the sequence $(u_n) * (u_n)$, then

$$u_{n+1} = v_n. \tag{5.59}$$

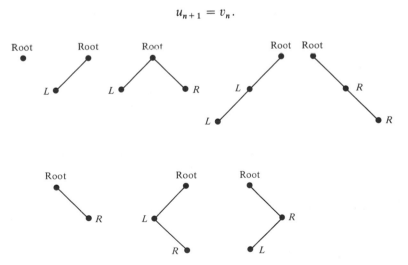

Figure 5.9 The distinct SOR trees of one, two, and three vertices.

†For the definition of rooted tree, see Section 3.6.1.

Let $U(x) = \sum_{n=0}^{\infty} u_n x^n$ and $V(x) = \sum_{n=0}^{\infty} v_n x^n$ be the ordinary generating functions for the sequences (u_n) and (v_n), respectively. Then by (5.59),

$$\sum_{n=0}^{\infty} u_{n+1} x^n = \sum_{n=0}^{\infty} v_n x^n.$$

We conclude that

$$\frac{1}{x}[U(x) - u_0] = V(x),$$

so

$$\frac{1}{x}[U(x) - 1] = V(x). \tag{5.60}$$

But $V(x) = U(x)U(x) = U^2(x)$, so (5.60) gives us

$$xU^2(x) - U(x) + 1 = 0. \tag{5.61}$$

Equation (5.61) is a functional equation for $U(x)$. We can solve this functional equation by treating the unknown $U(x)$ as a variable y. Then, assuming that $x \neq 0$, we apply the quadratic formula to the equation

$$xy^2 - y + 1 = 0$$

and solve for y to obtain

$$U(x) = \frac{1 \pm \sqrt{1 - 4x}}{2x}. \tag{5.62}$$

We can now solve for u_n by expanding out. In particular, we note that $\sqrt{1 - 4x}$ can be expanded out using the binomial theorem (Theorem 4.3), giving us

$$(1 - 4x)^{1/2} = 1 + \frac{1}{2}(-4x) + \frac{\frac{1}{2}(-\frac{1}{2})}{2!}(-4x)^2 + \frac{\frac{1}{2}(-\frac{1}{2})(-\frac{3}{2})}{3!}(-4x)^3 + \cdots$$

$$+ \binom{(\frac{1}{2})}{r}(-4x)^r + \cdots$$

For $n \geq 1$, the coefficient of x^n here can be written as

$$\binom{\frac{1}{2}}{n}(-4)^n = \frac{(\frac{1}{2})(-\frac{1}{2})(-\frac{3}{2}) \cdots (\frac{1}{2} - n + 1)}{n!}(-4)^n$$

$$= \frac{(\frac{1}{2})(-\frac{1}{2})(-\frac{3}{2}) \cdots \left[-\frac{(2n-3)}{2}\right]}{n!}(-4)^n$$

$$= \frac{(\frac{1}{2})(-1)^{n-1}(\frac{1}{2})(\frac{3}{2}) \cdots \left[\frac{2n-3}{2}\right]}{n!}(-1)^n 4^n$$

$$= \frac{\left(\frac{-1}{2^n}\right)[1 \cdot 3 \cdot 5 \cdot \ldots \cdot (2n-3)]4^n}{n!}$$

$$= \frac{-2^n[1 \cdot 3 \cdot 5 \cdot \ldots \cdot (2n-3)]}{n!}$$

$$= \frac{-2}{n} \frac{2^{n-1}}{(n-1)!}[1 \cdot 3 \cdot 5 \cdot \ldots \cdot (2n-3)]$$

$$= \frac{-2}{n} \frac{2^{n-1}}{(n-1)!} \frac{[1 \cdot 3 \cdot 5 \cdot \ldots \cdot (2n-3)](n-1)!}{(n-1)!}$$

$$= \frac{-2}{n} \frac{[1 \cdot 3 \cdot 5 \cdot \ldots \cdot (2n-3)][2 \cdot 4 \cdot 6 \cdot \ldots \cdot (2n-2)]}{(n-1)!(n-1)!}$$

$$= \frac{-2}{n} \frac{(2n-2)!}{(n-1)!(n-1)!}$$

$$= \frac{-2}{n} \binom{2n-2}{n-1}.$$

Thus,

$$(1-4x)^{1/2} = 1 - \sum_{n=1}^{\infty} \frac{2}{n} \binom{2n-2}{n-1} x^n. \tag{5.63}$$

Now (5.62) has two signs, that is, two possible solutions. If we take the solution of (5.62) with the $-$ sign, we have

$$U(x) = \frac{1}{2x}[1 - \sqrt{1-4x}],$$

so

$$U(x) = \frac{1}{2x}\left[\sum_{n=1}^{\infty} \frac{2}{n} \binom{2n-2}{n-1} x^n\right],$$

$$U(x) = \sum_{n=1}^{\infty} \frac{1}{n} \binom{2n-2}{n-1} x^{n-1},$$

$$U(x) = \sum_{n=0}^{\infty} \frac{1}{n+1} \binom{2n}{n} x^n. \tag{5.64}$$

We conclude from (5.64) that

$$u_n = \frac{1}{n+1} \binom{2n}{n}. \tag{5.65}$$

If we take the solution of (5.62) with the $+$ sign, we find similarly that

$$U(x) = \frac{1}{x} - \sum_{n=1}^{\infty} \frac{1}{n} \binom{2n-2}{n-1} x^{n-1}, \tag{5.66}$$

so for $n \geq 1$,

$$u_n = -\frac{1}{n+1}\binom{2n}{n}. \tag{5.67}$$

Now the coefficients u_n must be nonnegative (why?), so (5.65) must be the solution, not (5.67). We also see that we must take the $-$ sign in (5.62) to get $U(x)$. We can see this directly from (5.66) also, since if $U(x)$ were given by (5.66), $U(x)$ would have a term $1/x$, yet $U(x) = \sum_{n=0}^{\infty} u_n x^n$.

The numbers u_n defined by (5.65) are called the *Catalan numbers*, after Eugene Charles Catalan. For instance, we find

$$u_0 = \frac{1}{1}\binom{0}{0} = 1,$$

$$u_1 = \frac{1}{2}\binom{2}{1} = 1,$$

$$u_2 = \frac{1}{3}\binom{4}{2} = 2,$$

$$u_3 = \frac{1}{4}\binom{6}{3} = 5,$$

$$u_4 = \frac{1}{5}\binom{8}{4} = 14.$$

The first four results agree with our earlier computations and the fifth can readily be verified. We shall see that the Catalan numbers are very common in combinatorics.

5.4.2 The Ways to Multiply a Sequence of Numbers in a Computer

Suppose that we are given a sequence of n numbers, $x_1, x_2, \ldots, x_n$, and we wish to find their product. There are various ways in which we can find the product. For instance, suppose that $n = 4$. We can first multiply x_1 and x_2, then this product by x_3, and then this product by x_4. Alternatively, we can begin by multiplying x_1 and x_2, then multiply x_3 and x_4, and finally multiply the two products. We can distinguish these two and other approaches by inserting parentheses* as appropriate in the string $x_1 x_2 \cdots x_n$. Thus, the first method corresponds to

$$(((x_1 x_2)x_3)x_4)$$

and the second to

$$((x_1 x_2)(x_3 x_4)).$$

*The parentheses do not distinguish between first performing $x_1 x_2$, then $x_3 x_4$, and multiplying the product, and first performing $x_3 x_4$, then $x_1 x_2$, and multiplying the product. We are only concerned with what products will have to be calculated.

Let us assume that we must perform multiplications in the order given. For example, we do not allow multiplying x_1 by x_3 directly and so forth. Suppose that we are given a sequence of n numbers. How many different ways are there to instruct a computer to find the product? Suppose that P_n represents the number of ways in question. It is easy to see that finding the product corresponds to inserting $n-1$ left and $n-1$ right parentheses into the sequence $x_1 x_2 \cdots x_n$ in such a way that

1. one never has parentheses around a single x_i [for example, (x_i)], and
2. as we go from left to right, the number of right parentheses never exceeds the number of left parentheses.

Note that $P_1 = 1$, for there is only one way to insert 0 left and right parentheses. Also, $P_2 = 1$, $P_3 = 2$, and $P_4 = 5$. Table 5.4 demonstrates the parenthesizations corresponding to these numbers. It is easy to find a recurrence for P_n. Suppose that $n \geq 2$. Consider the last multiplication performed. This involves the product of two subproducts, $x_1 \cdots x_r$ and $x_{r+1} \cdots x_n$. That is, we have for $1 < r < n-1$, the multiplication

$$((x_1 \cdots x_r)(x_{r+1} \cdots x_n)).$$

If $r = 1$ or $n-1$, we have

$$(x_1(x_2 \cdots x_n)) \qquad \text{or} \qquad ((x_1 \cdots x_{n-1})x_n).$$

In either case, there are P_r ways to find the first subproduct and P_{n-r} ways to find the second subproduct, so we obtain the recurrence

$$P_n = \sum_{r=1}^{n-1} P_r P_{n-r}, \tag{5.68}$$

$n \geq 2$. Now if we let $P_0 = 0$, (5.68) becomes

$$P_n = \sum_{r=0}^{n} P_r P_{n-r}, \tag{5.69}$$

$n \geq 2$. Now let $P(x) = \sum_{n=0}^{\infty} P_n x^n$ be the ordinary generating function for the sequence (P_n). Equation (5.69) suggests that (P_n) is related to the convolution $(P_n) * (P_n)$. However, since (5.69) holds only for $n \geq 2$, we cannot conclude that $P(x) = P^2(x)$. To get around this

Table 5.4 The ways of performing a multiplication of two, three, or four numbers

$(x_1 x_2)$	$((x_1 x_2)x_3)$	$(((x_1 x_2)x_3)x_4)$
	$(x_1(x_2 x_3))$	$(x_1(x_2(x_3 x_4)))$
		$((x_1(x_2 x_3))x_4)$
		$(x_1((x_2 x_3)x_4))$
		$((x_1 x_2)(x_3 x_4))$

difficulty, suppose that we define the sequence (Q_n) to be the sequence $(P_n) * (P_n)$. Then note that

$$Q_n = \begin{cases} 0 = P_0 P_0 & \text{if } n = 0 \\ 0 = P_0 P_1 + P_1 P_0 & \text{if } n = 1 \\ P_n & \text{if } n \geq 2 \end{cases} \qquad (5.70)$$

and the ordinary generating function $Q(x) = \sum_{n=0}^{\infty} Q_n x^n$ satisfies

$$Q(x) = P^2(x).$$

Moreover, by (5.70),

$$Q(x) = P(x) - x,$$

since $P_n = Q_n$ for $n \neq 1$, and $P_1 = 1$, $Q_1 = 0$. Thus, we know that

$$P(x) - x = P^2(x).$$

This is a functional equation for $P(x)$. We solve it by rewriting it as a quadratic in the unknown $y = P(x)$ and using the quadratic formula, obtaining

$$P^2(x) - P(x) + x = 0,$$

$$P(x) = \frac{1 \pm \sqrt{1 - 4x}}{2}. \qquad (5.71)$$

To find P_n, we could expand out $P(x)$ using the binomial theorem. Alternatively, we recognize that $P(x)$ is $xU(x)$ for $U(x)$ of (5.62). Thus, $P_n = u_{n-1}$ for $n \geq 1$. We have defined $P_0 = 0$. By formula (5.65), we find that for $n \geq 1$,

$$P_n = \frac{1}{n}\binom{2n-2}{n-1}. \qquad (5.72)$$

The Catalan numbers have shown up again.

The close relation between the numbers u_n and P_n suggests that we might be able to find a direct relationship between SOR trees and order of multiplication. Following Even [1973], we shall describe such a relationship. Let us consider just a sequence of n left and n right parentheses. Such a sequence is called *well formed* if condition 2 above holds. Let K_n be the number of such sequences. Then clearly $P_n = K_{n-1}$. Given an SOR tree of n vertices, associate with each vertex of degree 1 the sequence of parentheses ( ). Associate with every other vertex the following sequence: (, followed by the sequence associated with its left son (if there is one), followed by the sequence associated with its right son (if there is one), followed by). This associates a unique well-formed sequence of parentheses with each SOR tree, the sequence assigned to its root. Figure 5.10 illustrates the procedure. Conversely, given a well-formed sequence of n left and n right parentheses, one can show that it comes from an SOR tree of n vertices. This is left as an exercise (Exercise 14). Thus, $K_n = u_n$, and we again have the conclusion $P_n = u_{n-1}$.

5.4.3 Secondary Structure in RNA

In Sections 2.12 and 2.13 we studied the linear chain of bases in an RNA molecule. This chain is sometimes said to define the *primary structure* of RNA. When RNA has only the bonds between neighbors in the chain, it is said to be a *random coil*. Now RNA does not

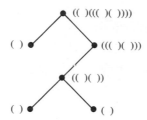

Figure 5.10 Next to each vertex is the corresponding well-formed sequence of parentheses.

remain a random coil. It folds back on itself and forms new bonds referred to as *Watson–Crick bonds*, creating helical regions. In such Watson–Crick bonding of an RNA chain $s = s_1 s_2 \cdots s_n$, each base can be bonded to at most one other nonneighboring base and if s_i and s_j are bonded, and $i < k < j$, then s_k can only be bonded with bases between s_{i+1} and s_{j-1}; that is, there is no crossover.* The new bonds define the *secondary structure* of the original RNA chain. Figure 5.11 shows one possible secondary structure for the RNA chain

AACGGGCGGGACCCUUCAACCCUU.

Watson–Crick bonds usually form between A and U bases or between G and C bases, but we shall, following Howell *et al.* [1980], find it convenient to allow all possible bonds in our discussion. In studying RNA chains, Howell *et al.* [1980] use recurrences to compute the number R_n of possible secondary structures for an RNA chain of length n. We briefly discuss their approach.†

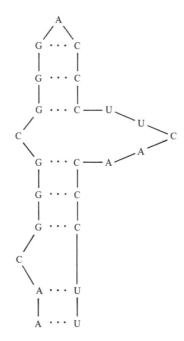

Figure 5.11 A secondary structure for the RNA chain AACGGGCGGGACC-CUUCAACCCUU. The Watson–Crick bonds are dotted. (Reproduced from Howell *et al.* [1980] with the permission of SIAM.)

*For a related bonding problem, see Nussinov *et al.* [1978].
†For related work, see Stein and Waterman [1978] and Waterman [1978].

Note that $R_1 = R_2 = 1$, for there can be no Watson–Crick bonds. Also, $R_0 = 1$ by convention. Consider an RNA chain $s_1 s_2 \cdots s_{n+1}$ of length $n + 1 \geq 3$. Now either s_{n+1} is not Watson–Crick bonded, or it is bonded with s_j, $1 \leq j \leq n - 1$. Thus, for $n \geq 2$,

$$R_{n+1} = R_n + \sum_{j=1}^{n-1} R_{j-1} R_{n-j}, \tag{5.73}$$

since in the first case, the chain $s_1 s_2 \cdots s_n$ is free to form any secondary structure, and in the second case, the subchains $s_1 s_2 \cdots s_{j-1}$ and $s_{j+1} s_{j+2} \cdots s_n$ are free to form any secondary structure. This is a recurrence for R_n. Let $R(x)$ be the ordinary generating function for R_n, that is,

$$R(x) = \sum_{n=0}^{\infty} R_n x^n.$$

To find $R(x)$, we note that the second term on the right-hand side of (5.73) almost arises from a convolution. If we define $(T_n) = (R_n) * (R_n)$, and we use the fact that $R_0 = 1$, we have for $n \geq 2$,

$$R_{n+1} = R_n + \sum_{j=1}^{n} R_{j-1} R_{n-j} - R_{n-1} R_0,$$

and hence for $n \geq 2$,

$$R_{n+1} = R_n - R_{n-1} + T_{n-1}. \tag{5.74}$$

It is easy to see that (5.74) still holds for $n = 1$. Furthermore, if we define $R_{-1} = T_{-1} = 0$, (5.74) holds for all $n \geq 0$. Now let $T(x) = \sum_{n=0}^{\infty} T_n x^n$ be the ordinary generating function for (T_n). By (5.74),

$$\sum_{n=0}^{\infty} R_{n+1} x^n = \sum_{n=0}^{\infty} R_n x^n - \sum_{n=0}^{\infty} R_{n-1} x^n + \sum_{n=0}^{\infty} T_{n-1} x^n.$$

Hence,

$$\frac{1}{x}[R(x) - R_0] = R(x) - x\left[\frac{R_{-1}}{x} + R(x)\right] + x\left[\frac{T_{-1}}{x} + T(x)\right],$$

so

$$\frac{1}{x}[R(x) - 1] = R(x) - xR(x) + xT(x).$$

Since (T_n) is a convolution of (R_n) with itself,

$$T(x) = R^2(x).$$

Thus, we have

$$\frac{1}{x} R(x) - \frac{1}{x} = R(x) - xR(x) + xR^2(x),$$

or

$$x^2 R^2(x) + (-x^2 + x - 1)R(x) + 1 = 0.$$

We find, for $x \neq 0$,

$$R(x) = \frac{x^2 - x + 1 \pm \sqrt{(-x^2 + x - 1)^2 - 4x^2}}{2x^2}$$

or

$$R(x) = \frac{1}{2x^2} \left[x^2 - x + 1 \pm \sqrt{1 - (2x + x^2 + 2x^3 - x^4)} \right]. \tag{5.75}$$

The square root in (5.75) can be expanded using the binomial theorem. Note that we can easily determine whether the $+$ or $-$ sign is to be used in (5.75) by noting that if the $+$ sign is used, there is a term $1/2x^2$. Thus the $-$ sign must be right. We can also see this by considering what happens as x approaches 0. Now $R(x)$ should approach $R(0) = R_0 = 1$. If we use the $+$ sign in (5.75), $R(x)$ approaches ∞ as x approaches 0. Thus, the $-$ sign must be right.

5.4.4 Organic Compounds Built Up from Benzene Rings

Harary and Read [1970] and Anderson [1974] point out that certain organic compounds built up from benzene rings can be represented by a configuration of hexagons joined together along a common edge, as for example in Figure 5.12. Let us assume that in such

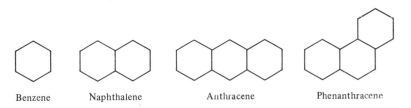

| Benzene | Naphthalene | Anthracene | Phenanthracene |

Figure 5.12 Some organic compounds built up from benzene rings.

a configuration, three hexagons may not meet in a point. Also, suppose that there is one fixed base hexagon, as shown in Figure 5.13. Onto this base, one can fit a hexagon only at the sides labeled 1, 2, and 3. Moreover, by the first assumption, one cannot fit a hexagon at both 1 and 2 or both 2 and 3. Finally, suppose that every hexagon is joined to another hexagon, and the entire configuration has one piece, so every hexagon is ultimately connected to the base hexagon by a chain of hexagons. Let h_k denote the number of possible configurations of k hexagons. We wish to compute h_k. Rather than derive a recurrence for h_k directly, we introduce two other sequences and use these to compute h_k. Let s_k denote the number of such configurations where only one hexagon is joined to the base, and d_k denote the number where exactly two hexagons are joined to the base. Obviously,

$$h_k = s_k + d_k, \tag{5.76}$$

$k \geq 2$. However, (5.76) fails for $k = 1$, since $h_1 = 1$ and $s_1 = d_1 = 0$. If we have a configur-

Figure 5.13 A base hexagon.

ation with $k + 1$ hexagons and only 1 joined to the base, there are three possible edges on which to join it, so

$$s_{k+1} = 3h_k, \qquad (5.77)$$

$k \geq 1$. If we have a configuration of $k + 1$ hexagons and two are joined to the base, they must use edges 1 and 3 of Figure 5.13, since three hexagons may not meet in a point. Thus, a configuration of r hexagons is joined to edge 1 and one of $k - r$ hexagons is joined to edge 3, with $1 \leq r \leq k - 1$. We conclude that

$$d_{k+1} = h_1 h_{k-1} + h_2 h_{k-2} + \cdots + h_{k-1} h_1, \qquad (5.78)$$

for $k \geq 2$. Note that the three recurrences (5.76), (5.77), and (5.78) can be used together to compute the desired numbers h_k iteratively. Knowing $h_1, h_2, \ldots, h_k$, we use (5.77) and (5.78) to compute s_{k+1} and d_{k+1}, respectively, and then obtain h_{k+1} from (5.76). This situation is analogous to the situation in Section 5.1.4 where we first encountered a system of recurrences.

Consider now the ordinary generating functions

$$H(x) = \sum_{k=0}^{\infty} h_k x^k, \quad S(x) = \sum_{k=0}^{\infty} s_k x^k, \quad D(x) = \sum_{k=0}^{\infty} d_k x^k. \qquad (5.79)$$

We shall compute $H(x)$. The technique for finding it will be somewhat different from that in the previous subsections. It will make use of the results of Section 4.2 on operating on generating functions. Note that h_k, s_k, and d_k are all defined only from $k = 1$ to ∞. To use the generating functions of (5.79), it is convenient to define $h_0 = s_0 = d_0 = 0$ and so to be able to take the sums from 0 to ∞. If we take $h_0 = s_0 = d_0 = 0$, then (5.76) holds for $k = 0$ as well as $k \geq 2$. Also, since $s_1 = 0$ and $h_0 = 0$, (5.77) now holds for all $k \geq 0$. Since $h_0 = 0$, we can add $h_0 h_k + h_k h_0$ to the right-hand side of (5.78), obtaining

$$d_{k+1} = h_0 h_k + h_1 h_{k-1} + h_2 h_{k-2} + \cdots + h_k h_0, \qquad (5.80)$$

for all $k \geq 2$. But it is easy to see that (5.80) holds for all $k \geq 0$, since $d_1 = 0 = h_0 h_0$ and $d_2 = 0 = h_0 h_1 + h_1 h_0$.

Using (5.76), we are tempted to conclude, by the methods of Section 4.2, that

$$H(x) = S(x) + D(x).$$

However, this is not true since (5.76) is false for $k = 1$. If we define

$$g_k = \begin{cases} h_k & \text{if } k \neq 1 \\ 0 & \text{if } k = 1 \end{cases},$$

then

$$g_k = s_k + d_k$$

holds for all $k \geq 0$. Moreover, if

$$G(x) = \sum_{k=0}^{\infty} g_k x^k,$$

then

$$G(x) = S(x) + D(x).$$

Finally,

$$H(x) = G(x) + x \qquad (5.81)$$

since the sequence (h_k) is the sum of the sequence (g_k) and the sequence $(0, 1, 0, 0, \ldots)$. Thus,

$$H(x) = S(x) + D(x) + x. \qquad (5.82)$$

Next, we have observed that (5.77) holds for $k \geq 0$. Hence,

$$\sum_{k=0}^{\infty} s_{k+1} \, x^k = 3 \sum_{k=0}^{\infty} h_k \, x^k,$$

$$\frac{1}{x} [S(x) - s_0] = 3H(x),$$

and, since $s_0 = 0$,

$$\frac{1}{x} S(x) = 3H(x).$$

Thus,

$$S(x) = 3xH(x). \qquad (5.83)$$

Finally, let us simplify $D(x)$. Letting $e_k = d_{k+1}$, $k \geq 0$, we see from (5.80) that (e_k) is the convolution of the sequence (h_k) with itself. Letting

$$E(x) = \sum_{k=0}^{\infty} e_k x^k,$$

we have

$$E(x) = H^2(x). \qquad (5.84)$$

Then

$$D(x) = \sum_{k=0}^{\infty} d_k x^k$$

$$= \sum_{k=1}^{\infty} d_k x^k$$

$$= \sum_{k=1}^{\infty} e_{k-1} x^k$$

$$= x \sum_{k=1}^{\infty} e_{k-1} x^{k-1}$$

$$= x \sum_{k=0}^{\infty} e_k x^k$$

$$= xE(x).$$

Thus, by (5.84),

$$D(x) = xH^2(x).\tag{5.85}$$

Using (5.83) and (5.85) in (5.82) gives

$$H(x) = 3xH(x) + xH^2(x) + x,$$

or

$$xH^2(x) + (3x - 1)H(x) + x = 0.\tag{5.86}$$

Equation (5.86) is a quadratic equation for the unknown function $H(x)$. We solve it by the quadratic formula, obtaining

$$H(x) = \frac{1}{2x}\left[1 - 3x \pm \sqrt{(3x - 1)^2 - 4x^2}\right]$$

or

$$H(x) = \frac{1}{2x}\left[1 - 3x \pm \sqrt{1 - (6x - 5x^2)}\right],\tag{5.87}$$

$x \neq 0$. $H(x)$ can be expanded out using the binomial theorem, and the proper sign, $+$ or $-$, can be chosen once the expansion has been obtained.

EXERCISES FOR SECTION 5.4

1. Check that the Catalan number $u_4 = 14$ does indeed count the number of SOR trees of four vertices.

2. Compute the Catalan numbers u_5 and u_6.

3. Use (5.72) to compute P_5 and check that it does indeed count the number of ways to multiply a sequence of 5 numbers.

4. Compute R_3, R_4, and R_5 from the recurrence (5.73) and the initial conditions, and check by drawing the appropriate secondary structures.

5. Compute h_2, h_3, and h_4 from the recurrences (5.76)–(5.78) and the initial conditions, and check your answers by drawing the appropriate configurations of hexagons.

6. A *rooted tree* is called *ordered* if a fixed ordering from left to right is assigned to all the sons of a given vertex. Put another way, the k sons of a given vertex are labeled with the integers $1, 2, \ldots,$ k. Two rooted, ordered trees are considered different if they are not isomorphic or if they have a different root or if they are isomorphic and have the same root, but the order of sons of two associated vertices differs. Suppose that r_n is the number of rooted, ordered trees of n vertices and $r_n(k)$ is the number of rooted, ordered trees of n vertices where the root has degree k.
   (a) Find an expression for r_n in terms of the $r_n(k)$.
   (b) Find an expression for $r_n(2)$ in terms of the other r_n's.

7. Suppose that

$$A(x) = \sum_{n=0}^{\infty} a_n x^n, \qquad B(x) = \sum_{n=0}^{\infty} b_n x^n, \qquad \text{and} \qquad C(x) = \sum_{n=0}^{\infty} c_n x^n$$

are the ordinary generating functions for the sequences (a_n), (b_n), and (c_n), respectively. Suppose that $a_0 = b_0 = c_0 = 0$, $a_1 = c_1 = 0$, $b_1 = 1$, and $a_2 = b_2 = 0$, $c_2 = 1$.

(a) Suppose that $c_n = a_n + b_n$, $n \geq 3$. Translate this into a statement in terms of generating functions.

(b) Suppose that $a_{n+1} = 4c_n$, $n \geq 0$. Translate this into a statement in terms of generating functions.

(c) Suppose that $b_{n+1} = c_1 c_{n-1} + c_2 c_{n-2} + \cdots + c_{n-1} c_1$, $n \geq 2$. Translate this into a statement using generating functions.

(d) Use your answers to parts (a)–(c) to derive an equation involving only $C(x)$.

8. Suppose that $A(x)$, $B(x)$, $C(x)$, a_0, b_0, c_0, a_1, b_1, c_1, a_2, and b_2 are as in Exercise 7, and $c_2 = 4$.

(a) Suppose that $c_n = a_n + 2b_n + 2$, $n \geq 3$. Translate this into a statement using generating functions.

(b) Suppose that $a_{n+1} = 3c_n$, $n \geq 0$. Translate this into a statement using generating functions.

(c) Suppose that $b_{n+1} = c_1 c_{n-1} + c_2 c_{n-2} + \cdots + c_{n-1} c_1$, $n \geq 2$. Translate this into a statement using generating functions.

(d) Use your answers to parts (a)–(c) to derive an equation involving only $C(x)$.

9. If $H(x)$ is given by (5.87), how can you tell whether to use the $+$ sign or the $-$ sign in computing $H(x)$?

10. (a) Use the formula for $H(x)$ [equation (5.87)] to compute the number h_1.

(b) Repeat for h_2.

(c) Repeat for h_3.

11. (Riordan [1975]) Suppose that $2n$ points are arranged on the circumference of a circle. Pair up these points and join corresponding points by chords of the circle. Show that the number C_n of ways of doing this pairing so that none of the chords cross is given by a Catalan number. (Maurer [1978] and Nussinov et al. [1978] relate intersection patterns of these chords to biochemical problems and Ko [1979], Read [1979], and Riordan [1975] study the number of ways to obtain exactly k overlaps of the chords.)

12. (Even [1973]) A *last in–first out* (*LIFO*) *stack* is a memory device which is like the stack of trays in a cafeteria: The last tray put on top of the stack is the first one that can be removed. A sequence of items labeled $1, 2, \ldots, n$ is waiting to be put into an empty LIFO stack. It must be put in in the order given. However, at any time, we may remove an item. Removed items are never returned to the stack or the sequence awaiting storage. At the end, we remove all items from the stack, and achieve a permutation of the labels $1, 2, \ldots, n$. For instance, if $n = 3$, we can first put in 1, then put in 2, then remove 2, then remove 1, then put in 3, and finally remove 3, obtaining the permutation 2, 1, 3. Let q_n be the number of permutations attainable.

(a) Find q_1, q_2, q_3, and q_4.

(b) Find q_n by obtaining a recurrence and solving.

13. In Section 5.4.4, suppose that we take $h_0 = \frac{3}{2}$ instead of $h_0 = 0$.*

(a) Show that

$$h_{k+1} = h_0 h_k + h_1 h_{k-1} + \cdots + h_k h_0$$

holds for all $k \geq 2$.

(b) Define (c_k) to be the sequence $(h_k) * (h_k)$ and let $w_k = h_{k+1}$. Let $C(x)$ be the ordinary generating function for (c_k) and $W(x)$ be the ordinary generating function for (w_k). Relate $W(x)$ and $C(x)$ to $H(x)$ and derive a functional equation for $H(x)$.

(c) Solve for $H(x)$.

(d) Why is the answer in part (c) different from the formula for $H(x)$ given in (5.87)? What is the relation of the new $H(x)$ to the old $H(x)$?

14. Show that each well-formed sequence of n left and n right parentheses comes from some SOR tree by the method described in Section 5.4.2.

*This idea is due to Martin Farber [personal communication].

15. (Anderson [1974]) A *simple, partly ordered, rooted tree* (*SPR tree*) is a simple rooted tree in which the labels L and R are placed on the sons of a vertex only if there are two sons. Figure 5.14 shows several SPR trees. Let u_n count the number of SPR trees of n vertices, let a_n count the number of SPR trees of n vertices in which the root has one son, and let b_n count the number of SPR trees of n vertices in which the root has two sons. Assume that $a_0 = b_0 = u_0 = 0$. Let $U(x)$, $A(x)$, and $B(x)$ be the ordinary generating functions for (u_n), (a_n), and (b_n), respectively.

  (a) Compute a_1, b_1, u_1, a_2, b_2, and u_2.
  (b) Derive a relation that gives u_n in terms of a_n and b_n and holds for all $n \neq 1$.
  (c) Derive a relation that gives a_{n+1} in terms of u_n and holds for all $n \geq 0$.
  (d) Derive a relation that gives b_{n+1} in terms of $u_1, u_2, \ldots, u_n$, and holds for all $n \geq 2$.
  (e) Derive a relation that gives b_{n+1} in terms of $u_0, u_1, u_2, \ldots, u_n$, and holds for all $n \geq 0$.
  (f) Find u_3, u_4, a_3, a_4, b_3, and b_4 from the answers to parts (b), (c), and (e), and check by drawing SPR trees.
  (g) Translate your answer to part (b) into a statement in terms of generating functions.
  (h) Do the same for part (c).
  (i) Do the same for part (e).
  (j) Show that

$$U(x) = \frac{1}{2x}\left[1 - x \pm \sqrt{(x-1)^2 - 4x^2}\right].$$

16. (Liu [1968]) Suppose that $A(x)$ is the ordinary generating function for the sequence (a_n) and $B(x)$ is the ordinary generating function for the sequence (b_n), and that

$$b_n = a_{n-1}b_0 + a_{n-2}b_1 + \cdots + a_0 b_{n-1}, \qquad n \geq 1.$$

  Find a relation involving $A(x)$ and $B(x)$.

17. Generalize the result in Exercise 16 to the case

$$b_n = a_{n-r}b_0 + a_{n-r-1}b_1 + \cdots + a_0 b_{n-r},$$

for $n \geq k$, where $k \geq r$.

18. (Liu [1968]) Recall the definition of pattern in a bit string introduced in Exercise 28, Section 5.1. Let a_n be the number of n-digit bit strings that have the pattern 010 occurring for the first time at the nth digit.

  (a) Show that

$$2^{n-3} = a_n + a_{n-2} + a_{n-3}2^0 + a_{n-4}2^1 + \cdots + a_3 2^{n-6},$$

$n \geq 6$.

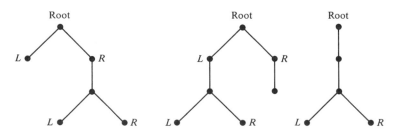

Figure 5.14 Several SPR trees.

(b) Let $b_0 = 1$, $b_1 = 0$, $b_2 = 1$, $b_3 = 2^0$, $b_4 = 2^1$, $b_5 = 2^2$, ..., and let $a_0 = a_1 = a_2 = 0$. Show that

$$2^{n-3} = a_n b_0 + a_{n-1} b_1 + a_{n-2} b_2 + \cdots + a_0 b_n,$$

$n \geq 3$.

(c) Letting $A(x)$ and $B(x)$ be the ordinary generating functions for the sequences (a_n) and (b_n), respectively, translate the equation obtained in part (b) into a statement involving $A(x)$ and $B(x)$. (See Exercise 17.)

(d) Solve for $A(x)$.

5.5 DIVIDE-AND-CONQUER ALGORITHMS*

Frequently, an algorithm for solving a large problem proceeds by *dividing* it into a number of smaller subproblems, which can then be more easily *conquered*. We have seen this approach, to give several examples, in the computation of the chromatic polynomial of a graph by reduction theorems (Section 3.4.2), in the process of sorting or placing a set of items in their natural order using the algorithm known as quik sort (Exercise 21, Section 3.6), and in the procedures for computing rook polynomials by reducing to smaller boards (Examples 4.14 and 5.9). A divide-and-conquer approach often can be implemented iteratively, by in turn dividing the subproblems into subproblems, and so on. In computing the computational complexity $f(n)$ of such an algorithm with input of size n, we will usually be able to reduce $f(n)$ to the complexity functions $f(m)$ for the smaller values of m corresponding to the subproblems, and thus find a recurrence for $f(n)$. We shall present one example of this approach in this section of the text, and other examples in the exercises.

Suppose that a divide-and-conquer algorithm proceeds by dividing a problem of size n as nearly as possible into a subproblems of size $\lfloor n/b \rfloor$ or $\lceil n/b \rceil$. Let us denote by $f(k)$ the cost of solving a problem of size k, and assume that $f(k)$ is a *monotone increasing function* of k, that is, if $k > k'$, then $f(k) > f(k')$. Let us also suppose that the cost of breaking a problem of size n into the subproblems plus the cost of combining the solutions of these subproblems into a solution of the whole problem is a constant c which is independent of n. Thus, if n is divisible by b, we have the recurrence

$$f(n) = af\left(\frac{n}{b}\right) + c. \tag{5.88}$$

We also assume that $f(1) = c$. To solve the recurrence (5.88) if $n = b^k$, $k \geq 1$, we note that

$$f(n) = af\left(\frac{n}{b}\right) + c,$$

$$f(n) = a^2 f\left(\frac{n}{b^2}\right) + ac + c,$$

---

*This section follows closely Stanat and McAllister [1977, pp. 248ff.] See also Aho *et al.* [1974, Sec. 2.6] and Tucker [1980, Sec. 4.5].

$$f(n) = a^3 f\left(\frac{n}{b^3}\right) + a^2 c + ac + c,$$

$$\vdots$$

$$f(n) = a^k f\left(\frac{n}{b^k}\right) + c \sum_{i=0}^{k-1} a^i. \tag{5.89}$$

Since $f(n/b^k) = f(1) = c$, we have

$$f(n) = c \sum_{i=0}^{k} a^i. \tag{5.90}$$

Assuming that $a \neq 1$ (a reasonable assumption), we recall that

$$1 + a + a^2 + \cdots + a^k = \frac{a^{k+1} - 1}{a - 1}.$$

Also, $k = \log_b n$, so

$$f(n) = c\left[\frac{a^{k+1} - 1}{a - 1}\right]$$

$$= c\left[\frac{aa^{\log_b n} - 1}{a - 1}\right]$$

$$= \frac{c(an^{\log_b a} - 1)}{a - 1},$$

using $a^{\log_b n} = n^{\log_b a}$. If α is the constant $ca/(a - 1)$, β is the constant $-c/(a - 1)$, and λ is the constant $\log_b a$, we have

$$f n) = \alpha n^\lambda + \beta. \tag{5.91}$$

This result holds whenever n is a power of b. If we are only interested in an upper bound on $f(n)$, then we can derive the inequality

$$f(n) \leq \alpha n^\lambda + \beta, \tag{5.92}$$

for n a power of b, if we simply assume

$$f(n) \leq af\left(\frac{n}{b}\right) + c \tag{5.93}$$

for n a power of b, and $f(1) \leq c$. The derivation of (5.92) is analogous to the derivation of (5.91), and is left as an exercise (Exercise 5).

What if n is not a power of b? Then there is a value of k such that

$$b^k < n \leq b^{k+1}.$$

Since the complexity function f is monotone increasing, if $m = b^{k+1}$, we have

$$f(n) \leq f(b^{k+1})$$
$$= f(m)$$
$$\leq \alpha m^{\lambda} + \beta$$
$$= \alpha b^{\lambda(k+1)} + \beta$$
$$= \alpha b^{\lambda} b^{\lambda k} + \beta.$$

Hence, $b^k < n$ implies

$$f(n) \leq \gamma n^{\lambda} + \beta, \tag{5.94}$$

for constants γ, λ, and β. Thus, $f(n)$ is bounded by a polynomial in n. In particular, since λ is $\log_b a$, then in the language of Section 2.18, $f(n)$ is $O(n^{\log_b a})$.

To illustrate this result, suppose that we consider the following algorithm for finding the maximum and minimum values of the entries $A(1)$, $A(2)$, ..., $A(n)$ of a vector $\mathbf{A}$. If $n = 1$, the algorithm gives $A(1)$ as both the maximum and minimum. If $n > 1$, the algorithm divides the entries into two groups of sizes as equal as possible, and finds the maximum and mininum in each half. These maxima and minima are compared to find the overall maximum and minimum. Suppose that $f(n)$ is the number of comparisons required between elements of $\mathbf{A}$ if there are n entries. Then clearly $f(1) = 0$, $f(n)$ is monotone increasing, and

$$f(n) = 2f\left(\frac{n}{2}\right) + 2$$

for n divisible by 2. (The constant 2 arises from comparisons of the two maxima and of the two minima.) Thus, (5.88) holds with $a = b = c = 2$. Here, we cannot conclude $f(1) = c$, only $f(1) \leq c$. Thus, we get (5.92) for n a power of 2, and (5.94) in general. We conclude that $f(n)$ is $O(n^{\log_2 2})$.

We shall consider other cases of divide-and-conquer algorithms in the exercises.

EXERCISES FOR SECTION 5.5

1. Solve the following recurrence relations for numbers n of the form indicated.
   (a) $a_n = 5a_{n/4} + 6$, $n = 4^k$, $a_1 = 6$;
   (b) $a_n = 8a_{n/7} + 3$, $n = 7^k$, $a_1 = 3$;
   (c) $a_n = 2a_{n/5} + n$, $n = 5^k$, $a_1 = 1$;
   (d) $a_n = a_{n/3} + 2n$, $n = 3^k$, $a_1 = 2$.

2. In an elimination tennis tournament with n players, there are $n/2$ matches in round 1. All winners of round 1 play in round 2, in an elimination tournament of $n/2$ players. And so on. Find a recurrence for the number $f(n)$ of rounds if there are n players, and solve for $f(n)$ assuming that $n = 2^k$.

3. (Tucker [1980]) We wish to arrange n numbers into increasing order, where n is a power of 2. First we pair off the numbers and order the numbers in each pair. Then we pair off the pairs into groups of four numbers, and order within each group of four. Then we pair up the ordered 4-tuples into groups of eight numbers, and order the eight numbers in each group. And so on. How many comparisons are needed to order all the n numbers?

4. Recurrences arising in divide-and-conquer problems can often be reduced to simpler recurrences by means of substitutions. For instance, suppose that in the recurrence (5.88), we substitute

$g(k) = f(b^k)$. Translate (5.88) into a recurrence for $g(k)$, solve for $g(k)$, and then use the solution to solve for $f(b^k)$. Compare to the solution in (5.91).

5. Derive (5.92) for n a power of b from (5.93) for n a power of b and $f(1) \leq c$.

6. Starting with Equation (5.88), assume $a = 1$. The next exercise will apply the results of this exercise.
   (a) Solve for $f(n)$ if $n = b^k$.
   (b) Find a bound on $f(n)$ if f is monotone increasing.

7. In the procedure for sorting a set of p elements into their natural order which was described by Steinhaus [1950], we pick two elements of the set at random and see which ranks higher. Having completely ordered n of the alternatives, we add an $(n + 1)$st alternative x to the ordering as follows. Compare x to the middle alternative y of the n alternatives already ordered (either of the middle alternatives if n is even). If x ranks below this middle alternative y, compare x next to the middle alternative of those ranking below y. If x ranks above y, compare x next to the middle alternative of those ranking above y. And so on. Eventually, the exact place of x in the ordering can be determined. Let the complexity $f(n)$ be the number of comparisons required for fitting in the $(n + 1)$st element in the worst possible case. Find a bound on $f(n)$. (Use the result in Exercise 6.)

8. Suppose that $A(1), A(2), \ldots, A(n)$ is an ordered list of index numbers in a file. We would like to know if a given argument or number α is on this list. The binary search algorithm (Stanat and McAllister [1977]) finds the middle element $\lceil \frac{n}{2} \rceil$ and compares α to $A(\lceil \frac{n}{2} \rceil)$. If α is $A(\lceil \frac{n}{2} \rceil)$, we are done. If $\alpha > A(\lceil \frac{n}{2} \rceil)$, we next search the list $A(\lceil \frac{n}{2} \rceil + 1), \ldots, A(n)$, again comparing α to A of the middle element. If $\alpha < A(\lceil \frac{n}{2} \rceil)$, we search the list $A(1), \ldots, A(\lceil \frac{n}{2} \rceil - 1)$, again comparing α to A of the middle element. And so on. If the computational complexity $f(n)$ is given by the number of comparisons required in the worst case, find a bound on $f(n)$.

9. (Tucker [1980]) To multiply two n-digit numbers, g and h, one normally performs n^2 digit-times-digit multiplications. However, a divide-and-conquer approach can work better. Suppose that n is even. Split both g and h into two $n/2$-digit parts:

$$g = g_1 10^{n/2} + g_2, \qquad h = h_1 10^{n/2} + h_2.$$

Note that

$$g \times h = (g_1 \times h_1 10^n) + (g_1 \times h_2 10^{n/2}) + (g_2 \times h_1 10^{n/2}) + (g_2 \times h_2). \qquad (5.95)$$

Since

$$(g_1 \times h_2) + (g_2 \times h_1) = [(g_1 + g_2) \times (h_1 + h_2)] - (g_1 \times h_1) - (g_2 \times h_2),$$

we only need to perform three multiplications to use (5.95) to obtain $g \times h$. Namely, we need only compute

$$g_1 \times h_1, \qquad g_2 \times h_2, \qquad \text{and} \qquad (g_1 + g_2) \times (h_1 + h_2).$$

The first two multiplications are $(n/2)$-digit multiplications, and the third is either an $(n/2)$-digit multiplication or an $[(n/2) + 1]$-digit multiplication.* Suppose that $a(n)$ represents the number of digit-times-digit multiplications needed to multiply two n-digit numbers by the procedure above. [Using $a(n)$ as the complexity function treats the time required to add as negligible compared to the time required to multiply.] For n even, we have

$$a(n) \leq 2a\left(\frac{n}{2}\right) + a\left(\frac{n}{2} + 1\right). \qquad (5.96)$$

*For further discussion of divide-and-conquer algorithms for multiplying two numbers, see Aho et al. [1974].

(a) Show from (5.96) that for $n \geq 4$, n even,

$$a(n) \leq 5a\left(\frac{n}{2}\right).$$

(b) Draw some conclusions about $a(n)$.

10. (Stanat and McAllister [1977]) Suppose that f is monotone increasing, $f(1) = c$, and

$$f(n) \leq af\left(\frac{n}{b}\right) + cn \qquad \text{for } n = b^k, \quad k \geq 1.$$

This situation arises in sorting algorithms such as quik sort (Exercise 21, Section 3.6), where the cost of splitting the problem and of combining the solutions of the subproblems increases proportionately to n.
(a) Find a bound for $f(n)$ if $n = b^k$.
(b) Show that if $a = b$, then for $n = b^k$,

$$f(n) \leq \alpha n \log_b n + \beta n,$$

for α, β constants and all $n \geq 0$. [Thus, $f(n)$ is $O(n \log_b n)$.]
(c) Show that if $a \neq b$, then for $n = b^k$,

$$f(n) \leq dbn - adn^{\log_b a}$$

for

$$d = \frac{c}{b - a}.$$

(d) Show that if $a < b$, then f is $O(n)$.
(e) Show that if $a > b$, then f is $O(n^{\log_b a})$.

11. Stanat and McAllister [1977] describe an algorithm for constructing a binary search tree of minimum height which contains a given set of elements as vertex values. (See Sections 3.6.1 and 3.6.2.) They show that for some constant c, the computational complexity $f(n)$, measured by the number of comparisons required to build the binary search tree, is monotone increasing and satisfies

$$f(1) \leq c + 1$$

and

$$f(n) \leq 2f\left(\frac{n}{2}\right) + (c + 1)n, \qquad n \text{ a power of 2.}$$

(a) Find a bound on $f(n)$ if n is a power of 2;
(b) Find a bound on $f(n)$ if n is arbitrary.
(c) Show that f is $O(n \log_2 n)$.

12. (Aho et al. [1974]) This exercise investigates a procedure for multiplying two $n \times n$ matrices. Suppose that two $n \times n$ matrices can be added in n^2 operations. Let us see how to multiply them. Suppose that $\mathbf{A}$ and $\mathbf{B}$ are both $n \times n$ matrices, with n a power of 2 for simplicity. We can partition $\mathbf{A}$ and $\mathbf{B}$ into four $n/2 \times n/2$ matrices, as

$$\mathbf{A} = \begin{bmatrix} \mathbf{A}_{11} & \mathbf{A}_{12} \\ \mathbf{A}_{21} & \mathbf{A}_{22} \end{bmatrix}, \qquad \mathbf{B} = \begin{bmatrix} \mathbf{B}_{11} & \mathbf{B}_{12} \\ \mathbf{B}_{21} & \mathbf{B}_{22} \end{bmatrix}.$$

Then we can express the product **AB** as

$$\mathbf{AB} = \begin{bmatrix} \mathbf{C}_{11} & \mathbf{C}_{12} \\ \mathbf{C}_{21} & \mathbf{C}_{22} \end{bmatrix},$$

where

$$\mathbf{C}_{11} = \mathbf{A}_{11}\mathbf{B}_{11} + \mathbf{A}_{12}\mathbf{B}_{21},$$

$$\mathbf{C}_{12} = \mathbf{A}_{11}\mathbf{B}_{12} + \mathbf{A}_{12}\mathbf{B}_{22},$$

$$\mathbf{C}_{21} = \mathbf{A}_{21}\mathbf{B}_{11} + \mathbf{A}_{22}\mathbf{B}_{21},$$

$$\mathbf{C}_{22} = \mathbf{A}_{21}\mathbf{B}_{12} + \mathbf{A}_{22}\mathbf{B}_{22}.$$

In other words, we can think of **A**, **B**, and **AB** as 2×2 matrices, each of whose entries is an $n/2 \times n/2$ matrix. Multiplication of **A** by **B** amounts to multiplication of two 2×2 matrices. Suppose that we can find an algorithm which multiplies two 2×2 matrices using m multiplications and a additions (or subtractions). (Strassen [1969] discovered a method in which $m = 7$, $a = 18$.) Suppose that $f(n)$ is the number of operations required to multiply two $n \times n$ matrices. If n is a power of 2, our procedure reduces the problem to m multiplications of $n/2 \times n/2$ matrices, and a additions of $n/2 \times n/2$ matrices. Since the latter require $a[(n/2) \times (n/2)] = a(n^2/4)$ operations, and the former require $mf(n/2)$ operations, we have

$$f(n) = mf\left(\frac{n}{2}\right) + a\,\frac{n^2}{4}.$$

For n a power of 2, show that if $m > 4$, $f(n)$ is $O(n^{\log_2 m})$. [Note that the usual method of multiplying matrices is $O(n^3)$. Since $\log_2 7 < 3$, Strassen's algorithm gives us a more efficient way to multiply matrices.]

REFERENCES FOR CHAPTER 5

ADBY, P. R., and DEMPSTER, M. A. H., *Introduction to Optimization Methods*, Chapman & Hall, London, 1974.

ADLER, I., "The Consequence of Constant Pressure in Phyllotaxis," *J. Theor. Biol.*, 65 (1977), 29–77.

AHO, A. V., HOPCROFT, J. E., and ULLMAN, J. D., *The Design and Analysis of Computer Algorithms*, Addison-Wesley, Reading, Mass., 1974.

ANDERSON, I., *A First Course in Combinatorial Mathematics*, Clarendon Press, Oxford, 1974.

BATSCHELET, E., *Introduction to Mathematics for Life Scientists*, Springer-Verlag, New York, 1971.

BRUALDI, R. A., *Introductory Combinatorics*, North-Holland, New York, 1977.

COXETER, H. S. M., *Introduction to Geometry*, Wiley, New York, 1961.

EVEN, S., *Algorithmic Combinatorics*, Macmillan, New York, 1973.

GAMOW, G., *One, Two, Three ... Infinity*, Mentor Books, New American Library, New York, 1954.

GOLDBERG, S., *Introduction to Difference Equations*, Wiley, New York, 1958.

HARARY, F., and PALMER, E. M., *Graphical Enumeration*, Academic Press, New York, 1973.

HARARY, F., and READ, R. C., "The Enumeration of Tree-like Polyhexes," *Proc. Edinb. Math. Soc.*, 17 (1970), 1–14.

HOLLINGDALE, S. H., "Methods of Operational Analysis," in J. Lighthill (ed.), *Newer Uses of Mathematics*, Penguin Books, Hammondsworth, Middlesex, England, 1978, pp. 176–280.

HOWELL, J. A., SMITH, T. F., and WATERMAN, M. S., "Computation of Generating Functions for Biological Molecules," *SIAM J. Appl. Math.*, *39* (1980), 119–133.

KIEFER, J., "Sequential Minimax Search for a Maximum," *Proc. Amer. Math. Soc.*, *4* (1953), 502–506.

KO, C.-S., "Broadcasting, Graph Homomorphisms, and Chord Intersections," Ph.D. thesis, Department of Mathematics, Rutgers University, New Brunswick, N.J., 1979.

LEVY, H., and LESSMAN, F., *Finite Difference Equations*, Macmillan, New York, 1961.

LIU, C. L., *Introduction to Combinatorial Mathematics*, McGraw-Hill, New York, 1968.

MAURER, S. B., "A Minimum Cycle Problem in DNA Research," unfinished manuscript, Swarthmore College, Swarthmore, Penn., 1978.

METZLER, L., "The Nature and Stability of Inventory Cycles," *Rev. Econ. Statist.*, *23* (1941), 113–129.

NUSSINOV, R. P., PIECZENIK, G., GRIGGS, J. R., and KLEITMAN, D., "Algorithms for Loop Matchings," *SIAM J. Appl. Math.*, *35* (1978), 68–82.

READ, R. C., "The Chord Intersection Problem," *Ann. N. Y. Acad. Sci.*, *319* (1979), 444–454.

RIORDAN, J., "The Distribution of Crossings of Chords Joining Pairs of $2n$ Points on a Circle," *Math. Comp.*, *29* (1975), 215–222.

RYSER, H. J., *Combinatorial Mathematics*, Carus Mathematical Monographs No. 14, Mathematical Association of America, Washington, D.C., 1963.

SAMUELSON, P. A., "Interactions between the Multiplier Analysis and the Principle of Acceleration," *Rev. Econ. Statist.*, *21* (1939), 75–78. (Reprinted in *Readings in Business Cycle Theory*, Blakiston, Philadelphia, 1944.)

SCHIPS, M., *Mathematik und Biologie*, Teubner, Leipzig, 1922.

SHANNON, C. E., "The Zero-Error Capacity of a Noisy Channel," *IRE Trans. Inf. Theory*, *IT-2* (1956), 8–19.

STANAT, D. F., and McALLISTER, D. F., *Discrete Mathematics in Computer Science*, Prentice-Hall, Englewood Cliffs, N.J., 1977.

STEIN, P. R., and WATERMAN, M. S., "On Some New Sequences Generalizing the Catalan and Motzkin Numbers," *Discrete Math.*, *26* (1978), 261–272.

STEINHAUS, H., *Mathematical Snapshots*, Oxford University Press, London, 1950.

STRASSEN, V., "Gaussian Elimination is Not Optimal," *Numerische Mathematik*, *13* (1969), 354–356.

TUCKER, A., *Applied Combinatorics*, Wiley, New York, 1980.

WATERMAN, M. S., "Secondary Structure of Single-Stranded Nucleic Acids," *Studies on Foundations and Combinatorics, Advances in Mathematics Supplementary Studies*, Vol. 1, Academic Press, New York, 1978, pp. 167–212.

6 The Principle of Inclusion and Exclusion

6.1 THE PRINCIPLE AND SOME OF ITS APPLICATIONS

6.1.1 Some Simple Examples

In this chapter we introduce still another basic counting tool, known as the principle of inclusion and exclusion. We introduce it with the following example.

Example 6.1   Job Applicants

Suppose that in a group of 18 job applicants, 10 have computer programming expertise, 5 have statistical expertise, and 2 have both programming and statistical expertise. How many of the group have neither expertise? To answer this question we draw a Venn diagram such as that shown in Figure 6.1.* There are 18 people altogether. To find out how many people have neither expertise, we want to subtract from 18 the number having programming expertise (10) and the number having statistical expertise (5). However, we may have counted several people twice. In particular, all people who have both kinds of expertise (the number of people in the intersection of the two sets programming expertise and statistical expertise in Figure 6.1) have been counted twice. There are 2 such people. Thus, we have to add these 2 back in to obtain the right count. Altogether, we conclude that

$$18 - 10 - 5 + 2 = 5$$

---

*The reader who is unfamiliar with Venn diagrams should consult any finite mathematics text, for example Kemeny *et al.* [1974].

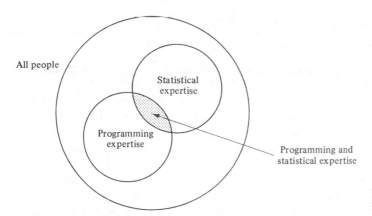

All people

Statistical expertise

Programming expertise

Programming and statistical expertise

Figure 6.1 A Venn diagram for Example 6.1.

is the number of people who have neither expertise. We shall generalize the reasoning we have just gone through.

Suppose that we have a set A of N objects. Let $a_1, a_2, \ldots, a_r$ be a set of properties the objects may have, and let $N(a_i)$ be the number of objects having property a_i. An object may have several (or none) of the properties in question. Let $N(a_i')$ count the number of objects not having the property a_i. Hence, we have

$$N = N(a_i) + N(a_i').$$

Since an object can have more than one property, it is useful to count the number of objects having both properties a_i and a_j. This will be denoted by $N(a_i a_j)$. The number of objects having neither of the properties a_i and a_j will be denoted $N(a_i' a_j')$ and the number of objects having property a_j but not property a_i will be denoted $N(a_i' a_j)$. We shall also use the following notation, which has the obvious interpretation:

$$N(a_i a_j \cdots a_k),$$

$$N(a_i' a_j' \cdots a_k'),$$

$$N(a_i' a_j \cdots a_k),$$

etc.

In Example 6.1 the objects in A are the 18 people, so $N = 18$. Let property a_1 be having programming expertise and property a_2 be having statistical expertise. Then

$$N(a_1) = 10, \qquad N(a_2) = 5, \qquad N(a_1 a_2) = 2.$$

By computation, we determined that

$$N(a_1' a_2') = 5.$$

It is also possible to compute $N(a_1' a_2)$ and $N(a_1 a_2')$. The former is the number of people who have statistical expertise but do not have programming expertise, and this is given by $5 - 2 = 3$.

Our computation in Example 6.1 used the following formula for $N(a_1'a_2')$:

$$N(a_1'a_2') = N - N(a_1) - N(a_2) + N(a_1a_2). \tag{6.1}$$

Note that certain objects are *included* too often, so some of these have to be *excluded*. Equation (6.1) is a special case of a principle known as the principle of inclusion and exclusion. Let us develop a similar principle for $N(a_1'a_2'a_3')$, the number of objects having neither property a_1 nor property a_2 nor property a_3. The principle is illustrated in the Venn diagram of Figure 6.2. We first include all the objects in A (all N of them). Then we exclude those having property a_1, those having property a_2, and those having property a_3. Since some objects have more than one of these properties, we need to add back in those objects which have been excluded more than once. We add back in (include) those objects having two of the properties, the objects corresponding to areas in Figure 6.2 that are colored in. Then we have added several objects back in too often, namely those which have all three of the properties, the objects in the one area of Figure 6.2 which has crosshatching. These objects must now be excluded. The result of this reasoning, which we shall formalize below, is the following formula:

$$N(a_1'a_2'a_3') = N - N(a_1) - N(a_2) - N(a_3)$$
$$+ N(a_1a_2) + N(a_1a_3) + N(a_2a_3) - N(a_1a_2a_3). \tag{6.2}$$

In general, the formula for the number of objects not having any of r properties is obtained by generalizing the formula (6.2). The general formula is called the *principle of inclusion and exclusion*. In the form we shall present it, the principle was discovered by Sylvester about 100 years ago. In another form, it was discovered by De Moivre some years earlier. The principle is given in the following theorem:

Theorem 6.1 (Principle of Inclusion and Exclusion). If N is the number of objects in a set A, the number of objects in A having none of the properties $a_1, a_2, \ldots, a_r$ is given by

$$N(a_1'a_2' \cdots a_r') = N - \sum_i N(a_i) + \sum_{i \ne j} N(a_i a_j) - \sum_{\substack{i, j, k \\ \text{different}}} N(a_i a_j a_k) + \cdots$$
$$+ (-1)^r N(a_1 a_2 \cdots a_r). \tag{6.3}$$

In (6.3), the first sum is over all i from $\{1, 2, \ldots, r\}$. The second sum is over all unordered

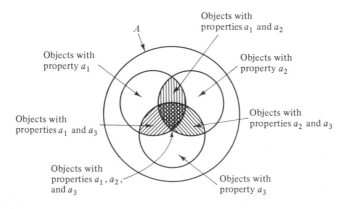

Figure 6.2 A Venn diagram.

A

Objects with property a_1

Objects with properties a_1 and a_2

Objects with property a_2

Objects with properties a_1 and a_3

Objects with properties a_2 and a_3

Objects with properties $a_1, a_2,$ and a_3

Objects with property a_3

pairs $\{i, j\}$, with i and j from $\{1, 2, \ldots, r\}$ and $i \neq j$. The third sum is over all unordered triples $\{i, j, k\}$, with $i, j,$ and k from $\{1, 2, \ldots, r\}$ and i, j, k distinct. The general term is $(-1)^t$ times a sum of terms of the form $N(a_{i_1} a_{i_2} \cdots a_{i_t})$, where the sum is over all unordered t-tuples $\{i_1, i_2, \ldots, i_t\}$ from $\{1, 2, \ldots, r\}$, with $i_1, i_2, \ldots, i_t$ distinct. In the remainder of this chapter, we present applications and variants of the principle of inclusion and exclusion. We present a proof in Section 6.1.2.

Example 6.2 Emissions Testing

Fifty cars are tested for pollutant emissions of nitrogen oxides (NO_x), hydrocarbons (HC), and carbon monoxide (CO). One of the cars exceeds the environmental standards for all three pollutants. Three cars exceed them for NO_x and HC, two for NO_x and CO, one for HC and CO, six for NO_x, four for HC, and three for CO. How many cars meet the environmental standards for all three pollutants? We let A be the set of cars, let a_1 be the property of exceeding the standards for NO_x and let a_2 and a_3 be the same property for HC and CO, respectively. We would like to calculate $N(a_1' a_2' a_3')$. We are given the following information:

$$N = 50,$$
$$N(a_1 a_2 a_3) = 1,$$
$$N(a_1 a_2) = 3,$$
$$N(a_1 a_3) = 2,$$
$$N(a_2 a_3) = 1,$$
$$N(a_1) = 6,$$
$$N(a_2) = 4,$$
$$N(a_3) = 3.$$

Thus, by the principle of inclusion and exclusion,

$$N(a_1' a_2' a_3') = 50 - 6 - 4 - 3 + 3 + 2 + 1 - 1 = 42.$$

6.1.2 Proof of Theorem 6.1*

The idea of the proof of Theorem 6.1 is a very simple one. The left-hand side of (6.3) counts the number of objects in A having none of the properties. We shall simply show that every object having none of the properties is counted exactly one time in the right-hand side of (6.3) and every object having at least one property is counted exactly zero times (in a net sense). Suppose that an object has none of the properties in question. Then it is counted once in computing N, but not in $\sum N(a_i)$, $\sum N(a_i a_j)$, and so on. Hence, it is counted exactly once in the right side of (6.3). Suppose that an object has exactly p of the properties $a_1, a_2, \ldots, a_r$, $p > 0$. Now the object is counted $1 = \binom{p}{0}$ times in computing N, the number of objects in A. It is counted once in each expression $N(a_i)$ for a property a_i it has, so exactly $p = \binom{p}{1}$ times in $\sum N(a_i)$. In how many terms $N(a_i a_j)$ is the object counted? The answer is it is the number of pairs of properties a_i and a_j which the object has, and

*This subsection can be omitted. However, the reader is urged to read it.

this number is given by the number of ways to choose two properties from p, that is, by $\binom{p}{2}$. Thus, the object is counted exactly $\binom{p}{2}$ times in $\sum N(a_i a_j)$. Similarly, in $\sum N(a_i a_j a_k)$ it is counted exactly $\binom{p}{3}$ times, and so on. All together, the number of times the object is counted in the right-hand side of (6.3) is given by

$$\binom{p}{0} - \binom{p}{1} + \binom{p}{2} - \binom{p}{3} \pm \cdots + (-1)^r\binom{p}{r}. \tag{6.4}$$

Since $p \le r$ and since by convention $\binom{p}{k} = 0$ if $p < k$, (6.4) becomes

$$\binom{p}{0} - \binom{p}{1} + \binom{p}{2} - \binom{p}{3} \pm \cdots + (-1)^p\binom{p}{p}. \tag{6.5}$$

Since $p > 0$, Theorem 2.9 implies that the expression (6.5) is 0, so the object contributes a net count of 0 to the right side of (6.3). This completes the proof.

6.1.3 The Sieve of Erastothenes

One of the earliest problems about numbers to interest mathematicians was the problem of identifying all *prime numbers*, integers greater than 1 whose only divisors are 1 and themselves. The Greek Erastothenes is credited with inventing the following procedure for identifying all prime numbers between 1 and N. First write out all the numbers between 1 and N. Cross out 1. Then cross out those numbers divisible by and larger than 2. Search for the first number larger than 2 not yet crossed out—it is 3—and cross out all those numbers divisible by and larger than 3. Then search for the first number larger than 3 not yet crossed out—it is 5—and cross out all numbers divisible by and larger than this number. And so on. The prime numbers are the ones remaining when the procedure ends. The following shows the steps in this procedure if $N = 25$:

1. Cross out 1 and cross out numbers divisible by and larger than 2:

   ̸1 2 3 ̸4 5 ̸6 7 ̸8 9 ̸10 11 ̸12 13 ̸14 15 ̸16 17 ̸18 19 ̸20 21 ̸22 23 ̸24 25

2. Cross out numbers divisible by and larger than 3 among those not yet crossed out:

   ̸1 2 3 ̸4 5 ̸6 7 ̸8 ̸9 ̸10 11 ̸12 13 ̸14 ̸15 ̸16 17 ̸18 19 ̸20 ̸21 ̸22 23 ̸24 25.

3. Cross out numbers divisible by and larger than 5 among those not yet crossed out:

   ̸1 2 3 ̸4 5 ̸6 7 ̸8 ̸9 ̸10 11 ̸12 13 ̸14 ̸15 ̸16 17 ̸18 19 ̸20 ̸21 ̸22 23 ̸24 ̸25.

Dividing by 7, 11, and so on, does not remove any more numbers, so the numbers remaining are the prime numbers between 1 and 25. This procedure used to be carried out on a wax tablet (Vilenkin [1971]), with numbers punched out rather than crossed out. The result was something like a sieve, and hence the procedure came to be known as the *Sieve of Erastothenes*. A basic question that arises is: How many primes are there between 1 and N? The answer is closely related to the answer to the following question: How many numbers between 1 and N (other than 1) are not divisible by 2, 3, 5, ... ? In the next example, we see how to answer questions of this type.

Example 6.3

How many integers between 1 and 1000 are

(a) not divisible by 2?

(b) not divisible by either 2 or 5?

(c) not divisible by 2, 5, or 11?

To answer this question, let us consider the set of 1000 integers between 1 and 1000 and let a_1 be the property of being divisible by 2, a_2 the property of being divisible by 5, and a_3 the property of being divisible by 11. We would like the following information:

(a) $N(a_1')$;   (b) $N(a_1'a_2')$;   (c) $N(a_1'a_2'a_3')$.

We are given $N = 1000$. Also,

$$N(a_1) = 500,$$

since every other integer is divisible by 2. Hence,

$$N(a_1') = N - N(a_1) = 500,$$

which gives us the answer to (a). Next,

$$N(a_2) = \tfrac{1}{5}(1000) = 200.$$

Also, every tenth integer is divisible by 2 and by 5, so

$$N(a_1 a_2) = \tfrac{1}{10}(1000) = 100.$$

Hence, by the principle of inclusion and exclusion,

$$N(a_1'a_2') = 1000 - 500 - 200 + 100 = 400,$$

which answers (b). Finally, we have

$$N(a_3) = \tfrac{1}{11}(1000) = 90.9.$$

Of course, since $N(a_3)$ is an integer, this means that

$$N(a_3) = 90.$$

In short,

$$N(a_3) = \lfloor 90.9 \rfloor = 90.$$

Also, every 22nd integer is divisible by 2 and by 11, so

$$N(a_1 a_3) = \left\lfloor \frac{1}{22}(1000) \right\rfloor = \lfloor 45.5 \rfloor = 45.$$

Similarly,

$$N(a_2 a_3) = \left\lfloor \frac{1}{55}(1000) \right\rfloor = \lfloor 18.2 \rfloor = 18.$$

Finally, every 110th integer is divisible by 2, 5, and 11, so

$$N(a_1 a_2 a_3) = \left\lfloor \frac{1}{110} (1000) \right\rfloor = \lfloor 9.1 \rfloor = 9.$$

Thus,

$$N(a_1' a_2' a_3') = 1000 - (500 + 200 + 90) + (100 + 45 + 18) - 9 = 364.$$

For more about prime numbers, see Exercises 29 and 30.

6.1.4 The Probabilistic Case

Suppose that an integer between 1 and 1000 is selected at random. What is the probability that it is not divisible by 2, 5, or 11? The answer is simple. Consider an experiment in which the outcome is one of the integers between 1 and 1000, and the outcomes are equally likely. The number of outcomes signalling the event "is not divisible by 2, 5, or 11" is 364, by the computation in the preceding section. Hence, the probability in question is $364/1000 = .364$.

More generally, suppose we consider an experiment that produces an outcome in a set S, the sample space. Let us consider the events $E_1, E_2, \ldots, E_r$. What is the probability that none of these events occur? To answer this, we shall assume, as we have throughout this book, that all outcomes in the sample space S are equally likely. (However, it can be shown that this assumption is not needed to obtain the main result of this subsection.) Let the set A be the set S, and let a_i be the property that an outcome signals event E_i. Let $P_{ijk\cdots}$ be the probability that events E_i and E_j and E_k and $\ldots$ occur. We conclude from Theorem 6.1 that the probability p that none of the events $E_1, E_2, \ldots, E_r$ occurs is given by

$$p = \frac{N(a_1' a_2' \cdots a_r')}{N(S)}$$

$$= 1 - \frac{\sum N(a_i)}{N(S)} + \frac{\sum N(a_i a_j)}{N(S)} - \frac{\sum N(a_i a_j a_k)}{N(S)} + \cdots + (-1)^r \frac{N(a_1 a_2 \cdots a_r)}{N(S)}.$$

Thus,

$$p = 1 - \sum p_i + \sum p_{ij} - \sum p_{ijk} + \cdots + (-1)^r p_{12 \cdots r}. \tag{6.6}$$

6.1.5 The Occupancy Problem with Distinguishable Balls and Cells

In Section 2.11, we considered the occupancy problem of placing n distinguishable balls into c distinguishable cells. Let us now ask: What is the probability that no cell will be empty? Let S be the set of distributions of balls to cells, and let E_i be the event that the ith cell is empty. Define A and a_i as above. Then $N(S) = c^n$, $N(a_i) = (c-1)^n$, $N(a_i a_j) = (c-2)^n$, $N(a_i a_j a_k) = (c-3)^n$, $\ldots$. Moreover, there are $\binom{c}{1}$ ways to choose property a_i, $\binom{c}{2}$ ways to choose properties a_i and a_j, and so on. Hence, the number of distributions of n

balls into c cells with no empty cell is given by

$$c^n - \binom{c}{1}(c-1)^n + \binom{c}{2}(c-2)^n - \binom{c}{3}(c-3)^n + \cdots + (-1)^c\binom{c}{c}(c-c)^n,$$

which equals

$$\sum_{t=0}^{c}(-1)^t\binom{c}{t}(c-t)^n. \tag{6.7}$$

Then the probability that no cell is empty is given by

$$\frac{c^n - \binom{c}{1}(c-1)^n + \binom{c}{2}(c-2)^n - \binom{c}{3}(c-3)^n + \cdots + (-1)^c\binom{c}{c}(c-c)^n}{c^n},$$

which equals

$$\sum_{t=0}^{c}(-1)^t\binom{c}{t}\left(1-\frac{t}{c}\right)^n, \tag{6.8}$$

since

$$\frac{(c-t)^n}{c^n} = \left(1-\frac{t}{c}\right)^n.$$

Example 6.4 Coupon Collecting

Suppose that a manufacturer gives away three different coupons in packages of breakfast cereal, one to a package. If we buy six packages of cereal, what is the probability of getting all three different coupons? We imagine placing $n = 6$ balls or coupons into $c = 3$ cells or types of coupons. The number of ways the coupons can be placed into types so that no cell (or type) is empty is given by Equation (6.7) as

$$3^6 - \binom{3}{1}\cdot 2^6 + \binom{3}{2}\cdot 1^6 - \binom{3}{3}\cdot 0^6 = 540.$$

The probability of this happening is $540/3^6 = .741$. This can also be computed directly by Equation (6.8) as

$$1 - \binom{3}{1}\cdot\left(\frac{2}{3}\right)^6 + \binom{3}{2}\cdot\left(\frac{1}{3}\right)^6 - \binom{3}{3}\cdot 0^6 = .741.$$

6.1.6 Chromatic Polynomials

In Section 3.4 we introduced the idea of a chromatic polynomial of a graph. The principle of inclusion and exclusion can be used to calculate chromatic polynomials. It is interesting to note how the same counting problem can be solved in more than one way. On several occasions in this chapter we shall be able to apply the principle of inclusion and exclusion to count a quantity that we have previously counted in a different way. The practical implication of this observation is that "there is more than one way to skin a cat." Consider, for example, the graph of Figure 6.3. We consider all possible colorings of the vertices of G in x or fewer colors. We shall even allow colorings where two vertices joined by an edge get the same color, but we shall call such colorings *improper*, and all others *proper*. Let us consider the set of all colorings, proper or improper, of the graph G in x or fewer colors. There are x^4 such colorings. We shall introduce one property a_i for each edge of the graph G. Thus,

a_1 is the property that a and b get the same color,
a_2 is the property that b and c get the same color,
a_3 is the property that c and d get the same color,
a_4 is the property that d and a get the same color.

To calculate $P(G, x)$, the number of (proper) colorings of G with x or fewer colors, we have to calculate $N(a_1' a_2' a_3' a_4')$. We have $N(a_1) = x^3$, since there are x choices for the color for a and b, then x choices for the color for c, and x for the color for d (recall that improper colorings are allowed). Similarly,

$$N(a_2) = N(a_3) = N(a_4) = x^3.$$

Next,

$$N(a_1 a_2) = x^2.$$

For a and b must receive the same color, and also b and c. Hence, there are x choices for the one color that a, b, and c receive, and then x choices for the color for d. Similar reasoning shows that

$$N(a_1 a_3) = N(a_1 a_4) = N(a_2 a_3) = N(a_2 a_4) = N(a_3 a_4) = x^2.$$

Similarly,

$$N(a_1 a_2 a_3) = N(a_1 a_2 a_4) = N(a_1 a_3 a_4) = N(a_2 a_3 a_4) = x,$$

since in all these cases all the vertices must receive the same color. This reasoning also leads us to conclude

$$N(a_1 a_2 a_3 a_4) = x.$$

Hence, by the principle of inclusion and exclusion,

$$P(G, x) = N(a_1' a_2' a_3' a_4')$$
$$= x^4 - 4x^3 + 6x^2 - 4x + x$$
$$= x^4 - 4x^3 + 6x^2 - 3x.$$

This computation can be checked by the methods of Chapter 3.

Figure 6.3   A graph.

Let us generalize this example as follows. Suppose that G is any graph and we wish to compute $P(G, x)$. Consider the set A of all colorings, proper or improper, of the vertices of G in x or fewer colors. For each edge i, let a_i be the property that the end vertices of edge i get the same color. Suppose that $|V(G)| = n$ and $|E(G)| = r$. Then $N = |A| = x^n$ and

$$P(G, x) = N(a_1' a_2' \cdots a_r').$$

Thus,

$$P(G, x) = x^n - \sum N(a_i) + \sum N(a_i a_j) \mp \cdots + (-1)^e \sum N(a_{i_1} a_{i_2} \cdots a_{i_e}) + \cdots \qquad (6.9)$$

Let us consider the term $N(a_{i_1} a_{i_2} \cdots a_{i_e})$. Suppose that H is the subgraph of G consisting of all the vertices of G and having edges $i_1, i_2, \ldots, i_e$. A subgraph H containing all the vertices of G was called a *spanning subgraph* in Chapter 3. Note that a coloring (proper or improper) of G satisfying properties $a_{i_1}, a_{i_2}, \ldots, a_{i_e}$ is equivalent to a coloring (proper or improper) of H satisfying properties $a_{i_1}, a_{i_2}, \ldots, a_{i_e}$. Now in such a coloring of H, any connected component of H must have all of its vertices colored the same. A color for a component of H can be chosen at random. Thus, the number of colorings of vertices of G in x or fewer colors and satisfying properties $a_{i_1}, a_{i_2}, \ldots, a_{i_e}$ is given by $x^{c(H)}$, where $c(H)$ is the number of connected components of H.

Each spanning subgraph H of e edges and c components corresponds to some set of properties $a_{i_1}, a_{i_2}, \ldots, a_{i_e}$ and will contribute a term $(-1)^e x^c$ to the right-hand side of (6.9). Thus, we have the following theorem.

Theorem 6.2* If G is a graph and $h(e, c)$ is the number of spanning subgraphs of e edges and c components, then

$$P(G, x) = \sum_{e, c} (-1)^e h(e, c) x^c.$$

In this theorem, note that $h(0, c)$ is 1 if $c = n$ and 0 otherwise. Note that the theorem gives a quick proof that $P(G, x)$ is a polynomial. In our example of Figure 6.3, we have the following results, which are illustrated in Figure 6.4:

$$h(4, 1) = 1,$$
$$h(3, 1) = 4,$$
$$h(2, 2) = 6,$$
$$h(1, 3) = 4,$$
$$h(0, 4) = 1,$$

and otherwise,

$$h(e, c) = 0.$$

*This theorem was discovered by Birkhoff [1912] (for graphs arising from maps) and first worked out by inclusion and exclusion by Whitney [1932].

Thus, we have

$$P(G, x) = (-1)^4 h(4, 1)x + (-1)^3 h(3, 1)x + (-1)^2 h(2, 2)x^2 + (-1)^1 h(1, 3)x^3$$
$$+ (-1)^0 h(0, 4)x^4$$
$$= x - 4x + 6x^2 - 4x^3 + x^4$$
$$= x^4 - 4x^3 + 6x^2 - 3x,$$

which agrees with our computation above.

6.1.7 Derangements

The reader will recall from Section 5.1.3 that a derangement is a permutation in which no object is put into its proper position. We shall show how to calculate the number of derangements D_n of a set of n objects by use of the principle of inclusion and exclusion. Consider the set A of all permutations of the n objects. Let a_i be the property that object i is placed in the ith position. Thus,

$$D_n = N(a_1' a_2' \cdots a_n').$$

We have

$$N = n!,$$

the number of permutations. Also,

$$N(a_i) = (n - 1)!,$$

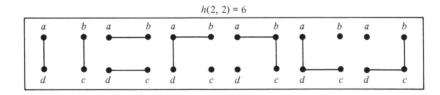

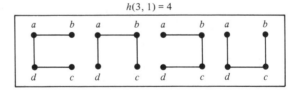

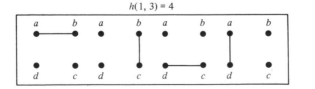

Figure 6.4 The spanning subgraphs of the graph of Figure 6.3.

for a permutation in which object i returns to its original position is equivalent to a permutation of the remaining objects. Similarly,

$$N(a_i a_j) = (n - 2)!$$

and

$$N(a_{i_1} a_{i_2} \cdots a_{i_t}) = (n - t)!$$

Hence,

$$\sum N(a_{i_1} a_{i_2} \cdots a_{i_t}) = \binom{n}{t}(n - t)!,$$

since there are $\binom{n}{t}$ choices for the properties $a_{i_1}, a_{i_2}, \ldots, a_{i_t}$. It follows by the principle of inclusion and exclusion that

$$D_n = N - \sum N(a_i) + \sum N(a_i a_j) - \sum N(a_i a_j a_k) + \cdots + (-1)^n N(a_1 a_2 \cdots a_n)$$

$$= n! - \binom{n}{1}(n - 1)! + \binom{n}{2}(n - 2)! - \binom{n}{3}(n - 3)! + \cdots + (-1)^n \binom{n}{n}(n - n)!$$

Simplifying, we have

$$D_n = n! - \frac{n!}{1!(n - 1)!}(n - 1)! + \frac{n!}{2!(n - 2)!}(n - 2)! - \frac{n!}{3!(n - 3)!}(n - 3)! + \cdots + (-1)^n$$

$$= n!\left[1 - \frac{1}{1!} + \frac{1}{2!} - \frac{1}{3!} + \cdots + (-1)^n \frac{1}{n!}\right],$$

as we have seen previously.

6.1.8 Counting Combinations

In Section 4.3 we studied a variety of counting problems which we solved by means of generating functions. Here we note that the principle of inclusion and exclusion can also be applied to such problems. We illustrate the method by means of an example. Suppose that we are doing a survey, and we have three teachers, four plumbers, and six autoworkers we are considering interviewing. Suppose that we consider two workers with the same job to be indistinguishable, and we seek the number of ways to choose 11 workers to interview. By the methods of Chapter 4, this can be computed by finding the coefficient of x^{11} in the generating function

$$(1 + x + x^2 + x^3)(1 + x + x^2 + x^3 + x^4)(1 + x + x^2 + x^3 + x^4 + x^5 + x^6). \quad (6.10)$$

However, we shall compute it a different way.

Consider the case where there are infinitely many teachers, plumbers, and autoworkers available, and consider the set A consisting of all the ways of choosing 11 workers to interview. For a particular element of the set A, say it satisfies property a_1 if it uses at least four teachers, property a_2 if it uses at least five plumbers, and property a_3 if it uses at least seven autoworkers. We seek to count all elements of the set A satisfying none of the properties a_i; thus we seek $N(a_1' a_2' a_3')$. To compute this, note that by Theorem 4.4,

$$N = |A| = \binom{3 + 11 - 1}{11} = \binom{13}{11} = 78.$$

What is $N(a_1)$? Note that a choice satisfies a_1 if and only if it has at least four teachers. Such a choice is equivalent to choosing seven (arbitrary) workers when there are infinitely many workers of each kind, so can be done, by Theorem 4.4, in

$$\binom{3 + 7 - 1}{7} = \binom{9}{7} = 36$$

ways. Thus,

$$N(a_1) = 36.$$

Similarly, a choice satisfying a_2 is equivalent to a choice of six workers when there are infinitely many of each kind, so

$$N(a_2) = \binom{3 + 6 - 1}{6} = \binom{8}{6} = 28.$$

Finally,

$$N(a_3) = \binom{3 + 4 - 1}{4} = \binom{6}{4} = 15.$$

Next, a choice satisfying both a_1 and a_2 has at least four teachers and at least five plumbers, and so is equivalent to a choice of two workers when each is in infinite supply. Thus,

$$N(a_1 a_2) = \binom{3 + 2 - 1}{2} = \binom{4}{2} = 6.$$

Similarly, a choice satisfying a_1 and a_3 is equivalent to a choice of no workers when each is in infinite supply; as there is exactly one way to choose no workers,

$$N(a_1 a_3) = 1.$$

Also, there is no way to choose 11 workers, at least 5 of whom are plumbers and at least 7 of whom are autoworkers, so

$$N(a_2 a_3) = 0.$$

Similarly,

$$N(a_1 a_2 a_3) = 0.$$

Thus, by the principle of inclusion and exclusion, the desired number of choices is

$$78 - (36 + 28 + 15) + (6 + 1 + 0) - 0 = 6.$$

It is easy to check this result by computing the coefficient of x^{11} in the generating function (6.10).

6.1.9 Rook Polynomials*

In Examples 4.10, 4.14, and 5.9, we studied the notion of rook polynomial of a board B consisting of acceptable (darkened) or unacceptable squares. If as in Figures 4.1 and 4.2, the board in B has a predominance of darkened squares, it is useful to consider the *complementary board* B' of B, the board obtained from B by interchanging acceptable and forbidden squares. Suppose that B is an $n \times m$ board, and we are interested in $r_n(B)$, the number of ways to place n nontaking rooks on acceptable squares of the board B. We shall show that we can obtain $r_n(B)$ from $R(x, B')$, the rook polynomial of the complementary board, rather than from $R(x, B)$. The former rook polynomial, based on a board with fewer darkened squares, will be easier to compute. We may assume that $n \leq m$. For if $n > m$, then $r_n(B) = 0$. Note that we shall not be computing $r_j(B)$ for $j < n$, only for the special case $j = n$. In Exercise 35 of Section 6.2 the reader will be asked to generalize the results to arbitrary $r_j(B), j \leq n$.

Let us say that an *assignment* of n nontaking rooks to an $n \times m$ board B means each rook is placed in a square, acceptable or not, with no two rooks in the same row and no two rooks in the same column. There are

$$P(m, n) = m(m - 1) \cdots (m - n + 1)$$

possible assignments. For we choose one of m positions in the first row, then one of $m - 1$ positions in the second row, ..., and finally one of $m - n + 1$ positions in the nth row. Let A be the set of all possible assignments for board B and let a_i be the property that an assignment has a rook in a forbidden square in the ith column. Then $r_n(B)$ is given by the number of assignments having none of the properties $a_1, a_2, \ldots, a_m$. We compute this number using the principle of inclusion and exclusion.

Given t, we shall see how to compute

$$\sum N(a_{i_1} a_{i_2} \cdots a_{i_t}),$$

where t must of course be at most m. Note that this sum is 0 if $t > n$ because there could be no assignment with rooks in t different columns, let alone in a forbidden square in t different columns. If $t \leq n$, then consider the complementary board B'. An assignment of n nontaking rooks to B with a rook in forbidden position in each of t columns of B corresponds to an assignment of t nontaking rooks to acceptable squares of the board B'—this can be done in $r_t(B')$ ways—and then an arbitrary placement of the remaining $n - t$ rooks in any of the remaining $m - t$ columns—this can be done in $P(m - t, n - t)$ ways. Thus, for $t \leq n$,

$$\sum N(a_{i_1} a_{i_2} \cdots a_{i_t}) = P(m - t, n - t) r_t(B').$$

By the principle of inclusion and exclusion, we conclude that

$$
\begin{aligned}
r_n(B) = P(m, n) - P(m - 1, n - 1) r_1(B') + P(m - 2, n - 2) r_2(B') \pm \cdots \\
+ (-1)^t P(m - t, n - t) r_t(B') \pm \cdots + (-1)^n P(m - n, 0) r_n(B').
\end{aligned}
\tag{6.11}
$$

Let us apply this theorem to the 5×5 board B of Figure 4.1 and compute $r_5(B)$. The

---

*This subsection can be omitted.

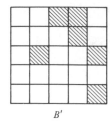

Figure 6.5 The complementary board B' of the board B of Figure 4.1.

B'

complementary board B' is shown in Figure 6.5. By the reduction results of Exercise 15, Section 4.1, and Example 4.14, one can show that

$$R(x, B') = 1 + 6x + 11x^2 + 6x^3 + x^4. \tag{6.12}$$

Using (6.12), noting that $P(a, a) = a!$, and applying (6.11), we have

$$r_5(B) = 5! - 4!(6) + 3!(11) - 2!(6) + 1!(1) - 0!(0) = 31.$$

EXERCISES FOR SECTION 6.1

1. Three newspapers, A, B, and C, are published in a city. The following results were obtained in a survey of the adult population of the city: 20 percent read A, 16 percent read B, 14 percent read C, 8 percent read both A and B, 5 percent read both A and C, 4 percent read both B and C, and 2 percent read all three. What percentage of the population of the city reads none of the papers?

2. In an experiment, there are two kinds of treatments, the controls and the noncontrols. There are 3 controls and 60 experimental units or blocks. Each control is used in 24 blocks, each pair of controls is used in the same block 12 times, and all three controls are used in the same block together 8 times. In how many blocks are none of the controls used? (We shall study similar conditions on experimental designs in detail in Chapter 9.)

3. A cigarette company surveys 100,000 people. Of these, 40,000 are males, according to the company's report. Also, 80,000 are smokers and 10,000 of those surveyed have cancer. However, of those surveyed, there are 1000 males with cancer, 2000 smokers with cancer, and 3000 male smokers. Finally, there are 100 male smokers with cancer. How many female nonsmokers without cancer are there? Is there something wrong with the cigarette company's report?

4. Thirty people were tested for immunity to the diseases tuberculosis, rubella, and smallpox. Of the 30 people, 13 were found to have immunity to tuberculosis, 17 to rubella, 13 to smallpox, 7 to tuberculosis and rubella, 8 to rubella and smallpox, 7 to tuberculosis and smallpox, and 1 to tuberculosis, rubella, but not smallpox. How many people were found to have immunity to none of the diseases?

5. Find an expression for the number of objects in a set A which have at least one of the properties $a_1, a_2, \ldots, a_r$.

6. One hundred water samples were tested for traces of three different types of chemicals, mercury, arsenic, and lead. Of the 100 samples, 7 were found to have mercury, 5 to have arsenic, 4 to have lead, 3 to have mercury and arsenic, 3 to have arsenic and lead, 2 to have mercury and lead, and 1 to have mercury, arsenic, but no lead. How many samples had a trace of at least one of the three chemicals?

7. Of 100 cars tested at an inspection station, 9 had defective headlights, 8 defective brakes, 7 defective horns, 2 defective windshield wipers, 4 defective headlights and brakes, 3 defective headlights and horns, 2 defective headlights and windshield wipers, 3 defective brakes and horns, none defective brakes and windshield wipers, 1 defective horn and windshield wipers, 1 had defective headlights, brakes and horn, 1 had defective headlights, horn, and windshield wipers, and none had any other combination of defects. Find the number of cars which had at least one of the defects in question.

8. A total of 77 students at a college were interviewed. Of these, 32 were taking a French course; 40 were taking a physics course; 30 a mathematics course; 23 a history course; 19 were taking French and physics; 13 French and mathematics; 15 physics and mathematics; 2 French and history; 15 physics and history; 14 mathematics and history; 8 French, physics, and mathematics; 8 French, physics, and history; 2 French, mathematics, and history; 6 physics, mathematics, and history; and 2 were taking courses in all four subjects. How many students were taking at least one course in the subjects in question?

9. An experimenter has 3 baboons and some gorillas. He has used each baboon in an experiment 12 times, each pair of baboons in the same experiment together 6 times, and all 3 baboons in the same experiment together 4 times. In 8 experiments, none of the baboons were used. How many experiments were performed altogether?

10. How many integers between 1 and 10,000 inclusive are divisible by none of 5, 7, and 11?

11. How many integers between 1 and 500 inclusive are divisible by none of 2, 3, and 5?

12. How many integers between 1 and 700 inclusive are divisible by none of 2, 3, 5, and 7?

13. Eight accidents occur during a week. Write an expression for computing the probability that there is at least one accident each day.

14. A total of five misprints occur on 4 pages of a book. What is the probability that each of these pages has at least one misprint?

15. Twenty light particles hit a section of the retina which has eight cells. What is the probability that at least one cell is not hit by a light particle?

16. Use the principle of inclusion and exclusion (not Theorem 6.2) to find the chromatic polynomial of each of the graphs of Figure 6.6.

17. Use Theorem 6.2 to find the chromatic polynomial of each of the graphs of Figure 6.6.

18. The *star* $S(1, n)$ is the graph consisting of one central vertex and n neighboring vertices, with no

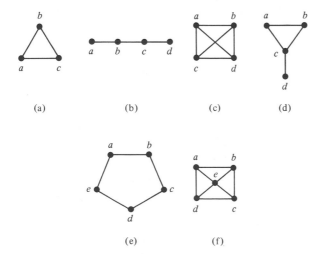

Figure 6.6 Graphs for exercises of Section 6.1.

other edges. Figure 6.7 shows some stars. Find the chromatic polynomial of $S(1, n)$ using the methods of Section 6.1.6.

19. Use the principle of inclusion and exclusion to count the number of ways to choose
    (a) 8 elements from a set of 3 a's, 3 b's, and 4 c's;
    (b) 9 elements from a set of 3 a's, 4 b's, and 5 c's;
    (c) 12 elements from a set of 5 a's, 5 b's, and 5 c's.

20. Verify Equation (6.12).

21. Use Equation (6.11) to compute $r_n(B)$ if B is the $n \times n$ board with all squares acceptable.

22. Use Equation (6.11) and earlier reduction results to compute $r_5(B)$ for board B of Figure 4.2.

23. Suppose that $r = 3$ and that an object has exactly two properties. How many times is the object counted in computing
    (a) $\sum N(a_i)$?   (b) $\sum N(a_i a_j)$?

24. Suppose that $r = 8$ and that an object has exactly three properties. How many times is the object counted in computing
    (a) $\sum N(a_i)$?   (b) $\sum N(a_i a_j)$?   (c) $\sum N(a_i a_j a_k)$?   (d) $\sum N(a_i a_j a_k a_l)$?

25. Suppose that d distinguishable computer decks are placed in n distinguishable input cells. More than one deck can go in a cell. We distinguish placing deck 5 in cell 1 from placing deck 5 in cell 2, and also distinguish placing deck 5 in cell 1 from placing deck 6 in cell 1, and so on. Suppose that the decks are placed so that no cells are empty. In how many ways can this be done?

26. Use inclusion and exclusion to find the number of solutions to the equation

$$x_1 + x_2 + x_3 = 16$$

in which each x_i is a nonnegative integer and $x_i \leq 7$.

27. Use inclusion and exclusion to find the number of solutions to the equation

$$x_1 + x_2 + x_3 + x_4 = 18$$

in which each x_i is a positive integer and $x_i \leq 9$.

28. Find the number of n-digit codewords from the alphabet $\{0, 1, 2, \ldots, 9\}$ in which the digits 1, 2, and 3 each appear at least once.

29. Exercises 29 and 30 are concerned with results of significance in number theory. For more on this subject, see such number theory books as Hardy and Wright [1960]. Two integers are *relatively prime* if the only divisor they have in common is the number 1. For instance, 14 and 15 are relatively prime, but 14 and 21 are not. Euler defined $\phi(n)$ to be the number of positive integers less than n that are relatively prime to n.
    (a) If the number n has as its distinct prime divisors the numbers $p_1, p_2, \ldots, p_k$, describe how to use the principle of inclusion and exclusion to compute $\phi(n)$.
    (b) Apply your method to compute $\phi(30)$ and check your answer by writing out the numbers less than 30 which are relatively prime to 30.
    (c) Repeat part (b) for $\phi(42)$.
    (d) Show that

$$\phi(n) = n - \frac{n}{p_1} - \frac{n}{p_2} - \cdots + \frac{n}{p_1 p_2} + \frac{n}{p_1 p_3} + \cdots + (-1)^k \frac{n}{p_1 p_2 \cdots p_k}.$$

$S(1, 2)$ $S(1, 3)$ $S(1, 4)$ **Figure 6.7**   Some stars.

(e) Conclude that

$$\phi(n) = n\left(1 - \frac{1}{p_1}\right)\left(1 - \frac{1}{p_2}\right) \cdots \left(1 - \frac{1}{p_k}\right).$$

(f) Use the formula in part (e) to check your result in part (b).
(g) Use the formula in part (e) to check your result in part (c).

30. Every positive integer n can be written in a unique way as the product of powers of primes,

$$n = p_1^{e_1} p_2^{e_2} \cdots p_r^{e_r},$$

where $p_1, p_2, \ldots, p_r$ are distinct primes and $e_i \geq 1$, all i. For example, $100 = 2^2 \cdot 5^2$, $12 = 2^2 \cdot 3^1$, and $30 = 2^1 \cdot 3^1 \cdot 5^1$. The *Moebius function* $\mu(n)$ is defined by

$$\mu(n) = \begin{cases} 1 & \text{if } n = 1 \\ 0 & \text{if } e_i > 1, \text{ any } i \\ (-1)^r & \text{if } e_1, e_2, \ldots, e_r \text{ all equal } 1. \end{cases}$$

Thus, $\mu(100) = 0$ since 2^2 is a factor of 100, and $\mu(30) = (-1)^3 = -1$.
(a) Show from the principle of inclusion and exclusion that

$$\sum_{d|n} \mu(d) = \begin{cases} 1 & \text{if } n = 1 \\ 0 & \text{if } n > 1, \end{cases} \tag{6.13}$$

where the sum in (6.13) is taken over all integers d that divide n. For example,

$$\sum_{d|12} \mu(d) = \mu(1) + \mu(2) + \mu(3) + \mu(4) + \mu(6) + \mu(12)$$

$$= 1 + (-1) + (-1) + 0 + (-1)^2 + 0$$

$$= 0.$$

(b) Suppose that f and g are functions such that

$$f(n) = \sum_{d|n} g(d).$$

Show from the result in part (a) that

$$g(n) = \sum_{d|n} f\left(\frac{n}{d}\right)\mu(d). \tag{6.14}$$

Equation (6.14) is called the *Moebius inversion formula*. For generalizations of this formula of significance in combinatorics, see Rota [1964] (see also Berge [1971], Hall [1967], and Liu [1972]).
(c) Show that if ϕ is the Euler function defined in Exercise 29, then

$$n = \sum_{d|n} \phi(d).$$

(d) Conclude that

$$\phi(n) = \sum_{d|n} \mu(d)\frac{n}{d}.$$

(e) Show that

$$\phi(p^c) = p^c\left(1 - \frac{1}{p}\right).$$

31. (Cohen [1978]) Each of n gentlemen checks both a hat and an umbrella. The hats are returned at random and then the umbrellas are independently returned at random. What is the probability that no man gets back both his hat and his umbrella?

32. Exercises 32 and 33 consider permutations with restrictions on certain patterns. To say that the *pattern uv does not appear* in a permutation $j_1 j_2 j_3 \cdots j_n$ of $\{1, 2, \ldots, n\}$ means that $j_i j_{i+1}$ is never uv. Similarly, to say that the *pattern uvw does not appear* means that $j_i j_{i+1} j_{i+2}$ is never uvw. Let b_n be the number of permutations of the set $\{1, 2, \ldots, n\}$ in which the patterns 12, 23, $\ldots, (n-1)n$ do not appear. Find b_n.

33. Find the number of permutations of $\{1, 2, 3, 4, 5, 6\}$ in which neither the pattern 124 nor the pattern 35 appears (see Exercise 32).

34. Find the number of ways the letters $a, a, b, b, c, c, d, d, d$ can be arranged so that two letters of the same kind never appear consecutively.

35. How many codewords of length 9 from the alphabet $\{0, 1, 2\}$ have three of each digit, but no three consecutive digits the same?

36. How many RNA chains have two A's, two U's, two C's, and two G's, and have no repeated base?

37. In our study of partitions of an integer (Exercises 11–15, Section 4.3), let $p^*(k)$ be the number of partitions of k with distinct integers and $p_0(k)$ be the number of partitions of k with odd integers.
(a) Define a set A and properties a_i and b_i for elements of A so that

$$p_0(k) = N(a_1' a_2' \cdots)$$

and

$$p^*(k) = N(b_1' b_2' \cdots).$$

(b) Show that A, a_i, and b_i in part (a) can be chosen so that

$$N(a_{i_1} a_{i_2} \cdots a_{i_k}) = N(b_{i_1} b_{i_2} \cdots b_{i_k})$$

for all k. Conclude that $p_0(k) = p^*(k)$. (This result was derived by generating functions in Exercise 14, Section 4.3.)

6.2 THE NUMBER OF OBJECTS HAVING EXACTLY m PROPERTIES

6.2.1 The Main Result and Its Applications

Let us return to the general situation of a set of N objects, each of which may or may not have each of r different properties, $a_1, a_2, \ldots, a_r$. There are situations where we want to know how many objects have exactly m of these properties. Let e_m be the number of objects having exactly m properties, $m \leq r$. To express a formula for e_m, suppose that for $t \geq 1$, we let

$$s_t = \sum N(a_{i_1} a_{i_2} \cdots a_{i_t}),$$

where the sum is taken over all choices of t distinct properties $a_{i_1}, a_{i_2}, \ldots, a_{i_t}$. Then we have the following theorem.

Theorem 6.3. The number of objects having exactly m properties if there are r properties and $m \leq r$ is given by

$$e_m = s_m - \binom{m+1}{1} s_{m+1} + \binom{m+2}{2} s_{m+2} - \binom{m+3}{3} s_{m+3} + \cdots$$

$$+ (-1)^p \binom{m+p}{p} s_{m+p} + \cdots + (-1)^{r-m} \binom{m+r-m}{r-m} s_r. \tag{6.15}$$

The reader should note that if s_0 is taken to be N, Theorem 6.3 yields the principle of inclusion and exclusion as a special case when $m = 0$.

We prove this theorem in Section 6.2.2. Here, let us apply it to several examples. In particular, let us return to Example 6.2. How many cars exceed the environmental standards on exactly one pollutant? We seek e_1. To compute e_1, we note that by the computation in the example,

$$s_1 = 6 + 4 + 3 = 13,$$

$$s_2 = 3 + 2 + 1 = 6,$$

$$s_3 = 1.$$

Thus, by Theorem 6.3,

$$e_1 = s_1 - \binom{2}{1} s_2 + \binom{3}{2} s_3$$

$$= 13 - 2(6) + 3(1)$$

$$= 4.$$

Example 6.5 The Hatcheck Problem

In Example 5.7, we considered a situation in which the hats of n gentlemen are returned at random. In this situation, let us compute the number of ways in which exactly one gentleman gets his hat back. Let us consider the set A of possible ways of returning hats to gentlemen if they are returned at random—these correspond to permutations—and let a_i be the property that the ith gentleman gets his own hat back. In Section 6.1.7, in dealing with derangements, we calculated

$$N(a_i) = (n-1)!, \quad \text{all } i,$$

$$N(a_i a_j) = (n-2)!, \quad \text{all } i \neq j,$$

and in general

$$N(a_{i_1} a_{i_2} \cdots a_{i_t}) = (n-t)!$$

Hence,

$$s_t = \binom{n}{t}(n-t)!,$$

since there are $\binom{n}{t}$ ways to pick the t properties $a_{i_1}, a_{i_2}, \ldots, a_{i_t}$. Then by Theorem 6.3, with $r = n$ and $m = 1$, we have

$$e_1 = s_1 - \binom{2}{1}s_2 + \binom{3}{2}s_3 - \binom{4}{3}s_4 \pm \cdots + (-1)^{n-1}\binom{n}{n-1}s_n$$

$$= \binom{n}{1}(n-1)! - \binom{2}{1}\binom{n}{2}(n-2)! + \binom{3}{2}\binom{n}{3}(n-3)! - \binom{4}{3}\binom{n}{4}(n-4)! \pm \cdots$$

$$+ (-1)^{n-1}\binom{n}{n-1}\binom{n}{n}(n-n)!$$

$$= \frac{n!}{1!(n-1)!}(n-1)! - \frac{2!}{1!1!}\frac{n!}{2!(n-2)!}(n-2)! + \frac{3!}{2!1!}\frac{n!}{3!(n-3)!}(n-3)!$$

$$- \frac{4!}{3!1!}\frac{n!}{4!(n-4)!}(n-4)! \pm \cdots + (-1)^{n-1}\frac{n!}{(n-1)!1!}$$

$$= n! - \frac{n!}{1!} + \frac{n!}{2!} - \frac{n!}{3!} \pm \cdots + (-1)^{n-1}\frac{n!}{(n-1)!}$$

$$= n!\left[1 - \frac{1}{1!} + \frac{1}{2!} - \frac{1}{3!} \pm \cdots + (-1)^{n-1}\frac{1}{(n-1)!} \right]$$

$$= nD_{n-1}.$$

This result is clear, for we pick one gentleman to get his hat back—this can be done in n ways—and then choose a derangement of the rest of the gentlemen—this can be done in D_{n-1} ways.

We conclude that the probability that exactly one gentleman gets his hat back is

$$\frac{nD_{n-1}}{n!} = \frac{D_{n-1}}{(n-1)!},$$

which approaches $1/e$ as n approaches infinity, since $D_n/n!$ does. Thus, in the long run, the probability that exactly one gentleman gets his hat back is the same as the probability that no gentlemen get their hats back.

Example 6.6 Testing Guessing Abilities

In some psychic experiments, we present a sequence of n elements in an order unknown to a person who claims to have psychic powers. That person predicts the order in advance. We count the number of correct elements, that is, the number of elements whose place in the sequence is predicted exactly right. Suppose that in a sequence of 10 elements, a person gets five right. Would we take this as evidence of psychic powers? To answer the question, we ask whether the observed number of successes is very unlikely if the person is only guessing. In particular, we ask what is the probability of guessing at least five elements correctly. (We are really interested in how likely it is the person did at least as well as he did.) The number of ways to guess exactly m elements correctly in a sequence of n elements can be computed from Theorem 6.3. We let A be the set of all permutations of the set $\{1, 2, \ldots, n\}$ and a_i be the property that i is in the ith position. Then $N(a_i)$, $N(a_i a_j)$, and so on, are exactly as in our analysis of the hatcheck problem, and so is s_t for every t. Thus, one can show that the probability of guessing exactly m positions correctly is given by

$$P_m^n = \frac{1}{m!}\left[1 - \frac{1}{1!} + \frac{1}{2!} - \frac{1}{3!} \pm \cdots + (-1)^{(n-m)}\,\frac{1}{(n-m)!}\right]. \tag{6.16}$$

The detailed verification of (6.16) is left as an exercise (Exercise 25). The probability of guessing at least five positions right out of a sequence of 10 is given by

$$P_5^{10} + P_6^{10} + P_7^{10} + P_8^{10} + P_9^{10} + P_{10}^{10}$$

$$= .00306 + .00052 + .00007 + .00001 + .00000 + .00000 = .00366.$$

(Notice that the next-to-last .00000 here is in fact exactly 0, while the last is actually $1/10!$.) We conclude that the probability of achieving this much success by guessing is *very* small. We would have to conclude that the person *does* seem to have psychic powers. For further references on tests of psychic powers and other applications of the notion of derangement of interest to psychologists, see Barton [1958] and Vernon [1936].

Example 6.7 RNA Chains

Let us find the number of RNA chains of length n with exactly two U's. We can calculate this number directly. For to get an RNA chain of length n with exactly two U's, we choose two positions out of n for the U's and then have three choices of base for each of the remaining $n - 2$ positions. This gives us

$$\binom{n}{2}3^{n-2}$$

chains. It is interesting to see how we can obtain this number from Theorem 6.3. Let A be the set of all n-digit sequences from the alphabet $\{U, A, C, G\}$, and let a_i be the property that there is a U in the ith position. Then we seek e_2. Note that

$$N(a_{i_1}a_{i_2} \cdots a_{i_t}) = 4^{n-t},$$

for we have four choices for the ith element in the chain if $i \neq i_1, i_2, \ldots, i_t$. Hence,

$$s_t = \binom{n}{t}4^{n-t},$$

since there are t properties to choose from n properties. We conclude by Theorem 6.3 that

$$e_2 = s_2 - \binom{3}{1}s_3 + \binom{4}{2}s_4 \mp \cdots + (-1)^p\binom{p+2}{p}s_{p+2} + \cdots + (-1)^{n-2}\binom{n}{n-2}s_n$$

$$= \binom{n}{2}4^{n-2} - \binom{3}{1}\binom{n}{3}4^{n-3} + \binom{4}{2}\binom{n}{4}4^{n-4} \mp \cdots$$

$$+ (-1)^p\binom{p+2}{p}\binom{n}{p+2}4^{n-p-2} + \cdots + (-1)^{n-2}\binom{n}{n-2}\binom{n}{n}4^{n-n}.$$

To evaluate this expression for e_2, note that

$$(-1)^p \binom{p+2}{p} \binom{n}{p+2} 4^{n-p-2} = (-1)^p \frac{(p+2)!}{p!2!} \frac{n!}{(p+2)!(n-p-2)!} 4^{n-p-2}$$

$$= (-1)^p \frac{n(n-1)}{2} \frac{(n-2)!}{p!(n-p-2)!} 4^{n-p-2}$$

$$= \binom{n}{2} (-1)^p \binom{n-2}{p} 4^{n-p-2}.$$

Thus

$$e_2 = \binom{n}{2} \sum_{p=0}^{n-2} \binom{n-2}{p} (-1)^p 4^{n-p-2}.$$

By the binomial expansion (Theorem 2.7),

$$e_2 = \binom{n}{2} (4-1)^{n-2} = \binom{n}{2} 3^{n-2}.$$

This result agrees with our initial computation. In this case, use of Theorem 6.3 was considerably more difficult!

Example 6.8 Legitimate Codewords

In Example 5.4 we defined a codeword from the alphabet $\{0, 1, 2, 3\}$ as *legitimate* if it had an even number of 0's and we let a_k be the number of legitimate codewords of length k. In Section 5.3.1 we used generating functions to show that

$$a_k = \tfrac{1}{2}(2)^k + \tfrac{1}{2}(4)^k.$$

Here, we shall derive the same result using Theorem 6.3. Let A be the set of all sequences of length k from $\{0, 1, 2, 3\}$ and let a_i be the property that the ith digit is 0, $i = 1, 2, \ldots, k = r$. We seek the number of elements of A having an even number of these properties, that is, we seek $e_0 + e_2 + e_4 + \cdots$. To compute this sum, note that

$$S_t = \binom{k}{t} 4^{k-t}$$

for

$$N(a_{i_1} a_{i_2} \cdots a_{i_t}) = 4^{k-t}.$$

From this and Theorem 6.3, one can show by algebraic manipulation that

$$e_0 + e_2 + e_4 + \cdots = \tfrac{1}{2}(2)^k + \tfrac{1}{2}(4)^k. \tag{6.17}$$

An easier way to show (6.17) is to use the following theorem, whose proof comes in Section 6.2.2.

Theorem 6.4. If there are r properties, then the number of objects having an even number of the properties is given by

$$e_0 + e_2 + e_4 + \cdots = \frac{1}{2}\left[s_0 + \sum_{t=0}^{r}(-2)^t s_t\right]$$

and the number of objects having an odd number of the properties is given by

$$e_1 + e_3 + e_5 + \cdots = \frac{1}{2}\left[s_0 - \sum_{t=0}^{r}(-2)^t s_t\right].$$

Applying Theorem 6.4 to Example 6.8, and recalling that s_0 is taken to be N, we find using the binomial expansion (Theorem 2.7) that

$$e_0 + e_2 + e_4 + \cdots = \frac{1}{2}\left[4^k + \sum_{t=0}^{k}(-2)^t \binom{k}{t}4^{k-t}\right]$$
$$= \tfrac{1}{2}[4^k + (-2 + 4)^k]$$
$$= \tfrac{1}{2}[4^k + 2^k],$$

which agrees with (6.17).

Example 6.9 Cosmic Rays and Occupancy Problems

Suppose that we have a Geiger counter with c cells which is exposed to a shower of cosmic rays, getting hit by n rays. What is the probability that exactly q counters will go off? To answer this question, we can follow the analysis in Section 6.1.4, where we introduced a sample space S and events E_i. Here, S consists of all distributions of n rays to c cells, and E_i is the event that counter i is not hit. We want the probability that exactly $m = c - q$ counters are not hit, that is, that exactly m of the events in question occur. We can introduce a set A and properties a_i exactly as in Section 6.1.4, and observe that among events $E_1, E_2, \ldots, E_r$, the probability that exactly m of them will occur can be computed from Theorem 6.3, by using $e_m/N(S)$. In our example, we can compute e_m by thinking of this as an occupancy problem and using the computations for $N(a_i)$, $N(a_i a_j)$, and so on, from Section 6.1.5. Then we find that

$$s_t = \sum N(a_{i_1}a_{i_2} \cdots a_{i_t}) = \binom{c}{t}(c - t)^n.$$

Thus, one can show from Theorem 6.3 that the probability that exactly m of the events $E_1, E_2, \ldots, E_r$ will occur is given by

$$\binom{c}{m}\sum_{p=0}^{c-m}(-1)^p\binom{c-m}{p}\left(1 - \frac{m+p}{c}\right)^n. \tag{6.18}$$

A detailed verification is left to the reader (Exercise 27). The result in (6.18) can also be derived directly from the case $m = 0$ (see Exercise 28).

For a variety of other applications of Theorem 6.3, see Feller [1968], Irwin [1955], or Parzen [1960].

6.2.2 Proofs of Theorems 6.3 and 6.4*

We close this section by presenting proofs of Theorems 6.3 and 6.4.

Proof of Theorem 6.3. The proof is similar to the proof of Theorem 6.1. As a preliminary, we note that

$$\binom{m+j}{m+p}\binom{m+p}{p} = \frac{(m+j)!}{(m+p)!(j-p)!} \frac{(m+p)!}{p!m!}$$

$$= \frac{(m+j)!}{m!p!(j-p)!}$$

$$= \frac{(m+j)!}{m!j!} \frac{j!}{p!(j-p)!}$$

$$= \binom{m+j}{m}\binom{j}{p}.$$

Thus,

$$\binom{m+j}{m+p}\binom{m+p}{p} = \binom{m+j}{m}\binom{j}{p}. \tag{6.19}$$

Let us now consider Equation (6.15). If an object has fewer than m properties a_i, then it is not counted in calculating e_m and it is not counted in any of the terms in the right-hand side of (6.15). Suppose that an object has exactly m of the properties. It is counted exactly once in calculating e_m, and counted exactly once in calculating the right-hand side of (6.15), namely in calculating s_m. Finally, suppose that an object has more than m properties, say $m + j$ properties. It is not counted in calculating e_m. We shall argue that the number of times it is counted in the right-hand side of (6.15) is 0. The object is counted $\binom{m+j}{m}$ times in calculating s_m: It is counted once for every m properties we can choose out of the $m + j$ properties the object has. It is counted $\binom{m+j}{m+1}$ times in calculating s_{m+1}. In general, it is counted $\binom{m+j}{m+p}$ times in calculating s_{m+p}, for $p \le j$. It is not counted otherwise. Hence, the total number of times the object is counted in the right-hand side of (6.15) is calculated by multiplying $\binom{m+j}{m+p}$ by $(-1)^p\binom{m+p}{p}$, the coefficient of s_{m+p}, and adding these terms for $p = 0$ up to j. We obtain

$$\binom{m+j}{m} - \binom{m+j}{m+1}\binom{m+1}{1} + \binom{m+j}{m+2}\binom{m+2}{2} - \cdots$$

$$+ (-1)^p\binom{m+j}{m+p}\binom{m+p}{p} + \cdots + (-1)^j\binom{m+j}{m+j}\binom{m+j}{j}. \tag{6.20}$$

Now by (6.19), Equation (6.20) becomes

$$\binom{m+j}{m} - \binom{m+j}{m}\binom{j}{1} + \binom{m+j}{m}\binom{j}{2} - \cdots + (-1)^j\binom{m+j}{m}\binom{j}{j},$$

*This subsection can be omitted.

which equals

$$\binom{m+j}{m}\left[\binom{j}{0}-\binom{j}{1}+\binom{j}{2}-\cdots+(-1)^{j}\binom{j}{j}\right]. \tag{6.21}$$

By Theorem 2.9, the bracketed material in (6.21) equals 0 [it arises by expanding $(1-1)^{j}$ using the binomial expansion], so (6.21) is 0. This completes the proof of Theorem 6.3.

Q.E.D.

Proof of Theorem 6.4. Let $E(x) = \sum e_m x^m$ be the ordinary generating function for the sequence $e_0, e_1, e_2, \ldots$. By Theorem 6.3,

$$E(x) = [s_0 - s_1 + s_2 - \cdots + (-1)^r s_r]$$

$$+\left[s_1 - \binom{2}{1}s_2 + \binom{3}{2}s_3 - \cdots + (-1)^{r-1}\binom{r}{r-1}s_r\right]x$$

$$+\left[s_2 - \binom{3}{1}s_3 + \binom{4}{2}s_4 - \cdots + (-1)^{r-2}\binom{r}{r-2}s_r\right]x^2$$

$$+\cdots$$

$$+\left[s_m - \binom{m+1}{1}s_{m+1} + \binom{m+2}{2}s_{m+2} - \cdots + (-1)^{r-m}\binom{m+r-m}{r-m}s_r\right]x^m$$

$$+\cdots$$

$$+ s_r x^r$$

$$= s_0$$

$$+ s_1[x-1]$$

$$+ s_2\left[x^2 - \binom{2}{1}x + 1\right]$$

$$+ s_3\left[x^3 - \binom{3}{1}x^2 + \binom{3}{2}x - 1\right]$$

$$+\cdots$$

$$+ s_m\left[x^m - \binom{m}{1}x^{m-1} + \binom{m}{2}x^{m-2} + \cdots + (-1)^{m-1}\binom{m}{m-1}x + (-1)^m\right]$$

$$+\cdots$$

$$+ s_r\left[x^r - \binom{r}{1}x^{r-1} + \binom{r}{2}x^{r-2} + \cdots + (-1)^{r-1}\binom{r}{r-1}x + (-1)^r\right].$$

Thus,

$$E(x) = \sum_{m=0}^{r} s_m(x-1)^m. \tag{6.22}$$

The first part of the theorem follows by noting that

$$e_0 + e_2 + e_4 + \cdots = \tfrac{1}{2}[E(1) + E(-1)]$$

and taking $x = 1$ and $x = -1$ in (6.22). The second part of the theorem follows by noting that

$$e_1 + e_3 + e_5 + \cdots = \tfrac{1}{2}[E(1) - E(-1)].$$ Q.E.D.

EXERCISES FOR SECTION 6.2

1. In Exercise 1, Section 6.1, what percentage of the population reads exactly one of the papers?
2. In Exercise 2, Section 6.1, how many blocks use exactly two controls?
3. In Exercise 4, Section 6.1, how many people were immune to exactly one of the diseases?
4. In Exercise 7, Section 6.1, find the number of cars having exactly two of the defects in question.
5. In Exercise 8, Section 6.1, find the number of students taking exactly three of the subjects in question.
6. A variant of Montmort's "problème des rencontres" discussed in Section 5.1.3 is the following. A deck of n cards is laid out in a row on the table. Cards of a second deck with n cards are placed one by one at random on top of the first set of cards. You get m points if there are m matches between the first and second deck.
   (a) How many ways are there to get 2 points if $n = 4$?
   (b) What is the probability of getting 6 points if $n = 8$?
7. The names on the files of 9 different job candidates appearing for an interview were unfortunately lost and a new receptionist placed the names on the files at random. In how many ways could this be done so that exactly 3 candidates' files were properly labeled?
8. In the hatcheck problem, use our formula for e_1 to determine the probability that exactly one gentleman gets his hat back if there are 3 gentlemen.
9. In the hatcheck problem, if there are 4 gentlemen, compute the number of ways that exactly 2 of them will get their hats back.
10. Compute e_m for the hatcheck problem for arbitrary m.
11. (a) If six fair coins are tossed, use Theorem 6.3 to compute the probability that there will be exactly 3 heads.
    (b) Check your answer by computing it directly.
12. Use Theorem 6.3 to compute the number of ways to get exactly m heads if a coin is tossed n times.
13. (a) Use Theorem 6.3 to find the number of permutations of $\{1, 2, 3, 4, 5, 6, 7\}$ in which exactly 3 integers are in their natural positions.
    (b) Check your answer by computing it directly.
14. (a) Use Theorem 6.3 to compute the number of legitimate codewords of length 6 from the alphabet $\{0, 1, 2\}$ if a codeword is legitimate if and only if it has exactly three 0's.
    (b) Check your answer by an alternative computation.
15. (a) Use Theorem 6.3 to compute the number of legitimate codewords of length n from the alphabet $\{0, 1, 2\}$ if a codeword is legitimate if and only if it has exactly four 1's.
    (b) Check your answer by an alternative computation.

16. **(a)** Suppose that n children are born to a family. Use Theorem 6.3 to compute the number of ways the family can have exactly 2 boys.

    (b) Check your answer by the methods of Chapter 2.

17. A psychic predicts a sequence of 4 elements, getting 2 right. What is the probability of getting at least this many right?

18. In a wine-tasting experiment, a taster is told there will be 5 different wines given to him. After each, he guesses which of the 5 it was, making sure never to repeat a guess. He gets 3 right. What is his probability of getting at least 3 right if he is guessing randomly?

19. Write an expression for the probability that in a sequence of 8 random digits chosen from 0, 1, 2, ..., 9, exactly 2 of the digits will not appear.

20. In a genetics experiment, each mouse in a litter of n mice is classified as belonging to one of M genotypes. What is the probability that exactly g genotypes will be represented among the n mice?

21. Use Theorem 6.4 to find the number of families of 8 children which have an even number of boys. Check your answer by direct computation.

22. Use Theorem 6.4 to find the number of 8-digit sequences from the alphabet $\{0, 1, 2\}$ which have an odd number of 1's. Check your answer by direct computation.

23. Find the number of 10-digit RNA chains which have no U's and an even number of G's.

24. Give an alternative proof of Theorem 6.3 by using mathematical induction.

25. Use Theorem 6.3 to verify Equation (6.16).

26. In Exercise 25, Section 6.1, show that the number of ways to place the decks so that exactly m cells are empty is given by

$$\binom{n}{m}\sum_{i=0}^{n-m}(-1)^i\binom{n-m}{i}(n-m-i)^d.$$

27. Use Theorem 6.3 to verify (6.18).

28. Suppose that $P_m(c, n)$ is the probability that exactly m cells will be empty if n distinguishable balls are distributed into c distinguishable cells.

    (a) Show that

$$P_m(c, n) = \binom{c}{m}\left(1 - \frac{m}{c}\right)^n P_0(c - m, n).$$

    (b) Derive (6.18) from the equation for $P_0(c - m, n)$.

29. Let e_m^* be the number of elements of the set A having at least m of the properties $a_1, a_2, \ldots, a_r$. Show that

$$e_m^* = S_m - \binom{m}{m-1}S_{m+1} + \binom{m+1}{m-1}S_{m+2} + \cdots + (-1)^p\binom{m+p-1}{m-1}S_{m+p} + \cdots$$

$$+ (-1)^{r-m}\binom{m+r-m-1}{m-1}S_r.$$

30. Suppose that $E_1, E_2, \ldots, E_r$ are events, that $p_{i_1 i_2 \cdots i_t}$ is the probability that events $E_{i_1}, E_{i_2}, \ldots, E_{i_t}$ all occur, and that $S_t = \sum p_{i_1 i_2 \cdots i_t}$, where the sum in question is taken over all t-element subsets $\{i_1, i_2, \ldots, i_t\}$ of $\{1\ 2, \ldots, r\}$. In terms of the S_t, derive expressions for

    (a) The probability that exactly m of the events occur;

    (b) The probability that at least m of the events occur.

31. In Exercise 7, Section 6.1, how many cars have at least 2 of the defects in question?

32. In Exercise 8, Section 6.1, how many students are taking at least 1 of the subjects in question?

33. Compute the number of RNA chains of length 8 with at least 2 U's.

34. Compute the number of permutations of $\{1, 2, 3, 4, 5\}$ in which at least three integers are in their natural position.

35. Suppose that B' is the complement of the $n \times m$ board B, $n \leq m$. If $j \leq n$, find a formula for $r_j(B)$ in terms of the numbers $r_k(B')$ which generalizes the result of Equation (6.11).

36. Use the result of Exercise 35 to show that

$$R(x, B) = x^n R\left(\frac{1}{x}, B'\right).$$

37. (a) If $E(x)$ is the ordinary generating function for the sequence $e_0, e_1, e_2, \ldots$ and the e_i are defined as in Example 6.8, what is $E(1)$?
(b) Find a formula for $E(1)$ that holds in general.

REFERENCES FOR CHAPTER 6

BARTON, D. E., "The Matching Distributions: Poisson Limiting Forms and Derived Methods of Approximation," *J. Roy. Statist. Soc.*, 20 (1958), 73–92.

BERGE, C., *Principles of Combinatorics*, Academic Press, New York, 1971.

BIRKHOFF, G. D., "A Determinant Formula for the Number of Ways of Coloring a Map," *Ann. Math.*, *14* (1912), 42–46.

COHEN, D. I. A., *Basic Techniques of Combinatorial Theory*, Wiley, New York, 1978.

FELLER, W., *An Introduction to Probability Theory and Its Applications*, 3d ed., Wiley, New York, 1968.

HALL, M., *Combinatorial Theory*, Blaisdell, Waltham, Mass., 1967.

HARDY, G. H., and WRIGHT, E. M., *An Introduction to the Theory of Numbers*, 4th ed., Oxford University Press, London, 1960.

IRWIN, J. O., "A Unified Derivation of Some Well-Known Frequency Distributions of Interest in Biometry and Statistics," *J. Roy. Statist. Soc., Ser. A*, *118* (1955), 389–404.

KEMENY, J. G., SNELL, J. L., and THOMPSON, G. L., *Introduction to Finite Mathematics*, Prentice-Hall, Englewood Cliffs, N.J., 1974.

LIU, C. L., *Topics in Combinatorial Mathematics*, Mathematical Association of America, Washington, D.C., 1972.

PARZEN, E., *Modern Probability Theory and Its Applications*, Wiley, New York, 1960.

ROTA, G.-C., "On the Foundations of Combinatorial Theory I. Theory of Möbius Functions," *Z. Wahrscheinlichkeitsth. verw. Geb.*, *2* (1964), 340–368.

VERNON, P. E., "The Matching Method Applied to Investigations of Personality," *Psychol. Bull.*, *33* (1936), 149–177.

VILENKIN, N. YA., *Combinatorics*, Academic Press, New York, 1971. (Translated from the Russian by A. Shenitzer and S. Shenitzer.)

WHITNEY, H., "A Logical Expansion in Mathematics," *Bull. Amer. Math. Soc.*, *38* (1932), 572–579.

7 The Polya Theory of Counting*

7.1 EQUIVALENCE RELATIONS

7.1.1 Definition and Examples

A major difficulty in counting the number of distinct configurations of a particular kind is to define clearly when two such configurations are the same. In this chapter we develop techniques for counting the number of distinct configurations of a certain kind. These techniques of course make heavy use of the ideas involved in determining whether or not two configurations are the same. Hence, we begin the chapter by studying what it means to say that two things are the same.

Suppose that V is a set and S is a set of ordered pairs of elements of V. Then S is called a (binary) *relation* on V. For instance, if $V = \{1, 2, 3\}$ and $S = \{(1, 2), (2, 3)\}$, then S is a relation on V. We write aSb if the ordered pair (a, b) is in S. Thus, in our example, $1S2$, but not $2S1$ and not $1S3$.

Suppose that V is a set of configurations and for a, b in V, we write aSb to mean that a and b are the same. Then the relation S should have the following properties:

1. *Reflexivity.*   For all a in V, aSa. (A configuration is the same as itself.)
2. *Symmetry.*   For all a, b in V, if aSb, then bSa. (If a is the same as b, then b is the same as a.)
3. *Transitivity.*   For all a, b, c in V, if aSb and bSc, then aSc. (If a is the same as b and b is the same as c, then a is the same as c.)

If S satisfies these three properties, it is called an *equivalence relation*.

*This chapter should be omitted in an elementary course.

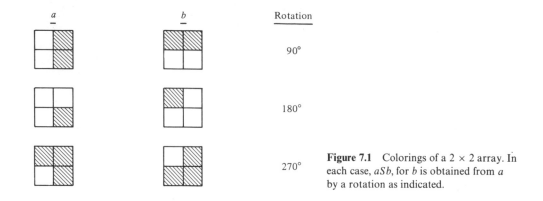

Figure 7.1 Colorings of a 2 × 2 array. In each case, aSb, for b is obtained from a by a rotation as indicated.

We now give several other examples of equivalence relations. Let V be the set of people in New Jersey, and let aSb mean that a and b have the same height. Then S defines an equivalence relation. Let V be the set of all people in the United States, and let aSb mean that a and b have the same birthdate. Then S is an equivalence relation. Let V be the set of all people in the United States and let aSb mean that a is the father of b. Then S does not define an equivalence relation: It is not reflexive, not symmetric, and not transitive. Let V be the set of real numbers and let aSb mean that $a \leq b$. Then S is not an equivalence relation: It is reflexive and transitive, but not symmetric.

Let us give some more complicated examples, which will illustrate the basic problems we shall discuss in this chapter.

Example 7.1 Coloring a 2 × 2 Array

Let us consider a 2 × 2 array in which each block is occupied or not. We color the block black if it is occupied and color it white or leave it uncolored otherwise. Figure 7.1 shows several such colorings. Let V be the collection of all such colorings. Let us suppose that we allow rotation* of the array by 0°, 90°, 180°, or 270°. We consider colored arrays a and b the same, and write aSb, if b can be obtained from a by one of the rotations in question. Then S defines an equivalence relation. To illustrate, note that Figure 7.1 shows some pairs of arrays that are considered in the relation S. To see why S is an equivalence relation, note that it is reflexive because a can be obtained from a by a 0° rotation. It is symmetric because if b can be obtained from a by a rotation, then a can be obtained from b by a rotation. Finally, it is transitive because if b can be obtained from a by a rotation and c from b by a rotation, then c can be obtained from a by following the first rotation by the second. (It is assumed that a rotation by $360 + x$ degrees is equivalent to a rotation by x degrees.)

Example 7.2 Necklaces

Suppose that an open necklace consists of a string of k beads, each being either blue or red. Thus, a typical necklace of three beads can be represented by a string like *bbr* or *brb*. A necklace is not considered to have a designated front end, so two such

---

*All rotations in this chapter are counterclockwise unless noted otherwise.

Table 7.1 Two Switching Functions T and U

Bit string x	$T(x)$	$U(x)$
00	1	1
01	0	1
10	1	0
11	1	1

necklaces x and y are considered the same, and we write xSy, if x equals y or if y can be obtained from x by reversing. Thus, *bbr* is the same as *rbb*. S defines an equivalence relation. The verification is left to the reader (Exercise 4).

Example 7.3   Switching Functions

Recall from Example 2.4 that a switching function of n variables is a function that assigns to every bit string of length n a number 0 or 1. These functions arise in computer engineering. Recall from our discussion in Example 2.4 that certain switching functions are considered equivalent or the same. To make this precise, suppose that T and U are the two switching functions defined by Table 7.1. It is easy to see that $T(x_1 x_2) = U(x_2 x_1)$ for all bit strings $x_1 x_2$. Thus, T can be obtained from U by simply reordering the input, interchanging the two positions. In this sense, T and U can be considered equivalent. Indeed, for all practical purposes they are. For suppose that we can design an electronic circuit which computes U. Then we can design one to compute T, as shown in Figure 7.2, where the circuit computing U is shown as a black box. In general, we consider two switching functions T and U of two variables the same, and write TSU, if either $T = U$ or $T(x_1 x_2) = U(x_2 x_1)$ for all bit strings $x_1 x_2$. Then S is an equivalence relation. We leave the proof to the reader (Exercise 5). In what follows, we will generalize this concept of equivalence to switching functions of more than two variables. In Section 2.1, we noted that there were many switching functions, even of four variables. Hence, it is impractical to compile a manual listing, for each switching function of n variables, the best corresponding electronic circuit. However, it is not necessary to include every switching function in such a manual, but only enough switching functions so that every switching function of n variables is equivalent to one of the included ones. Counting the number of switching functions required was an historically important problem in computer science (see Section 2.1) and we shall show how to make this computation.

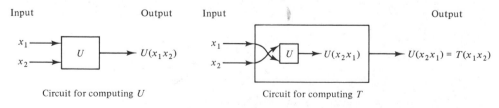

Circuit for computing U Circuit for computing T

Figure 7.2   A circuit for computing T can be obtained from a circuit for computing U.

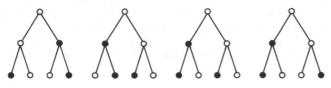

Figure 7.3 Four equivalent colorings of the binary tree of seven vertices.

Example 7.4   Coloring Trees*

Let T be a fixed tree, for instance the binary tree of seven vertices shown in Figure 7.3. Color each vertex of T black or white, and do not distinguish left from right. Let V be the collection of all colorings of T. Let aSb mean that a and b are considered the same, that is, if b can be obtained from a by interchanging left and right subtrees. We shall define this more precisely below. However, since we do not distinguish left from right, it should be clear that all colorings in Figure 7.3 are considered the same. It follows from our general results below that S defines an equivalence relation.

Example 7.5   Organic Molecules†

One of the historically important motivations for the theory developed in this chapter was the desire to count distinct organic molecules in chemistry. Consider the set V of molecules of the form shown in Figure 7.4, where C is a carbon atom and X can be either CH_3 (methyl), C_2H_5 (ethyl), H (hydrogen), or Cl (chlorine). A typical such molecule is CH_2Cl_2, which has two hydrogen atoms and two chlorine atoms. We can model such a molecule using a regular tetrahedron, a figure consisting of four equilateral triangles that meet at six edges and four corners, as in Figure 7.5. The carbon atom is thought of as being at the center of this tetrahedron and the four components labeled X are at the corners labeled a, b, c, and d. Two such molecules x and y are considered the same, and we write xSy, if y can be obtained from x by one of the following 12 symmetries of the tetrahedron: no change; a rotation by 120° or 240° around a line connecting a vertex and the center of its opposite face (there are eight of these rotations); or a rotation by 180° around a line connecting the midpoints of opposite edges (there are three of these rotations). Figures 7.6 and 7.7 illustrate the second and third kinds of symmetries.

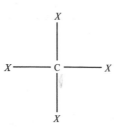

Figure 7.4 An organic molecule.

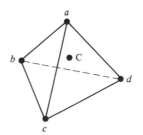

Figure 7.5 A regular tetrahedron.

*This example is from Reingold *et al.* [1977].
†This example is from Liu [1968].

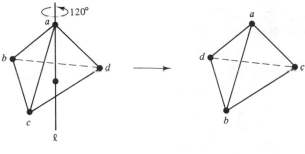

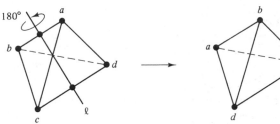

Figure 7.6 Rotation by 120° around a line ℓ connecting vertex a and the center of the opposite face.

Figure 7.7 Rotation by 180° around a line ℓ connecting the midpoints of opposite edges.

7.1.2 Equivalence Classes

An equivalence relation S on V divides the elements of V into classes called *equivalence classes*. Specifically, if a is any element of V, the *equivalence class containing a*, $C(a)$, consists of all elements b such that aSb. By reflexivity, aSa, so every element of V is in some equivalence class; in particular $a \in C(a)$. Moreover, for all a, b in V, either $C(a) = C(b)$ or $C(a)$ and $C(b)$ are disjoint. For suppose that x is in both $C(a)$ and $C(b)$. Then aSx and bSx. By symmetry, aSx and xSb. Transitivity now implies that aSb. This shows that $C(a) = C(b)$. For if y is in $C(b)$, then bSy. Now aSb and bSy imply aSy, so y is in $C(a)$. Thus, $C(b) \subseteq C(a)$. Similarly, if y is in $C(a)$, then aSy, so ySa. Thus, ySa and aSb, which implies ySb, which implies bSy. Thus, y is in $C(b)$, and $C(a) \subseteq C(b)$. Thus,

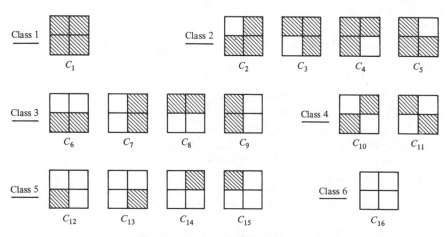

Figure 7.8 Equivalence classes of black-white colorings of the 2×2 array.

$C(a) = C(b)$. If we now think of $C(a)$ and $C(b)$ as being the same if they have the same members, we have the following theorem.

Theorem 7.1. If S is an equivalence relation, then every element is in one and only one equivalence class.

To illustrate this result, note that in Example 7.2, if the necklaces have length 2, there are four kinds of necklaces, bb, br, rb, and rr. The second and third are equivalent. Thus, for instance, $C(bb) = \{bb\}$ and $C\{br\} = \{br, rb\}$. There are three distinct equivalence classes, $\{bb\}$, $\{br, rb\}$, and $\{rr\}$.

In Example 7.1, there are six equivalence classes. These are shown in Figure 7.8.

EXERCISES FOR SECTION 7.1

1. In each of the following cases, is S an equivalence relation on V? If not, determine which of the properties of an equivalence relation hold.
   (a) $V =$ real numbers, aSb iff $a = b$.
   (b) $V =$ real numbers, aSb iff $a \neq b$.
   (c) $V =$ real numbers, aSb iff $a > b$.
   (d) $V =$ all subsets of $\{1, 2, \ldots, n\}$, aSb iff a and b have the same number of elements.
   (e) V as in part (d), aSb iff a and b are disjoint.
   (f) $V =$ all people in the world, aSb iff a is the brother of b.
   (g) $V =$ all people in New York City, aSb iff a and b have the same blood type.
   (h) $V = \{1, 2, 3, 4\}$, $S = \{(1, 1), (2, 2), (3, 4), (4, 3)\}$.
   (i) $V = \{x, y, z\}$, $S = \{(x, x), (y, y), (z, z), (x, z), (z, x)\}$.
   (j) $V =$ all residents of California, aSb iff a and b live within 10 miles of each other.

2. Suppose that V is the set of bit strings of length 10, and aSb holds if and only if a and b have the same number of 1's. Is (V, S) an equivalence relation?

3. For each of the following equivalence relations, identify all equivalence classes.
   (a) $V = \{1, 2, 3, 4\}$, $S = \{(1, 1), (2, 2), (3, 3), (4, 4), (1, 2), (2, 1), (3, 4), (4, 3)\}$.
   (b) $V = \{L, M, N\}$, $S = \{(L, L), (M, M), (N, N), (L, M), (M, L), (M, N), (N, M), (L, N), (N, L)\}$.
   (c) $V = \{\alpha, \beta, \gamma, \delta\}$, $S = \{(\alpha, \alpha), (\beta, \beta), (\gamma, \gamma), (\delta, \delta), (\alpha, \beta), (\beta, \alpha), (\beta, \gamma), (\gamma, \beta), (\alpha, \gamma) (\gamma, \alpha)\}$.
   (d) $V =$ the set of all positive integers, aSb iff $|a - b|$ is an even number.

4. Show that S of Example 7.2 is an equivalence relation.

5. Show that S of Example 7.3 is an equivalence relation among switching functions of two variables.

6. In Example 7.2, identify the equivalence classes of necklaces of length 3.

7. In Example 7.2, identify the equivalence classes of necklaces of length 2 if each bead can be one of three colors, blue, red, or purple.

8. In Example 7.1, suppose that we can use any of three different colors, black (b), white (w), or red (r). Describe all equivalence classes of colorings.

9. In Example 7.1, suppose that we allow not only rotations, but also reflections in either a vertical, a horizontal, or a diagonal line. (The latter would switch the colors assigned to two diagonally opposite cells.) Identify all equivalence classes of colorings. (Only two colors are used, black and white.)

10. In Example 7.3, identify all equivalence classes of switching functions of two variables.

11. For each tree of Figure 7.9, draw all trees that are equivalent to it in the sense of Example 7.4.

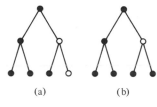

Figure 7.9 Trees for Exercise 11, Section 7.1.

Figure 7.10 Tree for Exercise 13, Section 7.1.

12. The *complement* x' of a bit string x is obtained from x by interchanging all 0's and 1's. For instance, if $x = 00110$, then $x' = 11001$. Suppose that we consider two switching functions T and U of n variables the same if $T = U$ or for every bit string x, $T(x) = U(x')$. Describe all equivalence classes of switching functions under this sameness relation if $n = 3$.

13. Suppose that V is the set of all colorings of the binary tree of Figure 7.10 in which each vertex gets one of the colors black or white. Find all equivalence classes of colorings if two colorings are considered the same if one can be obtained from the other by interchanging the colors of the vertices labeled 1 and 2.

14. In Example 7.5, suppose that a molecule has three hydrogen atoms and one nonhydrogen. How many other molecules with three hydrogens and one nonhydrogen are considered the same as this one?

15. Suppose that V is the set of unlabeled graphs of n vertices and that aSb iff a and b are isomorphic.
    (a) Show that S is an equivalence relation on V.
    (b) Find one unlabeled graph from each equivalence class if $n = 3$.

7.2 PERMUTATION GROUPS

7.2.1 Definition of a Permutation Group

In studying examples such as Examples 7.1–7.5, we shall make heavy use of the notion of a permutation. Recall that a permutation of a set $A = \{1, 2, \ldots, n\}$ is an ordering of the elements of A. The permutation that sends 1 to a_1, 2 to a_2, and so on, can be written as

$$\begin{pmatrix} 1 & 2 & 3 & \cdots & n \\ a_1 & a_2 & a_3 & \cdots & a_n \end{pmatrix},$$

or as $a_1 a_2 \cdots a_n$ for short. Thus, the permutation 132 stands for

$$\begin{pmatrix} 1 & 2 & 3 \\ 1 & 3 & 2 \end{pmatrix}.$$

Similarly, the permutation

$$\begin{pmatrix} 1 & 2 & 3 & 4 \\ 3 & 1 & 4 & 2 \end{pmatrix}$$

is written as 3142.

A permutation of A can also be thought of as a function from A into itself. This function must be one-to-one. Thus, the permutation

$$\begin{pmatrix} 1 & 2 & 3 \\ 1 & 3 & 2 \end{pmatrix}$$

can be thought of as the function $\pi\colon \{1, 2, 3\} \to \{1, 2, 3\}$ defined by $\pi(1) = 1$, $\pi(2) = 3$, $\pi(3) = 2$. Similarly, if A is any finite set, any one-to-one function from A into A can be thought of as a permutation of A; we simply identify elements of A with the integers 1, 2, ..., n. For instance, suppose that $A = \{a, b, c, d\}$ and $f(a) = b, f(b) = c, f(c) = d, f(d) = a$. If $a = 1, b = 2, c = 3, d = 4, f$ can be thought of as the permutation

$$\begin{pmatrix} 1 & 2 & 3 & 4 \\ 2 & 3 & 4 & 1 \end{pmatrix}.$$

Suppose that π_1 and π_2 are permutations of the set A. We can define the *product* or *composition* $\pi_1 \circ \pi_2$ of the permutations π_1 and π_2 as the permutation that first permutes by the permutation π_2 and then permutes the resulting arrangement by the permutation π_1. For instance, if

$$\pi_1 = \begin{pmatrix} 1 & 2 & 3 & 4 \\ 4 & 2 & 1 & 3 \end{pmatrix} \quad \text{and} \quad \pi_2 = \begin{pmatrix} 1 & 2 & 3 & 4 \\ 2 & 1 & 4 & 3 \end{pmatrix},$$

then

$$\pi_1 \circ \pi_2 = \begin{pmatrix} 1 & 2 & 3 & 4 \\ 2 & 4 & 3 & 1 \end{pmatrix}.$$

For 1 is sent to 2 by π_2, which is sent to 2 by π_1, so the composition sends 1 to 2. Similarly, 2 is sent to 1 by π_2 and 1 to 4 by π_1, so the composition sends 2 to 4. And so on.

Let X be the collection of all permutations of the set A. Note that the product of permutations satisfies the following conditions:

Condition **G1** (*Closure*).   If π_1 and π_2 are in X, then so is $\pi_1 \circ \pi_2$.

Condition **G2** (*Associativity*).   If π_1, π_2, and π_3 are in X, then

$$\pi_1 \circ (\pi_2 \circ \pi_3) = (\pi_1 \circ \pi_2) \circ \pi_3.$$

Condition **G3** (*Identity*).   There is an element I in X, called the *identity*, so that for all π in X,

$$I \circ \pi = \pi \circ I = \pi.$$

Condition **G4** (*Inverse*).   For all π in X, there is π^{-1} in X, called the *inverse* of π, so that

$$\pi \circ \pi^{-1} = \pi^{-1} \circ \pi = I.$$

To verify these conditions, note for example that **G3** follows by taking I to be the permutation

$$\begin{pmatrix} 1 & 2 & \cdots & n \\ 1 & 2 & \cdots & n \end{pmatrix}.$$

Also, **G4** holds if we take π^{-1} to be the permutation that reverses what π does. **G1** has been tacitly assumed above. Its verification and that of **G2** are straightforward.

If X is any set and $\circ$ defines a product* on elements of X, then the pair $G = (X, \circ)$ is called a *group* if these four properties hold. Let us give some examples of groups. If X is the positive real numbers and $a \circ b$ means $a \times b$, then the pair $(X, \circ)$ is a group. Axiom **G1** holds because $a \times b$ is always a positive real number if a and b are positive reals. Axiom **G2** holds because $a \times (b \times c) = (a \times b) \times c$. Axiom **G3** holds because we take I to be 1. Axiom **G4** holds because we take a^{-1} to be $1/a$.

Another example of a group is $(X, \circ)$, where X is all the real numbers and $a \circ b$ is defined to be $a + b$. The identity element for Axiom **G3** is the number 0 and the inverse of element a is $-a$. Note that the real numbers where $a \circ b$ is defined to be $a \times b$ do not define a group. For the only possible identity is 1. But then the number 0 does not have an inverse: There is no number 0^{-1} so that $0 \times 0^{-1} = 1$.

We have observed that the collection of all permutations of $A = \{1, 2, \ldots, n\}$ defines a group. This group is called the *symmetric group*. We shall be interested in groups of permutations, or *permutation groups*. Another example of a permutation group consists of the following three permutations of the set $\{1, 2, 3\}$:

$$\pi_1 = \begin{pmatrix} 1 & 2 & 3 \\ 1 & 2 & 3 \end{pmatrix}, \qquad \pi_2 = \begin{pmatrix} 1 & 2 & 3 \\ 2 & 3 & 1 \end{pmatrix}, \qquad \pi_3 = \begin{pmatrix} 1 & 2 & 3 \\ 3 & 1 & 2 \end{pmatrix}. \tag{7.1}$$

It is left to the reader (Exercise 4) to verify that the group axioms are satisfied here.

Often the symmetries of physical objects or configurations define groups, and hence the theory of groups is very important in modern physics. To give an example, consider the symmetries of the 2×2 array studied in Example 7.1, namely the rotations by $0°$, $90°$, $180°$, and $270°$. These symmetries define a group if we take $a \circ b$ to mean first perform symmetry b and then perform symmetry a. For instance, if a is rotation by $90°$ and b is rotation by $180°$, then $a \circ b$ is rotation by $270°$.

This group of symmetries can be thought of as a permutation group, each symmetry permuting $\{1, 2, 3, 4\}$. To see why, let us label the four cells in the 2×2 array as in the first part of Figure 7.11. Then Figure 7.11 shows the resulting labeling from the different symmetries. We can think of this labeling as corresponding to a permutation that takes the label i into the label j. For example, we can think of the $90°$ rotation as the permutation that takes 1 into 4, 2 into 1, 3 into 2, and 4 into 3, that is, the permutation

$$\begin{pmatrix} 1 & 2 & 3 & 4 \\ 4 & 1 & 2 & 3 \end{pmatrix}.$$

The permutations corresponding to the other rotations are also shown in Figure 7.11.

Suppose that A is any finite set and f is any one-to-one function from A into A.

---

*Technically, a product $\circ$ is a function that assigns to each pair of elements a and b of X, another element (of X) denoted $a \circ b$. (Note that we can either define $a \circ b$ to always be an element of X, or make explicit condition **GI**.)

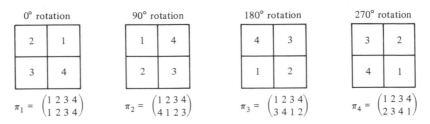

Figure 7.11 The permutations corresponding to the rotations of the 2 × 2 array.

Then as we have observed before, f can be thought of as a permutation of A. If X is a collection of such functions and $\circ$ is the composition of functions and $G = (X, \circ)$ is a group, we can think of G as a permutation group. For instance, suppose that $A = \{a, b, c\}$ and f, g, and h are defined as follows:

$$f(a) = a, \qquad f(b) = b, \qquad f(c) = c;$$

$$g(a) = b, \qquad g(b) = c, \qquad g(c) = a;$$

$$h(a) = c, \qquad h(b) = a, \qquad h(c) = b.$$

Then f, g, and h are one-to-one functions. It is easy to show that if $X = \{f, g, h\}$, then $(X, \circ)$ is a group. It is a permutation group. Indeed, if we take $a = 1$, $b = 2$, and $c = 3$, then f, g, and h are the permutations π_1, π_2, and π_3 of (7.1), so this is exactly the permutation group we have encountered earlier using different notation.

7.2.2 The Equivalence Relation Induced by a Permutation Group

Suppose that G is a permutation group on a set A. We can define a sameness relation S on A by saying that aSb iff there is a permutation π in G such that $\pi(a) = b$; that is, π takes a into b. For instance, if $A = \{1, 2, 3\}$ and G consists of the three permutations of (7.1), then $1S2$ because $\pi_2(1) = 2$, and $3S2$ because $\pi_3(3) = 2$. It is easy to see that for this S, aSb for all a, b.

Theorem 7.2. If G is a permutation group, then S defines an equivalence relation.

Proof. We have to show that S satisfies reflexivity, symmetry, and transitivity. Since the identity permutation I is in G, $I(a) = a$ for all a, and so aSa for all a. Thus, reflexivity holds. If aSb, then there is π in G so that $\pi(a) = b$. Now π^{-1} is in G and $\pi^{-1}(b) = a$. We conclude that bSa. Thus, symmetry holds. Finally, suppose that aSb and bSc. Then there are π_1 and π_2 in G so that $\pi_1(b) = c$ and $\pi_2(a) = b$. Then $\pi_1 \circ \pi_2$ maps a into c, so aSc follows. Q.E.D.

The relation S will be called the *equivalence relation induced by the permutation group* G.

Let us give several more examples. If G is the group of permutations of $\{1, 2, 3, 4\}$ shown in Figure 7.11, that is, the group of rotations of the 2 × 2 array, then aSb for all a, b in $\{1, 2, 3, 4\}$. Thus, S has one equivalence class, $\{1, 2, 3, 4\}$. Next, suppose that $A = \{1, 2, 3\}$ and G consists of the permutations

$$\pi_1 = \begin{pmatrix} 1 & 2 & 3 \\ 1 & 2 & 3 \end{pmatrix} \quad \text{and} \quad \pi_2 = \begin{pmatrix} 1 & 2 & 3 \\ 3 & 2 & 1 \end{pmatrix}.$$

Then G is a group (Exercise 3). Moreover, the equivalence classes under S are $\{1, 3\}$ and $\{2\}$. Note that $C(a)$, the equivalence class containing a, consists of all b in A such that aSb, equivalently all b in A such that $\pi(a) = b$ for some π in G. Thus,

$$C(a) = \{\pi(a): \pi \in G\}.$$

In the special case of a permutation group, $C(a)$ is sometimes called the *orbit* of a. In the example we have just given,

$$C(1) = \{\pi_1(1), \pi_2(1)\} = \{1, 3\}$$

is the orbit of 1.

In counting the number of distinct configurations, we shall be interested in counting the number of (distinct) equivalence classes under a sameness relation. One way to count is simply to compute all the equivalence classes and enumerate them. But this is often impractical. In the next section we present a method for counting the number of equivalence classes.

EXERCISES FOR SECTION 7.2

1. Write each of the following permutations in the form

$$\begin{pmatrix} 1 & 2 & 3 & \cdots & n \\ a_1 & a_2 & a_3 & \cdots & a_n \end{pmatrix}.$$

   (a) 1246753   (b) 53421   (c) 67854312.

2. Find $\pi_1 \circ \pi_2$ if π_1 and π_2 are as follows.

   (a) $\pi_1 = \begin{pmatrix} 1 & 2 & 3 & 4 \\ 4 & 3 & 2 & 1 \end{pmatrix}$, $\pi_2 = \begin{pmatrix} 1 & 2 & 3 & 4 \\ 2 & 3 & 1 & 4 \end{pmatrix}$

   (b) $\pi_1 = \begin{pmatrix} 1 & 2 & 3 & 4 \\ 2 & 1 & 4 & 3 \end{pmatrix}$, $\pi_2 = \begin{pmatrix} 1 & 2 & 3 & 4 \\ 4 & 3 & 1 & 2 \end{pmatrix}$

   (c) $\pi_1 = \begin{pmatrix} 1 & 2 & 3 & 4 & 5 \\ 1 & 3 & 2 & 4 & 5 \end{pmatrix}$, $\pi_2 = \begin{pmatrix} 1 & 2 & 3 & 4 & 5 \\ 2 & 1 & 3 & 5 & 4 \end{pmatrix}$

   (d) $\pi_1 = \begin{pmatrix} 1 & 2 & 3 & 4 & 5 & 6 \\ 6 & 1 & 5 & 2 & 4 & 3 \end{pmatrix}$, $\pi_2 = \begin{pmatrix} 1 & 2 & 3 & 4 & 5 & 6 \\ 2 & 3 & 4 & 5 & 6 & 1 \end{pmatrix}$.

3. Suppose that $A = \{1, 2, 3\}$ and X is the set of permutations

$$\begin{pmatrix} 1 & 2 & 3 \\ 1 & 2 & 3 \end{pmatrix} \quad \text{and} \quad \begin{pmatrix} 1 & 2 & 3 \\ 3 & 2 & 1 \end{pmatrix}.$$

   If $\circ$ is composition, show that $(X, \circ)$ is a group.

4. Suppose that $A = \{1, 2, 3\}$ and X is the set of the three permutations given in Equation (7.1). Show that X defines a group under composition.

5. For each of the following X and $\circ$, check which of the four axioms for a group hold.

   (a) X = the permutations $\begin{pmatrix} 1 & 2 & 3 & 4 & 5 \\ 1 & 2 & 3 & 4 & 5 \end{pmatrix}$ and $\begin{pmatrix} 1 & 2 & 3 & 4 & 5 \\ 5 & 4 & 3 & 2 & 1 \end{pmatrix}$, and $\circ$ = composition.

(b) X = the permutations $\begin{pmatrix} 1 & 2 & 3 & 4 & 5 \\ 1 & 2 & 3 & 4 & 5 \end{pmatrix}$ and $\begin{pmatrix} 1 & 2 & 3 & 4 & 5 \\ 1 & 3 & 2 & 4 & 5 \end{pmatrix}$, and $\circ$ = composition.

(c) $X = \{0, 1\}$ and $\circ$ is defined by the following rules: $0 \circ 0 = 0, 0 \circ 1 = 1, 1 \circ 0 = 1, 1 \circ 1 = 1$.

(d) X = rational numbers, $\circ$ = addition.

(e) X = rational numbers, $\circ$ = multiplication.

(f) X = negative real numbers, $\circ$ = addition.

(g) X = all 2×2 matrices of real numbers, $\circ$ = matrix multiplication.

6. Show that the set of functions $\{f, g\}$ is a group of permutations on $A = \{x, y, u, v\}$, if $f(x) = v$, $f(y) = u, f(u) = y, f(v) = x$, $g(x) = x$, $g(y) = y$, $g(u) = u$, $g(v) = v$.

7. Is the conclusion of Exercise 6 still true if f is redefined by $f(x) = y, f(y) = x, f(u) = v, f(v) = u$?

8. Suppose that $A = \{1, 2, 3, 4, 5, 6\}$ and G is the following group of permutations:

$$\left\{ \begin{pmatrix} 1 & 2 & 3 & 4 & 5 & 6 \\ 1 & 2 & 3 & 4 & 5 & 6 \end{pmatrix}, \begin{pmatrix} 1 & 2 & 3 & 4 & 5 & 6 \\ 6 & 5 & 4 & 3 & 2 & 1 \end{pmatrix} \right\}.$$

If S is the equivalence relation induced by G,
(a) is $1S2$?   (b) is $3S4$?   (c) is $5S6$?

9. In Exercise 8, find the orbits $C(1)$ and $C(4)$.

10. If A and G are as follows, find the equivalence classes under the equivalence relation S induced by G.

(a) $A = \{1, 2, 3, 4, 5\}$,

$$G = \left\{ \begin{pmatrix} 1 & 2 & 3 & 4 & 5 \\ 1 & 2 & 3 & 4 & 5 \end{pmatrix}, \begin{pmatrix} 1 & 2 & 3 & 4 & 5 \\ 5 & 4 & 3 & 2 & 1 \end{pmatrix} \right\}$$

(b) $A = \{1, 2, 3, 4, 5, 6\}$,

$$G = \left\{ \begin{pmatrix} 1 & 2 & 3 & 4 & 5 & 6 \\ 1 & 2 & 3 & 4 & 5 & 6 \end{pmatrix}, \begin{pmatrix} 1 & 2 & 3 & 4 & 5 & 6 \\ 2 & 1 & 3 & 4 & 5 & 6 \end{pmatrix}, \begin{pmatrix} 1 & 2 & 3 & 4 & 5 & 6 \\ 1 & 2 & 4 & 3 & 5 & 6 \end{pmatrix}, \right.$$

$$\begin{pmatrix} 1 & 2 & 3 & 4 & 5 & 6 \\ 1 & 2 & 3 & 4 & 6 & 5 \end{pmatrix}, \begin{pmatrix} 1 & 2 & 3 & 4 & 5 & 6 \\ 2 & 1 & 4 & 3 & 5 & 6 \end{pmatrix}, \begin{pmatrix} 1 & 2 & 3 & 4 & 5 & 6 \\ 2 & 1 & 3 & 4 & 6 & 5 \end{pmatrix},$$

$$\left. \begin{pmatrix} 1 & 2 & 3 & 4 & 5 & 6 \\ 1 & 2 & 4 & 3 & 6 & 5 \end{pmatrix}, \begin{pmatrix} 1 & 2 & 3 & 4 & 5 & 6 \\ 2 & 1 & 4 & 3 & 6 & 5 \end{pmatrix} \right\}$$

(c) $A = \{1, 2, 3, 4, 5\}$,

$$G = \left\{ \begin{pmatrix} 1 & 2 & 3 & 4 & 5 \\ 1 & 2 & 3 & 4 & 5 \end{pmatrix}, \begin{pmatrix} 1 & 2 & 3 & 4 & 5 \\ 3 & 2 & 1 & 4 & 5 \end{pmatrix}, \begin{pmatrix} 1 & 2 & 3 & 4 & 5 \\ 1 & 2 & 3 & 5 & 4 \end{pmatrix}, \begin{pmatrix} 1 & 2 & 3 & 4 & 5 \\ 3 & 2 & 1 & 5 & 4 \end{pmatrix} \right\}.$$

11. Consider the collection of all symmetries of the 2×2 array described in Exercise 9, Section 7.1. If π_1 and π_2 are the following symmetries, find $\pi_1 \circ \pi_2$.

(a) π_1 = 90° rotation, π_2 = reflection in a horizontal line.

(b) π_1 = reflection in a vertical line, π_2 = rotation by 180°.

(c) π_1 = rotation by 270°, π_2 = reflection in a vertical line.

(d) π_1 = rotation by 180°, π_2 = reflection in the diagonal going from lower left to upper right.

12. Continuing with Exercise 11, describe the following symmetries as permutations of $\{1, 2, 3, 4\}$:

(a) reflection in a horizontal line;

(b) reflection in a vertical line;

(c) reflection in a diagonal going from lower left to upper right;

(d) reflection in a diagonal going from upper left to lower right.

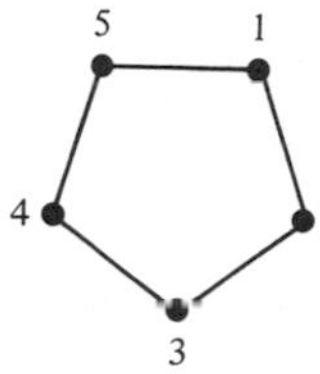

Figure 7.12 An automorphism is given by $\pi(1) = 4$, $\pi(2) = 5$, $\pi(3) = 1$, $\pi(4) = 2$, $\pi(5) = 3$.

13. Continuing with Exercise 11, is the collection of all the symmetries (rotations and reflections) a group?

14. If π_1 and π_2 are permutations, $\pi_1 \circ \pi_2$ may not equal $\pi_2 \circ \pi_1$. (Thus, we say that the product of permutations is not necessarily *commutative*.)

  (a) Demonstrate this with $\pi_1 = \begin{pmatrix} 1 & 2 & 3 & 4 \\ 1 & 4 & 2 & 3 \end{pmatrix}$ and $\pi_2 = \begin{pmatrix} 1 & 2 & 3 & 4 \\ 2 & 1 & 3 & 4 \end{pmatrix}$.

  (b) Find two symmetries π_1 and π_2 of the 2×2 array (π_1 and π_2 can be rotations or reflections) such that $\pi_1 \circ \pi_2 \ne \pi_2 \circ \pi_1$.

15. Show that for all prime numbers p, the set of integers $\{1, 2, \ldots, p-1\}$ with $\circ$ equal to multiplication modulo p forms a group.

16. In Exercise 15, do we still get a group if p is not a prime? Why?

17. Suppose that G is a permutation group. Fix a permutation σ in G. If π_1 and π_2 are in G, we say that $\pi_1 S \pi_2$ if $\pi_1 = \sigma^{-1} \circ (\pi_2 \circ \sigma)$. Show that S is an equivalence relation.

18. Let G be a fixed, unlabeled graph of n vertices. An *automorphism* of G is a permutation π of the vertices of G so that if $\{x, y\} \in E(G)$, then $\{\pi(x), \pi(y)\} \in E(G)$. In Figure 7.12 we can define an automorphism by labeling the vertices as 1, 2, 3, 4, 5 as shown and by taking $\pi(1) = 4$, $\pi(2) = 5$, $\pi(3) = 1$, $\pi(4) = 2$, $\pi(5) = 3$. To use the terminology of Section 3.1.3, we can think of an automorphism as an isomorphism of a graph into itself. In any graph G, the collection of all automorphisms defines a permutation group. It is called the *automorphism group* of G, and is denoted $\Gamma(G)$.

  (a) Find $\Gamma(L_4)$, where L_4 is the chain of four vertices.

  (b) Find $\Gamma(Z_4)$, where Z_4 is the circuit of four vertices.

  (c) Find $\Gamma[K(1, 3)]$, for $K(1, 3)$ the graph shown in Figure 7.13.

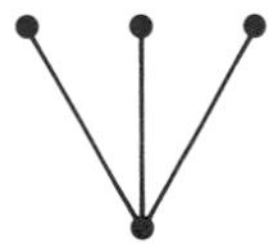

$K(1, 3)$

Figure 7.13 Graph for Exercise 18, Section 7.2.

7.3 BURNSIDE'S LEMMA

In this section we present a method for counting the number of (distinct) equivalence classes under the equivalence relation induced by a permutation group. Suppose that G is a group of permutations of a set A. An element a in A is said to be *invariant* under a permutation π of G if $\pi(a) = a$. Let Inv (π) be the number of elements of A that are invariant under π.

Theorem 7.3 (Burnside's Lemma).* Let G be a group of permutations of a set A and let S be the equivalence relation on A induced by G. Then the number of equivalence classes in S is given by

$$\frac{1}{|G|} \sum_{\pi \in G} \text{Inv } (\pi).$$

To illustrate this theorem, let us first consider the set $A = \{1, 2, 3\}$ and the group G of permutations of A defined by Equation (7.1). Then Inv $(\pi_1) = 3$ since 1, 2, and 3 are invariant under π_1, and Inv $(\pi_2) = $ Inv $(\pi_3) = 0$, since no element is invariant under either π_2 or π_3. Hence, the number of equivalence classes under the induced equivalence relation S is given by $\frac{1}{3}(3 + 0 + 0) = 1$. This is correct, since aSb holds for all a, b. There is just one equivalence class, $\{1, 2, 3\}$.

To give a second example, suppose that $A = \{1, 2, 3, 4\}$ and G consists of the following permutations:

$$\pi_1 = \begin{pmatrix} 1 & 2 & 3 & 4 \\ 1 & 2 & 3 & 4 \end{pmatrix}, \qquad \pi_2 = \begin{pmatrix} 1 & 2 & 3 & 4 \\ 2 & 1 & 3 & 4 \end{pmatrix},$$

$$\pi_3 = \begin{pmatrix} 1 & 2 & 3 & 4 \\ 1 & 2 & 4 & 3 \end{pmatrix}, \qquad \pi_4 = \begin{pmatrix} 1 & 2 & 3 & 4 \\ 2 & 1 & 4 & 3 \end{pmatrix}. \tag{7.2}$$

It is easy to check that G is a group. Now Inv $(\pi_1) = 4$, Inv $(\pi_2) = 2$, Inv $(\pi_3) = 2$, Inv $(\pi_4) = 0$, and the number of equivalence classes under the induced equivalence relation S is $\frac{1}{4}(4 + 2 + 2 + 0) = 2$. This is correct since the two equivalence classes are $\{1, 2\}$ and $\{3, 4\}$ (Exercise 1).

In Section 7.4 we shall see how to apply Burnside's Lemma to examples such as Examples 7.1–7.5.

We now present a proof of Burnside's Lemma.† Suppose that G is a group of permutations on a set A. For each $a \in A$, let St (a), the *stabilizer* of a, be the set of all permutations in G under which a is invariant. Let $C(a)$ be the orbit of a, the equivalence class containing a under the induced equivalence relation S, that is, the set of all b such that $\pi(a) = b$ for some π in G. To illustrate, suppose that $A = \{1, 2, 3\}$ and G is defined by Equation (7.1). Then $C(2) = \{\pi_1(2), \pi_2(2), \pi_3(2)\} = \{1, 2, 3\}$. Also, St $(2) = \{\pi_1\}$.

Lemma 1. Suppose that G is a group of permutations on a set A and a is in A. Then

$$|\text{St } (a)| \cdot |C(a)| = |G|.$$

Proof. Suppose that $C(a) = \{b_1, b_2, \ldots, b_r\}$. Then there is a permutation π_1 which sends a to b_1. (There may be other permutations that send a to b_1, but we pick one such.) There is also a permutation π_2 that sends a to b_2, a permutation π_3 that sends a to b_3, and so on. Let $P = \{\pi_1, \pi_2, \ldots, \pi_r\}$. Note that $|P| = |C(a)|$. The key idea is that every

---

*This version of the lemma is a simple consequence of the crucial lemma given by Burnside [1911], and is usually called Burnside's Lemma.

†The remainder of Section 7.3 is optional.

permutation π in G can be written in exactly one way as the product of a permutation in P and a permutation in St (a). It follows by the product rule that $|G| = |P| \cdot |\text{St}(a)| = |C(a)| \cdot |\text{St}(a)|$.

Given π in G, note that $\pi(a) = b_k$, some k. Thus, $\pi(a) = \pi_k(a)$, so $\pi_k^{-1} \circ \pi$ leaves a invariant. Thus, $\pi_k^{-1} \circ \pi$ is in St (a). But

$$\pi_k \circ (\pi_k^{-1} \circ \pi) = (\pi_k \circ \pi_k^{-1}) \circ \pi = I \circ \pi = \pi,$$

so π is the product of a permutation in P and a permutation in St (a).

Next, suppose that π can be written in two ways as a product of a permutation in P and a permutation in St (a), that is, suppose that $\pi = \pi_k \circ \gamma = \pi_l \circ \delta$, where γ, δ are in St (a). Now $(\pi_k \circ \gamma)(a) = b_k$ and $(\pi_l \circ \gamma)(a) = b_l$. Since $\pi_k \circ \gamma = \pi_l \circ \delta$, b_k must equal b_l, so $k = l$. Thus, $\pi_k \circ \gamma = \pi_k \circ \delta$, and by multiplying by π_k^{-1}, we conclude that $\gamma = \delta$.   Q.E.D.

To illustrate this lemma, let $A = \{1, 2, 3\}$ and let G be defined by (7.1). By our computation above, $C(2) = \{1, 2, 3\}$ and St $(2) = \{\pi_1\}$. Thus,

$$|G| = 3 = (1) \cdot (3) = |\text{St}(2)| \cdot |C(2)|.$$

To complete the proof of Burnside's Lemma, we show that if $A = \{1, 2, \dots\}$ and $G = \{\pi_1, \pi_2, \dots\}$, then

$$\text{Inv}(\pi_1) + \text{Inv}(\pi_2) + \cdots = |\text{St}(1)| + |\text{St}(2)| + \cdots.$$

This is because both sides of this equation count the number of ordered pairs (a, π) such that $\pi(a) = a$. It follows by Lemma 1 that

$$\frac{1}{|G|}[\text{Inv}(\pi_1) + \text{Inv}(\pi_2) + \cdots] = \frac{1}{|C(1)|} + \frac{1}{|C(2)|} + \cdots. \tag{7.3}$$

Note that x is always in $C(x)$, since $I(x) = x$. Thus, by Theorem 7.1, $C(x) = C(y)$ iff x is in $C(y)$. Hence, if $C(x) = \{b_1, b_2, \dots, b_k\}$, there are exactly k equivalence classes $C(b_1)$, $C(b_2)$, $\dots$, $C(b_k)$ that equal $C(x)$. It follows that we may split the equivalence classes up into groups such as $\{C(b_1), C(b_2), \dots, C(b_k)\}$, each group being a list of identical equivalence classes. Note that $|C(b_i)| = k$. Thus,

$$\frac{1}{|C(b_1)|} + \frac{1}{|C(b_2)|} + \cdots + \frac{1}{|C(b_k)|} = \frac{1}{k} + \frac{1}{k} + \cdots + \frac{1}{k} = 1.$$

It follows that the sum in the right-hand side of (7.3) will count the number of distinct equivalence classes, so this number is also given by the left-hand side of (7.3). Burnside's Lemma follows.

To illustrate the proof, suppose that $A = \{1, 2, 3, 4\}$ and G is given by the four permutations of (7.2). Then $C(1) = \{1, 2\}$, $C(2) = \{1, 2\}$, $C(3) = \{3, 4\}$, and $C(4) = \{3, 4\}$. Thus,

$$\frac{1}{|C(1)|} + \frac{1}{|C(2)|} = 1,$$

$$\frac{1}{|C(3)|} + \frac{1}{|C(4)|} = 1,$$

and

$$\frac{1}{|C(1)|} + \frac{1}{|C(2)|} + \frac{1}{|C(3)|} + \frac{1}{|C(4)|} = 2,$$

the number of equivalence classes.

EXERCISES FOR SECTION 7.3

1. Verify that the four permutations of Equation (7.2) define a group and that the equivalence classes under the induced equivalence relation are $\{1, 2\}$ and $\{3, 4\}$.

2. In Exercise 8, Section 7.2, use Burnside's Lemma to find the number of equivalence classes under S.

3. In each case of Exercise 10, Section 7.2, use Burnside's Lemma to find the number of equivalence classes under S and check by computing the equivalence classes.

4. For each case of Exercise 10, Section 7.2, take $a = 1$ and **(i)** find St (a), **(ii)** find $C(a)$, and **(iii)** verify Lemma 1.

5. For each case of Exercise 10, Section 7.2, check that

$$\frac{1}{|C(1)|} + \frac{1}{|C(2)|} + \cdots$$

gives the number of equivalence classes under S.

6. Use Burnside's Lemma to compute the number of distinct ways to seat 5 negotiators in fixed chairs around a circular table if rotating seat assignments around the circle is not considered to change the seating arrangement.

7. Suppose that

$$A = \{1, 2, 3, 4\}, \qquad G = \left\{ \begin{pmatrix} 1 & 2 & 3 & 4 \\ 1 & 2 & 3 & 4 \end{pmatrix}, \begin{pmatrix} 1 & 2 & 3 & 4 \\ 2 & 1 & 3 & 4 \end{pmatrix}, \begin{pmatrix} 1 & 2 & 3 & 4 \\ 1 & 2 & 4 & 3 \end{pmatrix} \right\}.$$

Is $|\text{St}(1)| \cdot |C(1)| = |G|$? Explain what happened.

8. Suppose we label the vertices of a graph G of n vertices with the labels $1, 2, \ldots, n$. Any labeling of G can be thought of as a permutation of $\{1, 2, \ldots, n\}$, if we start with a fixed labeling. For instance, if we start with the labeling shown in Figure 7.12, the new labeling shown in Figure 7.14 corresponds to the permutation

$$\begin{pmatrix} 1 & 2 & 3 & 4 & 5 \\ 2 & 1 & 5 & 3 & 4 \end{pmatrix}.$$

Figure 7.14 A new labeling of the graph of Figure 7.12.

In Exercise 18 of Section 7.2, we defined the automorphism group of G. Now every automorphism π of G can be thought of as taking any labeling σ of G into another labeling: We simply use the labeling $\pi \circ \sigma$. Thus, the number of distinct labelings of G corresponds to the number of equivalence classes in the equivalence relation induced on the set of permutations of $\{1, 2, \ldots, n\}$ by the automorphism group $\Gamma(G)$.

(a) Show that the number of distinct labelings is given by $n!/|\Gamma(G)|$.
(b) If G is the chain of four vertices, L_4, find the number of distinct labelings from the result in part (a) and check by enumerating the labelings.
(c) Repeat for Z_4, the circuit of length 4.
(d) Repeat for $K(1, 3)$, the graph of Figure 7.13.

9. Using the methods of this section, do Exercise 21 from the Additional Exercises for Chapter 2 at the end of that chapter.

7.4 DISTINCT COLORINGS

7.4.1 Definition of a Coloring

Suppose that D is a collection of objects. A *coloring* of D assigns a color to each object in D. In this sense, if D is the vertex set of a graph, a coloring simply assigns a color to each vertex, independent of the rule used in Chapter 3 that if x and y are joined by an edge, they must get different colors. A coloring can be thought of as a function $f: D \rightarrow R$, where R is the set of colors. If D has n elements and R has m elements, then there are m^n colorings of D.

In Example 7.1, the set D is the set of four boxes in the 2×2 array, and the set R is the set {black, white}. In Example 7.2, the set D can be thought of as the integers 1, 2, ..., k, representing the k spaces for beads, and R is the set $\{b, r\}$. In Example 7.3, D is the set of bit strings of length n, and R is the set $\{0, 1\}$. In Example 7.4, D is the set of vertices of the tree of seven vertices, and $R = \{$black, white$\}$. Finally, in Example 7.5, the set D consists of the vertices a, b, c, d of the regular tetrahedron, and the set R is the set $\{CH_3, C_2H_5, H, Cl\}$.

Every graph G on the vertex set $V = \{1, 2, \ldots, p\}$ can be thought of as a coloring. Take D to be the set of all 2-element subsets of V, R to be $\{0, 1\}$, and let $f(\{i, j\})$ be 1 if $\{i, j\} \in E(G)$ and 0 otherwise. Exercises 16 and 17 will exploit this idea to compute the number of distinct (nonisomorphic) graphs of p vertices.

In all of our examples, we also allow certain permutations of the elements of D. These permutations define a group G. In particular, in Example 7.1, we allow the four rotations given in Figure 7.11, which define a group G of permutations. In Example 7.2, in the case of two beads, the permutations are

$$\pi_1 = \begin{pmatrix} 1 & 2 \\ 1 & 2 \end{pmatrix} \quad \text{and} \quad \pi_2 = \begin{pmatrix} 1 & 2 \\ 2 & 1 \end{pmatrix}.$$

Note that π_1 and π_2 define a group—this is the group G. More generally, if there are k beads, the group G is the group of the two permutations

$$\begin{pmatrix} 1 & 2 & \cdots & k \\ 1 & 2 & \cdots & k \end{pmatrix} \quad \text{and} \quad \begin{pmatrix} 1 & 2 & \cdots & k \\ k & k-1 & \cdots & 1 \end{pmatrix}.$$

In Example 7.5 (the molecules), the permutations in the group G correspond to the symmetries of the regular tetrahedron which were described in Section 7.1. We shall return to this example in Section 7.5, where we describe a simple way to represent these permutations.

What is the group in Example 7.4? Suppose that we start with the first labeled tree shown in Figure 7.15, that with π_1 under it. Then any other labeling of this tree corresponds to a permutation of $\{1, 2, \ldots, 7\}$. Not every permutation of $\{1, 2, \ldots, 7\}$ corresponds to a labeling which is considered equivalent in the sense that left and right have been interchanged. Figure 7.15 shows all labeled trees obtained from the first one by interchanging left and right. For instance, the labeled tree with π_2 under it is obtained by interchanging vertices 4 and 5, and the labeled tree with π_3 under it is obtained by interchanging vertices 6 and 7. The labeled tree with π_4 under it is obtained by interchanging the subtree T_1 generated by vertices 2, 4, 5 with the subtree T_2 generated by vertices 3, 6, 7. The labeled tree with π_5 under it is obtained by interchanging both 4 and 5 and 6 and 7. The labeled tree with π_6 under it is obtained by first interchanging subtrees T_1 and T_2 and then interchanging vertices 6 and 7. And so on. These eight trees correspond to the legitimate permutations of the elements of D, the members of the group G. The permutation corresponding to each labeled tree is also shown in Figure 7.15. Note that

$$\pi_4 = \begin{pmatrix} 1 & 2 & 3 & 4 & 5 & 6 & 7 \\ 1 & 3 & 2 & 6 & 7 & 4 & 5 \end{pmatrix},$$

because vertex 1 stays unchanged, vertex 2 is changed to vertex 3 and vertex 3 to vertex 2, and so on. Verification that G is a group is left to the reader (Exercise 15).

We shall discuss Example 7.3, the switching functions, shortly. In Examples 7.1–7.5,

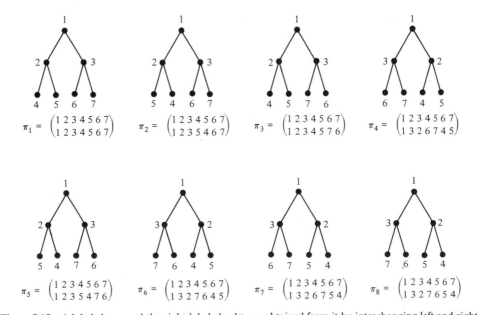

$$\pi_1 = \begin{pmatrix} 1234567 \\ 1234567 \end{pmatrix} \quad \pi_2 = \begin{pmatrix} 1234567 \\ 1235467 \end{pmatrix} \quad \pi_3 = \begin{pmatrix} 1234567 \\ 1234576 \end{pmatrix} \quad \pi_4 = \begin{pmatrix} 1234567 \\ 1326745 \end{pmatrix}$$

$$\pi_5 = \begin{pmatrix} 1234567 \\ 1235476 \end{pmatrix} \quad \pi_6 = \begin{pmatrix} 1234567 \\ 1327645 \end{pmatrix} \quad \pi_7 = \begin{pmatrix} 1234567 \\ 1326754 \end{pmatrix} \quad \pi_8 = \begin{pmatrix} 1234567 \\ 1327654 \end{pmatrix}$$

Figure 7.15 A labeled tree and the eight labeled subtrees obtained from it by interchanging left and right.

we are interested in determining whether or not two colorings are distinct, and in counting the number of equivalence classes of colorings. However, the equivalence relation of two colorings being the same is not the same as the equivalence relation S induced by the permutation group G. For S is a relation on the set D, not on the set of colorings of D. Thus, a direct use of Burnside's Lemma would not help us to count the number of equivalence classes of colorings. In Section 7.4.2 we shall discuss how to define the appropriate equivalence relation.

7.4.2 Equivalent Colorings

Suppose that $C(D, R)$ is the set of all colorings of D using colors in R and that G is a group of permutations of the set D and π is in G. Corresponding to π is a permutation π^* of $C(D, R)$. π^* takes each coloring in $C(D, R)$ into another coloring. If f is a coloring, then the new coloring $\pi^* f$ is defined by taking $(\pi^* f)(a)$ to be $f(\pi(a))$. That is $\pi^* f$ assigns to a the same color f assigns to $\pi(a)$. In Example 7.1, π^* takes a given coloring C_i of the 2×2 array into another one. For instance, if π_2 is the $90°$ rotation of the array, as shown in Figure 7.11, let us compute π_2^*. If $C_1, C_2, \dots$ are as in Figure 7.8, first note that $C_1(x) =$ black, all x, so $(\pi_2^* C_1)(a) = C_1(\pi_2(a)) =$ black. That means that $(\pi_2^* C_1)(a) =$ black, all a, so $\pi_2^* C_1$ is the coloring C_1. Next, since $4 = \pi_2(1)$, $(\pi_2^* C_2)(1) = C_2(\pi_2(1)) = C_2(4) =$ black. Also, $(\pi_2^* C_2)(2) = C_2(\pi_2(2)) = C_2(1) =$ black, $(\pi_2^* C_2)(3) = C_2(\pi_2(3)) = C_2(2) =$ white, and $(\pi_2^* C_2)(4) = C_2(\pi_2(4)) = C_2(3) =$ black. Thus, $\pi_2^* C_2$ is the same as the coloring C_3. Similarly, $\pi_2^* C_3 = C_4$, $\pi_2^* C_4 = C_5$, and so on. In sum, the permutation π_2^* is given by

$$\pi_2^* = \begin{pmatrix} C_1 & C_2 & C_3 & C_4 & C_5 & C_6 & C_7 & C_8 & C_9 & C_{10} & C_{11} & C_{12} & C_{13} & C_{14} & C_{15} & C_{16} \\ C_1 & C_3 & C_4 & C_5 & C_2 & C_7 & C_8 & C_9 & C_6 & C_{11} & C_{10} & C_{13} & C_{14} & C_{15} & C_{13} & C_{16} \end{pmatrix} \quad (7.4)$$

Similarly, if π_1 is the $0°$ rotation, π_3 is the $180°$ rotation, and π_4 is the $270°$ rotation, then

$$\pi_1^* = \begin{pmatrix} C_1 & C_2 & C_3 & C_4 & C_5 & C_6 & C_7 & C_8 & C_9 & C_{10} & C_{11} & C_{12} & C_{13} & C_{14} & C_{15} & C_{16} \\ C_1 & C_2 & C_3 & C_4 & C_5 & C_6 & C_7 & C_8 & C_9 & C_{10} & C_{11} & C_{12} & C_{13} & C_{14} & C_{15} & C_{16} \end{pmatrix}, \quad (7.5)$$

$$\pi_3^* = \begin{pmatrix} C_1 & C_2 & C_3 & C_4 & C_5 & C_6 & C_7 & C_8 & C_9 & C_{10} & C_{11} & C_{12} & C_{13} & C_{14} & C_{15} & C_{16} \\ C_1 & C_4 & C_5 & C_2 & C_3 & C_8 & C_9 & C_6 & C_7 & C_{10} & C_{11} & C_{14} & C_{15} & C_{12} & C_{13} & C_{16} \end{pmatrix}, \quad (7.6)$$

and

$$\pi_4^* = \begin{pmatrix} C_1 & C_2 & C_3 & C_4 & C_5 & C_6 & C_7 & C_8 & C_9 & C_{10} & C_{11} & C_{12} & C_{13} & C_{14} & C_{15} & C_{16} \\ C_1 & C_5 & C_2 & C_3 & C_4 & C_9 & C_6 & C_7 & C_8 & C_{11} & C_{10} & C_{15} & C_{12} & C_{13} & C_{14} & C_{16} \end{pmatrix}. \quad (7.7)$$

Thus, the group G of permutations of D corresponds to a group G^* of permutations of $C(D, R)$; $G^* = \{\pi^* : \pi \in G\}$. (Why is G^* a group?) Note that G and G^* have the same number of elements. Moreover, if S^* is the equivalence relation induced by G^*, then S^* is the sameness relation in which we are interested. Under S^*, two colorings f and g are considered the same if for some permutation π of D, $g = \pi^* f$, that is, for all a in D, $g(a) = f(\pi(a))$. In Example 7.2, suppose that

$$\pi = \begin{pmatrix} 1 & 2 \\ 2 & 1 \end{pmatrix}.$$

Let $f(1) = r$, $f(2) = b$, $g(1) = b$, $g(2) = r$. Then $g(a) = f(\pi(a))$ for all a, so $f S^* g$. This just says that the two colorings rb and br are equivalent. Henceforth, to distinguish equivalence classes under S from those under S^*, we shall refer to the latter equivalence classes as *patterns*. We are interested in computing the number of distinct patterns. This can be done by applying Burnside's Lemma to G^*.

In Example 7.1, Inv $(\pi_1^*) = 16$, Inv $(\pi_2^*) = 2$, Inv $(\pi_3^*) = 4$, and Inv $(\pi_4^*) = 2$, since π_1^* leaves all 16 colorings C_i invariant, π_2^* and π_4^* leave only C_1 and C_{16} invariant, and π_3^* leaves C_1, C_{10}, C_{11}, and C_{16} invariant. Thus, the number of equivalence classes under S^* is given by $\frac{1}{4}(16 + 2 + 4 + 2) = 6$, which agrees with Figure 7.8. (We could have computed Inv (π_i^*) directly without first computing π_i^*. For instance, π_3^* leaves invariant only those colorings which agree in boxes 1 and 3 and agree in boxes 2 and 4. Since in such a coloring there are 2 choices for the color for boxes 1 and 3 and 2 choices for the color for boxes 2 and 4, there are $2^2 = 4$ choices for the coloring. Similarly, π_2^* leaves invariant only those colorings which agree in all four boxes, since each box must get the same color as the one $90°$ away in a clockwise direction. Thus, there are only 2 such colorings.)

In the case of the necklaces (Example 7.2) if

$$\pi_1 = \begin{pmatrix} 1 & 2 \\ 1 & 2 \end{pmatrix} \quad \text{and} \quad \pi_2 = \begin{pmatrix} 1 & 2 \\ 2 & 1 \end{pmatrix},$$

then

$$\pi_1^* = \begin{pmatrix} bb & br & rb & rr \\ bb & br & rb & rr \end{pmatrix} \quad \text{and} \quad \pi_2^* = \begin{pmatrix} bb & br & rb & rr \\ bb & rb & br & rr \end{pmatrix}. \tag{7.8}$$

Note that Inv $(\pi_1^*) = 4$ and Inv $(\pi_2^*) = 2$, so that the number of equivalence classes or patterns under S^* is given by $\frac{1}{2}(4 + 2) = 3$. This agrees with our earlier observation that the patterns are $\{bb\}$, $\{br, rb\}$, and $\{rr\}$.

In the tree colorings (Example 7.4), note that there are 2^7 tree colorings in all: Each vertex of the tree can get one of the two colors. Suppose that π_i^* is the permutation of tree colorings that corresponds to the permutation π_i of labelings shown in Figure 7.15. It is impractical to write out π_i^*. However, note that $\pi_1^* = I^*$ leaves invariant all 2^7 tree colorings, so Inv $(\pi_1^*) = 2^7 = 128$. Also, permutation π_2 interchanges vertices 4 and 5. Thus, π_2^* leaves invariant exactly those colorings which color vertices 4 and 5 the same, that is, 2^6 colorings. Thus, Inv $(\pi_2^*) = 2^6 = 64$. Similarly, Inv $(\pi_3^*) = 2^6 = 64$, Inv $(\pi_4^*) = 2^4 = 16$, Inv $(\pi_5^*) = 2^5 = 32$, Inv $(\pi_6^*) = 2^3 = 8$, Inv $(\pi_7^*) = 2^3 = 8$, and Inv $(\pi_8^*) = 2^4 = 16$. Thus, the number of patterns or the number of distinct colorings is

$$\frac{1}{8}(128 + 64 + 64 + 16 + 32 + 8 + 8 + 16) = 42.$$

7.4.3 The Case of Switching Functions*

Let us now apply the theory we have been developing to the case of switching functions, Example 7.3. If there are two variables, we considered two such functions T and U the same if $T = U$ or $T(x_1 x_2) = U(x_2 x_1)$. This idea generalizes as follows: Two switching functions T and U of n variables are considered the same if there is a permutation π of

---

*This subsection can be omitted.

$\{1, 2, \ldots, n\}$ so that

$$T(x_1 x_2 \cdots x_n) = U(x_{\pi(1)} x_{\pi(2)} \cdots x_{\pi(n)}). \tag{7.9}$$

In the case $n = 2$, the two possible π are

$$\begin{pmatrix} 1 & 2 \\ 1 & 2 \end{pmatrix} \quad \text{and} \quad \begin{pmatrix} 1 & 2 \\ 2 & 1 \end{pmatrix}.$$

If $n = 3$, an example of two switching functions satisfying (7.9) with

$$\pi = \begin{pmatrix} 1 & 2 & 3 \\ 2 & 3 & 1 \end{pmatrix}$$

is given in Table 7.2. That Equation (7.9) should correspond to sameness or equivalence makes sense. For if (7.9) holds, then a circuit design for T can be obtained from one for U, in a manner analogous to Figure 7.2. Alternative sameness relations also make sense for computer engineering. We explore them in the exercises.

How does this sameness relation fit into the formal structure we have developed? D here is the set B_n of bit strings of length n. Let π be any permutation of $\{1, 2, \ldots, n\}$, and let H be the group of all permutations of $\{1, 2, \ldots, n\}$. Then a bit string $x_1 x_2 \cdots x_n$ can be looked at as a coloring of $\{1, 2, \ldots, n\}$ using the colors 0 and 1. The corresponding group of permutations of colorings is H^*. This is the group G of our theory. The group G^* is the group $(H^*)^*$. $G^* = (H^*)^*$ consists of permutations $(\pi^*)^*$ for all π in H. How does $(\pi^*)^*$ work? First, note that $\pi^*(x_1 x_2 \cdots x_n) = x_{\pi(1)} x_{\pi(2)} \cdots x_{\pi(n)}$. Note that H^* is a group of permutations of the collection of bit strings. $(H^*)^*$ consists of permutations of colorings of bit strings. But a coloring of bit strings using colors 0, 1 is a switching function. Note that by definition, if U is a switching function,

$$[(\pi^*)^* U](x_1 x_2 \cdots x_n) = U[\pi^*(x_1 x_2 \cdots x_n)] = U(x_{\pi(1)} x_{\pi(2)} \cdots x_{\pi(n)}).$$

Thus, if $T = (\pi^*)^* U$, (7.9) follows.

Let $n = 2$. Then $H = \{\pi_1, \pi_2\}$, where

$$\pi_1 = \begin{pmatrix} 1 & 2 \\ 1 & 2 \end{pmatrix} \quad \text{and} \quad \pi_2 = \begin{pmatrix} 1 & 2 \\ 2 & 1 \end{pmatrix}.$$

Table 7.2 Switching Functions T and U Satisfying Equation (7.9) with $\pi = \begin{pmatrix} 1 & 2 & 3 \\ 2 & 3 & 1 \end{pmatrix}$

Bit string x	$T(x)$	$U(x)$
000	1	1
001	0	0
010	1	0
011	1	0
100	0	1
101	0	1
110	1	1
111	0	0

Table 7.3 The 16 Switching Functions of Two Variables

Bit string x	T_1	T_2	T_3	T_4	T_5	T_6	T_7	T_8	T_9	T_{10}	T_{11}	T_{12}	T_{13}	T_{14}	T_{15}	T_{16}
00	0	0	0	0	0	0	0	0	1	1	1	1	1	1	1	1
01	0	0	0	0	1	1	1	1	0	0	0	0	1	1	1	1
10	0	0	1	1	0	0	1	1	0	0	1	1	0	0	1	1
11	0	1	0	1	0	1	0	1	0	1	0	1	0	1	0	1

The permutations in $G = H^*$ are

$$\pi_1^* = \begin{pmatrix} 00 & 01 & 10 & 11 \\ 00 & 01 & 10 & 11 \end{pmatrix} \quad \text{and} \quad \pi_2^* = \begin{pmatrix} 00 & 01 & 10 & 11 \\ 00 & 10 & 01 & 11 \end{pmatrix}. \tag{7.10}$$

There are $2^{2^2} = 16$ switching functions of two variables. These are shown as $T_1, T_2, \ldots,$ T_{16} of Table 7.3. Then the permutations in $G^* = (H^*)^*$ are

$$(\pi_1^*)^* = \begin{pmatrix} T_1 & T_2 & T_3 & T_4 & T_5 & T_6 & T_7 & T_8 & T_9 & T_{10} & T_{11} & T_{12} & T_{13} & T_{14} & T_{15} & T_{16} \\ T_1 & T_2 & T_3 & T_4 & T_5 & T_6 & T_7 & T_8 & T_9 & T_{10} & T_{11} & T_{12} & T_{13} & T_{14} & T_{15} & T_{16} \end{pmatrix} \tag{7.11}$$

and

$$(\pi_2^*)^* = \begin{pmatrix} T_1 & T_2 & T_3 & T_4 & T_5 & T_6 & T_7 & T_8 & T_9 & T_{10} & T_{11} & T_{12} & T_{13} & T_{14} & T_{15} & T_{16} \\ T_1 & T_2 & T_5 & T_6 & T_3 & T_4 & T_7 & T_8 & T_9 & T_{10} & T_{13} & T_{14} & T_{11} & T_{12} & T_{15} & T_{16} \end{pmatrix}. \tag{7.12}$$

Note that $(\pi_2^*)^*U$ is the function T which does on 01 what U does on 10, and on 10 what U does on 01, and otherwise agrees with U. We have Inv $((\pi_1^*)^*) = 16$ and Inv $((\pi_2^*)^*) = 8$, and the number of equivalence classes or patterns of switching functions of two variables is given by $\frac{1}{2}(16 + 8) = 12$. The number of patterns of switching functions of three variables can similarly be shown to be 80, the number of patterns of switching functions of four variables can be shown to be 3984, and the number of patterns of switching functions of five variables can be shown to be 37,333,248 (see Harrison [1965] or Prather [1976]). By allowing other symmetries (such as interchange of 0 and 1 in the domain or range) of a switching function—see Exercises 18 and 19—we can further reduce the number of equivalence classes. In fact, the number can be reduced to 222 if $n = 4$ (Harrison [1965], Stone [1973]). This gives a small enough number so that for $n = 4$ it is reasonable to prepare a catalog of optimal circuit designs for realizing switching functions which contains a representative of each equivalence class.

EXERCISES FOR SECTION 7.4

1. Suppose that $D = \{a, b, c\}$ and $R = \{1, 2\}$. Find all colorings in $C(D, R)$.

2. How many colorings (not necessarily distinct) are there for the vertices of a cube if the set of allowable colors is {red, green, blue}?

3. How many allowable colorings (not necessarily distinct) are there for the vertices of a regular tetrahedron if there are five colors available?

4. In Example 7.1, check (7.4), (7.6), and (7.7) by computing
   (a) $(\pi_2^* C_4)(2)$ (b) $(\pi_3^* C_5)(4)$ (c) $(\pi_4^* C_{11})(2)$.
5. In Example 7.2, check (7.8) by computing
   (a) $(\pi_2^* br)(1)$ (b) $(\pi_2^* br)(2)$ (c) $(\pi_2^* rr)(1)$.
6. Suppose that $D = \{1, 2, 3, 4\}$, $R = \{1, 2\}$, and G consists of the permutations in Equation (7.2).
   (a) Suppose that f and g are the following colorings: $f(a) = 1$, all a, and $g(1) = g(2) = 2$, $g(3) = g(4) = 1$. Is $f\,S*g$?
   (b) Suppose that $f(1) = f(3) = 2, f(2) = f(4) = 1$, $g(1) = g(2) = 2$, $g(3) = g(4) = 1$. Is $f\,S*g$?
   (c) Find π_2^*.   (d) Find π_3^*.   (e) Find π_4^*.
   (f) Find Inv (π_2^*).   (g) Find Inv (π_3^*).   (h) Find Inv (π_4^*)
   (i) Find S.   (j) Find $S*$.
7. Repeat Exercise 6 [except parts (e) and (h)] if G consists of

$$\pi_1 = \begin{pmatrix} 1 & 2 & 3 & 4 \\ 1 & 2 & 3 & 4 \end{pmatrix}, \qquad \pi_2 = \begin{pmatrix} 1 & 2 & 3 & 4 \\ 2 & 3 & 1 & 4 \end{pmatrix}, \qquad \pi_3 = \begin{pmatrix} 1 & 2 & 3 & 4 \\ 3 & 1 & 2 & 4 \end{pmatrix}.$$

8. In Example 7.2, suppose that $k = 2$ and there are three different colors available, red (r), blue (b), and purple (p). Find π_2^* if

$$\pi_2 = \begin{pmatrix} 1 & 2 \\ 2 & 1 \end{pmatrix}.$$

9. In Example 7.2, suppose that $k = 3$ and there are two different colors available, red (r) and blue (b).
   (a) Find $G*$.
   (b) Find the number of distinct necklaces using Burnside's Lemma.
   (c) Check your answer by enumerating the distinct necklaces.
10. In Exercise 13, Section 7.1, find
   (a) D (b) R (c) G (d) $G*$
   (e) the number of distinct colorings.
11. In Example 7.4, verify that
   (a) Inv $(\pi_3^*) = 2^6$ (b) Inv $(\pi_4^*) = 2^4$
   (c) Inv $(\pi_5^*) = 2^5$ (d) Inv $(\pi_6^*) = 2^3$.
12. In Example 7.3, verify (7.12) by computing
   (a) $(\pi_2^*)*T_4$ (b) $(\pi_2^*)*T_{11}$ (c) $(\pi_2^*)*T_{14}$.
13. In the situation of Exercise 9, Section 7.1,
   (a) Find $G*$.
   (b) Use Burnside's Lemma to compute the number of distinct colorings.
   (c) Check your answer by comparing the enumeration of equivalence classes you gave as your answer in Section 7.1.
14. Find the number of distinct ways to 2-color a 4×4 array that can rotate by $0°$ or $180°$.
15. Show that the eight permutations in Figure 7.15 define a group.
16. Suppose that $V = \{1, 2, \ldots, p\}$. Recall that there is a one-to-one correspondence between graphs (V, E) on the vertex set V and functions f which assign 0 or 1 to each 2-element subset of V. The idea is that

$$f(\{i, j\}) = 1 \quad \text{iff} \quad \{i, j\} \in E.$$

The function f is a coloring of the set D of all 2-element subsets of V, using the colors 0 and 1.
   (a) If $p = 3$, find all such functions f and their corresponding graphs.
   (b) If G and G' are two graphs on V and f and f' are their corresponding functions, show that G

and G' are isomorphic iff there is a permutation π on D so that for all $\{i, j\}$ in D, $f(\{i, j\}) = f'(\pi(\{i, j\}))$, that is, so that f and f' are equivalent.

(c) Let G be the group of all permutations π of D. If $p = 3$, write down all the elements of G and compute Inv (π_i^*) for all π_i in G.

(d) Use Burnside's Lemma to determine the number of distinct (nonisomorphic) graphs of three vertices, and verify your result by identifying the classes of equivalent (isomorphic) graphs.

17. Repeat parts (a), (c), and (d) of Exercise 16 for $p = 4$.

18. Suppose that D is the collection of all bit strings of length 3 and let G be the group that consists of the identity permutation and the permutation that complements a string by interchanging 0 and 1 (see Exercise 12, Section 7.1). Find the number of distinct switching functions (number of distinct colorings using the colors 0 and 1), that is, the number of equivalence classes in the equivalence relation induced by G^*. Note that in this case we have no H.

19. Suppose that two switching functions are considered equivalent if one can be obtained from the other by permuting or complementing the variables as in Exercise 18 or both. Find the number of distinct switching functions of two variables.

20. How many distinct ways are there to 3-color the vertices of a square if we allow rotation of the square by 0°, 90°, 180°, or 270°?

21. (Reingold *et al.* [1977]) A manufacturer of integrated circuits makes chips that have 16 elements arranged in a 4×4 array. These elements are interconnected between some adjacent horizontal or vertical elements. Figure 7.16 shows some sample interconnection patterns. A photomask of the interconnection pattern is used to deposit interconnections on a chip. Two patterns are considered the same if the same photomask could be used for each. For instance, by flipping the photomask over on a diagonal, it can be used for both the interconnection patterns shown in Figure 7.16. Thus, they are considered the same. How many photomasks are required in order to lay out all possible interconnection patterns? Formulate this problem as a coloring problem by defining an appropriate D, R, and G. However, do not attempt to compute G^* or to solve the problem completely with the tools developed so far.

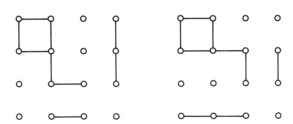

Figure 7.16   Sample interconnection patterns for chips.

7.5 THE CYCLE INDEX

7.5.1 Permutations as Products of Cycles

It gets rather messy to apply Burnside's Lemma to many counting problems. For instance, it gets rather long and complicated to compute the permutations in the group G^*. We shall develop alternative procedures, procedures that will also allow us to get more information than provided by Burnside's Lemma.

The permutation

$$\begin{pmatrix} 1 & 2 & 3 & 4 & 5 \\ 2 & 3 & 4 & 5 & 1 \end{pmatrix}$$

cycles the numbers around, sending 1 to 2, 2 to 3, 3 to 4, 4 to 5, and 5 to 1. It can be abbreviated by simply writing (12345). More generally, $(a_1 a_2 \cdots a_{m-1} a_m)$ will represent the permutation that takes a_1 to a_2, a_2 to a_3, ..., a_{m-1} to a_m, and a_m to a_1. This is called a *cyclic permutation*. For instance,

$$\begin{pmatrix} 1 & 2 & 3 \\ 3 & 1 & 2 \end{pmatrix}$$

is a cyclic permutation (132). Now consider the permutation

$$\begin{pmatrix} 1 & 2 & 3 & 4 & 5 & 6 \\ 5 & 1 & 6 & 3 & 2 & 4 \end{pmatrix}.$$

This consists of two cycles, (152) and (364). We say that

$$\begin{pmatrix} 1 & 2 & 3 & 4 & 5 & 6 \\ 5 & 1 & 6 & 3 & 2 & 4 \end{pmatrix}$$

is the *product* of these two cycles, and write it as (152)(364). It is the product of these cycles in the same sense as taking the product of two permutations, if we think of a cycle like (152) as leaving 3, 4, and 6 fixed. In the same way,

$$\begin{pmatrix} 1 & 2 & 3 & 4 & 5 & 6 \\ 2 & 1 & 3 & 5 & 6 & 4 \end{pmatrix}$$

is the *product* of three cycles, (12)(3)(456), where (3) means that 3 is mapped into itself. In this product, the three cycles are *disjoint* in the sense that no two of them involve the same element.

We now show that every permutation of $\{1, 2, ..., n\}$ can be written as the product of *disjoint* cycles, with each element i of 1, 2, ..., n appearing in some cycle. To see why, let us take the permutation

$$\begin{pmatrix} 1 & 2 & 3 & 4 & 5 & 6 & 7 & 8 \\ 5 & 4 & 8 & 6 & 7 & 2 & 1 & 3 \end{pmatrix}.$$

Take the element 1. It goes to 5. In turn, 5 goes to 7, and 7 to 1, so we have a cycle (157). Take the first element not in this cycle, 2. It goes to 4, which goes to 6, which goes to 2, so we have a cycle (246). Take the first element not in either of these cycles. It is 3. Now 3 goes to 8, which goes to 3, so we have a cycle (38). Thus, the original permutation is the product of the three cycles (157)(246)(38). Similar reasoning applies to any permutation.

Could a permutation π be written in two different ways as the product of disjoint cycles (with every element in some cycle)? The answer is yes if we consider $(a_1 a_2 \cdots a_m)$ and $(a_i a_{i+1} \cdots a_m a_1 a_2 \cdots a_{i-1})$ as different. However, we consider them the same, since they correspond to the same permutation. Thus, (123) and (231) and (312) are the same. Suppose that

$$\pi = (xyz \cdots) \cdots (abc \cdots) = (uvw \cdots) \cdots (\alpha\beta\gamma \cdots).$$

If these two ways of writing the permutation are different, there must be a number k such that the cycle containing k on the left is different from the cycle containing k on the right. We can write these two cycles with k first. Then whatever π takes k into must be next in each cycle, and whatever π takes this into must be third in each cycle, and so on. Thus, the two cycles must be the same. To summarize, we have the following result.

Theorem 7.4. Every permutation of $\{1, 2, \ldots, n\}$ can be written in exactly one way as the product of disjoint cycles with every element of $\{1, 2, \ldots, n\}$ appearing in some cycle.

We shall call the unique way of writing a permutation described in Theorem 7.4 the *cycle decomposition* of the permutation.

7.5.2 A Special Case of Polya's Theorem

We are now ready to present another result about counting equivalence classes, which is a special case of the main theorem we are aiming for. Suppose that cyc (π) counts the number of cycles in the unique cycle decomposition of the permutation π. For instance, if $\pi = (12)(3)(456)$, then cyc $(\pi) = 3$.

Theorem 7.5 (Special Case of Polya's Theorem). Suppose that G is a group of permutations of the set D and $C(D, R)$ is the set of colorings of elements of D using colors in R, a set of m elements. Then the number of distinct colorings in $C(D, R)$ (the number of equivalence classes or patterns in the equivalence relation S^* induced by G^*) is given by

$$\frac{1}{|G|} [m^{\text{cyc} \,(\pi_1)} + m^{\text{cyc} \,(\pi_2)} + \cdots],$$

where $G = \{\pi_1, \pi_2, \ldots\}$.

Note that this theorem allows us to compute the number of distinct colorings without first computing G^*. We prove the theorem in Section 7.5.5.

To illustrate this theorem, let us reconsider the case of the 2×2 arrays, Example 7.1. There are four permutations in G, the four rotations $\pi_1, \pi_2, \pi_3, \pi_4$ shown in Figure 7.11. These have the following cycle decompositions: $\pi_1 = (1)(2)(3)(4)$, $\pi_2 = (1432)$, $\pi_3 = (13)(24)$, $\pi_4 = (1234)$. Thus, cyc $(\pi_1) = 4$, cyc $(\pi_2) = 1$, cyc $(\pi_3) = 2$, cyc $(\pi_4) = 1$. The number of distinct colorings (number of patterns) is given by $\frac{1}{4}(2^4 + 2^1 + 2^2 + 2^1) = 6$, which agrees with our earlier computation.

In Example 7.2, with necklaces of k beads, we can write the two permutations in G as

$$\pi_1 = (1)(2) \cdots (k)$$

and as

$$\pi_2 = (1 \quad k)(2 \quad k - 1)(3 \quad k - 2) \cdots \left(\frac{k}{2} \quad \frac{k}{2} + 1\right)$$

if k is even and as

$$\pi_2 = (1 \quad k)(2 \quad k - 1)(3 \quad k - 2) \cdots \left(\frac{k-1}{2} \quad \frac{k+3}{2}\right)\left(\frac{k+1}{2}\right)$$

if k is odd. Thus, for instance, if k is 4, $\pi_2 = (14)(23)$. If k is 5, $\pi_2 = (15)(24)(3)$. It follows that cyc $(\pi_1) = k$ and cyc $(\pi_2) = k/2$ if k is even and $(k + 1)/2$ if k is odd. Hence, the number of distinct necklaces is $\frac{1}{2}(2^k + 2^{k/2})$ if k is even and $\frac{1}{2}(2^k + 2^{(k+1)/2})$ if k is odd. For instance, the case $k = 2$ gives us 3, which agrees with our earlier computation. The case $k = 3$ gives us 6, which is left to the reader to check (Exercise 8). If there are three different colors of beads and k is even, we would have $\frac{1}{2}(3^k + 3^{k/2})$ distinct necklaces. For instance, for $k = 2$, we would have 6 distinct necklaces. If the colors of beads are r, b, and p, then the 6 equivalence classes are $\{rb, br\}$, $\{rp, pr\}$, $\{bp, pb\}$, $\{rr\}$, $\{bb\}$, $\{pp\}$.

In Example 7.4, the tree colorings, G is given by the eight permutations $\pi_1, \pi_2, \ldots, \pi_8$ of Figure 7.15. We have

$$\pi_1 = (1)(2)(3)(4)(5)(6)(7), \qquad \pi_2 = (1)(2)(3)(45)(6)(7),$$
$$\pi_3 = (1)(2)(3)(4)(5)(67), \qquad \pi_4 = (1)(23)(46)(57),$$
$$\pi_5 = (1)(2)(3)(45)(67), \qquad \pi_6 = (1)(23)(4756),$$
$$\pi_7 = (1)(23)(4657), \qquad \pi_8 = (1)(23)(47)(56).$$

Then the number of distinct tree colorings is given by

$$\tfrac{1}{8}(2^7 + 2^6 + 2^6 + 2^4 + 2^5 + 2^3 + 2^3 + 2^4) = 42,$$

which is what we computed earlier.

7.5.3 The Case of Switching Functions*

In Example 7.3, the switching functions, we have to consider the elements of $G = H^*$. In case we consider switching functions of two variables, we have $G = \{\pi_1^*, \pi_2^*\}$, where

$$\pi_1 = \begin{pmatrix} 1 & 2 \\ 1 & 2 \end{pmatrix} \quad \text{and} \quad \pi_2 = \begin{pmatrix} 1 & 2 \\ 2 & 1 \end{pmatrix}.$$

By our computation in Section 7.4.3, π_1^* and π_2^* are given by (7.10). Now we can think of π_1^* and π_2^* as acting on $\{1, 2, 3, 4\}$, in which case $\pi_1^* = (1)(2)(3)(4)$ and $\pi_2^* = (1)(23)(4)$, so cyc $(\pi_1^*) = 4$, cyc $(\pi_2^*) = 3$, and the number of distinct switching functions is $\frac{1}{2}(2^4 + 2^3) = 12$, which agrees with our earlier computation.

7.5.4 The Cycle Index of a Permutation Group

It will be convenient to summarize the cycle structure of the permutations in a permutation group in a manner analogous to generating functions. Suppose that π is a permutation with b_1 cycles of length 1, b_2 cycles of length 2, ... in its unique cycle decomposition. Then if $x_1, x_2, \ldots$ are placeholders and k is at least the length of the longest cycle in the cycle decomposition of π, we can encode π by using the expression $x_1^{b_1} x_2^{b_2} \cdots x_k^{b_k}$. Moreover, we can encode an entire permutation group G by taking the sum of these expressions for members of G divided by the number of permutations of G. That is, if k is

*This subsection can be omitted.

the length of the longest cycle in the cycle decomposition of any π of G^*, we write

$$P_G(x_1, x_2, \ldots, x_k) = \frac{1}{|G|} \sum_{\pi \in G} x_1^{b_1} x_2^{b_2} \cdots x_k^{b_k}$$

and call $P_G(x_1, x_2, \ldots, x_k)$ the *cycle index* of G. For instance, consider Example 7.4. Then, to use the notation of Figure 7.15, $\pi_6 = (1)(23)(4756)$, and its corresponding code is $x_1 x_2 x_4$. Also, $\pi_4 = (1)(23)(46)(57)$, and it is encoded as $x_1 x_2^3$. By a similar analysis, the cycle index for the group of permutations is

$$P_G(x_1, x_2, \ldots, x_8) = \tfrac{1}{8}[x_1^7 + x_1^5 x_2 + x_1^5 x_2 + x_1 x_2^3 + x_1^3 x_2^2 + x_1 x_2 x_4 + x_1 x_2 x_4 + x_1 x_2^3].$$

(7.13)

Note that if π is a permutation with corresponding code $x_1^{b_1} x_2^{b_2} \cdots x_k^{b_k}$, then cyc $(\pi) = b_1 + b_2 + \cdots + b_k$ and $x_1^{b_1} x_2^{b_2} \cdots x_k^{b_k}$ is $m^{\text{cyc } (\pi)}$ if all x_i are taken to be m. Hence, Theorem 7.5 can be restated as follows:

Corollary 7.5.1. Suppose that G is a group of permutations of the set D and that $C(D, R)$ is the set of colorings of elements of D using colors in R, a set of m elements. Then the number of distinct colorings in $C(D, R)$ is given by $P_G(m, m, \ldots m)$.

It is this version of the result that we will generalize in Section 7.6.

To make use of this result here, we return to Example 7.4. Here $m = 2$. We let $x_1 = x_2 = \cdots = x_8 = 2$ in (7.13), obtaining

$$P_G(2, 2, \ldots, 2) = \tfrac{1}{8}(2^7 + 2^6 + 2^6 + 2^4 + 2^5 + 2^3 + 2^3 + 2^4) = 42,$$

which agrees with our earlier result about the number of distinct colorings.

Let us now consider Example 7.5. In Section 7.1, we identified 12 different symmetries of the tetrahedron. These can be thought of as permutations of the letters a, b, c, d of Figure 7.5. The identity symmetry is the permutation $(a)(b)(c)(d)$, which has cycle structure code x_1^4. The 120° rotation about the line joining vertex a to the middle of the face determined by b, c, and d corresponds to the permutation $(a)(bdc)$, which can be encoded as $x_1 x_3$. (See Figure 7.6.) All eight 120° and 240° rotational symmetries have similar structure and coding. Finally, the rotation by 180° about the line connecting the midpoints of edges ab and cd corresponds to the permutation $(ab)(cd)$, which has the coding x_2^2. The other two 180° rotations have similar coding. Thus, the cycle index is given by

$$P_G(x_1, x_2, x_3) = \tfrac{1}{12}(x_1^4 + 8x_1 x_3 + 3x_2^2).$$

We seek a coloring of the set $D = \{a, b, c, d\}$ using the four colors contained in the set $R = \{CH_3, C_2H_5, H, Cl\}$. Hence, $m = 4$, and the number of distinct colorings (number of distinct molecules) is given by

$$P_G(m, m, m) = \tfrac{1}{12}(4^4 + 8(4)(4) + 3(4)^2) = 36.$$

---

*We can also take $k = |D|$ and note that the length of the longest cycle in the cycle decomposition of any π of G is at most k.

7.5.5 Proof of Theorem 7.5*

We shall apply Burnside's Lemma (Theorem 7.3) to G^* in order to prove Theorem 7.5. Since $|G| = |G^*|$, it suffices to show that $m^{\text{cyc}\,(\pi)} = \text{Inv}\,(\pi^*)$. Let π be in G. We try to compute $\text{Inv}\,(\pi^*)$. Note that an element of $C(D, R)$ is left invariant by π^* iff in the corresponding permutation π of D, all the elements of D in each cycle of π receive the same color. For instance, suppose that $\pi = (12)(345)(67)(8)$. Let f be the coloring such that $f(1) = f(2) = \text{black}$, $f(3) = f(4) = f(5) = \text{white}$, $f(6) = f(7) = \text{red}$, and $f(8) = \text{blue}$. Then clearly π^*f is the same coloring.

In sum, to find a coloring that is left invariant by π^*, we compute the cycle decomposition of π and color each element in a cycle with the same color. Now π has cyc (π) different cycles in its cycle decomposition, and we have m choices for the common color of each cycle. Hence, there are $m^{\text{cyc}\,(\pi)}$ different colorings left invariant by π^*. In short, $\text{Inv}\,(\pi^*) = m^{\text{cyc}\,(\pi)}$.

EXERCISES FOR SECTION 7.5

1. Find the cycle decomposition of each permutation of Exercise 1, Section 7.2.
2. Find the cycle decomposition for all permutations π_i arising from the situation of Exercise 9, Section 7.1.
3. Compute cyc (π) for every permutation in parts (a), (b), and (c) of Exercise 10, Section 7.2.
4. For every permutation of parts (a), (b), (c) of Exercise 10, Section 7.2, encode the permutation as $x_1^{b_1} x_2^{b_2} \cdots x_k^{b_k}$.
5. For each group of permutations in Exercise 10, Section 7.2, compute the cycle index.
6. Given D, R, and G as in Exercise 6, Section 7.4:
   (a) Find the number of distinct colorings by Theorem 7.5.
   (b) Repeat using Corollary 7.5.1.
7. Repeat Exercise 6 for D, R, and G as in Exercise 7, Section 7.4.
8. In Example 7.2, check that if $k = 3$, there are six distinct necklaces.
9. In Example 7.2, use Theorem 7.5 to find the number of distinct necklaces if
   (a) the number of colors is 2 and k is 4;
   (b) the number of colors is 2 and k is 5;
   (c) the number of colors is 3 and k is 3;
   (d) the number of colors is 3 and k is 4.
10. Repeat Exercise 9 using Corollary 7.5.1.
11. (a) In Exercise 9, Section 7.1, use Theorem 7.5 to find the number of equivalence classes.
    (b) Repeat using Corollary 7.5.1.
12. (a) In Exercise 13, Section 7.1, use Theorem 7.5 to find the number of distinct colorings.
    (b) Repeat using Corollary 7.5.1.
13. (a) In Exercise 8, Section 7.1, use Theorem 7.5 to find the number of distinct colorings.
    (b) Repeat using Corollary 7.5.1.
14. In Example 7.1, for each π_i, $i = 1, 2, 3, 4$, verify that $\text{Inv}\,(\pi_i^*) = m^{\text{cyc}\,(\pi_i)}$.
15. (a) Use Theorem 7.5 to find the number of nonisomorphic graphs of $p = 3$ vertices. (See Exercise 16, Section 7.4.)
    (b) Repeat using Corollary 7.5.1.
16. Repeat Exercise 15 for $p = 4$.

*This subsection can be omitted.

17. **(a)** In the situation of Exercise 18, Section 7.4, use Theorem 7.5 to find the number of distinct switching functions.
   (b) Repeat using Corollary 7.5.1.

18. Repeat Exercise 17 for the situation of Exercise 19, Section 7.4.

19. If the definition of sameness of Exercise 19, Section 7.4, is adopted, find the number of distinct switching functions of three variables given that the cycle index for the appropriate group of permutations is

$$\tfrac{1}{12}[x_1^8 + 4x_2^4 + 2x_1^2x_3^2 + 2x_2\,x_6 + 3x_1^4x_2^2].$$

20. Consider a cube in 3-space. There are eight vertices. The following symmetries correspond to permutations of these vertices. Encode each of these symmetries in the form $x_1^{b_1}x_2^{b_2} \cdots x_k^{b_k}$ and compute the cycle index of the group G of all the permutations corresponding to these symmetries.
   (a) The identity symmetry.
   (b) Rotations by $180°$ around lines connecting the centers of opposite faces (there are three such).
   (c) Rotations by $90°$ or $270°$ around lines connecting the centers of opposite faces (there are six such).
   (d) Rotations by $180°$ around lines connecting the midpoints of opposite edges (there are six such).
   (e) Rotations by $120°$ around lines connecting opposite vertices (there are eight such).

21. In Exercise 20, find the number of distinct ways of coloring the vertices of the cube with two colors, red and blue.

22. Complete the solution of Exercise 21, Section 7.4.

23. A *transposition* is a cycle (ij). Show that every permutation is the product of transpositions. (*Hint:* It suffices to show that every cycle is the product of transpositions.)

24. Continuing with Exercise 23, write (123456) as a product of transpositions.

25. Write each permutation of Exercise 1, Section 7.2, as the product of transpositions.

26. Show that a permutation can be written as a product of transpositions in more than one way.

27. Although a permutation can be written in more than one way as a product of transpositions, it turns out that every way of writing the permutation as such a product either includes an even number of transpositions or an odd number. (For a proof, see Exercise 29.) A permutation, therefore, can be called *even* if every way of writing it as a product of transpositions uses an even number of transpositions, and *odd* otherwise.
   (a) Identify all even permutations of $\{1, 2, 3\}$.
   (b) Show that the collection of even permutations of $\{1, 2, \ldots, n\}$ forms a group.
   (c) Does the collection of odd permutations of $\{1, 2, \ldots, n\}$ form a group?

28. The number of permutations of $\{1, 2, \ldots, n\}$ with code $x_1^{b_1}x_2^{b_2} \cdots x_n^{b_n}$ is given by the formula

$$\frac{n!}{b_1!b_2! \cdots b_n!1^{b_1}2^{b_2} \cdots n^{b_n}}.$$

   This is called Cauchy's formula.
   (a) Verify this formula for $n = 5$, $b_1 = 3$, $b_2 = 1$, $b_3 = b_4 = b_5 = 0$.
   (b) Verify this formula for $n = 3$ and all possible codes.

29. Suppose that

$$D_n = (2 - 1)(3 - 2)(3 - 1)(4 - 3)(4 - 2)(4 - 1) \cdots (n - 1). \qquad (7.14)$$

If π is a permutation of $\{1, 2, \ldots, n\}$, define πD_n from D_n by replacing the term $(i - j)$ in (7.14) by the term $(\pi(i) - \pi(j))$.

(a) Find D_5.

(b) Find πD_5 if

$$\pi = \begin{pmatrix} 1 & 2 & 3 & 4 & 5 \\ 4 & 3 & 5 & 2 & 1 \end{pmatrix}.$$

(c) Show that if π is a transposition, then $\pi D_n = -D_n$.

(d) Conclude from part (c) that if π is the product of an even number of transpositions, then $\pi D_n = D_n$, and if π is the product of an odd number of transpositions, then $\pi D_n = -D_n$.

(e) Conclude that a permutation cannot both be written as the product of an even number of transpositions and the product of an odd number of transpositions.

7.6 POLYA'S THEOREM

7.6.1 The Inventory of Colorings

We may be interested in counting not just the number of distinct colorings, but the number of distinct colorings of a certain kind. For instance, in Example 7.1, we might be interested in counting the number of distinct 2-colorings of the 2×2 array in which exactly two black colors are used. In Example 7.5, we might be interested in counting the number of distinct molecules with at least one hydrogen atom. And so on. We shall now present a general result for answering questions of this type.

Let $D = \{a_1, a_2, \ldots, a_n\}$ be the set of objects to be colored and $R = \{r_1, r_2, \ldots, r_m\}$ be the set of colors. We shall distinguish colorings by assigning a *weight* $w(r)$ to each color r. This weight can be either a symbol or a number.

If we have assigned weights to the colors, we can assign a *weight to a coloring*. It is defined to be the product of the weights of the colors assigned to the elements of D. To illustrate this, suppose that $R = \{x, y, z\}$ and $w(x) = 1$, $w(y) = 5$, $w(z) = 7$. Suppose that the objects being colored are the seven vertices of the first binary tree of Figure 7.15, and we color vertices 1, 2, 4, 6 with color x, vertices 3, 7 with color y, and vertex 5 with color z. Then the weight of this coloring is $w(x)^4 w(y)^2 w(z) = (1)^4(5)^2 7 = 175$. If $w(x) = r$, $w(y) = g$, and $w(z) = b$, then the weight of the coloring is $r^4 g^2 b$. We shall see below how the weight of a coloring encodes the coloring in a very useful way.

Suppose now that K is a set of colorings. The sum of the weights of colorings in K is called the *inventory* of K. For instance, suppose that $D = \{a, b, c, d\}$, $R = \{x, y, z\}$, and $w(x) = r$, $w(y) = g$, $w(z) = b$. Let colorings f_1, f_2, and f_3 in $C(D, R)$ be defined as follows:

$$f_1(a) = x, \quad f_1(b) = y, \quad f_1(c) = y, \quad f_1(d) = z,$$

$$f_2(a) = z, \quad f_2(b) = z, \quad f_2(c) = x, \quad f_2(d) = z,$$

$$f_3(a) = x, \quad f_3(b) = z, \quad f_3(c) = y, \quad f_3(d) = x.$$

Let $W(f_i)$ be the weight of coloring f_i. Then $W(f_1) = w(x)w(y)w(y)w(z) = rg^2 b$, and similarly $W(f_2) = rb^3$ and $W(f_3) = r^2 gb$. The inventory of the set $K = \{f_1, f_2, f_3\}$ is given by

$rg^2b + rb^3 + r^2gb$. If all the weights of colors are different symbols, then the weight of a coloring represents the distribution of colors used. For instance, $W(f_1) = rg^2b$ shows that f_1 used color x once, color y twice, and color z once. The inventory of a set of colorings summarizes the distribution of colors in the different colorings in the set. This is like a generating function.

Now suppose that G is a group of permutations of the set D and that f and g are two equivalent colorings in $C(D, R)$. Then as observed in Section 7.4.2, there is a π in G so that for all a in D, $g(a) = f(\pi(a))$. If $D = \{a_1, a_2, \ldots, a_n\}$, then

$$W(f) = w[f(a_1)]w[f(a_2)] \cdots w[f(a_n)] \tag{7.15}$$

and

$$W(g) = w[g(a_1)]w[g(a_2)] \cdots w[g(a_n)]. \tag{7.16}$$

Since π is a permutation, the set $\{a_1, a_2, \ldots, a_n\}$ has exactly the same elements as the set $\{\pi(a_1), \pi(a_2), \ldots, \pi(a_n)\}$. Thus, (7.15) implies that

$$W(f) = w[f(\pi(a_1))]w[f(\pi(a_2))] \cdots w[f(\pi(a_n))]. \tag{7.17}$$

But since $g(a) = f(\pi(a))$, (7.16) and (7.17) imply that $W(f) = W(g)$. Thus, we have shown the following.

Theorem 7.6. If colorings f and g are equivalent, they have the same weight.

As a result of this theorem, we can speak of the *weight of an equivalence class of colorings* or, what is the same, the *weight of a pattern*. This is the weight of any coloring in this class. We shall also be able to speak of the *inventory of a set of patterns* or of a set of equivalence classes, the *pattern inventory*, as the sum of the weights of the patterns in the set. For instance, let us consider Example 7.1, the colorings of the 2×2 arrays. There are six patterns of colorings, as shown in Figure 7.8. Let color black have weight b and color white weight w. Then the coloring of class 1 of Figure 7.8 has weight b^4, all the colorings of class 2 have weight b^3w, all of class 3 have weight b^2w^2, all of class 4 have weight b^2w^2, all of class 5 have weight bw^3, and the coloring of class 6 has weight w^4. Note that two different equivalence classes can have the same weight. The pattern inventory is given by

$$b^4 + b^3w + 2b^2w^2 + bw^3 + w^4. \tag{7.18}$$

We find that there is one equivalence class using four black colors, one using three black and one white, two using two blacks and two whites, and so on. This information can be read directly from the pattern inventory. If we simply wanted to find the number of patterns, we would proceed as we did with generating functions in Chapter 4, and take all the weights to be 1. Here, setting $b = w = 1$ in (7.18) gives us 6, the number of patterns. If we wanted to find the number of patterns using no black, we would set $w(\text{black}) = 0$ and $w(\text{white}) = 1$, or equivalently let $b = 0$ and $w = 1$ in (7.18). The result is 1. There is only one pattern with no black, that corresponding to the term w^4 in the pattern inventory. We shall now seek a method for computing the pattern inventory without knowing the equivalence classes.

7.6.2 Computing the Pattern Inventory

Theorem 7.7 (Polya's Theorem).* Suppose that G is a group of permutations on a set D and $C(D, R)$ is the collection of all colorings of D using colors in R. If w is a weight assignment on R, then the pattern inventory of colorings in $C(D, R)$ is given by

$$P_G\left(\sum_{r \in R} w(r), \sum_{r \in R} [w(r)]^2, \sum_{r \in R} [w(r)]^3, \ldots, \sum_{r \in R} [w(r)]^k\right),$$

where $P_G(x_1, x_2, x_3 \ldots, x_k)$ is the cycle index of the group G.

Note that Corollary 7.5.1 is a special case of this theorem in which $w(r) = 1$ for all r in R. To illustrate the theorem, let us return to Example 7.1, the 2×2 arrays, one more time. Note that G consists of the permutations $\pi_1 = (1)(2)(3)(4)$, $\pi_2 = (1432)$, $\pi_3 = (13)(24)$, and $\pi_4 = (1234)$. Thus, the cycle index is given by

$$P_G(x_1, x_2, x_3, x_4) = \tfrac{1}{4}(x_1^4 + 2x_4 + x_2^2).$$

Now let us assign a weight b to a black coloring and a weight w to a white coloring. Then $R = \{\text{black, white}\}$ and

$$\sum_{r \in R} w(r) = b + w, \qquad \sum_{r \in R} [w(r)]^2 = b^2 + w^2, \qquad \sum_{r \in R} [w(r)]^3 = b^3 + w^3,$$

$$\sum_{r \in R} [w(r)]^4 = b^4 + w^4.$$

By Polya's Theorem, the pattern inventory is given by taking $P_G(x_1, x_2, x_3, x_4)$ and substituting $\sum_{r \in R} w(r)$ for x_1, $\sum_{r \in R} [w(r)]^2$ for x_2, and so on. Thus, the pattern inventory is

$$\tfrac{1}{4}[(b + w)^4 + 2(b^4 + w^4) + (b^2 + w^2)^2]. \tag{7.19}$$

By using the binomial expansion (Theorem 2.7), we can expand out (7.19) and obtain (7.18), which was our previous description of the pattern inventory.

Suppose that we allow three different colors in coloring the 2×2 array, black, white, and red. If we let $w(\text{red}) = r$, then we find the pattern inventory is

$$\tfrac{1}{4}[(b + w + r)^4 + 2(b^4 + w^4 + r^4) + (b^2 + w^2 + r^2)^2] = b^4 + w^4 + r^4 + b^3w + w^3b$$

$$+ b^3r + r^3b + w^3r + r^3w + 2b^2w^2 + 2b^2r^2 + 2w^2r^2 + 3b^2wr + 3w^2br + 3r^2wb.$$

We see, for instance, that there are three patterns with two blacks, one white, and one red. One example of each of these patterns is shown in Figure 7.17. Any other pattern using two black, one white, and one red can be obtained from one of these by rotation. The number of patterns is obtained by substituting $b = w = r = 1$ into the pattern inventory. Notice how, once having computed the cycle index, we can apply it easily to do a great many different counting procedures without having to repeat the computation of this index.

---

*Polya's fundamental theorem was first presented in his classic paper (Polya [1937]). The result was anticipated by Redfield [1927], but few people understood Redfield's results and Polya was unaware of them. A generalization of Polya's Theorem can be found in de Bruijn [1959]—for an exposition of this, see, for example, Liu [1968]. A well-known exposition of Polya theory is in the paper by de Bruijn [1964].

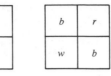

Figure 7.17 Examples of the three different patterns of colorings of the 2 × 2 array using two blacks, one white, and one red.

Let us next consider Example 7.5, the organic molecules. We have already noted that

$$P_G(x_1, x_2, x_3) = \frac{1}{12}(x_1^4 + 8x_1x_3 + 3x_2^2).$$

Suppose that we want to find the number of distinct molecules (patterns) containing at least one chlorine atom. It is a little easier to compute first the number of patterns having no chlorine atoms. This can be obtained by assigning the weight of 1 to each color CH_3, C_2H_5, and H, and the weight of 0 to the color Cl. Then for all $k \geq 1$,

$$\sum_{r \in R} [w(r)]^k = [w(CH_3)]^k + [w(C_2H_5)]^k + [w(H)]^k + [w(Cl)]^k = 1 + 1 + 1 + 0 = 3.$$

It follows that the pattern inventory is given by $\frac{1}{12}(3^4 + 8(3)(3) + 3(3)^2) = 15$. Since we have previously calculated that there are 36 patterns in all, the number with at least one chlorine atom is $36 - 15 = 21$.

Continuing with this example, suppose that we assign a weight of 1 to each color except Cl and a weight of c to Cl. Then the pattern inventory is given by

$$\tfrac{1}{12}[(c + 3)^4 + 8(c + 3)(c^3 + 3) + 3(c^2 + 3)^2] = c^4 + 3c^3 + 6c^2 + 11c + 15.$$

We conclude that there is one pattern consisting of four chlorine atoms, while three patterns consist of three chlorine atoms, six of two chlorine atoms, eleven of one chlorine atom, and fifteen of no chlorine atoms.

7.6.3 The Case of Switching Functions*

Next let us turn to Example 7.3, the switching functions, and take $n = 2$. Then G consists of the permutations π_1^* and π_2^* given by (7.10). As before, it is natural to think of π_1^* as (1)(2)(3)(4) and π_2^* as (1)(23)(4). Hence,

$$P_G(x_1, x_2) = \tfrac{1}{2}(x_1^4 + x_1^2x_2).$$

Setting $w(0) = a$ and $w(1) = b$, we find that $\sum_{r \in R} [w(r)]^k = a^k + b^k$. Thus, the pattern inventory is given by

$$\tfrac{1}{2}[(a + b)^4 + (a + b)^2(a^2 + b^2)] = a^4 + 3a^3b + 4a^2b^2 + 3ab^3 + b^4.$$

The term $3a^3b$ indicates that there are three patterns of switching functions which assign three 0's and one 1. The reader might wish to identify these patterns.

*This subsection can be omitted.

7.6.4 Proof of Polya's Theorem*

We now present a proof of Polya's Theorem. We proceed by a series of lemmas. Throughout, let us assume that $R = \{1, 2, \ldots, m\}$.

Lemma 1. Suppose that D is divided up into disjoint sets $D_1, D_2, \ldots, D_p$. Let C be the subset of $C(D, R)$ that consists of all colorings f with the property that if a and b are both in D_i, some i, then $f(a) = f(b)$. Then the inventory of the set C is given by

$$[w(1)^{|D_1|} + w(2)^{|D_1|} + \cdots + w(m)^{|D_1|}] \times [w(1)^{|D_2|} + w(2)^{|D_2|} + \cdots + w(m)^{|D_2|}]$$

$$\times \cdots \times [w(1)^{|D_p|} + w(2)^{|D_p|} + \cdots + w(m)^{|D_p|}]. \tag{7.20}$$

Proof. Multiplying out (7.20), we get terms such as

$$w(i_1)^{|D_1|} w(i_2)^{|D_2|} \cdots w(i_p)^{|D_p|}.$$

This is the weight of the coloring that gives color i_1 to all elements of D_1, color i_2 to all elements of D_2, and so on. Thus, (7.20) gives the sum of the weights of colorings which color all of D_i the same color. Q.E.D.

Lemma 2. Suppose that $G^* = \{\pi_1^*, \pi_2^*, \ldots\}$ is a group of permutations of $C(D, R)$. For each π^* in G^*, let $\overline{w}(\pi^*)$ be the sum of the weights of all colorings f in $C(D, R)$ left invariant by π^*. Suppose that $C_1, C_2, \ldots$ are the equivalence classes of colorings and $w(C_i)$ is the common weight of all f in C_i. Then

$$w(C_1) + w(C_2) + \cdots = \frac{1}{|G^*|} [\overline{w}(\pi_1^*) + \overline{w}(\pi_2^*) + \cdots]. \tag{7.21}$$

Note that if all weights are 1, Lemma 2 reduces to Burnside's Lemma.

Proof of Lemma 2. The sum on the right-hand side of (7.21) adds up for each π^* the weights of all colorings f left fixed by π^*. Thus, $w(f)$ is added in here exactly the number of times it is left invariant by some π^*. This is, to use the terminology of Section 7.3, the number of elements in the stabilizer of f, St (f). By Lemma 1 of Section 7.3, $|\text{St } (f)| = |G^*|/|C(f)|$, where $C(f)$ is the equivalence class containing f. Therefore, if $C(D, R) = \{f_1, f_2, \ldots\}$, the right-hand side of (7.21) is given by

$$\frac{1}{|G^*|} [w(f_1) \cdot |\text{St } (f_1)| + w(f_2) \cdot |\text{St } (f_2)| + \cdots] =$$

$$\frac{1}{|G^*|} \left[w(f_1) \frac{|G^*|}{|C(f_1)|} + w(f_2) \frac{|G^*|}{|C(f_2)|} + \cdots \right],$$

which equals

$$\frac{w(f_1)}{|C(f_1)|} + \frac{w(f_2)}{|C(f_2)|} + \cdots. \tag{7.22}$$

If we add up the terms $w(f_i)/|C(f_i)|$ for f_i in equivalence class C_j, we get $w(C_j)$, since each $w(f_i) = w(C_j)$ and since $|C(f_i)| = |C_j|$. Thus, (7.22) equals $w(C_1) + w(C_2) + \cdots$. Q.E.D.

*This subsection can be omitted.

We are now ready to complete the proof of Polya's Theorem. In (7.21) of Lemma 2, the left-hand side is the pattern inventory. Recall that $\overline{w}(\pi^*)$ is the sum of the weights of the colorings f left invariant by π^*. Suppose that the permutation π has cycles D_1, $D_2, \ldots, D_p$ in its cycle decomposition. Note that a coloring f is left invariant by π^* iff $f(a) = f(b)$ whenever a and b are in the same D_i. Thus, by Lemma 1, (7.20) gives the inventory or the sum of the weights of the set of colorings left invariant by π^*, i.e., (7.20) gives $\overline{w}(\pi^*)$. Each term in (7.20) is of the form

$$[w(1)]^j + [w(2)]^j + \cdots + [w(m)]^j = \sum_{r \in R} [w(r)]^j, \tag{7.23}$$

where $j = |D_i|$. Thus, a term (7.23) occurs in (7.20) as many times as $|D_i|$ equals j, that is, as many times as π has a cycle of length j. This we denoted b_j in Section 7.5.4, when we defined the cycle index. Hence, $\overline{w}(\pi^*)$ or (7.20) can be rewritten as

$$\left[\sum_{r \in R} [w(r)]^1 \right]^{b_1} \left[\sum_{r \in R} [w(r)]^2 \right]^{b_2} \cdots.$$

Therefore, the right-hand side of (7.21) becomes

$$P_G \left(\sum_{r \in R} [w(r)]^1, \sum_{r \in R} [w(r)]^2, \ldots \right).$$

This proves Polya's Theorem.

EXERCISES FOR SECTION 7.6

1. Find the weight of each coloring in column a in Figure 7.1 if $w(\text{black}) = 2$ and $w(\text{white}) = 3$.
2. If $w(1) = u$ and $w(2) = v$, find the weights of colorings f and g in parts (a) and (b) of Exercise 6, Section 7.4.
3. Suppose that K consists of the colorings C_2, C_7, C_{10}, and C_{15} of Figure 7.8. If $w(\text{black}) = b$ and $w(\text{white}) = w$, find the inventory of the collection K.
4. Suppose that K consists of the switching functions T_2, T_3, T_8, T_9, and T_{15} of Table 7.3. Find the inventory of K if $w(0) = a$ and $w(1) = b$.
5. In the situation of Exercise 9, Section 7.1, find the pattern inventory if $w(\text{black}) = b$ and $w(\text{white}) = w$.
6. In the situation of Exercise 7, Section 7.4, find the pattern inventory if $w(1) = \alpha$ and $w(2) = \beta$.
7. Use Polya's Theorem to compute the number of distinct four-bead necklaces, where each bead has one of three colors.
8. In Example 7.4, suppose that we have four possible colors for the vertices. Use Polya's Theorem to find the number of distinct colorings of the tree.
9. In Example 7.5, find the number of distinct molecules with no CH_3's.
10. How many four-bead necklaces are there in which each bead is one of the colors b, r, or p, and there is at least one p?
11. Find the number of distinct switching functions of two variables that have at least one 1 in the range, that is, which assign 1 to at least one bit string.
12. In the situation of Exercise 18, Section 7.4, find the number of distinct switching functions of three variables that have at least one 0 in the range.

13. In Example 7.5, find the number of distinct molecules that have at least one Cl atom and at least one H atom.

14. Use Polya's Theorem to compute the number of nonisomorphic graphs with
   (a) three vertices;
   (b) three vertices and two edges;
   (c) three vertices and at least one edge;
   (d) four vertices;
   (e) four vertices and three edges;
   (f) four vertices and at least two edges.
   (See Exercises 16 and 17, Section 7.4). For further applications of Polya's Theorem to graph theory, see Behzad *et al.* [1979], Harary [1969], or Harary and Palmer [1973].

15. The vertices of a cube are to be colored and five colors are available, red, blue, green, yellow, and purple. Count the number of distinct colorings in which at least one green and one purple are used (see Exercises 20 and 21, Section 7.5).

16. Let $D = \{a, b, c, d\}$, $R = \{0, 1\}$, let G consist of the permutations $(1)(2)(3)(4)$, $(12)(34)$, $(13)(24)$, $(14)(23)$, and take $w(0) = 1$, $w(1) = x$.
   (a) Find $C(D, R)$.
   (b) Find G^*.
   (c) Find all equivalence classes of colorings under G^*.
   (d) Find the weights of all equivalence classes under G^*.
   (e) Let e_i be the number of colors of weight x^i and let $e(x)$ be the ordinary generating function of the e_i, that is, $\sum_{i=0}^{\infty} e_i x^i$. Compute $e(x)$.
   (f) Let E_j be the number of patterns of weight x^j and let $E(x)$ be the ordinary generating function of the E_j, that is, $E(x) = \sum_{j=0}^{\infty} E_j x^j$. Compute $E(x)$.
   (g) Show that for $e(x)$ and $E(x)$ as computed in parts (e) and (f),

$$E(x) = P_G[e(x), e(x^2), e(x^3), \dots].\tag{7.24}$$

17. Repeat Exercise 16 for $D = \{a, b, c\}$, $R = \{0, 1\}$, G the set of permutations $(1)(2)(3)$ and $(12)(3)$, and $w(0) = x^2$, $w(1) = x^7$.

18. Generalizing the results of Exercises 16 and 17, suppose that for every r, $w(r) = x^p$ for some nonnegative integer p. Let $e(x)$ and $E(x)$ be defined as in Exercise 16. Show that, in general, (7.24) holds.

REFERENCES FOR CHAPTER 7

BEHZAD, M., CHARTRAND, G., and LESNIAK-FOSTER, L., *Graphs and Digraphs*, Wadsworth, Belmont, Calif., 1979.

BURNSIDE, W., *Theory of Groups of Finite Order*, 2d. ed., Cambridge University Press, Cambridge, 1911. (Reprinted by Dover, New York, 1955.)

DE BRUIJN, N. G., "Generalization of Polya's Fundamental Theorem in Enumerative Combinatorial Analysis," *Ned. Akad. Wet., Proc. Ser. A 62, Indag. Math., 21* (1959), 59–79.

DE BRUIJN, N. G., "Polya's Theory of Counting," in E. F. Beckenbach (ed.), *Applied Combinatorial Mathematics*, Wiley, New York, 1964, pp. 144–184.

HARARY, F., *Graph Theory*, Addison-Wesley, Reading, Mass., 1969.

HARARY, F., and PALMER, E. M., *Graphical Enumeration*, Academic Press, New York, 1973.

HARRISON, M. A., *Introduction to Switching and Automata Theory*, McGraw-Hill, New York, 1965.

LIU, C. L., *Introduction to Combinatorial Mathematics*, McGraw-Hill, New York, 1968.

POLYA, G., "Kombinatorische Anzahlbestimmungen für Gruppen, Graphen und Chemische Verbindungen," *Acta Math.*, *68* (1937), 145–254.

PRATHER, R. E., *Discrete Mathematical Structures for Computer Science*, Houghton Mifflin, Boston, 1976.

REDFIELD, J. H., "The Theory of Group-Reduced Distributions," *Amer. J. Math.*, *49* (1927), 433–455.

REINGOLD, E. M., NIEVERGELT, J., and DEO, N., *Combinatorial Algorithms: Theory and Practice*, Prentice-Hall, Englewood Cliffs, N.J., 1977.

STONE, H. S., *Discrete Mathematical Structures and Their Applications*, Science Research Associates,

PART III The Existence Problem

8 The Pigeonhole Principle and its Generalizations*

8.1 PIGEONS IN HOLES

8.1.1 The Simplest Version of the Pigeonhole Principle

In this part of the book, we study the second major problem in combinatorics, the existence problem. In combinatorics, one of the most widely used tools for proving that a certain kind of arrangement or pattern *exists* is the *pigeonhole principle*. Stated informally, this principle says the following: If there are "many" pigeons and "few" pigeonholes, then there must be two or more pigeons occupying the same pigeonhole. This principle is also called the *Dirichlet drawer principle†*, the *shoebox principle*, and other names. It says that if there are many objects (shoes) and few drawers (shoeboxes), then some drawer (shoebox) must have two or more objects (shoes). In this chapter we present several variants of this basic combinatorial principle and several applications of it. We then discuss a deep combinatorial generalization of the principle, known as Ramsey's Theorem. Note that the pigeonhole principle simply states that there must *exist* two or more pigeons occupying the same pigeonhole. It does not help us to identify such pigeons.

Let us start by stating the pigeonhole principle more formally.

Theorem 8.1 (Pigeonhole Principle). If $k + 1$ pigeons are placed into k pigeonholes, then at least one pigeonhole will contain two or more pigeons.

*A short version of this chapter, accessible at a relatively elementary level, would cover Section 8.1 and, if time permits, as much of Sections 8.2.1 and 8.2.2 as possible. The rest of the chapter should be accessible to more advanced students.

†Although the origin of the pigeonhole principle is not clear, it was widely used by the nineteenth-century mathematician Peter Dirichlet.

To illustrate Theorem 8.1, we note that if there are 13 people in a room, at least two of them are sure to have a birthday in the same month. Similarly, if there are 677 people chosen from the telephone book, then there will be at least two whose first and last names begin with the same letter. The next few examples are somewhat deeper.

Example 8.1   Clique Number and Chromatic Number

Suppose that G is a graph. A *clique* in G is a collection of vertices, each joined to the other by an edge. For instance, in the graph G of Figure 8.1, $\{a, b, c\}$, $\{d, e\}$, and $\{a, b, c, d\}$ are cliques. The *clique number* of G, $\omega(G)$, is the size of the largest clique of G. In our example, $\omega(G) = 4$. Recall that the chromatic number of G, $\chi(G)$, is the size of the smallest set of colors so that the vertices of G can be colored with these colors, with two vertices joined by an edge getting different colors. Since all vertices in a clique must receive different colors, the pigeonhole principle implies that $\chi(G) \geq \omega(G)$. To see why, let the vertices of a clique be the pigeons and the colors the pigeonholes.

Example 8.2   Manufacturing Furniture

A manufacturer of custom furniture makes at least one table every day over a period of 30 days, and he averages no more than $1\frac{1}{2}$ tables per day. Then there must be a period of consecutive days during which he makes *exactly* 14 tables. To see why, let a_i be the number of tables he has made through the end of the ith day. Since he makes at least one table each day, and at most 45 tables in 30 days, we have

$$a_1 < a_2 < \cdots < a_{30},$$

$$a_1 \geq 1,$$

$$a_{30} \leq 45.$$

Also,

$$a_1 + 14 < a_2 + 14 < \cdots < a_{30} + 14 \leq 45 + 14 = 59.$$

Now consider the following numbers:

$$a_1, a_2, \ldots, a_{30}, a_1 + 14, a_2 + 14, \ldots, a_{30} + 14.$$

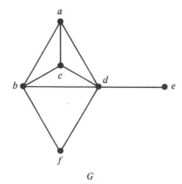

G

Figure 8.1   A graph of clique number 4.

These are 60 numbers, each between 1 and 59. By the pigeonhole principle, two of these numbers are equal. Since $a_1, a_2, \ldots, a_{30}$ are all different and $a_1 + 14$, $a_2 + 14$, $\ldots, a_{30} + 14$ are all different, there are i and j so that $a_i = a_j + 14$. Thus, between days j and i, the manufacturer makes exactly 14 tables.

8.1.2 Some Generalizations and Applications of the Pigeonhole Principle

Let us now state some stronger versions of the pigeonhole principle. In particular, if $2k + 1$ pigeons are placed into k pigeonholes, then at least one pigeonhole will contain more than two pigeons. This result follows since if every pigeonhole contains at most two pigeons, then since there are k pigeonholes, there would be at most $2k$ pigeons in all. By the same line of reasoning, if $3k + 1$ pigeons are placed into k pigeonholes, then at least one pigeonhole will contain more than three pigeons.

Speaking generally, we have the following theorem.

Theorem 8.2. If m pigeons are placed into k pigeonholes, then at least one pigeonhole will contain more than

$$\left\lfloor \frac{m-1}{k} \right\rfloor$$

pigeons.

Proof. If the largest number of pigeons in a pigeonhole is at most $\lfloor (m-1)/k \rfloor$, then the total number of pigeons is at most

$$k \left\lfloor \frac{m-1}{k} \right\rfloor \leq m - 1 < m. \qquad \text{Q.E.D.}$$

To illustrate this theorem, note that if there are 40 people in a room, a group of more than three will have a common birth month, for $\lfloor 39/12 \rfloor = 3$.

To give another application of this result, suppose we know that a computer memory has a capacity of 8000 bits in eight storage locations. Then we know that we have room in at least one location for at least 1000 bits. For $m = 8000$ and $k = 8$, so $\lfloor (m-1)/k \rfloor = 999$. Similarly, if a factory has 40 electrical outlets with a total 9600-volt capacity, we know there must be at least one outlet with a capacity of 240 or more volts.

These last two examples illustrate the following corollary of Theorem 8.2, whose formal proof is left to the reader (Exercise 22).

Corollary 8.2.1. The average value of a set of numbers is between the smallest and the largest of the numbers.

An alternative way to state this corollary is:

Corollary 8.2.2. Given a set of numbers, there is always a number in the set whose value is at least as large (at least as small) as the average value of the numbers in the set.

Example 8.3   The Chromatic Number and the Independence Number of a Graph

To give one further application of Corollary 8.2.2, suppose that G is a graph and W is a set of vertices in G. W is called an *independent set* of G if no two vertices of W are joined by an edge. The *independence number* $\alpha(G)$ is the size of a largest independent set of G. For instance, in the graph of Figure 8.1, $\{a, f\}$ and $\{e, c, f\}$ are independent sets. There is no independent set of four vertices, so $\alpha(G) = 3$. Suppose that $\chi(G)$ is the chromatic number of G. Let us suppose the vertices of G have been colored. There can be no edges between vertices of the same color, so all vertices of a given color define an independent set. Thus, a coloring of the $n = |V(G)|$ vertices of G in $\chi(G)$ colors partitions the vertices into $k = \chi(G)$ "color classes," each defining an independent set. The average size of such an independent set is $n/k = |V(G)|/\chi(G)$. Thus, there is at least one independent set of size at least n/k, that is,

$$\alpha(G) \geq \frac{|V(G)|}{\chi(G)}$$

or

$$\chi(G)\alpha(G) \geq |V(G)|. \tag{8.1}$$

This well-known formula is very useful in graph theory.

Example 8.4   Minicomputers and Printers*

There are 15 minicomputers and 10 printers in a workroom. We never expect more than 10 computers to be in use at one time. Every 5 minutes, some subset of the computers requests printers. We wish to connect each computer to some of the printers, in such a way that we use as few connections as possible, but we are always sure that a computer will have a printer to use. (A printer can be used by at most one computer at a time.) How many connections are needed? To answer this question, note that if there are fewer than 60 connections, then the average printer will have fewer than six connections, so by Corollary 8.2.2, there will be some printer that will be connected to five or fewer computers. If the remaining 10 computers were used at one time, there would be only 9 printers left for them. Thus, at least 60 connections are required. It is left to the reader (Exercise 21) to show that there is an arrangement with 60 connections that has the desired properties.

Another application of Theorem 8.2 is a result about increasing and decreasing subsequences of a sequence of numbers. Consider the sequence of numbers $x_1, x_2, \ldots, x_p$. A *subsequence* is any sequence $x_{i_1}, x_{i_2}, \ldots, x_{i_q}$ such that $1 \leq i_1 < i_2 < \cdots < i_q \leq p$. For instance, if $x_1 = 9$, $x_2 = 6$, $x_3 = 14$, $x_4 = 8$, and $x_5 = 17$, then we have the sequence 9, 6, 14, 8, 17; the subsequence x_2, x_4, x_5 is the sequence 6, 8, 17; the subsequence x_1, x_3, x_4, x_5 is the sequence 9, 14, 8, 17; and so on. A subsequence is *increasing* if its entries go successively up in value, and *decreasing* if its entries go successively down in value. In our

---

*This example is due to Tucker [1980].

example, a longest increasing subsequence is 9, 14, 17, and a longest decreasing subsequence is 14, 8.

To give another example, consider the sequence

$$12, 5, 4, 3, 8, 7, 6, 11, 10, 9.$$

A longest increasing subsequence is 5, 8, 11 and a longest decreasing subsequence is 12, 11, 10, 9. These two examples illustrate the following theorem, whose proof depends on Theorem 8.2.

Theorem 8.3 (Erdös and Szekeres [1935]). Given a sequence of $n^2 + 1$ *distinct* integers, either there is an increasing subsequence of $n + 1$ terms or a decreasing subsequence of $n + 1$ terms.

Note that $n^2 + 1$ is required for this theorem; that is, the conclusion can fail for a sequence of fewer than $n^2 + 1$ integers. For example, consider the sequence

$$4, 3, 2, 1, 8, 7, 6, 5, 12, 11, 10, 9, 16, 15, 14, 13.$$

This is a sequence of 16 integers arranged so that the longest increasing subsequences and the longest decreasing subsequences are four terms long.

Proof of Theorem 8.3. Let the sequence be

$$x_1, x_2, \ldots, x_{n^2+1}.$$

Let t_i be the number of terms in the longest increasing subsequence beginning at x_i. If any t_i is at least $n + 1$, the theorem is proved. Thus, assume that each t_i is between 1 and n. We therefore have $n^2 + 1$ pigeons (the $n^2 + 1$ t_i's) to be placed into n pigeonholes (the numbers $1, 2, \ldots, n$). By Theorem 8.2, there is a pigeonhole containing at least

$$\left\lfloor \frac{(n^2 + 1) - 1}{n} \right\rfloor + 1 = n + 1$$

pigeons. That is, there are at least $n + 1$ t_i's which are equal. We shall show that the x_i's associated with these t_i's form a decreasing subsequence. For suppose that $t_i = t_j$, with $i < j$. We shall show that $x_i > x_j$. If $x_i \leq x_j$, then $x_i < x_j$ because of the hypothesis that the $n^2 + 1$ integers are all distinct. Then x_i followed by the longest increasing subsequence beginning at x_j forms an increasing subsequence of length $t_j + 1$. Thus, $t_i \geq t_j + 1$, which is a contradiction. $\qquad$ Q.E.D.

To illustrate this proof, let us consider the following sequence:

$$10, 3, 2, 1, 6, 5, 4, 9, 8, 7.$$

Here $n = 3$, and we have

$$t_1 = 1, \quad t_2 = 3, \quad t_3 = 3, \quad t_4 = 3, \quad t_5 = 2, \quad t_6 = 2,$$
$$t_7 = 2, \quad t_8 = 1, \quad t_9 = 1, \quad t_{10} = 1.$$

Hence, there are four 1's among the t_i's, and the corresponding x_i's, namely x_1, x_8, x_9, x_{10}, form a decreasing subsequence, 10, 9, 8, 7.

We close this section by stating one more generalization of the pigeonhole principle, whose proof we leave to the reader (Exercise 23). In Section 8.2 we shall obtain a very general result of this form.

Theorem 8.4. Suppose that $p_1, p_2, \ldots, p_k$ are positive integers. If

$$p_1 + p_2 + \cdots + p_k - k + 1$$

pigeons are put into k pigeonholes, then either the first pigeonhole contains at least p_1 pigeons, or the second pigeonhole contains at least p_2 pigeons, or $\ldots$, or the kth pigeonhole contains at least p_k pigeons.

EXERCISES FOR SECTION 8.1

1. How many people must be chosen to be sure that at least two have
   (a) the same first initial?
   (b) a birthday on the same day of the year?
   (c) the same last four digits in their social security numbers?
   (d) the same first three digits in their telephone numbers?
2. Repeat Exercise 1 if we ask for at least three people to have the desired property.
3. If five different pairs of socks are put unsorted into a drawer, how many individual socks must be chosen before we can be sure of finding a pair?
4. (a) How many three-digit bit strings must we choose to be sure that two of them agree on at least one digit?
   (b) How many n-digit bit strings?
5. Give examples of graphs G so that
   (a) $\chi(G) = \omega(G)$; (b) $\chi(G) > \omega(G)$.
6. Let $\theta(G)$ be the size of the smallest set S of cliques of G so that every vertex is in some clique of S. What is the relation between $\theta(G)$ and $\alpha(G)$?
7. If a factory has 95 worktables with a total of 465 seats, can we be sure that there is a table with at least 7 seats?
8. If a school has 400 courses with an average of 40 students, what conclusion can you draw about the largest course?
9. If a graph has 100 vertices and seven connected components, what can you say about the largest component? The smallest?
10. If a telephone switching network of 20 switching stations averages 65,000 connections for each station, what can you say about the number of connections in the smallest station?
11. For each graph of Figure 3.43:
    (a) Find $\alpha(G)$.
    (b) Find $\chi(G)$ and verify the inequality (8.1).
12. A social worker has 77 days to make his rounds. He wants to make at least one visit a day, and has 132 visits to make. Is there a period of consecutive days in which he makes exactly 21 visits? Why?
13. A tennis player preparing for a tournament wants to practice by playing at least one match a day over a period of 50 days, but no more than 75 matches in all.
    (a) Show that during those 50 days, there is a period of consecutive days during which the player plays exactly 24 matches.
    (b) Is the conclusion still true for 30 matches?

14. Find the longest increasing and longest decreasing subsequences of each of the following sequences and check that your conclusions verify the Erdös–Szekeres Theorem.
 (a) 5, 3, 8, 2, 1
 (b) 7, 6, 8, 1, 2, 5, 9, 11, 14, 4
 (c) 6, 11, 3, 9, 4, 13, 15, 18, 10, 8.

15. Give an example of a sequence of 25 distinct integers which has neither an increasing nor a decreasing subsequence of 6 terms.

16. If the vertices of the graph Z_{11}, the circuit of length 11, are colored in four colors, what can you say about the size of the largest set of vertices each of which gets the same color?

17. An employee's time clock shows that he worked 81 hours over a period of 10 days. Show that on some pair of consecutive days, the employee worked at least 17 hours.

18. A computer is used for 300 hours over a period of 15 days. Show that on some period of 3 consecutive days, the computer was used at least 60 hours.

19. There are 25 executives in a corporation sharing 12 secretaries. Every hour, some group of the executives needs secretarial help. We never expect more than 12 executives to require secretarial help at any given time. We give each secretary a list of the executives he or she is working for, and make sure that each executive is on at least one secretary's list. If the number of names on each of the lists is added up, the total is 95. Show that it is possible that at some hour some executive might not be able to obtain secretarial help.

20. Consider the following sequence:

$$9, \ 8, \ 4, \ 3, \ 2, \ 7, \ 6, \ 5, \ 10, \ 1.$$

Find the numbers t_l as defined in the proof of Theorem 8.3, and use these t_i's to find a decreasing subsequence of at least four terms.

21. In Example 8.4, show that there is an arrangement with 60 connections which has the desired properties.

22. Prove Corollary 8.2.1 from Theorem 8.2.

23. Prove Theorem 8.4.

24. Show that if $n + 1$ numbers are selected from the set $\{1, 2, 3, \ldots, 2n\}$, then one of these will divide a second one of them.

25. Prove that in a group of at least 2 people, there are always 2 people who have the same number of acquaintances in the group.

26. Given a sequence of p integers $a_1, a_2, \ldots, a_p$, show that there exist consecutive terms in the sequence whose sum is divisible by p. That is, show that there are i and j, with $1 \le i \le j \le p$, such that $a_i + a_{i+1} + \cdots + a_j$ is divisible by p.

8.2 RAMSEY THEORY

8.2.1 Ramsey's Theorem

In this section we shall state a deep generalization of the pigeonhole principle, which is due to Ramsey [1930]. We shall mention a variety of applications of the theorem in Section 8.3. For a further discussion of Ramsey's famous theorem and related results, the reader is referred to Graham [1981] and Graham, et al. [1980]. See also Chung and Grinstead [1983] for a recent survey article and the special issue on Ramsey theory of the *Journal of Graph Theory*, Volume 7, Number 1, Spring 1983.

To begin with, let us give one more relatively simple application of the pigeonhole principle.

Theorem 8.5. Assume that among 6 persons, each pair of persons are either friends or enemies. Then either there are 3 persons who are mutual friends or 3 persons who are mutual enemies.

Proof. Let a be any person. By the pigeonhole principle, of the remaining 5 people, either 3 or more are friends of a or 3 or more are enemies of a. Suppose first that b, c, and d are friends of a. If any 2 of these persons are friends, these 2 and a form a group of 3 mutual friends. If none of b, c, and d are friends, then b, c, and d form a group of 3 mutual enemies. The argument is similar if we suppose that b, c, and d are enemies of a.   Q.E.D.

We can restate Theorem 8.5 as follows.

Theorem 8.6. Suppose that S is any set of 6 elements. If we divide the 2-element subsets of S into two classes, X and Y, then either

1. there is a 3-element subset of S all of whose 2-element subsets are in X,

or

2. there is a 3-element subset of S all of whose 2-element subsets are in Y.

Generalizing these conclusions, suppose that p and q are integers with p, $q \geq 2$. We say that a positive integer N has the (p, q) *Ramsey property* if the following holds: Given any set S of N elements, if we divide the 2-element subsets of S into two classes X and Y, then either

1. there is a p-element subset of S all of whose 2-element subsets are in X,

or

2. there is a q-element subset of S all of whose 2-element subsets are in Y.

For instance, by Theorem 8.6, the number 6 has the $(3, 3)$ Ramsey property. However, the number 3 does not have the $(3, 3)$ Ramsey property. For consider the set $S = \{a, b, c\}$ and the division of 2-element subsets of S into $X = \{\{a, b\}, \{b, c\}\}$ and $Y = \{\{a, c\}\}$. Then clearly there is no 3-element subset of S all of whose 2-element subsets are in X or 3-element subset of S all of whose 2-element subsets are in Y. Similarly, the number 5 does not have the $(3, 3)$ Ramsey property. We shall show this below.

Note that if the number N has the (p, q) Ramsey property and $M > N$, then the number M has the (p, q) Ramsey property. (Why?) We shall make use of this observation throughout.

We can now state our main theorem, which is a generalization of Theorem 8.5 and, as we shall see, of the pigeonhole principle.

Theorem 8.7 (Ramsey's Theorem).* If p and q are integers with p, $q \geq 2$, then there is a positive integer N which has the (p, q) Ramsey property.

Theorem 8.7 is proved in Section 8.2.3.

---

*This theorem is essentially contained in the original paper by Ramsey [1930], which was mainly concerned with applications to formal logic. The basic results were rediscovered and popularized by Erdös and Szekeres [1935]. See Graham *et al.* [1980] for an account.

8.2.2 The Ramsey Numbers $R(p, q)$

By Ramsey's Theorem, it makes sense to speak of the smallest integer N which has the (p, q) Ramsey property. We shall denote this smallest number $R(p, q)$ and call it a *Ramsey number*. Note that by Theorem 8.6, $R(3, 3) \leq 6$. We shall show that $R(3, 3) = 6$. The problem of computing a Ramsey number is an example of an optimization problem, and so in trying to find Ramsey numbers, we are working on the third basic type of combinatorics problem, the optimization problem. The computation of the Ramsey numbers is in general a difficult problem. Very few Ramsey numbers are known explicitly.

To study the Ramsey numbers $R(p, q)$, it is convenient to look at them using graph theory. If G is a graph, its *complement* G^c is the graph with the same vertex set as G, and so that for all $a \neq b$ in $V(G)$, $\{a, b\} \in E(G^c)$ if and only if $\{a, b\} \notin E(G)$. Figure 8.2 shows a graph G and its complement G^c. In studying the Ramsey numbers $R(p, q)$, we shall think of a set S as the vertex set of a graph, and of a set of 2-element subsets of S as the edge set of this graph. Then, to say that a number N has the (p, q) Ramsey property means that whenever S is a set of N elements and we have a graph G with vertex set S, then if we divide the 2-element subsets of S into edges of G and edges of G^c (edges not in G), either there are p vertices all of which are joined by edges in G or there are q vertices all of which are joined by edges in G^c. Put another way, we have the following theorem.

Theorem 8.8. A number N has the (p, q) Ramsey property if and only if for every graph G of N vertices, either G has a complete p-gon (complete subgraph of p vertices) or G^c has a complete q-gon.

Corollary 8.8.1. $R(p, 2) = p$ and $R(2, q) = q$.

Proof. For every graph of p vertices, either it is complete or its complement has an edge. This shows that $R(p, 2) \leq p$. Certainly $R(p, 2) \geq p$. (Why?)   Q.E.D.

In the language of Theorem 8.8, Theorem 8.5 says that if G is a graph of 6 (or more) vertices, then either G has a triangle or G^c has a triangle. (The reader should try this out on a number of graphs of his or her choice.) More generally, every graph of (at least) $R(p, q)$ vertices has either a complete p-gon or its complement has a complete q-gon.

Consider now the graph $G = Z_5$, the circuit of length 5. Now G^c is again (isomorphic to) Z_5. Thus, neither G nor G^c has a triangle. We conclude that the number 5 does not have the $(3, 3)$ Ramsey property, so $R(3, 3) > 5$. This completes the proof of the following theorem.

Theorem 8.9. $R(3, 3) = 6$.

To use the terminology of Section 8.1, a complete p-gon in G is a clique of p vertices. A complete q-gon in G^c corresponds to an independent set of q vertices in G. Thus, we can restate Theorem 8.8 as follows.

Theorem 8.10. A number N has the (p, q) Ramsey property if and only if for every graph G of N vertices, either G has a clique of p vertices or G has an independent set of q vertices.

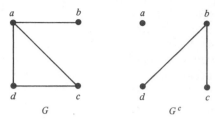

Figure 8.2 A graph G and its complement G^c.

There is still another way to restate Theorem 8.8, which is as follows.

Theorem 8.11. A number N has the (p, q) Ramsey property if and only if whenever we color the *edges* of K_N, the complete graph on N vertices, with each edge being colored either blue or red, then either K_N has a complete red p-gon or K_N has a complete blue q-gon.

Proof. Given an edge coloring, let G be the graph whose vertices are the same as those of K_N and whose edges are the red edges. Q.E.D.

We have already observed that Ramsey numbers are difficult to compute, and that very few Ramsey numbers are known. Indeed, the only known Ramsey numbers $R(p, q)$ are given in Table 8.1. We shall verify some parts of this table here and in Section 8.2.3. (We already know the first two lines to be true.)

To verify (at least partly) some of the numbers in Table 8.1, we derive some bounds on the Ramsey numbers. Consider the graph of Figure 8.3. This graph has 8 vertices. It also has no triangle (3-gon) and no independent set of 4 vertices. Thus, by Theorem 8.10, the number 8 does not have the (3, 4) Ramsey property. It follows that $R(3, 4) \geq 9$.

Table 8.1 The known Ramsey numbers $R(p, q)$

$R(p, 2) = p, R(2, q) = q$
$R(3, 3) = 6^a$
$R(3, 4) = R(4, 3) = 9^a$
$R(3, 5) = R(5, 3) = 14^a$
$R(3, 6) = R(6, 3) = 18^b$
$R(3, 7) = R(7, 3) = 23^c$
$R(3, 9) = R(9, 3) = 36^d$
$R(4, 4) = 18^a$
Note: $R(3, 8) = R(8, 3) = 28$ or 29^d

[a] Greenwood and Gleason [1955].
[b] Kalbfleisch [1966], Kéry [1964].
[c] Graver and Yackel [1968].
[d] Grinstead and Roberts [1982].

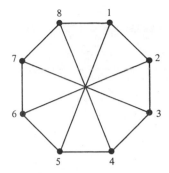

Figure 8.3 Graph that demonstrates $R(3, 4) \geq 9$.

Similarly, the graphs of Figures 8.4 and 8.5 show that $R(3, 5) \geq 14$ and $R(4, 4) \geq 18$ (Exercises 20 and 21).

8.2.3 Bounds on Ramsey Numbers*

We now derive some general upper bounds on Ramsey numbers, and these results will help us verify some of the entries in Table 8.1.

Theorem 8.12 (Erdös and Szekeres [1935] and Greenwood and Gleason [1955]). For all integers $p, q \geq 3$,†

$$R(p, q) \leq R(p, q - 1) + R(p - 1, q).$$

Proof. By Theorem 8.10, it suffices to show the following. Suppose that G is a graph of $R(p, q - 1) + R(p - 1, q)$ vertices. Then either G has a clique of p vertices or G has an independent set of q vertices. We demonstrate this fact.

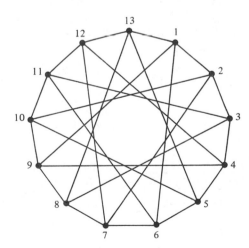

Figure 8.4 Graph that demonstrates $R(3, 5) \geq 14$.

*This subsection can be omitted without loss of continuity.
†The hypothesis $p, q \geq 3$ is necessary for $R(p, q - 1)$ and $R(p - 1, q)$ to be defined.

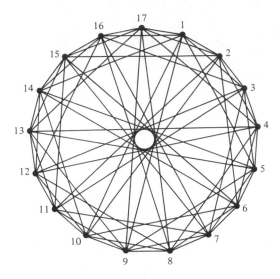

Figure 8.5 Graph that demonstrates $R(4, 4) \geq 18$.

Fix a vertex v in G. Let C be the set of neighbors of v in G, and D be all other vertices of G other than v. Since the number of vertices different from v is $R(p, q-1) + R(p-1, q) - 1$, the pigeonhole principle (Theorem 8.4) says that either C has at least $R(p-1, q)$ vertices or D has at least $R(p, q-1)$ vertices. We consider the second possibility. The argument in the case of the first possibility is the same. By definition of the Ramsey number $R(p, q-1)$, the subgraph of G generated by vertices in D either has a clique of p vertices or an independent set of $q-1$ vertices. Since v is not adjacent to any of the vertices of D, we conclude that the subgraph of G generated by vertices of $D \cup \{v\}$ either has a clique of p vertices or an independent set of q vertices. Thus, either G has a clique of p vertices or G has an independent set of q vertices. Q.E.D.

Note. This proof really gives a proof of Ramsey's Theorem, Theorem 8.7. The argument is by induction on $k = p + q$.

Corollary 8.12.1. Suppose that $p, q \geq 3$. If $R(p, q-1)$ and $R(p-1, q)$ are both even, then

$$R(p, q) \leq R(p, q-1) + R(p-1, q) - 1.$$

Proof. Let G have $R(p, q-1) + R(p-1, q) - 1$ vertices, which by assumption is an odd number. Note that in any graph, the number of vertices of odd degree is even. For, as we observed in Theorem 3.12, the sum of the degrees of all the vertices is twice the number of edges, an even number. Thus,

$$\sum \{\text{degree } v: \quad \text{degree } v \text{ even}\} + \sum \{\text{degree } v: \quad \text{degree } v \text{ odd}\}$$

is an even number. This implies that the latter sum is even, so there is an even number of vertices of odd degree. Since G has an odd number of vertices, it must have a vertex v of even degree. Now either D as defined in the proof of the theorem has at least $R(p, q-1)$

vertices or C as defined in the proof of the theorem has at least $R(p-1, q) - 1$ vertices. Since C must have an even number of vertices, the latter implies that C has at least $R(p-1, q)$ vertices. Then the proof of Theorem 8.12 implies that either G has a clique of p vertices or G has an independent set of q vertices.   Q.E.D.

Corollary 8.12.2.   If $p \geq 2$ and $q \geq 2$, then

$$R(p, q) \leq \binom{p + q - 2}{p - 1}.$$

Proof. We first observe that the result follows easily from Corollary 8.8.1 if $p = 2$ or $q = 2$. We now prove the result by induction on $p + q$. If $p + q \leq 5$, then $p, q \geq 2$ implies that $p = 2$ or $q = 2$, and the result holds. Thus, we may assume that $p + q \geq 6$, that $p, q \geq 3$, and that the result holds for p' and q' with $p' + q' = p + q - 1$. By Theorem 8.12,

$$R(p, q) \leq R(p, q-1) + R(p-1, q).$$

By the inductive assumption,

$$R(p, q) \leq \binom{p + q - 3}{p - 1} + \binom{p + q - 3}{p - 2}. \tag{8.2}$$

By Theorem 2.2, it follows that the right-hand side of (8.2) is equal to

$$\binom{p + q - 2}{p - 1},$$

which completes the proof.   Q.E.D.

We now apply these results. We know that $R(3, 3) = 6$ and, by Corollary 8.8.1, that $R(2, 4) = 4$. Since both numbers are even, Corollary 8.12.1 implies that

$$R(3, 4) \leq R(3, 3) + R(2, 4) - 1 = 6 + 4 - 1 = 9.$$

Thus, $R(3, 4) \leq 9$. Since we have already demonstrated $R(3, 4) \geq 9$ at the end of Section 8.2.2, we conclude that $R(3, 4) = 9$. Clearly, $R(4, 3) = R(3, 4)$. This verifies the third line of Table 8.1.

Next, note that $R(2, 5) = 5$, by Corollary 8.8.1. Thus, by Theorem 8.12,

$$R(3, 5) \leq R(3, 4) + R(2, 5) = 9 + 5 = 14,$$

so $R(3, 5) \leq 14$. We have already observed that $R(3, 5) \geq 14$, so we conclude that $R(3, 5) = 14$. Clearly, $R(5, 3) = R(3, 5)$, which verifies the fourth line of Table 8.1.

Other entries of Table 8.1 are studied by the results of this subsection in Exercises 22 and 23.

8.2.4 Generalizations of Ramsey's Theorem

In this subsection we note two generalizations of Ramsey's Theorem as stated earlier. The first arises if we consider not just 2-element subsets of an N-element set, but r-element subsets. Suppose that p, q, and r are integers with $p \geq r$, $q \geq r$, and $r \geq 1$. We say that a positive integer N has the $(p, q; r)$ *Ramsey property* if the following holds: Given any set S of N elements, if we divide the r-element subsets of S into two classes X and Y, then either

1. there is a p-element subset of S all of whose r-element subsets are in X,

or

2. there is a q-element subset of S all of whose r-element subsets are in Y.

Thus, the (p, q) Ramsey property is the same as the $(p, q; 2)$ Ramsey property.

Theorem 8.13 (Ramsey's Theorem—First Generalization).   If p, q, and r are integers with $p \geq r$, $q \geq r$, and $r \geq 1$, then there is a positive integer N which has the $(p, q; r)$ Ramsey property.

For a proof of Theorem 8.13, we refer the reader to Graham *et al.* [1980] or Liu [1972].

The *Ramsey number* $R(p, q; r)$ is defined to be the smallest integer N with the $(p, q; r)$ Ramsey property. Thus, $R(3, 3; 2) = R(3, 3) = 6$. Except for the few Ramsey numbers $R(p, q; 2)$ given by Table 8.1, very few Ramsey numbers are known explicitly. Theorems 8.14 and 8.15 give the only known exceptions to this statement.

Theorem 8.14.   Suppose that p and q are integers with $p, q \geq 1$. Then

$$R(p, q; 1) = p + q - 1.$$

Proof. First, the number $p + q - 1$ has the $(p, q; 1)$ Ramsey property. To show this, suppose that S is a set of $p + q - 1$ elements, and the 1-element subsets of S are divided into two classes, X and Y. Then by Theorem 8.4, either $|X| \geq p$ or $|Y| \geq q$. Hence, either (1) or (2) of the definition of the $(p, q; r)$ Ramsey property holds. We conclude that $p + q - 1$ has the $(p, q; 1)$ Ramsey property, and hence that $R(p, q; 1) \leq p + q - 1$. However, clearly, $R(p, q; 1) \geq p + q - 1$. To see this, it suffices to show that $p + q - 2$ does not have the $(p, q; 1)$ Ramsey property. This follows by splitting the 1-element subsets of a $(p + q - 2)$-element set S into two sets, X and Y, with $|X| = p - 1$ and $|Y| = q - 1$.

Q.E.D.

It should be clear from the proof of Theorem 8.14 why Ramsey's Theorem is thought of as a generalization of the pigeonhole principle.

Theorem 8.15.   If p, q, and r are integers, with $p \geq r$, $q \geq r$, $r \geq 1$, then

$$R(p, r; r) = p \qquad \text{and} \qquad R(r, q; r) = q.$$

The proof of this theorem is left as an exercise (Exercise 12).

To generalize Ramsey's Theorem even further, suppose that $p_1, p_2, \ldots, p_t, r$ are integers, with $p_i \geq r$, all i, with $r \geq 1$, and $t \geq 2$. We say that a positive integer N has the $(p_1, p_2, \ldots, p_t; r)$ *Ramsey property* if whenever S is a set of N elements and we divide the r-element subsets of S into t sets $X_1, X_2, \ldots, X_t$, then for some i, there is a p_i-element subset of S all of whose r-element subsets are in X_i.

Theorem 8.16 (Ramsey's Theorem—Second Generalization). Suppose that $p_1, p_2,$ $\ldots, p_t, r$ are integers, with $p_i \geq r$, all i, with $r \geq 1$, and $t \geq 2$. Then there is a finite number N which has the $(p_1, p_2, \ldots, p_t; r)$ Ramsey property.

For a proof of Theorem 8.16, we refer the reader to Graham *et al.* [1980] or Liu [1972]. Now we can define the *Ramsey number* $R(p_1, p_2, \ldots, p_t; r)$ to be the smallest number N that has the $(p_1, p_2, \ldots, p_t; r)$ Ramsey property.

To illustrate these ideas, let us consider the complete graph K_N of N vertices. Let us color the edges of K_N, coloring each edge in exactly one of the three colors, red, white, or blue. If $N \geq R(3, 3, 3; 2)$, we conclude that there is a monochromatic triangle in K_N, that is, a triangle all of whose edges are red, or all of whose edges are white, or all of whose edges are blue. (This result is closely related to Theorem 8.11.) Exercise 25 asks the reader to show that $R(3, 3, 3; 2) \leq 17$. In fact, Greenwood and Gleason [1955] show, by an argument using algebra, that $R(3, 3, 3; 2) \geq 17$. Berge [1973, p. 442] shows a coloring of the edges of K_{16} in the three colors red, blue, and green, so that there is no monochromatic triangle. This again shows that $R(3, 3, 3; 2) \geq 17$. We conclude that $R(3, 3, 3; 2) = 17$. Very few other numbers $R(p_1, p_2, \ldots, p_t; r)$ are known. See Exercises 13 and 14 for exceptions.

8.2.5 Graph Ramsey Numbers

Let $G_1, G_2, \ldots, G_t$ be graphs. An integer N is said to have the *graph Ramsey property* $(G_1, G_2, \ldots, G_t)$ if every coloring of the edges of the complete graph K_N in the t colors $1, 2, \ldots, t$ gives rise, for some i, to a subgraph that is (isomorphic to) G_i and is colored all in color i, that is, to a monochromatic G_i. The *graph Ramsey number* $R(G_1, G_2, \ldots, G_t)$ is the smallest N with the graph Ramsey property $(G_1, G_2, \ldots, G_t)$. Thus, $R(K_p, K_q)$ is $R(p, q)$.

Theorem 8.17. If $t \geq 2$ and $G_1, G_2, \ldots, G_t$ are arbitrary graphs of more than one vertex, then some integer N has the graph Ramsey property $(G_1, G_2, \ldots, G_t)$, so the graph Ramsey number $R(G_1, G_2, \ldots, G_t)$ is well defined.

Proof. Suppose that G_i has p_i vertices and that $N = R(p_1, p_2, \ldots, p_t; 2)$, and consider an edge-coloring of K_N in t colors. Then for some i, K_{p_i} is monochromatic. But G_i is a subgraph of K_{p_i}. Q.E.D.

Corollary 8.17.1. $R(G_1, G_2, \ldots, G_t) \leq R(|V(G_1)|, |V(G_2)|, \ldots, |V(G_t)|; 2)$.

To illustrate these ideas, note that by the corollary, if L_3 is the chain of three vertices, then $R(L_3, L_3) \leq R(3, 3) = 6$. In fact, it is easy to see that every coloring of the edges of K_3 in two colors, red and blue, gives rise to a red L_3 or a blue L_3. Thus, $R(L_3, L_3) \leq 3$. However, $R(L_3, L_3) \geq 3$ since it is easy to color the edges of K_2 using the two colors red and blue so that there is no monochromatic L_3. Thus, $R(L_3, L_3) = 3$.

The numbers $R(Z_4, Z_4, \ldots, Z_4)$ have had applications in network design in the Bell system (see Section 8.3.3). For further discussion of graph Ramsey numbers, see Exercises 26–28 and see the books by Graham [1981], Graham *et al.* [1980], and Behzad *et al.* [1979].

EXERCISES FOR SECTION 8.2

1. Show by exhibiting a division X and Y that
   (a) the number 4 does not have the (3, 3) Ramsey property;
   (b) the number 5 does not have the (3, 4) Ramsey property;
   (c) the number 6 does not have the (4, 4) Ramsey property;
   (d) the number 3 does not have the (3, 4; 1) Ramsey property;
   (e) the number 4 does not have the (8, 7; 1) Ramsey property;
   (f) the number 4 does not have the (8, 3; 3) Ramsey property.

2. Find the following Ramsey numbers.
   (a) $R(2, 2)$ (b) $R(2, 8)$ (c) $R(7, 2)$
   (d) $R(3, 3; 3)$ (e) $R(3, 5; 3)$.

3. Show that if the number N has the (p, q) Ramsey property and $M > N$, then the number M has the (p, q) Ramsey property.

4. For each of the graphs of Figure 8.6, either find a clique of 3 vertices or an independent set of 3 vertices, or conclude that neither of these can be found.

5. Let G be any graph of 11 vertices and chromatic number 3.
   (a) Does G necessarily have either a clique of 3 vertices or an independent set of 3 vertices?
   (b) Does G necessarily have either a clique of 4 vertices or an independent set of 3 vertices?

6. Let G be a graph of 16 vertices and largest clique of size 3.
   (a) Does G necessarily have an independent set of 4 vertices?
   (b) Of 5 vertices?

7. Let G be a tree of 20 vertices.
   (a) Does G necessarily have an independent set of 5 vertices?
   (b) Of 6 vertices?

8. Let G be a complete graph of 20 vertices and let the edges of G be colored either green or brown. If there is no brown triangle, what is the largest complete green m-gon you can be sure G has?

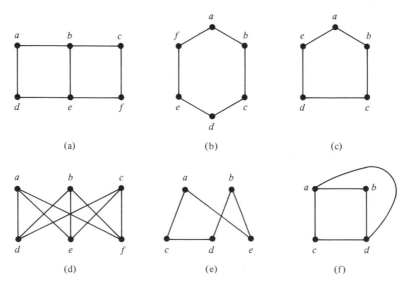

(a)   (b)   (c)

(d)   (e)   (f)

Figure 8.6 Graphs for Exercise 4, Section 8.2.

9. Color the edges of K_{10} either red or blue.
   (a) Show that if there are at least 4 red edges from one vertex, then there are 3 vertices all joined by red edges or 4 vertices all joined by blue edges.
   (b) Similarly, if there are at least 6 blue edges from a vertex, show that either there are 3 vertices all joined by red edges or 4 vertices all joined by blue edges.
   (c) Show that by parts (a) and (b), K_{10} has in any coloring of its edges with red or blue colors either 3 vertices all joined by red edges or 4 vertices all joined by blue edges.
   (d) Does part (c) tell you anything about a Ramsey number?

10. An interviewer wants to assign each job applicant interviewed to a rating of P (pass) or F (fail). She finds that no matter how she assigns the ratings, there are 3 people who receive the same rating. What is the least number of applicants she could have interviewed?

11. Repeat Exercise 10 if she always finds 4 people who receive the same rating.

12. Prove Theorem 8.15.

13. If $p \geq r \geq 1$, find $R(p, r, r, \ldots, r; r)$.

14. If $p_i \geq 1$, all i, and $t \geq 2$, find $R(p_1, p_2, \ldots, p_t; 1)$.

15. If $p = \max \{p_1, p_2, \ldots, p_t\}$, show that $R(p, p, \ldots, p; r) \geq R(p_1, p_2, \ldots, p_t; r)$.

16. Show that if the edges of a complete graph are colored in the colors red, white, blue, brown, green, or purple, then if the graph has sufficiently many vertices, there is a complete 4-gon all of whose edges have the same color.

17. The edges of a complete graph are colored red, yellow, brown, or purple. How many vertices must there be in the graph to guarantee that there will be 4 vertices each joined to the other by an edge of the same color? (Answer in general terms.)

18. Show that if the edges of the complete graph are colored in the colors red, white, blue, and green, then if there are sufficiently many vertices, and there is no red, white, or blue triangle, we can be sure there is a complete 12-gon all of whose edges are colored green.

19. Show that given a sequence of $R(n + 1, n + 1)$ distinct integers, either there is an increasing subsequence of $n + 1$ terms or a decreasing subsequence of $n + 1$ terms.

20. Use Figure 8.4 to show that $R(3, 5) \geq 14$.

21. Use Figure 8.5 to show that $R(4, 4) \geq 18$.

22. Use the theorems of Section 8.2.3 to obtain an upper bound for $R(4, 4)$ and verify the entry for $R(4, 4)$ in Table 8.1.

23. Use the theorems of Section 8.2.3 to obtain an upper bound for $R(3, 6)$. Does this upper bound agree with the value of $R(3, 6)$ in Table 8.1?

24. Show that
   (a) $R(5, 5) \leq 70$; (b) $R(5, 6) \leq 126$.

25. Color the edges of the graph K_{17} in red, white, or blue. This exercise will argue that there are 3 vertices all joined by edges of the same color, thus showing that $R(3, 3, 3; 2) \leq 17$.
   (a) Fix one vertex a. Show that of the edges joining this vertex, at least 6 must have the same color.
   (b) Suppose that the 6 edges in (a) are all red. These lead from a to 6 vertices, b, c, d, e, f, and g. Argue from here that either K_{17} has a red triangle, a blue triangle, or a white triangle.

26. Suppose that L_p is the chain of p vertices.
   (a) Find $R(L_3, L_4)$.
   (b) Find $R(L_3, Z_4)$.
   (c) Show that $R(L_4, L_4) = 5$.
   (d) Find $R(L_4, Z_4)$.
   (e) Find $R(Z_4, Z_4)$.

27. (Chvátal and Harary [1972]) Let $c(G)$ be the size of the largest connected component of G. Show that

$$R(G, H) \geq (\chi(G) - 1)(c(H) - 1) + 1.$$

28. (Chvátal [1977]) If T_m is a tree of m vertices, show that

$$R(T_m, K_n) = 1 + (m - 1)(n - 1).$$

29. We say that the Ramsey number $R(p_1, p_2, \ldots, p_t; r)$ *exists* if there is a number N with the $(p_1, p_2, \ldots, p_t; r)$ Ramsey property. Assuming that $R(p_1, p_2; 2)$ exists for all $p_1, p_2 \geq 2$, show directly that $R(p_1, p_2, p_3; 2)$ exists for all $p_1, p_2, p_3 \geq 2$ by showing that

$$R(p_1, p_2, p_3; 2) \leq R(p_1, R(p_2, p_3; 2); 2).$$

30. Argue analogously to Exercise 29, and using induction, that $R(p_1, p_2, \ldots, p_t; 2)$ exists for all $p_1, p_2, \ldots, p_t \geq 2$, and $t \geq 2$.

31. (a) Let $p(t) = R(3, 3, \ldots, 3; 2)$, with t 3's. Show that

$$p(t + 1) \leq (t - 1)[p(t) - 1] + 2,$$

all $t \geq 2$.

(b) Use the result of part (a) to derive an upper bound for $R(3, 3, 3; 2)$.

32. Show that

$$R(p_1, p_2, \ldots, p_t; 2) \leq R(p_1 - 1, p_2, \ldots, p_t; 2) + R(p_1, p_2 - 1, \ldots, p_t; 2) + \cdots$$

$$+ R(p_1, p_2, \ldots, p_t - 1; 2).$$

33. Show that

$$R(p_1 + 1, p_2 + 1, \ldots, p_t + 1; 2) \leq \binom{p_1 + p_2 + \cdots + p_t}{p_1, p_2, \ldots, p_t},$$

where

$$\binom{M}{p_1, p_2, \ldots, p_t}$$

is the multinomial coefficient of Section 2.12.

8.3 APPLICATIONS OF RAMSEY THEORY

In this section we present five applications of Ramsey Theory. The first application is a geometric one, the second and third involve communications, the fourth is to a problem of information retrieval in computer science, and the fifth is to a problem in decisionmaking.*

8.3.1 Convex m-gons

In this subsection, we present an application that was crucial in the development of Ramsey Theory. (For a historical account, see Graham *et al.* [1980].)

An m-gon in the plane is called *convex* if whenever two points are within the m-gon,

---

*The five applications are independent, and may be read in any order.

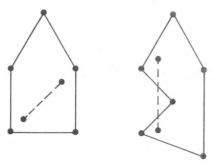

Figure 8.7 A convex 5-gon and a concave 6-gon.

then so is the straight line joining the two points. All other *m*-gons are called *concave*. (See Figure 8.7.) Let us say that a set of points in the plane is *in general position* if no three are collinear. Suppose that we have *n* points in general position in the plane. Can we guarantee that four of these points determine a convex quadrilateral (4-gon)? This is clearly false if $n = 4$ (see Figure 8.8). But we shall see that it is true for higher *n*. This is the observation with which we start.

The *convex hull* of any finite set of points is the smallest convex set containing the points. Note that *n* points in the plane determine a convex *n*-gon if and only if none of the points is in the interior of the convex hull of the remaining $n - 1$ points. Figure 8.8 illustrates this observation: There is one point in the interior of the convex hull of the remaining three points.

Theorem 8.18. If five points in the plane are in general position, then four of the points are the vertices of a convex quadrilateral.

Proof. Find a minimal subset of the five points such that their convex hull contains all five points: If this set consists of all five points as in Figure 8.9(a), then any four of the points determine a convex quadrilateral. If this set consists of four points, with one additional point inside, as in Figure 8.9(b), then these four points determine a convex quadrilateral. Finally, suppose that the set consists of three points, with two additional points inside, as in Figure 8.9(c). Then the two interior points determine a straight line and two of the remaining points lie on one side of this straight line. These two points and the two interior points determine a convex quadrilateral. Q.E.D.

In what follows, we shall ask if this theorem generalizes. Specifically, we shall ask this: If *n* is sufficiently large, and *n* points are located in general position in the plane, can we be sure that five of them determine a convex 5-gon? Can we be sure that *m* of them determine a convex *m*-gon? We shall answer these questions in the affirmative, by using Ramsey's Theorem. First, we prove the following preliminary result.

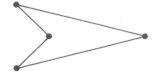

Figure 8.8 Four points in general position in the plane which do not determine a convex quadrilateral.

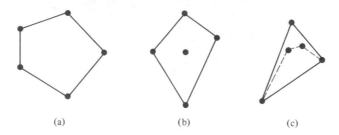

Figure 8.9   The possible cases of five points in general position in the plane.

Theorem 8.19.   If m points are located in general position in the plane, and all quadrilaterals formed from the m points are convex, then the m points are vertices of a convex m-gon.

Proof. As in the proof of Theorem 8.18, let q of the points, $V_1, V_2, \ldots, V_q$, form a convex q-gon with the rest of the points inside, as in Figure 8.10. Assume that $q < m$. Then one of the m points V must lie inside the convex q-gon. Now consider the triangles $V_1 V_2 V_3$, $V_1 V_3 V_4$, $V_1 V_4 V_5, \ldots, V_1 V_{q-1} V_q$. Now V must be inside one of these triangles, say $V_1 V_r V_{r+1}$ (again see Figure 8.10). Then V, V_1, V_r, V_{r+1} do not form a convex quadrilateral, which is a contradiction.   Q.E.D.

Theorem 8.20   **(Erdös and Szekeres [1935]).**   Given $m \geq 4$, there is a number $N(m)$ such that if n points are in general position in the plane and $n \geq N(m)$, then m of these points determine a convex m-gon.

Proof. Let $N(m)$ be the Ramsey number $R(5, m; 4)$. Suppose that $n \geq N(m)$ and n points are located in general position in the plane. Divide the subsets of four points into two sets X and Y, the first set containing all those four-point subsets that determine a concave quadrilateral and the second set containing all those four-point subsets that determine a convex quadrilateral. (Since the points are in general position, these sets cover all possibilities.) By Ramsey's Theorem, there are five points with all four-point

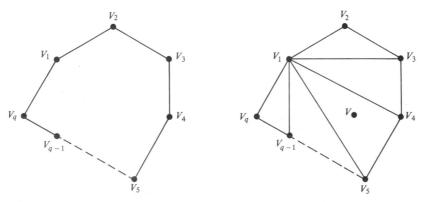

Figure 8.10   The point V inside the convex q-gon determined by the points $V_1, V_2, \ldots, V_q$ must be inside one of the triangles $V_1 V_r V_{r+1}$.

subsets forming concave quadrilaterals or m points with all four-point subsets forming convex quadrilaterals. The former contradicts Theorem 8.18. Hence, the latter holds. Theorem 8.19 now implies that the m points form a convex m-gon.   Q.E.D.

For an alternative proof of this result, see Exercise 5.

8.3.2 Confusion Graphs for Noisy Channels

In communication theory, a *noisy channel* gives rise to a *confusion graph*, a graph whose vertices are elements of a transmission alphabet T and which has an edge between two letters of T if and only if, when sent over the channel, they can be received as the same letter. Given a noisy channel, we would like to make errors impossible by choosing a set of signals that can be unambiguously received, that is, so that no signal in the set is confusable with another signal in the set. This corresponds to choosing an independent set in the confusion graph G. In the confusion graph G of Figure 8.11 the largest independent set consists of two vertices. Thus, we may choose two letters (vertices), say a and c, and use these as an *unambiguous code alphabet* for sending messages. In general, the largest unambiguous code alphabet has $\alpha(G)$ elements, where $\alpha(G)$ is the size of the largest independent set in G.

To see whether we can find a better unambiguous code alphabet, we shall introduce the notion of *normal product* $G \cdot H$ of two graphs G and H. This is defined as follows. The vertices are the pairs in the Cartesian product $V(G) \times V(H)$. There is an edge between (a, b) and (c, d) if and only if one of the following holds:

1. $\{a, c\} \in E(G)$ and $\{b, d\} \in E(H)$,
2. $a - c$ and $\{b, d\} \in E(II)$,
3. $b = d$ and $\{a, c\} \in E(G)$.

(The term *normal product* is used by Berge [1973]; another term in use for this is *strong product*.) Figure 8.12 shows a normal product.

We can find a larger unambiguous code alphabet by allowing combinations of letters from the transmission alphabet to form the code alphabet. For example, suppose that we consider all possible ordered pairs of elements from the transmission alphabet T, or strings of two elements from T. Then under the confusion graph of Figure 8.11, we can find four such ordered pairs, aa, ac, ca, and cc, none of which can be confused with any of the others. In general, two strings of letters from the transmission alphabet can be con-

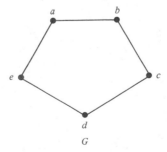

G

Figure 8.11   Confusion graph.

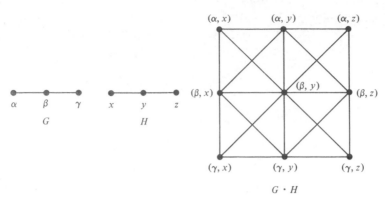

Figure 8.12   The normal product of two graphs.

fused if and only if they can be received as the same string. In this sense, strings *aa* and *ac* cannot be confused, since *a* and *c* cannot be received as the same letter. We can draw a new confusion graph whose vertices are strings of length 2 from *T*. This graph has the following property: Strings *xy* and *uv* can be confused if and only if one of the following holds:

1. *x* and *u* can be confused and *y* and *v* can be confused.
2. *x* = *u* and *y* and *v* can be confused.
3. *y* = *v* and *x* and *u* can be confused.

In terms of the original confusion graph *G*, the new confusion graph is the normal product $G \cdot G$.

If *G* is the confusion graph of Figure 8.11, we have already observed that one independent set or unambiguous code alphabet in $G \cdot G$ can be found by using the strings *aa*, *ac*, *ca*, and *cc*. However, there is a larger independent set that consists of the strings *aa*, *bc*, *ce*, *db*, and *ed*. What is the largest independent set in $G \cdot G$? The following theorem can be used to help answer this question.

Theorem 8.21   (Hedrlín [1966]).   If *G* and *H* are any graphs, then

$$\alpha(G \cdot H) \le R(\alpha(G) + 1, \alpha(H) + 1) - 1.$$

Proof. Let $N = R(\alpha(G) + 1, \alpha(H) + 1)$. Suppose that $\alpha(G \cdot H) \ge N$. We reach a contradiction. Let *I* be an independent set of $G \cdot H$ with *N* vertices. Suppose that (a, b) and (c, d) are two distinct vertices in *I*. Since *I* is independent, either

(1) $a \ne c$ and $\{a, c\} \notin E(G)$,

or

(2) $b \ne d$ and $\{b, d\} \notin E(H)$.

Consider a complete graph with vertex set the vertices of *I*. Color an edge (a, b) to (c, d) of this graph blue if (1) holds and red otherwise. This is a coloring of the edges of the complete graph K_N in two colors, blue and red. By Theorem 8.11, either there is a blue clique *C* with $\alpha(G) + 1$ vertices or a red clique *D* with $\alpha(H) + 1$ vertices. In the former case, note that $(a, b) \in C$ and $(c, d) \in C$ implies that (1) holds, and hence

$$\{a: a \in V(G) \text{ and } (a, b) \in C \text{ for some } b\}$$

is an independent set of G with $\alpha(G) + 1$ vertices. This is a contradiction. In the latter case,

$$\{b: b \in V(H) \text{ and } (a, b) \in D \text{ for some } a\}$$

is an independent set of H with $\alpha(H) + 1$ vertices, again a contradiction. We conclude that $\alpha(G \cdot H) \leq N - 1$.   Q.E.D.

As a corollary of this result, note that if $G = Z_5$ is the confusion graph of Figure 8.11 then

$$\alpha(G \cdot G) \leq R(3, 3) - 1 = 5.$$

Hence, we have found a largest independent set here.

Going beyond strings of length 2, we can seek strings of length k from the transmission alphabet, and seek independent sets in the graph $G^k = G \cdot G \cdot \ldots \cdot G$, where there are k terms in the product. We obtain larger and larger unambiguous code alphabets this way, but at a cost in efficiency: We use longer strings. This observation led Shannon [1956] to compensate by considering the number $\sqrt[k]{\alpha(G^k)}$ as a measure of the capacity of the channel to build an unambiguous code alphabet of strings of length k, and to consider the number $c(G) = \sup_k \sqrt[k]{\alpha(G^k)}$. The number $c(G)$ is called the *capacity of the graph* or the *zero-error capacity of the channel.** Computation of the capacity of a graph is a difficult problem. Indeed, even the capacity of the graph $G = Z_5$ which we discussed above was not known precisely until Lovász [1979] showed that it equals $\sqrt{5}$. Meanwhile, as of this writing, $c(Z_7)$ remains unknown. For some bounds on $c(G)$, see Lovász [1979], Haemers [1979], Schrijver [1979], and Rosenfeld [1967].

8.3.3 Design of Packet-Switched Networks

Stephanie Boyles and Geoff Exoo (personal communication, June 7, 1981) have found an application of Ramsey theory in the design of a packet-switched network, the Bell System signaling network. We describe the application in this subsection.†

Consider a graph in which vertices represent communications equipment joined by communications links or edges. The graph is assumed to be complete; that is, every pair of vertices is joined by a link. In some applications, vertices are paired up, and we would like to guarantee that in case of outages of some links, there will always remain at least

---

*We have encountered this idea in another setting in Example 5.5. To show the relation with the concept defined there, note that it can be shown that

$$\sup_k \sqrt[k]{\alpha(G^k)} = \lim_{k \to \infty} \sqrt[k]{\alpha(G^k)}$$

and that

$$\log_2 \sqrt[k]{\alpha(G^k)} = \frac{\log_2 \alpha(G^k)}{k}.$$

†The author thanks Drs. Boyles and Exoo for directing him to this application and for permission to present it here.

one link joining every paired set of vertices. For instance, consider the graph shown in Figure 8.13. The vertices labeled x_1 and x_2 are paired, the vertices labeled y_1 and y_2 are paired, and the vertices labeled z_1 and z_2 are paired. Outages occur at intermediate facilities such as microwave towers, trunk groups, and so on. An outage at such a facility will affect all links sharing this facility. Let us color the intermediate facilities and hence the corresponding links. Figure 8.13 shows such a coloring. Note that in case the red intermediate facility goes out, there will be no operative links between the pair of vertices x_1 and x_2 and the pair of vertices z_1 and z_2. This corresponds to the fact that the four edges $\{x_i, z_j\}$ form a monochromatic (red) Z_4. In general, designing a network involves a decision as to the number of intermediate facilities and which links will use which intermediate facilities. We would like to design the network so that if any intermediate facility is destroyed, there will remain at least one link for each paired set of vertices. If the vertex pairing may change after the network is constructed, we want to avoid all monochromatic Z_4's.

It turns out that $R(Z_4, Z_4) = 6$ (Faudree and Schelp [1974] or Rosta [1973 a, b]). Thus, if there are just two intermediate facilities, there is a network with five vertices which has an assignment of links to intermediate facilities so that there is no monochromatic Z_4. Chung and Graham [1975] show that $R(Z_4, Z_4, Z_4) \geq 8$. Thus, there is a network with three different intermediate facilities and seven vertices and no monochromatic Z_4.

As we have said, designing a network involves a decision as to the number of intermediate facilities and which links will use which intermediate facilities. Intermediate facilities are expensive, and it is desirable to minimize the number of them. Thus, one is led to ask the following. If we have a network of n vertices, what is the least number of colors or intermediate facilities so that there is some network of n vertices and some coloring of edges (assignment of links to intermediate facilities) with no monochromatic Z_4. In other words, what is the least r so that if there are rZ_4's, $R(Z_4, Z_4, \ldots, Z_4) > n$? If $n = 6$, as in our example of Figure 8.13, then since $R(Z_4, Z_4) = 6$, and $R(Z_4, Z_4, Z_4) \geq 8$, we have $r = 3$. We need three intermediate facilities. Boyles and Exoo point out that for their purposes, it is enough to estimate the number r using a result of Erdös (see Graham et al. [1980]) that a graph of n vertices always contains Z_4 if it has at least $\frac{1}{2}n^{3/2} + \frac{1}{4}n$ edges. If the $\binom{n}{2}$ edges of an n-vertex graph are divided into r color classes, the average

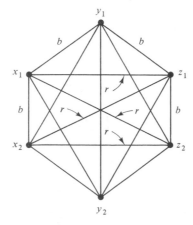

Figure 8.13 Links are colored red (r), blue (b), or white (unlabeled).

	Sorted table structure			Cyclic table structure			Permuted sorted table structure	
Set S	Corresponding table		Set S	Corresponding table		Set S	Corresponding table	
$\{1, 2\}$	1	2	$\{1, 2\}$	1	2	$\{1, 2\}$	2	1
$\{2, 3\}$	2	3	$\{2, 3\}$	2	3	$\{2, 3\}$	3	2
$\{1, 3\}$	1	3	$\{1, 3\}$	3	1	$\{1, 3\}$	3	1

Figure 8.14 Three table structures for storing two keys from a three-element key space $M = \{1, 2, 3\}$.

class will have $\binom{n}{2}/r$ edges, and so, by the pigeonhole principle (Corollary 8.2.2), some class will have at least $\binom{n}{2}/r$ edges. We want to pick r so that no class has $\frac{1}{2}n^{3/2} + \frac{1}{4}n$ edges, so we must pick r so that

$$\binom{n}{2}\bigg/r < \frac{1}{2}n^{3/2} + \frac{1}{4}n.$$

8.3.4 Information Retrieval

Yao [1978, 1981] uses Ramsey theory in the study of information retrieval.* Suppose that a table or a file has n different entries, chosen from a *key space* $M = \{1, 2, \ldots, m\}$, whose elements are called *keys*. We wish to find a way to store all subsets S of n elements from M in a table so that it is easy to answer queries of the form: Is x in S? A rule for telling us how to store the n-element subsets S of M is called a *table structure* or an (m, n) *table structure*. The simplest table structure is called a *sorted table structure*: We just list all elements of S in increasing order. For instance, if $m = 3$ and $n = 2$, a sorted table structure is shown in Figure 8.14. The second table structure in Figure 8.14 is called *cyclic*. Note that if we have the sorted table structure of Figure 8.14, if we want to know if x is in S, we need to ask two questions. However, in the cyclic table structure, we need to ask only one question, since by the cyclic nature of the table structure, the first entry in the row corresponding to S determines the second entry. A variant of the sorted table structure is the *permuted sorted table structure*. Here, we fix a permutation σ of $\{1, 2, \ldots, n\}$, and list elements of S in order according to this permutation. For instance, the third table structure of Figure 8.14 is a permuted sorted table structure corresponding to the permutation which interchanges the first and second elements. Again, to determine if x is in S, two questions are needed with this table structure.

The computational complexity of information retrieval depends on the table structure and the search strategy, that is, the kinds of questions asked. It is measured by the number of queries needed to determine if $x \in S$ in the worst case. For instance, for a sorted table structure, the number of queries required is $\lceil \log_2 (n + 1) \rceil$ if a binary search tree is used (see Section 3.6.1 and Exercise 16, Section 3.6). Let the complexity $f(n, m)$ be defined to be the minimum worst case complexity over all conceivable (m, n) table structures and search strategies for determining if $x \in S$.

*See Chandra *et al.* [1983] for a different use of Ramsey theory in information exchange.

Theorem 8.22 (Yao). For every n, there exists a number $N(n)$ so that $f(n, m) = \lceil \log_2 (n + 1) \rceil$ for all $m \geq N(n)$.

It follows from Theorem 8.22 that for m sufficiently large, using a sorted table structure is the most efficient method as far as information retrieval is concerned. There are two crucial ideas in proving this result:

Lemma 1. If $m \geq 2n - 1$ and $n \geq 2$, then for a permuted sorted table structure, $\lceil \log_2 (n + 1) \rceil$ probes are needed to determine if $x \in S$ in the worst case by any search strategy.

Lemma 2. Given n, there is a number $N(n)$ with the following property. If $m \geq N(n)$ and we are given an (m, n) table structure, then there is a set K of $2n - 1$ keys so that the tables corresponding to the n-element subsets of K form a permuted sorted table structure.

Theorem 8.22 follows from these lemmas. For given an (m, n) table structure and search strategy and a number $m \geq N(n)$, find the set K of Lemma 2. Then by Lemma 1, $\lceil \log_2 (n + 1) \rceil$ probes are needed in the worst case, just restricting the problem to subsets of K. Thus, the complexity is at least $\lceil \log_2 (n + 1) \rceil$, so $f(n, m) \geq \lceil \log_2 (n + 1) \rceil$. But we know that binary search on a sorted table structure has complexity $\lceil \log_2 (n + 1) \rceil$. Thus, $f(n, m) = \lceil \log_2 (n + 1) \rceil$.

We shall omit the proof of Lemma 1, referring the reader to Yao [1981]. We shall present the proof of Lemma 2.

Proof of Lemma 2. Let us note that a set $S = \{j_1, j_2, \ldots, j_n\}$ of n keys is stored in the table structure in some order. If $j_1 < j_2 < \cdots < j_n$ and j_i is stored in the u_ith box of the table, then the set S corresponds to the permutation $u_1, u_2, \ldots, u_n$ of the integers 1, 2, ..., n. For instance, in the cyclic table structure of Figure 8.14, if $S = \{1, 3\}$, then $j_1 = 1$, $j_2 = 3$, $u_1 = 2$, and $u_2 = 1$. In a permuted sorted table structure, each set of n keys corresponds to the same permutation $u_1, u_2, \ldots, u_n$. Given an (m, n) table structure, let $\sigma(u_1, u_2, \ldots, u_n)$ consist of all sets S of n keys whose corresponding permutation is $u_1, u_2, \ldots, u_n$. For instance, in the cyclic table structure of Figure 8.14, $\sigma(1, 2)$ consists of the sets $\{1, 2\}$ and $\{2, 3\}$ and $\sigma(2, 1)$ consists of the set $\{1, 3\}$.

We shall use the general version of Ramsey's Theorem given in Theorem 8.16 to prove Lemma 2. Let $p_i = 2n - 1$, all i, let $t = n!$, and let $r = n$. Let $N(n)$ be the Ramsey number $R(p_1, p_2, \ldots, p_t; r)$. Suppose that $m \geq N(n)$ and that we divide the r-element subsets (the n-element subsets) of the key space M into $t = n!$ parts, with each part consisting of the set $\sigma(u_1, u_2, \ldots, u_n)$ of all n-element subsets S of M which are stored in the permutation $u_1, u_2, \ldots, u_n$. By the definition of $R(p_1, p_2, \ldots, p_t; r)$, there is for some i, a p_i-element subset [$(2n - 1)$-element subset] K of M all of whose n-element subsets belong to a given $\sigma(u_1, u_2, \ldots, u_n)$. This proves Lemma 2.   Q.E.D.

To illustrate this proof, consider the table structure of Figure 8.15. Here, $m = 6$ and $n = 2$. The set $\sigma(1, 2)$ is given by the elements labeled $*$ and the set $\sigma(2, 1)$ is given by the remaining elements. Note that there is a 3-element subset $K = \{1, 2, 5\}$ all of whose

Set S	Corresponding table		Set S	Corresponding table
*{1, 2}	1 2		*{2, 5}	2 5
{1, 3}	3 1		{2, 6}	6 2
{1, 4}	4 1		*{3, 4}	3 4
*{1, 5}	1 5		*{3, 5}	3 5
*{1, 6}	1 6		*{3, 6}	3 6
*{2, 3}	2 3		{4, 5}	5 4
*{2, 4}	2 4		{4, 6}	6 4
			{5, 6}	6 5

Figure 8.15 A (6, 2) table structure with elements of the set $\sigma(1, 2)$ represented by *.

2-element subsets belong to $\sigma(1, 2)$. We know K exists because

$$m \geq 6 = R(3, 3; 2) = R(p_1, p_2, \ldots, p_t; r).$$

We next note that if $n = 2$, there is an alternative proof of Lemma 2 which gives a better value of $N(n)$. If $n = 2$, then any table structure can be represented as a digraph whose vertex set is $M = \{1, 2, \ldots, m\}$, and which has an arc from i to j if the set $\{i, j\}$ is stored as $\boxed{i\,j}$. For example, the table structure of Figure 8.16 yields the digraph shown in that figure. This digraph is a tournament, an object to be defined in Section 11.6.1. For $m \geq 4$, such a tournament always has a *transitive triple*, a triple of vertices $\{i, j, k\}$, with arcs (i, j), (j, k), (i, k) (see Exercise 23 (c), Section 11.6). This means that all 2-element subsets of the 3-element set $\{i, j, k\}$ will be stored in a manner corresponding to the same permutation i, j, k. Thus, if we relabel the elements of M so that i becomes 1, j becomes 2, and k becomes 3, the 2-element subsets of $\{1, 2, 3\}$ will appear in the sorted table structure shown in Figure 8.14. Hence, if $n = 2$, $N(n) = 4$ will suffice to give us the conclusion of Lemma 2. In the example of Figure 8.16, one transitive triple is $\{4, 1, 2\}$ and all 2-element subsets of this triple are stored in the same permutation 4, 1, 2. If we relabel the elements of M so that 4 becomes 1, 1 becomes 2, and 2 becomes 3, we have a sorted table structure in which the sets $\{1, 2\}$, $\{2, 3\}$, and $\{1, 3\}$ are stored as in the sorted table of Figure 8.14. Note that the conclusion of the lemma does not hold here; that is, for this table structure, there is no 3-element subset K and no permutation u_1, u_2 of $\{1, 2\}$ so that all 2-element

Set S	Corresponding table	Associated digraph
{1, 2}	1 2	
{1, 3}	3 1	
{1, 4}	4 1	
{2, 3}	2 3	
{2, 4}	4 2	
{3, 4}	3 4	

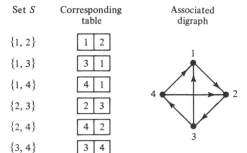

Figure 8.16 A (4, 2) table structure and the associated digraph.

subsets of K are stored in the order u_1, u_2. If m were at least $R(3, 3; 2) = 6$, we would be able to find such a subset and permutation.

8.3.5 The Dimension of Partial Orders: A Decisionmaking Application

A digraph $D = (V, A)$ is *asymmetric* if $(u, v) \in A$ implies that $(v, u) \notin A$. An asymmetric digraph is *transitive* if $(u, v) \in A$, $(v, w) \in A$ imply that $(u, w) \in A$.* For instance, in Figure 8.17 all the digraphs are asymmetric, except for D_2, in which $(a, b) \in A$ and $(b, a) \in A$. Of the asymmetric digraphs, D_1 is not transitive, since $(a, b) \in A$ and $(b, c) \in A$, but $(a, c) \notin A$. But digraphs D_3 and D_4 are transitive. A digraph that is both asymmetric and transitive is called a *(strict) partial order*. Partial orders arise in many contexts in decisionmaking. For instance, if V is a set of alternatives being considered, and $(u, v) \in A$ means that u is preferred to v, we get a partial order if preference satisfies the following conditions. If you prefer u to v, you do not prefer v to u; if you prefer u to v and prefer v to w, then you prefer u to w. Partial orders arise similarly if $(u, v) \in A$ means u is judged more important than v, u is judged more qualified than v, and so on. We shall use preference as a concrete example.

Suppose that $D = (V, A)$ is a digraph representing preference. If we are judging our alternatives a on the basis of one characteristic, say monetary value $f(a)$, we would have

$$(u, v) \in A \iff f(u) > f(v);$$

that is, we would prefer u to v if and only if the value of u is greater than the value of v. If we judge on the basis of several characteristics, say monetary value $f_1(a)$, quality $f_2(a)$, beauty $f_3(a), \ldots, f_t(a)$, we might only express preference for u over v if we are sure that u is better than v on every characteristic. Thus, we would have

$$(u, v) \in A \iff [f_1(u) > f_1(v)] \& [f_2(u) > f_2(v)] \& [f_3(u) > f_3(v)] \& \cdots \& [f_t(u) > f_t(v)]. \quad (8.3)$$

If A is defined using (8.3), then it is easy to show that (V, A) is a partial order.

The converse problem is of importance in preference theory. Suppose that we are given a partial order $D = (V, A)$. Can we find functions $f_1, f_2, \ldots, f_t$, each f_i assigning a real number to each a in V, so that (8.3) holds? It is not hard to prove that for every partial order (with V finite), we can find such functions for sufficiently large t. (The proof uses Szpilrajn's [1930] extension theorem. See Baker *et al.* [1972].) The smallest t such that there are t such functions is called the *dimension* of the partial order.† This notion is originally due to Dushnik and Miller [1941], and has been widely studied. See Baker *et al.* [1972], Kelly and Trotter [1982] and Trotter and Moore [1976] for surveys, and Roberts [1972] for some applications.

To illustrate the idea of dimension, let us consider the partial orders of Figure 8.18.

---

*An arbitrary digraph is *transitive* if $(u, v) \in A$, $(v, w) \in A$, and $u \neq w$ imply that $(u, w) \in A$. Asymmetry implies $u \neq w$.

†Strictly speaking, the dimension of the partial order is usually defined to be the smallest t such that the partial order is the intersection of t linear orders. However, our definition of dimension agrees with the more common one except for dimensions 1 and 2: The so-called (strict) weak orders can have dimension 1 by our definition, but not by the more common definition (see Baker *et al.* [1970]).

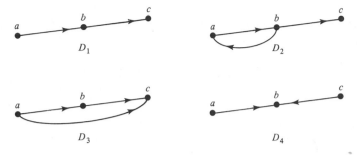

Figure 8.17 Digraphs D_1, D_3, and D_4 are asymmetric, and D_3 and D_4 are transitive.

The partial order D_1 has dimension 1. To see this, take $f_1(a) = 4$, $f_1(b) = 3$, $f_1(c) = 2$, $f_1(d) = 1$. The partial order D_2 does not have dimension 1, since we would have to have $f_1(b) = f_1(d)$, since neither (b, d) nor (d, b) is in A, yet we would have to have $f_1(b) > f_1(c)$ while $f_1(d) = f_1(c)$. This partial order has dimension 2. To see this, take $f_1(a) = 3$, $f_1(b) = 2$, $f_1(c) = 1$, $f_1(d) = 2$, $f_2(a) = 3$, $f_2(b) = 2$, $f_2(c) = 1$, $f_2(d) = 1$.

The dimension of many important partial orders has been computed. Here we shall study the dimension of one very important class of partial orders, the interval orders. To get an interval order, imagine that for each alternative a that you are considering, you do not know its exact value, but you estimate a range of possible values, given by a closed interval $J(a) = [\alpha(a), \beta(a)]$. Then you prefer a to b if and only if you are sure that the value of a is greater than the value of b, that is, if and only if $\alpha(a) > \beta(b)$. It is easy to show (Exercise 27) that the corresponding digraph gives a partial order; that is it is asymmetric and transitive. (In this digraph, the vertices are a family of closed real intervals, and there is an arc from an interval $[a, b]$ to an interval $[c, d]$ if and only if $a > d$.) Any partial order that arises this way is called an *interval order*. The notion of interval order is due to Fishburn [1970].

In studying interval orders, which are somehow one-dimensional in nature, it came as somewhat of a surprise that their dimension as partial orders could be arbitrarily large. That is the content of the main theorem of this subsection. It implies that if preferences arise in the very natural way that defines interval orders, we might need very many dimensions or characteristics to explain preference in the sense of Equation (8.3).

Theorem 8.23 (Bogart, Rabinovitch and Trotter [1976]). There are interval orders of arbitrarily high dimension.

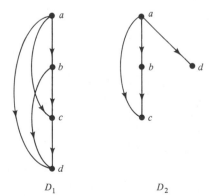

Figure 8.18 The partial order D_1 has dimension 1, while the partial order D_2 has dimension 2.

Proof. Suppose that $I(0, n)$ is the interval order defined by taking all closed intervals $[a, b]$ with a, b integers between 0 and n inclusive, and by taking an arc from $[a, b]$ to $[c, d]$ if and only if $a > d$. We shall show that given $t \geq 2$, there is a number $N(t)$ so that if $n \geq N(t)$, interval order $I(0, n)$ has dimension greater than t. In particular, let $N(t) = R(p_1, p_2, \ldots, p_t; r) - 1$ with $p_1 = p_2 = \cdots = p_t = 4$ and $r = 3$. Now suppose that $n \geq N(t)$ and that $I(0, n)$ has dimension less than or equal to t. Then there are functions $f_1, f_2, \ldots, f_t$ so that (8.3) holds. Now consider the set of all integers between 0 and n and consider the 3-element subsets $\{u, v, w\}$. Suppose that $u < v < w$. Then neither $([u, v], [v, w]) \in A$ nor $([v, w], [u, v]) \in A$. It follows by (8.3) that there are i and j so that

$$f_j([v, w]) \geq f_j([u, v]) \qquad \text{and} \qquad f_i([u, v]) \geq f_i([v, w]).$$

Place the triple $\{u, v, w\}$ in the ith class, $i = 1, 2, \ldots, t$, if i is the smallest integer so that $f_i([u, v]) \geq f_i([v, w])$. Since $n \geq N(t)$, we have $n + 1 \geq R(4, 4, \ldots, 4; 3)$. Thus, we know that for some i, there is a 4-element subset $\{x, y, z, t\}$ of $\{0, 1, \ldots, n\}$ all of whose 3-element subsets are in the ith class. Therefore, if $x < y < z < t$, we have

$$f_i([x, y]) \geq f_i([y, z]) \qquad \text{and} \qquad f_i([y, z]) \geq f_i([z, t]).$$

Hence,

$$f_i([x, y]) \geq f_i([z, t]).$$

But

$$([z, t], [x, y]) \in A,$$

so we should have

$$f_i([z, t]) > f_i([x, y]).$$

Hence, we have reached a contradiction, which implies that $I(0, n)$ has dimension larger than t. Q.E.D.

EXERCISES FOR SECTION 8.3

1. Show that five points not necessarily in general position in the plane do not necessarily determine a convex quadrilateral.

2. For each set of points in Figure 8.19, find a subset of four points that does not form a convex quadrilateral.

3. Suppose that n points are in general position in the plane and each point is joined to each other point by a line. Show that if the lines are colored either red or blue, then if n is sufficiently large, we can be sure of finding five points each pair of which are joined by a red line or five points each pair of which are joined by a blue line.

4. Suppose that n points are in general position in the plane. Show that if n is sufficiently large, then either there are six points all of whose five-point subsets form concave 5-gons or 10 points all of whose five-point subsets form convex 5-gons.

5. An alternative proof of Theorem 8.20, discovered by M. Tarsy (Graham *et al.* [1980]), is outlined here. Let $N(m) = R(m, m; 3)$. Given n points in general position, with $n \geq N(m)$, number the points $1, 2, \ldots, n$ arbitrarily. Consider the triple $\{i, j, k\}$ with $i < j < k$. Color the triple red if

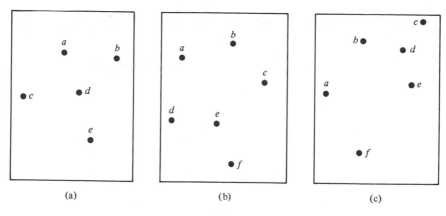

Figure 8.19 Sets of points for Exercise 2, Section 8.3.

traveling from i to j to k to i is in a clockwise direction and blue if it is in a counterclockwise direction. Complete the proof of Theorem 8.20 by using the observation that if m points have every triple with the same orientation (clockwise or counterclockwise), then the m points form a convex m-gon.

6. For each pair of graphs G and H of Figure 8.20, draw the normal product $G \cdot H$.

7. If G is Z_4, find $\alpha(G^2)$.

8. Use Theorem 8.21 to find upper bounds for $\alpha(G \cdot H)$ if
   (a) $G = Z_4, H = Z_5$ (b) $G = K_5, H = Z_6$
   (c) $G = L_5$, the chain of five vertices, and $H = Z_4$ (d) $G = L_5, H = L_6$.

9. Use Theorem 8.21 to find an upper bound for $\alpha(G^2)$ if
   (a) $G = Z_4$ (b) $G = K_5$ (c) $G = L_5$.

10. Show that $\alpha(G \cdot H)$ can be less than $R(\alpha(G) + 1, \alpha(H) + 1) - 1$.

11. Show that $\alpha(G \cdot H) \geq \alpha(G)\alpha(H)$.

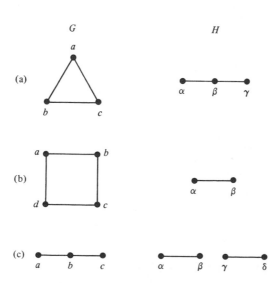

Figure 8.20 Graphs G and H for Exercise 6, Section 8.3.

12. Suppose that we are given an edge coloring of a graph G. Show that it is possible that one vertex pairing has some pairs joined by links of only one color, while another vertex pairing does not.

13. Find an edge coloring of the following graphs with the following number of colors so that there is no monochromatic Z_4.
    (a) K_4 with three colors
    (b) K_5 with three colors
    (c) K_6 with three colors
    (d) K_6 with four colors.

14. Chung and Graham [1975] prove that if $r - 1$ is a power of a prime number and there are r Z_4's, then $R(Z_4, Z_4, \ldots, Z_4) \geq r^2 - r + 2$. What is the implication of this result for the size of networks with four intermediate facilities and no possibility of having an outage in one of the facilities totally disconnect two paired sets of vertices?

15. For each table structure of Figure 8.21, find the corresponding digraph.

16. For the table structure of Figure 8.16:
    (a) Find the sets $\sigma(1, 2)$ and $\sigma(2, 1)$.
    (b) Show that there is no subset K of M of 4 elements all of whose 3-element subsets belong to $\sigma(1, 2)$ or all of whose 3-element subsets belong to $\sigma(2, 1)$.

17. For each table structure of Figure 8.21, find the sets $\sigma(1, 2)$ and $\sigma(2, 1)$.

18. (a) For the table structure (a) of Figure 8.22, find the sets $\sigma(1, 2, 3)$, $\sigma(1, 3, 2)$, $\sigma(2, 1, 3)$, $\sigma(2, 3, 1)$, $\sigma(3, 1, 2)$, and $\sigma(3, 2, 1)$.
    (b) For the table structure (b) of Figure 8.22, find the set $\sigma(4, 3, 2, 1)$.

19. For each table structure of Figure 8.21, find a set K of 3 keys and a permutation u_1, u_2 such that every 2-element subset S of K belongs to $\sigma(u_1, u_2)$, or show that no such set K exists.

20. Given n, if we are given an (m, n) table structure, is it true that for m sufficiently large, there is a set K of $25n$ keys and a fixed permutation $u_1, u_2, \ldots, u_n$ so that every n-element subset S of K belongs to $\sigma(u_1, u_2, \ldots, u_n)$? Why?

21. Which of the digraphs of Figure 8.23 are partial orders?

22. Do the data of Table 8.2 define a partial order?

$M = \{1, 2, 3, 4, 5\}$

Set S	Corresponding table	
$\{1, 2\}$	1	2
$\{1, 3\}$	3	1
$\{1, 4\}$	1	4
$\{1, 5\}$	5	1
$\{2, 3\}$	2	3
$\{2, 4\}$	4	2
$\{2, 5\}$	5	2
$\{3, 4\}$	3	4
$\{3, 5\}$	5	3
$\{4, 5\}$	4	5

(a)

$M = \{1, 2, 3, 4\}$

Set S	Corresponding table	
$\{1, 2\}$	2	1
$\{1, 3\}$	3	1
$\{1, 4\}$	4	1
$\{2, 3\}$	3	2
$\{2, 4\}$	4	2
$\{3, 4\}$	3	4

(b)

Figure 8.21 Table structures for exercises of Section 8.3.

$M - \{1, 2, 3, 4, 5\}$

Set S	Corresponding table
$\{1, 2, 3\}$	1 2 3
$\{1, 2, 4\}$	2 4 1
$\{1, 2, 5\}$	5 2 1
$\{1, 3, 4\}$	1 3 4
$\{1, 3, 5\}$	3 1 5
$\{1, 4, 5\}$	5 1 4
$\{2, 3, 4\}$	4 2 3
$\{2, 3, 5\}$	2 3 5
$\{2, 4, 5\}$	2 5 4
$\{3, 4, 5\}$	3 5 4

(a)

$M = \{1, 2, 3, 4, 5\}$

Set S	Corresponding table
$\{1, 2, 3, 4\}$	1 2 4 3
$\{1, 2, 3, 5\}$	1 5 3 2
$\{1, 2, 4, 5\}$	5 4 2 1
$\{1, 3, 4, 5\}$	5 4 3 1
$\{2, 3, 4, 5\}$	3 4 5 2

(b)

Figure 8.22 Table structures for exercises of Section 8.3.

23. Show that the partial orders of Figure 8.24 have dimension 1.
24. Show that the partial orders of Figure 8.25 do not have dimension 1.
25. Show that the partial orders of Figure 8.26 have dimension at most 2.
26. Find the dimension of the partial orders of Figure 8.27.
27. Show that if V is any set of closed real intervals and there is an arc from $[a, b]$ to $[c, d]$ if and only if $a > d$, then the resulting digraph is a partial order.

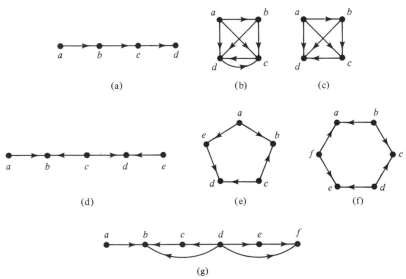

Figure 8.23 Digraphs for Exercise 21, Section 8.3.

Table 8.2 Taste preference
for vanilla puddings[a]

	1	2	3	4	5
1	0	0	1	1	0
2	1	0	0	1	0
3	0	1	0	0	0
4	0	0	1	0	0
5	1	1	0	1	0

[a]Entry i, j is 1 if and only if pudding i is (strictly) preferred to pudding j by a group of judges. *Source*: Data obtained from an experiment of Davidson and Bradley [1969].

28. Draw the digraphs corresponding to the interval orders
(a) $I(0, 2)$ (b) $I(0, 3)$ (c) $I(0, 4)$ (d) $I(0, 5)$.

29. Find a partial order of dimension n for every n.

30. Draw the digraph of the partial order whose vertex set is the set of all subsets of $\{1, 2, 3\}$ and which has an arc from subset X to subset Y if $X \subsetneqq Y$. Compute the dimension of this partial order.

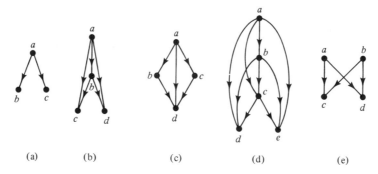

(a)   (b)   (c)   (d)   (e)

Figure 8.24 Partial orders for Exercises 23 and 31, Section 8.3.

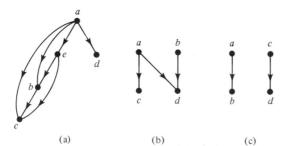

(a)   (b)   (c)

Figure 8.25 Partial orders for Exercises 24 and 31, Section 8.3.

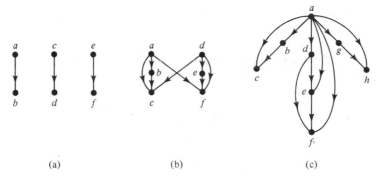

(a)   (b)   (c)

Figure 8.26 Partial orders for Exercises 25 and 31, Section 8.3.

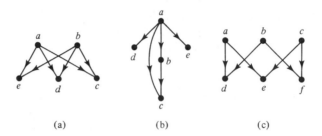

(a)   (b)   (c)

Figure 8.27 Partial orders for Exercises 26 and 31, Section 8.3.

31. Fishburn [1970] shows that a digraph $D = (V, A)$ is an interval order if and only if (D has no loops and) the following condition C holds: Whenever $(a, b) \in A$ and $(c, d) \in A$, then either $(a, d) \in A$ or $(c, b) \in A$.
 (a) Show that if a digraph is defined as in Exercise 27, then condition C holds.
 (b) Use Fishburn's theorem to determine which of the digraphs of Figures 8.24–8.27 are interval orders.

32. If $n \geq R(4, 4; 3) - 1$, what can you say about the dimension of the interval order $I(0, n)$?

33. Repeat Exercise 32 if $n \geq R(4, 4, 4; 3) - 1$.

34. Show that if a partial order has at least $R(a + 1, b + 1)$ vertices, then it either has a path of $a + 1$ vertices or a set of $b + 1$ vertices, no two of which are joined by arcs. (A famous theorem of Dilworth [1950] says that the same conclusion holds as long as the partial order has at least $ab + 1$ vertices.)

REFERENCES FOR CHAPTER 8

BAKER, K. A., FISHBURN, P. C., and ROBERTS, F. S., "Partial Orders of Dimension 2, Interval Orders, and Interval Graphs," Paper P-4376, The RAND Corporation, Santa Monica, Calif., 1970.

BAKER, K. A., FISHBURN, P. C., and ROBERTS, F. S., "Partial Orders of Dimension 2," *Networks, 2* (1972), 11–28.

BEHZAD, M., CHARTRAND, G., and LESNIAK-FOSTER, L., *Graphs and Digraphs*, Wadsworth, Belmont, Calif., 1979.

BERGE, C., *Graphs and Hypergraphs*, American Elsevier, New York, 1973.

BOGART, K. P., RABINOVITCH, I., and TROTTER, W. T., "A Bound on the Dimension of Interval Orders," *J. Comb. Theory, 21* (1976), 319–328.

CHANDRA, A. K., FURST, M. L., and LIPTON, R. J., "Multi-party Protocols," *Proceedings of the Fifteenth Annual ACM Symposium on the Theory of Computing*, Association for Computing Machinery, New York, 1983, pp. 94–99.

CHUNG, F. R. K., and GRAHAM, R. L., "On Multicolor Ramsey Numbers for Complete Bipartite Graphs," *J. Comb. Theory B, 18* (1975), 164–169.

CHUNG, F. R. K., and GRINSTEAD, C., "A Survey of Bounds for Classical Ramsey Numbers," *J. Graph Theory, 7* (1983), 25–37.

CHVÁTAL, V., "Tree-Complete Graph Ramsey Numbers," *J. Graph Theory, 1* (1977), 93.

CHVÁTAL, V., and HARARY, F., "Generalized Ramsey Theory for Graphs. III. Small Off-Diagonal Numbers," *Pacific J. Math., 41* (1972), 335–345.

DAVIDSON, R. R., and BRADLEY, R. A., "Multivariate Paired Comparisons: The Extension of a Univariate Model and Associated Estimation and Test Procedures," *Biometrika, 56* (1969), 81–95.

DILWORTH, R. P., "A Decomposition Theorem for Partially Ordered Sets," *Ann. Math., 51* (1950), 161–166.

DUSHNIK, B., and MILLER, E. W., "Partially Ordered Sets," *Amer. J. Math., 63* (1941), 600–610.

ERDÖS, P., and SZEKERES, G., "A Combinatorial Problem in Geometry," *Composito Math., 2* (1935), 464–470.

FAUDREE, R. J., and SCHELP, R. H., "All Ramsey Numbers for Cycles in Graphs," *Discrete Math., 8* (1974), 313–329.

FISHBURN, P. C., "Intransitive Indifference with Unequal Indifference Intervals," *J. Math. Psychol., 7* (1970), 144–149.

GRAHAM, R. L., *Rudiments of Ramsey Theory*, CBMS Regional Conference Series in Mathematics, No. 45, American Mathematical Society, Providence, R.I., 1981.

GRAHAM, R. L., ROTHSCHILD, B. L., and SPENCER, J. H., *Ramsey Theory*, Wiley, New York, 1980.

GRAVER, J. E., and YACKEL, J., "Some Graph Theoretic Results Associated with Ramsey's Theorem," *J. Comb. Theory, 4* (1968), 125–175.

GREENWOOD, R. E., and GLEASON, A. M., "Combinatorial Relations and Chromatic Graphs," *Canad. J. Math., 7* (1955), 1–7.

GRINSTEAD, C. M., and ROBERTS, S. M., "On the Ramsey Numbers $R(3, 8)$ and $R(3, 9)$," *J. Comb. Theory B, 33* (1982), 27–51.

HAEMERS, W., "On Some Problems of Lovász Concerning the Shannon Capacity of a Graph," *IEEE Trans. Inf. Theory, IT-25* (1979), 231–232.

HEDRLÍN, Z., "An Application of the Ramsey Theorem to the Topological Product," *Bull. Acad. Pol. Sci., 14* (1966), 25–26.

KALBFLEISCH, J. G., "Chromatic Graphs and Ramsey's Theorem," Ph.D. Thesis, University of Waterloo, Waterloo, Ontario, 1966.

KELLY, D., and TROTTER, W. T., "Dimension Theory for Ordered Sets," in I. Rival (ed.), *Ordered Sets*, Reidel, Boston, 1982, pp. 171–211.

KÉRY, G., "Ramsey egy Gráfelméleti Tételéröl, *Mat. Lapok, 15* (1964), 204–224.

LIU, C. L., *Topics in Combinatorial Mathematics*, Mathematical Association of America, Washington, D. C., 1972.

LOVÁSZ, L., "On the Shannon Capacity of a Graph," *IEEE Trans. Inf. Theory, IT-25* (1979), 1–7.

RAMSEY, F. P., "On a Problem of Formal Logic," *Proc. Lond. Math. Soc., 30* (1930), 264–286.

ROBERTS, F. S., "What If Utility Functions Do Not Exist?" *Theory and Decision, 3* (1972), 126–139.

ROSENFELD, M., "On a Problem of C. E. SHANNON in Graph Theory," *Proc. Amer. Math. Soc., 18* (1967), 315–319.

ROSTA, V., "On a Ramsey Type Problem of J. A. BONDY and P. ERDÖS, I," *J. Comb. Theory B, 15* (1973a), 94–104.

ROSTA, V., "On a Ramsey Type Problem of J. A. BONDY and P. ERDÖS, II," *J. Comb. Theory B, 15* (1973b), 105–120.

SCHRIJVER, A., "A Comparison of the Delsarte and Lovász Bounds," *IEEE Trans. Inf. Theory, IT-25* (1979), 425–429.

SHANNON, C. E., "The Zero-Error Capacity of a Noisy Channel," *IRE Trans. Inf. Theory, IT-2* (1956), 8–19.

SZPILRAJN, E., "Sur l'extension de l'ordre partiel," *Fund. Math., 16* (1930), 386–389.

TROTTER, W. T., and MOORE, J. I., "Characterization Problems for Graphs, Partially Ordered Sets, Lattices, and Families of Sets," *Discrete Math., 15* (1976), 361–368.

TUCKER, A., *Applied Combinatorics*, Wiley, New York, 1980.

YAO, A. C., "Should Tables Be Sorted? (Preliminary Version)," *Proc. 1978 IEEE Symposium on Foundations of Computer Science*, Ann Arbor, Mich., October, 1978.

YAO, A. C., "Should Tables Be Sorted?", *J. ACM, 28* (1981), 615–628.

9 Experimental Design

9.1 BLOCK DESIGNS

In the history of attempts to perform scientifically sound experiments, combinatorics has played an important role. We have already encountered problems of experimental design in Section 1.1, where we discussed the design of an experiment to study the effects of different drugs, and used this design problem to introduce the notion of Latin squares. In this chapter we study the combinatorial questions that arise from issues in experimental design, and discuss the role of combinatorial analysis in the theory of experimental design.

The theory of design of experiments came into being largely through the work of R. A. Fisher, F. Yates, and others, motivated by questions of design of careful field experiments in agriculture. Although the applicability of this theory is now very widespread, much of the terminology still bears the stamp of its origin.

We shall be concerned with experiments aimed at comparing effects of different *treatments* or *varieties*, for instance different types of fertilizers, different doses of a drug, or different brands of shoes or tires. Each treatment is applied to a number of *experimental units* or *plots*. In agriculture, the experimental unit may be an area in which a crop is grown. However, the experimental unit may be a human subject on a given day, a piece of animal tissue, or the site on an animal or plant where an injection or chemical treatment is applied, or it may in other cases be a machine used in a certain location for some purpose.

Certain experimental units are grouped together in *blocks*. These are usually chosen because they have some inherent features in common, for example because they are all on the same human subject or all in the same horizontal row in a field or all on the skin of the same animal, or all on the same machine.

Table 9.1 An experimental design for testing tread wear[a]

| | | Car | | | |
		A	B	C	D
	Left front	1	2	3	4
Wheel	Right front	1	2	3	4
position	Left rear	1	2	3	4
	Right rear	1	2	3	4

[a]The i, j entry is the brand of tire used in position i on car j.

To be concrete, let us consider the problem of comparing the tread wear of four different brands of tires.* The treatments we are comparing are the different brands of tires. Clearly, individual tires of a given brand may differ. Hence, we certainly want to try out more than one tire of each brand. A particular tire is an experimental unit. Now suppose that the tires are to be tested under real driving conditions. Then we naturally group four tires or experimental units together, since a car used to test the tires takes four of them. The test cars define the blocks.

It is natural to try to let each brand of tire or treatment be used as often as any other. Suppose that each is used r times. Then we need $4r$ experimental units in all, since there are four treatments or tire brands. Since the experimental units are split into blocks of size 4, $4r$ must be divisible by 4. In this case, r could be any positive integer. If there were five brands of tires, we would need $5r$ experimental units in all, and then r could only be chosen to be an integer so that $5r$ is divisible by 4.

If we take r to be 4, then we could have a very simple experimental design. Find four cars, say A, B, C, D, and place four tires of brand 1 on car A, four tires of brand 2 on car B, four tires of brand 3 on car C, and four tires of brand 4 on car D. This design is summarized in Table 9.1. This is clearly an unsatisfactory experimental design. Different cars (and different drivers) may lead to different amounts of tire wear, and the attempt to distinguish brands of tires as to wear will be confused by extraneous factors.

Much of the theory of experimental design has been directed at eliminating the biasing or confusing effect caused by variations in particular experimental units. One often tries to eliminate the effect by randomizing, and assigning treatments to experimental units in a random way. For instance, we could start with four tires of each brand, and assign tires to each car completely at random. This might lead to a design such as the one shown in Table 9.2. Unfortunately, as the table shows, we could end up with a tire brand such as 4 never being used on a particular car such as A, or one brand such as 3 used several times in a particular car such as A. The results might still be biased by car effects. We can avoid this situation if we require that each treatment or brand be used in each block or car, and then make the assignment of tires to wheels of the car randomly. A major question in the theory of experimental design is what we have called in Chapter 1 the existence question. Here, we ask this question as follows: Does there exist a design in which there are four brands and four cars, each brand is used four times, and it is used at least once, equivalently exactly once, in each car? The answer is yes. Table 9.3 gives such a design.

*Our treatment follows Hicks [1973].

Table 9.2 A randomized design for testing tread wear[a]

| | | Car | | | |
		A	B	C	D
Wheel position	Left front	3	4	2	2
	Right front	1	1	4	4
	Left rear	3	4	1	3
	Right rear	2	3	2	1

[a]The i, j entry is the brand of tire used in position i on car j.

The design in Table 9.3 still has some defects. For instance, suppose that the position of a tire on a car affects its tread life. For instance, rear tires get different wear than front tires, and even the side of a car a tire is on could affect its tread life. If we wish also to eliminate the biasing effect of wheel position, we could require that each brand or treatment be used exactly once on each car and also exactly once in each of the possible positions. Then, we ask for an assignment of the numbers 1, 2, 3, 4 in a 4 × 4 array with each number appearing exactly once in each row and in each column. That is, we ask for a *Latin square* (see Section 1.1). Table 9.4 shows such a design. Among all possible 4 × 4 Latin square designs, we might still want to pick the particular one to use randomly.

In some experiments, it may not be possible to apply all treatments to every block. For instance, if there were five brands of tires, we could only use four of them in each block. How would we design an experiment now? If each brand of tire is used r times, we have $5r$ tires in all to distribute into groups of four, so as we observed above, $5r$ must be divisible by 4. For example, r must be 4, 8, 12, and so on. Note that we could not do the experiment with six cars; that is, there does not exist an experimental design using five brands and six cars, with each brand used the same number of times, and four (different) brands assigned to each car. For there are 24 tire locations in all, and $5r = 24$ is impossible. Suppose that we take $r = 4$. Then there are $5r = 20$ tire locations in all. If s is the number of cars, $4s$ should be 20, so s should be 5. One possible design is given in Table 9.5. Here there are four different brands of tires on each car, each brand is used exactly once in each position, and each brand is used the same number of times, 5. There are various additional requirements we can place on such a design. We shall discuss some of them below.

Table 9.3 A complete block design for testing tread wear[a]

| | | Car | | | |
		A	B	C	D
Wheel position	Left front	1	1	3	4
	Right front	2	3	4	2
	Left rear	3	2	1	1
	Right rear	4	4	2	3

[a]The i, j entry is the brand of tire used in position i on car j.

Table 9.4 A Latin square design for testing tread wear[a]

		Car			
		A	B	C	D
Wheel position	Left front	1	2	3	4
	Right front	2	3	4	1
	Left rear	3	4	1	2
	Right rear	4	1	2	3

[a]The i, j entry is the brand of tire used in position i on car j.

Let us now introduce some general terminology. Suppose that P is a set of experimental units or plots, and V is a set of treatments or varieties. Certain subsets of P will be called *blocks*. Given P and V, a *block design* is defined by giving the collection of blocks and assigning to each experimental unit in P a treatment in V. Thus, corresponding to each block is a set (possibly with repetitions) of treatments. Speaking abstractly, we shall be able to disregard the experimental units and think of a block design as simply consisting of a set V of treatments and a collection of subsets of V (possibly with repetitions) called blocks. Thus, the block design corresponding to Table 9.2 has $V = \{1, 2, 3, 4\}$ and has the following blocks:

$$\{3, 1, 3, 2\}, \quad \{4, 1, 4, 3\}, \quad \{2, 4, 1, 2\}, \quad \{2, 4, 3, 1\}.$$

If order counts, as in the case of Latin squares, we can think of the blocks as sequences rather than subsets. A block design is called *complete* if each block is all of V, and *incomplete* otherwise. Tables 9.3 and 9.4 define complete block designs, and Table 9.5 defines an incomplete block design. A block design is called *randomized* if elements within each block are ordered by some random device, such as a random number table or a computer program designed to pick out random permutations.

We shall study two types of block designs in this chapter, the complete designs that come from Latin squares and families of Latin squares, and the incomplete designs which are called balanced. We shall also relate experimental design to the study of the finite geometries known as finite projective planes. In Chapter 10 we shall apply our results about experimental design to the design of error-correcting codes.

Table 9.5 An incomplete block design for testing tread wear[a]

		Car				
		A	B	C	D	E
Wheel position	Left front	1	2	3	4	5
	Right front	2	3	4	5	1
	Left rear	3	4	5	1	2
	Right rear	4	5	1	2	3

[a]The i, j entry is the brand of tire used in position i on car j.

9.2 LATIN SQUARES

9.2.1 Some Examples

A Latin square design is appropriate if there are two factors, for example subject and day, wheel position and car, or *row* and *column*, and we want to control for both factors. In agricultural experiments, the rows and columns are literally rows and columns in a rectangular field. Latin squares were introduced by Fisher [1926] to deal with such experiments. Suppose, for example, that there are k different row effects and k different column effects, and we wish to test k different treatments. We wish to arrange things so that each treatment appears once and only once in a given row and in a given column, for example, in a given position and on a given car. Clearly, there is such an arrangement or $k \times k$ Latin square for every k. Table 9.6 shows a $k \times k$ Latin square. Thus, for Latin squares, the existence problem is very simple. The existence problem will not be so simple for the other designs we consider in this chapter.

We now turn to a series of examples of the use of Latin square designs.

Example 9.1 Prosthodontics

Cox [1958]* discusses an experiment in prosthodontics which compares seven treatments, which are commercial dentures of different materials and set at different angles. It is desirable to eliminate as much as possible of the variation due to differences between patients. Hence, each patient wears dentures of one type for a month, then dentures of another type for another month, and so on. After seven months, each patient has worn each type of denture, that is, has been subjected to each treatment.

In this experiment, it seems likely that the results in later months will be different from those in earlier months, and hence it is sensible to arrange that each treatment be used equally often in each time position. Thus there are two types of variation, namely between-patient and between-time variation. The desire to balance out both types suggests the use of a Latin square. The rows correspond to the months and the columns to the patients. Each patient defines a block, and the experimental unit is the jth patient in the ith month.

Example 9.2 Cardiac Drugs

Chen *et al.* [1942] tested the effects of 12 different cardiac drugs on cats. The experiment required an observer to measure carefully the effect over a period of time, so a given observer could only observe four different cats in a day. The experimenters desired to eliminate the effects of the day an observation was made, the observer who made the observation, and the time of day (early A.M., late A.M., early P.M., late P.M.) the observation was made. Thus, there were three factors, which is inappropriate for a Latin square design. However, a Latin square design could be carried out by taking as one factor the day on which the observation was made and as a second factor the observer and the time of day of the observation. A 12×12

---

*Examples 9.2, 9.3, 9.5, 9.11, and 9.12 below are also discussed by Cox [1958]. These and other examples can also be found in Box *et al.* [1978], Cochran and Cox [1957], Finney [1960], or Hicks [1973].

Table 9.6 A $k \times k$ Latin square

1	2	3	$\cdots$	$k-1$	k
2	3	4	$\cdots$	k	1
3	4	5	$\cdots$	1	2
			$\vdots$		
$k-1$	k	1	$\cdots$	$k-3$	$k-2$
k	1	2	$\cdots$	$k-2$	$k-1$

Latin square experiment was performed, carried out over 12 days, with each of three observers observing four cats per day, two in the morning and two in the afternoon. The design used had 12 rows, coded by observer and time of observation, and 12 columns, coded by date. The i, j entry was the drug used on date j at the time of day and by the observer encoded by i. The dates defined the blocks.

Example 9.3 Market Research

Brunk and Federer [1953] discuss some investigations in market research. One of these studied the effect on the sale of apples of varying practices of pricing, displaying and packaging. In each experiment of a series, four merchandising practices (treatments), 1, 2, 3, and 4, were compared and four supermarkets took part. It was clearly desirable that each treatment should be used in each store, so it was sensible to arrange for the experiment to continue for a multiple of four time periods. The experimenters wanted to eliminate systematic differences between stores, and between different periods. Since there were two types of variations, a Latin square design, in particular a 4×4 Latin square, was appropriate. In fact, however, the week was divided into two parts, Monday through Thursday, and Friday and Saturday, and one 4×4 Latin square was built up for each part of the week. This was a good idea because the grocery order per customer was larger over the weekend and it was quite possible that the treatment differences would not be the same in the two parts of the experiment. For an experiment lasting one week and comparing four treatments, the design of Table 9.7 was used.

Example 9.4 Spinning Synthetic Yarn

Box *et al.* [1978] discuss an experiment dealing with the breaking strength of synthetic yarn and how this is affected by changes in draw ratio, the tension applied to yarn as it is spun. The three treatments tested were (1) the usual draw ratio, (2) a 5 per cent increase in draw ratio, and (3) a 10 per cent increase in draw ratio. One spinning machine was used, with three different spinnerets supplying yarn to three different bobbins under different draw ratios. When all the bobbins were completely wound with yarn, they were each replaced with an empty bobbin spool and the experiment was continued. The experimenter wished to control for the two factors: The effect of the three different spinnerets and the effect of the time (order) in which the spinnerets were used. This called for a 3×3 Latin square design, with columns labeled I, II, III corresponding to order of production of the yarn, and rows labeled A, B, C corresponding to which spinneret was used. The i, j entry was the treatment or draw ratio (1, 2, or 3) used in producing yarn from the ith spinneret in the jth

Table 9.7 Two Latin square designs for the two different parts of the week in the market research experiment[a]

First Part of the Week

Store		Time			
		Mon.	Tues.	Wed.	Thur.
	A	2	1	4	3
	B	3	2	1	4
	C	4	3	2	1
	D	1	4	3	2

Second Part of the Week

Store		Time			
		Fri. A.M.	Fri. P.M.	Sat. A.M.	Sat. P.M.
	A	2	3	1	4
	B	1	4	2	3
	C	3	2	4	1
	D	4	1	3	2

[a]The i, j entry gives the treatment used in store i in period j.

production run. When small Latin squares are used, it is often desirable to replicate them, and so in fact the experiment was replicated four times, using different 3×3 Latin square designs. Table 9.8 shows the designs.

9.2.2 Orthogonal Latin Squares

Let us return to the example of the differing effects on tire wear of four tire brands, which we discussed in Section 9.1. Let us imagine that we are also interested in the effect of brake linings on tire wear. Suppose for simplicity that we also have four different brands of brake linings. Thus, we would like to arrange, in addition to having each brand of tire tested exactly once on each car and exactly once in each tire position, that each tire brand be tested exactly once in combination with each brand of brake lining. We can accomplish this by building a 4×4 array, with rows corresponding to wheel position and columns to cars, and placing in each box *both* a tire brand and a brake lining brand to be used in the corresponding position and on the corresponding car. If a_{ij} is the tire brand in entry i, j of the array and b_{ij} is the brake lining brand in this entry, we require that every possible ordered pair (a, b) of tire brands a and brake lining brands b appear if we list all

Table 9.8 Latin square designs for the synthetic yarn experiment[a]

Order of production

Spinneret		I	II	III		I	II	III		I	II	III		I	II	III
	A	1	2	3	A	2	1	3	A	3	1	2	A	1	2	3
	B	2	3	1	B	3	2	1	B	1	2	3	B	2	3	1
	C	3	1	2	C	1	3	2	C	2	3	1	C	3	1	2
		First replication				Second replication				Third replication				Fourth replication		

[a]The i, j entry is the draw ratio used with the ith spinneret in the jth production run.

ordered pairs (a_{ij}, b_{ij}). Equivalently, since there are $4 \times 4 = 16$ possible ordered pairs (a, b) and exactly 16 spots in the array, we require that all the pairs (a_{ij}, b_{ij}) be different. Can we accomplish this? We certainly can. If the brake lining brands are denoted 1, 2, 3, 4, simply test brake lining brand i on every wheel of the ith car. Combining this design with the tire brand design of Table 9.4 gives us the array of ordered pairs of Table 9.9. All the ordered pairs in this table are different.

Unfortunately, the array of Table 9.9 is not a very satisfactory design if we consider just brake linings. For we only use brake linings of brand 1 on car A, of brand 2 on car B, and so on. It would be good to have the brake linings tested by a Latin square design, not just the tires. Thus we would like to find two Latin square experiments, $A = (a_{ij})$ and $B = (b_{ij})$, one for tire brands and one for brake lining brands, both using the same row and column effects. Moreover, we want the ordered pairs (a_{ij}, b_{ij}) all to be different. Can this be done? In our case, it can. Table 9.10 shows a pair of Latin square designs and the corresponding array of ordered pairs, which is easily seen to have each ordered pair (a, b), with $1 \leq a \leq 4$ and $1 \leq b \leq 4$, appearing exactly once. Equivalently, the ordered pairs are all different.

We shall say that two distinct $n \times n$ Latin squares $A = (a_{ij})$ and $B = (b_{ij})$ are *orthogonal* if the n^2 ordered pairs (a_{ij}, b_{ij}) are all different. Thus, the two 4×4 Latin squares of Table 9.10 are orthogonal. However, the two Latin squares of Table 9.7 are not, as the ordered pair (2, 4) appears twice, once in the 2, 2 position and once in the 3, 3 position. More generally, if $A^{(1)}, A^{(2)}, \ldots, A^{(r)}$ are distinct $n \times n$ Latin squares, they are said to form an *orthogonal family* if every pair of them is orthogonal.

The main question we shall address in this section is the fundamental existence question: If we want to design an experiment using a pair of $n \times n$ orthogonal Latin squares, can we always be sure that such a pair exists? More generally, we shall ask: When does an orthogonal family of r different $n \times n$ Latin squares exist?

Before addressing these questions, we give several examples of the use of orthogonal Latin square designs.

Example 9.5 Fuel Economy

Davies [1945] has used a pair of orthogonal Latin squares in the comparison of fuel economy in miles per gallon achieved with different grades of gasoline. There were

Table 9.9 Design for testing the combined effects of tire brand and brake lining brand on tread wear[a]

		Car			
		A	B	C	D
Wheel position	Left front	(1, 1)	(2, 2)	(3, 3)	(4, 4)
	Right front	(2, 1)	(3, 2)	(4, 3)	(1, 4)
	Left rear	(3, 1)	(4, 2)	(1, 3)	(2, 4)
	Right rear	(4, 1)	(1, 2)	(2, 3)	(3, 4)

[a]The i, j entry is an ordered pair, giving first the tire brand used in position i on car j, and then the brake lining brand used there.

Table 9.10 Two orthogonal Latin square designs for testing the combined effects of tire brand and brake lining brand on tread wear[a]

		Car			
		A	B	C	D
Wheel position	Left front	1	2	3	4
	Right front	2	1	4	3
	Left rear	3	4	1	2
	Right rear	4	3	2	1

Latin Square
Design for
Tire Brands

		Car			
		A	B	C	D
Wheel position	Left front	4	1	2	3
	Right front	3	2	1	4
	Left rear	1	4	3	2
	Right rear	2	3	4	1

Latin Square
Design for
Brake Lining
Brands

		Car			
		A	B	C	D
Wheel position	Left front	(1, 4)	(2, 1)	(3, 2)	(4, 3)
	Right front	(2, 3)	(1, 2)	(4, 1)	(3, 4)
	Left rear	(3, 1)	(4, 4)	(1, 3)	(2, 2)
	Right rear	(4, 2)	(3, 3)	(2, 4)	(1, 1)

Combined Design

[a]The combined array lists in the i, j entry the ordered pair consisting of the tire brand and then the brake lining brand used in the two Latin squares in tire position i on car j.

seven grades of gasoline tested. One car was used throughout. Each test involved driving the test car over a fixed route of 20 miles, including various gradients. To remove possible biases connected with the driver, seven drivers were used; and to remove possible effects connected with the traffic conditions, the experiment was run on different days and at seven different times of the day. Thus, in addition to the seven treatments under comparison, there are three classifications of the experimental units: by drivers, by days, and by times of the day. A double classification of the experimental units suggests the use of a Latin square, a triple classification a pair of orthogonal Latin squares. The latter allows for an experiment in which each grade of gasoline is used once on each day, once by each driver, and once at each time of day, ensuring a balanced comparison. The design assigns to each day (row) and each time of day (column) one grade of gasoline (in the first Latin square) and one driver (in the second square). (In our tire wear example of Section 9.1, we could not control for the driver in the same way, that is, it would not make sense to use a pair of orthogonal Latin square experiments, the first indicating tire brand at position i on car j and the second indicating driver at position i on car j. For the same driver must be assigned to all positions of a given car!)

Example 9.6   Testing Cloth for Wear

Box *et al.* [1978] describe an experiment involving a Martindale wear tester, a machine used to test the wearing quality of materials such as cloth. In one run of a Martindale wear tester of the type considered, four pieces of cloth could be rubbed simultaneously, each against a sheet of emery paper, and then the weight loss could be measured. There were four different specimen holders, labeled A, B, C, D, and each could be used in one of four positions, P_1, P_2, P_3, P_4, on the machine. In a particular experiment, four types of cloth or treatments, labeled 1, 2, 3, 4, were compared. The experimenters wanted to control for the effects of the four different specimen holders, the four positions of the machine, which run the cloth was tested in, and which sheet of emery paper the cloth was rubbed against. A quadruple classification of experimental units suggests an orthogonal family of three 4×4 Latin squares. It was decided to use four sheets of emery paper, labeled α, β, γ, δ, cut each into four quarters, and use each quarter in one experimental unit. There were four runs in all, R_1, R_2, R_3, R_4, each testing four specimens of cloth with different holders in varying positions and with different quarter-pieces of emery paper. Table 9.11 shows the three Latin square designs used. The reader can check that these are pairwise orthogonal. (In fact, the experiment was replicated, using four more runs and four more sheets of emery paper, again in a design involving an orthogonal family of three 4×4 Latin squares.)

9.2.3 Existence Results for Orthogonal Families

Let the *order* of an $n \times n$ Latin square be n. In what follows, we shall usually assume that the entries in a Latin square of order n are the integers $1, 2, \ldots, n$. We shall now discuss the question: Does there exist an orthogonal family of r Latin squares of order n? We shall assume that $n > 1$, for there is only one 1×1 Latin square. There does not exist a pair of orthogonal 2×2 Latin squares. For the only Latin squares of order 2 are shown in Table 9.12. They are not orthogonal since the pair (1, 2) appears twice. We have seen in

Table 9.11   An orthogonal family of three Latin squares for testing cloth for wear[a]

Run

	R_1	R_2	R_3	R_4
P_1	1	3	4	2
P_2	2	4	3	1
P_3	3	1	2	4
P_4	4	2	1	3

Position

Latin Square Design for Treatments

	R_1	R_2	R_3	R_4
P_1	A	D	B	C
P_2	B	C	A	D
P_3	C	B	D	A
P_4	D	A	C	B

Latin Square Design for Holders

	R_1	R_2	R_3	R_4
P_1	α	β	γ	δ
P_2	β	α	δ	γ
P_3	γ	δ	α	β
P_4	δ	γ	β	α

Latin Square Design for Emery Paper Sheets

[a]The i, j entry shows the treatment (cloth type), holder, and emery paper sheet, respectively, used in run R_j in position P_i.

Table 9.12 The two
Latin squares of order 2

1	2
2	1

2	1
1	2

Table 9.10 that there is a pair of orthogonal Latin squares of order 4, and in Table 9.11 that there is an orthogonal family of three Latin squares of order 4. It is easy enough to give a pair of orthogonal Latin squares of order 3. (Try it.)

The first theorem gives necessary conditions for the existence of an orthogonal family of r Latin squares of order n.

Theorem 9.1. If there is an orthogonal family of r Latin squares of order n, then $r \leq n - 1$.

Proof. Suppose that $A^{(1)}$, $A^{(2)}$, ..., $A^{(r)}$ form an orthogonal family of $n \times n$ Latin squares. Let $a_{ij}^{(p)}$ be the i, j entry of $A^{(p)}$. Relabel the entries in the first square so that 1 comes in the 1, 1 spot, that is, so that $a_{11}^{(1)} = 1$. Do this as follows. If $a_{11}^{(1)}$ was k, switch 1 with k and k with 1 throughout $A^{(1)}$. This does not change $A^{(1)}$ from being a Latin square and it does not change the orthogonality. For if the pair

$$(a_{ij}^{(1)}, a_{ij}^{(p)}) \quad \text{was} \quad (k, l),$$

it is now $(1, l)$, and if it was $(1, l)$, it is now (k, l).

By the same reasoning, without affecting the fact that we have an orthogonal family of $n \times n$ Latin squares, we can arrange matters so that each 1, 1 entry in each square is 1, and more generally so that

$$a_{11}^{(1)} = a_{11}^{(2)} = \cdots = a_{11}^{(r)} = 1,$$

$$a_{12}^{(1)} = a_{12}^{(2)} = \cdots = a_{12}^{(r)} = 2,$$

$$a_{13}^{(1)} = a_{13}^{(2)} = \cdots = a_{13}^{(r)} = 3,$$

$$\vdots$$

$$a_{1n}^{(1)} = a_{1n}^{(2)} = \cdots = a_{1n}^{(r)} = n.$$

That is, we can arrange matters so that each $A^{(p)}$ has the same first row:

$$1 \quad 2 \quad 3 \quad \cdots \quad n.$$

Let us consider the 2, 1 entry in each square. Since $A^{(p)}$ is a Latin square, and since $a_{11}^{(p)}$ is 1 and 1 can appear only once in a column, $a_{21}^{(p)}$ must be different from 1. Moreover, by orthogonality,

$$a_{21}^{(p)} \neq a_{21}^{(q)}$$

if $p \neq q$. For otherwise,

$$\left(a_{21}^{(p)}, a_{21}^{(q)}\right) = (i, i)$$

for some i, so

$$\left(a_{21}^{(p)}, a_{21}^{(q)}\right) = \left(a_{1i}^{(p)}, a_{1i}^{(q)}\right),$$

which violates orthogonality. Thus, the numbers

$$a_{21}^{(1)}, a_{21}^{(2)}, \ldots, a_{21}^{(r)}$$

are all different and all different from 1. It follows that there are at most $n - 1$ of these numbers, and $r \leq n - 1$. (Formally, this reasoning uses the pigeonhole principle of Section 8.1.) Q.E.D.

We illustrate the proof of this theorem by starting with an orthogonal family of three Latin squares of order 4 as shown in Table 9.13. The procedure to arrange that all the first rows are 1 2 3 4 is illustrated in the table. Note that the 2, 1 entries in the three Latin squares in the last row of Table 9.13 are 2, 3, and 4, respectively.

Theorem 9.1 says that we can never find an orthogonal family of $n \times n$ Latin squares consisting of more than $n - 1$ squares. Let us say that an orthogonal family of Latin squares of order n is *complete* if it has $n - 1$ Latin squares in it. Thus, the three Latin squares of order 4 shown in Table 9.13 form a complete orthogonal family. It will be convenient to think of a single 2×2 Latin square as constituting an orthogonal family.

Theorem 9.2 gives sufficient conditions for the existence of a complete orthogonal family of Latin squares.

Table 9.13 The procedure of changing an orthogonal family of Latin squares into one where each square has first row 1 2 3 $\cdots$ n

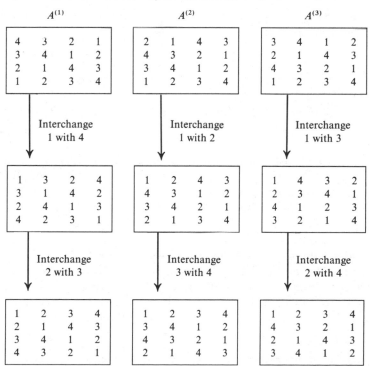

Theorem 9.2. If $n > 1$ and $n = p^k$, where p is a prime number* and k is a positive integer, then there is a complete orthogonal family of Latin squares of order n.

We omit a proof of Theorem 9.2 at this point. In Section 9.3.3 we shall describe a constructive procedure for finding a complete orthogonal family of Latin squares of order n if n is a power of a prime. In particular, Theorem 9.2 says that there exists a pair of orthogonal 3×3 Latin squares, since $3 = 3^1$. It also says that there exist three pairwise orthogonal 4×4 Latin squares, since $4 = 2^2$. (We have already seen three such squares in Tables 9.11 and 9.13.) There also exists a family of four pairwise orthogonal 5×5 Latin squares, since $5 = 5^1$. Since 6 is not a power of a prime, Theorem 9.2 does not tell us whether or not there is a set of five pairwise orthogonal 6×6 Latin squares, or indeed whether there is even a pair of such squares.

Any integer $n > 1$ can be written *uniquely* as the product of (integer) powers of primes:

$$p_1^{t_1} p_2^{t_2} \cdots p_s^{t_s}. \tag{9.1}$$

This product is called the *prime power decomposition*. For example,

$$6 = 2^1 3^1,$$

$$12 = 3 \times 4 = 3^1 2^2,$$

$$80 = 16 \times 5 = 2^4 5^1,$$

$$60 = 4 \times 15 = 4 \times 3 \times 5 = 2^2 3^1 5^1.$$

Theorem 9.3. Suppose that $n = p_1^{t_1} p_2^{t_2} \cdots p_s^{t_s}$ is the prime power decomposition of n, $n > 1$, and r is the smallest of the quantities

$$(p_1^{t_1} - 1), \quad (p_2^{t_2} - 1), \quad \ldots, \quad (p_s^{t_s} - 1).$$

Then there is an orthogonal family of r Latin squares of order n.

We shall prove this theorem below. To illustrate it, we recall that $12 = 2^2 3^1$. Then

$$2^2 - 1 = 3, \qquad 3^1 - 1 = 2,$$

so $r = 2$. It follows that there are two orthogonal Latin squares of order 12. This does not say that there is not a larger orthogonal family of 12×12's. Note that Theorem 9.2 does not apply, since 12 is not a power of a prime.

Let us try to apply Theorem 9.3 to $n = 6$. We have $6 = 2^1 3^1$. Since

$$2^1 - 1 = 1, \qquad 3^1 - 1 = 2,$$

$r = 1$, and we do not even know from Theorem 9.3 if there exists a pair of orthogonal 6×6 Latin squares. The famous mathematician Euler conjectured in 1782 that there was no such pair. For more than 100 years, the conjecture could neither be proved nor disproved. Around 1900, Tarry looked systematically at all possible pairs of 6×6 Latin squares. (There are 812,851,200; but by making the first row 123456, it is only necessary

---

*Recall that a *prime number* n is an integer > 1 whose only divisors are 1 and n.

to consider 9408 pairs.) He succeeded in proving that Euler was right. Thus, there does not exist a pair of orthogonal Latin squares of order 6. (See Tarry [1900, 1901].)

The following is a corollary of Theorem 9.3.

Corollary 9.3.1. Suppose that $n > 1$ and either 2 does not divide n or the prime power decomposition of n is

$$n = 2^{t_1} p_2^{t_2} p_3^{t_3} \cdots,$$

with $t_1 > 1$. Then there is a pair of orthogonal Latin squares.

Proof. If $t_1 > 1$,

$$2^{t_1} - 1 \geq 3.$$

Each other p_i is greater than 2, so

$$p_i^{t_i} - 1 \geq 2.$$

It follows that $r \geq 2$.   Q.E.D.

Corollary 9.3.1 leaves open the question of the existence of pairs of orthogonal Latin squares of order $n = 2k$ where 2 does not divide k. Euler, also in 1782, conjectured that there does not exist a pair of orthogonal Latin squares of order n for all such n. He was right for $n = 2$ and $n = 6$. However, contrary to his usual performance, he was wrong otherwise. It was not until 1960 that he was proved wrong.

Theorem 9.4 (Bose, Shrikhande, and Parker [1960]).   If $n > 6$, $n = 2k$, and 2 does not divide k, then there is a pair of orthogonal Latin squares of order n.

We can now summarize what we know about the existence of pairs of orthogonal Latin squares.

Theorem 9.5.   There is a pair of orthogonal Latin squares of order n for all $n > 1$ except $n = 2$ or 6.

Thus, the existence of pairs of orthogonal Latin squares is completely settled. This is not the case for larger families of orthogonal Latin squares. For $n = 2, 3, \ldots, 9$, the size of the largest orthogonal family of $n \times n$ Latin squares is known. For by Theorems 9.2 and 9.5, it is $n - 1$ for $n \neq 2$, 6, and it is 1 for $n = 2$, 6. However, for $n = 10$, it is not even known if there is a family of three pairwise orthogonal $n \times n$ Latin squares.

Example 9.7   The Problem of the 36 Officers

Euler encountered the notion of orthogonal Latin squares not in connection with experimental design, but in connection with the following problem. There are 36 officers, six officers of six different ranks in each of six regiments. Find an arrangement of the 36 officers in a 6×6 square formation such that each row and each column contains one and only one officer of each rank and one and only one officer from each regiment and there is only one officer from each regiment of each rank. Can this be done? The officers must be arranged so that their ranks form a Latin

Table 9.14 Orthogonal Latin squares $A^{(1)}$ and $A^{(2)}$ and $B^{(1)}$ and $B^{(2)}$; matrices $(a_{12}^{(1)}, B^{(1)})$ and $(a_{32}^{(2)}, B^{(2)})$

$$A^{(1)} = \begin{array}{|ccc|} \hline 1 & 2 & 3 \\ 2 & 3 & 1 \\ 3 & 1 & 2 \\ \hline \end{array}, \qquad A^{(2)} = \begin{array}{|ccc|} \hline 1 & 2 & 3 \\ 3 & 1 & 2 \\ 2 & 3 & 1 \\ \hline \end{array}$$

$$B^{(1)} = \begin{array}{|cccc|} \hline 4 & 3 & 2 & 1 \\ 3 & 4 & 1 & 2 \\ 2 & 1 & 4 & 3 \\ 1 & 2 & 3 & 4 \\ \hline \end{array}, \qquad B^{(2)} = \begin{array}{|cccc|} \hline 2 & 1 & 4 & 3 \\ 4 & 3 & 2 & 1 \\ 3 & 4 & 1 & 2 \\ 1 & 2 & 3 & 4 \\ \hline \end{array}$$

$$(a_{12}^{(1)}, B^{(1)}) = \begin{bmatrix} (2,4) & (2,3) & (2,2) & (2,1) \\ (2,3) & (2,4) & (2,1) & (2,2) \\ (2,2) & (2,1) & (2,4) & (2,3) \\ (2,1) & (2,2) & (2,3) & (2,4) \end{bmatrix}$$

$$(a_{32}^{(2)}, B^{(2)}) = \begin{bmatrix} (3,2) & (3,1) & (3,4) & (3,3) \\ (3,4) & (3,3) & (3,2) & (3,1) \\ (3,3) & (3,4) & (3,1) & (3,2) \\ (3,1) & (3,2) & (3,3) & (3,4) \end{bmatrix}$$

square and also so that their regiments form a Latin square. Moreover, the pairs of rank and regiment appear once and only once, so the two squares must be orthogonal. We now know that this cannot be done.

We close this subsection by presenting a proof of Theorem 9.3.* We first prove the following result.

Theorem 9.6 (MacNeish [1922]). Suppose that there is an orthogonal family of r Latin squares of order m and another orthogonal family of r Latin squares of order n. Then there is an orthogonal family of r Latin squares of order mn.

Proof. Let $A^{(1)}, A^{(2)}, \ldots, A^{(r)}$ be pairwise orthogonal Latin squares of order m and $B^{(1)}, B^{(2)}, \ldots, B^{(r)}$ be pairwise orthogonal Latin squares of order n. For $e = 1, 2, \ldots, r$, let $(a_{ij}^{(e)}, B^{(e)})$ represent the $n \times n$ matrix whose u, v entry is the ordered pair $(a_{ij}^{(e)}, b_{uv}^{(e)})$. For instance, suppose that $A^{(1)}, A^{(2)}, B^{(1)}$, and $B^{(2)}$ are as in Table 9.14. Then $(a_{12}^{(1)}, B^{(1)})$ and $(a_{32}^{(2)}, B^{(2)})$ are shown in the table. Let $C^{(e)}$ be the matrix that can be represented schematically as follows:

$$C^{(e)} = \begin{bmatrix} (a_{11}^{(e)}, B^{(e)}) & (a_{12}^{(e)}, B^{(e)}) & \cdots & (a_{1m}^{(e)}, B^{(e)}) \\ (a_{21}^{(e)}, B^{(e)}) & (a_{22}^{(e)}, B^{(e)}) & \cdots & (a_{2m}^{(e)}, B^{(e)}) \\ & \cdots & & \\ (a_{m1}^{(e)}, B^{(e)}) & (a_{m2}^{(e)}, B^{(e)}) & \cdots & (a_{mm}^{(e)}, B^{(e)}) \end{bmatrix}.$$

Then $C^{(e)}$ is an $mn \times mn$ matrix. We shall show that $C^{(1)}, C^{(2)}, \ldots, C^{(r)}$ is an orthogonal family of Latin squares of order mn.

*The rest of this subsection can be omitted.

To see that $C^{(e)}$ is a Latin square, note first that in a given row, two entries in different columns are given by $(a_{ij}^{(e)}, b_{uv}^{(e)})$ and $(a_{ik}^{(e)}, b_{uw}^{(e)})$, and so are distinct since $A^{(e)}$ and $B^{(e)}$ are Latin squares. In a given column, two entries in different rows are given by $(a_{ij}^{(e)}, b_{uv}^{(e)})$ and $(a_{kj}^{(e)}, b_{wv}^{(e)})$, and so again are distinct because $A^{(e)}$ and B^e are Latin squares.

To see that $C^{(e)}$ and $C^{(f)}$ are orthogonal, suppose that

$$\langle (a_{ij}^{(e)}, b_{uv}^{(e)}), (a_{ij}^{(f)}, b_{uv}^{(f)}) \rangle = \langle (a_{pq}^{(e)}, b_{st}^{(e)}), (a_{pq}^{(f)}, b_{st}^{(f)}) \rangle.$$

Then it follows that

$$(a_{ij}^{(e)}, a_{ij}^{(f)}) = (a_{pq}^{(e)}, a_{pq}^{(f)}),$$

so by orthogonality of $A^{(e)}$ and $A^{(f)}$, $i = p$ and $j = q$. Similarly, orthogonality of $B^{(e)}$ and $B^{(f)}$ implies that $u = s$ and $v = t$.   Q.E.D.

Proof of Theorem 9.3. By Theorem 9.2, for $i = 1, 2, \ldots, s$, there is an orthogonal family of $p_i^{t_i} - 1$ Latin squares of order $p_i^{t_i}$. Thus, for $i = 1, 2, \ldots, s$, there is an orthogonal family of r Latin squares of order $p_i^{t_i}$. The result follows from Theorem 9.6 by mathematical induction on s.   Q.E.D.

EXERCISES FOR SECTION 9.2

1. For each pair of Latin squares in Table 9.15, determine if they are orthogonal.
2. Check that the three Latin squares of Table 9.11 form an orthogonal family.
3. For each family of Latin squares of Table 9.16, determine if it is orthogonal.
4. Suppose that A is an $n \times n$ Latin square. For each of the following operations, determine if it results in a new Latin square.
   (a) Interchange the entries 2 and 3 whenever they occur.
   (b) Replace each row by going from last to first.
   (c) Replace A by its transpose.

Table 9.15 Pairs of Latin squares for exercises of Section 9.2

(a)

```
1 2 3     1 2 3
2 3 1     3 1 2
3 1 2     2 3 1
```

(b)

```
1 2 3 4     1 2 3 4
2 3 4 1     3 4 1 2
3 4 1 2     2 3 4 1
4 1 2 3     4 1 2 3
```

(c)

```
1 2 3 4 5     5 1 2 3 4
2 3 4 5 1     4 5 1 2 3
3 4 5 1 2     3 4 5 1 2
4 5 1 2 3     2 3 4 5 1
5 1 2 3 4     1 2 3 4 5
```

(d)

```
1 2 3 4 5 6     1 2 3 4 5 6
6 1 2 3 4 5     2 3 4 5 6 1
2 3 4 5 6 1     5 6 1 2 3 4
5 6 1 2 3 4     3 4 5 6 1 2
3 4 5 6 1 2     4 5 6 1 2 3
4 5 6 1 2 3     6 1 2 3 4 5
```

Table 9.16 Families of Latin squares
for exercises of Section 9.2

1	3	2	4
3	1	4	2
2	4	1	3
4	2	3	1

3	1	4	2
2	4	1	3
1	3	2	4
4	2	3	1

2	4	1	3
1	3	2	4
3	1	4	2
4	2	3	1

(a)

1	2	3	4	5
2	3	4	5	1
3	4	5	1	2
4	5	1	2	3
5	1	2	3	4

1	2	3	4	5
3	4	5	1	2
5	1	2	3	4
2	3	4	5	1
4	5	1	2	3

1	2	3	4	5
5	1	2	3	4
4	5	1	2	3
3	4	5	1	2
2	3	4	5	1

(b)

5. For which of the following numbers n can you be sure that there is an orthogonal family of 3 Latin squares of order n? Why?
   (a) $n = 12$ **(b)** $n = 13$ **(c)** $n = 21$ **(d)** $n = 25$
   (e) $n = 35$ **(f)** $n = 36$ **(g)** $n = 39$ **(h)** $n = 75$
   (i) $n = 140$ **(j)** $n = 185$ **(k)** $n = 369$ **(l)** $n = 539$.

6. If $n = 275$, show that there is a set of 10 pairwise orthogonal Latin squares of order n.

7. If $n = 370$, does there exist a pair of orthogonal Latin squares of order n? Why?

8. Does there exist a pair of orthogonal Latin squares of order 54? Why?

9. If n is divisible by 4, can you be sure (with the theorems we have stated) whether or not there is a set of three pairwise orthogonal Latin squares of order n? Give a reason for your answer.

10. For the orthogonal family of Latin squares of order 5 in Table 9.17, use the procedure in the proof of Theorem 9.1 to rearrange elements so that the first row of each Latin square is 1 2 3 4 5.

11. Suppose that two orthogonal 7×7 Latin squares both have   7 6 5 4 3 2 1   as the first row.
    (a) Is it possible for them to have the same 2, 4 entry?
    (b) What does your answer tell you about the number of possible 7×7 pairwise orthogonal Latin squares each of which has   7 6 5 4 3 2 1   as the first row?

12. Suppose that two orthogonal 5×5 Latin squares both have   1 2 3 4 5   as the last row.
    (a) Is it possible for them to have the same 1, 3 entry?
    (b) What does your answer tell you about the number of possible 5×5 pairwise orthogonal Latin squares each of which has   1 2 3 4 5   as the last row?

Table 9.17 An orthogonal family of Latin squares

5	1	2	3	4
4	5	1	2	3
3	4	5	1	2
2	3	4	5	1
1	2	3	4	5

4	5	1	2	3
2	3	4	5	1
5	1	2	3	4
3	4	5	1	2
1	2	3	4	5

3	4	5	1	2
5	1	2	3	4
2	3	4	5	1
4	5	1	2	3
1	2	3	4	5

2	3	4	5	1
3	4	5	1	2
4	5	1	2	3
5	1	2	3	4
1	2	3	4	5

13. Suppose that two orthogonal 4 × 4 Latin squares both have 1 2 3 4 as the main diagonal.

    (a) Is it possible for them to have the same 2, 3 entry?

    (b) What does your answer tell you about the number of possible 4 × 4 pairwise orthogonal Latin squares each of which has 1 2 3 4 as the main diagonal?

14. Use the Latin squares of Table 9.14 to find a pair of orthogonal Latin squares of order 12.

15. Find a pair of orthogonal Latin squares of order 9.

16. If there exists a pair of orthogonal Latin squares of order n, and A is a Latin square of order n, A is not necessarily a member of an orthogonal pair of Latin squares. Give an example to illustrate this.

9.3 FINITE FIELDS AND COMPLETE ORTHOGONAL FAMILIES OF LATIN SQUARES*

In this section we aim to present a constructive proof of Theorem 9.2, namely, that if $n > 1$ and $n = p^k$ for p prime, then there is a complete orthogonal family of Latin squares of order n. We begin with some mathematical preliminaries.

9.3.1 Modular Arithmetic

Arithmetics with only finitely many numbers underlie the construction of combinatorial designs. They are also vitally important in computing, where there are practical bounds on the size of the sets of integers that can be considered. In this subsection we introduce a simple example of an arithmetic with only finitely many elements, modular arithmetic. In the next subsection we shall introduce a general notion of an arithmetic with only finitely many elements, namely a finite field. Then, we shall use finite fields to construct complete orthogonal families of Latin squares. Modular arithmetic and finite fields underlie the operation of the shift registers which operate in a computer to take a bit string and produce another one. For a discussion of this application, see, for example, Fisher [1977].

   Let us consider the remainders left when integers are divided by the number 3. We find

$$0 = 0 \cdot 3 + 0, \qquad 1 = 0 \cdot 3 + 1, \qquad 2 = 0 \cdot 3 + 2,$$
$$3 = 1 \cdot 3 + 0, \qquad 4 = 1 \cdot 3 + 1, \qquad 5 = 1 \cdot 3 + 2,$$
$$6 = 2 \cdot 3 + 0, \qquad 7 = 2 \cdot 3 + 1, \qquad 8 = 2 \cdot 3 + 2,$$
$$9 = 3 \cdot 3 + 0, \qquad 10 = 3 \cdot 3 + 1, \qquad 11 = 3 \cdot 3 + 2.$$

The remainder is always one of the three integers 0, 1, or 2. We say that two integers a and b are *congruent modulo* 3, and write $a \equiv b$ (mod 3), if they leave the same remainder on division by 3. For instance, $2 \equiv 5$ (mod 3) and $1 \equiv 7$ (mod 3). In general, if a, b, and n are integers, we say that a is *congruent to b modulo n*, and write $a \equiv b$ (mod n), if a and b leave the same remainder on division by n. For instance, $29 \equiv 17$ (mod 4), since $29 = 7 \cdot 4 + 1$ and $17 = 4 \cdot 4 + 1$. Congruence modulo 12 is used every day when we look

---

*This section can be omitted without loss of continuity. As an alternative, the reader might wish to read all but Section 9.3.4.

at the clock. The hands of the clock indicate the hour modulo 12. Similarly, the mileage indicator on a car gives the mileage the car has traveled modulo 100,000.

Let us now fix a number n and consider the set of integers $Z_n = \{0, 1, 2, \ldots, n - 1\}$. Every integer is congruent modulo n to one of the integers in the set Z_n. If we add two integers in the set Z_n, their sum is not necessarily in the set. However, their sum is congruent to an element in Z_n. In performing *modular addition*, if a and b are in Z_n, we *define* $a + b$ to be that number in Z_n which is congruent to the ordinary sum of a and b, modulo n. For instance, suppose that $n = 3$. Then $2 + 2$ is 4 in ordinary arithmetic, which is congruent to 1, modulo 3. Hence, in addition modulo 3, $2 + 2$ is 1. Similarly, $1 + 2$ is 0 and $1 + 1$ is 2. In addition modulo 4, $3 + 3$ is 2 and $2 + 2$ is 0. Modular multiplication works similarly. If a, b are in Z_n, we define $a \times b$ to be that integer in Z_n congruent to the ordinary product of a and b, modulo n. For instance, in multiplication modulo 3, 2×2 is 1, since the ordinary product of 2 and 2 is 4, which is congruent to 1 modulo 3. Similarly, in multiplication modulo 4, 3×3 is 1 and 2×2 is 0.

We can summarize the *operations* $+$ and $\times$ on Z_n by giving addition and multiplication tables. Tables 9.18 and 9.19 give these tables for the cases $n = 2$ and $n = 3$, respectively. In these tables, the elements of Z_n are listed in the same order along both rows and columns and the entry in the row corresponding to a and the column corresponding to b is $a + b$ or $a \times b$, depending on the table.

9.3.2 The Finite Fields GF(p^k)

Suppose that X is a set. A *binary operation* $\circ$ on X is a function that assigns to each ordered pair of elements of X another element of X, usually denoted $a \circ b$. For example, if X is the set of integers, then $+$ and $\times$ define binary operations on X. If X is a finite set, we can define a binary operation $\circ$ by giving a table such as those in Tables 9.18 or 9.19.

A *field* $\mathscr{F}$ is a triple $(F, +, \times)$, where F is a set and $+$ and $\times$ are two binary operations on F (not necessarily the usual operations of $+$ and $\times$), with certain conditions holding. These are the following:*

Condition **F1** (*Closure*).† For all a, b in F, $a + b$ is in F and $a \times b$ is in F.

Condition **F2** (*Associativity*). For all a, b, c in F,

$$a + (b + c) = (a + b) + c,$$

$$a \times (b \times c) = (a \times b) \times c.$$

Condition **F3** (*Commutativity*). For all a, b in F,

$$a + b = b + a,$$

$$a \times b = b \times a.$$

---

*Our treatment of fields is necessarily brief. The reader without a background in this subject might consult such elementary treatments as Birkhoff and MacLane [1953], Dornhoff and Hohn [1978], Fisher [1977], or Zariski and Samuel [1958].

†Condition **F1** is actually implicit in our definition of operation.

Table 9.18 The operations
+ and × of addition and
multiplication modulo 2 on Z_2

+	0	1		×	0	1
0	0	1		0	0	0
1	1	0		1	0	1

Condition **F4** (*Identity*).
(a) There is an element in F, which is denoted 0 and called the *additive identity*, so that for all a in F,

$$a + 0 = a.$$

(b) There is an element in F different from 0, which is denoted 1 and called the *multiplicative identity*, so that for all a in F,

$$a \times 1 = a.$$

Condition **F5** (*Inverse*).
(a) For all a in F, there is an element b in F, called an *additive inverse* of a, so that

$$a + b = 0.$$

(b) For all a in F with $a \neq 0$, there is an element b in F, called a *multiplicative inverse* of a, so that

$$a \times b = 1.$$

Condition **F6** (*Distributivity*). For all a, b, c in F,

$$a \times (b + c) = (a \times b) + (a \times c).$$

(Conditions **F1, F2, F4**, and **F5** say that $(F, +)$ is a group in the sense of Chapter 7. Also, if $F' = F$ less the element 0, these conditions also say that $(F', \times)$ is a group, since we can show that $a \times b$ is never 0 if $a \neq 0$ and $b \neq 0$ and we can show that the multiplicative inverse of a is never 0.) Note that it is possible to prove from conditions **F1–F6** that the additive and multiplicative inverses of an element a are, respectively, unique; they are denoted, respectively, as $-a$ and a^{-1}.

Table 9.19 The operations + and ×
of addition and multiplication modulo 3 on Z_3

+	0	1	2		×	0	1	2
0	0	1	2		0	0	0	0
1	1	2	0		1	0	1	2
2	2	0	1		2	0	2	1

The following are examples of fields.

1. $(Re, +, \times)$, where Re is the set of real numbers and $+$ and $\times$ are the usual addition and multiplication operations.
2. $(Q, +, \times)$, where Q is the set of rational numbers and $+$ and $\times$ are the usual addition and multiplication operations.
3. $(C, +, \times)$, where C is the set of complex numbers and $+$ and $\times$ are the usual addition and multiplication operations.

However, $(Z, +, \times)$, where Z is the set of integers and $+$ and $\times$ are the usual addition and multiplication operations, is not a field. Conditions **F1**–**F4** and **F6** hold. However, part (b) of condition **F5** fails. There is no b in Z so that $2 \times b = 1$.

We now consider some examples of *finite fields*, fields where F is a finite set.

Consider $(Z_2, +, \times)$, where $+$ and $\times$ are modulo 2. Then this is a field, as is easy to check. Note that 0 and 1 are, respectively, the additive and multiplicative identities. The additive inverse of 1 is 1, since $1 + 1 = 0$, and the multiplicative inverse of 1 is 1, since $1 \times 1 = 1$.

Similarly, $(Z_3, +, \times)$, where $+$ and $\times$ are modulo 3, is a field. The additive and multiplicative inverses are again 0 and 1, respectively. Note that the additive inverse of 2 is 1, since $2 + 1 = 0$. The multiplicative inverse of 2 is 2, since $2 \times 2 = 1$.

Is Z_n under modular addition and multiplication always a field? The answer is no. Consider Z_6. We have $3 \times 2 = 0$. If Z_6 under addition and multiplication modulo 6 is a field, then let 2^{-1} denote the multiplicative inverse of 2. We have

$$(3 \times 2) \times 2^{-1} = 0 \times 2^{-1} = 0.$$

However,

$$(3 \times 2) \times 2^{-1} = 3 \times (2 \times 2^{-1}) = 3 \times 1 = 3.$$

We conclude that $0 = 3$, a contradiction.

Theorem 9.7. For $n \geq 2$, Z_n under addition and multiplication modulo n is a field if and only if n is a prime number.

The proof of Theorem 9.7 is left as an exercise (Exercise 12).

We close this subsection by asking: For what values of n does there exist a finite field of n elements? We shall be able to give an explicit answer. Note that by Theorem 9.7, Z_4 does not define a field under modular addition and multiplication. However, it is possible to define addition and multiplication operations on $\{0, 1, 2, 3\}$ which define a field. Such operations are shown in Table 9.20. Verification that these define a field is left to the reader (Exercise 13). (The reader familiar with algebra will be able to derive these tables by letting $2 = w$ and $3 = 1 + w$ and performing addition and multiplication modulo the irreducible polynomial $1 + w + w^2$ over GF[2], the finite field of 2 elements to be defined below.) The arithmetic of binary numbers which is actually used in many large computers is based on Z_{2^n} for some n. (Here, $n = 2$.) For a discussion of this particular arithmetic, see Dornhoff and Hohn [1978], Kapur [1970, Chapter 6], and Vickers [1971, Chapter 4].

Table 9.20 Addition and multiplication tables
for a field $GF(2^2)$ of four elements

+	0	1	2	3		×	0	1	2	3
0	0	1	2	3		0	0	0	0	0
1	1	0	3	2		1	0	1	2	3
2	2	3	0	1		2	0	2	3	1
3	3	2	1	0		3	0	3	1	2

If $n = 6$, it is not possible to define addition and multiplication on a set of n elements, such as $\{0, 1, 2, 3, 4, 5\}$, so that we get a finite field. The situation is summarized in Theorem 9.8.

Theorem 9.8. If $(F, +, \times)$ is a finite field, then there is a prime number p and a positive integer k so that F has p^k elements. Conversely, for all prime numbers p and positive integers k, there is a finite field of p^k elements.

The proof of this theorem can be found in most books on modern algebra, for instance any of the references in the first footnote on page 374. It turns out that there is essentially only one field of p^k elements for p a prime and k a positive integer, in the sense that any two of these fields are isomorphic.* The unique field of p^k elements will be denoted $GF(p^k)$. (The letters GF stand for Galois field, and are in honor of the famous mathematician Galois, who made fundamental contributions to modern algebra.)

9.3.3 Construction of a Complete Orthogonal Family of $n \times n$ Latin Squares if n is a Power of a Prime

We now present a construction of a complete orthogonal family of $n \times n$ Latin squares that applies whenever $n = p^k$, for p prime and k a positive integer, and $n > 1$. This will prove Theorem 9.2. Let $b_1, b_2, \ldots, b_n$ be elements of a finite field $GF(n)$ of $n = p^k$ elements. Let b_1 be the multiplicative identity of this field and b_n be the additive identity. For $e = 1, 2, \ldots, n - 1$, define the $n \times n$ array $A^{(e)} = (a_{ij}^{(e)})$ by taking

$$a_{ij}^{(e)} = (b_e \times b_i) + b_j, \tag{9.2}$$

where $+$ and $\times$ are the operations of the field $GF(n)$. In Section 9.3.4 we shall show that $A^{(e)}$ is a Latin square and that $A^{(1)}, A^{(2)}, \ldots, A^{(n-1)}$ is an orthogonal family. Thus, if $n > 1$, we get a complete orthogonal family of $n \times n$ Latin squares. For instance, if $n = 3$, we use $GF(3)$, whose addition and multiplication operations are addition and multiplication modulo 3, as defined by Table 9.19. We let $b_1 = 1$, $b_2 = 2$, and $b_3 = 0$. (Remember that b_1 is to be chosen as the multiplicative identity and b_n as the additive identity.) Then we find that $A^{(1)}$ and $A^{(2)}$ are given by Table 9.21. To see how the 1, 2 entry of $A^{(2)}$ was computed,

---

*Two fields $\mathscr{F}$ and $\mathscr{G}$ are *isomorphic* if there is a one-to-one mapping from $\mathscr{F}$ onto $\mathscr{G}$ which preserves addition and multiplication.

$$A^{(1)} = \begin{array}{ccc} 2 & 0 & 1 \\ 0 & 1 & 2 \\ 1 & 2 & 0 \end{array} \qquad A^{(2)} = \begin{array}{ccc} 0 & 1 & 2 \\ 2 & 0 & 1 \\ 1 & 2 & 0 \end{array}$$

for instance, note that

$$(b_2 \times b_1) + b_2 = (2 \times 1) + 2 = 2 + 2 = 1.$$

It is easy to check directly that $A^{(1)}$ and $A^{(2)}$ of Table 9.21 are Latin squares and that they are orthogonal.

To give another example, suppose that $n = 4$. Then we use GF(4) = GF(2^2), whose addition and multiplication operations are given in Table 9.20. Taking $b_1 = 1$, $b_2 = 2$, $b_3 = 3$, and $b_4 = 0$, and using (9.2), we get the three pairwise orthogonal Latin squares of Table 9.22. To see how these entries are obtained, note, for example, that

$$a_{23}^{(3)} = (b_3 \times b_2) + b_3$$
$$= (3 \times 2) + 3$$
$$= 1 + 3$$
$$= 2,$$

where we have used the addition and multiplication rules of Table 9.20.

9.3.4 Justification of the Construction of a Complete Orthogonal Family if $n = p^k$†

To justify the construction of Section 9.3.3, we first show in general that if $A^{(e)}$ is defined by (9.2), then it is a Latin square. Suppose that $a_{ij}^{(e)} = a_{ik}^{(e)}$. Then

$$(b_e \times b_i) + b_j = (b_e \times b_i) + b_k. \tag{9.3}$$

By adding the additive inverse c of $(b_e \times b_i)$ to both sides of (9.3) and using associativity and commutativity of addition, we find that

$$c + [(b_e \times b_i) + b_j] = c + [(b_e \times b_i) + b_k],$$
$$[c + (b_e \times b_i)] + b_j = [c + (b_e \times b_i)] + b_k,$$
$$0 + b_j = 0 + b_k,$$
$$b_j = b_k,$$
$$j = k.$$

Thus, all elements of the same row are different.

†This subsection can be omitted.

Table 9.22 The orthogonal family of 4 × 4 Latin squares obtained from the finite field $GF(2^2)$ of Table 9.20

$$A^{(1)} = \begin{array}{|cccc|} \hline 0 & 3 & 2 & 1 \\ 3 & 0 & 1 & 2 \\ 2 & 1 & 0 & 3 \\ 1 & 2 & 3 & 0 \\ \hline \end{array} \qquad A^{(2)} = \begin{array}{|cccc|} \hline 3 & 0 & 1 & 2 \\ 2 & 1 & 0 & 3 \\ 0 & 3 & 2 & 1 \\ 1 & 2 & 3 & 0 \\ \hline \end{array} \qquad A^{(3)} = \begin{array}{|cccc|} \hline 2 & 1 & 0 & 3 \\ 0 & 3 & 2 & 1 \\ 3 & 0 & 1 & 2 \\ 1 & 2 & 3 & 0 \\ \hline \end{array}$$

Next, suppose that $a_{ji}^{(e)} = a_{ki}^{(e)}$. Then

$$(b_e \times b_j) + b_i = (b_e \times b_k) + b_i. \tag{9.4}$$

By adding the additive inverse of b_i to both sides of (9.4) and using associativity of addition, we obtain

$$b_e \times b_j = b_e \times b_k. \tag{9.5}$$

We now multiply (9.5) by the multiplicative inverse a of b_e, which exists since $b_e \neq 0$, and use commutativity and associativity of the $\times$ operation. We obtain

$$a \times (b_e \times b_j) = a \times (b_e \times b_k),$$

$$(a \times b_e) \times b_j = (a \times b_e) \times b_k,$$

$$1 \times b_j = 1 \times b_k,$$

$$b_j = b_k,$$

$$j = k.$$

Thus, all elements of the same column are different, and we conclude that $A^{(e)}$ is a Latin square.

Finally, we verify orthogonality of $A^{(e)}$ and $A^{(f)}$, for $e \neq f$. Suppose that

$$(a_{ij}^{(e)}, a_{ij}^{(f)}) = (a_{kl}^{(e)}, a_{kl}^{(f)}).$$

Then

$$a_{ij}^{(e)} = a_{kl}^{(e)} \qquad \text{and} \qquad a_{ij}^{(f)} = a_{kl}^{(f)},$$

so

$$(b_e \times b_i) + b_j = (b_e \times b_k) + b_l \tag{9.6}$$

and

$$(b_f \times b_i) + b_j = (b_f \times b_k) + b_l. \tag{9.7}$$

Using freely the properties of fields, we subtract (9.7) from (9.6), that is, we add the additive inverse of both sides of (9.7) to both sides of (9.6). This yields

$$(b_e \times b_i) - (b_f \times b_i) = (b_e \times b_k) - (b_f \times b_k),$$

where $-$ means add the additive inverse. Thus, again using freely the properties of fields, we find that

$$(b_e - b_f) \times b_i = (b_e - b_f) \times b_k. \tag{9.8}$$

Finally, since $e \neq f$, it follows that $(b_e - b_f) \neq 0$, so $(b_e - b_f)$ has a multiplicative inverse. Multiplying (9.8) by this multiplicative inverse, we derive the equation

$$b_i = b_k,$$

whence

$$i = k.$$

Now (9.6) gives us

$$(b_e \times b_i) + b_j = (b_e \times b_i) + b_l,$$

from which we derive

$$b_j = b_l,$$
$$j = l.$$

Hence, $i = k$ and $j = l$, and we conclude that $A^{(e)}$ and $A^{(f)}$ are orthogonal. This completes the proof that $A^{(1)}, A^{(2)}, \ldots, A^{(n-1)}$ is an orthogonal family of $n \times n$ Latin squares.

EXERCISES FOR SECTION 9.3

1. For each of the following values of a and n, find a number b in $\{0, 1, \ldots, n-1\}$ so that $a \equiv b \pmod{n}$.
   (a) $a = 27, n = 5$ (b) $a = 36, n = 3$ (c) $a = 11, n = 10$
   (d) $a = 117, n = 51$ (e) $a = 1025, n = 7$.

2. For each of the following values of a, b, and n, compute $a + b$ and $a \times b$ using addition and multiplication modulo n.
   (a) $a = 3, b = 4, n = 6$ (b) $a = 2, b = 3, n = 12$
   (c) $a = 5, b = 6, n = 8$ (d) $a = 4, b = 8, n = 15$
   (e) $a = 3, b = 11, n = 14$.

3. Verify the following facts about congruence.
   (a) If $a \equiv b \pmod{n}$, then $b \equiv a \pmod{n}$.
   (b) If $a \equiv b \pmod{n}$ and $b \equiv c \pmod{n}$, then $a \equiv c \pmod{n}$.
   (c) If $a \equiv a' \pmod{n}$ and $b \equiv b' \pmod{n}$, then $a + b \equiv a' + b' \pmod{n}$.
   (d) If $a \equiv a' \pmod{n}$ and $b \equiv b' \pmod{n}$, then $a \times b \equiv a' \times b' \pmod{n}$.

4. Suppose that $c_1 c_2 \cdots c_n$ is any permutation of $0, 1, 2, \ldots, n-1$. Build a matrix A as follows. The first row of A is the permutation $c_1 c_2 \cdots c_n$. Each successive row of A is obtained from the previous row by adding 1 to each element and using addition modulo n.
   (a) Build A for the permutation 3201.
   (b) Show that A is always a Latin square.

5. (Williams [1949]) Let $n = 2m$ and let

$$0 \quad 1 \quad 2m-1 \quad 2 \quad 2m-2 \quad 3 \quad 2m-3 \quad \cdots \quad m+1 \quad m$$

be a permutation of $\{0, 1, 2, \ldots, n-1\}$. Let A be the Latin square constructed from that permutation by the method of Exercise 4.
   (a) Build A for $m = 2$.
   (b) Show that for every value of $m \geq 1$, A is *horizontally complete* in the sense that whenever $1 \leq \alpha \leq n$ and $1 \leq \beta \leq n$ and $\alpha \neq \beta$ then there is a row of A in which α is followed

immediately by β. (Such Latin squares are important in agricultural experiments where we wish to minimize interaction of treatments applied to adjacent plots.)

(c) For arbitrary $m \geq 1$, is A vertically complete (in the obvious sense)?

6. Write down the addition and multiplication tables for the following fields.
   (a) GF(5)   (b) GF(7)   (c) GF(9).

7. (a) Write down the tables for the binary operations of addition and multiplication modulo 4 on the set Z_4.
   (b) Find an element in Z_4 that does not have a multiplicative inverse.

8. Repeat Exercise 7 for addition and multiplication modulo 10 on the set Z_{10}.

9. Find the additive and multiplicative inverse of 8 in each of the following fields.
   (a) GF(11)   (b) GF(13)   (c) GF(17).

10. Repeat Exercise 9 for 6 in place of 8.

11. Which of the following triples $(F, +, \times)$ define fields?
    (a) F is the *positive* reals, $+$ and $\times$ are ordinary addition and multiplication.
    (b) F is the reals with an additional element ∞. The operations $+$ and $\times$ are the usual addition and multiplication operations on the reals, and in addition we have for all real numbers a,

    $$a + \infty = a \times \infty = \infty + a = \infty \times a = \infty + \infty = \infty \times \infty = \infty.$$

    (c) F is Re, $a + b$ is ordinary addition, and $a \times b = 1$ for all a, b in F.
    (d) F is Re, and $a + b = a \times b = 0$ for all a, b in F.

12. Consider Z_n under addition and multiplication modulo n and consider the conditions for a field.
    (a) Show that condition **F1** holds.
    (b) Show that condition **F2** holds.
    (c) Show that condition **F3** holds.
    (d) Show that condition **F4** holds by showing that 0 and 1 are the additive and multiplicative identities, respectively.
    (e) Show that condition **F5**(a) holds by showing that $n - a$ is the additive inverse of a.
    (f) Show that condition **F6** holds.
    (g) Show that condition **F5**(b) fails if n is not a prime number.
    (h) Show that condition **F5**(b) holds if n is a prime number. Do this as follows. Use induction and the binomial expansion to show that $a^n \equiv a \pmod{n}$ if n is a prime. Then show that $a \neq 0$ implies that $a^{n-1} \equiv 1 \pmod{n}$. Conclude that $a^{-1} = a^{n-2}$.

13. Verify that Table 9.20 defines a field.

14. *Use the method of Section 9.3.3* to find a complete orthogonal family of Latin squares of order (a) 5 (b) 7 (c) 8 (d) 9. Parts (c) and (d) are for the reader with knowledge of modern algebra.

9.4 BALANCED INCOMPLETE BLOCK DESIGNS

9.4.1 (b, v, r, k, λ)-designs

In Section 9.1 we pointed out that in a block design, it is not always possible to test each treatment in each block. For instance, in testing tire wear, if there are five brands of tires, then, as we observed, only four of these can be tested in any one block. Thus, it is necessary to use an incomplete block design. The basic incomplete block design we shall study is called a balanced incomplete block design. A *balanced block design* consists of a set X of $v \geq 2$ elements called *varieties* or *treatments*, and a collection of $b > 0$ subsets of X, called *blocks*, such that the following conditions are satisfied:

each block consists of exactly the same number k of
varieties, $k > 0$, (9.9)

each variety appears in exactly the same number r of
blocks, $r > 0$, (9.10)

each pair of varieties appears simultaneously in exactly
the same number λ of blocks, $\lambda > 0$. (9.11)

A balanced block design with $k < v$ is called a *balanced incomplete block design* since each block has fewer than the total number of varieties. Such a design is also called a *BIBD*, a (b, v, r, k, λ)-*design*, or a (b, v, r, k, λ)-*configuration*. The basic ideas behind BIBD's were introduced by Yates [1936]. Note that if $k = v$, then (as long as no block has repeated varieties) conditions (9.9), (9.10) and (9.11) hold trivially, with $k = v$, $r = b$, and $\lambda = b$. That is why we usually assume that $k < v$.

Example 9.8   A (7, 7, 3, 3, 1)-design

If $b = 7$, $v = 7$, $r = 3$, $k = 3$, and $\lambda = 1$, there is a (b, v, r, k, λ)-design. It is given by taking the set of varieties X to be $\{1, 2, 3, 4, 5, 6, 7\}$ and using the following blocks:

$$B_1 = \{1, 2, 4\}, \qquad B_2 = \{2, 3, 5\}, \qquad B_3 = \{3, 4, 6\},$$

$$B_4 = \{4, 5, 7\}, \qquad B_5 = \{5, 6, 1\}, \qquad B_6 = \{6, 7, 2\}, \qquad B_7 = \{7, 1, 3\}.$$

It is easy to see that each block consists of exactly 3 varieties, that each variety appears in exactly 3 blocks, and that each pair of varieties appears simultaneously in exactly 1 block (for example, 3 and 6 appear together in B_3 and nowhere else).

Example 9.9   A (4, 4, 3, 3, 2)-design

If $b = 4$, $v = 4$, $r = 3$, $k = 3$, and $\lambda = 2$, a (b, v, r, k, λ)-design is given by

$$X = \{1, 2, 3, 4\}$$

and the blocks

$$\{1, 2, 3\}, \{2, 3, 4\}, \{3, 4, 1\}, \{4, 1, 2\}.$$

In the tire wear example of Section 9.1, the varieties are tire brands and the blocks are the sets of tire brands corresponding to the tires used in a given test car. The conditions (9.9), (9.10), and (9.11) correspond to the following reasonable requirements:

each car uses the same number k of tire brands,
each brand appears on the same number r of cars,
each pair of brands is tested together on the same car the same number λ of times.

It should be clear from our experience with orthogonal families of Latin squares that (b, v, r, k, λ)-designs may not exist for all combinations of the parameters $b, v, r, k,$ and λ. Indeed, the basic combinatorial question of the subject of balanced incomplete block designs is the existence question: For what values of b, v, r, k, λ does a (b, v, r, k, λ)-design exist? We shall address this question below. In general, it is an unsolved problem to state complete conditions on the parameters b, v, r, k, λ necessary and sufficient for existence of a (b, v, r, k, λ)-design. A typical reference book on practical experimental design will list, for reasonable values of the parameters, those (b, v, r, k, λ)-designs that do exist. For now, let us give some examples of the use of (b, v, r, k, λ)-designs. Then we shall study the basic existence question in some detail.

Example 9.10   Testing Cloth for Wear Revisited

Suppose that we have a Martindale wear tester as described in Example 9.6, and we wish to use it to compare seven different types of cloth. Since only four pieces of cloth can be tested in one run of the machine, an incomplete block design must be used. The number v of varieties is 7 and the blocks will all be of size $k = 4$. Box *et al.* [1978] describe a (b, v, r, k, λ)-design for this situation in which there are 7 blocks $(b = 7)$, each type of cloth is run $r = 4$ times, and each pair of cloth types is used together in a run $\lambda = 2$ times. If the cloth types are labeled 1, 2, 3, 4, 5, 6, 7, the blocks used can be described as

$$B_1 = \{2, 4, 6, 7\}, \qquad B_2 = \{1, 3, 6, 7\}, \qquad B_3 = \{3, 4, 5, 7\},$$

$$B_4 = \{1, 2, 5, 7\}, \qquad B_5 = \{2, 3, 5, 6\}, \qquad B_6 = \{1, 4, 5, 6\}, \qquad B_7 = \{1, 2, 3, 4\}.$$

Since there were four positions in which to place the cloth, and since the design could be chosen so that each cloth type was used in 4 runs, it was also possible to arrange to place each type of cloth exactly once in each position. Thus, it was possible to control for differences due to machine position. Such an incomplete block design which is balanced for 2 different sources of block variation is called a *Youden square*, after its inventor W. J. Youden (see Youden [1937]). In this case, the Youden square can be summarized in Table 9.23, where $p, q, r,$ and s represent the 4 positions, and the i, j entry gives the position of variety i in block j.

Example 9.11   Tuberculosis in Cattle

Wadley [1948] used balanced incomplete block designs in work on diagnosing tuberculosis in cattle. The disease can be diagnosed by injecting the skin of a cow with an appropriate allergen and observing the thickening produced. In an experiment to compare allergens, the observation for each allergen was the log concentration required to produce a 3-millimeter thickening. This concentration was being estimated by observing the thickenings at four different concentrations and interpolating. Thus, each test of an allergen required a number of injections of the allergen at different concentrations.

In Wadley's experiment, 16 allergens were under comparison. Thus, $v = 16$. On each cow there were four main regions and in each region about 16 injections could be made. This suggests using each region as a block, with four allergen preparations in each block, each used four times at different concentrations, making 16 injections in all. Thus, $k = 4$. There is a design with $k = 4$, $v = 16$, $b = 20$, and $r = 5$. This information is available from a typical reference book. Since there are 20

Table 9.23 A Youden square for the wear testing experiment[a]

Block (run)

Variety (cloth type)	A	B	C	D	E	F	G
1		p		q		r	s
2	p			s	r		q
3		q	p		s		r
4	q		r			s	p
5			s	r	q	p	
6	r	s			p	q	
7	s	r	q	p			

[a]The i, j entry gives the position of variety i in block j.

blocks, and four blocks per cow, this calls for five cows. (Or, repeating the experiment, some multiple of five cows. Wadley's experiment used 10 cows.) If no suitable design had been available with $k = 4$, $v = 16$, it would have been natural to have considered whether five preparations could have been included in each region (20 injections per region).

Example 9.12   Comparing Dishwashing Detergents

In experiments such as that by Pugh [1953] to compare detergents used for domestic dishwashing, the following procedure has been used. In order to obtain a series of homogeneous experimental units, a pile of plates from one course in a canteen is divided into groups. Each group of plates is then washed with water at a standard temperature and with a controlled amount of one detergent. The experimenter records the (logarithm of the) number of plates washed before the foam is reduced to a thin surface layer. The detergents form the varieties. Each group of plates from the one course forms an experimental unit and the different groups of plates from the same course form a block. The washing for one block is done by one person. Each group of plates within a course is assigned to a variety. The experimental conditions are as constant as possible within one block. Different blocks consist of plates soiled in different ways and washed by different people.

Now the number of plates available in one block is limited. It frequently allows only four tests to be completed; that is, there is a restriction to four experimental units and hence varieties per block. If eight varieties are to be compared, not every variety can be tried out in every block. This calls for an incomplete block design, with $v = 8$, $k = 4$. There is such a design with $r = 7$ and $b = 14$.

In sum, the experimenter takes a set of dishes from a given course, divides it into four groups, and applies a different detergent to each of the groups. The experiment is repeated 14 times, each time with a collection of four detergents chosen to make up the four varieties in the corresponding block.

9.4.2 Necessary Conditions for the Existence of (b, v, r, k, λ)-designs

Our first theorem states some necessary conditions which the parameters for a balanced incomplete block design must satisfy.

Theorem 9.9. In a (b, v, r, k, λ)-design,

$$bk = vr \tag{9.12}$$

and

$$r(k - 1) = \lambda(v - 1). \tag{9.13}$$

To illustrate this theorem, we note that no $(12, 9, 4, 3, 2)$-design exists, for $bk = 36$, $vr = 36$, $r(k - 1) = 8$, and $\lambda(v - 1) = 16$. Hence, although (9.12) is satisfied, (9.13) is not. If $b = 12$, $v = 9$, $r = 4$, $k = 3$, and $\lambda = 1$, $bk = vr = 36$ and $r(k - 1) = \lambda(v - 1) = 8$, so (9.12) and (9.13) are satisfied. This says that a $(12, 9, 4, 3, 1)$-design *could* exist; it does not guarantee that such a design *does* exist. [Conditions (9.12) and (9.13) are necessary, but not sufficient.]

Proof of Theorem 9.9. The product bk is the product of the number of blocks (b) by the number of varieties in each block (k), and hence gives the total number of elements which are listed if the blocks are written out as

$$B_1: \quad \ldots \qquad B_2: \quad \ldots \qquad \cdots \qquad B_b: \quad \ldots$$

The product vr is the product of the number of varieties (v) by the number of replications of each variety (r) and hence also gives the number of elements listed above. Hence, $bk = vr$, and (9.12) holds.

The product $r(k - 1)$ is the product of the number of blocks in which a variety i appears (r) by the number of other varieties in each block in which i appears, and hence gives the number of pairs $\{i, j\}$ appearing in a common block (counting a pair once for each time it occurs). The product $\lambda(v - 1)$ is the product of the number of times each pair $\{i, j\}$ appears in a block (λ) by the number of possible j's $(v - 1)$, and hence gives the number of pairs $\{i, j\}$ appearing in a common block (counting a pair once for each time it occurs). Thus, $r(k - 1) = \lambda(v - 1)$, and (9.13) holds.   Q.E.D.

Corollary 9.9.1. Suppose that in an incomplete block design with $v \geq 2$ varieties and b blocks,

1. each block consists of the same number k of varieties, and
2. each pair of varieties appears simultaneously in exactly the same number λ of blocks, $\lambda > 0$.

Then each variety appears in the same number r of blocks, r is given by $\lambda(v - 1)/(k - 1)$, and the block design is balanced.

Proof. The proof of (9.13) above actually shows this. For suppose that a given variety i appears in r_i blocks. The proof above shows that

$$r_i(k - 1) = \lambda(v - 1).$$

Note that since $v \geq 2$ and $\lambda > 0$, k could not be 1. Thus,

$$r_i = \frac{\lambda(v-1)}{k-1}.$$

This is the same number r_i for each i. $\hspace{6cm}$ Q.E.D.

Corollary 9.9.1 shows that the definition of balanced incomplete block design is redundant: One of the conditions in the definition, namely the condition (9.10) that every variety appears in the same number of blocks, follows from the other conditions, (9.9) and (9.11).

Theorem 9.10 **(Fisher's Inequality).*** $\hspace{0.5cm}$ In a (b, v, r, k, λ)-design, $b \geq v$.

We shall prove this result in Section 9.4.3. To prove it, it will be helpful to introduce a concept that will also be very useful in our study of error-correcting codes in Chapter 10. This is the notion of an *incidence matrix* $\mathbf{A}$ of a block design. If the design has varieties x_1, $x_2, \ldots, x_v$ and blocks $B_1, B_2, \ldots, B_b$, then $\mathbf{A}$ is a $v \times b$ matrix of 0's and 1's, with the i, j entry of $\mathbf{A}$ being 1 if x_i is in B_j and 0 otherwise. (This is the point-set incidence matrix of Section 3.7.) For example, in the (b, v, r, k, λ)-design of Example 9.8, we have the following incidence matrix:

$$
\mathbf{A} =
\begin{array}{c}
 \\
1 \\
2 \\
3 \\
4 \\
5 \\
6 \\
7
\end{array}
\begin{array}{c}
\begin{array}{ccccccc}
B_1 & B_2 & B_3 & B_4 & B_5 & B_6 & B_7
\end{array} \\
\left[
\begin{array}{ccccccc}
1 & 0 & 0 & 0 & 1 & 0 & 1 \\
1 & 1 & 0 & 0 & 0 & 1 & 0 \\
0 & 1 & 1 & 0 & 0 & 0 & 1 \\
1 & 0 & 1 & 1 & 0 & 0 & 0 \\
0 & 1 & 0 & 1 & 1 & 0 & 0 \\
0 & 0 & 1 & 0 & 1 & 1 & 0 \\
0 & 0 & 0 & 1 & 0 & 1 & 1
\end{array}
\right]
\end{array}.
$$

Designs can be defined by giving $v \times b$ matrices of 0's and 1's. An arbitrary $v \times b$ matrix of 0's and 1's with $v \geq 2$ is the incidence matrix of a (b, v, r, k, λ)-design, $b, v, r, k, \lambda > 0$, if and only if the following conditions hold:

$$\text{each column has the same number of 1's,} \quad k \text{ of them,} \quad k > 0, \tag{9.14}$$

$$\text{each row has the same number of 1's, } r \text{ of them,} \quad r > 0, \tag{9.15}$$

$$\begin{array}{l}\text{each pair of rows has the same number of columns with} \\ \text{a common 1, } \lambda \text{ of them,} \quad \lambda > 0.\end{array} \tag{9.16}$$

We have seen (in Corollary 9.9.1) that (9.14) and (9.16) imply (9.15). A natural analog of (9.16) is the following:

$$\begin{array}{l}\text{each pair of columns has the same number of rows with} \\ \text{a common 1.}\end{array} \tag{9.17}$$

Exercise 35 investigates the relations among conditions (9.14)–(9.17).

*This theorem is due to Fisher [1940].

9.4.3 Proof of Fisher's Inequality*

To prove Fisher's Inequality, we shall first prove a result about incidence matrices of (b, v, r, k, λ)-designs.

Theorem 9.11. If $\mathbf{A}$ is the incidence matrix of a (b, v, r, k, λ)-design, then

$$\mathbf{AA}^T = (r - \lambda)\mathbf{I} + \lambda\mathbf{J}, \qquad (9.18)$$

where $\mathbf{A}^T$ is the transpose of $\mathbf{A}$, $\mathbf{I}$ is a $v \times v$ identity matrix, and $\mathbf{J}$ is the $v \times v$ matrix of all 1's.

Proof. Let b_{ij} be the i, j entry of $\mathbf{AA}^T$. Then b_{ij} is the *inner product* of the ith row of $\mathbf{A}$ with the jth row of $\mathbf{A}$, that is,

$$b_{ij} = \sum_{k=1}^{b} a_{ik} a_{jk}.$$

If $i = j$, we see that $a_{ik} a_{ik}$ is 1 if the ith variety belongs to the kth block, and it is 0 otherwise. Thus, b_{ii} counts the number of blocks that i belongs to, that is, r. If $i \neq j$, then $a_{ik} a_{jk}$ is 1 if the ith and jth varieties both belong to the kth block, and it is 0 otherwise. Thus, b_{ij} counts the number of blocks that the ith and jth varieties both belong to, that is, λ. Translating these conclusions into matrix language gives us (9.18).   Q.E.D.

Proof of Fisher's Inequality (Theorem 9.10). We shall suppose that $b < v$ and reach a contradiction. Let $\mathbf{A}$ be the incidence matrix. Since $b < v$, we can add $v - b$ columns of 0's to $\mathbf{A}$, giving us a square $v \times v$ matrix $\mathbf{B}$. Now $\mathbf{AA}^T = \mathbf{BB}^T$, since the inner product of two rows of $\mathbf{A}$ is the same as the inner product of two rows of $\mathbf{B}$. Taking determinants, we conclude that

$$\det(\mathbf{AA}^T) = \det(\mathbf{BB}^T) = (\det \mathbf{B})(\det \mathbf{B}^T).$$

But $\det \mathbf{B} = 0$ since $\mathbf{B}$ has a column of 0's. Thus, $\det(\mathbf{AA}^T) = 0$. Now by Theorem 9.11,

$$\det(\mathbf{AA}^T) = \det \begin{bmatrix} r & \lambda & \lambda & \lambda & \cdots & \lambda \\ \lambda & r & \lambda & \lambda & \cdots & \lambda \\ \lambda & \lambda & r & \lambda & \cdots & \lambda \\ \lambda & \lambda & \lambda & r & \cdots & \lambda \\ \vdots & \vdots & \vdots & \vdots & \vdots & \vdots \\ \lambda & \lambda & \lambda & \lambda & \cdots & \lambda \\ \lambda & \lambda & \lambda & \lambda & \cdots & r \end{bmatrix}. \qquad (9.19)$$

Subtracting the first column from each of the others in the matrix in the right-hand side of (9.19) does not change the determinant. Hence,

$$\det(\mathbf{AA}^T) = \det \begin{bmatrix} r & \lambda - r & \lambda - r & \lambda - r & \cdots & \lambda - r \\ \lambda & r - \lambda & 0 & 0 & \cdots & 0 \\ \lambda & 0 & r - \lambda & 0 & \cdots & 0 \\ \lambda & 0 & 0 & r - \lambda & \cdots & 0 \\ \vdots & \vdots & \vdots & \vdots & \vdots & \vdots \\ \lambda & 0 & 0 & 0 & \cdots & 0 \\ \lambda & 0 & 0 & 0 & \cdots & r - \lambda \end{bmatrix}. \qquad (9.20)$$

*This subsection can be omitted.

Adding to the first row of the matrix on the right-hand side of (9.20) all the other rows does not change the determinant. Hence,

$$\det\left(\mathbf{A}\mathbf{A}^T\right) = \det \begin{bmatrix} r + (v-1)\lambda & 0 & 0 & 0 & \cdots & 0 \\ \lambda & r - \lambda & 0 & 0 & \cdots & 0 \\ \lambda & 0 & r - \lambda & 0 & \cdots & 0 \\ \lambda & 0 & 0 & r - \lambda & \cdots & 0 \\ \vdots & \vdots & \vdots & \vdots & \vdots & \vdots \\ \lambda & 0 & 0 & 0 & \cdots & 0 \\ \lambda & 0 & 0 & 0 & \cdots & r - \lambda \end{bmatrix}. \qquad (9.21)$$

Since the matrix in the right-hand side of (9.21) has all 0's above the diagonal, its determinant is the product of the diagonal elements, so

$$\det\left(\mathbf{A}\mathbf{A}^T\right) = [r + (v-1)\lambda](r - \lambda)^{v-1}.$$

Since we have concluded that $\det\left(\mathbf{A}\mathbf{A}^T\right) = 0$, we have

$$[r + (v-1)\lambda](r - \lambda)^{v-1} = 0. \qquad (9.22)$$

But since r, v, and λ are all assumed positive,

$$[r + (v-1)\lambda] > 0.$$

Also, by Equation (9.13) of Theorem 9.9, since $k < v$, it follows that $r > \lambda$. Hence,

$$(r - \lambda)^{v-1} > 0.$$

We conclude that the left-hand side of (9.22) is positive, which is a contradiction.   Q.E.D.

9.4.4 Steiner Triple Systems

So far our results have given necessary conditions for the existence of (b, v, r, k, λ)-designs, but have not given us sufficient conditions for their existence, or constructive procedures for finding them. We shall describe several such procedures. We begin by considering special cases of (b, v, r, k, λ)-designs.

In particular, suppose that $k = 2$ and $\lambda = 1$. In this case, each block consists of two varieties. Equation (9.13) implies that $r = v - 1$, so (9.12) implies that

$$2b = v(v-1),$$

or

$$b = \frac{v(v-1)}{2}.$$

Now

$$\frac{v(v-1)}{2} = \binom{v}{2}$$

is the number of two-element subsets of a set of v elements. Hence, the number of blocks is the number of two-element subsets of the set of varieties. If, for example, $v = 3$, such a

design with $X = \{1, 2, 3\}$ has as blocks the subsets
$$\{1, 2\}, \{1, 3\}, \{2, 3\}.$$

In this subsection we shall concentrate on another special case of (b, v, r, k, λ)-designs, that where $k = 3$ and $\lambda = 1$. Such a design is a set of triples in which each pair of varieties appears in exactly one triple. These designs are called *Steiner triple systems*. Some authors define Steiner triple systems as block designs in which the blocks are triples from a set X and each pair of varieties appears in exactly one triple. This definition allows inclusion of the complete block design where $k = v$. This is the trivial design where $X = \{1, 2, 3\}$ and there is only one block, $\{1, 2, 3\}$. For the purposes of this subsection, we shall include this design as a Steiner triple system. A more interesting example of a Steiner triple system occurs when $v = 7$. Example 9.8 is such an example.

We shall now discuss the existence problem for Steiner triple systems. Note that in a Steiner triple system, (9.13) implies that

$$r(2) = v - 1, \tag{9.23}$$

so

$$r = \frac{v - 1}{2}. \tag{9.24}$$

Equation (9.12) now implies that

$$3b = \frac{v(v - 1)}{2},$$

so

$$b = \frac{v(v - 1)}{6}. \tag{9.25}$$

Equation (9.24) implies that $v - 1$ is even and v is odd. Also, $v \geq 2$ implies that v is at least 3. Equation (9.25) implies that $v(v - 1) = 6b$, so $v(v - 1)$ is a multiple of 6. These are necessary conditions. Let us begin to tabulate what values of v satisfy the two necessary conditions: v odd and at least 3, $v(v - 1)$ a multiple of 6. If $v = 3$, then $v(v - 1) = 6$, so there could be a Steiner triple system with $v = 3$, that is, the necessary conditions are satisfied. However, with $v = 5$, $v(v - 1) = 20$, which is not divisible by 6, so there is no Steiner triple system with $v = 5$. Similar calculations are summarized in Table 9.24. In general, Steiner triple systems are possible for $v = 3, 7, 9, 13, 15, 19, 21, \ldots$, that is, for $v = 6n + 1$ or $6n + 3$, $n \geq 1$, and $v = 3$. In fact, these systems do exist for all of these values of v.

Theorem 9.12 (Kirkman [1847]). There is a Steiner triple system of v varieties if and only if $v = 3$ or $v = 6n + 1$ or $v = 6n + 3$, $n \geq 1$.

We have already proved the necessity of the conditions in Theorem 9.12. Rather than prove sufficiency, we shall prove a simpler theorem, which gives us the existence of some of these Steiner triple systems, for instance those with $3 \cdot 3 = 9$ varieties, $3 \cdot 7 = 21$ varieties, $7 \cdot 7 = 49$ varieties, $9 \cdot 7 = 63$ varieties, and so on. For a proof of sufficiency, see for instance Hall [1967].

Table 9.24 Values of v for which a Steiner triple system is possible

Value of v (odd, $\neq 1$)	$v(v-1)$	Possible Steiner triple system?
3	$3(2) = 6$	Yes
5	$5(4) = 5 \cdot 2^2$	No
7	$7(6)$	Yes
9	$9(8) = 12 \cdot 6$	Yes
11	$11(10) = 11 \cdot 5 \cdot 2$	No
13	$13(12) = 26 \cdot 6$	Yes
15	$15(14) = 35 \cdot 6$	Yes
17	$17(16) = 17 \cdot 2^4$	No

Theorem 9.13. If there is a Steiner triple system S_1 of v_1 varieties and a Steiner triple system S_2 of v_2 varieties, then there is a Steiner triple system S of $v_1 v_2$ varieties.

Proof.* The proof provides a construction for building a Steiner triple system S given Steiner triple systems S_1 and S_2. Suppose that the varieties of S_1 are $a_1, a_2, \ldots, a_{v_1}$ and those of S_2 are $b_1, b_2, \ldots, b_{v_2}$. Let S consist of the $v_1 v_2$ elements c_{ij}, $i = 1, 2, \ldots, v_1$, $j = 1, 2, \ldots, v_2$. A triple $\{c_{ir}, c_{js}, c_{kt}\}$ is in S if and only if one of the following conditions holds:

(1) $r = s = t$ and $\{a_i, a_j, a_k\} \in S_1$,
(2) $i = j = k$ and $\{b_r, b_s, b_t\} \in S_2$,

or

(3) $\{a_i, a_j, a_k\} \in S_1$ and $\{b_r, b_s, b_t\} \in S_2$.

Then it is easy to prove that S forms a Steiner triple system. Q.E.D.

Let us illustrate the construction in the proof of Theorem 9.13. Suppose that $v_1 = v_2 = 3$ and S_1 has the one triple $\{a_1, a_2, a_3\}$, S_2 the one triple $\{b_1, b_2, b_3\}$. Then S has the triples shown in Table 9.25 and forms a Steiner triple system of 9 varieties and 12 blocks.

If an experimental design is to be a Steiner triple system on v varieties, then the specific choice of design is simple if $v = 3$, 7, or 9, for there is (up to relabeling of varieties) only one Steiner triple system of v varieties in these cases. However, for $v = 13$ there are two essentially different Steiner triple systems and for $v = 15$, there are 80. Presumably, one of these will be chosen at random if a Steiner triple system of 13 or 15 varieties is required. The number of distinct Steiner triple systems of v varieties is in general unknown.

9.4.5 Symmetric Balanced Incomplete Block Designs

A balanced incomplete block design or (b, v, r, k, λ)-design is called *symmetric* if $b = v$ (the number of blocks is the same as the number of varieties) and if $r = k$ (the number of times a variety occurs is the same as the number of varieties in a block). A symmetric BIBD is

*The proof and the illustration of it can be omitted.

Table 9.25 Construction of a Steiner triple system S of $v_1 v_2 = 9$ varieties if S_1 has only the triple $\{a_1, a_2, a_3\}$ and S_2 only the triple $\{b_1, b_2, b_3\}$

By (1):	$r = s = t = 1$	$r = s = t = 2$	$r = s = t = 3$
	$\{c_{11}, c_{21}, c_{31}\}$	$\{c_{12}, c_{22}, c_{32}\}$	$\{c_{13}, c_{23}, c_{33}\}$
By (2):	$i = j = k = 1$	$i = j = k = 2$	$i = j = k = 3$
	$\{c_{11}, c_{12}, c_{13}\}$	$\{c_{21}, c_{22}, c_{23}\}$	$\{c_{31}, c_{32}, c_{33}\}$
By (3):	$\{c_{11}, c_{22}, c_{33}\}$	$\{c_{12}, c_{23}, c_{31}\}$	$\{c_{13}, c_{21}, c_{32}\}$
	$\{c_{11}, c_{23}, c_{32}\}$	$\{c_{12}, c_{21}, c_{33}\}$	$\{c_{13}, c_{22}, c_{31}\}$

sometimes called a (v, k, λ)-*design* or a (v, k, λ)-*configuration*. By Equation (9.12) of Theorem 9.9,

$$b = v \quad \text{iff} \quad k = r.$$

Hence, the two conditions in the definition are redundant. Example 9.9 is an example of a symmetric BIBD: We have $b = v = 4$ and $r = k = 3$. So is Example 9.8: We have $b = v = 7$ and $r = k = 3$. The Steiner triple system of Table 9.25 is an example of a BIBD that is not symmetric.

Theorem 9.14 (Bruck–Ryser–Chowla Theorem).* The following conditions are necessary for the existence of a (v, k, λ)-design.

1. If v is even, then $k - \lambda$ is the square of an integer.
2. If v is odd, then the following equation has a solution in integers x, y, z, not all of which are 0:

$$x^2 = (k - \lambda)y^2 + (-1)^{(v-1)/2}\lambda z^2. \tag{9.26}$$

We shall omit the proof of Theorem 9.14. For a proof, see Ryser [1963] or Hall [1967]. To illustrate the theorem, suppose that $v = 8$, $k = 7$, and $\lambda = 3$. Then v is even and $k - \lambda = 4$ is a square, so condition 1 says that an $(8, 7, 3)$-design *could* exist. However, it also implies that an $(8, 7, 4)$-design *could not* exist, since $k - \lambda = 3$ is not a square. Suppose that $v = 5$, $k = 3$, and $\lambda = 1$. Then v is odd and (9.26) becomes

$$x^2 = 2y^2 + z^2.$$

This has a solution $x = z = 1$, $y = 0$. Hence, a $(5, 3, 1)$-design could exist.

The conditions for existence of a symmetric BIBD given in Theorem 9.14 are not sufficient. Some specific sufficient conditions are given by the following theorem, whose proof we leave to Section 10.5.2. (That section can be read at this point.)

Theorem 9.15. For arbitrarily large values of m, and in particular for $m = 2^k$, $k \geq 1$, there is a $(4m - 1, 2m - 1, m - 1)$-design.

A $(4m - 1, 2m - 1, m - 1)$-design is called a *Hadamard design of dimension m*. The case

---

*This theorem was proved for $\lambda = 1$ by Bruck and Ryser [1949] and in generality by Chowla and Ryser [1950].

$m = 2$ gives a (7, 3, 1)-design, an example of which is given in Example 9.8. That Hadamard designs of dimension m *could* exist for all $m \geq 2$ follows from Theorem 9.14. For $v = 4m - 1$ is odd and (9.26) becomes

$$x^2 = my^2 - (m - 1)z^2,$$

which has the solution $x = y = z = 1$. Hadamard designs will be very important in the theory of error-correcting codes in Section 10.5.

A second theorem giving sufficient conditions for the existence of symmetric BIBD's is the following, which will be proved in Section 9.5.2 in our study of projective planes.

Theorem 9.16. If $m \geq 1$ is a power of a prime, then there is an $(m^2 + m + 1, m + 1, 1)$-design.

To illustrate this theorem, note that taking $m = 1$ gives us a (3, 2, 1)-design. We have seen such a design at the beginning of Section 9.4.4. Taking $m = 2$ gives us a (7, 3, 1)-design. We have seen such a design in Example 9.8. Taking $m = 3$ gives us a (13, 4, 1)-design, which is something new.

9.4.6 Building New (b, v, r, k, λ)-designs from Existing Ones

Theorem 9.13 gives us a way of building new (b, v, r, k, λ)-designs from old ones. Here we shall present other such ways. The most trivial way to get one design from another is to simply repeat blocks. If we take p copies of each block in a (b, v, r, k, λ)-design, we get a $(pb, v, pr, k, p\lambda)$-design. For example, from the (4, 4, 3, 3, 2)-design of Example 9.9, we get an (8, 4, 6, 3, 4)-design by repeating each block twice. To describe more interesting methods of obtaining new designs from old ones, we need one preliminary result.

Theorem 9.17. In a (v, k, λ)-design, any two blocks have exactly λ elements in common.

Proof. Exercise 35.

If U and V are sets, $U - V$ will denote the set $U \cap V^c$.

Theorem 9.18. Suppose that $B_1, B_2, \ldots, B_v$ are the blocks of a (v, k, λ)-design with $X = \{x_1, x_2, \ldots, x_v\}$ the set of varieties. Then for any i,

$$B_1 - B_i, B_2 - B_i, \ldots, B_{i-1} - B_i, B_{i+1} - B_i, \ldots, B_v - B_i$$

are the blocks of a $(v - 1, v - k, k, k - \lambda, \lambda)$-design on the set of varieties $X - B_i$.

Proof. There are clearly $v - 1$ blocks and $v - k$ varieties. By Theorem 9.17, each block $B_j - B_i$ has $k - \lambda$ elements. Each variety in $X - B_i$ appears in k blocks of the original design and hence in k blocks of the new design. Similarly, each pair of varieties in $X - B_i$ appear in common in λ blocks of the original design and hence in λ blocks of the new design.   Q.E.D.

To illustrate this construction, suppose that we start with the (7, 3, 1)-design of Example 9.8 and let $B_i = \{3, 4, 6\}$. Then the following blocks form a (6, 4, 3, 2, 1)-design

Table 9.26 The blocks of a (15, 7, 3)-design on the set of varieties $X = \{1, 2, \ldots, 15\}$

$\{2, 4, 6, 8, 10, 12, 14\}$,	$\{1, 4, 5, 8, 9, 12, 13\}$,	$\{3, 4, 7, 8, 11, 12, 15\}$,
$\{1, 2, 3, 8, 9, 10, 11\}$,	$\{2, 5, 7, 8, 10, 13, 15\}$,	$\{1, 6, 7, 8, 9, 14, 15\}$,
$\{3, 5, 6, 8, 11, 13, 14\}$,	$\{1, 2, 3, 4, 5, 6, 7\}$,	$\{2, 4, 6, 9, 11, 13, 15\}$,
$\{1, 4, 5, 10, 11, 14, 15\}$,	$\{3, 4, 7, 9, 10, 13, 14\}$,	$\{1, 2, 3, 12, 13, 14, 15\}$,
$\{2, 5, 7, 9, 11, 12, 14\}$,	$\{1, 6, 7, 10, 11, 12, 13\}$,	$\{3, 5, 6, 9, 10, 12, 15\}$

---

on the set of varieties $\{1, 2, 5, 7\}$:

$$\{1, 2\}, \quad \{2, 5\}, \quad \{5, 7\}, \quad \{1, 5\}, \quad \{2, 7\}, \quad \{1, 7\}.$$

Theorem 9.19. Suppose that $B_1, B_2, \ldots, B_v$ are the blocks of a (v, k, λ)-design with $X = \{x_1, x_2, \ldots, x_v\}$ the set of varieties. Then for any i,

$$B_1 \cap B_i, \quad B_2 \cap B_i, \ldots, B_{i-1} \cap B_i, \quad B_{i+1} \cap B_i, \ldots, B_v \cap B_i$$

are the blocks of a $(v - 1, k, k - 1, \lambda, \lambda - 1)$-design on the set of varieties B_i.

Proof. There are clearly $v - 1$ blocks and k varieties. By Theorem 9.17, each block $B_j \cap B_i$ has λ elements. Moreover, a given variety in B_i appears in the original design in blocks

$$B_{j_1}, \quad B_{j_2}, \quad \ldots, \quad B_{j_{k-1}}, \quad B_i.$$

Then it appears in the new design in $k - 1$ blocks,

$$B_{j_1} \cap B_i, \quad B_{j_2} \cap B_i, \quad \ldots, \quad B_{j_{k-1}} \cap B_i.$$

Moreover, any pair of varieties in B_i appear in common in the original design in λ blocks,

$$B_{j_1}, \quad B_{j_2}, \quad \ldots, \quad B_{j_{\lambda-1}}, \quad B_i,$$

and hence appear in common in the new design in $\lambda - 1$ blocks,

$$B_{j_1} \cap B_i, \quad B_{j_2} \cap B_i, \quad \ldots, \quad B_{j_{\lambda-1}} \cap B_i. \qquad \text{Q.E.D.}$$

To illustrate this theorem, we note that by Theorem 9.15, there is a (15, 7, 3)-design. Hence, Theorem 9.19 implies that there is a (14, 7, 6, 3, 2)-design. To exhibit such a design, we note that the blocks in Table 9.26 define a (15, 7, 3)-design on $X = \{1, 2, \ldots, 15\}$. We show how to construct this design in Section 10.5.2. Taking $B_i = \{1, 2, 3, 8, 9, 10, 11\}$, we get the (14, 7, 6, 3, 2)-design of Table 9.27 on the set of varieties B_i. Note that this design has repeated blocks. If we take only one copy of each of these blocks, we get a (7, 3, 1)-design.

Table 9.27 The blocks of a (14, 7, 6, 3, 2)-design obtained from the design of Table 9.26 by intersecting blocks with $B_i = \{1, 2, 3, 8, 9, 10, 11\}$

$\{2, 8, 10\}$,	$\{1, 8, 9\}$,	$\{3, 8, 11\}$,	$\{2, 8, 10\}$,	$\{1, 8, 9\}$,	$\{3, 8, 11\}$,	$\{1, 2, 3\}$,
$\{2, 9, 11\}$,	$\{1, 10, 11\}$,	$\{3, 9, 10\}$,	$\{1, 2, 3\}$,	$\{2, 9, 11\}$,	$\{1, 10, 11\}$,	$\{3, 9, 10\}$

---

1. For each of the following block designs, determine if the design is a BIBD and if so, determine its parameters $b, v, r, k,$ and λ.
   (a) Varieties: $\{1, 2, 3\}$
       Blocks: $\{1, 2\},\ \{1, 3\},\ \{2, 3\},\ \{1, 2, 3\}$
   (b) Varieties: $\{1, 2, 3, 4, 5\}$
       Blocks: $\{1, 2, 3\},\ \{2, 3, 4\},\ \{3, 4, 5\},\ \{1, 4, 5\},\ \{1, 2, 5\}$
   (c) Varieties: $\{1, 2, 3, 4, 5\}$
       Blocks: $\{1, 2, 3, 4\},\ \{1, 3, 4, 5\},\ \{1, 2, 4, 5\},\ \{1, 2, 3, 5\},\ \{2, 3, 4, 5\}$
   (d) Varieties: $\{1, 2, 3, 4, 5, 6, 7, 8, 9\}$
       Blocks: $\{1, 2, 3\},\ \{4, 5, 6\},\ \{7, 8, 9\},\ \{1, 4, 7\},\ \{2, 5, 8\},\ \{3, 6, 9\},$
       $\{1, 5, 9\},\ \{2, 6, 7\},\ \{3, 4, 8\},\ \{1, 6, 8\},\ \{2, 4, 9\},\ \{3, 5, 7\}.$

2. (a) A BIBD has parameters $v = 15, k = 10,$ and $\lambda = 9$. Find b and r.
   (b) A BIBD has parameters $v = 47, b = 47,$ and $r = 23$. Find k and λ.
   (c) A BIBD has parameters $b = 14, k = 3,$ and $\lambda = 2$. Find v and r.

3. Show that there is no (b, v, r, k, λ)-design with the following parameters.
   (a) $b = 7, v = 5, r = 4, k = 3, \lambda = 2$
   (b) $b = 22, v = 22, r = 7, k = 7, \lambda = 1$.

4. Show that there is no (b, v, r, k, λ)-design with the following parameters:
   $b = 4, v = 9, r = 4, k = 9, \lambda = 4$.

5. Could there be a $(12, 6, 8, 7, 1)$-design?

6. Could there be a $(12, 6, 12, 6, 1)$-design?

7. Could there be a $(5, 13, 5, 13, 5)$-design?

8. In Wadley's experiment (Example 9.11), in the case where there are five cows, find λ.

9. For each of the block designs of Exercise 1, find its incidence matrix.

10. For each of the following block designs, compute $\mathbf{AA}^T$ for $\mathbf{A}$ the incidence matrix of the design.
    (a) The design of Example 9.9.
    (b) The design of Example 9.10.
    (c) A $(15, 15, 7, 7, 3)$-design.

11. In a Steiner triple system with $v = 9$, find b and r.

12. The following nine blocks form part of a Steiner triple system with nine varieties:

    $\{a, b, c\},\ \{d, e, f\},\ \{g, h, i\},\ \{a, d, g\},\ \{c, e, h\},\ \{b, f, i\},\ \{a, e, i\},\ \{c, f, g\},\ \{b, d, h\}.$

    (a) How many missing blocks are there?
    (b) Add additional blocks that will lead to a Steiner triple system.

13. If a Steiner triple system has 63 varieties, how many blocks does it have?

14. Compute $\mathbf{AA}^T$ for $\mathbf{A}$ the incidence matrix of a Steiner triple system of 13 varieties.

15. Given a design, the incidence matrix of the *complementary design* is obtained by interchanging 0 and 1 in the incidence matrix of the original design. In general, if one starts with a (b, v, r, k, λ)-design, the complementary design is a $(b', v', r', k', \lambda')$-design.
    (a) Find formulas for $b', v', r',$ and k'.
    (b) Show that $\lambda' = b + \lambda - 2r$.
    (c) Find a $(7, 7, 4, 4, 2)$-design.
    (d) Find a $(12, 9, 8, 6, 5)$-design.

16. Suppose that the complementary design (Exercise 15) of a Steiner triple system with 13 varieties is a (b, v, r, k, λ)-design. Find $b, v, r, k,$ and λ.

17. Construct a Steiner triple system of 21 varieties.

18. Complete the proof of Theorem 9.13 by showing that the set S is in fact a Steiner triple system.

19. Four of the blocks of a $(7, 3, 1)$-design are

$$\{1, 2, 3\}, \quad \{2, 4, 6\}, \quad \{3, 4, 5\}, \quad \text{and} \quad \{3, 6, 7\}.$$

Find the remaining blocks.

20. Show by construction that there is a $(v, v - 1, v - 2)$-design.

21. Show that each of the following designs exists.
  (a) A $(31, 15, 7)$-design
  (b) A $(63, 31, 15)$-design
  (c) A $(21, 5, 1)$-design
  (d) A $(31, 6, 1)$-design.

22. Compute $\mathbf{AA}^T$ for $\mathbf{A}$ the incidence matrix of a $(31, 15, 7)$-design.

23. Show that if $m \geq 1$ is a power of a prime, there is a $(2m^2 + 2m + 2, m^2 + m + 1, 2m + 2, m + 1, 2)$-design.

24. Use the Bruck–Ryser–Chowla Theorem to show that a $(17, 2, 1)$-design could exist.

25. Which of the following (v, k, λ)-designs could possibly exist?
  (a) $(16, 9, 1)$ **(b)** $(18, 9, 5)$ **(c)** $(21, 4, 2)$.

26. Show that a $(22, 22, 7, 7, 2)$-design does not exist.

27. Show by construction that there is a $(14, 8, 7, 4, 3)$-design. (*Hint:* Use Theorem 9.15 and another theorem.)

28. Show by construction that there is a $(30, 16, 15, 8, 7)$-design. (*Hint:* Use Theorem 9.15 and another theorem.)

29. Show that there is a $(30, 15, 14, 7, 6)$-design.

30. Suppose that there is a (v, k, λ)-design.
  (a) Show that there is a $(2v, v, 2k, k, 2\lambda)$-design.
  (b) Show that for any positive integer p, there is a $(pv, v, pk, k, p\lambda)$-design.

31. Show that there is a $(62, 31, 30, 15, 14)$-design.

32. If $t \geq 2$, a t-(b, v, r, k, λ)-*design* consists of a set X of $v \geq 2$ varieties, and a collection of $b > 0$ subsets of X called blocks, such that (9.9) and (9.10) hold, such that

$$\text{every } t\text{-element subset of } X \text{ is a subset of exactly } \lambda \text{ blocks}, \quad \lambda > 0, \tag{9.27}$$

and such that $k < v$. Obviously, a 2-(b, v, r, k, λ)-design is a (b, v, r, k, λ)-design.
  (a) Suppose that $x_{i_1}, x_{i_2}, \ldots, x_{i_t}$ are t distinct varieties of a t-(b, v, r, k, λ)-design. For $1 \leq j \leq t$, let λ_j be the number of blocks containing $x_{i_1}, x_{i_2}, \ldots, x_{i_j}$. Let $\lambda_0 = b$. Show that for $0 \leq j \leq t$,

$$\lambda_j = \frac{\lambda \dbinom{v - j}{t - j}}{\dbinom{k - j}{t - j}}, \tag{9.28}$$

and conclude that λ_j is independent of the choice of $x_{i_1}, x_{i_2}, \ldots, x_{i_j}$. Hence, conclude that for all $1 \leq j \leq t$, a t-(b, v, r, k, λ)-design is also a j-(b, v, r, k, λ)-design.
  (b) Show that for $t \geq 2$, (9.9) and (9.27) imply (9.10).
  (c) Note that if a t-(b, v, r, k, λ)-design exists, then the numbers λ_j defined by (9.28) are integers for all j with $0 \leq j \leq t$.

(d) The results of Section 9.5.2 imply that there is no (43, 7, 1)-design. Use this result to prove that even if all λ_j are integers, this is not sufficient for the existence of a t-(b, v, r, k, λ)-design.

33. Suppose that the square matrix $\mathbf{A}$ is the incidence matrix of a BIBD. Show that $\mathbf{A}^{-1}$ exists.

34. If $\mathbf{A}$ is the incidence matrix of a (b, v, r, k, λ)-design, show that $\mathbf{AJ} = k\mathbf{J}$, where $\mathbf{J}$ is a square matrix of all 1's.

35. Suppose $\mathbf{A}$ is a $v \times v$ matrix of 0's and 1's, $v \geq 2$, and that there are $k > 0$ and $\lambda > 0$ with $k < \lambda$ and so that:

   (1) Any row of $\mathbf{A}$ contains exactly k 1's

and

   (2) Any pair of rows of $\mathbf{A}$ have 1's in common in exactly λ columns.

This exercise asks the reader to prove that:

   (3) Any column of $\mathbf{A}$ contains exactly k 1's

and

   (4) Any pair of columns of $\mathbf{A}$ have 1's in common in exactly λ rows.

[In particular, it follows that (3) and (4) hold for incidence matrices of (v, k, λ)-designs, and Theorem 9.17 follows.]
   (a) Show that $\mathbf{AJ} = k\mathbf{J}$, where $\mathbf{J}$ is a square matrix of all 1's.
   (b) Show that $\mathbf{AA}^T = (k - \lambda)\mathbf{I} + \lambda\mathbf{J}$.
   (c) Show that $\mathbf{A}^{-1}$ exists.
   (d) Show that $\mathbf{A}^{-1}\mathbf{J} = k^{-1}\mathbf{J}$.
   (e) Show that $\mathbf{A}^T\mathbf{A} = (k - \lambda)\mathbf{I} + \lambda k^{-1}\mathbf{JA}$.
   (f) Show that if $\mathbf{JA} = k\mathbf{J}$, then (3) and (4) follow.
   (g) Show that $\mathbf{JA} = k^{-1}(k - \lambda + \lambda v)\mathbf{J}$.
   (h) Show that $k - \lambda + \lambda v = k^2$.
   (i) Show that $\mathbf{JA} = k\mathbf{J}$ and hence that (3) and (4) hold.

36. In an experiment, there are two kinds of treatments or varieties, the controls and the non-controls. There are three controls and 120 blocks. Each control is used in 48 blocks. Each pair of controls is used in the same block 24 times. All three controls are used in the same block together 16 times. In how many blocks are none of the controls used?

9.5 FINITE PROJECTIVE PLANES

9.5.1 Basic Properties

It is interesting that experimental designs have geometric applications, and conversely that geometry has played an important role in the analysis of experimental designs. Let us consider the design of Example 9.8. This is a Steiner triple system and a symmetric BIBD. It can be represented geometrically by letting the varieties be points and representing a block by a "line" (not necessarily straight) through the points it contains. Figure 9.1 shows this geometric representation. All but one "line" is straight. This representation is known as a *projective plane*, the *Fano plane*. It has the following properties:

   (P_1)   Two distinct points lie on one and only one common line.
   (P_2)   Two distinct lines pass through one and only one common point.

In general, a *projective plane* consists of a set of objects called *points*, a second set of objects called *lines*, and a notion of when a *point lies on a line*, or equivalently when a *line passes through a point*, so that conditions (P_1) and (P_2) hold. A projective plane is *finite* if the set of points is finite. Projective planes are important not only in combinatorial design but also in art, where they arise in the study of perspective. They are also important in geometry, for they define a geometry where Euclid's parallel postulate is violated: By (P_2), there is no line that passes through a given point and has no points in common with (and hence is "parallel" to) a given line. The development of projective geometry had its roots in the work of Pappas of Alexandria in the fourth century. It led in the 1840's to the algebraic theory of invariance, developed by the famous mathematicians Boole, Cayley, and Sylvester. This in turn led to the tensor calculus, and eventually to ideas of fundamental importance in physics, in particular to the work of Einstein in the theory of gravitation.

The basic existence question that dominated the theory of experimental design arises also for projective planes: For what values of n is there a projective plane of n points? If $n = 2$, we can take two points a and b and one line L that passes through the two points. The postulates (P_1) and (P_2) for a projective plane are trivially satisfied. They are also trivially satisfied if there are n points, any n, and just one line, which passes through all n points. Finally, they are trivially satisfied if there are three points, a, b, and c, and three lines, L_1, L_2, and L_3, with a and b lying on L_1, b and c on L_2, and a and c on L_3. To rule out these dull examples, one usually adds one additional postulate:

(P_3)   There are four distinct points, no three of which lie on the same line.

A finite projective plane satisfying (P_3) is called *nondegenerate* and we shall assume (without making the assumption explicit every time) that *all finite projective planes are nondegenerate*. Any theorem about these planes will be proved using the postulates (P_1), (P_2), and (P_3).

The smallest possible projective plane would now have at least four points. Is there such a plane with exactly four points? Suppose that a, b, c, and d are four points, and that no three lie on a line. By (P_1), there must be a line L_1 passing through a and b and a line L_2 passing through c and d. Since no three of these points lie on a line, c and d are not on L_1 and a and b are not on L_2. Then if a, b, c, and d are all the points of the projective plane, L_1 and L_2 do not have a common point, which violates (P_2). Thus, there is no projective plane of four points. We shall see below that there is no projective plane of five or six points either. However, the Fano plane of Figure 9.1 is a projective plane of seven points, for (P_3) is easy to verify.

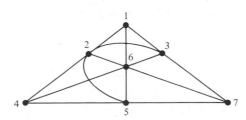

Figure 9.1   The Fano plane.

The reader will note that the postulates (P_1) and (P_2) have a certain *duality*: We obtain (P_2) from (P_1) by interchanging the words point and line and interchanging the expressions point lying on a line and line passing through a point. We obtain (P_1) from (P_2) by the same interchanges. If (P) is any *statement* about finite projective planes, the *dual* of (P) is the statement obtained from (P) by making these interchanges. The dual of postulate (P_3) turns out to be true, and we formulate this result as a theorem.

Theorem 9.20.   In a finite projective plane, the following holds:

(P_4)   There are four distinct lines, no three of which pass through the same point.

Proof. By (P_3), there are points a, b, c, d no three of which lie on a line. By (P_1), there are lines L_1 through a and b, L_2 through b and c, L_3 through c and d, and L_4 through d and a. Now these four lines are distinct, because c and d are not in L_1, a and d are not in L_2, and so on. Moreover, no three of these lines pass through a common point. We prove this by contradiction. Suppose without loss of generality that L_1, L_2, and L_3 have the point x in common. Then x could not be b, for b is not on L_3. Now L_1 and L_2 have two distinct points in common, b and x. Since $L_1 \neq L_2$, postulate (P_2) is violated, which is a contradiction.   Q.E.D.

Now conditions (P_1) and (P_2) are duals and conditions (P_3) and (P_4) are duals. Any theorem (provable statement) about finite projective planes must be proved from the postulates (P_1), (P_2), and (P_3). Any such theorem will have a *dual theorem*, obtained by interchanging the words point and line and interchanging the expressions point lying on a line with line passing through a point. A proof of the dual theorem can be obtained from a proof of the theorem by replacing (P_1), (P_2), and (P_3) by their appropriate dual statements, which we know to be true. Thus, we have the following result.

Theorem 9.21   (Duality Principle).   For every statement about finite projective planes which is a theorem, the dual statement is also a theorem.

The next basic theorem about finite projective planes is the following.

Theorem 9.22.   In a finite projective plane, every point lies on the same number of lines, and every line passes through the same number of points.

To illustrate this theorem, we note that the projective plane of Figure 9.1 has three points on a line and three lines through every point.

Proof of Theorem 9.22. We first show that every line passes through the same number of points. The basic idea of the proof is to set up a one-to-one correspondence between points on two distinct lines, L and L', which shows that the two lines have the same number of points.

We first show that there is a point x not on either L or L'. By postulate (P_3), there are four points a, b, c, and d, no three of which lie on a line. If any one of these is not on either L or L', we can take that as x. If all of these are on L or L', we must have two points (say a and b) on L and two (say c and d) on L'. By (P_1), there are lines K through b and c and K' through a and d. By (P_2), the lines K and K' have a point x in common (see

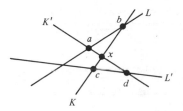

Figure 9.2 The point x is not on either line L or line L'.

Figure 9.2). If x lies on L, then x and b lie on two distinct lines, violating (P$_1$). If x lies on L', then x and c lie on two distinct lines, again violating (P$_1$). Thus, x is the desired point.

Now given a point p on line L, the line through p and x [which exists by (P$_1$)] must meet L' in exactly one point p' [by (P$_2$)]. We say p' is the *projection* of p through x onto L' (see Figure 9.3). If q is any other point on L, let q' be its projection through x onto L'. Now if $q \neq p$, q' must be different from p'. For otherwise q' and x are on two distinct lines, violating (P$_1$). We conclude that projection defines a one-to-one correspondence between points of L and points of L'. We know that it is one-to-one. To see it is a correspondence, note that every point r of L' is obtained from some point of L by this procedure. To see that, simply project back from r through x onto L. Thus, L and L' have the same number of points. This proves that every line passes through the same number of points.

By using the duality principle (Theorem 9.21), we conclude that every point lies on the same number of lines. Q.E.D.

Theorem 9.23. In a finite projective plane, the number of lines through each point is the same as the number of points on each line.

Proof. Pick an arbitrary line L. By (P$_3$), there is a point x not on line L. By (P$_1$), for any point y on L, there is one and only one line $L(y)$ passing through x and y. Moreover, any line L' through x cannot be L, and hence by (P$_2$) it must pass through a point y of L, so L' is $L(y)$. Thus, $L(y)$ defines a one-to-one correspondence between points of L and lines through x. Thus, there are the same number of lines through x as there are points on L. The theorem follows from Theorem 9.22. Q.E.D.

It follows from Theorem 9.23 that the projective plane with $m + 1$ points on each line has $m + 1$ lines through each point.

Corollary 9.23.1. A projective plane with $m + 1$ points on each line and $m + 1$ lines through each point has $m^2 + m + 1$ points and $m^2 + m + 1$ lines.

Proof. Let x be any point. There are $m + 1$ lines through x. Each such line has m points other than x. Every point lies on one and only one line through x. Hence, we count

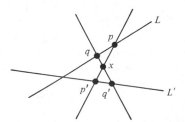

Figure 9.3 The point p' is the projection of p through x onto L'.

Table 9.28 A Projective Plane of Order 3 (having 13 Points and 13 Lines)

Points: 1, 2, 3, 4, 5, 6, 7, 8, 9, 10, 11, 12, 13

Lines: {1, 2, 8, 12}, {3, 9, 10, 12}, {4, 5, 11, 12}, {1, 4, 6, 9}, {3, 6, 8, 11}
{2, 5, 6, 10}, {1, 10, 11, 13}, {5, 8, 9, 13}, {2, 3, 4, 13}, {1, 3, 5, 7},
{4, 7, 8, 10}, {2, 7, 9, 11}, {6, 7, 12, 13}

all the points by counting the number of points other than x on the lines through x and adding x; we get

$$(m + 1)(m) + 1 = m^2 + m + 1$$

points. The rest of the corollary follows by a duality argument. Q.E.D.

In our example of Figure 9.1, $m + 1 = 3$, $m = 2$, and $m^2 + m + 1 = 7$. We know now that projective planes of n points can only exist for n of the form $m^2 + m + 1$. In particular, no such planes can exist for $n = 4$, 5, or 6. Postulate (P_3) rules out $n = 3$ (even though $3 = 1^2 + 1 + 1$). Thus, $n = 7$ corresponds to the first possible projective plane. The next possible one is obtained by taking $m = 3$, obtaining $n = 3^2 + 3 + 1 = 13$. We will see that projective planes exist whenever m is a power of a prime. Thus, the 13-point plane ($m = 3$) will exist. (Table 9.28 shows such a plane.) The number m will be called the *order* of the projective plane. Note that the order is different from the number of points.

9.5.2 Projective Planes, Latin Squares, and (v, k, λ)-designs

In this section we investigate the relations among projective planes and certain kinds of experimental designs.

Theorem 9.24 (Bose [1938]). Suppose that $m \geq 2$. A finite projective plane of order m exists if and only if a complete orthogonal family of $m \times m$ Latin squares exists.*

The proof of this theorem is constructive, that is, it shows how to go back and forth between finite projective planes and sets of orthogonal Latin squares. We sketch the proof in Exercises 19 and 20.

Corollary 9.24.1. If $m = p^k$ for p prime and k a positive integer, then there is a projective plane of order m.

Proof. The result follows by Theorem 9.2. Q.E.D.

It follows from Theorem 9.24 and Corollary 9.24.1 that the first possible order m for which there does not exist a finite projective plane of that order (that is, of $m^2 + m + 1$ points) is $m = 6$. In fact, for $m = 6$, we have seen in Section 9.2 that there does not exist a set of five orthogonal 6×6 Latin squares (or indeed a pair of such squares), and hence there is no finite projective plane of order 6 (that is, of $6^2 + 6 + 1 = 43$ points). There are projective planes of orders 7, 8, and 9, since $7 = 7^1$, $8 = 2^3$, and $9 = 3^2$. However, $m = 10$ is not a power of a prime. No one has been able to find a finite projective plane of order

*Recall that by our convention, a single 2×2 Latin square constitutes an orthogonal family.

10 ($10^2 + 10 + 1 = 111$ points), or prove that there can be no such plane. (In Section 9.2, we observed that there were pairs of orthogonal 10×10 Latin squares, but that it was not known whether or not there exists a set of three orthogonal 10×10 Latin squares, let alone a set of nine. The latter is, of course, equivalent to the existence of a finite projective plane of order 10.)

The next theorem takes care of some of the cases not covered by Corollary 9.24.1. We omit the proof.

Theorem 9.25 (Bruck and Ryser [1949]). Let $m \equiv 1$ or 2 (mod 4). Suppose that the largest square dividing into m is d and that $m = m'd$. If m' is divisible by a prime number p which is $\equiv 3$ (mod 4), then there does not exist a projective plane of order m.

To illustrate this theorem, suppose that $m = 6$. Note that $6 \equiv 2$ (mod 4). Then $d = 1$ and $m' = m$. Since m' is divisible by 3, there is no projective plane of order 6, as we have previously observed. Next, suppose that $m = 54$. Note that $54 \equiv 2$ (mod 4), that $d = 9$, and that $m' = 6$. Since 3 divides m', there is no projective plane of order 54. It follows by Theorem 9.24 that there is no complete orthogonal family of Latin squares of order 54.

A projective plane gives rise to a (b, v, r, k, λ)-design by taking the varieties as the points and the blocks as the lines. Then $b = v = m^2 + m + 1$, $k = r = m + 1$, and $\lambda = 1$ since each pair of points lies on one and only one line. Hence, we have a symmetric balanced incomplete block design or a (v, k, λ)-design. Conversely, for every $m \geq 2$, an $(m^2 + m + 1, m + 1, 1)$-design gives rise to a projective plane by taking the varieties as the points and the blocks as the lines. To see why Axiom (P3) holds, let a and b be any two points. There is a unique block B_1 containing a and b. Since the design is incomplete, there is a point x not in block B_1. Now there is a unique block B_2 containing a and x and there is a unique block B_3 containing b and x. Each block has $m + 1$ points. Thus, other than a, b, and x, B_1, B_2, and B_3 each have at most $m - 1$ other points. In short, the number of points in B_1, B_2, or B_3 is at most $3 + 3(m - 1) = 3m$. Since $m \geq 2$, we have $m^2 + m + 1 > 3m$. Thus, there is a point y not in B_1, B_2, or B_3. No three of the points a, b, x, and y lie in a block. Thus, we have the following result.

Theorem 9.26. If $m \geq 2$, a finite projective plane of order m exists if and only if an $(m^2 + m + 1, m + 1, 1)$-design exists.

Corollary 9.26.1. There are $(m^2 + m + 1, m + 1, 1)$-designs whenever m is a power of a prime.

Corollary 9.26.2. Suppose that $m \geq 2$. Then the following are equivalent.

(a) There exists a finite projective plane of order m.
(b) There exists a complete orthogonal family of Latin squares of order m.
(c) There exists an $(m^2 + m + 1, m + 1, 1)$-design.

EXERCISES FOR SECTION 9.5

1. In each of the following, P is a set of points and Q a set of lines. A point x lies on a line L if $x \in L$ and a line L passes through a point x if $x \in L$. Determine which of axioms (P1), (P2), and (P3) hold.

(a) P = all points in 3-space, Q = all lines in 3-space.
(b) $P = \{1, 2, 3\}$, $Q = \{\{1, 2\}, \{1, 3\}, \{2, 3\}\}$.
(c) P = any set, Q = all ordered pairs from P.
(d) P = all lines in 3-space, Q = all planes in 3-space.
(e) P = all lines through the origin in 3-space, Q = all planes through the origin in 3-space.

2. State the dual of each of the following (not necessarily true) statements about finite projective planes.
   (a) There are seven distinct lines, no three of which pass through the same point.
   (b) There is a point that lies on every line.
   (c) There are three distinct lines so that every point lies on one of these lines.
   (d) There are four distinct points, no three of which lie on the same line, and so that every line passes through one of these points.

3. If a projective plane has four points on every line, how many points does it have in all?

4. If a projective plane has five lines through every point, how many points does it have in all?

5. If a projective plane has 31 points, how many points lie on each line?

6. If a projective plane has 57 points, how many lines pass through each point?

7. If a projective plane has 73 lines, how many points lie on each line?

8. Is there a projective plane of
   (a) 20 points?   (b) 73 points?   (c) 41 lines?   (d) 91 lines?

9. Suppose that a projective plane has n points and a (v, k, λ)-design is defined from the plane with the points as the varieties and the lines as the blocks. For each of the following values of n, compute v, k, and λ.
   (a) 31   (b) 57   (c) 133.

10. Could there be a finite projective plane of order (not number of points) equal to
    (a) 11?   (b) 25?   (c) 36?
    Justify your answer.

11. Show that there could be no finite projective plane of order 245.

12. Show that there could be no finite projective plane of order 150.

13. Could there be a finite projective plane of order
    (a) 60?   (b) 81?   (c) 93?
    Justify your answer.

14. Show that there is no complete orthogonal family of Latin squares of order 378.

15. In the Fano plane (Figure 9.1):
    (a) Find the projection of the point 6 on the line $\{3, 4, 6\}$ through the point 5 onto the line $\{1, 2, 4\}$.
    (b) Find the projection of the point 7 on the line $\{2, 6, 7\}$ through the point 3 onto the line $\{1, 2, 4\}$.
    (c) Find the projection of the point 3 on the line $\{3, 4, 6\}$ through the point 5 onto the line $\{1, 2, 4\}$.

16. In the projective plane of Table 9.28:
    (a) Find the projection of the point 2 on the line $\{1, 2, 8, 12\}$ through the point 9 onto the line $\{4, 5, 11, 12\}$.
    (b) Find the projection of the point 4 on the line $\{1, 4, 6, 9\}$ through the point 11 onto the line $\{1, 3, 5, 7\}$.

17. (Bogart [1983]).   Take the points of an $(n^2 + n, n^2, n + 1, n, 1)$-design as the points and the blocks of this design as the lines. Which of axioms (P_1), (P_2), (P_3) are satisfied?

18. Show that in the previous exercise, the following "parallel postulate" is satisfied: Given a point

x and a line L not passing through x, there is one and only one line L' passing through x and which has no points in common with L.

19. The next two exercises sketch a proof of Theorem 9.24. Let P be a finite projective plane of order m.

    (a) Pick a line L from P arbitrarily and call L the *line at infinity*. Let L have points

$$u, v, w_1, w_2, \ldots, w_{m-1}.$$

Through each point of L there are $m + 1$ lines, hence m lines in addition to L. Let these be listed as follows:

$$\text{lines through } u: \quad U_1, U_2, \ldots, U_m,$$
$$\text{lines through } v: \quad V_1, V_2, \ldots, V_m,$$
$$\text{lines through } w_j: \quad W_{j1}, W_{j2}, \ldots, W_{jm}.$$

Every point x not on line L is joined by a unique line to each point on L. Suppose that U_h is the line containing x and u, V_i the line containing x and v, and W_{jk_j} the line containing x and w_j. Thus we can associate with the point x the $(m + 1)$-tuple $(h, i, k_1, k_2, \ldots, k_{m-1})$. Show that the correspondence between points x not on L and ordered pairs (h, i) is one-to-one.

    (b) Illustrate this construction with the projective plane of Table 9.28. Write out all of the lines and the associations $x \to (h, i)$ and $x \to (h, i, k_1, k_2)$, assuming $\{1, 2, 8, 12\}$ is the line at infinity.

    (c) Let $a_{hi}^{(j)} = k_j$ if the point x corresponding to ordered pair (h, i) gives rise to the $(m + 1)$-tuple $(h, i, k_1, k_2, \ldots, k_{m-1})$, and let $A^{(j)} = (a_{hi}^{(j)})$, $j = 1, 2, \ldots, m - 1$. Find $A^{(1)}$ and $A^{(2)}$ for the projective plane of Table 9.28.

    (d) Show that $A^{(j)}$ is a Latin square.

    (e) Show that $A^{(p)}$ and $A^{(q)}$ are orthogonal if $p \neq q$.

20. Suppose that $A^{(1)}, A^{(2)}, \ldots, A^{(m-1)}$ is a family of pairwise orthogonal Latin squares of order m.

    (a) Consider m^2 "finite" points (h, i), $h = 1, 2, \ldots, m$, $i = 1, 2, \ldots, m$. Given the point (h, i), associate with it the $(m + 1)$-tuple

$$(h, i, k_1, k_2, \ldots, k_{m-1}),$$

where k_j is $a_{hi}^{(j)}$. Find these $(m + 1)$-tuples given the two orthogonal Latin squares $A^{(1)}$ and $A^{(2)}$ of order 3 shown in Table 9.14. (This will be our running example.)

    (b) Form $m^2 + m = m(m + 1)$ lines W_{jk}, $j = -1, 0, 1, 2, \ldots, m - 1$, $k = 1, 2, \ldots, m$, by letting W_{jk} be the set of all finite points (h, i) where the $(j + 2)$th entry in the $(m + 1)$-tuple corresponding to (h, i) is k. (These lines will be extended by one point later on.) Identify the lines W_{jk} in our example.

    (c) Note that for fixed j,

$$W_{j1}, W_{j2}, \ldots, W_{jm} \tag{9.29}$$

as we have defined them is a collection of m lines, no two of which intersect. We say that two of these are *parallel* in the sense of having no finite points in common. Show that W_{jk} has m finite points.

    (d) Show that if $j \neq j'$, then W_{jk} and $W_{j'k'}$ as we have defined them, have one and only one common point.

    (e) We now have $m + 1$ sets of m parallel lines, and any two nonparallels intersect in one point. To each set of parallels (9.29), we now add a distinct "point at infinity," w_j, lying on each line in the set. Let $w_{-1} = u$ and $w_0 = v$. We have added $m + 1$ infinite points in all. We then add one more line L, the "line at infinity," defined to be the line consisting of u, v, w_1, w_2, ..., w_{m-1}. Complete and update the list of lines begun for our example in part (b). Also list all points (finite or infinite) in this example.

(f) Find in general the number of points and lines constructed.

(g) Find the number of points on each line and the number of lines passing through each point.

(h) Verify that postulates (P1), (P2), and (P_3) hold with the collection of all points (finite or infinite) and the collection of all lines W_{jk} as augmented plus the line L at infinity. [*Hint:* Verify (P_2) first.]

REFERENCES FOR CHAPTER 9

BIRKHOFF, G., and MACLANE, S., *A Survey of Modern Algebra*, Macmillan, New York, 1953.

BOGART, K. P., *Introductory Combinatorics*, Pitman, Marshfield, Mass., 1983.

BOSE, R. C., "On the Application of the Properties of Galois Fields to the Problem of Construction of Hyper-Graeco-Latin Squares," *Sankhyā, 3* (1938), 323–338.

BOSE, R. C., SHRIKHANDE, S. S., and PARKER, E. T., "Further Results on the Construction of Mutually Orthogonal Latin Squares and the Falsity of Euler's Conjecture," *Canad. J. Math., 12* (1960), 189–203.

BOX, G. E. P., HUNTER, W. G., and HUNTER, J. S., *Statistics for Experimenters: An Introduction to Design, Data Analysis, and Model Building*, Wiley, New York, 1978.

BRUCK, R. H., and RYSER, H. J., "The Nonexistence of Certain Finite Projective Planes," *Canad. J. Math, 1* (1949), 88–93.

BRUNK, M. E., and FEDERER, W. T., "Experimental Designs and Probability Sampling in Marketing Research," *J. Am. Statist. Assoc., 48* (1953), 440–452.

CHEN, K. K., BLISS, C. I., and ROBBINS, E. B., "The Digitalis-like Principles of *Calotropis* Compared with Other Cardiac Substances," *J. Pharmacol. Exp. Ther., 74* (1942), 223–234.

CHOWLA, S., and RYSER, H. J. "Combinatorial Problems," *Canad. J. Math., 2* (1950), 93–99.

COCHRAN, W. G., and COX, G. M., *Experimental Designs*, 2d ed., Wiley, New York, 1957.

COX, D. R., *Planning of Experiments*, Wiley, New York, 1958.

DAVIES, H. M., "The Application of Variance Analysis to Some Problems of Petroleum Technology," Technical paper, Institute of Petroleum, London, 1945.

DORNHOFF, L. L., and HOHN, F. E., *Applied Modern Algebra*, Macmillan, New York, 1978.

FINNEY, D. J., *The Theory of Experimental Design*, University of Chicago Press, Chicago, 1960.

FISHER, J. L., *Application-Oriented Algebra*, Harper & Row, New York, 1977.

FISHER, R. A., "The Arrangement of Field Experiments," *J. Minist. Agric., 33* (1926), 503–513.

FISHER, R. A., "An Examination of the Different Possible Solutions of a Problem in Incomplete Blocks," *Ann. Eugen., 10* (1940), 52–75.

HALL, M., *Combinatorial Theory*, Blaisdell, Waltham, Mass., 1967. (Second printing, Wiley, New York, 1980.)

HICKS, C. R., *Fundamental Concepts in the Design of Experiments*, Holt, Rinehart and Winston, New York, 1973.

KAPUR, G. K., *IBM 360 Assembler Language Programming*, Wiley, New York, 1970.

KIRKMAN, T. A., "On a Problem in Combinatorics," *Camb. Dublin Math J., 2* (1847), 191–204.

MACNEISH, H. F., "Euler Squares," *Ann. Math., 23* (1922), 221–227.

PUGH, C., "The Evaluation of Detergent Performance in Domestic Dishwashing," *Appl. Statist., 2* (1953), 172–179.

RYSER, H. J., *Combinatorial Mathematics*, Carus Mathematical Monographs, No. 14, Mathematical Association of America, Washington, D.C., 1963.

TARRY, G., "Le Problème de 36 officieurs," *C.R. Assoc. Fr. Avance. Sci. Nat.*, *1* (1900), 122–123.

TARRY, G., "Le Problème de 36 officieurs, " *C. R. Assoc. Fr. Avance. Sci. Nat.*, *2* (1901), 170–203.

VICKERS, F. D., *Introduction to Machine and Assembly Langauge: Systems 360/370*, Holt, Rinehart and Winston, New York, 1971.

WADLEY, F. M., "Experimental Design in the Comparison of Allergens on Cattle," *Biometrics*, *4* (1948), 100–108.

WILLIAMS, E. J., "Experimental Designs Balanced for the Estimation of Residual Effects of Treatments, *Aust. J. Sci. Res.*, *A2* (1949), 149–168.

YATES, F., "Incomplete Randomized Blocks," *Ann. Eugen.*, *7* (1936), 121–140.

YOUDEN, W. J., "Use of Incomplete Block Replications in Estimating Tobacco-Mosaic Virus," *Contrib. Boyce Thompson Inst.*, *9* (1937), 41–48.

ZARISKI, O., and SAMUEL, P., *Commutative Algebra*, Vol. 1, D. Van Nostrand, Princeton, N.J., 1958.

10 Coding Theory*

10.1 INFORMATION TRANSMISSION

In this chapter we provide a brief overview of coding theory. Our concern will be with two aspects of the use of codes: to ensure the secrecy of transmitted messages and to detect and correct errors in transmission. The methods we shall discuss have application to communications with computers, with distant space probes, and with missiles in launching pads; to electronic banking; to genetic codes; and so on. For many references on the applications of coding theory, and in particular to those having to do with communication with computers, see MacWilliams and Sloane [1977, p. 34]. For more detailed treatments of coding theory as a whole, see, in addition to MacWilliams and Sloane [1977], Berlekamp [1968], Blake and Mullin [1975], Peterson [1961], Peterson and Weldon [1972], or van Lint [1982]. For good short treatments, see Dornhoff and Hohn [1978] or Fisher [1977].

The basic steps in information transmission are modeled in Figure 10.1. We imagine that we start with a "word," an English word or a word already in some code, for example a bit string. In step (a) we encode it, usually into a bit string. We then transmit the encoded word over a transmission channel in step (b). Finally, the received word is decoded in step (c). This model applies to transmission over physical communication paths such as telegraph lines or across space via radio waves. However, as MacWilliams and Sloane [1977] point out, a similar analysis applies, for example, to the situation when data are stored in a computer and later retrieved. As Fisher [1977] points out, the

*Sections 10.1–10.3 form the basis for this chapter. The reader interested in a brief treatment can then go directly to Section 10.5 (which depends on Chapter 9) or sample part of Section 10.4 (which does not depend on Chapter 9.)

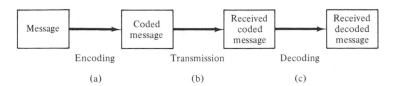

| Message | → | Coded message | → | Received coded message | → | Received decoded message |

Encoding   Transmission   Decoding

(a)   (b)   (c)

Figure 10.1 The basic steps in information transmission.

analysis also applies to communication of visual input into the retina (patterns of photons). The input is encoded into electrical impulses in some of the cells in the retina, and these impulses are transmitted through neurons to a visual area of the brain, where they are "decoded" as a visual pattern. There are many other applications as well.

We shall assume that the only errors which occur are in step (b), and are caused by the presence of noise or by weak signals. In Section 10.2 we discuss steps (a) and (c), the encoding and decoding steps, without paying attention to errors in transmission. In Section 10.3 we begin to see how to deal with such errors. In Section 10.4 we demonstrate how to use the encoding and decoding process to detect and correct errors in transmission. Finally in Section 10.5 we discuss the use of block designs to obtain error-detecting and error-correcting codes. That section should be omitted by the reader who has skipped Chapter 9.

Much of the emphasis in this chapter is on the existence question: Is there a code of a certain kind? However, we also deal with the optimization question: What is the best or richest or largest code of a certain kind?

10.2 ENCODING AND DECODING

The field called *cryptography* is concerned with coding and decoding messages, and also with deciphering encoded messages if the code is not known. We shall discuss the encoding and decoding problems of cryptography here but not the deciphering problem.

A message to be encoded involves a sequence of symbols from some *message alphabet A*. The encoded message will be a sequence of symbols from a *code alphabet B*, possibly the same as the message alphabet. A simple encoding rule will encode each symbol a of A as a symbol $E(a)$ of B.

Example 10.1 Caesar Cyphers

If A and B are both the 26 letters of the alphabet, a simple encoding rule $E(a)$ might take $E(a)$ to be the letter following a, with $E(Z) = A$. Thus, the message

DEPOSIT SIX MILLION DOLLARS   (10.1)

would be encoded as

EFQPTJU TJY NJMMJPO EPMMBST.

If A is as above but $B = \{0, 1, 2, \ldots, 25\}$, we could take $E(a) =$ the position of a in the alphabet $+ 2$, where Z is assumed to have position 0. Here $24 + 2$ is interpreted as 0 and $25 + 2$ as 1. (Addition is modulo 26, to use the terminology of

Section 9.3.1.) Thus, $E(D) = 6$. Similar encodings were used by Julius Caesar about 2000 years ago. They are now called *Caesar cyphers*.

Any function $E(a)$ from A into B will suffice for coding provided we can adequately decode, that is, find a from $E(a)$. To be able to do so unambiguously, we cannot have $E(a) = E(b)$ if $a \neq b$; that is, $E(a)$ must be one-to-one.

It is frequently useful to break a long message up into blocks of symbols rather than just one symbol, and encode the message in blocks. For instance, if we use blocks of length 2, the message (10.1) becomes

$$\text{DE \quad PO \quad SI \quad TS \quad IX \quad MI \quad LL \quad IO \quad ND \quad OL \quad LA \quad RS.} \tag{10.2}$$

We then encode each sequence of two symbols, that is, each block.* One way to encode blocks uses matrices, as the next example shows.

Example 10.2   Matrix Codes

Let us replace each letter of the alphabet by a number representing its position in the alphabet (with Z having position 0). Then a block corresponds to a vector. For instance, the block DE above corresponds to the vector $(4, 5)$. Suppose that all blocks have length m. Let $\mathbf{M}$ be an $m \times m$ matrix, for instance

$$\mathbf{M} = \begin{pmatrix} 2 & 3 \\ 1 & 2 \end{pmatrix}.$$

Then we can encode a block a as a block $a\mathbf{M}$. In our case, we encode a block (i, j) as a block $E(i, j) = (i, j)\mathbf{M}$. Hence, DE or $(4, 5)$ gets encoded as

$$(4, 5) \begin{pmatrix} 2 & 3 \\ 1 & 2 \end{pmatrix} = (13, 22).$$

The reader can check that the message (10.1), when broken up as in (10.2), now gets encoded as

13, 22, 47, 78, 47, 75, 59, 98, 42, 75, 35, 57, 36, 60, 33, 57, 32, 50, 42, 69, 25, 38, 55, 92.

In general, the procedure we have defined is called *matrix encoding*. It is highly efficient. Moreover, decoding is easy. To decode, we break a coded message into blocks b, and find a so that $a\mathbf{M} = b$. We can unambiguously decode provided that $\mathbf{M}$ has an inverse. For then $a = b\mathbf{M}^{-1}$. In our example,

$$\mathbf{M}^{-1} = \begin{pmatrix} 2 & -3 \\ -1 & 2 \end{pmatrix}.$$

Thus, for instance, if we have the encoded block $(6, 11)$, we find that it came from

$$(6, 11)\mathbf{M}^{-1} = (6, 11) \begin{pmatrix} 2 & -3 \\ -1 & 2 \end{pmatrix} = (1, 4) = \text{AD}.$$

Note that if we have the encoded block $(3, 4)$, then decoding gives us

$$(3, 4)\mathbf{M}^{-1} = (2, -1).$$

---

*To guarantee that a message will always break up into blocks of the same size, we can always lengthen the message by adding a recognizable "closing" string of copies of the same letter, for example, Z or ZZZ.

Since $(2, -1)$ does not correspond to any pair of letters, we can only conclude that an error was made in sending us the message that included the block $(3, 4)$. Much of the emphasis in this chapter will be on ways to detect errors, as we have done here.

In general, suppose that a message is broken up into blocks of length k. Let A^k consist of all blocks (sequences) of length k from the message alphabet A and B^n consist of all blocks (sequences) of length n from the code alphabet B. Let $\mathcal{A}$ be a subset of A^k called the set of *message blocks*. In most practical applications, $\mathcal{A} = A^k$, and we shall assume this unless stated explicitly otherwise. A *block code* or $k \to n$ *block code* is a one-to-one function $E: \mathcal{A} \to B^n$. The set C of all $E(a)$ for a in $\mathcal{A}$ is defined to be the set of *codewords*. Sometimes C alone is called the *code*. In Example 10.2, $k = n = 2$,

$$A = \{A, B, \cdots, Z\},$$

$$B = \{0, 1, \cdots, 25\},$$

and $\mathcal{A} = A^k$. The blocks $(13, 22)$ and $(6, 11)$ are codewords. However, the block $(3, 4)$ is not a codeword. In most practical examples, A and B will both be $\{0, 1\}$, messages will be bit strings, and the encoding will take bit strings of length k into bit strings of length n. We shall see in Section 10.4 that taking $n > k$ will help us in error detection and correction.

We close this section by giving one more example of a block code called a *permutation code*.

Example 10.3   Permutation Codes

Suppose that $A = B = \{0, 1\}$ and π is a permutation of $\{1, 2, \ldots, k\}$. Then we can define $E: A^k \to B^k$ by taking $E(a_1 a_2 \cdots a_k) = a_{\pi(1)} a_{\pi(2)} \cdots a_{\pi(k)}$. For instance, suppose that $k = 3$ and $\pi(1) = 2$, $\pi(2) = 3$, $\pi(3) = 1$. Then $E(a_1 a_2 a_3) = a_2 a_3 a_1$, so $E(011) = 110$, $E(101) = 011$, and so on. It is easy to see (Exercise 12) that every such permutation code E is a matrix encoding.

EXERCISES FOR SECTION 10.2

1. Suppose that

$$\mathbf{M} = \begin{pmatrix} 1 & 1 \\ 2 & 3 \end{pmatrix}.$$

   Encode the following expressions using the matrix code defined by $\mathbf{M}$.
   (a) AX
   (b) UV
   (c) DE
   (d) ASSEMBLE AT NOON
   (e) INVEST TWO MILLION
   (f) ABORT THE MISSIONZ
   (*Note*: Z has been added at the end because there is an odd number of letters.)

2. Suppose that

$$\mathbf{M} = \begin{bmatrix} 1 & 2 & 3 \\ 2 & 3 & 4 \\ 1 & 2 & 1 \end{bmatrix}.$$

Encode the following expressions using the matrix code defined by **M**.

(a) ABC

(b) XAT

(c) TTU

(d) BUY TWELVE SHARES

(e) RUN THE PROGRAM AT TEN

(f) OBSERVE THE TRANSFERSZZ. (*Note:* Two *Z*'s have been added to make the number of letters divisible by 3.)

3. Repeat Exercise 1 with

$$\mathbf{M} = \begin{pmatrix} 1 & 2 \\ 3 & 4 \end{pmatrix}.$$

4. Repeat Exercise 2 with

$$\mathbf{M} = \begin{bmatrix} 1 & 1 & 0 \\ 0 & 3 & 1 \\ 2 & 3 & 3 \end{bmatrix}.$$

5. If **M** of Exercise 1 is used in a matrix code, decode the following or show that an error was made; that is, this is not proper code.

(a) (5, 7)   (b) (8, 12)   (c) (1, 2)

(d) (2, 5)   (e) 18, 23, 44, 60   (f) 5, 7, 51, 69, 20, 20.

6. If **M** of Exercise 2 is used in a matrix code, decode the following or show that an error was made; that is, this is not proper code.

(a) (1, 2, 3)   (b) (4, 7, 8)   (c) (1, 0, 0)

(d) 26, 47, 44, 50, 80, 100   (e) 59, 100, 105, 51, 84, 117.

7. Repeat Exercise 5 for **M** of Exercise 3.

8. Repeat Exercise 6 for **M** of Exercise 4.

9. For each of the following permutations of $\{1, 2, \ldots, k\}$, find the indicated values of E in the corresponding permutation code.

(a) $k = 3$, $\pi(1) = 1$, $\pi(2) = 3$, $\pi(3) = 2$.

   (i) $E(110)$ (ii) $E(001)$ (iii) $E(101)$

(b) $k = 4$, $\pi(1) = 2$, $\pi(2) = 3$, $\pi(3) = 4$, $\pi(4) = 1$.

   (i) $E(1001)$ (ii) $E(0011)$ (iii) $E(1000)$

(c) $k = 5$, $\pi(1) = 5$, $\pi(2) = 4$, $\pi(3) = 3$, $\pi(4) = 2$, $\pi(5) = 1$.

   (i) $E(11110)$ (ii) $E(01010)$ (iii) $E(00100)$.

10. Find C if $\mathscr{A}$ consists of the blocks 010, 111, 001, and 110 and $E(abc)$ is $a + b, a + c, b + c$, where addition is modulo 2.

11. Make up an example of a $1 \to 2$ block code.

12. Show that every permutation code is a matrix encoding.

10.3 ERROR-CORRECTING CODES

10.3.1 Error Correction and Hamming Distance

In this section we study the use of codes to detect and correct errors in transmission. In particular, we study step (b) of Figure 10.1. We assume that we start with an encoded message, which has been encoded using a block code, with blocks in the code alphabet having length n. For concreteness, we assume that the encoded message is a bit string, so all blocks are bit strings of length n. We speak of *binary codes* or *binary block codes* or *binary n-codes*. In Exercise 3 we modify the assumption and take encoded messages which are strings from an alphabet $\{0, 1, \ldots, q - 1\}$. We then speak of *q-ary codes*. The message is sent by blocks. Since we are assuming there are no errors in encoding, the only blocks that are ever sent are codewords. Let us suppose that the only possible errors which can take place in transmission are interchanges of the digits 0 and 1. Other errors, for example deletion or addition of a digit, will be disregarded. They are easily detected since all blocks have the same length. Following common practice, we shall assume that the probability of changing 1 to 0 is the same as the probability of changing 0 to 1, and that the probability of an error is the same at each digit, independent of any previous errors which may have occurred. In this case, we speak of a *binary symmetric channel*. For the implications of these assumptions, see Section 10.3.3.

Recall that we will be sending only codewords. If we ever receive a block that is not a codeword, we have *detected* an error in transmission. For instance, if the codeword 10010 is sent and the block 10110 is received, and if 10110 is not a codeword, we know there was an error. There are situations where we wish to *correct* the error, that is, guess what codeword was sent. This is especially the case where we cannot ask for re-transmission, for instance in transmission of a photograph from a space probe (such as occurred on the Mariner 9 Mars probe, whose code we shall discuss below), or if the transmission is based on an old magnetic tape. Error-detecting and error-correcting codes have many applications. For instance, simple error-detecting codes called parity check codes (see Example 10.5) were used on the earliest computers: UNIVAC, Whirlwind I, and IBM 650 (Wakerly [1978]). They are now used extensively in designing fault-tolerant computers. (For references on the applications of error-detecting and error-correcting codes to computing, see MacWilliams and Sloane [1977], Sellers *et al.* [1968], Wakerly [1978], and the survey articles by Carter and Bouricius [1971] and Avizenius [1976].)

There are good ways of designing error-correcting codes. Suppose we choose the code so that in the set of codewords, no two codewords are too close or too similar. For example, suppose that the only codewords are

$$000000, \quad 010101, \quad 101010, \quad \text{and} \quad 111111. \tag{10.3}$$

Then we would be very unlikely to confuse two of these, and a message that is received which is different from one of the codewords could readily be interpreted as the codeword to which it is closest.

Let us make this notion of closest more precise. The *Hamming distance* between two bit strings of the same length is the number of digits on which they differ.* Hence if $d(\cdot, \cdot)$ denotes this distance, we have

$$d(000000, 010101) = 3.$$

Our aim is to find a set of codewords with no two words too close in terms of the Hamming distance.

Suppose that we have found a set of codewords (all of the same length, according to our running assumption). Suppose that the smallest distance between two codewords is d. Then we can *detect* all errors of $d - 1$ or fewer digits. For if $d - 1$ or fewer digits are

---

*The Hamming distance was named after R. W. Hamming, who wrote the pioneering paper (Hamming [1950]) on error-detecting and error-correcting codes.

interchanged, the resulting bit string will not be a codeword and we can recognize or detect that there was an error. For example, if the possible codewords are those of (10.3), then d is 3 and we can detect errors in up to two digits. Suppose we use the strategy that if we receive a word which is not a codeword, we interpret it as the codeword to which it is closest in terms of Hamming distance. (In case of a tie, choose arbitrarily.) This is called the *nearest-neighbor rule*. For example, if the codewords are those of (10.3), and we receive the word 010000, we would interpret it as 000000, since

$$d(010000, 000000) = 1$$

while for every other codeword α,

$$d(010000, \alpha) > 1.$$

Using the nearest-neighbor rule, we can *correct* all errors that involve fewer than $d/2$ digits. For if fewer than $d/2$ digits are interchanged, the resulting bit string is closest to the correct codeword (the one transmitted), and so is interpreted as that codeword by the nearest neighbor rule. Hence, we have an *error-correcting code*. A code that can correct up to t errors is called a *t-error-correcting code*. We summarize our results as follows.

Theorem 10.1. Suppose that d is the minimum (Hamming) distance between two codewords in the binary code C. Then, the code C can detect up to $d - 1$ errors and, using the nearest-neighbor rule, can correct up to $\lceil (d/2) - 1 \rceil$ errors.

Our next theorem says that no error-correcting rule can do better than the nearest-neighbor rule.

Theorem 10.2. Suppose that d is the minimum Hamming distance between two codewords in the binary code C. Then no error-detecting rule can detect more than $d - 1$ errors and no error-correcting rule can correct more than $t = \lceil (d/2) - 1 \rceil$ errors.

Proof. The error-detecting conclusion is obvious. Next, note that an error-correcting rule assigns a codeword $R(\lambda)$ to each word λ (each bit string λ of length n). Now there are two codewords α and β with $d(\alpha, \beta) = d$. Let γ be a word different from α and β with $d(\alpha, \gamma) \le t + 1$ and $d(\beta, \gamma) \le t + 1$. Then $R(\gamma)$ must be some codeword. This cannot be both α and β. Say that it is not α without loss of generality. Then α could be sent, and with at most $t + 1$ errors, γ could be received. This would be "corrected" as $R(\gamma)$, which is not α. $\hspace{2cm}$ Q.E.D.

Since the nearest-neighbor rule (and any other reasonable rule) does not allow us to correct as many errors as we could detect, we sometimes do not attempt to correct errors but only to detect them and ask for a retransmission if errors are detected. This is the case in such critical situations as sending a command to fire a missile or to change course in a space shot.

Suppose that we know how likely errors are to occur. If they are very likely to occur, we would like to have a code with minimum Hamming distance d fairly large. If errors are unlikely, this minimum distance d could be smaller, perhaps as small as 3 if no more than 1 error is likely to occur. In general, we would like to be able to construct sets of codewords of given length n with minimum Hamming distance a given number d. Such a set of codewords will be called an (n, d)-*code*.

There always is an (n, d)-code if $d \le n$. Let the set of codewords be

$$000 \cdots 0 \qquad \text{and} \qquad \underbrace{111 \cdots 1}_{} 00 \cdots 0.$$

$$d \text{ places}$$

The trouble with this code is that there are very few codewords. We could only encode sets $\mathcal{A}$ of two different message blocks. We want ways of constructing richer codes, with more possible codewords.

In Section 10.4 we shall see how to use clever encoding methods $E \colon \mathcal{A} \to B^n$ to find richer (n, d) codes. In Section 10.5 we shall see how to use block designs, and in particular their incidence matrices, to define very rich (n, d)-codes. In the next subsection, we obtain an upper bound on the size of an (n, d)-code.

10.3.2 The Hamming Bound

Fix a given bit string s in an (n, d)-code C. Let the collection of bit strings of length n which have distance exactly t from string s be denoted $B_t(s)$. Then the number of bit strings in $B_t(s)$ is given by $\binom{n}{t}$, for we choose t positions out of n in which to change digits. Let the collection of bit strings of length n which have distance at most t from s be denoted by $B'_t(s)$. Thus, the number of bit strings in $B'_t(s)$ is given by

$$\alpha = \binom{n}{0} + \binom{n}{1} + \cdots + \binom{n}{t}. \tag{10.4}$$

If $t = \lceil (d/2) - 1 \rceil$, then $B'_t(s) \ne B'_t(s^*)$ for every $s \ne s^*$ in C. Thus every bit string of length n is in at most one set $B'_t(s)$. Since there are 2^n bit strings of length n, we have

$$|C|\alpha = \sum_{s \in C} |B'_t(s)| = \left| \bigcup_{s \in C} B'_t(s) \right| \le 2^n.$$

Thus, we get the following theorem.

Theorem 10.3 (Hamming Bound).* If C is an (n, d)-code and $t = \lceil (d/2) - 1 \rceil$, then

$$|C| \le \frac{2^n}{\binom{n}{0} + \binom{n}{1} + \cdots + \binom{n}{t}}. \tag{10.5}$$

To illustrate this theorem, let us take $n = d = 3$. Then $t = 1$. We find that any (n, d)-code has at most

$$\frac{2^3}{\binom{3}{0} + \binom{3}{1}} = 2$$

*This result is due to Hamming [1950]. It is sometimes called the *sphere packing bound*.

codewords. There is a (3, 3)-code of two codewords: Use the bit strings 111 and 000. We investigate an alternative upper bound on the size of (n, d)-codes in Section 10.5.3.

10.3.3 The Probability of Error

Recall that we are assuming that we have a binary symmetric channel: The probability of switching 1 to 0 is the same as the probability of switching 0 to 1, and this common probability p is the same at each digit, independent of any previous errors that may have occurred. Then we have the following theorem.

Theorem 10.4. In a binary symmetric channel, the probability that exactly r errors will be made in transmitting a bit string of length n is given by

$$\binom{n}{r} p^r (1 - p)^{n-r}.$$

*Proof.** The student familiar with probability will recognize that we are in a situation of Bernoulli trials as described in Example 4.30. We seek the probability of r successes in n independent, repeated trials, if the probability of success on any one trial is p. The formula given in the theorem is the well-known formula for this probability (see Example 4.30.)   Q.E.D.

For instance, suppose that $p = .01$. Then in a 4-digit message, the probability of no errors is

$$(1 - .01)^4 = .960596,$$

the probability of one error is

$$\binom{4}{1} (.01)(1 - .01)^3 = .038812,$$

the probability of two errors is

$$\binom{4}{2} (.01)^2 (1 - .01)^2 = .000588,$$

and the probability of more than two errors is

$$1 - .960596 - .038812 - .000588 = .000004.$$

EXERCISES FOR SECTION 10.3

1. In each of the following cases, find the Hamming distance between the two codewords x and y.
   (a) $x = 1110010$, $y = 0101010$
   (b) $x = 10001000$, $y = 10100101$
   (c) $x = 111111000$, $y = 001001001$
   (d) $x = 111000111000$, $y = 101010101010$.

*The proof should be omitted by the student who is not familiar with probability.

2. For each of the following codes C, find the number of errors that could be detected and the number that could be corrected using the nearest-neighbor rule.
    (a) $C = \{0000000, 1111110, 1010100, 0101010\}$
    (b) $C = \{0000000, 1111111, 1111000, 0000111\}$
    (c) $C = \{000000000, 111111111, 111100000, 000011111, 101010101, 010101010\}$.

3. A *q-ary block n-code* (*q-ary code*) is a collection of strings of length n chosen from the alphabet $\{0, 1, \ldots, q - 1\}$. The *Hamming distance* between two strings from this alphabet is again defined to be the number of digits on which they differ. For instance, if $q = 4$, then $d(0123, 1111) = 3$.
    (a) Find the Hamming distance between the following strings x and y.
        (i) $x = 0113456$, $y = 1013546$
        (ii) $x = 000111222333$, $y = 001112223334$
        (iii) $x = 1010101010$, $y = 1020304050$.
    (b) If the minimum distance between two codewords in a q-ary block n-code is d, how many errors can the code detect?
    (c) How many errors can the code correct under the nearest-neighbor rule?

4. Find an upper bound on the number of codewords in a code where each codeword has length 8 and the minimum distance between any two codewords is 5.

5. Repeat Exercise 4 for length 9 and minimum distance 4.

6. Find an upper bound on the number of codewords in a code C whose codewords have length 7 and which corrects up to 2 errors.

7. If codewords are bit strings of length 100, we have a binary symmetric channel, and the probability of error is .1, find the probability that if a codeword is sent there will be
    (a) no errors;   (b) exactly one error;
    (c) exactly two errors;   (d) more than two errors.

8. Repeat Exercise 7 if the length of codewords is 5 and the probability of error is .001.

9. Suppose we assume that under transmission of a q-ary codeword (Exercise 3), the probability p of error is the same at each digit, independent of any previous errors. Moreover, if there is an error at the ith digit, the digit is equally likely to be changed into any one of the remaining $q - 1$ symbols. We then refer to a *q-ary symmetric channel*.
    (a) In a q-ary symmetric channel, what is the probability of exactly t errors if a string of length n is sent?
    (b) If $q = 3$, what is the probability of receiving 22222 if 11111 is sent?

10. Suppose that in a binary symmetric channel, codewords have length $n = 3$ and that the probability of error is .1. If we want a code which with probability $\lambda \geq .95$, will correct a received message if it is in error, what must be the minimum distance between two codewords?

11. Repeat Exercise 10 if $n = 5$ and $\lambda \geq .90$.

12. Exercises 12-17 deal with q-ary codes (Exercise 3). Find a bound on the number of codewords in a q-ary code which is analogous to Theorem 10.3.

13. Suppose that a code is built up of strings of symbols using 0's, 1's, and 2's, and that S is a set of codewords with the following properties:

    (i) Each word in S has length n.
    (ii) Each word in S has a 2 in the first and last places and nowhere else.
    (iii) $d(a, b) = d$ for each pair of words a and b in S, where $d(a, b)$ is the Hamming distance.

    Let T be defined from S by changing 0 to 1 and 1 to 0, but leaving 2 unchanged. For example, the word 2102 becomes the word 2012.
    (a) What is the distance between two words of T?
    (b) What is the distance between a word of S and a word of T?

(c) Suppose that R consists of the words of S plus the words of T. How many errors can R detect? Correct?

14. Let A be a $q \times q$ Latin square on the set of symbols $\{0, 1, 2, \ldots, q - 1\}$. Consider the q-ary code C consisting of all strings of length 3 of the form i, j, a_{ij}.
   (a) If 5 is replaced by 0, find C corresponding to the Latin square of Table 1.4.
   (b) Given the code C for an arbitrary $q \times q$ Latin square, how many errors can the code detect?
   (c) How many errors can it correct?

15. Suppose that we have an orthogonal family of Latin squares of order q, $A^{(1)}, A^{(2)}, \ldots, A^{(r)}$, each using the symbols $0, 1, \ldots, q - 1$. Consider the q-ary code S consisting of all strings of length $r + 2$ of the form

$$i, j, a_{ij}^{(1)}, a_{ij}^{(2)}, \ldots, a_{ij}^{(r)}.$$

   (a) Find S corresponding to the orthogonal family of Table 9.22.
   (b) For arbitrary q, how many errors can the code S detect?
   (c) How many errors can it correct?

16. Consider codewords of length 4, with each digit chosen from the alphabet $\{0, 1, 2, 3, 4, 5\}$. Show that there cannot be a set of 36 codewords, each one of which has Hamming distance at least 3 from each other one.

17. Let A be a Latin square of order q on the set of symbols $\{0, 1, \ldots, q - 1\}$ which is horizontally complete in the sense of Section 9.3, Exercise 5. Consider the q-ary code T consisting of all strings of length 4 of the form

$$i, j, a_{ij}, a_{i, j+1},$$

for $i = 0, 1, \ldots, q - 1$, $j = 0, 1, \ldots, q - 2$.
   (a) Find T corresponding to the Latin square constructed in Exercise 5, Section 9.3, in the case $n = 4$.
   (b) For arbitrary q, how many errors can the code T detect?
   (c) How many errors can it correct?

10.4 LINEAR CODES*

10.4.1 Generator Matrices

In this section we shall see how to use the encoding step to build error-detecting and error-correcting codes. Moreover, the encoding we present will have associated with it very efficient encoding and decoding procedures. The approach we describe was greatly influenced by the work of R. W. Hamming and D. E. Slepian in the 1950's. See, for example, the papers of Hamming [1950] and Slepian [1956a, b, 1960].

Example 10.4   A Repetition Code

Perhaps the simplest way to check for errors is to repeat the message. For instance, suppose that $k = 4$. A code that repeats a block three times, a $k \rightarrow 3k$ code, takes $E(a) = aaa$. For example,

$$E(1100) = 110011001100.$$

*This section can be omitted if the reader wants to go directly to Section 10.5.

Using such a code, we can easily detect errors and correct errors. For the set C of codewords consists of blocks of length $3k$ which have a block of length k repeated three times. To detect errors, we compare the ith digits in the three repetitions of the k-block. Thus, suppose that $k = 4$ and we receive 001101110110, which breaks into k-blocks as 0011/0111/0110. Then we know that there has been an error, since, for example, the second digits of the first two k-blocks differ. We can detect up to two transmission errors in this way. However, three errors might go undetected if they were all in the same digit of the original k-block. Of course, the minimum distance between two codewords aaa is $d = 3$, so this observation agrees with the result of Theorem 10.2. We could try to correct the error as follows. Note that the second digit of the k-blocks is 1 twice, and 0 once. If we assume that errors are unlikely, we could use the majority rule and assume that the correct digit was 1. Using similar reasoning, we would decode 001101110110 as 0111. This error-correcting procedure can correct up to one error. However, two errors would be "corrected" incorrectly if they were both on the same digit. This observation agrees with the results of Theorem 10.2, as we have a code with $d = 3$. If we want to correct more errors, we simply use more repetitions. Thus, for example, a five-repetitions code $E(a) = aaaaa$ has $d = 5$ and can detect up to four errors and correct up to two. Unfortunately, the repetition method of designing error-correcting codes is expensive in terms of both time and space.

Example 10.5 Parity Check Codes

A simple way to detect a single error is to add one digit to a block, a digit that will always bring the number of 1's to an even number. The extra digit added is called a *parity check digit*. The corresponding code is a $k \to (k + 1)$ code, and it lets $E(0011) = 00110$, $E(0010) = 00101$, and so on. We can represent E by

$$E(a_1 a_2 \cdots a_k) = a_1 a_2 \cdots a_k \sum_{i=1}^{k} a_i,$$

where $\sum_{i=1}^{k} a_i$ is interpreted as the summation modulo 2, as defined in Section 9.3.1. Throughout this section and the next, addition will be interpreted this way. Thus, $1 + 1 = 0$, which is why $E(0011) = 00110$. In such a *parity check code* E, the minimum distance between two codewords is $d = 2$, so one error can be detected. No errors can be corrected. As we have pointed out before, such parity check codes were used on the earliest computers: UNIVAC, Whirlwind I, and IBM 650. Large and small computers of the 1960s had memory parity check as an optional feature, and by the 1970s this was standard on a number of machines.

It is useful to compare the two codes of Examples 10.4 and 10.5. Suppose that we wish to send a message of 1000 digits, with the probability of error equal to $p = .01$. If there is no encoding, the probability of making no errors is $(1 - .01)^{1000}$, which is approximately .000043. Suppose by way of contrast that we use the $k \to 3k$ triple repetition code of Example 10.4, with $k = 1$. Suppose that we want to send a digit a. We encode it as aaa. The probability of correctly sending the block aaa is $(.99)^3$, or approximately .970299. By Theorem 10.4, the probability of a single error is $\binom{3}{1}(.01)(.99)^2$, or approximately .029403. Thus, since we can correct single errors, the probability of correctly interpreting the message aaa and hence of correctly decoding the single digit a is

.970299 + .029403 = .999702. There are 1000 digits in the original message, so the probability of decoding the whole message correctly is $(.999702)^{1000}$, or approximately .742268. This is much higher than .000043. Note that the greatly increased likelihood of error-free transmission is bought at a price: we need to send 3000 digits in order to receive 1000.

Let us compare the parity check code. Suppose we break the 1000 digit message into blocks of length 10; there are 100 such in all. Each such block is encoded as an 11 digit block by adding a parity check digit. The probability of sending the 11-digit block without making any errors is $(.99)^{11}$, or approximately .895338. Also, by Theorem 10.4, the probability of sending the block with exactly one error is $\binom{11}{1}(.01)(.99)^{10}$, or approximately .099482. Now if a single error is made, we can detect it and ask for a retransmission. Thus, it is reasonable to assume that we can eliminate single errors. Hence, the probability of receiving the 11-digit block correctly is

$$.895338 + .099482 = .994820.$$

Now the original 1000-digit message has 100 blocks of 10 digits, so the probability of eventually decoding this entire message correctly is $(.994820)^{100}$, or approximately .594909. The probability of correct transmission is less than with a triple repetition code, but much greater than without a code. The price is much less than the triple repetition code: We need to send 1100 digits to receive 1000.

To generalize Example 10.5, we can consider a $k \to n$ code E as adding $n - k$ *parity check digits* to a block of length k. In general, a *message block* $a_1 a_2 \cdots a_k$ is encoded as $x_1 x_2 \cdots x_n$, where $x_1 = a_1$, $x_2 = a_2$, ..., $x_k = a_k$ and the parity check digits are determined from the k *message digits* a_1, a_2, ..., a_k. The easiest way to obtain such an encoding is to generalize the matrix encoding method of Example 10.2. We let $\mathbf{M}$ be a $k \times n$ matrix, called the *generator matrix*, and we define $E(\mathbf{a})$ to be $\mathbf{aM}$.

Example 10.6

Suppose that $k = 3$, $n = 6$, and

$$\mathbf{M} = \begin{bmatrix} 1 & 0 & 0 & 1 & 1 & 0 \\ 0 & 1 & 0 & 0 & 1 & 1 \\ 0 & 0 & 1 & 1 & 0 & 1 \end{bmatrix}.$$

Then the message word 110 is encoded as $110\mathbf{M} = 110101$. The fifth digit is 0 because $1 + 1 = 0$: we are using addition modulo 2. Note that $\mathbf{M}$ begins with a 3×3 identity matrix. Since we want $a_i = x_i$ for $i = 1, 2, ..., k$, every generator matrix will begin with the $k \times k$ identity matrix $\mathbf{I}_k$.

To give some other examples, the parity check codes of Example 10.5 are defined by taking

$$\mathbf{M} = \begin{bmatrix} 1 & 0 & 0 & \cdots & 0 & 1 \\ 0 & 1 & 0 & \cdots & 0 & 1 \\ 0 & 0 & 1 & \cdots & 0 & 1 \\ \cdot & \cdot & \cdot & \cdots & \cdot & \cdot \\ \cdot & \cdot & \cdot & \cdots & \cdot & \cdot \\ \cdot & \cdot & \cdot & \cdots & \cdot & \cdot \\ 0 & 0 & 0 & \cdots & 1 & 1 \end{bmatrix} \tag{10.6}$$

We add a column of all 1's to the $k \times k$ identity matrix $\mathbf{I}_k$. The triple repetition code of Example 10.4 is obtained by taking $\mathbf{M}$ to be three copies of the $k \times k$ identity matrix.

If a code is defined by a generator matrix, decoding is trivial (when the transmission is correct). We simply decode $x_1 x_2 \cdots x_n$ as $x_1 x_2 \cdots x_k$, that is, we drop the $n - k$ parity check digits.

In general, codes definable by such generator matrices are called *linear codes*. This is because if $\mathbf{x} = x_1 x_2 \cdots x_n$ and $\mathbf{y} = y_1 y_2 \cdots y_n$ are codewords, then so is the word $\mathbf{x} + \mathbf{y}$ whose ith digit is $x_i + y_i$, where addition is modulo 2. To see this, simply observe that if $\mathbf{aM} = x$ and $\mathbf{bM} = y$, then $(\mathbf{a} + \mathbf{b})\mathbf{M} = \mathbf{x} + \mathbf{y}$.*

It follows that $\mathbf{0} = 00 \cdots 0$ is a codeword of every (nonempty) linear code. For if $\mathbf{x}$ is any codeword, then since addition is modulo 2, $\mathbf{x} + \mathbf{x} = \mathbf{0}$. Indeed, it turns out that the set of all codewords under the operation $+$ defines a group, to use the language of Section 7.2 (see Exercise 29). Thus, linear codes are sometimes called *binary group codes*.

We next note that in a linear code, the minimum distance between two codewords is easy to find. Let the *Hamming weight* of a bit string $\mathbf{x}$, denoted wt $(\mathbf{x})$, be the number of nonzero digits of $\mathbf{x}$.

Theorem 10.5. In a linear code C, the minimum distance d between two codewords is equal to the minimum Hamming weight w of a codeword other than the codeword $\mathbf{0}$.

Proof. Note that if $d(\mathbf{x}, \mathbf{y})$ is Hamming distance, then since C is linear, $\mathbf{x} + \mathbf{y}$ is in C, so

$$d(\mathbf{x}, \mathbf{y}) = \text{wt } (\mathbf{x} + \mathbf{y}).$$

This conclusion uses the fact that addition is modulo 2. Suppose that the minimum distance d is given by $d(\mathbf{x}, \mathbf{y})$ for a particular $\mathbf{x}, \mathbf{y}$ in C. Then

$$d = d(\mathbf{x}, \mathbf{y}) = \text{wt } (\mathbf{x} + \mathbf{y}) \geq w.$$

Next, suppose that $\mathbf{u} \neq \mathbf{0}$ in C has minimum weight w. Then since $\mathbf{0} = \mathbf{u} + \mathbf{u}$ is in C, we have

$$w = \text{wt } (\mathbf{u}) = \text{wt } (\mathbf{u} + \mathbf{0}) = d(\mathbf{u}, \mathbf{0}) \geq d. \qquad \text{Q.E.D.}$$

It follows by Theorem 10.5 that in the code of Example 10.5, the minimum distance between two codewords is 2, since the minimum weight of a codeword other than $\mathbf{0}$ is 2. This weight is attained, for example, in the string 11000. The minimum distance is attained, for example, as

$$d(11000, 00000).$$

10.4.2 Error Correction Using Linear Codes

If $\mathbf{M}$ is a generator matrix, it can be represented schematically as $[\mathbf{I}_k \mathbf{G}]$, where $\mathbf{G}$ is a $k \times (n - k)$ matrix. Let $\mathbf{G}^T$ be the transpose of matrix $\mathbf{G}$ and let $\mathbf{H} = [\mathbf{G}^T \mathbf{I}_{n-k}]$ be an $(n - k) \times n$ matrix, called the *parity check matrix*. The matrix $\mathbf{G}^T$ in Example 10.5 is a row

---

*We use the assumption, which we shall make throughout this section, that $\mathcal{A} = A^k$.

vector of k 1's, and the matrix $\mathbf{H}$ is a row vector of $k + 1$ 1's. In Example 10.4, with three repetitions and message blocks of length $k = 4$, we have

$$\mathbf{G}^T = \begin{bmatrix} 1 & 0 & 0 & 0 \\ 0 & 1 & 0 & 0 \\ 0 & 0 & 1 & 0 \\ 0 & 0 & 0 & 1 \\ 1 & 0 & 0 & 0 \\ 0 & 1 & 0 & 0 \\ 0 & 0 & 1 & 0 \\ 0 & 0 & 0 & 1 \end{bmatrix}$$

and

$$\mathbf{H} = \begin{bmatrix} 1 & 0 & 0 & 0 & 1 & 0 & 0 & 0 & 0 & 0 & 0 & 0 \\ 0 & 1 & 0 & 0 & 0 & 1 & 0 & 0 & 0 & 0 & 0 & 0 \\ 0 & 0 & 1 & 0 & 0 & 0 & 1 & 0 & 0 & 0 & 0 & 0 \\ 0 & 0 & 0 & 1 & 0 & 0 & 0 & 1 & 0 & 0 & 0 & 0 \\ 1 & 0 & 0 & 0 & 0 & 0 & 0 & 0 & 1 & 0 & 0 & 0 \\ 0 & 1 & 0 & 0 & 0 & 0 & 0 & 0 & 0 & 1 & 0 & 0 \\ 0 & 0 & 1 & 0 & 0 & 0 & 0 & 0 & 0 & 0 & 1 & 0 \\ 0 & 0 & 0 & 1 & 0 & 0 & 0 & 0 & 0 & 0 & 0 & 1 \end{bmatrix}. \qquad (10.7)$$

In Example 10.6,

$$\mathbf{H} = \begin{bmatrix} 1 & 0 & 1 & 1 & 0 & 0 \\ 1 & 1 & 0 & 0 & 1 & 0 \\ 0 & 1 & 1 & 0 & 0 & 1 \end{bmatrix}. \qquad (10.8)$$

Note that a linear code can be defined by giving either the generator matrix $\mathbf{G}$ or the parity check matrix $\mathbf{H}$. For we can derive $\mathbf{H}$ from $\mathbf{G}$ and $\mathbf{G}$ from $\mathbf{H}$. We shall see that parity check matrices will provide a way to detect and correct errors. The basic ideas of this approach are due to Slepian [1956a,b, 1960].

First, we note that the parity check matrix can be used to identify codewords.

Theorem 10.6. In a linear code, a block $\mathbf{a} = a_1 a_2 \cdots a_k$ is encoded as $\mathbf{x} = x_1 x_2 \cdots x_n$ if and only if $a_i = x_i$ for $i \leq k$ and $\mathbf{H}\mathbf{x}^T = \mathbf{0}$.

*Proof.** By definition of the encoding, if block $\mathbf{a} = (a_1 a_2 \cdots a_k)$ is encoded as $\mathbf{x} = (x_1 x_2 \cdots x_n)$, then

$$(a_1 a_2 \cdots a_k)[\mathbf{I}_k \mathbf{G}] = (x_1 x_2 \cdots x_n).$$

Now clearly $a_i = x_i$, $i \leq k$. Also,

$$\mathbf{H}\mathbf{x}^T = \mathbf{H}(\mathbf{a}[\mathbf{I}_k\mathbf{G}])^T$$
$$= \mathbf{H}[\mathbf{I}_k\mathbf{G}]^T\mathbf{a}^T$$
$$= [\mathbf{G}^T\mathbf{I}_{n-k}]\begin{bmatrix} \mathbf{I}_k \\ \mathbf{G}^T \end{bmatrix}\mathbf{a}^T$$

*The proof can be omitted.

$$= (\mathbf{G}^T + \mathbf{G}^T)\mathbf{a}^T$$
$$= \mathbf{0}\mathbf{a}^T$$
$$= \mathbf{0},$$

where $\mathbf{G}^T + \mathbf{G}^T = \mathbf{0}$ since addition is modulo 2.

Conversely, suppose that $a_i = x_i$, $i \le k$, and $\mathbf{Hx}^T = \mathbf{0}$. Now suppose that $\mathbf{a}$ is encoded as $\mathbf{y} = y_1 y_2 \cdots y_n$. Then $a_i = y_i$, $i \le k$, so $x_i = y_i$, $i \le k$. Also, by the first part of the proof, $\mathbf{Hy}^T = \mathbf{0}$. It follows that $x = y$. To see why, note that the equations $\mathbf{Hx}^T = \mathbf{0}$ and $\mathbf{Hy}^T = \mathbf{0}$ each give rise to a system of equations. The jth equation in the first case involves at most the variables $x_1, x_2, \ldots, x_k$ and x_{k+j}. Thus, since $x_1, x_2, \ldots, x_k$ are given by $a_1, a_2, \ldots, a_k$, the jth equation defines x_{k+j} uniquely in terms of the k message digits $a_1, a_2, \ldots, a_k$. The same is true of y_{k+j}. Thus $x_{k+j} = y_{k+j}$.　　　Q.E.D.

Corollary 10.6.1.　A bit string $\mathbf{x} = x_1 x_2 \cdots x_n$ is a codeword if and only if $\mathbf{Hx}^T = \mathbf{0}$.

Corollary 10.6.2.　There is a unique bit string $\mathbf{x}$ so that $x_i = a_i$, $i \le k$, and $\mathbf{Hx}^T = \mathbf{0}$. This bit string $\mathbf{x}$ is the encoding of $a_1 a_2 \cdots a_k$. An expression for the entry x_{k+j} of $\mathbf{x}$ in terms of $a_1, a_2, \ldots, a_k$ is obtained by multiplying the jth row of $\mathbf{H}$ by $\mathbf{x}^T$.

Proof. This is a corollary of the proof.　　　Q.E.D.

To illustrate these results, note that in the parity check codes, $\mathbf{H} = (11 \cdots 1)$ with $n - k + 1$ 1's. Then $\mathbf{Hx}^T = \mathbf{0}$ if and only if

$$x_1 + x_2 + \cdots + x_{k+1} = 0. \tag{10.9}$$

But since addition is modulo 2, (10.9) is exactly equivalent to the condition that

$$x_{k+1} = x_1 + x_2 + \cdots + x_k,$$

which defines codewords. This equation is called the *parity check equation*. Similarly, in the triple repetition codes with $k = 4$, we see from (10.7) that $\mathbf{Hx}^T = \mathbf{0}$ if and only if

$$x_1 + x_5 = 0, \qquad x_2 + x_6 = 0, \qquad x_3 + x_7 = 0, \qquad x_4 + x_8 = 0,$$
$$x_1 + x_9 = 0, \qquad x_2 + x_{10} = 0, \qquad x_3 + x_{11} = 0, \qquad x_4 + x_{12} = 0. \tag{10.10}$$

Since addition is modulo 2, Equations (10.10) are equivalent to

$$x_1 = x_5, \qquad x_2 = x_6, \qquad x_3 = x_7, \qquad x_4 = x_8,$$
$$x_1 = x_9, \qquad x_2 = x_{10}, \qquad x_3 = x_{11}, \qquad x_4 = x_{12},$$

which define a codeword. These equations are called the *parity check equations*. Finally, in Example 10.6, by (10.8), $\mathbf{Hx}^T = \mathbf{0}$ if and only if

$$x_1 + x_3 + x_4 = 0,$$
$$x_1 + x_2 + x_5 = 0,$$
$$x_2 + x_3 + x_6 = 0.$$

Using the fact that addition is modulo 2, these equations are equivalent to the following parity check equations, which determine the parity check digits x_{k+j} in terms of the message digits $x_i, i \leq k$:

$$x_4 = x_1 + x_3,$$

$$x_5 = x_1 + x_2,$$

$$x_6 = x_2 + x_3.$$

Thus, as we have seen before, 110 is encoded as 110101, since $x_4 = 1 + 0 = 1$, $x_5 = 1 + 1 = 0$, and $x_6 = 1 + 0 = 1$.

In general, by Corollary 10.6.2, the equations $\mathbf{Hx}^T = \mathbf{0}$ define x_{k+j} in terms of a_1, $a_2, \ldots, a_k$. The equations giving x_{k+j} in these terms are the *parity check equations*.

Theorem 10.6 and Corollary 10.6.1 provide a way to detect errors. We simply note that if $\mathbf{x}$ is received and if $\mathbf{Hx}^T \neq \mathbf{0}$, then an error occurred.

The parity check matrix allows us not only to detect errors, but to correct them, as the next theorem shows.

Theorem 10.7. Suppose that the columns of the parity check matrix $\mathbf{H}$ are all nonzero and all distinct. Suppose that a codeword $\mathbf{y}$ is transmitted and a word $\mathbf{x}$ is received. If $\mathbf{x}$ differs from $\mathbf{y}$ only on the ith digit, then $\mathbf{Hx}^T$ is the ith column of $\mathbf{H}$.

Proof. Note that by Corollary 10.6.1, $\mathbf{Hy}^T = \mathbf{0}$. Now $\mathbf{x}$ can be written as $\mathbf{y} + \mathbf{e}$, where $\mathbf{e}$ is an error string, a bit string with 1's in the digits that differ from $\mathbf{y}$. (Recall that addition is modulo 2.) We conclude that

$$\mathbf{Hx}^T = \mathbf{H}(\mathbf{y} + \mathbf{e})^T = \mathbf{H}(\mathbf{y}^T + \mathbf{e}^T) = \mathbf{Hy}^T + \mathbf{He}^T = \mathbf{He}^T.$$

Now if $\mathbf{e}$ is the vector with all 0's except a 1 in the ith digit, then $\mathbf{He}^T$ is the ith column of $\mathbf{H}$. Q.E.D.

To illustrate this result, suppose that we have a triple repetition code and $k = 4$. Then the parity check matrix $\mathbf{H}$ is given by (10.7). Suppose that a codeword $\mathbf{y} = 110111011101$ is sent and a codeword $\mathbf{x} = 110011011101$, which differs on the fourth digit, is received. Then note that

$$\mathbf{Hx}^T = \begin{bmatrix} 0 \\ 0 \\ 0 \\ 1 \\ 0 \\ 0 \\ 0 \\ 1 \end{bmatrix},$$

which is the fourth column of $\mathbf{H}$.

In general, error correction using the parity check matrix $\mathbf{H}$ will proceed as follows, assuming $\mathbf{H}$ has distinct columns and all are nonzero. Having received a block $\mathbf{x}$, compute $\mathbf{Hx}^T$. If this is $\mathbf{0}$, and errors in transmission are unlikely, it is reasonable to assume that $\mathbf{x}$ is correct. If it is not $\mathbf{0}$, but is the ith column of $\mathbf{H}$, and if errors are unlikely, it is reasonable to assume that only one error was made, so the correct word differs from $\mathbf{x}$ on the ith digit. If $\mathbf{Hx}^T$ is not $\mathbf{0}$ and not a column of $\mathbf{H}$, then at least two errors occurred in transmission and error correction cannot be carried out this way.

10.4.3 Hamming Codes

Recall that for error correction, we want the $(n - k) \times n$ parity check matrix $\mathbf{H}$ to have columns that are nonzero and distinct. Now if $p = n - k$, the columns of $\mathbf{H}$ are bit strings of length p. There are 2^p such strings, $2^p - 1$ nonzero ones. The *Hamming code* $\mathscr{H}_p$ arises when we take $\mathbf{H}$ to be a matrix whose columns are all of these $2^p - 1$ nonzero bit strings of length p, arranged in any order. Technically, to conform with our definition, the last $n - k$ columns should form the identity matrix. The resulting code is a $k \rightarrow n$ code, where $n = 2^p - 1$ and $k = n - p = 2^p - 1 - p$. For instance, if $p = 2$, then $n = 2^2 - 1 = 3$ and a typical $\mathbf{H}$ is given by

$$\mathbf{H} = \begin{bmatrix} 1 & 1 & 0 \\ 1 & 0 & 1 \end{bmatrix}.$$

It is easy to see that $\mathbf{H}$ defines a $1 \rightarrow 3$ triple repetition code. If $p = 3$, then $n = 2^3 - 1 = 7$, and a typical $\mathbf{H}$ is given by

$$\mathbf{H} = \begin{bmatrix} 0 & 1 & 1 & 1 & 1 & 0 & 0 \\ 1 & 0 & 1 & 1 & 0 & 1 & 0 \\ 1 & 1 & 0 & 1 & 0 & 0 & 1 \end{bmatrix}.$$

We get a $4 \rightarrow 7$ code. The Hamming codes were introduced by Hamming [1950] and Golay [1949]. (Note that if we change the order of columns in H, we get (essentially) the same code or set of codewords. For any two such codes are seen to be equivalent (in the sense to be defined in Exercise 28) by changing the order of the parity check digits. That is why we speak of *the* Hamming code $\mathscr{H}_p$.)

Theorem 10.8. In the Hamming codes $\mathscr{H}_p$, $p \geq 2$, the minimum distance d between two codewords is 3.

Proof. Since the columns of the parity check matrix $\mathbf{H}$ are nonzero and distinct, single errors can be corrected (see the discussion following Theorem 10.7). Thus, by Theorem 10.2, $d \geq 3$. Now it is easy to show that for $p \geq 2$, there are always three nonzero bit strings of length p whose sum (under addition modulo 2) is zero (Exercise 26). If these occur as columns u, v, and w of the matrix $\mathbf{H}$, we take $\mathbf{x}$ to be the vector that is 1 in positions u, v, and w, and 0 otherwise. Then $\mathbf{Hx}^T$ is the sum of the uth, vth, and wth columns of $\mathbf{H}$, and so is $\mathbf{0}$. Thus, $\mathbf{x}$ is a codeword of weight 3. By Theorem 10.5, $d \leq 3$.
Q.E.D.

It follows from Theorem 10.8 that the Hamming codes *always* detect up to two errors and correct up to one. In Exercise 32 we shall ask the reader to show that the Hamming codes can *never* correct more than one error. We shall call them *perfect* single-error-correcting codes.

1. In the parity check code, find $E(\mathbf{a})$ if $\mathbf{a}$ is
   (a) 1111   (b) 100111   (c) 0011001   (d) 0101011.
2. Suppose that

$$\mathbf{M} = \begin{bmatrix} 1 & 0 & 1 & 1 \\ 0 & 1 & 1 & 0 \end{bmatrix}.$$

   Using $\mathbf{M}$, find the codeword $x_1 x_2 \cdots x_n$ corresponding to each of the following message words $a_1 a_2 \cdots a_k$.
   (a) 11   (b) 10   (c) 01   (d) 00.
3. Suppose that

$$\mathbf{M} = \begin{bmatrix} 1 & 0 & 0 & 1 & 0 & 0 \\ 0 & 1 & 0 & 0 & 1 & 0 \\ 0 & 0 & 1 & 0 & 0 & 1 \end{bmatrix}.$$

   Repeat Exercise 2 for the following message words.
   (a) 111   (b) 101   (c) 000.
4. Suppose that

$$\mathbf{M} = \begin{bmatrix} 1 & 0 & 0 & 0 & 0 & 1 & 1 \\ 0 & 1 & 0 & 0 & 1 & 0 & 1 \\ 0 & 0 & 1 & 0 & 1 & 1 & 0 \\ 0 & 0 & 0 & 1 & 1 & 1 & 1 \end{bmatrix}.$$

   Repeat Exercise 2 for the following message words.
   (a) 1111   (b) 1000   (c) 0001.
5. Find the linear code C generated by the matrices $\mathbf{M}$ of
   (a) Exercise 2   (b) Exercise 3   (c) Exercise 4.
6. Find a generator matrix for the code that translates 0 into 000000 and 1 into 111111.
7. Suppose that we have a message of 10 digits and that $p = .1$ is the probability of an error.
   (a) What is the probability of no errors in transmission if we use no code?
   (b) What is the probability of correctly transmitting the message if we use the $k \to 3k$ triple repetition code with $k = 1$?
   (c) What is this probability if we use a parity check code with blocks of size 2?
8. Suppose that we have a message of 1000 digits and that $p = .01$ is the probability of error. Suppose that we use the $k \to 5k$ five repetitions code with $k = 1$. What is the probability of correctly transmitting the message? (*Hint:* How many errors can be corrected?)
9. Each of the following codes is linear. For each, find the minimum distance between two codewords.
   (a) 000000, 001001, 010010, 100100, 011011, 101101, 110110, 111111;
   (b) 0000, 0011, 0101, 1001, 0110, 1010, 1100, 1111;
   (c) 00000, 00011, 00101, 01001, 10001, 00110, 01010, 01100, 10010, 10100, 11000, 01111, 11011, 10111, 11101, 11110;
   (d) 11111111, 10101010, 11001100, 10011001, 11110000, 10100101, 11000011, 10010110, 00000000, 01010101, 00110011, 01100110, 00001111, 01011010, 00111100, 01101001.
10. Find the parity check matrix $\mathbf{H}$ corresponding to $\mathbf{M}$ of
    (a) Exercise 2   (b) Exercise 3   (c) Exercise 4.

11. Find the generator matrix **M** corresponding to the following parity check matrices:

(a) $\mathbf{H} = \begin{bmatrix} 1 & 1 & 0 & 1 & 0 & 0 \\ 1 & 0 & 1 & 0 & 1 & 0 \\ 0 & 1 & 1 & 0 & 0 & 1 \end{bmatrix}$

(b) $\mathbf{H} = \begin{bmatrix} 1 & 1 & 1 & 0 & 0 \\ 1 & 0 & 0 & 1 & 0 \\ 1 & 1 & 0 & 0 & 1 \end{bmatrix}$

(c) $\mathbf{H} = \begin{bmatrix} 1 & 0 & 1 & 0 & 0 & 0 \\ 1 & 1 & 0 & 1 & 0 & 0 \\ 0 & 1 & 0 & 0 & 1 & 0 \\ 0 & 0 & 0 & 0 & 0 & 1 \end{bmatrix}.$

12. Find the parity check equations in each of the following exercises:
   (a) 2; (b) 3; (c) 4; (d) 11(a); (e) 11(b); (f) 11(c).

13. Suppose that we use the $k \to 5k$ five repetitions code with $k = 2$. Find the parity check matrix **H** and derive the parity check equations.

14. For the matrix **H** of Exercise 11(a), suppose that a word **x** is received over a noisy channel. Assume that errors are unlikely. For each of the following **x**, determine if an error was made in transmission.
   (a) $\mathbf{x} = 111000$ (b) $\mathbf{x} = 111100$ (c) $\mathbf{x} = 000100.$

15. Repeat Exercise 14 for **H** of Exercise 11(b) and
   (a) $\mathbf{x} = 11101$ (b) $\mathbf{x} = 01101$ (c) $\mathbf{x} = 00011.$

16. Repeat Exercise 14 for **M** of Exercise 2 and
   (a) $\mathbf{x} = 1111$ (b) $\mathbf{x} = 1000$ (c) $\mathbf{x} = 0100.$

17. In each part of Exercise 14, if possible, correct an error if there was one.

18. In each part of Exercise 15, if possible, correct an error if there was one.

19. Consider the parity check matrix

$$\mathbf{H} = \begin{bmatrix} 1 & 0 & 0 & 1 & 0 & 0 \\ 1 & 1 & 0 & 0 & 1 & 0 \\ 1 & 1 & 1 & 0 & 0 & 1 \end{bmatrix}.$$

Find all the codewords in the corresponding code. Does the code correct all single errors?

20. In the Hamming code $\mathcal{H}_p$ with $p = 3$, encode the following words.
   (a) 1001   (b) 0001   (c) 1110.

21. For the Hamming code $\mathcal{H}_p$ with $p = 4$, find a parity check matrix and encode the word 10000000000.

22. Find all codewords in the Hamming code $\mathcal{H}_2$.

23. Find all codewords in the Hamming code $\mathcal{H}_3$.

24. Find a word of weight 3 in the Hamming code $\mathcal{H}_p$, for every $p \geq 2$.

25. Consider a codeword $x_1 x_2 \cdots x_n$ in the Hamming code $\mathcal{H}_p$, $p \geq 2$. Let $k = n - p$.
   (a) Show that if $x_i = 0$, all $i \leq k$, then $x_i = 0$, all i.
   (b) Show that it is impossible to have exactly one x_i equal to 1 and all other x_i equal to 0.

26. Show that if $p \geq 2$, there are always three bit strings of length p whose sum (under addition modulo 2) is zero.

27. Let C be a linear code with codewords of length n in which some codewords have odd weight. Form a new code C' by adding 0 at the end of each word of C of even weight and 1 at the end of each word of C of odd weight.

(a) If **M** is the parity check matrix for C, show that the parity check matrix for C' is given by

$$\begin{bmatrix} 1 & 1 & \cdots & 1 \\ & & & 0 \\ & \mathbf{M} & & 0 \\ & & & \vdots \\ & & & 0 \end{bmatrix}$$

(b) Show that the distance between any two codewords of C' is even.

(c) Show that if the minimum distance d between two codewords of C was odd, then the minimum distance between two codewords of C' is $d + 1$.

(d) For all $p \geq 2$, find a linear code with codewords of length $n = 2^p$ and with minimum distance between two codewords equal to 4.

28. Two n-codes C and C' are *equivalent* if they differ only in the order of symbols, that is, if one can be obtained from the other by applying the same permutation to each element. For example, the following C and C' are equivalent:

C	C'
0000	0000
1111	1111
0011	0101
1100	1010

Show that if **M** is a generator matrix giving rise to a code C, and **M'** is obtained from **M** by interchanging rows in the matrix **G** corresponding to **M**, and **M'** gives rise to a code C', then C and C' are equivalent.

29. Show that under a linear code, the set of codewords under the operation $+$ defines a group.

30. An (n, d)-code C is called a *perfect t-error-correcting code* if $t = \lceil (d/2) - 1 \rceil$ and inequality (10.5) is an equality. Show that such a C *never* corrects more than t errors.

31. Show that the $1 \rightarrow 2t + 1$ repetition codes (Example 10.4) are perfect t-error-correcting codes.

32. Show that the Hamming codes $\mathcal{H}_p$, $p \geq 2$, are perfect 1-error correcting codes.

33. It turns out (Tietäväinen [1973], van Lint [1971]) that there is only one perfect t-error-correcting code other than the repetition codes and the Hamming codes. This is a $12 \rightarrow 23$ perfect 3-error correcting code due to Golay [1949]. How many codewords does this code have?

34. We can extend the theory of linear codes from binary codes to q-ary codes (Exercises 3, 9, and 12–17, Section 10.3), where q is a power of a prime. We continue to define a codeword from a message word by adding parity check digits, and specifically use a generator matrix **M**, but with addition in the finite field $GF(q)$, rather than modulo 2. Note that most of our results hold in this general setting, with the major exceptions being noted in Exercises 35 and 36. Suppose that $q = 5$.

(a) If **M** is as in Exercise 2, find the codeword corresponding to the message words
  (i) 14   **(ii)** 03   **(iii)** 13.

(b) If **M** is as in Exercise 3, find the codeword corresponding to the message words
  (i) 124   **(ii)** 102   **(iii)** 432.

35. Continuing with Exercise 34:

(a) Note that $d(\mathbf{x}, \mathbf{y})$ is not necessarily wt $(\mathbf{x} + \mathbf{y})$.

(b) What is the relation between distance and weight?

36. Continuing with Exercise 34, show that Theorem 10.6 may not be true.

37. Continuing with Exercise 34, show that Theorem 10.5 still holds.

38. Suppose that C is a q-ary block n-code of minimum distance d. Generalize the Hamming bound (Theorem 10.3) to find an upper bound on $|C|$.

10.5 THE USE OF BLOCK DESIGNS TO FIND ERROR-CORRECTING CODES*

10.5.1 Hadamard Codes

One way to find error-correcting codes is to concentrate first on finding a rich set C of codewords, and then perform an encoding of messages into C. This is the opposite of the approach we have taken so far, which first defined the encoding, and defined the set C to be the set of all the words arising from the encoding. Our goal will be to find a set C of codewords which corrects a given number of errors or equivalently has a given minimum distance d between codewords, and which has many codewords in it, thus allowing more possible message blocks to be encoded.

Recall that an (n, d)-*code* has codewords of length n and the minimum distance between two codewords is d. A useful way to build (n, d)-codes C is to use the incidence matrix of a (v, k, λ)-design (see Section 9.4.2). Each row of this incidence matrix has $r = k$ 1's and the rest 0's. The rows have length $b = v$. Any two rows have 1's in common in a column exactly λ times. The rows can define codewords for an (n, d)-code, with $n = v$. What is d? Let us consider two rows, the ith and jth. There are λ columns where there is a 1 in both rows. There are k 1's in each row, and hence there are $k - \lambda$ columns where row i has 1 and row j has 0, and there are $k - \lambda$ columns where row i has 0 and row j has 1. All other columns have 0's in both rows. It follows that the two rows differ in $2(k - \lambda)$ places. This is true for every pair of rows, so

$$d = 2(k - \lambda).$$

Theorem 9.15 says that, for arbitrarily large m, there are Hadamard designs of dimension m, that is, $(4m - 1, 2m - 1, m - 1)$-designs. It follows that we can find (v, k, λ)-designs with arbitrarily large $k - \lambda$, and hence (n, d)-codes for arbitrarily large d. For given a Hadamard design of dimension m, we have

$$k - \lambda = (2m - 1) - (m - 1) = m,$$

$$d = 2m.$$

Hence, if there are Hadamard designs of dimension m for arbitrarily large m, there are error-correcting codes which will detect up to $d - 1 = 2m - 1$ errors and correct up to $\lceil (d/2) - 1 \rceil = m - 1$ errors, for arbitrarily large m. These codes are $(4m - 1, 2m)$-codes, since each codeword has length $4m - 1$. We call them *Hadamard codes*. We shall first set out to prove the existence of Hadamard designs of arbitrarily large dimension m, that is, to prove Theorem 9.15. Then we shall ask how rich are the codes constructed from the

*This section depends on Section 9.4.

incidence matrices of Hadamard designs; that is, how do these codes compare to the richest possible (n, d)-codes?

10.5.2 Constructing Hadamard Designs*

The basic idea in constructing Hadamard designs is that certain kinds of matrices will give rise to the incidence matrices of these designs. An $n \times n$ matrix $\mathbf{H} = (h_{ij})$ is called a *Hadamard matrix of order n* if h_{ij} is $+1$ or -1 for every i and j and if

$$\mathbf{HH}^T = n\mathbf{I},$$

where $\mathbf{H}^T$ is the transpose of $\mathbf{H}$ and $\mathbf{I}$ is the $n \times n$ identity matrix.† The matrix $n\mathbf{I}$ has n's down the diagonal, and 0's otherwise. To give an example, suppose that

$$\mathbf{H} = \begin{bmatrix} 1 & 1 \\ -1 & 1 \end{bmatrix}. \tag{10.11}$$

Then

$$\mathbf{H}^T = \begin{bmatrix} 1 & -1 \\ 1 & 1 \end{bmatrix}$$

and

$$\mathbf{HH}^T = \begin{bmatrix} 2 & 0 \\ 0 & 2 \end{bmatrix}.$$

Hence, the matrix $\mathbf{H}$ of (10.11) is a Hadamard matrix of order 2. A Hadamard matrix of order 4 is given by

$$\mathbf{H} = \begin{bmatrix} 1 & 1 & -1 & 1 \\ 1 & -1 & -1 & -1 \\ -1 & -1 & -1 & 1 \\ 1 & -1 & 1 & 1 \end{bmatrix}. \tag{10.12}$$

For it is easy to check that

$$\mathbf{HH}^T = \begin{bmatrix} 4 & 0 & 0 & 0 \\ 0 & 4 & 0 & 0 \\ 0 & 0 & 4 & 0 \\ 0 & 0 & 0 & 4 \end{bmatrix}.$$

An equivalent definition of Hadamard matrix is the following. Recall that if

$$(x_1, x_2, \ldots, x_n)$$

and

$$(y_1, y_2, \ldots, y_n)$$

are two vectors, their *inner product* is defined to be

$$\sum_i x_i y_i.$$

*This subsection can be omitted.

†Hadamard matrices were introduced by Hadamard [1893] and, in an earlier use, by Sylvester [1867]. Plotkin [1960] and Bose and Shrikhande [1959] constructed binary codes from Hadamard matrices.

Then an $n \times n$ matrix of $+1$'s and -1's is a Hadamard matrix if for all i, the inner product of the ith row with itself is n, and for all $i \neq j$, the inner product of the ith row with the jth is 0. This is just a restatement of the definition. To illustrate, let us consider the matrix $\mathbf{H}$ of (10.11). The first row is the vector $(1, 1)$ and the second the vector $(-1, 1)$. The inner product of the first row with itself is

$$1 \cdot 1 + 1 \cdot 1 = 2.$$

The inner product of the first row with the second is

$$1 \cdot (-1) + 1 \cdot 1 = 0.$$

And so on.

A Hadamard matrix is called *normalized* if the first row and first column consist of just $+1$'s. For example, the matrix

$$\begin{bmatrix} 1 & 1 \\ 1 & -1 \end{bmatrix} \tag{10.13}$$

is a normalized Hadamard matrix. So is the matrix

$$\begin{bmatrix} 1 & 1 & 1 & 1 \\ 1 & -1 & 1 & -1 \\ 1 & 1 & -1 & -1 \\ 1 & -1 & -1 & 1 \end{bmatrix}. \tag{10.14}$$

Some of the most important properties of Hadamard matrices are summarized in the following theorem.

Theorem 10.9. If $\mathbf{H}$ is a normalized Hadamard matrix of order $n > 2$, then $n = 4m$, for some m. Moreover, each row (column) except the first has exactly $2m$ $+1$'s and $2m$ -1's, and for any two rows (columns) other than the first, there are exactly m columns (rows) in which both rows (columns) have $+1$.

We shall prove Theorem 10.9 later in this subsection. However, let us see how Theorem 10.9 gives rise to a proof of Theorem 9.15. Given a normalized Hadamard matrix, we can define a (v, k, λ)-design. We do so by deleting the first row and column and, in what is left, replacing every -1 by a 0. As we shall see, this gives an incidence matrix $\mathbf{A}$ of a (v, k, λ)-design. Performing this procedure on the normalized Hadamard matrix of (10.14) gives

$$\begin{bmatrix} -1 & 1 & -1 \\ 1 & -1 & -1 \\ -1 & -1 & 1 \end{bmatrix}$$

and hence the incidence matrix

$$
\begin{array}{c c}
 & \begin{array}{c c c} B_1 & B_2 & B_3 \end{array} \\
\mathbf{A} = \begin{array}{c} 1 \\ 2 \\ 3 \end{array} & \begin{bmatrix} 0 & 1 & 0 \\ 1 & 0 & 0 \\ 0 & 0 & 1 \end{bmatrix}.
\end{array}
$$

This gives rise to the blocks

$$B_1 = \{2\}, \qquad B_2 = \{1\}, \qquad B_3 = \{3\},$$

and to a design with $v = 3$, $k = 1$, and $\lambda = 0$. (This is, technically, not a design, since we have required $\lambda > 0$. However, it illustrates the procedure.)

To show that this procedure always gives rise to a (v, k, λ)-design, let us note that by Theorem 10.9, the incidence matrix $\mathbf{A}$ has $4m - 1$ rows and $4m - 1$ columns, so $b = v = 4m - 1$. Also, one 1 (the first one) has been removed from each row, so each row of $\mathbf{A}$ has $2m - 1$ 1's and $r = 2m - 1$. By a similar argument, each column of $\mathbf{A}$ has $2m - 1$ 1's and $k = 2m - 1$. Finally, any two rows had a pair of 1's in the first column in common, so now have one fewer pair, namely $m - 1$, in common. Hence, $\lambda = m - 1$. Thus, we have a (v, k, λ)-design with $v = 4m - 1$, $k = 2m - 1$, $\lambda = m - 1$, provided that $m \geq 2$.*

Let us next show how to construct normalized Hadamard matrices of order $4m$ for arbitrarily large m. This will prove Theorem 9.15. Suppose that $\mathbf{H}$ is an $n \times n$ Hadamard matrix. Let $\mathbf{K}$ be the following matrix:

$$\mathbf{K} = \begin{bmatrix} \mathbf{H} & \mathbf{H} \\ \mathbf{H} & -\mathbf{H} \end{bmatrix},$$

where $-\mathbf{H}$ is the matrix obtained from $\mathbf{H}$ by multiplying each entry by -1. For example, if $\mathbf{H}$ is the matrix of (10.13), then $\mathbf{K}$ is the matrix

$$\begin{bmatrix} 1 & 1 & 1 & 1 \\ 1 & -1 & 1 & -1 \\ 1 & 1 & -1 & -1 \\ 1 & -1 & -1 & 1 \end{bmatrix}.$$

$\mathbf{K}$ is always a Hadamard matrix of order $2n$. For each entry is $+1$ or -1. Moreover, it is easy to show that the inner product of a row with itself is always $2n$, and the inner product of two different rows is always 0. Finally, we observe that if $\mathbf{H}$ is normalized, then so is $\mathbf{K}$. Thus, there are normalized Hadamard matrices of arbitrarily large orders, and in particular of all orders 2^p for $p \geq 1$. A Hadamard matrix of order $4m$ for $m \geq 2$ gives rise to a $(4m - 1, 2m - 1, m - 1)$-design. Since $4m = 2^p$, $m \geq 2$ is equivalent to $m = 2^k$, $k \geq 1$, this completes the proof of Theorem 9.15.

It is interesting to go through the construction we have just gone through if the matrix $\mathbf{H}$ is the matrix shown in (10.14). Then the corresponding matrix $\mathbf{K}$ is given by

$$\begin{bmatrix} 1 & 1 & 1 & 1 & 1 & 1 & 1 & 1 \\ 1 & -1 & 1 & -1 & 1 & -1 & 1 & -1 \\ 1 & 1 & -1 & -1 & 1 & 1 & -1 & -1 \\ 1 & -1 & -1 & 1 & 1 & -1 & -1 & 1 \\ 1 & 1 & 1 & 1 & -1 & -1 & -1 & -1 \\ 1 & -1 & 1 & -1 & -1 & 1 & -1 & 1 \\ 1 & 1 & -1 & -1 & -1 & -1 & 1 & 1 \\ 1 & -1 & -1 & 1 & -1 & 1 & 1 & -1 \end{bmatrix}. \qquad (10.15)$$

---

*If $m = 1$, λ turns out to be 0, as we have seen.

The corresponding incidence matrix for a (v, k, λ)-design is given by deleting the first row and column and changing the remaining -1's to 0's. We get

$$\begin{bmatrix} 0 & 1 & 0 & 1 & 0 & 1 & 0 \\ 1 & 0 & 0 & 1 & 1 & 0 & 0 \\ 0 & 0 & 1 & 1 & 0 & 0 & 1 \\ 1 & 1 & 1 & 0 & 0 & 0 & 0 \\ 0 & 1 & 0 & 0 & 1 & 0 & 1 \\ 1 & 0 & 0 & 0 & 0 & 1 & 1 \\ 0 & 0 & 1 & 0 & 1 & 1 & 0 \end{bmatrix}. \tag{10.16}$$

This is the incidence matrix for a symmetric BIBD with $v = 7$, $k = 3$, and $\lambda = 1$. It defines a design different from that of Example 9.8. Repeating the construction one more time gives a 16×16 Hadamard matrix and hence a $(15, 7, 3)$-design. It is this design whose blocks are shown in Table 9.26. A proof is left to the reader as Exercise 8.

The construction procedure we have outlined certainly leads to normalized Hadamard matrices of order $4m$ for arbitrarily large m, in particular for $m = 1, 2, 4, 8, \ldots$, and so of orders $4, 8, 16, 32, \ldots$. Notice that we do not claim to construct normalized Hadamard matrices of order $4m$ for every m. It is conjectured that for every m, a normalized Hadamard matrix of order $4m$ exists. Whether or not this conjecture is true is another open question of the mathematics of experimental design and coding. Note that if there is any Hadamard matrix of order $4m$, there is a normalized one (Exercise 6). (For other methods of constructing Hadamard matrices, see, for example, Hall [1967] or MacWilliams and Sloane [1977].)

We conclude this subsection by filling in the one missing step in the proof of Theorem 9.15, namely, by proving Theorem 10.9.* We need one preliminary result.

Theorem 10.10. If $\mathbf{H}$ is a Hadamard matrix, so is $\mathbf{H}^T$.

Proof. If $\mathbf{H}\mathbf{H}^T = n\mathbf{I}$, then

$$\frac{\mathbf{H}}{\sqrt{n}} \frac{\mathbf{H}^T}{\sqrt{n}} = \mathbf{I},$$

where $\mathbf{H}/\sqrt{n}$ is obtained from $\mathbf{H}$ by dividing each entry by $\sqrt{n}$, and similarly for $\mathbf{H}^T/\sqrt{n}$. Since $\mathbf{A}\mathbf{B} = \mathbf{I}$ implies that $\mathbf{B}\mathbf{A} = \mathbf{I}$ for square matrices $\mathbf{A}$ and $\mathbf{B}$, we have

$$\frac{\mathbf{H}^T}{\sqrt{n}} \frac{\mathbf{H}}{\sqrt{n}} = \mathbf{I},$$

or

$$\mathbf{H}^T\mathbf{H} = n\mathbf{I}.$$

Since $(\mathbf{H}^T)^T = \mathbf{H}$, it follows that $\mathbf{H}^T$ is Hadamard. Q.E.D.

Proof of Theorem 10.9. Let $\mathbf{H}$ be a normalized Hadamard matrix of order n. The results for columns follow from the results for rows by applying Theorem 10.10. Hence,

---

*The rest of this subsection can be omitted.

we need only prove the row results. Since the first row of **H** is

$$1 \quad 1 \quad 1 \quad \cdots \quad 1$$

and since the inner product with any other row is 0, any other row must have an equal number of $+1$'s and -1's. Thus, n is even, and there are $n/2$ $+1$'s and $n/2$ -1's. Interchange columns so that the second row has $+1$'s coming first, and then -1's. Thus, the first two rows look like this:

$$1 \quad 1 \quad \cdots \quad 1 \quad 1 \quad 1 \quad \cdots \quad 1$$
$$1 \quad 1 \quad \cdots \quad 1 \quad -1 \quad -1 \quad \cdots \quad -1.$$

This interchange does not affect the size of the matrix or the number of $+1$'s in each row or the number of columns where two rows share a $+1$. Consider the ith row, $i \neq 1, 2$. The ith row has u digits which are $+1$ and $n/2 - u$ digits which are -1 in the first half (the first $n/2$ entries), and v digits which are $+1$ and $n/2 - v$ digits which are -1 in the second half. The first row and ith row have inner product 0. Hence, the ith row has $n/2$ $+1$'s, and

$$u + v = \frac{n}{2}. \tag{10.17}$$

The second and ith rows have inner product 0 if $i \neq 1, 2$. Hence,

$$u - \left(\frac{n}{2} - u\right) - v + \left(\frac{n}{2} - v\right) = 0. \tag{10.18}$$

(See Figure 10.2.) It follows that

$$u - v = 0. \tag{10.19}$$

Equations (10.17) and (10.19) give us

$$2u = \frac{n}{2}$$

or

$$n = 4u.$$

Thus, n is a multiple of 4, which proves the first part of Theorem 10.9. Moreover, there are $n/2 = 2u$ $+1$'s in each row other than the first, proving another part of Theorem 10.9. Finally, the second and the ith rows, $i \neq 1, 2$, have exactly u columns in which 1's appear in common. The same can be shown for rows j and i, $j \neq i$, $j, i \neq 1$, by repeating the

Second row	1	1	-1	-1
ith row, $i \neq 1, 2$	1	-1	1	-1
	u digits	$\frac{n}{2} - u$ digits	v digits	$\frac{n}{2} - v$ digits

Figure 10.2 Schematic for computing the inner product of the second and ith rows $(i \neq 1, 2)$ in a normalized Hadamard matrix.

argument above, interchanging columns so that the 1's in row j come first and then the -1's. This completes the proof of Theorem 10.9. Q.E.D.

10.5.3 The Richest (n, d)-Codes

We consider next the question: How "rich" are the codes obtained from Hadamard matrices? That is, compared to other possible codes with the same n (length of codewords) and the same d (minimum distance between codewords), does such a code have many or few codewords?

Let $A(n, d)$ be the maximum number of codewords in an (n, d) code. Equation (10.5) of Theorem 10.3 gives an upper bound for $A(n, d)$. Our next result gives a better bound in most cases.

Theorem 10.11 (Plotkin [1960]). Suppose that an (n, d)-code has N codewords. If $d > n/2$, then

$$N \le \frac{2d}{2d - n}.$$

*Proof.** Form the $N \times n$ matrix $\mathbf{M} = (m_{ij})$ whose rows are codewords. Consider

$$S = \sum d(u, v), \tag{10.20}$$

where the sum is taken over all (unordered) pairs of words u, v from the set of codewords. We know that $d(u, v) \ge d$ for all u, v, so

$$S \ge \frac{N(N - 1)}{2} d. \tag{10.21}$$

Let $t_0^{(i)}$ be the number of times 0 appears in the ith column of $\mathbf{M}$, and $t_1^{(i)}$ be the number of times 1 appears in the ith column.

Note that

$$S = \sum_{\{i, k\}} \sum_j |m_{ij} - m_{kj}| = \sum_j \sum_{\{i, k\}} |m_{ij} - m_{kj}|. \tag{10.22}$$

Now

$$\sum_{\{i, k\}} |m_{ij} - m_{kj}|$$

is the number of times that we have a pair of rows i and k with one having a 1 in the jth column and the other a 0 in the jth column, and this is given by $t_0^{(j)} t_1^{(j)}$. Thus, by (10.22),

$$S = \sum_j t_0^{(j)} t_1^{(j)}. \tag{10.23}$$

Now

$$t_1^{(j)} = N - t_0^{(j)},$$

so

$$t_0^{(j)} t_1^{(j)} = t_0^{(j)}[N - t_0^{(j)}].$$

*The proof can be omitted.

We seek an upper bound on $t_0^{(j)}t_1^{(j)}$. Let $f(x)$ be the function defined by

$$f(x) = x(N - x),$$

$0 \leq x \leq N$. Note that $f(x)$ is maximized when $x = N/2$ and the maximum value of $f(x)$ is $(N/2)(N/2) = N^2/4$. Thus,

$$t_0^{(j)}t_1^{(j)} \leq \frac{N^2}{4},$$

so by (10.23),

$$S \leq n\,\frac{N^2}{4}. \tag{10.24}$$

Then by (10.21) and (10.24),

$$\frac{N(N - 1)d}{2} \leq \frac{nN^2}{4},$$

$$(N - 1)d \leq \frac{nN}{2},$$

$$N\left(d - \frac{n}{2}\right) \leq d.$$

Since $d > n/2$, we have

$$N \leq \frac{d}{d - n/2} = \frac{2d}{2d - n}. \qquad \text{Q.E.D.}$$

Corollary 10.11.1. $A(n, d) \leq \dfrac{2d}{2d - n}$, if $d > \dfrac{n}{2}$.

A normalized Hadamard matrix of order $4m$ gives rise to an (n, d)-code with $n = 4m - 1$, $d = 2m$, and $N = 4m - 1$ codewords. Thus,

$$A(4m - 1, 2m) \geq 4m - 1. \tag{10.25}$$

By Corollary 10.11.1,

$$A(4m - 1, 2m) \leq \frac{2(2m)}{2(2m) - (4m - 1)} = \frac{4m}{1} = 4m. \tag{10.26}$$

Thus, the code obtained from a normalized Hadamard matrix is close to best possible. One gets the best possible code in terms of number of codewords by adding one code-word

$$1 \quad 1 \quad \cdots \quad 1.$$

Adding this word is equivalent to modifying our earlier construction and not deleting the first row of the normalized Hadamard matrix of order $4m$. The distance of this word to any codeword obtained from the Hadamard matrix is $2m$, because any such codeword has $2m - 1$ 1's and $2m$ 0's. It follows that we have a code with $4m$ words, each of length

$4m - 1$, and having minimum distance between two codewords equal to $2m$. This code will be called the $(4m - 1)$-*Hadamard code*. It follows that

$$A(4m - 1, 2m) \geq 4m. \tag{10.27}$$

Equations (10.26) and (10.27) give us the following theorem.

Theorem 10.12.   For all m for which there is a (normalized)* Hadamard matrix of order $4m$,

$$A(4m - 1, 2m) = 4m.$$

The bound is attained by using a $(4m - 1)$-Hadamard code.

Let $B(n, d)$ be the maximum number of words of length n each of which has distance exactly d from the other. Clearly, $B(n, d) \leq A(n, d)$.

Corollary 10.12.1.   For all m for which there is a (normalized) Hadamard matrix of order $4m$,

$$B(4m - 1, 2m) = 4m.$$

Proof. The code we have constructed has all distances equal to $2m$.   Q.E.D.

It is interesting to note that sometimes we can get much richer codes by increasing the length of codewords by 1. We shall prove the following theorem.

Theorem 10.13.   For all m for which there is a (normalized) Hadamard matrix of order $4m$,

$$A(4m, 2m) = 8m.$$

This theorem shows that in cases where there is a (normalized) Hadamard matrix of order $4m$, adding one digit leads to a doubling in the number of possible codewords.

We first show that $A(4m, 2m) \geq 8m$ by constructing a $(4m, 2m)$-code with $8m$ codewords.

Starting with a $4m \times 4m$ normalized Hadamard matrix **H**, we have constructed a $(4m - 1, 2m - 1, m - 1)$-design. Let **A** be the incidence matrix of this design. Use the rows of **A** and the rows of **B**, the matrix obtained from **A** by interchanging 0's and 1's. This gives us $8m - 2$ words in all. The distance between two words in **A** is $2m$, the distance between two words in **B** is $2m$, and the distance between a word in **A** and a word in **B** is $2m - 1$ if one is the ith word in **A** and the second is the jth word in **B** with $i \neq j$, and it is $4m - 1$ if one is the ith word in **A** and the second the ith word in **B**. Now add the digit 1 before words of **A** and the digit 0 before words of **B**, obtaining two sets of words **A**′ and **B**′. The distance between two words of **A**′ or of **B**′ is now still $2m$, while the distance between the ith word of **A**′ and the jth word of **B**′ is $2m$ if $i \neq j$ and $4m$ if $i = j$. There are still only $8m - 2$ words. Add the $4m$-digit words

*Recall that if there is any Hadamard matrix of order $4m$, there is a normalized one.

$$0 \quad 0 \quad \cdots \quad 0$$

and

$$1 \quad 1 \quad \cdots \quad 1.$$

We now have $8m$ words. The distance between these two new words is $4m$. Any word in $\mathbf{A}'$ or $\mathbf{B}'$ has $2m$ 1's and $2m$ 0's (since words of $\mathbf{A}$ had $k = 2m - 1$ 1's). Thus, the new words have distance $2m$ from any word of $\mathbf{A}'$ or of $\mathbf{B}'$. We now have $8m$ codewords, each of which has length $4m$ and with the minimum distance between two such codewords equal to $2m$. This set of codewords defines the *$4m$-Hadamard code*. It is equivalent to the set of codewords obtained by taking the normalized Hadamard matrix $\mathbf{H}$ of order $4m$, not deleting the first row or column, considering the matrix

$$\begin{bmatrix} \mathbf{H} \\ -\mathbf{H} \end{bmatrix},$$

changing -1 to 0, and using the rows in the resulting matrix as codewords. The $4m$-Hadamard code with $m = 2$ derived from the normalized Hadamard matrix of (10.15) is given in Table 10.1. This completes the proof of the fact that if there is a (normalized) Hadamard matrix of order $4m$, then

$$A(4m, 2m) \geq 8m. \tag{10.28}$$

We now prove that for all $m > 0$,

$$A(4m, 2m) \leq 8m, \tag{10.29}$$

which will prove Theorem 10.13.

Theorem 10.14. If $0 < d < n$, then $A(n, d) \leq 2A(n - 1, d)$.

Proof. Consider an (n, d)-code C of $A(n, d)$ words. Let C' be obtained from C by choosing all codewords beginning with 1 and deleting the first digit of these codewords. Then C' defines an $(n - 1, d)$-code, so

$$|C'| \leq A(n - 1, d).$$

Similarly, if C'' is obtained by choosing all codewords of C beginning with 0 and deleting

Table 10.1 The $8m = 16$ Codewords of the $4m$-Hadamard Code Derived from the Normalized Hadamard Matrix $\mathbf{H}$ of (10.15)

1	1	1	1	1	1	1	1		0	0	0	0	0	0	0	0
1	0	1	0	1	0	1	0		0	1	0	1	0	1	0	1
1	1	0	0	1	1	0	0		0	0	1	1	0	0	1	1
1	0	0	1	1	0	0	1		0	1	1	0	0	1	1	0
1	1	1	1	0	0	0	0		0	0	0	0	1	1	1	1
1	0	1	0	0	1	0	1		0	1	0	1	1	0	1	0
1	1	0	0	0	0	1	1		0	0	1	1	1	1	0	0
1	0	0	1	0	1	1	0		0	1	1	0	1	0	0	1

| From $\mathbf{H}$ | From $-\mathbf{H}$ |

the first digit, then

$$|C''| \le A(n - 1, d).$$

Thus,

$$A(n, d) = |C| = |C'| + |C''| \le 2A(n - 1, d). \qquad \text{Q.E.D.}$$

Now (10.29) follows immediately from Theorem 10.14 and (10.26). For we have

$$A(4m, 2m) \le 2A(4m - 1, 2m) \le 2(4m) = 8m.$$

(Note that Theorem 10.12 could not be used directly to give this result. Why?)

The $4m$-Hadamard codes we have constructed are also called *Reed–Muller codes* (of the first kind), after Reed [1954] and Muller [1954].

The results in Theorems 10.12 and 10.13 are due to Levenshtein [1964], who obtains more general results as well.

Example 10.7 The Mariner 9 Space Probe

The Reed–Muller codes were used in the 1972 Mariner 9 space probe, which returned photographs of Mars. The specific code used was the $4m$-Hadamard code for $m = 8$, that is, the code based on a 32×32 normalized Hadamard matrix. This code has 64 codewords of 32 bits each. A photograph of Mars was broken into very small dots, and each dot was assigned 1 of 64 levels of grayness, which was encoded in one of the 32-bit codewords. (See Posner [1968] or MacWilliams and Sloane [1977] for more details.)

Example 10.8 "Reading" DNA to Produce Protein

Golomb [1962] speculated that error-correcting codes might be at work in genetic coding, specifically in the process by which DNA strands are "read" in order to produce proteins. Error correction would be used whenever a sequence of three bases does not correspond to a code for an amino acid (see Example 2.2). Golomb speculated that a $4m$-Hadamard code is used. The smallest such code which would encode for the 20 different amino acids has 24 codewords and results from a 12×12 normalized Hadamard matrix. (That such a matrix exists does not follow from our theorems.) Codewords have length 12. A DNA chain would be encoded into a string of 0's and 1's in two steps. First, one of the letters A, C, G, and T would be encoded as 00, one as 01, one as 10, and one as 11. This represents a DNA chain of length m as a bit string of length $2m$, the message word. Every message word would be broken into blocks of length $k = 2$ and encoded by a $2 \to 12$ code. Every six bits of the message word (every three letters of the chain) or every 36 bits of the code would correspond to an amino acid.

In closing this section we note that the $(4m - 1)$-Hadamard codes and the $4m$-Hadamard codes are in fact linear codes, if 0's and 1's are interchanged and $m = 2^p$ for some p, as in our construction. However, the 24-Hadamard code is an example of one that is not linear. We shall not present a proof of these facts here, but instead refer the reader to the coding theory literature, for example to MacWilliams and Sloane [1977], for a proof. (See also Exercises 23 to 26).

EXERCISES FOR SECTION 10.5

1. A Hadamard design includes the following blocks. Find the missing blocks.

$$\{1, 2, 3\}, \{2, 4, 5\}, \{1, 4, 6\}, \{3, 4, 7\}.$$

2. Suppose that $\mathbf{A}$ is the incidence matrix of a (b, v, r, k, λ)-design and that a code is made up of the rows of this matrix.
   (a) What is the distance between two words in this code?
   (b) How many errors will the code detect?
   (c) How many errors will the code correct?
   (d) Suppose that $\mathbf{B}$ is the incidence matrix of the complementary design (Exercise 15, Section 9.4). What is the distance between the ith row of $\mathbf{A}$ and the jth row of $\mathbf{B}$ if $i \neq j$?
   (e) If rows of $\mathbf{B}$ form a code, how many errors will the code detect?
   (f) How many will it correct?

3. Given a Steiner triple system of nine varieties, build a code by using the rows of the incidence matrix. Find the minimum distance between two codewords.

4. (a) Could there be a 3×3 Hadamard matrix?
   (b) A 6×6?

5. If $\mathbf{H}$ is a Hadamard matrix and some row is multiplied by -1, is the resulting matrix still Hadamard? Why?

6. Show that if there is a Hadamard matrix of order $4m$, there is a normalized Hadamard matrix of order $4m$.

7. Suppose that

$$\mathbf{H} = \begin{bmatrix} 1 & 1 & 1 & 1 \\ 1 & 1 & -1 & -1 \\ 1 & -1 & 1 & -1 \\ 1 & -1 & -1 & 1 \end{bmatrix}$$

   Use $\mathbf{H}$ to find an 8×8 Hadamard matrix and a $(7, 3, 1)$-design.

8. Derive the $(15, 7, 3)$-design of Table 9.26 from a 16×16 normalized Hadamard matrix.

9. (a) Find a $(4m - 1)$-Hadamard code for $m = 4$.
   (b) How many errors can this code detect?
   (c) How many errors can it correct?

10. (a) Find a $4m$-Hadamard code for $m = 4$.
    (b) How many errors can this code detect?
    (c) How many errors can it correct?

11. Repeat Exercise 9 for $m = 8$.

12. Repeat Exercise 10 for $m = 8$.

13. The following is an $(11, 5, 2)$-design. Use it to find a code of 12 words that can detect up to five errors and a code of 24 words that can detect up to five errors.

$$\{1, 3, 4, 5, 9\}, \{2, 4, 5, 6, 10\}, \{3, 5, 6, 7, 11\}, \{1, 4, 6, 7, 8\}, \{2, 5, 7, 8, 9\}, \{3, 6, 8, 9, 10\},$$
$$\{4, 7, 9, 10, 11\}, \{1, 5, 8, 10, 11\}, \{1, 2, 6, 9, 11\}, \{1, 2, 3, 7, 10\}, \{2, 3, 4, 8, 11\}.$$

14. Suppose that $\mathbf{H}$ is the incidence matrix of a $(4m - 1, 2m - 1, m - 1)$-design and $\mathbf{K}$ is the incidence matrix of the complementary design (Exercise 15, Section 9.4).
    (a) What is the distance between the ith row of $\mathbf{H}$ and the jth row of $\mathbf{H}$?
    (b) What is the distance between the ith row of $\mathbf{H}$ and the jth row of $\mathbf{K}$?

15. Suppose that S is a set of binary codewords of length n with Hamming distance $d(a, b)$ equal to d for each pair of words a and b in S. Let T be defined from S by taking complements of words in S (interchanging 0 and 1).
    (a) What is the distance between two words of T?
    (b) What is the distance between a word of S and a word of T?
    (c) If n is 12 and d is 5, how many errors will the code T detect? How many will it correct?
    (d) Suppose that a code is defined from the words in S and the words in T. If n is 12 and d is 5, how many errors will this code detect? How many will it correct?

16. Suppose that the $m \times n$ matrix $\mathbf{M}$ is the incidence matrix of a block design, and that the ith row and the jth row have 1's in common in u columns.
    (a) What is the inner product of these two rows?
    (b) What is the inner product in the complementary design (Exercise 15, Section 9.4)?

17. Find a 2×4 matrix of 1's and -1's such that the inner product of rows i and j is 0 if $i \neq j$ and 4 if $i = j$.

18. For what values of k does there exist a $2 \times k$ matrix of 1's and -1's such that the inner product of rows i and j is 0 if $i \neq j$ and k if $i = j$?

19. Show that $A(n, n - 1) = 2$ for n sufficiently large.

20. (a) Show that if $n > m$, then $A(n, d) \geq A(m, d)$.
    (b) If $n > m$, is it necessarily the case that $A(n, d) > A(m, d)$? Why?

21. (a) Find an upper bound for $A(4m - 2, 2m)$.
    (b) Find $A(4m - 2, 2m)$ exactly for arbitrarily large values of m.

22. Using the results of Exercise 15, show that

$$A(2d, d) \geq 2B(2d, d).$$

23. Show that the $(4m - 1)$-Hadamard code as we have defined it is not linear. (*Hint:* Consider sums of codewords.)

24. If $m = 2^p$ and if 0's and 1's are interchanged in the $(4m - 1)$-Hadamard code as we have defined it, then the code becomes linear. Show this for the case $m = 1$ by finding a generator matrix.

25. Continuing with Exercise 24, show linearity for the case $m = 2$ as follows:
    (a) Using the generator matrix obtained by taking

$$\mathbf{G} = \begin{bmatrix} 0 & 1 & 1 & 1 \\ 1 & 0 & 1 & 1 \\ 1 & 1 & 0 & 1 \end{bmatrix},$$

    generate the corresponding code.
    (b) Show that this code is equivalent (Exercise 28, Section 10.4) to the $(4m - 1)$-Hadamard code with $m = 2$ and 0's and 1's interchanged.

26. If $m \neq 2^p$, show that even if 0's and 1's are interchanged in the $(4m - 1)$-Hadamard code, the code is not linear. (*Hint:* How many codewords does a linear code have?)

27. Let $A(n, d, w)$ be the maximum number of codewords in an (n, d)-code in which each word has the same weight w. Show that
    (a) $A(n, 2d - 1, w) = A(n, 2d, w)$.
    (b) $A(n, 2d, w) = A(n, 2d, n - w)$.
    (c) $A(n, 2d, w) = 1$ if $w < d$.
    (d) $A(n, 2d, d) = \lfloor n/d \rfloor$.

REFERENCES FOR CHAPTER 10

AVIZENIUS, A., "Fault-Tolerant Systems," *IEEE Trans. Comput.*, *C-25* (1976), 1304–1312.

BERLEKAMP, E. R., *Algebraic Coding Theory*, McGraw-Hill, New York, 1968.

BLAKE, I. F., and MULLIN, R. C., *The Mathematical Theory of Coding*, Academic Press, New York, 1975. (Abridged as *An Introduction to Algebraic and Combinatorial Coding Theory*, Academic Press, New York, 1976.)

BOSE, R. C., and SHRIKHANDE, S. S., "A Note on a Result in the Theory of Code Construction," *Inf. Control*, *2* (1959), 183–194.

CARTER, W. C., and BOURICIUS, W. G., "A Survey of Fault-Tolerant Architecture and Its Evaluation," *Computer*, *4* (1971), 9–16.

DORNHOFF, L. L., and HOHN, F. E., *Applied Modern Algebra*, Macmillan, New York, 1978.

FISHER, J. L., *Application-Oriented Algebra*: *An Introduction to Discrete Mathematics*, Harper & Row, New York, 1977.

GOLAY, M. J. E., "Notes on Digital Coding," *Proc. IEEE*, *37* (1949), 657.

GOLOMB, S. W., "Efficient Coding for the Deoxyribonucleic Channel," in *Mathematical Problems in the Biological Sciences*, Proceedings of Symposia in Applied Mathematics, Vol. 14, American Mathematical Society, Providence, R. I., 1962, pp. 87–100.

HADAMARD, J., "Résolution d'une question relative aux déterminants," *Bull. Sci. Math.*, *17* (1893), 240–248.

HALL, M., *Combinatorial Theory*, Ginn (Blaisdell), Boston, 1967. (Second printing, Wiley, New York, 1980.)

HAMMING, R. W., "Error Detecting and Error Correcting Codes," *Bell Syst. Tech. J.*, *29* (1950), 147–160.

LEVENSHTEIN, V. I., "The Application of Hadamard Matrices to a Problem in Coding," *Probl. Kibern.*, *5* (1961), 123–136. [English transl.: *Probl. Cybern.*, *5* (1964), 166–184.]

MACWILLIAMS, F. J., and SLOANE, N. J. A., *The Theory of Error-Correcting Codes*, Vols. 1 and 2, North-Holland, Amsterdam, 1977.

MULLER, D. E., "Application of Boolean Algebra to Switching Circuit Design and to Error Detection," *IEEE Trans. Comput.*, *3* (1954), 6–12.

PETERSON, W. W., *Error Correcting Codes*, MIT Press, Cambridge, Mass., 1961.

PETERSON, W. W., and WELDON, E. J., *Error Correcting Codes*, MIT Press, Cambridge, Mass., 1972. (Second edition of Peterson [1961].)

PLOTKIN, M., "Binary Codes with Specified Minimum Distances," *IEEE Trans. Inf. Theory*, *6* (1960), 445–450.

POSNER, E. C., "Combinatorial Structures in Planetary Reconnaissance," in H. B. Mann (ed.), *Error Correcting Codes*, Wiley, New York, 1968.

REED, I. S., "A Class of Multiple-Error-Correcting Codes and the Decoding Scheme," *IEEE Trans. Inf. Theory*, *4* (1954), 38–49.

SELLERS, F. F., HSIAO, M.-Y., and BEARNSON, L. W., *Error Detecting Logic for Digital Computers*, McGraw-Hill, New York, 1968.

SLEPIAN, D., "A Class of Binary Signaling Alphabets," *Bell Syst. Tech. J.*, *35* (1956a), 203–234.

SLEPIAN, D., "A Note on Two Binary Signaling Alphabets," *IEEE Trans. Inf. Theory*, *2* (1956b), 84–86.

SLEPIAN, D., "Some Further Theory of Group Codes," *Bell Syst. Tech. J.*, *39* (1960), 1219–1252.

SYLVESTER, J. J., "Thoughts on Inverse Orthogonal Matrices, Simultaneous Sign Successions, and

Tesselated Pavements in Two or More Colors, with Applications to Newton's Rule, Ornamental Tile-Work, and the Theory of Numbers," *Philos. Mag.*, *34* (1867), 461–475.

TIETÄVÄINEN, A., "On the Nonexistence of Perfect Codes over Finite Fields," *SIAM J. Appl. Math.*, *24* (1973), 88–96.

VAN LINT, J. H., "Nonexistence Theorems for Perfect Error-Correcting Codes," in *Computers in Algebra and Number Theory*, Vol. 4 (SIAM-AMS Proceedings), 1971.

VAN LINT, J. H., *Introduction to Coding Theory*, Springer-Verlag, New York, 1982.

WAKERLY, J., *Error Detecting Codes, Self-Checking Circuits, and Applications*, Elsevier North-Holland, New York, 1978.

11 Existence Problems in Graph Theory

Topics in graph theory underlie much of combinatorics. This chapter begins a sequence of three chapters dealing mostly with graphs and networks. The modern approach to graph theory is heavily influenced by computer science, and it emphasizes algorithms for solving problems. Among other things, we give an introduction to graph algorithms in these chapters. We also talk about applications of graph theory, many of which have already been mentioned in Section 3.1. This chapter emphasizes existence questions in graph theory and Chapter 13 emphasizes optimization questions. Chapter 12 is a transitional chapter, which begins with existence questions and ends with optimization questions. Of course, it is hard to make such a rigid partition. The same techniques and concepts of graph theory that are used on existence problems are usually useful for optimization problems, and vice versa.

In this chapter we examine four basic existence questions and discuss their applications. The existence questions are:

1. Is a given graph G connected; that is, does there exist, for every pair of vertices x and y of G, a chain between x and y?
2. Does a given graph G have a strongly connected orientation?
3. Does a given graph G have an eulerian chain, a chain that uses each edge exactly once?
4. Does a given graph G have a hamiltonian chain, a chain that uses each vertex exactly once?

We present theorems that help us to answer these questions. With practical use in mind, we shall be concerned with describing good procedures or algorithms for answering them.

The results will have applications to such subjects as traffic flow, RNA codes, street sweeping, and telecommunications.

11.1 DEPTH-FIRST SEARCH: A TEST FOR CONNECTEDNESS

Suppose that $G = (V, E)$ is a graph. The first question we shall ask is this: Is G connected, that is, does there exist, for every pair of vertices x and y of G, a chain between x and y?

Given a graph, it is easy to see from the corresponding diagram whether or not it is connected. However, for large graphs, drawing such diagrams is infeasible. Moreover, diagrams are not readily amenable to use as input in a computer. Instead, one has to input a graph into a computer by, for example, listing its edges. (See Section 3.7 for a more complete discussion of ways to input a graph into a computer.) In any case, it is now not so easy to check if a graph is connected. Thus, we would like an algorithm for testing connectedness. In Section 11.1.1, we shall present such an algorithm. The algorithm will play a crucial role in solving the traffic flow problem we study in Section 11.2 and can also be utilized in finding eulerian chains, the problem we discuss in Section 11.3. This algorithm is also important for solving optimization problems in graph theory, not just existence problems.

11.1.1 Depth-First Search

Suppose that $G = (V, E)$ is any graph. The method we describe for testing if G is connected is based on the *depth-first search procedure*, a highly efficient procedure which is the basis for a number of important computer algorithms. (See Tarjan [1972], Even [1979], Hopcroft and Tarjan [1973], Aho *et al.* [1974, Ch. 5], Baase [1978], Reingold *et al.* [1977], or Golumbic [1980], for a discussion. See Exercise 9 for a related search procedure known as breadth-first search.)

In the depth-first search procedure, we start with a graph G with n vertices, and aim to label the vertices with the integers $1, 2, \ldots, n$. We choose an arbitrary vertex and label it 1. Having just labeled a given vertex x with the integer k, we search among all neighbors of x and find an unlabeled neighbor, say y. We give vertex y the next label, $k + 1$. We also keep track of the edges used in the labeling procedure by *marking* the edge $\{x, y\}$ traversed from x to y. We then continue the search among neighbors of y. In this way, we progress along chains of G leading away from x. The complication arises if we have just labeled a vertex z which has no unlabeled neighbors. We then go back to the neighbor u of z from which z was labeled— u is called the *father* of z—and continue the search from u. We can keep track of the father of a vertex since we have marked the edges traversed from a vertex to the next one labeled. The procedure continues until all the labels $1, 2, \ldots, n$ have been used (equivalently all vertices have been labeled) or it is impossible to continue because we have returned to a labeled vertex with no father, namely the vertex labeled 1.

We illustrate the labeling procedure in the graph of Figure 11.1. Vertex a is chosen first and labeled 1. The labels are shown in the figure by circled numbers next to the vertices. Vertex b is an unlabeled neighbor of vertex a which is chosen next and labeled 2.

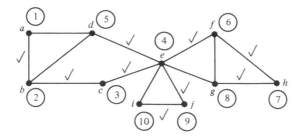

Figure 11.1 A labeling of vertices obtained using the depth-first search procedure. A check mark indicates marked edges.

The edge $\{a, b\}$ is marked (we put a check mark in the figure). Next, an unlabeled neighbor of vertex b is found, and labeled 3. This is vertex c. The edge $\{b, c\}$ is marked. An unlabeled neighbor of c, vertex e, is labeled 4, and the edge $\{c, e\}$ is marked. An unlabeled neighbor of e, vertex d, is labeled 5, and the edge $\{e, d\}$ is marked. Now after vertex d is labeled 5, all neighbors of d are labeled. Thus, we go back to e, the father of d, and seek an unlabeled neighbor. We find one, vertex f, and label it 6. We also mark edge $\{e, f\}$. We now label vertex h with integer 7 and mark edge $\{6, 7\}$ and label vertex g with integer 8 and mark edge $\{h, g\}$. Now all neighbors of g are labeled, so we go back to the father of g, namely h. All neighbors of h are also labeled, so we go back to the father of h, namely f. Finally, since all neighbors of f are labeled, we go back to the father of f, namely e. From e, we next label vertex j with integer 9, and mark edge $\{e, j\}$. Then from j, we label vertex i with integer 10 and mark edge $\{j, i\}$. Now all vertices are labeled, so we stop.

It is easy to show that the labeling procedure can be completed if and only if the graph is connected. Thus, the depth-first search procedure provides an algorithm for testing if a graph is connected. Figure 11.2 shows the procedure on a disconnected graph. After having labeled vertex d with label 4, we find that the father of d, namely c, has no unlabeled neighbors. Moreover, the same is true of the father of c, namely a. Finally, a has no father, so we cannot continue the labeling.

It should also be noted that if the labeling procedure is completed and T is the collection of marked edges, then there are $n - 1$ edges in T, for one edge is added to T each time we assign a new label. Moreover, suppose that H is the spanning subgraph of G whose edges are the edges in T. Then H has no circuits, because every edge added to T goes from a vertex labeled i to a vertex labeled j with $i < j$. Hence, H is a subgraph of G with $n = e + 1$ and no circuits. By Theorem 3.16, H is a tree. It is called the *depth-first search spanning tree*. (The reader should observe that the checked edges of Figure 11.1 form a tree.)

11.1.2 The Computational Complexity of Depth-First Search

The computational complexity of this depth-first search algorithm can be computed as follows. Suppose that the graph has n vertices and e edges. Each step of the procedure involves assigning a label or traversing an edge. The traversal is either forward from a labeled vertex to a neighbor or backward from a vertex to its father. By marking not only edges that are *used* in a forward direction, but also edges that are *investigated* in a forward direction but lead to already labeled vertices, we can be sure no edge is used or investigated in a forward direction more than once. Also, it is not hard to see that no edge can be used backward more than once. Thus, the entire procedure investigates at most $2e$ edges. Since at most n labels are assigned, the procedure terminates in at most $n + 2e$ steps. [In the notation of Section 2.18, this is an $O(n + e)$ algorithm.] Since $2e$ is at most

$$2\binom{n}{2} = 2\,\frac{n(n-1)}{2} = n^2 - n,$$

this is a polynomial bound on the complexity in terms of the number of vertices. [It is an $O(n^2)$ algorithm.*] The algorithm is very efficient.

It should be noted that a similar directed depth-first search procedure can be defined for digraphs. We simply label as before, but only go from a given vertex to vertices reachable from it by arcs, not edges. This method can be used to provide an efficient test for strong connectedness. For details, the reader is referred to Tarjan [1972], Hopcroft and Tarjan [1973], Aho *et al.* [1974, Ch. 5], or Reingold *et al.* [1977]. (See also Exercise 8.)

11.1.3 A Formal Statement of the Algorithm†

For the reader who is familiar with recursive programming, we close this section with a more formal statement of the depth-first search algorithm. We first state a subroutine called DFSEARCH (v, u). In this subroutine, as well as in the main algorithm, k will represent the current value of the label being assigned and T will be the set of marked edges. (We disregard the separate marking of edges that are investigated but lead to already labeled vertices.) The control is the vertex whose neighbors are currently being searched. Some vertices will already bear labels. The vertex v is the one that is just to be labeled and u is the father of v.

Algorithm 11.1.   DFSEARCH (v, u)

> **Input:** A graph G with some vertices labeled and some edges in T, a vertex v that is just to be labeled, a vertex u that is the father of v, and a current label k to be assigned.
>
> **Output:** Some additional label assignments and a (possibly) new set T and a new current label k.

Step 1. Set the control equal to v, mark v labeled, assign v the label k, and replace k by $k + 1$.

Step 2. For each neighbor w of v, if w is unlabeled, add the edge $\{v, w\}$ to T, mark the edge $\{v, w\}$, call v the father of w, and perform the algorithm DFSEARCH (w, v).

Step 3. If v has no more unlabeled neighbors and all vertices have been labeled, stop and output the labeling. If some vertex is still unlabeled, set the control equal to u and stop.

---

*Some authors call the algorithm *linear* because its complexity is linear in $n + e$. Others call it *quadratic* because its complexity is quadratic in n.

†This subsection can be omitted.

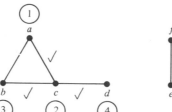

Figure 11.2 Depth-first search labeling and marking on a disconnected graph.

We can now summarize the entire depth-first search algorithm as follows.

> **Algorithm 11.2.   Depth-First Search**
>
> **Input:** A graph G of n vertices.
>
> **Output:** A labeling of the vertices of G using the integers 1, 2, ..., n and an assignment of $n - 1$ edges of G to a set T, or the message that the procedure cannot be completed.
>
> *Step 1.* Set $T = \emptyset$ and $k = 1$ and let no vertex be labeled.
>
> *Step 2.* Pick any vertex v, introduce a (dummy) vertex α, call α the father of v, and perform DFSEARCH (v, α).
>
> *Step 3.* If the control is ever set equal to α, output the message that the procedure cannot be completed.

Algorithm 11.2 terminates either because all vertices have been labeled or because the labeling procedure has returned to the vertex labeled 1, all neighbors of that vertex have labels, and there are some unlabeled vertices. In this case, it is easy to see that the graph is disconnected.

EXERCISES FOR SECTION 11.1

1. For each graph of Figure 11.3, perform a depth-first search beginning with the vertex labeled a. Indicate all vertex labels and all marked edges.

2. Given the adjacency structure of Table 11.1 for a graph of G of vertices a, b, c, d, e, f, g, use depth-first search to determine if G is connected.

3. From each of the following adjacency matrices for a graph G, use depth-first search to determine if G is connected.

$$
\textbf{(a) } A =
\begin{bmatrix}
0 & 1 & 0 & 0 & 0 & 0 & 1 \\
1 & 0 & 0 & 1 & 1 & 0 & 0 \\
0 & 0 & 0 & 1 & 0 & 1 & 0 \\
0 & 1 & 1 & 0 & 0 & 0 & 1 \\
0 & 1 & 0 & 0 & 0 & 0 & 0 \\
0 & 0 & 1 & 0 & 0 & 0 & 0 \\
1 & 0 & 0 & 1 & 0 & 0 & 0
\end{bmatrix}
$$

$$\textbf{(b) A} = \begin{bmatrix} 0 & 1 & 0 & 0 & 0 & 0 & 1 & 0 \\ 1 & 0 & 0 & 1 & 0 & 0 & 0 & 0 \\ 0 & 0 & 0 & 0 & 0 & 1 & 0 & 0 \\ 0 & 1 & 0 & 0 & 0 & 0 & 1 & 1 \\ 0 & 0 & 0 & 0 & 0 & 0 & 1 & 0 \\ 0 & 0 & 1 & 0 & 0 & 0 & 0 & 0 \\ 1 & 0 & 0 & 1 & 1 & 0 & 0 & 0 \\ 0 & 0 & 0 & 1 & 0 & 0 & 0 & 0 \end{bmatrix}.$$

4. For each graph of Figure 11.3, find a depth-first search spanning tree.

5. Is every spanning tree attainable as a depth-first search spanning tree? Why?

6. Suppose that the depth-first search algorithm is modified so that if further labeling is impossible, but there is still an unlabeled vertex, some unlabeled vertex is chosen, given a new label, and the

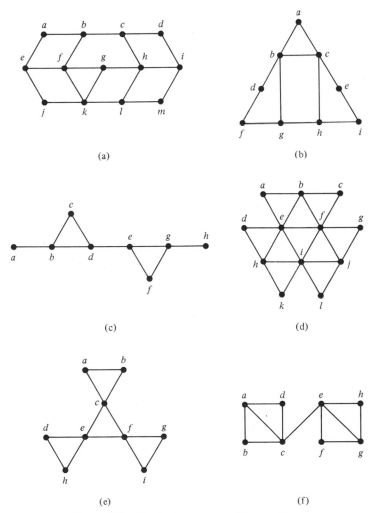

Figure 11.3 Graphs for exercises of Section 11.1.

Table 11.1 Adjacency Structure for a Graph

Vertex x	a	b	c	d	e	f	g
Vertices adjacent to x	b, c, g	a, c	a, b	e, f	d, f	d, e	a

process starts again. What can you say about the subgraph determined by the marked edges (the edges in T)?

7. How can the depth-first search algorithm as described in the text be modified to count the number of connected components of a graph G?

8. (a) Is the following test for strong connectedness of a digraph D correct? Pick any vertex of the digraph to receive label 1 and perform a directed depth-first search until no more vertices can be labeled. Then D is strongly connected if and only if all vertices have received a label.
   (b) If this is not a correct test, how can it be modified?

9. In the algorithm known as *breadth-first search*, we start with an arbitrarily chosen vertex x and place it at the head of an ordered list or *queue*. At each stage of the algorithm, we pick the vertex y at the head of the queue, delete y from the queue, and add to the tail of the queue one at a time all vertices adjacent to y which have never been on the queue. We continue until the queue is empty. If not all vertices have been on the queue, we start with another arbitrarily chosen vertex which has never been on the queue. Breadth-first search fans out in all directions from a starting vertex, whereas depth-first search follows one chain at a time from this starting vertex to an end. We have already employed breadth-first search in Algorithm 3.1 of Section 3.3.4. For a detailed description of breadth-first search, see, for example, Baase [1978], Even [1979], or Golumbic [1980]. Perform a breadth-first search on each graph of Figure 11.3, beginning with the vertex a. Indicate the order of vertices encountered by labeling each with an integer from 1 to n in the order they reach the head of the queue. (*Note:* In many applications, breadth-first search is inefficient or unwieldy compared to depth-first search. This is true, for example, in algorithms for testing if a given graph is planar. However, in many network optimization problems such as most of those studied in Chapter 13, breadth-first search is the underlying procedure.)

10. Can breadth-first search (Exercise 9) be used to test a graph for connectedness? If so, how?

11. What is the computational complexity of breadth-first search?

11.2 THE ONE-WAY STREET PROBLEM

11.2.1 Robbins' Theorem

In this section we discuss an application of graph theory to a traffic flow problem. Imagine that a city has a number of locations, some of which are joined by two-way streets. The number of cars on the road has markedly increased, resulting in traffic jams and increased air pollution, and it has been suggested that the city should make all its streets one way. This would, presumably, cut down on traffic congestion. The question is: Can this always be done? If not, when? The answer is: Of course, it can always be done. Just put a one-way sign on each street! However, it is quite possible that we will get into trouble if we make the assignment arbitrarily, for example by ending up with some places that we can get into and never leave. (See, for example, Figure 11.4, which shows an

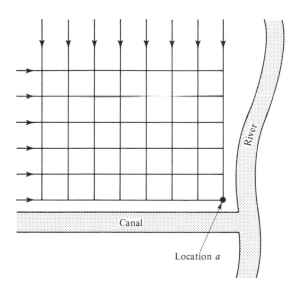

Figure 11.4 A one-way street assignment for the streets of a city which leaves travelers stuck at location *a*.

assignment of directions to the streets of a city that is satisfactory only for someone who happens to own a parking lot at location *a*.) We would like to make every street one-way in such a manner that for every pair of locations *x* and *y*, it is possible (legally) to reach *x* from *y* and reach *y* from *x*. Can this always be done?

To solve this problem, we assume for simplicity that all streets are two-way at the beginning. See Boesch and Tindell [1980] or Roberts [1978] for a treatment of the problem where some streets are one-way before we start (see also Exercise 12). We represent the transportation network of a city by a graph *G*. Let the locations in question be the vertices of *G* and draw an edge between two locations *x* and *y* if and only if *x* and *y* are joined by a two-way street. A simple example of such a graph is shown in Figure 11.5. In terms of the graph, our problem can now be restated as follows: Is it possible to put a direction or arrow (a one-way sign) on each edge of the graph *G* in such a way that by following arrows in the resulting figure, which is a digraph, one can always get from any point *x* to any other point *y*? If it is not always possible, when is it possible? Formally, we define an *orientation* of a graph *G* as an assignment of a direction to each edge of *G*. We seek an orientation of *G* which, to use the terminology of Section 3.2, is strongly connected.

In the graph of Figure 11.5, we can certainly assign a direction to each edge to obtain a strongly connected digraph. We have done this in Figure 11.6. However, it is not always possible to obtain a strongly connected orientation. For example, if our graph has two components, as in the graph of Figure 11.7, there is no way of assigning directions to

Figure 11.5 Two-way street graph for a city.

Figure 11.6 A strongly connected orientation for the graph of Figure 11.5.

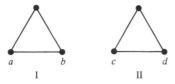

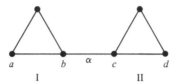

Figure 11.7 Graph representing a disconnected city.

Figure 11.8 α is a bridge.

edges which will make it possible to go from vertices in section I of the graph to vertices in section II. To have a strongly connected orientation, our graph must certainly be connected. However, even a connected graph may not have a strongly connected orientation. Consider the graph of Figure 11.8. This is connected. We must put a direction on edge α. If α is directed from section I to section II, then from no vertex in II can we reach any vertex in I. If α is directed from II to I, then from no vertex in I can we reach any vertex in II. What is so special about the edge α? The answer is that it is the only edge joining two separate pieces of the graph. Put more formally, removal of α (but not the two vertices it joins) results in a disconnected graph. Such an edge in a connected graph is called a *bridge*. Figure 11.9 gives another example of a bridge α. It is clear that for a graph G to have a strongly connected orientation, G must not only be connected, but it must also have no bridges.

In Figure 11.9, suppose that we add another bridge β joining the two separate components which are joined by the bridge α, obtaining the graph of Figure 11.10. Does this graph have a strongly connected orientation? The answer is yes. A satisfactory assignment is shown in Figure 11.11. Doesn't this violate the observation we have just made, namely that if a graph G has a strongly connected orientation, it can have no

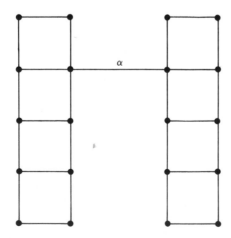

Figure 11.9 α is again a bridge.

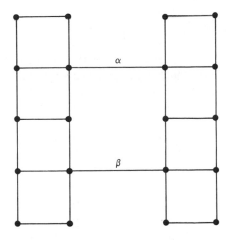

Figure 11.10 Edge α is no longer a bridge.

bridges? The answer here is no. For we were too quick to call β a bridge. In the sense of graph theory, neither β nor α is a bridge in the graph of Figure 11.10. A graph-theoretical bridge has the nasty habit that if we have a bridge and build another bridge, then neither is a bridge! In applying combinatorics, if we use suggestive terminology such as the term *bridge*, we have to be careful to be consistent in our usage.

Suppose now that G has the following properties: It is a connected graph and has no bridges. Are these properties sufficient to guarantee that G has a strongly connected orientation? The answer turns out to be yes, as we summarize in the following theorem.

Theorem 11.1 (Robbins [1939]). A graph G has a strongly connected orientation if and only if G is connected and has no bridges.

We omit the proof of Theorem 11.1. For two different proofs, see Roberts [1976, 1978] and see Boesch and Tindell [1980]. For a sketch of one of the proofs, see Exercise 12.

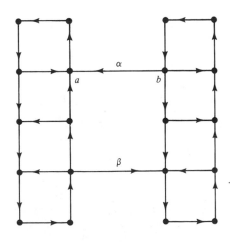

Figure 11.11 A satisfactory one-way assignment for the graph of Figure 11.10.

11.2.2 A Depth-First Search Algorithm

Robbins' Theorem in some sense completely solves the problem we have stated. However, it is not, by itself, a very useful result. For the theorem states that there is a one-way street assignment for a connected, bridgeless graph, but it does not tell us how to find such an assignment. In this section we present an efficient algorithm for finding such an assignment if one exists.

Suppose that $G = (V, E)$ is any graph. We shall describe a procedure for orienting G which can be completed if and only if G is connected. If G has no bridges, the resulting orientation is strongly connected. For a proof of the second assertion, see Roberts [1976]. The procedure begins by using the depth-first search procedure described in Section 11.1.1 to label the vertices and mark the edges. If the labeling procedure cannot be completed, G is disconnected and hence has no strongly connected orientation. The next step is to use the labeling and the set of marked edges T to define an orientation. Suppose that an edge is labeled with the labels i and j. If the edge is in T, that is, if it is marked, we orient from lower number to higher number, that is, from i to j if i is less than j and from j to i otherwise. If the edge is not T, that is, if it is unmarked, we orient from higher number to lower number. We illustrate this procedure using the labeling and marking of Figure 11.1. The corresponding orientation is shown in Figure 11.12. Note that we orient from a to b because edge $\{a, b\}$ is in T and a gets the label 1, which is less than the label 2 given to b. However, we orient from d to b because edge $\{d, b\}$ is not in T and the label of d, that is, 5, is more than the label of b, that is, 2.

This algorithm always leads to an orientation of G if the labeling procedure can be completed, that is, if G is connected. However, it is only guaranteed to lead to a strongly connected orientation if G has no bridges. If we assume that the decision of how to orient an edge, given a labeling and a record of edges in T, takes about the same amount of time as each step in the depth-first search procedure which is used to label and record edges in T, then the computational complexity $g(n, e)$ of this algorithm for an n vertex, e edge graph is at most the computational complexity $f(n, e)$ of the depth-first search procedure plus the number of edges in the graph G. In Section 11.1.2 we showed that $f(n, e)$ is at most $n + 2e$. Thus, $g(n, e)$ is at most $n + 3e$. [It is again an $O(n + e)$ algorithm.] In terms of n, the complexity is at most $\frac{3}{2}n^2 - \frac{3}{2}n + n$, since $e \leq \binom{n}{2}$. Thus, we have a polynomially bounded algorithm. [It is $O(n^2)$.]

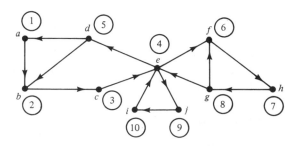

Figure 11.12 The strongly connected orientation for the graph G of Figure 11.1 obtained from the labeling and marking in that figure.

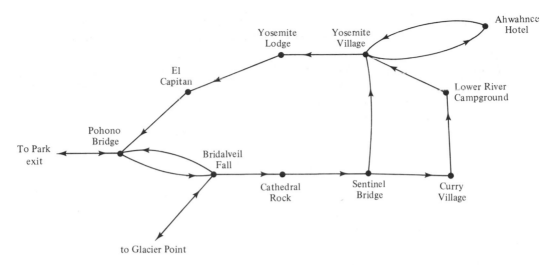

Figure 11.13 One-way street assignment for Yosemite Valley automobile traffic, summer of 1973. (*Note:* Buses may go both ways between Yosemite Lodge and Yosemite Village and between Yosemite Village and Sentinel Bridge.)

11.2.3 Efficient One-way Street Assignments

Not every one-way street assignment for a city's streets is very efficient. Consider, for example, the one-way street assignment shown in the graph of Figure 11.11. This one-way street assignment is obtainable by the algorithm we have described (Exercise 4). If a person at location *a* wanted to reach location *b*, he or she used to be able to do it very quickly, but now has to travel a long way around. In short, this assignment meets the criterion we have set up, of being able to go from any point to any other point, but it does not give a very efficient solution to the traffic flow problem. In general, the most efficient one-way street assignment is one in which "on the whole" distances traveled are not too great. See Roberts [1978] for several attempts to make this notion precise. See also Exercises 14–17. Unfortunately, it turns out that for several reasonable definitions of efficiency, no one has been able to find a good algorithm for obtaining an efficient (the most efficient) one-way street assignment or strongly connected orientation. Indeed, there is some evidence that no such algorithm exists.* Again, see Roberts [1978] for a discussion.

It should be remarked that efficiency is not always the goal of a one-way street assignment. In the National Park System throughout the United States, traffic congestion has become a serious problem. A solution being implemented by the U.S. National Park Service is to try to discourage people from driving during their visits to national parks. This can be done by designing very inefficient one-way street assignments, which make it hard to get from one place to another by car, and by encouraging people to use alternatives, for example bicycles or buses. Figure 11.13 shows approximately the one-way

*Specifically, Chvátal and Thomassen [1978] have shown that for at least one definition of efficiency, the problem of finding the most efficient strongly connected orientation is NP-hard, to use the terminology of Section 2.18 (see Exercise 14).

street assignment used in the Yosemite Valley section of Yosemite National Park during the summer of 1973.* (Note that there is a two-way street; the idea of making *every* street one-way is obviously too simple.†) This is a highly inefficient system. For example, to get from Yosemite Lodge to Yosemite Village, a distance of less than 1 mile by road, it is necessary to drive all the way around via Pohono Bridge, a distance of over 8 miles! However, the Park's buses are allowed to go directly from Yosemite Lodge to Yosemite Village. Many people are riding them!

From a mathematical point of view, it would be nice to find algorithms for finding an inefficient (the most inefficient) one-way street assignment, just as to find an efficient (the most efficient) such assignment. Unfortunately, no algorithms for inefficient assignments are known either.

EXERCISES FOR SECTION 11.2

1. In each graph of Figure 11.3, find all bridges.
2. Apply the procedure described in the text for finding one-way street assignments to the graphs of Figure 11.3. In each case, check if the assignment obtained is strongly connected.
3. Repeat Exercise 2 for the graphs of Figure 3.25.
4. Show that the one-way street assignment shown in Figure 11.11 is attainable using the algorithm described in the text.
5. Find conditions for a graph to have a weakly connected orientation (Exercise 10, Section 3.2).
6. If a graph has a weakly connected orientation, how many weakly connected orientations does it have?
7. Give an example of a graph that has a weakly connected orientation but not a unilaterally connected orientation (Exercise 9, Section 3.2).
8. Let G be a connected graph. A *cut vertex* of G is a vertex with the following property: When you remove it and all edges to which it belongs, the result is a disconnected graph. In each graph of Figure 11.3, find all cut vertices.
9. Can a graph with a cut vertex (Exercise 8) have a strongly connected orientation?
10. Prove or disprove: A connected graph with no cut vertices (Exercise 8) has a strongly connected orientation.
11. Prove that an edge $\{u, v\}$ in a connected graph G is a bridge if and only if every chain from u to v in G includes edge $\{u, v\}$.
12. In this exercise we consider the case of cities in which some streets are one-way to begin with. Our approach is based on Boesch and Tindell [1980]. A *mixed graph* consists of a set of vertices, some of which are joined by one-way arcs and some of which are joined by undirected edges. (The arcs correspond to one-way streets, the edges to two-way streets.) A mixed graph G can be translated into a digraph $D(G)$ by letting each edge be replaced by two arcs, one in each direction. G will be called *strongly connected* if $D(G)$ is strongly connected, and *connected* if $D(G)$ is weakly connected. An edge α in a connected mixed graph is called a *bridge* if removal of α but not its end vertices results in a mixed graph that is not connected.
    (a) Suppose that G is a strongly connected mixed graph and that $\{u, v\}$ is an edge of G. Let D' be the digraph obtained from $D(G)$ by omitting arcs (u, v) and (v, u) but not vertices u and v.

*In addition to making roads one-way, the Park Service has closed off others to cars entirely.

†In this situation, there is no satisfactory assignment in which every street is one-way. Why?

Let A be the set of all vertices reachable from u by a path in D', less the vertex u. Let B be defined similarly from v. Suppose that u is not in B and v is not in A. Prove that the edge $\{u, v\}$ must be a bridge of G.

(b) Use the result of part (a) to prove the following theorem of Boesch and Tindell [1980]: If G is a strongly connected mixed graph and $\{u, v\}$ is an edge of G that is not a bridge, then there is an orientation of $\{u, v\}$ so that the resulting mixed graph is still strongly connected.

(c) Prove from part (b) that every connected graph without bridges has a strongly connected orientation.

(d) Translate your proof of part (c) into an algorithm for finding a strongly connected orientation of a connected, bridgeless graph.

(e) Illustrate your algorithm in part (d) on the graphs of Figure 11.3.

(f) What is the computational complexity of your algorithm in part (d)? How does this compare to the algorithm described in the text?

13. Exercises 14–17 consider alternative definitions of efficiency of a strongly connected orientation. In this exercise we present some mathematical preliminaries. If u and v are two vertices in a digraph D, the *distance* $d_D(u, v)$ from u to v is the length of the shortest path from u to v in D. Distance is undefined if there is no path from u to v. Similarly, if u and v are two vertices in a graph G, the *distance* $d_G(u, v)$ from u to v is the length of the shortest chain from u to v in G. Again, distance is undefined if there is no chain.

(a) Show that $d_D(u, v)$ may be different from $d_D(v, u)$.

(b) Show that $d_G(u, v)$ always equals $d_G(v, u)$.

(c) Show that if v is reachable from u and w from v, then $d_D(u, w) \le d_D(u, v) + d_D(v, w)$.

14. One notion of efficiency of a strongly connected orientation is the following: Among all strongly connected orientations D of G, a most efficient one is one in which the maximum distance $d_D(u, v)$ over all u, v in V is as small as possible.

(a) For each graph of Figure 11.14, find a most efficient strongly connected orientation in this sense.

(b) Comment on the efficiency in this sense of the one-way street assignment for Yosemite Valley which is shown in Figure 11.13.

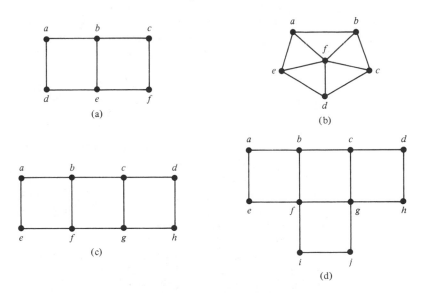

Figure 11.14   Graphs for exercises of Section 11.2.

(c) Consider a graph that arises from a city with m north-south streets and n east-west avenues. Comment, at least for the case $n = 2$, on the efficiency of the one-way street assignment which alternates one-way street signs, first north, then south; and similarly, first east, then west.

(*Note:* Chvátal and Thomassen [1978] show that finding a most efficient strongly connected orientation in the sense of this exercise is a combinatorial problem that may have no efficient algorithm to solve it. Specifically, they show that it is NP-hard. Even for the specific graphs of part (c), no one has been able to determine the most efficient strongly connected orientation in the case of arbitrary n.)

15. Repeat Exercise 14 for the following notion of efficiency: Among all strongly connected orientations D of G, a most efficient one is one in which the average distance $d_D(u, v)$ over all u, v in V is as small as possible.

16. Repeat Exercise 14 for the following notion of efficiency: Among all strongly connected orientations D of G, a most efficient one is one in which the maximum of the differences between the distances $d_D(u, v)$ and $d_G(u, v)$ is as small as possible.

17. Repeat Exercise 14 for the following notion of efficiency: Among all strongly connected orientations D of G, a most efficient one is one in which the difference between the distances $d_D(u, v)$ and $d_G(u, v)$ is on the average as small as possible.

18. One of the major concerns with transportation networks, communication networks, telephone networks, electrical networks, and so on, is to construct them so that they are not very vulnerable to disruption. Motivated by the notion of bridge, we shall be interested in Exercises 18–26 in the disruption of connectedness that arises from the removal of arcs or vertices. We say that a digraph D is in *connectedness category 3* if it is strongly connected, in *category 2* if it is unilaterally connected but not strongly connected, in *category 1* if it is weakly connected but not unilaterally connected, and in *category 0* if it is not weakly connected. Find the connectedness category of every digraph of Figure 3.6.

19. If digraph D is in connectedness category i (Exercise 18), we say that arc (u, v) is an (i, j)-*arc* if removal of (u, v) (but not vertices u and v) results in a digraph of category j. For each pair (i, j), $i = 0, 1, 2, 3$, and $j = 0, 1, 2, 3$, either give an example of a digraph with an (i, j)-arc or prove there is no such digraph.

20. If digraph D is in connectedness category i, we say that vertex u is an (i, j)-*vertex* if the subgraph generated by vertices of D other than u is in category j. For each pair (i, j), $i = 0, 1, 2, 3$ and $j = 0, 1, 2, 3$, either give an example of a digraph with an (i, j)-vertex or prove that there is no such digraph.

21. Let us define the *arc vulnerability* of a digraph in connectedness categories 1, 2, or 3 as the minimum number of arcs whose removal results in a digraph of lower connectedness category. For each digraph of Figure 11.15, determine its arc vulnerability.

22. Give an example of a digraph D with arc vulnerability equal to 4.

23. For every k, give an example of a digraph D with arc vulnerability equal to k.

24. Give an example of a digraph with n vertices and arc vulnerability equal to $n - 1$. Could there be a strongly connected digraph with n vertices and arc vulnerability equal to n?

25. (a) If D is strongly connected, what is the relation between the arc vulnerability of D and the minimum *indegree* of a vertex, the minimum number of incoming arcs of a vertex?
    (b) Show that the arc vulnerability of a strongly connected digraph is at most a/n, where a is the number of arcs and n the number of vertices.

26. (Whitney [1932])   If there is a set of vertices in a digraph D whose removal results in a digraph in a lower connectedness category than D, then we define the *vertex vulnerability* of D to be the size of the smallest such set of vertices. Otherwise, vertex vulnerability is undefined.
    (a) Show that if D is weakly connected and there is at least one pair $u \neq v$ such that neither

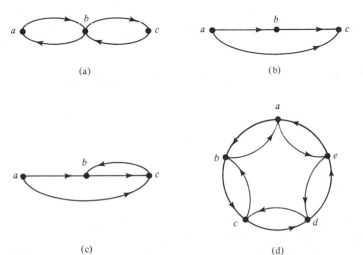

(a)

(b)

(c)

(d)

Figure 11.15 Digraphs for exercises of Section 11.2.

(u, v) nor (v, u) is an arc of D, then the vertex vulnerability of D is less than or equal to the minimum total degree of any vertex of D. [The *total degree* of a vertex x is the sum of the *indegree* (the number of incoming arcs) of x and the *outdegree* (the number of outgoing arcs) of x.]

(b) Show that the vertex vulnerability and the minimum total degree can be diffcrent.

(c) Why do we need to assume that there is at least one pair $u \neq v$ such that neither (u, v) nor (v, u) is an arc of D?

11.3 EULERIAN CHAINS AND PATHS

11.3.1 The Königsberg Bridge Problem

Graph theory was invented by the famous mathematician Leonhard Euler [1736] in the process of settling the famous Königsberg bridge problem.* Euler's techniques have found modern applications in the study of street sweeping, mechanical plotting by computer, RNA chains, coding, telecommunications, and other subjects, and we shall survey some of these applications in Section 11.4. Here, we shall present the Königsberg bridge problem and then present general techniques arising from its solution.

The city of Königsberg had seven bridges linking islands in the River Pregel to the banks and to each other, as shown in Figure 11.16. The residents wanted to know if it was possible to take a walk that starts at some point, crosses each bridge exactly once, and returns to the starting point. Euler translated this into a graph theory problem by letting the various land areas be vertices and joining two vertices by one edge for each bridge joining them. The resulting object, shown in Figure 11.17, is called a *multigraph*. We shall use the terms *multigraph* and *multidigraph* when more than one edge or arc is allowed between two vertices x and y or from vertex x to vertex y. The concepts of connectedness, such as chain, circuit, path, cycle, component, and so on, are defined just as for graphs

*See "The Königsberg Bridges," *Scientific American, 189* (1953), 66–70.

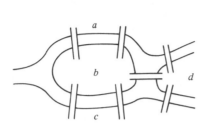
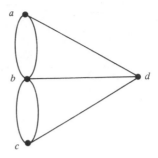

Figure 11.16   The Königsberg bridges.

Figure 11.17   Multigraph obtained
from Figure 11.16.

and digraphs. (Note, however, that in a multigraph, we can have a circuit from a to b to a, if there are two edges between a and b.) Loops will be allowed here.

We say that a *chain* in a multigraph G or *path* in a multidigraph D is *eulerian* if it uses every edge of G or arc of D once and only once. Euler observed that the citizens of Königsberg were seeking an *eulerian closed chain* in the multigraph of Figure 11.17. He asked the following general question: When does a multigraph G have an eulerian closed chain? We begin by attempting to answer this question. Clearly, if G has an eulerian closed chain, G must be *connected up to isolated* vertices*, that is, at most one component has an edge. Moreover, an eulerian closed chain must leave each vertex x as often as it enters x, and hence each vertex x must have even degree, where the degree of x, the reader will recall, is the number of neighbors of x. Euler discovered the following result.

Theorem 11.2   (Euler).   A multigraph G has an eulerian closed chain if and only if G is connected up to isolated vertices and every vertex of G has even degree.

Before proving the sufficiency of the condition stated in Theorem 11.2, we observe that the multigraph of Figure 11.17 does not have an eulerian closed chain. For vertex a has odd degree. Thus, the citizens of Königsberg could not complete their walk. To further illustrate the theorem, we note that the multigraph G of Figure 11.18 is connected and every vertex has even degree. An eulerian closed chain is given by $a, b, c, d, e, f, h, f, a, g, e, d, b, a$. If every vertex of a multigraph G has even degree, we say that G is *even*.

In applying Theorem 11.2 to the situation where there are loops, we have to be careful to add 2 to the degree of vertex x for every loop from x to x. (Why?)

11.3.2 An Algorithm for Finding an Eulerian Closed Chain

A good way to prove sufficiency in Theorem 11.2 is to describe a procedure for finding an eulerian closed chain in an even multigraph. Start with any vertex x that has a neighbor and choose any edge joining x, say $\{x, y\}$. Next, choose any edge $\{y, z\}$ joining y, making sure not to choose a previously used edge. Continue at any vertex by choosing a previously unused edge which joins that vertex. Now each time we pass through a vertex, we use up two adjoining edges. Thus, the number of unused edges joining any vertex other

---

*A vertex is called *isolated* if it has no neighbors.

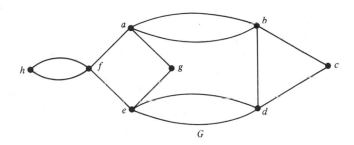

Figure 11.18 Multigraph with an eulerian closed chain.

than x remains even at all times. It follows that any time we enter such a vertex other than x, we can leave it. Since we started at x, the number of unused edges joining x remains odd at all times. Continue the procedure until it is impossible to continue. Now since every vertex except x can be left whenever it is entered, the only way the procedure can end is back at x. Let us illustrate it on the multigraph G of Figure 11.18. Suppose that x is c. We can use edge $\{c, b\}$ to get to b, then edge $\{b, a\}$ to get to a, then edge $\{a, g\}$ to get to g, then edge $\{g, e\}$ to get to e, then edge $\{e, d\}$ to get to d, and then edge $\{d, c\}$ to return to c. At this point, we can go no further. Note that we have found a closed chain c, b, a, g, e, d, c starting and ending at c which uses each edge of G at most once. However, the chain does not use up all the edges of G. Let us call the procedure so far "finding a closed chain from x to x," or CL CHAIN (x, x) for short.

The algorithm can now continue. Note that at this stage, we have a chain C from x to x. Moreover, every vertex has an even number of unused joining edges. Since the graph is connected up to isolated vertices, if there is any unused edge, there must be at least one unused edge joining a vertex u on the chain C. In the multigraph of unused edges, we apply the procedure CL CHAIN (u, u) to find a closed chain D from u to u which uses each previously unused edge at most once. We can now modify our original closed chain C from x to x by inserting the "detour" D when we first hit u. We get a new closed chain C' from x to x which uses each edge of G at most once. If there are still unused edges, we repeat the procedure. We must eventually use up all the edges of the original multigraph. Continuing with our example, note that we have so far obtained the closed chain $C = c, b, a, g, e, d, c$. One unused edge joining this chain is the edge $\{a, f\}$. Since a is on C, we look for a closed chain of unused edges from a to a. Such a chain is found by our earlier procedure CL CHAIN (a, a). One example is $D = a, f, e, d, b, a$. Note that we used the second edges $\{e, d\}$ and $\{b, a\}$ here. The first ones have already been used. We insert the detour D into C, obtaining the closed chain $C' = c, b, a, f, e, d, b, a, g, e, d, c$. Since there are still unused edges, we repeat the process again. We find a vertex f on C' which joins an unused edge, and we apply CL CHAIN (f, f) to find a closed chain f, h, f from f to f. We insert this detour into C' to find $C'' = c, b, a, f, h, f, e, d, b, a, g, e, d, c$, which uses each edge once and only once.

The algorithm we have described can now be formalized. We have a subroutine called CL CHAIN (x, x) which is described as follows.

Algorithm 11.3.   CL CHAIN (x, x).

> *Input:* A multigraph G, a set U of unused edges of G, with each vertex appearing in an even number of edges in U, and a designated vertex x which appears in some edge of U.

Output: A closed chain from x to x which uses each edge of U at most once.

Step 1. Set $v = x$ and output x. (Here, v is the last vertex visited.)

Step 2. If there is an edge $\{v, y\}$ in U, set $v = y$, output y, remove $\{v, y\}$ from U, and repeat this step.

Step 3. If there is no edge $\{v, y\}$ in U, stop. The outputs in order give us the desired chain from x to x.

Using Algorithm 11.3, we can now summarize the entire procedure.

Algorithm 11.4.   Finding An Eulerian Closed Chain

Input: An even multigraph G which is connected up to isolated vertices.

Output: An eulerian closed chain.

Step 1. Find any vertex x that has a neighbor. (If there is none, every vertex is isolated and any vertex x alone defines an eulerian closed chain. Output this and stop.) Let $U = E(G)$.

Step 2. Apply CL CHAIN (x, x) to obtain a chain C.

Step 3. Remove all edges in C from U. (Technically, this is already done in step 2.)

Step 4.

    Step 4.1. If $U = \varnothing$, stop and output C.

    Step 4.2. If $U \neq \varnothing$, find a vertex u on C that has a joining edge in U, and go to step 5.

Step 5. Apply CL CHAIN (u, u) to obtain a chain D.

Step 6. Redefine C by inserting the detour D at the first point u is visited, remove all edges of D from U, and go to step 4.

For a more formal statement of this algorithm, see Even [1979]. Even shows that the algorithm can be completed in the worst case in a number of steps that is a constant k times the number of edges e. Hence, if G is a graph, i.e., if there are no multiple edges, the computational complexity of the algorithm we have described is

$$ke \leq k \frac{n(n-1)}{2} = \frac{k}{2} n^2 - \frac{k}{2} n,$$

a polynomial bound in terms of n. [In the terminology of Section 2.18, this is an $O(e)$ or, for graphs, an $O(n^2)$ algorithm.] Since it is clear that the algorithm works, Theorem 11.2 is proved.

An alternative algorithm for finding an eulerian closed chain would use the depth-first search procedure of Section 11.1. We leave the description of such an algorithm to the reader (Exercise 16).

11.3.3 Further Results about Eulerian Chains and Paths

The next theorem tells when a multigraph has an eulerian chain (not necessarily closed). We leave the proof to the reader.

Figure 11.19 Multigraph with an eulerian chain.

Theorem 11.3 (Euler). A multigraph G has an eulerian chain if and only if G is connected up to isolated vertices and the number of vertices of odd degree is either 0 or 2.

According to this theorem, the multigraph of Figure 11.17 does not have an eulerian chain. However, the multigraph of Figure 11.19 does, since there are exactly two vertices of odd degree.

We now state analogous results for multidigraphs, leaving the proofs to the reader. In these theorems, the *indegree (outdegree)* of a vertex is the number of incoming (outgoing) arcs. A digraph (multidigraph) is called *weakly connected* if, when all directions on arcs are disregarded, the resulting graph (multigraph) is connected.

Theorem 11.4 (Good [1946]). A multidigraph D has an eulerian closed path if and only if D is weakly connected up to isolated vertices* and for every vertex, indegree equals outdegree.

Theorem 11.5 (Good [1946]). A multidigraph D has an eulerian path if and only if D is weakly connected up to isolated vertices and for all vertices with the possible exception of two, indegree equals outdegree, and for at most two vertices, indegree and outdegree differ by one.

Theorem 11.5 is illustrated by the two multidigraphs of Figure 11.20. Neither has an eulerian path. In the first example, there are four vertices where indegree is different from outdegree. In the second example, there are just two such vertices, but for each the indegree and outdegree differ by more than 1. Note that the hypotheses of Theorem 11.5 imply that if indegree and outdegree differ for any vertex, then exactly one vertex has an excess of one indegree and exactly one vertex has an excess of one outdegree (see Exercise 9). These vertices turn out to correspond to the first and last vertices of the eulerian path.

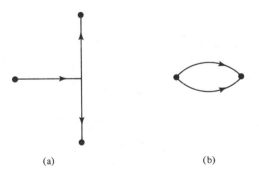

(a) (b)

Figure 11.20 Two multidigraphs without eulerian paths.

*That is, when directions on arcs are disregarded, the underlying multigraph is connected up to isolated vertices.

Note that Theorems 11.2–11.5 hold if there are loops. Note also that in a multi-graph, a loop adds 2 to the degree of a vertex, while in a multidigraph, it adds 1 to indegree and 1 to outdegree. Thus, loops do not affect the existence of eulerian (closed) chains or paths.

EXERCISES FOR SECTION 11.3

1. For each river with bridges shown in Figure 11.21, build a corresponding multigraph and determine if it is possible to take a walk that starts at a given location, crosses each bridge once and only once, and returns to the starting location.

2. Which of the multigraphs of Figure 11.22 have an eulerian closed chain? For those that do, find one.

3. Of the multigraphs of Figure 11.22 that do not have an eulerian closed chain, which have an eulerian chain?

4. Which of the multidigraphs of Figure 11.23 have an eulerian closed path? For those that do, find one.

5. Of the multidigraphs of Figure 11.23 that do not have an eulerian closed path, which have an eulerian path?

6. For each multigraph G of Figure 11.24, apply the subroutine CL CHAIN (a, a) to the vertex a with $U = E(G)$.

7. For each multigraph of Figure 11.24, apply Algorithm 11.4 to find an eulerian closed chain.

8. How would you modify Algorithm 11.4 to find an eulerian chain from x to y?

9. Show that in a multidigraph with an eulerian path but no eulerian closed path, exactly one vertex has an excess of one indegree and exactly one vertex has an excess of one outdegree.

10. (a) Can the drawing of Figure 11.25(a) be made without taking your pencil off the paper or retracing?
    (b) What about Figure 11.25(b)?
    (c) What about Figure 11.25(c)?

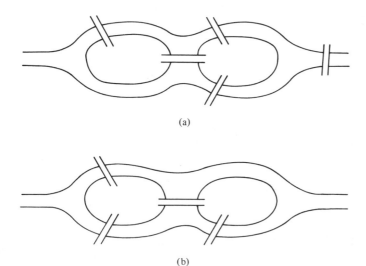

(a)

(b)

Figure 11.21 Rivers and bridges for Exercise 1, Section 11.3.

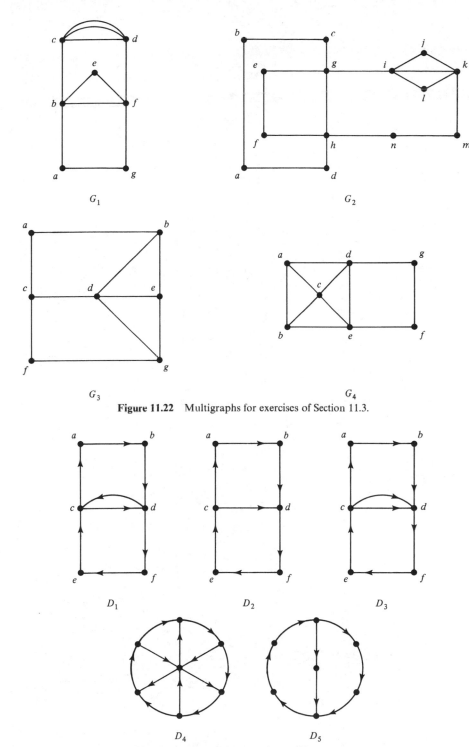

Figure 11.22 Multigraphs for exercises of Section 11.3.

Figure 11.23 Multidigraphs for exercises of Section 11.3.

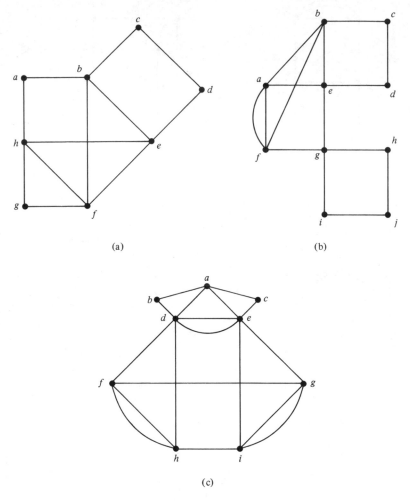

(a)   (b)

(c)

Figure 11.24   Multigraphs for exercises of Section 11.3.

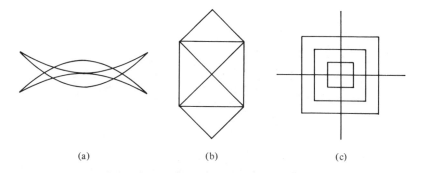

(a)   (b)   (c)

Figure 11.25   Drawings for Exercise 10, Section 11.3.

11. (Harary and Palmer [1973]) Show that the number of labeled even graphs of n vertices equals the number of labeled graphs of $n - 1$ vertices. (*Hint:* To a labeled graph G of $n - 1$ vertices, add a vertex labeled n and join it to the vertices of G of odd degree.)

12. Show that every digraph without isolated vertices and with an eulerian closed path is strongly connected.

13. Show that every digraph without isolated vertices and with an eulerian path is unilaterally connected.

14. Does every strongly connected digraph have an eulerian closed path? Why?

15. Does every unilaterally connected digraph have an eulerian path? Why?

16. Describe an algorithm for finding an eulerian closed chain that uses the depth-first search procedure.

17. Prove the necessity part of Theorem 11.3.

18. Prove the sufficiency part of Theorem 11.3.

19. Prove Theorem 11.4.

20. Prove Theorem 11.5.

11.4 APPLICATIONS OF EULERIAN CHAINS AND PATHS

In this section we present various applications of the ideas presented in Section 11.3. The different subsections of this section are independent, except Sections 11.4.2 and 11.4.3 depend on Section 11.4.1. Other than this, the subsections may be used in any order. The reader without the time to cover all of these subsections might sample Sections 11.4.1, 11.4.2, and 11.4.4, or 11.4.6.

11.4.1 The "Chinese Postman" Problem

A mail carrier starting out from a post office must deliver letters to each block in a territory and return to the post office. What is the least amount of walking the mail carrier can do? This problem was originally studied by Kwan [1962], and has traditionally been called the *Chinese postman problem*, or just the *postman problem*. A similar problem is faced by many delivery people, by farmers who have fields to seed, by street repair crews, and so on.

We can represent the mail carrier's problem by building a graph G with each vertex representing a street corner and each edge a street.* Assuming for simplicity that the post office is near a street corner, the mail carrier then seeks a closed chain in this graph, which starts and ends at the vertex x corresponding to the corner where the post office is located, and which uses each edge of the graph *at least once*.

If the graph G has an eulerian closed chain, we can pick up this chain at x and follow it back to x. No chain can give a shorter route. If there is no such eulerian closed chain, we can formulate the problem as follows. Any mail carrier's route will use all of the edges of G once and possibly some of them more than once. Suppose that we modify G by replacing each edge by as many copies of it as are in the mail carrier's route, or equiva-

---

*This already simplifies the problem. A street between x and y with houses on both sides should really be represented by two edges between x and y. A mail carrier can walk up one side of the street first, and later walk up (or down) the other side.

lently adding to G enough copies of each edge to exactly achieve the mail carrier's route. Then in the resulting multigraph, the mail carrier's route corresponds to an eulerian closed chain from x to x. For instance, consider the graph G of Figure 11.26, which represents a four-square-block area in a city. There is no eulerian closed chain in G, because for instance vertex d has degree 3. A possible mail carrier's route would be the closed chain $x, h, g, d, e, f, c, b, a, d, e, b, e, h, x, f, x$. This corresponds to the first multigraph also shown in Figure 11.26. (Added edges are dashed.) An alternative route would be $x, h, e, h, g, d, e, d, g, d, a, b, e, b, c, f, e, f, x$. This corresponds to the second multigraph shown in Figure 11.26. The first of these routes is the shorter; equivalently, it requires the addition of fewer copies of edges of G. (The second route can be shortened by omitting the d, g, d part.)

The problem of trying to find the shortest mail carrier's route from x to x in G is equivalent to the problem of determining the smallest number of copies of edges of G to add to G to obtain a multigraph which has an eulerian closed chain, that is, one in which all vertices have even degree. A general method for solving this combinatorial optimization problem is to translate it into the maximum weight matching problem of the type we will discuss in Chapter 12. We will discuss this translation specifically in Section 13.2.4. In our example it is easy to see that since there are four vertices of odd degree, and no two are adjacent, at least four edges must be added. Hence, the first multigraph of Figure 11.26 corresponds to a shortest mail carrier's route.

In Section 11.4.3 we will generalize this problem. For a good general discussion of the "Chinese postman" problem, see Lawler [1976] or Minieka [1978].

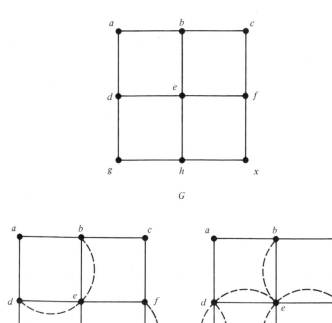

Figure 11.26 Graph representing a mail carrier's territory and two multigraphs corresponding to mail carriers' routes.

11.4.2 Computer Graph Plotting

Reingold and Tarjan [1981] point out that the "Chinese postman" problem of Section 11.4.1 arises in mechanical plotting by computer of a graph with prespecified vertex locations. Applications of mechanical graph plotting described by Reingold and Tarjan include shock-wave propagation problems, where meshes of thousands of vertices must be plotted, map drawing, electrical networks, and activity charts (see Reingold and Tarjan's paper for references).

Much time is wasted in mechanical graph plotting when the plotter pen moves with the pen off the paper. Thus, we seek to minimize the number of moves from vertex to vertex with the pen off the paper. This is exactly the problem of minimizing the number of edges to be added to the graph being plotted so as to obtain a multigraph with an eulerian closed chain.

11.4.3 Street Sweeping

A large area for applications of combinatorial techniques is the area of urban services. Cities spend billions of dollars a year providing such services. Combinatorics has been applied to problems involving the location and staffing of fire and police stations, design of rapid transit systems, assignments of shifts for municipal workers, routing of street-sweeping and snow-removal vehicles, and so on. See Beltrami [1977] and Helly [1975] for a variety of examples. We have previously discussed in Example 3.12 the problem of routing garbage trucks to pick up garbage. Here we discuss the problem of determining optimal routes for street-sweeping or snow-removal equipment. We follow Tucker and Bodin [1983] and Roberts [1978]. See also Liebling [1970].

Consider the street-sweeping problem for concreteness. Let the street corners in the neighborhood to be swept be the vertices of a multidigraph. Include an arc from x to y if there is a curb that can be traveled from x to y. In general, a one-way street will give rise to two arcs from x to y, which is why we get a multidigraph. Call this multidigraph the *curb multidigraph*.

During a given period of time, certain curbs are to be swept. The corresponding arcs define a subgraph of the curb multidigraph called the *sweep subgraph*. (The curbs to be swept in a large city such as New York are kept free of cars during the period in question by parking regulations.)

Now any arc in the sweep subgraph has associated with it a number indicating the length of time required to sweep the corresponding curb. Also, any arc in the curb multidigraph has associated with it a number indicating the length of time required to follow the corresponding curb without sweeping it. This is called the *deadheading time*.

Figure 11.27 shows a curb multidigraph. Some arcs are solid—these define the sweep subgraph. Each arc has a number in a square; this is the deadheading time. The solid arcs have in addition a number in a circle; this is the sweep time.

We would like to find a way to start from a particular location (the garage), sweep all the curbs in the sweep subgraph, and return to the start, and use as little time as possible. This is a generalization of the "Chinese postman" problem studied in Section 11.4.1, and our approach to it is similar. We seek a closed path in the curb multidigraph which includes all arcs of the sweep subgraph. The time associated with any acceptable path is the sum of the sweeping times for arcs swept plus the sum of the deadheading

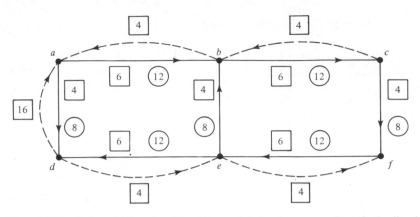

Figure 11.27 Curb multidigraph, with solid arcs defining the sweep subgraph, deadheading times in squares, and sweep times in circles.

times for arcs in the path which are not swept. Note that an arc can appear several times in the path. Even if it is a solid arc, we count the sweeping time only once. It is swept only the first time it is traversed. For each other appearance of the arc, we count the deadheading time.

If the sweep subgraph has an eulerian closed path, then as in the "Chinese postman" problem, this path must be the minimal time street-sweeping route. In our example, Theorem 11.4 implies that there could be no such path. For there are a number of vertices in the sweep subgraph for which indegree is different from outdegree. Thus, a solution to the street-sweeping problem will be a closed path P of the curb multidigraph using some arcs not in the sweep subgraph. Suppose that we add to the sweep subgraph all the arcs used in P for deadheading, either because they are not in the sweep subgraph or because they have previously been swept in P. Add these arcs as many times as they are used. Adding these arcs gives rise to a multidigraph in which the path P is an eulerian closed path. For instance, one closed path in the curb multidigraph of Figure 11.27 which uses all arcs of the sweep subgraph is given by

$$f, e, b, c, f, e, d, a, d, a, b, c, f. \tag{11.1}$$

(Assume that the garage is at f.) Path (11.1) corresponds to adding to the sweep subgraph the dashed arcs shown in Figure 11.28.

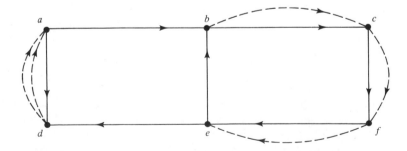

Figure 11.28 Multidigraph with an eulerian closed path (corresponding to (11.1)) obtained from the sweep subgraph of Figure 11.27 by adding the dashed arcs.

In general, we want the sum of the deadheading times on the added arcs to be minimized,* and we want to add deadheading arcs so that in the resulting multidigraph, there is an eulerian closed path; that is, every vertex has indegree equal to outdegree. Tucker and Bodin [1983] show how to solve the resulting combinatorial optimization problem by setting it up as a so-called transportation problem of the type to be discussed in Section 13.4.1. The approach is similar to the approach used in the "Chinese postman" problem in Section 13.2.4. In our example, an optimal route is the path $f, e, d, e, b, a, b, a, d, e, b, c, f$. The total time required for this route is 72 for sweeping plus 20 for deadheading. The total time required for the route (11.1) is 72 for sweeping plus 48 for deadheading, which is considerably worse.

11.4.4 RNA Chains and Complete Digests

In Section 2.13 we discussed the problem of finding an RNA chain given knowledge of its fragments after applying two enzymes, one of which breaks up the chain after each G link and the other of which breaks up the chain after each U or C link. In this subsection we show how to use the notion of eulerian closed path to find all RNA chains with given G and U, C fragments.

If there is only one G fragment or only one U, C fragment, the chain is determined. Hence, we shall assume that there are at least two G fragments and at least two U, C fragments.

We shall illustrate the procedure with the following G and U, C fragments:

G fragments: CCG, G, UCACG, AAAG, AA

U, C fragments: C, C, GGU, C, AC, GAAAGAA.

We first break down each fragment after each G, U, or C, for instance breaking fragment GAAAGAA into G·AAAG·AA, GGU into G·G·U, and UCACG into U·C·AC·G. Each piece is called an *extended base*, and all extended bases in a fragment except the first and last are called *interior extended bases*. Using this further breakup of the fragments, we shall first observe how to find the beginning and end of the desired RNA chain. We make two lists, one giving all interior extended bases of all fragments from both digests, and one giving all fragments consisting of one extended base. In our example, we obtain the following lists:

Interior extended bases: C, C, AC, G, AAAG

Fragments consisting of one extended base: G, AAAG, AA, C, C, C, AC.

It is not hard to show that every entry on the first list is on the second list (Exercise 11). Moreover, the first and last extended bases in the entire chain are on the second list but not the first (Exercise 11). Since we are assuming that there are at least two G fragments and at least two U, C fragments, it is not hard to show that there will always be exactly two entries on the second list which are not on the first list (Exercise 11). One of these will

---

*We are omitting sources of delay other than deadheading, for example, delays associated with turns. These delays can be introduced by defining a system of penalties associated with different kinds of turns (see Tucker and Bodin [1983]).

be the first extended base of the RNA chain and one will be the last. In our example, these are AA and C. How do we tell which is last? We do it by observing that one of these entries will be from an *abnormal fragment*, that is, it will be the last extended base of a G fragment not ending in G or a U, C fragment not ending in U or C. In our example, AA is the last extended base of two abnormal fragments AA and GAAAGAA. This implies that the chain we are seeking begins in C and ends in AA.

To find all possible chains, we build a multidigraph as follows. First identify all normal fragments with more than one extended base. From each such fragment, use the first and last extended bases as vertices and draw an arc from the first to the last, labeling it with the corresponding fragment. (We shall add one more arc shortly.) Figure 11.29 illustrates this construction. For example, we have included an arc from U to G, labeled with the name of the corresponding fragment UCACG. Similarly, we add arcs from C to G and G to U. There might be several arcs from a given extended base to another, if there are several normal fragments with the same first and last extended base. Finally, we add one additional arc to our multidigraph. This is obtained by identifying the longest abnormal fragment—here this is GAAAGAA—and drawing an arc from the first (and perhaps only) extended base in this abnormal fragment to the first extended base in the chain. Here, we add an arc from G to C. We label this arc differently, by marking it $X * Y$, where X is the longest abnormal fragment, $*$ is a symbol marking this as a special arc, and Y is the first extended base in the chain. Hence, in our example, we label the arc from G to C by GAAAGAA $*$ C. Every possible RNA chain with the given G and U, C fragments can now be identified from the multidigraph we have built. It turns out that each such chain corresponds to an eulerian closed path which ends with the special arc $X * Y$ (Exercise 14). In our example, the only such eulerian closed path goes from C to G to U to G to C. By using the corresponding arc labeling, we obtain the chain

$$CCGGUCACGAAAGAA.$$

It is easy to check that this chain has the desired G and U, C fragments. For more details on this graph-theoretic approach, see Hutchinson [1969].

11.4.5 A Coding Application

Hutchinson and Wilf [1975] study codes on an alphabet of n letters, under the following assumption: All the information in a codeword is carried in the number of letters of each type and in the frequency of ordered pairs of letters, that is, the frequency with which one letter follows a second. For instance, Hutchinson and Wilf treat a DNA or RNA molecule as a word, with bases (not extended bases) as the letters, and make the assumption above about the genetic code. Specifically, Hutchinson and Wilf ask the following question: Given nonnegative integers v_i, v_{ij}, $i, j = 1, 2, \ldots, n$, is there a word from an alphabet of n letters so that the ith letter occurs exactly v_i times and so that the ith letter is followed by

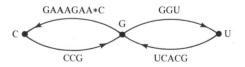

Figure 11.29 Multidigraph for reconstruction of RNA chains from complete enzyme digest.

the jth exactly v_{ij} times? If so, what are all such words? We shall present Hutchinson and Wilf's solution, which uses the notion of eulerian path.

To give an example, suppose that $v_1 = 2$, $v_2 = v_3 = 1$ and v_{ij} is given by the following matrix:

$$(v_{ij}) = \begin{pmatrix} 0 & 1 & 0 \\ 0 & 0 & 1 \\ 1 & 0 & 0 \end{pmatrix}. \tag{11.2}$$

Then one word that has the prescribed pattern is $ABCA$, if A corresponds to the first letter, B to the second, and C to the third. To give a second example, suppose that $v_1 = 2$, $v_2 = 4$, $v_3 = 3$, and

$$(v_{ij}) = \begin{pmatrix} 0 & 0 & 2 \\ 2 & 1 & 1 \\ 0 & 2 & 0 \end{pmatrix}. \tag{11.3}$$

One word that has the prescribed pattern is $BBCBACBAC$.

To analyze our problem, let us draw a multidigraph D with vertices the n letters A_1, A_2, ..., A_n, and with v_{ij} arcs from A_i to A_j. Loops are allowed. The multidigraphs corresponding to the matrices of (11.2) and (11.3) are shown in Figures 11.30 and 11.31, respectively. Let us suppose that $w = A_{i_1}, A_{i_2}, ..., A_{i_q}$ is a solution word. Then it is clear that w corresponds in the multidigraph D to an eulerian path which begins at A_{i_1} and ends at A_{i_q}. It is easy to see this for the two solution words we have given for our two examples. In what follows, we use the observation that a solution word corresponds to an eulerian path to learn more about solution words. The reader who wishes to may skip the rest of this subsection.

Since a solution word corresponds to an eulerian path, it follows that if there is a solution word, then D must be weakly connected up to isolated vertices. We consider first the case where $i_1 \neq i_q$. For every $i \neq i_1, i_q$, we have indegree at A_i equal to outdegree. For $i = i_1$, we have outdegree one higher than indegree, and for $i = i_q$, we have indegree one higher than outdegree. Thus, we have

$$\sum_{k=1}^{n} v_{ik} = \begin{cases} \displaystyle\sum_{k=1}^{n} v_{ki}, & i \neq i_1, i_q \\[2ex] \displaystyle\sum_{k=1}^{n} v_{ki} + 1, & i = i_1 \\[2ex] \displaystyle\sum_{k=1}^{n} v_{ki} - 1, & i = i_q. \end{cases} \tag{11.4}$$

This condition says that in the matrix (v_{ij}), the row sums equal the corresponding column sums, except in two places where they are off by one in the indicated manner. We also

$A = A_1$ $B = A_2$ $C = A_3$

Figure 11.30 Multidigraph corresponding to (11.2).

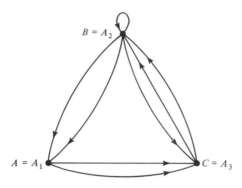

Figure 11.31 Multidigraph corresponding to (11.3).

have a consistency condition, which relates the v_i to the v_{ij}:

$$
v_i = \begin{cases} \sum_{k=1}^{n} v_{ki}, & i \neq i_1 \\ \\ \sum_{k=1}^{n} v_{ki} + 1, & i = i_1. \end{cases}
\tag{11.5}
$$

It is easy to see, using Theorem 11.5, that if conditions (11.4) and (11.5) hold for some i_1 and i_q, $i_1 \neq i_q$, and if D is weakly connected up to isolated vertices, then there is a solution word, and every solution word corresponds to an eulerian path that begins in A_{i_1} and ends in A_{i_q}. In our second example, conditions (11.4) and (11.5) hold with $i_1 = 2$, $i_q = 3$. There are a number of eulerian paths from $B = A_{i_1}$ to $C = A_{i_q}$, each giving rise to a solution word. A second example is $BACBBCBAC$.

What if the solution word begins and ends in the same letter, that is, if $i_1 = i_q$? Then there is an eulerian closed path, and we have

$$
\sum_{k=1}^{n} v_{ik} = \sum_{k=1}^{n} v_{ki}, \qquad i = 1, 2, \ldots, n.
\tag{11.6}
$$

Also, given i_1, (11.5) holds (for all i). Condition (11.6) says that in (v_{ij}), every row sum equals its corresponding column sum. Conversely, if (11.6) holds and for i_1, (11.5) holds (for all i), and if D is weakly connected up to isolated vertices, then there is a solution and every solution word corresponds to an eulerian closed path in the multidigraph D, beginning and ending at A_{i_1}. This is the situation in our first example with $i_1 = 1$.

In sum, if there is a solution word, then D is weakly connected up to isolated vertices and (11.5) holds for some i_1. Moreover, either (11.6) holds, or for i_1 and some i_q, $i_1 \neq i_q$, (11.4) holds. Conversely, suppose that D is weakly connected up to isolated vertices. If (11.5) holds for some i_1 and (11.4) holds for i_1 and some i_q, $i_1 \neq i_q$, then there is a solution and all solution words correspond to eulerian paths beginning at A_{i_1} and ending at A_{i_q}. If (11.5) holds for some i_1 and (11.6) holds, then there is a solution and all solution words correspond to eulerian closed paths beginning and ending at A_{i_1}.

11.4.6 De Bruijn Sequences and Telecommunications

In this subsection we consider another coding problem and its application in telecommunications. Let

$$\Sigma = \{0, 1, \ldots, p - 1\}$$

be an alphabet of p letters and consider all words of length n from Σ. A (p, n) *de Bruijn sequence* (named after N. G. de Bruijn) is a sequence

$$a_0 a_1 \cdots a_{L-1} \tag{11.7}$$

with each a_i in Σ such that every word w of length n from Σ is realized as

$$a_i a_{i+1} \cdots a_{i+n-1} \tag{11.8}$$

for exactly one i, where addition in the subscripts of (11.8) is modulo L.* For instance,

$$01110100$$

is a $(2, 3)$ de Bruijn sequence over the alphabet $\Sigma = \{0, 1\}$ if $n = 3$. For starting with the beginning, the three-letter words $a_i a_{i+1} a_{i+2}$ obtained are, respectively,

$$011, \quad 111, \quad 110, \quad 101, \quad 010, \quad 100, \quad 000, \quad 001.$$

The latter two are the words $a_6 a_7 a_0$ and $a_7 a_0 a_1$. De Bruijn sequences are of great significance in coding theory. They are implemented by shift registers and are sometimes called *shift register sequences*. For a detailed treatment of this topic, see Golomb [1967].

Clearly, if there is a (p, n) de Bruijn sequence (11.7), then L must be p^n where $p = |\Sigma|$. We shall show that for every positive p and n, there is a (p, n) de Bruijn sequence (11.7).

Given p and n, build a digraph $D_{p,n}$, called a *de Bruijn diagram*, as follows. Let $V(D_{p,n})$ consist of all words of length $n - 1$ from alphabet Σ, and include an arc from word $b_1 b_2 \cdots b_{n-1}$ to every word of the form $b_2 b_3 \cdots b_n$. Label such an arc with word $b_1 b_2 \cdots b_n$. Figures 11.32 and 11.33 show the digraph $D_{p,n}$ for cases $p = 2$, $n = 3$ and $p = 3$, $n = 2$. Suppose that the de Bruijn diagram $D_{p,n}$ has an eulerian closed path. We then use successively the first letter from each arc in this path to obtain a de Bruijn sequence (11.7), where $L = p^n$. To see this is a de Bruijn sequence, note that each word w of length n from Σ is realized. For if $w = b_1 b_2 \cdots b_n$, we know that the eulerian path covers the arc from $b_1 b_2 \cdots b_{n-1}$ to $b_2 b_3 \cdots b_n$. Thus, the eulerian path must go from $b_1 b_2 \cdots b_{n-1}$ to $b_2 b_3 \cdots b_n$. From there it must go to $b_3 b_4 \cdots$ to $b_4 b_5 \cdots$ to $b_n \cdots$. Thus, the first letters of the arcs in this path are $b_1 b_2 \cdots b_n$. It is easy to see, using Theorem 11.4, that the de Bruijn diagram of Figure 11.32 has an eulerian closed path. For this digraph is weakly connected and every vertex has indegree equal to outdegree. The de Bruijn sequence 01110100 corresponds to the eulerian closed path that goes from 01 to 11 to 11 to 10 to 01 to 10 to 00 to 00 to 01.

We shall prove the following theorem, which implies that for every pair of positive integers p and n, there is a (p, n) de Bruijn sequence. This theorem was discovered for the case $p = 2$ by de Bruijn [1946], and it was discovered for arbitrary p by Good [1946].

*For a discussion of modular arithmetic, see Section 9.3.1.

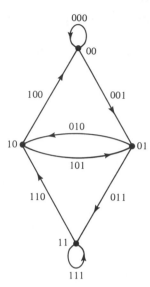

Figure 11.32　The digraph $D_{2,3}$.

Theorem 11.6. For all positive integers p and n, the de Bruijn diagram $D_{p,n}$ has an eulerian closed path.

Proof. We first show that $D_{p,n}$ is weakly connected, indeed, strongly connected. Let $b_1 b_2 \cdots b_{n-1}$ and $c_1 c_2 \cdots c_{n-1}$ be any two vertices of $D_{p,n}$. Since we have a path

$$b_1 b_2 \cdots b_{n-1}, \quad b_2 b_3 \cdots b_{n-1} c_1, \quad b_3 b_4 \cdots b_{n-1} c_1 c_2, \quad \cdots, \quad c_1 c_2 \cdots c_{n-1},$$

$D_{p,n}$ is strongly connected. Next, note that every vertex has indegree and outdegree equal to p. The result follows by Theorem 11.4. Q.E.D.

We close this section by applying our results to a problem in telecommunications. We follow Liu [1968]. A rotating drum has eight different sectors. The question is: Can we tell the position of the drum without looking at it? One approach is by putting conducting material in some of the sectors and nonconducting material in others of the

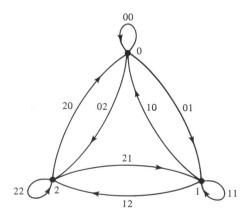

Figure 11.33　The digraph $D_{3,2}$.

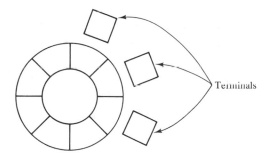

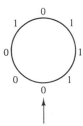

Figure 11.34 Rotating drum with eight sectors and three adjacent terminals.

Figure 11.35 Arrangement of 0's and 1's which solves the rotating drum problem.

sectors. Place three terminals adjacent to the drum so that in any position of the drum, the terminals adjoin three consecutive sectors, as shown in Figure 11.34. A terminal will be activated if it adjoins a sector with conducting material. If we are clever, then the pattern of conducting and nonconducting material will be so chosen that the pattern of activated and nonactivated terminals will tell us the position of the drum.

We can reformulate this as follows. Let each sector receive a 1 or a 0. We wish to arrange eight 0's and 1's around a circle so that every sequence of three consecutive digits is different. This is accomplished by finding a (2, 3) de Bruijn sequence, in particular the sequence 01110100. If we arrange these digits around a circle as shown in Figure 11.35, the following sequences of consecutive digits occur going counterclockwise beginning from the arrow: 011, 111, 110, 101, 010, 100, 000, 001. These are all distinct, as desired. Thus, each position of the drum can be uniquely encoded.

Related problems concerned with teleprinting and cryptography are described in Exercises 16 and 17.

EXERCISES FOR SECTION 11.4

1. In the graph G of Figure 11.26, find another way to add four copies of edges of G to obtain a multigraph with an eulerian closed chain.

2. In each graph of Figure 11.36, find a shortest mail carrier's route from x to x by finding the smallest number of copies of edges of G which give rise to a multigraph with an eulerian closed chain.

3. Given the sweep subgraph and curb multidigraph of Figure 11.37, determine a set of deadheading arcs to add to the sweep subgraph to obtain a multidigraph with an eulerian closed path. Determine the total time required to traverse this path, including sweeping and deadheading.

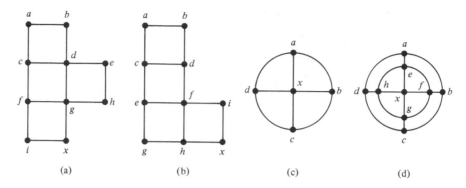

Figure 11.36 Graphs for Exercise 2 of Section 11.4.

4. Draw the de Bruijn diagram $D_{p, n}$ and find a (p, n) de Bruijn sequence for
   (a) $p = 2, n = 4$;   (b) $p = 3, n = 3$.

5. For the G and U, C fragments given in the following exercises of Section 2.13, find an RNA chain using the method of Section 11.4.4.
   (a) Exercise 1   (b) Exercise 3   (c) Exercise 5.

6. In each of the following cases, determine if there is a codeword with the ith letter occurring exactly v_i times and the ith followed by the jth exactly v_{ij} times, and find such a word if there is one.
   (a) $v_1 = 3, v_2 = 4, v_3 = 5$,

$$(v_{ij}) = \begin{pmatrix} 1 & 1 & 0 \\ 1 & 2 & 1 \\ 0 & 1 & 4 \end{pmatrix}.$$

   (b) $v_1 = 3, v_2 = 4, v_3 = 5$,

$$(v_{ij}) = \begin{pmatrix} 1 & 1 & 1 \\ 1 & 2 & 1 \\ 0 & 1 & 3 \end{pmatrix}.$$

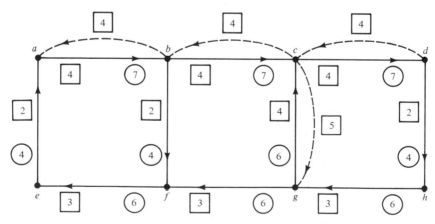

Figure 11.37 Curb multidigraph with sweep subgraph and sweeping and deadheading times shown as in Figure 11.27.

(c) $v_1 = 3$, $v_2 = 4$, $v_3 = 5$,

$$(v_{ij}) = \begin{pmatrix} 1 & 1 & 1 \\ 1 & 2 & 1 \\ 1 & 1 & 3 \end{pmatrix}.$$

(d) $v_1 = 3$, $v_2 = 3$, $v_3 = 3$, $v_4 - 3$,

$$(v_{ij}) = \begin{pmatrix} 1 & 1 & 0 & 1 \\ 1 & 1 & 1 & 0 \\ 1 & 0 & 1 & 1 \\ 0 & 1 & 1 & 1 \end{pmatrix}.$$

(e) $v_1 = 3$, $v_2 = 3$, $v_3 = 2$, $v_4 = 3$ and (v_{ij}) as in (d).

7. In the "Chinese postman" problem, show that if x and y are any two vertices of G, the shortest mail carrier's route from x to x has the same length as the shortest mail carrier's route from y to y.

8. Notice that in the second multigraph of Figure 11.26, the dashed edges can be divided into two chains, each joining two vertices of odd degree in G. These are the chains d, e, b and h, x, f.
   (a) Can two such chains be found in the third multigraph of Figure 11.26?
   (b) In general, if H is a multigraph obtained from G by using edges in a shortest mail carrier's route, show that the dashed edges in H can be grouped into several (not necessarily two) chains, each joining two vertices of odd degree in G. Moreover, every vertex of odd degree in G is an end vertex of exactly one of these chains. (This result will be important in discussing the last part of the solution to the "Chinese postman" problem in Section 13.2.4.)

9. In a shortest mail carrier's route, can any edge be used more than twice? Why?

10. Suppose that we have an optimal solution to the street-sweeping problem, that is, a closed path in the curb multidigraph which includes all arcs of the sweep subgraph and uses as little total time as possible. Is it possible that some arc is traversed more than twice?

11. Show the following about the list of interior extended bases of the G and U, C fragments and the list of fragments consisting of one extended base.
   (a) Every entry on the first list is on the second list.
   (b) The first and last extended bases in the entire RNA chain are on the second list, but not the first.
   (c) If there are at least two G fragments and at least two U, C fragments, there will always be exactly two entries on the second list which are not on the first.

12. Under what circumstances does an RNA chain have two abnormal fragments?

13. Find the length of the smallest possible ambiguous RNA chain, that is, one whose G and U, C fragments are the same as that of another RNA chain.

14. This exercise sketches a proof* of the claim S: Suppose we are given G and U, C fragments from an RNA chain with at least two G fragments and at least two U, C fragments. Then every RNA chain with given G and U, C fragments corresponds to an eulerian closed path which ends with the arc $X * Y$ in the multidigraph D constructed in Section 11.4.4. Let us say that an extended base has *type* G if it ends in G and *type* U, C if it ends in U or C. Note that any RNA chain can be written as

$$A_0 A_1 \cdots A_k A_{k+1} B,$$

where each A_i consists of extended bases of one type, the A_i alternate in type, and B is the last extended base if the chain ends in A and is the empty chain otherwise. For $i = 0, 1, \ldots, k$, let $\bar{A}_i$ be $A_i a_{i+1}$, where a_j is the first extended base in A_j. Let $\bar{A}_{k+1}$ be $A_{k+1}B$. Finally, say a fragment of one extended base is *trivial*. Show the following.

*The author thanks Michael Vivino for the idea of this proof.

(a) $\bar{A}_0, \bar{A}_1, \ldots, \bar{A}_k$ are all nontrivial normal fragments and there are no other nontrivial normal fragments.

(b) $\bar{A}_{k+1}$ is the longest abnormal fragment.

(c) $a_0, a_1, \ldots, a_{k+1}$ are the vertices of the multidigraph D, $\bar{A}_i$ is the label on the arc from a_i to a_{i+1}, and $\bar{A}_{k+1} * a_0$ is the label on the arc from a_{k+1} to a_0.

(d) The arcs labeled $\bar{A}_0, \bar{A}_1, \ldots, \bar{A}_k, \bar{A}_{k+1} * a_0$ in this order define an eulerian closed path.

(e) Any other RNA chain which produces the same set of nontrivial normal fragments and the same longest abnormal fragment produces the same multidigraph D.

(f) Any RNA chain built from D produces the same set of nontrivial normal fragments and the same longest abnormal fragment.

(g) The nontrivial normal fragments and the longest abnormal fragment uniquely determine all of the fragments.

(h) Statement S holds.

15. In the problem discussed in Section 11.4.5, if there is a solution word, then every eulerian path determines exactly one solution word. However, how many different eulerian paths are determined by each solution word?

16. The following problem arises in cryptography: Find a word from a given m-letter alphabet in which each arrangement of r letters appears exactly once. Find the solution to this problem if $r = 3$ and $m = 4$.

17. An important problem in communications known as the teleprinter's problem is the following: How long is the longest circular sequence of 0's and 1's such that no sequence of r consecutive bits appears more than once in the sequence? (The sequence of r bits is considered to appear if it starts near the end and finishes at the beginning.) Solve the teleprinter's problem. (It was first solved by Good [1946].)

11.5 HAMILTONIAN CHAINS AND PATHS

11.5.1 Definitions

Analogous to an eulerian chain or path in a graph or digraph is a *hamiltonian chain* or *path*, a chain or path that uses each vertex once and only once. The notion of hamiltonian chain goes back to Sir William Rowan Hamilton, who in 1857 described a puzzle that led to this concept (see below). In this section we discuss the question of existence of hamiltonian chains and paths. We discuss some applications here, and others in Section 11.6.

Note that a hamiltonian chain or path is automatically simple, and hence by our conventions cannot be closed. However, we shall call a *circuit* or a *cycle* $u_1, u_2, \ldots, u_t, u_1$ *hamiltonian* if $u_1, u_2, \ldots, u_t$ is, respectively, a hamiltonian chain or path.

Although the notions of hamiltonian chain, path, and so on, are analogous to the comparable eulerian notions, it is very hard to tell if a graph or digraph has such an object. Indeed, it is an NP-complete problem to determine if a graph or digraph has a hamiltonian chain (path, circuit, cycle). Some results are known about existence of these objects, and we present some here. First we mention some applications of hamiltonian chains, paths, circuits, or cycles.

Example 11.1   Following the Edges of a Dodecahedron

We begin with an example. Hamilton's puzzle was to determine a way to follow the edges of a dodecahedron that visited each corner exactly once and returned to the start. The vertices and edges of a dodecahedron can be drawn as a graph as in

Figure 11.38. The problem becomes: Find a hamiltonian circuit in this graph. There is one: $a, b, c, d, e, f, g, h, i, j, k, l, m, n, o, p, q, r, s, t, a$.

Example 11.2   The Traveling Salesman Problem Revisited

The notion of hamiltonian cycle arises in the traveling salesman problem, which we discussed in Section 2.4. A salesman wishes to visit n different cities, each exactly once, and return to his starting point, in such a way as to minimize cost. (In Section 2.4 we mentioned other applications of the traveling salesman problem, for example to finding optimal routes for a bank courier or a robot in an automated warehouse.) Suppose that we let the cities be the vertices of a complete symmetric digraph, a digraph with all pairs of vertices joined by two arcs, and suppose that we put a weight c_{ij} on the arc (i, j) if c_{ij} is the cost of traveling from city i to city j. Since the complete symmetric digraph has a hamiltonian cycle, the existence question is not of interest. Rather, we seek a hamiltonian cycle in this digraph which has minimum sum of weights. As we have previously pointed out, this traveling salesman problem is hard, that is, it is NP-complete.

Example 11.3   Scheduling Industrial Processes

A manufacturing plant has a single processing facility. A number of items are to be processed there. If item j is processed immediately after item i, there is a cost c_{ij} of resetting the processing facility from its configuration for processing item i to its configuration for processing item j. If no resetting is necessary, the cost is of course zero. Assuming the costs of processing items are independent of the order in which items are processed, the problem is to choose the order so as to minimize the sum of the resetting costs c_{ij}. This problem arises in scheduling computer runs, as we have already observed in Example 2.10. It also arises in the chemical and pharmaceutical industries, where the processing facility might be a reaction vessel and resetting means cleaning. Costs are of course minimized if we can find a production schedule

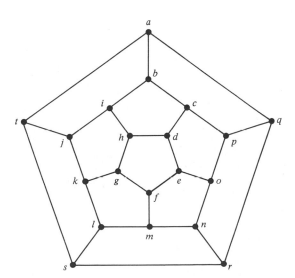

Figure 11.38   A graph representing the vertices and edges of a dodecahedron.

Figure 11.39 A graph without a hamiltonian circuit.

in which no resetting is necessary. To see if such a production schedule exists, we build a digraph D whose vertices are the items to be processed and which has an arc from i to j if j can follow i without resetting. If there is a hamiltonian path in D, then such a production schedule exists. For more information on this problem, and for a treatment of the case where there is no hamiltonian path, see Christofides [1975]. See also the related discussion in Section 11.6.3.

11.5.2 Sufficient Conditions for the Existence of a Hamiltonian Circuit in a Graph

Not every graph has a hamiltonian circuit. Consider the graph of Figure 11.39. Note that every edge joins one of the vertices in $A = \{a, b, c\}$ to one of the vertices in $B = \{x, y\}$. Thus, a hamiltonian circuit would have to pass alternately from one of the vertices in A to one of the vertices in B. This could only happen if $|A| = |B|$.

In this section we present sufficient conditions for the existence of a hamiltonian circuit.

Theorem 11.7 **(Ore [1960]).** Suppose that G is a graph with $n \geq 3$ vertices and whenever vertices $x \neq y$ are not joined by an edge, then the degree of x plus the degree of y is at least n. Then G has a hamiltonian circuit.

Proof.* Suppose that G has no hamiltonian circuit. We shall show that for some nonadjacent x, y in $V(G)$,

$$\deg_G (x) + \deg_G (y) < n, \tag{11.9}$$

where $\deg_G (a)$ means degree of a in G. If we add edges to G, we eventually obtain a complete graph, which has a hamiltonian circuit. Thus, in the process of adding edges, we must eventually hit a graph H with the property that H has no hamiltonian circuit, but adding any edge to H gives us a graph with a hamiltonian circuit. We shall show that in H, there are nonadjacent x and y so that

$$\deg_H (x) + \deg_H (y) < n. \tag{11.10}$$

But $\deg_G (a) \leq \deg_H (a)$ for all a, so (11.10) implies (11.9).

Pick any nonadjacent x and y in H. Then H plus the edge $\{x, y\}$ has a hamiltonian circuit. Since H does not, this circuit must use the edge $\{x, y\}$. Hence, it can be written as

$$x, y, a_1, a_2, \ldots, a_{n-2}, x.$$

*The proof can be omitted.

480 Existence Problems in Graph Theory Chap. 11

Now $V(H) = \{x, y, a_1, a_2, \ldots, a_{n-2}\}$. Moreover, we note that for $i > 1$,

$$\{y, a_i\} \in E(H) \Rightarrow \{x, a_{i-1}\} \notin E(H). \tag{11.11}$$

For if not, then

$$y, a_i, a_{i+1}, \ldots, a_{n-2}, x, a_{i-1}, a_{i-2}, \ldots, a_1, y$$

is a hamiltonian circuit in H, which is a contradiction. Now (11.11) and $\{x, y\} \notin E(H)$ imply (11.10). $\qquad$ Q.E.D.

To illustrate Ore's Theorem, note that the graph (a) of Figure 11.40 has a hamiltonian circuit, for any two vertices have sum of degrees at least 5. Note that Ore's Theorem does not give necessary conditions. For consider the circuit of length 5, Z_5. There is a hamiltonian circuit, but any two vertices have sum of degrees equal to 4, which is less than n.

The next result, originally proved independently, follows immediately from Theorem 11.7.

Corollary 11.7.1 (Dirac [1952]). Suppose that G is a graph with $n \geq 3$ vertices and each vertex has degree at least $n/2$. Then G has a hamiltonian circuit.

To illustrate Corollary 11.7.1, note that the graph (b) of Figure 11.40 has a hamiltonian circuit because every vertex has degree 3 and there are six vertices.

Corollary 11.7.2 (Bondy and Chvátal [1976]). Suppose that G is a graph with $n \geq 3$ vertices and that x and y are nonadjacent vertices in G so that

$$\deg(x) + \deg(y) \geq n.$$

Then G has a hamiltonian circuit if and only if G plus the edge $\{x, y\}$ has a hamiltonian circuit.

Proof. If G has a hamiltonian circuit, then certainly G plus edge $\{x, y\}$ does. The converse follows from the proof of Theorem 11.7. $\qquad$ Q.E.D.

Let us see what happens if we try to repeat the construction in Corollary 11.7.2. That is, we start with a graph $G = G_1$. We find a pair of nonadjacent vertices x_1 and y_1 in G_1 so that in G_1, the degrees of x_1 and y_1 sum to at least n. We let G_2 be G_1 plus edge $\{x_1, y_1\}$. We now find a pair of nonadjacent vertices x_2 and y_2 in G_2 so that in G_2, the degrees of x_2 and y_2 sum to at least n. We let G_3 be G_2 plus edge $\{x_2, y_2\}$. We continue this procedure until we obtain a graph $H = G_i$ with no nonadjacent x_i and y_i whose

(a)

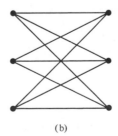

(b)

Figure 11.40 Two graphs with hamiltonian circuits.

degrees in G_i sum to at least n. It is not hard to show (Exercise 18) that no matter in what order we perform this construction, we always obtain the same graph H. We call this graph H the *closure* of G, and denote it by $c(G)$. We illustrate the construction of $c(G)$ in Figure 11.41. The next result follows from Corollary 11.7.2.

Corollary 11.7.3 (Bondy and Chvátal [1976]). G has a hamiltonian circuit if and only if $c(G)$ has a hamiltonian circuit.

Note that in Figure 11.41, $c(G)$ is a complete graph in parts (a) and (c), but not in part (b). Parts (a) and (c) illustrate the following theorem. [Note that part (c) could not be handled by Ore's Theorem.]

Theorem 11.8 (Bondy and Chvátal [1976]). Suppose that G is a graph with at least three vertices. If $c(G)$ is complete, then G has a hamiltonian circuit.

Proof. A complete graph with at least three vertices has a hamiltonian circuit. Q.E.D.

Note that Ore's Theorem is a corollary of Theorem 11.8.

We have given conditions sufficient for the existence of a hamiltonian circuit, but

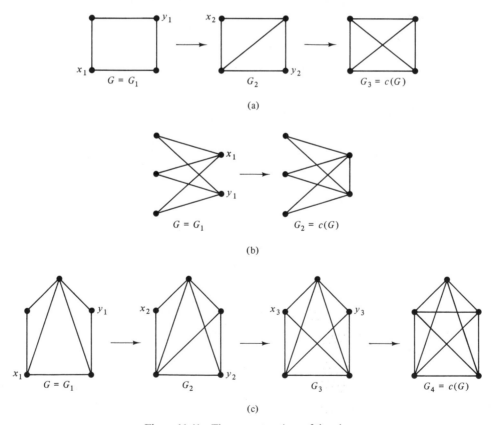

Figure 11.41 Three constructions of the closure.

not for a hamiltonian chain. For some conditions sufficient for the latter, see, for example, the text and exercises of Behzad *et al.* [1979].

11.5.3 Sufficient Conditions for the Existence of a Hamiltonian Cycle in a Digraph

Ore's and Dirac's results (Theorem 11.7 and Corollary 11.7.1) have the following analogues for digraphs. Here, od (u) is the outdegree of u and id (u) is the indegree of u.

Theorem 11.9 (Woodall [1972]). Suppose that D is a digraph with $n \geq 3$ vertices and whenever $x \neq y$ and there is no arc from x to y, then

$$\text{od } (x) + \text{id } (y) \geq n. \tag{11.12}$$

Then D has a hamiltonian cycle.

Theorem 11.10 (Ghouila-Houri [1960]). Suppose that D is a strongly connected digraph with n vertices and for every vertex x,

$$\text{od } (x) + \text{id } (x) \geq n. \tag{11.13}$$

Then D has a hamiltonian cycle.

Corollary 11.10.1. Suppose that D is a digraph with n vertices and for every vertex x,

$$\text{od } (x) \geq \frac{n}{2} \quad \text{and} \quad \text{id } (x) \geq \frac{n}{2}. \tag{11.14}$$

Then D has a hamiltonian cycle.

Proof. One shows that (11.14) implies that D is strongly connected. The proof is left to the reader (Exercise 19).

We have given conditions for the existence of a hamiltonian cycle, but not for a hamiltonian path. For a summary of conditions sufficient for the latter, see, for example, Behzad *et al.* [1979].

EXERCISES FOR SECTION 11.5

1. In each graph of Figure 11.42, find a hamiltonian circuit.
2. In each graph of Figure 11.43, find a hamiltonian chain.
3. In each digraph of Figure 11.44, find a hamiltonian cycle.
4. In each digraph of Figure 11.45, find a hamiltonian path.
5. Show that the graph of Figure 11.46 can have no hamiltonian circuit. (*Hint:* The a's and x's should tell you something.)
6. Can the graph of Figure 11.47 have a hamiltonian circuit? Why?
7. Give examples of graphs that
   (a) have both eulerian and hamiltonian circuits;

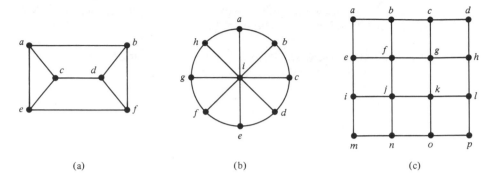

(a) (b) (c)

Figure 11.42 Graphs for exercises of Section 11.5.

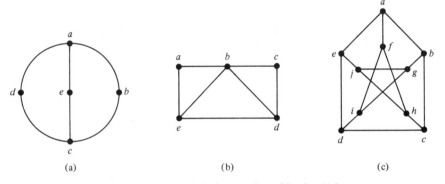

(a) (b) (c)

Figure 11.43 Graphs for exercises of Section 11.5.

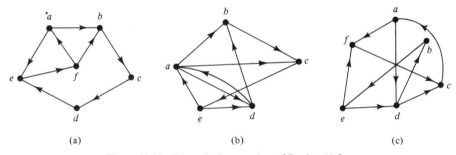

(a) (b) (c)

Figure 11.44 Digraphs for exercises of Section 11.5.

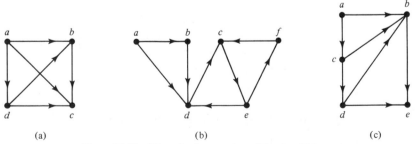

(a) (b) (c)

Figure 11.45 Digraphs for exercises of Section 11.5.

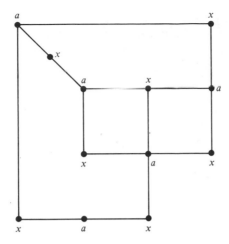

Figure 11.46 A graph with no hamilton-ian circuit.

(b) have a hamiltonian, but not an eulerian circuit;
(c) have an eulerian, but not a hamiltonian circuit;
(d) have neither an eulerian nor a hamiltonian circuit.

8. Give an example of
  (a) a graph with no hamiltonian chain;
  (b) a digraph with no hamiltonian path.

9. Suppose that G is a graph in which there are 10 vertices, a, b, c, d, e and x, y, u, v, w, and each of the first five vertices is joined to each of the last five.
  (a) Does Ore's Theorem (Theorem 11.7) imply that G has a hamiltonian circuit? Why?
  (b) Does Dirac's Theorem (Corollary 11.7.1) imply that G has a hamiltonian circuit? Why?
  (c) Does G have one?

10. Find the closure $c(G)$ of each graph of Figures 11.42 and 11.43.

11. For which graphs in Figures 11.42 and 11.43 does Theorem 11.8 imply that there is a hamiltonian circuit?

12. Give an example of a graph whose closure does not have a hamiltonian circuit.

13. Show that if G has a hamiltonian circuit, $c(G)$ is not necessarily complete.

14. For which digraphs of Figures 11.44 and 11.45 does Theorem 11.9 imply that there is a hamiltonian cycle?

15. For which digraphs of Figures 11.44 and 11.45 does Theorem 11.10 imply that there is a hamiltonian cycle?

16. A graph is *regular of degree k* if every vertex has the same degree, k. Show that if G has 11 vertices and is regular of degree 6, then G has a hamiltonian circuit.

17. Suppose that $n \geq 4$ and that of n people, any two of them together know all the remaining $n - 2$. Show that the people can be seated around a circular table so that everyone is seated between two friends.

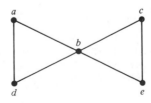

Figure 11.47 Graph for Exercise 6, Section 11.5.

18. Suppose that by successively adding edges joining nonadjacent vertices whose degrees sum to at least n, we eventually obtain a graph H and then are unable to continue. Suppose that by doing the construction in a different order, we eventually obtain a graph K and then are unable to continue. Show that $H = K$. (*Hint*: If not, find the first edge added to G in the first construction which is not an edge of K.)

19. Show that Equation (11.14) implies that D is strongly connected.

20. Suppose that the hypotheses of Theorem 11.9 hold with n in Equation (11.12) replaced by $n - 1$. Use Theorem 11.9 to show that D has a hamiltonian path.

21. (Pósa [1962]).   Use Theorem 11.8 to show that if G is a graph of at least three vertices and if for every integer j with $1 \leq j < n/2$, the number of vertices of degree not exceeding j is less than j, then G has a hamiltonian circuit.

11.6 APPLICATIONS OF HAMILTONIAN CHAINS AND PATHS

In this section we present several applications of the ideas of hamiltonian chains and paths. Sections 11.6.2 and 11.6.3 depend in small ways on Section 11.6.1. Otherwise, these sections are independent, and can be read in any order. In particular, a quick reading could include just a glance at Section 11.6.1, followed by Section 11.6.3, or it could consist of just Section 11.6.4.

11.6.1 Tournaments

Let (V, A) be a digraph and assume that for all $u \neq v$ in V, (u, v) is in A or (v, u) is in A, but not both. Such a digraph is called a *tournament*. Tournaments arise in many different situations. There are the obvious ones, the round-robin competitions in tennis, baseball, and so on.* In a round-robin competition, every pair of players (pair of teams) competes and one and only one member of each pair beats the other. (We assume that no ties are allowed.) Tournaments also arise from *pair comparison experiments*, where a subject or a decisionmaker is presented with each pair of alternatives from a set and is asked to say which of the two he or she prefers, which is more important, which is more qualified, and so on. Tournaments also arise in nature, where certain individuals in a given species develop dominance over others of the same species. In such situations, for every pair of animals of the species, one and only one is dominant over the other. The dominance relation defines what is called a *pecking order* among the individuals concerned.

Sometimes we want to *rank* the players of a tournament. This might be true if we are giving out second prize, third prize, and so on. It might also be true, for example, if the "players" are alternative candidates for a job and the tournament represents preferences among candidates. Then our first-choice job candidate might not accept and we might want to have a second choice, third choice, and so on, chosen ahead of time. One way to find a ranking of the players is to observe that every tournament has a hamiltonian path. This result, which we shall prove shortly, implies that we can label the players as $u_1, u_2, \ldots, u_n$ in such a way that u_1 beats u_2, u_2 beats u_3, $\ldots$, and u_{n-1} beats u_n. Such a labeling gives us a ranking of the players: u_1 is ranked first, u_2 second, and so on. To illustrate, consider the tournament of Figure 11.48. Here, a hamiltonian path and hence a

*The reader should distinguish a round-robin competition from the elimination-type competition which is more common.

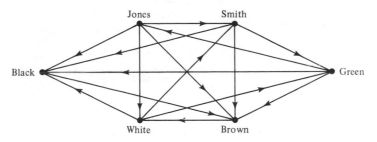

Figure 11.48 A round-robin Ping-Pong tournament.

ranking is given by: Jones, Smith, Green, Brown, White, Black. Another way to find a ranking of the players is to use the score sequence defined in Exercise 12. (See also Exercises 15 and 20.)

Theorem 11.11   (Rédei [1934]).   Every tournament (V, A) has a hamiltonian path.

*Proof.** We proceed by induction on the number n of vertices. The result is trivial if $n = 2$. Assuming the result for tournaments of n vertices, let us consider a tournament (V, A) with $n + 1$ vertices. Let u be an arbitrary vertex and consider the subgraph generated by $V - \{u\}$. It is easy to verify that this subgraph is still a tournament. Hence, by inductive assumption, it has a hamiltonian path $u_1, u_2, \ldots, u_n$. If there is an arc from u to u_1, then $u, u_1, u_2, \ldots, u_n$ is a hamiltonian path of (V, A). If there is no arc from u to u_1, then, since (V, A) is a tournament, there is an arc from u_1 to u. Let i be the largest integer such that there is an arc from u_i to u. If i is n, then there is an arc from u_n to u, and we conclude that $u_1, u_2, \ldots, u_n, u$ is a hamiltonian path of (V, A). If $i < n$, then there is an arc from u_i to u and, moreover, by definition of i, there is no arc from u_{i+1} to u. Since (V, A) is a tournament, there is an arc from u to u_{i+1}. Thus, $u_1, u_2, \ldots, u_i, u, u_{i+1}, \ldots, u_n$ is a hamiltonian path of (V, A).   Q.E.D.

If $u_1, u_2, \ldots, u_n$ is a hamiltonian path in a tournament (V, A), then, as we have already observed, we can use it to define a ranking of the players. Unfortunately, there can be other hamiltonian paths in the tournament. In our example, Smith, Brown, White, Green, Jones, Black is another. In this situation, then, there are many possible rankings of the players. The situation can get even worse. In the tournament of Figure 11.49, in fact, for each vertex a and each possible rank r, there is some hamiltonian path for which a has rank r.

In general, given a set of possible rankings, we might want to try to choose a "consensus" ranking. The problem of finding a consensus among alternative possible rankings is a rather difficult one, which deserves a longer discussion than we can give it here. (See Roberts [1976, Ch. 7].)

We next ask whether there are any circumstances where the problem of having many different rankings does not occur. That is, are there circumstances when a tournament has a unique hamiltonian path? The answer is given by saying that a digraph is *transitive* if whenever (u, v) is an arc and (v, w) is an arc and $u \neq w$, then (u, w) is an arc. (Note that in a tournament, we do not have to add the condition $u \neq w$, since there could

*The proof can be omitted.

Figure 11.49 For each vertex a and rank r, there is some hamiltonian path for which a has rank r.

never be arcs (u, v) and (v, u).) The tournament of Figure 11.48 is not transitive, since, for example, Jones beats Smith and Smith beats Green, but Jones does not beat Green.

Theorem 11.12. A tournament has a unique hamiltonian path if and only if the tournament is transitive.

Proof. See Exercise 19.

At this point, let us consider applications of Theorem 11.12 to a decisionmaker's preferences among alternatives, or to his or her ratings of relative importance of alternatives, and so on. The results are illustrated by pair comparison data for preference, for example, that shown in Figure 11.50. Here, transitivity holds, and so there is a unique hamiltonian path. The path is Northwest, Northeast, Southwest, Southeast, Central.

In studying preference, it is often reasonable to assume (or demand) that the decisionmaker's preferences define a transitive tournament, that is, that if he or she prefers u to v and v to w, then he or she prefers u to w. This turns out to be equivalent to assuming that the decisionmaker can uniquely rank the alternatives among which he or she is expressing preferences.

Performing a pair comparison experiment to elicit preferences can be a tedious job. If there are n alternatives, $\binom{n}{2} = n(n-1)/2$ comparisons are required, and this can get to be a large number even for moderate n. However, if we believe that a subject is transitive, we know that a pair comparison experiment amounts to a transitive tournament and

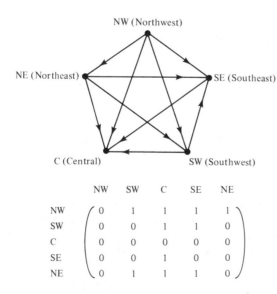

	NW	SW	C	SE	NE
NW	0	1	1	1	1
SW	0	0	1	1	0
C	0	0	0	0	0
SE	0	0	1	0	0
NE	0	1	1	1	0

Figure 11.50 Results of a pair comparison experiment for preference among geographical areas. The x, y entry in the matrix is 1 iff x is preferred to y. In the corresponding digraph, an arc from x to y indicates that x is preferred to y.

hence to a unique ranking of the alternatives. Thus, we are trying to order a set of n items. How many comparisons are really required to recover this ranking and hence the original tournament? This is the problem of sorting which we discussed in Section 3.6. There, we pointed out that good sorting algorithms can take on the order of $n \log_2 n$ steps, which is a smaller number than $n(n-1)/2$.

11.6.2 Topological Sorting

It turns out that a tournament is transitive if and only if it is *acyclic*, that is, has no cycles (Exercise 18). Finding the unique hamiltonian path in a transitive or acyclic tournament is a special case of the following more general problem. Suppose that we are given a digraph D of n vertices. Label the vertices of D with the integers $1, 2, \ldots, n$ so that every arc of D leads from a vertex with a smaller label to a vertex with a larger label. Such a labeling is called a *topological order* for D and the process of finding such an order is called *topological sorting*.

Theorem 11.13. A digraph D has a topological order if and only if D is acyclic.

Proof. Clearly, if there is a topological order $1, 2, \ldots, n$, then D could not have a cycle. Conversely, suppose that D is acyclic. By Exercise 22, there is a vertex x_1 with no incoming arcs. Label x_1 with the label 1, and delete it from D. Now the resulting digraph is still acyclic and hence has a vertex x_2 with no incoming arcs. Label x_2 with label 2. And so on. This clearly gives rise to a topological order. Q.E.D.

We illustrate the construction in the proof of Theorem 11.13 by finding a topological order in the digraph D shown in Figure 11.51(a). We first choose vertex a. After eliminating a, we choose vertex e. After eliminating e, we have a choice of either b or d; say we choose b for concreteness. Then after eliminating b, we choose d. Finally, we are left with c. This gives the labeling shown in Figure 11.45(b). One of the problems with the procedure described is that at each stage, we must search through the entire remaining digraph for the next vertex to choose. For a description of a more efficient algorithm, based on the depth-first search procedure described in Section 11.1, see Deo [1974], Golumbic [1980], or Reingold *et al.* [1977].

Topological sorting has a variety of applications. For instance, it arises in the analysis of *activity networks* in project planning (Deo [1974]). A large project is divided into individual tasks called *activities*. Some activities must be completed before others can be started—for example, an item must be sanded before it can be painted. Build a digraph D as follows. The vertices are the activities, and there is an arc from activity x to activity y if x must precede y. We seek to find an order in which to perform the activities so that if there is an arc from x to y in D, then x comes before y. This requires a topological sorting

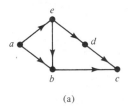

(a)

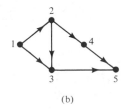

(b)

Figure 11.51 Part (b) shows a topological order for the digraph of part (a).

of D. It can be done if and only if D has no cycles. A similar situation arises, for example, if we wish to arrange words in a glossary and guarantee that no word is used before it is defined. This too requires a topological sorting.

11.6.3 Scheduling Problems in Operations Research*

Many scheduling problems in operations research, for instance those involving activity networks discussed in Section 11.6.2, involve finding an order in which to perform a certain number of operations. We often seek an optimal order. Sometimes such problems can be solved by finding hamiltonian paths. We have already seen an example of this in Example 11.3. Here we present another example. Related problems are discussed in Section 13.2.3.

Suppose that a printer has n different books to publish. He has two machines, a printing machine and a binding machine. A book must be printed before it can be bound. The binding machine operator earns much more money than the printing machine operator, and must be paid from the time the binding machine is first started until all the books have been bound. In what order should the books be printed so that the total pay of the binding machine operator is minimized?

Let p_k be the time required to print the kth book and b_k the time required to bind the kth book. Let us make the special assumption that for all i and j, either $p_i \le b_j$ or $p_j \le b_i$. Note that it is now possible to find an ordering of books so that if books are printed and bound in that order, the binding machine will be kept busy without idle time after the first book is printed. This clearly will minimize the pay of the binding machine operator. To find the desired ordering, draw a digraph with an arc from i to j if and only if $b_i \ge p_j$. Then this digraph contains a tournament and so, by Theorem 11.11, has a hamiltonian path. This path provides the desired ordering. More general treatment of this problem, without the special assumption, can be found in Johnson [1954]. See also Exercise 28.

11.6.4 Facilities Design†

Graph theory is finding a variety of applications in the design of such physical facilities as manufacturing plants, hospitals, schools, golf courses, and so on. In such design problems, we have a number of areas which need to be located and it is desired that certain of them be next to each other. We draw a graph G, the *relationship graph*, whose vertices are the areas in question, and which has an edge between two areas if they should be next to each other. If G is planar, then it comes from a map (see Example 3.13), and the corresponding map represents a layout plan where two areas (countries) which are joined by an edge in G in fact share a common wall (boundary). Figure 11.52 shows a planar graph and the corresponding facilities layout.

If the relationship graph G is not planar, then we seek to eliminate some requirements, i.e., eliminate some edges of G, to find a spanning subgraph G' of G which is planar. Then G' can be used to build the layout plan. One method for finding G' which is used in facilities planning is the following. (It is due to Demourcron *et al.* [1964].)

*This subsection is based on Berge [1962], Johnson [1954], and Liu [1972].
†This subsection is based on Chachra *et al.* [1979].

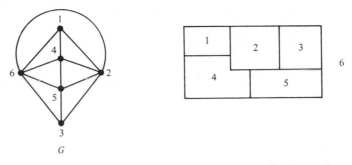

Figure 11.52 A relationship graph G and a corresponding layout plan.

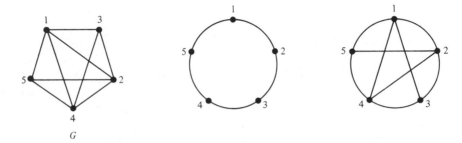

Figure 11.53 A relationship graph G, the vertices of a hamiltonian circuit arranged around a circle, and the remaining edges of G represented as chords of the circle.

Determine if G has a hamiltonian circuit. If so, find such a circuit $C = u_1, u_2, \ldots, u_n$. Locate the vertices u_i around a circle in this order. For instance, consider graph G of Figure 11.53. Then $C = 1, 2, 3, 4, 5$ is a hamiltonian circuit and it is represented around a circle in Figure 11.53. Build a new graph H as follows. The vertices of H are the edges in G which are not edges of C. Two edges of H are adjacent if and only if when the corresponding chords are drawn in the circle determined by C, these chords intersect. See Figure 11.53 for the chords in our example, and see Figure 11.54 for the graph H. Suppose that H is 2-colorable, say using the colors red and blue. Then G is planar. To see this, simply note that all red chords can be drawn in the interior of the circle determined by C and all blue chords in the exterior, with no edges crossing. For instance, in graph H of Figure 11.54, we can color vertices $\{1, 3\}$ and $\{1, 4\}$ red and vertices $\{2, 4\}$ and $\{2, 5\}$ blue. Then we obtain the planar drawing of G shown in Figure 11.55. If H is not 2-colorable, then we seek a maximal subgraph K of H which is 2-colorable. The edges of K can be added to C to get a planar graph G' which is a spanning subgraph of G. G' is used for finding a layout plan. For instance, suppose that G is obtained from G of Figure 11.53 by adding edge $\{3, 5\}$. Then, using the same hamiltonian circuit C, we see that H is as shown in Figure 11.56. This H is not 2-colorable. We have to eliminate some vertex to get K, for instance vertex $\{3, 5\}$. The resulting graph G' is the graph G of Figure 11.53, which is planar.

$\{1, 4\}$ $\{2, 5\}$ $\{1, 3\}$ $\{2, 4\}$

Figure 11.54 Graph H obtained from Figure 11.53.

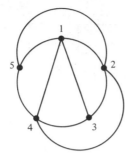

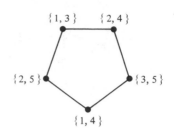

Figure 11.55 Planar drawing of graph G of Figure 11.53.

Figure 11.56 Graph H obtained if edge $\{3, 5\}$ is added to G of Figure 11.53.

EXERCISES FOR SECTION 11.6

1. In each tournament of Figure 11.57, find all hamiltonian paths.
2. For the tournament defined by the preference data of Figure 11.58, find all hamiltonian paths.
3. Which of the tournaments with four or fewer vertices is transitive?
4. Draw a digraph of a transitive tournament on five players.
5. Find a topological order for each of the digraphs of Figure 11.59.
6. In the book printing problem of Section 11.6.3, suppose $b_1 = 3$, $b_2 = 5$, $b_3 = 8$, $p_1 = 6$, $p_2 = 2$, $p_3 = 9$. Find an optimal order in which to produce books.
7. Repeat Exercise 6 if $b_1 = 10$, $b_2 = 7$, $b_3 = 5$, $p_1 = 11$, $p_2 = 4$, $p_3 = 8$.
8. Design a layout plan for the planar graphs (a) and (b) of Figure 11.43.
9. If G is graph (a) of Figure 11.42, find a hamiltonian circuit C and construct H as in Section 11.6.4. Use H to determine a planar diagram for G, if one exists. Otherwise, find a planar G'.
10. Repeat Exercise 9 if G is graph (b) of Figure 11.42.
11. Repeat Exercise 9 if G is graph (b) of Figure 11.40.

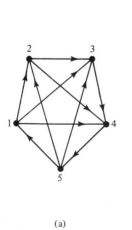

(a)

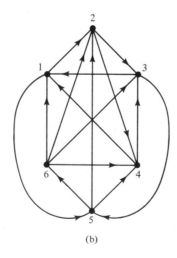

(b)

Figure 11.57 Tournaments for exercises of Section 11.6.

	New York	Boston	San Francisco	Los Angeles	Houston
New York	0	0	0	0	1
Boston	1	0	0	1	1
San Francisco	1	1	0	1	1
Los Angeles	1	0	0	0	0
Houston	0	0	0	1	0

Figure 11.58 Results of a pair comparison experiment for preference among cities as a place to live. Entry i, j is 1 iff i is preferred to j.

12. In a tournament, the *score* of vertex u is the outdegree of u. (It is the number of players u beats.) If the vertices are labeled $1, 2, \ldots, n$, with $s(1) \le s(2) \le \cdots \le s(n)$, then the sequence $(s(1), s(2), \ldots, s(n))$ is called the *score sequence* of the tournament. Exercises 12–16 and 20(b) investigate the score sequence. Find the score sequence of the tournaments of
    (a) Figure 11.48   **(b)** Figure 11.50.

13. Show that if $(s(1), s(2), \ldots, s(n))$ is the score sequence of a tournament, then $\sum_{i=1}^{n} s(i) = \binom{n}{2}$.

14. **(a)** Could $(1, 1, 2, 3)$ be the score sequence of a tournament? Why?
    (b) What about $(0, 0, 0, 2, 7)$?
    (c) What about $(0, 0, 3, 3)$?

15. Show that in a tournament, the ranking of players using the score sequence can be different from the ranking of players using a hamiltonian path.

16. Draw the digraph of a tournament with score sequence $(0, 1, 2, 3)$.

17. Show that a tournament is transitive if and only if it has no cycles of length 3.

18. Show that a tournament is transitive if and only if it is acyclic.

19. Prove Theorem 11.12 as follows.
    (a) Show that if D is transitive, then it has a unique hamiltonian path, by observing that if there are two such paths, there must be u and v with u following v in one, but v following u in the other.
    (b) Prove the converse by observing that in the unique hamiltonian path $u_1, u_2, \ldots, u_n$, we have $(u_i, u_j) \in A$ iff $i < j$.

20. **(a)** Use the result of Exercise 19(b) to find the score sequence of a transitive tournament of n vertices.

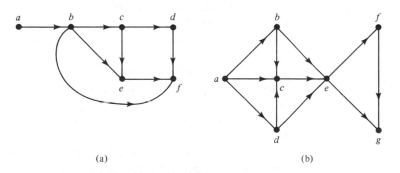

(a)   (b)

Figure 11.59 Digraphs for exercises of Section 11.6.

(b) Show that for a transitive tournament, the ranking obtained by the score sequence is the same as the ranking obtained by the unique hamiltonian path.

21. Use the result of Exercise 20(a) to determine if the preference data of Figure 11.58 are transitive.

22. Show that every acyclic digraph has a vertex with no incoming arcs.

23. In a tournament, a triple $\{x, y, z\}$ of vertices is called *transitive* if the subgraph generated by x, y, and z is transitive, equivalently if one of these three vertices beats the other two. This one vertex is called the *transmitter* of the transitive triple.
    (a) How would you find the number of transitive triples of which a given vertex u is the transmitter?
    (b) Show that in a tournament, if $s(x)$ is the score of x, there are $\sum_x \binom{s(x)}{x}$ transitive triples.
    (c) Show that every tournament of at least four vertices has to have a transitive triple.

24. If you have not already done so, do Exercise 25, Section 3.6.

25. (Harary *et al.* [1965]) If D has a level assignment (Exercise 25, Section 3.6) and r is the length of the longest simple path of D, show that $r + 1$ is the smallest number of distinct levels (values L_i) in a level assignment for D.

26. If you have not already done so, do Exercise 26, Section 3.6.

27. (Camion [1959] and Foulkes [1960]) Prove that every strongly connected tournament has a hamiltonian cycle. (*Hint*: Show that there is a cycle of length k, for $k = 3, 4, \ldots, n$, where n is the number of vertices.)

28. An alternative criterion for optimality in the book production problem of Section 11.6.3 is to finish printing and binding all the books in as short a time as possible. Show that even under the special assumption of Section 11.6.3, there can be an optimal solution which does not correspond to a hamiltonian path in the digraph constructed in that section.

REFERENCES FOR CHAPTER 11

AHO, A. V., HOPCROFT, J. E., and ULLMAN, J. D., *The Design and Analysis of Computer Algorithms*, Addison-Wesley, Reading, Mass., 1974.

BAASE, S., *Computer Algorithms: Introduction to Design and Analysis*, Addison-Wesley, Reading, Mass., 1978.

BEHZAD, M., CHARTRAND, G., and LESNIAK-FOSTER, L., *Graphs and Digraphs*, Wadsworth, Belmont, Calif., 1979.

BELTRAMI, E. J., *Models for Public Systems Analysis*, Academic Press, New York, 1977.

BERGE, C., *The Theory of Graphs and Its Applications*, Wiley, New York, 1962.

BOESCH, F., and TINDELL, R., "Robbins' Theorem for Mixed Graphs," *Amer. Math. Monthly*, *87* (1980), 716–719.

BONDY, J. A., and CHVÁTAL, V., "A Method in Graph Theory," *Discrete Math.*, *15* (1976), 111–136.

CAMION, P., "Chemins et circuits hamiltoniens des graphes complets," *C. R. Acad. Sci. Paris*, *249* (1959), 2151–2152.

CHACHRA, V., GHARE, P. M., and MOORE, J. M., *Applications of Graph Theory Algorithms*, Elsevier North Holland, New York, 1979.

CHRISTOFIDES, N., *Graph Theory: An Algorithmic Approach*, Academic Press, New York, 1975.

CHVÁTAL, V., and THOMASSEN, C., "Distances in Orientations of Graphs," *J. Comb. Theory B*, *24* (1978), 61–75.

DE BRUIJN, N. G., "A Combinatorial Problem," *Nedl. Akad. Wet., Proc., 49* (1946), 758–764; *Indag. Math., 8* (1946), 461–467.

DEMOURCRON, G., MALGRANCE, V., and PERTUISET, R., "Graphes planaires: reconnaissance et construction de representations planaires topologiques," *Recherche Operationelle, 30* (1964), 33.

DEO, N., *Graph Theory with Applications to Engineering and Computer Science*, Prentice-Hall, Englewood Cliffs, N.J., 1974.

DIRAC, G. A., "Some Theorems on Abstract Graphs," *Proc. Lond. Math. Soc., 2* (1952), 69–81.

EULER, L., "Solutio Problematis ad Geometriam Situs Pertinentis," *Comment. Acad. Sci. I. Petropolitanae, 8* (1736), 128–140. [Reprinted in *Opera Omnia*, Series I-7 (1766), 1–10.]

EVEN, S., *Graph Algorithms*, Computer Science Press, Potomac, Md., 1979.

FOULKES, J. D., "Directed Graphs and Assembly Schedules," *Proc. Symp. Appl. Math, Amer. Math. Soc., 10* (1960), 281–289.

GHOUILA-HOURI, A., "Une condition suffisante d'existence d'un circuit hamiltonien," *C. R. Acad. Sci. Paris, 156* (1960), 495–497.

GOLOMB, S. W., *Shift Register Sequences*, Holden-Day, San Francisco, 1967.

GOLUMBIC, M. C., *Algorithmic Graph Theory and Perfect Graphs*, Academic Press, New York, 1980.

GOOD, I. J., "Normal Recurring Decimals," *J. Lond. Math. Soc., 21* (1946), 167–169.

HARARY, F., NORMAN, R. Z., and CARTWRIGHT, D., *Structural Models: An Introduction to the Theory of Directed Graphs*, Wiley, New York, 1965.

HARARY, F., and PALMER, E. M., *Graphical Enumeration*, Academic Press, New York, 1973.

HELLY, W., *Urban Systems Models*, Academic Press, New York, 1975.

HOPCROFT, J. E., and TARJAN, R. E., "Algorithm 447: Efficient Algorithms for Graph Manipulation," *Commun. ACM, 16* (1973), 372–378.

HUTCHINSON, G., "Evaluation of Polymer Sequence Fragment Data Using Graph Theory," *Bull. Math. Biophys., 31* (1969), 541–562.

HUTCHINSON, J. P., and WILF, H. S., "On Eulerian Circuits and Words with Prescribed Adjacency Patterns," *J. Comb. Theory A, 18* (1975), 80–87.

JOHNSON, S. M., "Optimal Two- and Three-Stage Production Schedules with Setup Times Included," *Naval Res. Logist. Quart., 1* (1954), 61–68.

KWAN, M.-K., "Graphic Programming Using Odd or Even Points," *Chin. Math., 1* (1962), 273–277.

LAWLER, E. L., *Combinatorial Optimization: Networks and Matroids*, Holt, Rinehart and Winston, New York, 1976.

LIEBLING, T. M., *Graphentheorie in Planungs-und Tourenproblemen*, Lecture Notes in Operations Research and Mathematical Systems No. 21, Springer-Verlag, New York, 1970.

LIU, C. L., *Introduction to Combinatorial Mathematics*, McGraw-Hill, New York, 1968.

LIU, C. L., *Topics in Combinatorial Mathematics*, Mathematical Association of America, Washington, D.C., 1972.

MINIEKA, E., *Optimization Algorithms for Networks and Graphs*, Dekker, New York, 1978.

ORE, O., "Note on Hamilton Circuits," *Amer. Math Monthly, 67* (1960), 55.

PÓSA, L., "A Theorem Concerning Hamiltonian Lines," *Magyar Tud. Akad. Mat. Kutató Int. Közl., 7* (1962), 225–226.

RÉDEI, L., "Ein kombinatorischer Satz," *Acta Litt. Sci. (Sect. Sci. Math.)*, Szeged, *7* (1934), 39–43.

REINGOLD, E. M., NIEVERGELT, J., and DEO, N., *Combinatorial Algorithms: Theory and Practice*, Prentice-Hall, Englewood Cliffs, N.J., 1977.

REINGOLD, E. M., and TARJAN, R. E., "On a Greedy Heuristic for Complete Matching," *SIAM J. Comput.*, *10* (1981), 676–681.

ROBBINS, H. E., "A Theorem on Graphs, with an Application to a Problem of Traffic Control," *Amer. Math. Monthly*, *46* (1939), 281–283.

ROBERTS, F. S., *Discrete Mathematical Models, with Applications to Social, Biological, and Environmental Problems*, Prentice-Hall, Englewood Cliffs, N.J., 1976.

ROBERTS, F. S., *Graph Theory and Its Applications to Problems of Society*, NSF–CBMS Monograph No. 29, SIAM, Philadelphia, 1978.

TARJAN, R. E., "Depth-First Search and Linear Graph Algorithms," *SIAM J. Comput.*, *1* (1972), 146–160.

TUCKER, A. C., and BODIN, L., "A Model for Municipal Street-Sweeping Operations," in W. F. Lucas, F. S. Roberts, and R. M. Thrall (eds.), *Discrete and System Models*, Vol. 3 of *Modules in Applied Mathematics*, Springer-Verlag, New York, 1983, pp 76–111.

WHITNEY, H., "*Congruent Graphs and the Connectivity of Graphs*," *Amer. J. Math.*, *54* (1932), 150–168.

WOODALL, D. R., "*Sufficient Conditions for Circuits in Graphs*," *Proc. Lond. Math. Soc.*, *24* (1972), 739–755.

PART IV Combinatorial Optimization

12 Matching and Covering

12.1 SOME MATCHING PROBLEMS

In this chapter we study a variety of problems that fall into two general categories called matching problems and covering problems. We shall look at these problems in two ways, first as existence problems and then as optimization problems. Thus, this chapter will serve as a transition from our emphasis on the second basic problem of combinatorics, the existence problem, to the third basic problem, the optimization problem. We now give a variety of examples.

Example 12.1 Job Assignments Revisited

In Example 4.10, we discussed a job assignment problem. In general, one can formulate this problem as follows. There are n workers and m jobs. Each worker is suited for some of the jobs. Assign each worker to one job, making sure that it is a job for which he or she is suited, and making sure that no two workers get the same job. In Example 4.10 we were concerned with counting the number of ways to make such an assignment. We used rook polynomials to do the counting. Here we shall ask an existence question: Is there *any* assignment that assigns each worker to one job to which he or she is suited, making sure that no job has two workers? Later, we shall ask an optimization question: What is the best assignment?

It will be convenient to formulate the existence question graph-theoretically. Build a graph G as follows. There are $m + n$ vertices in G, one for each worker and one for each job. Join each worker by an edge to each job for which he or she is suited. There are no other edges. Figure 12.1 shows the resulting graph G for one specific job assignment problem. Graph G is a *bipartite graph* (X, Y, E), a graph

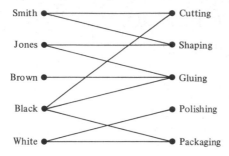

Smith — Cutting
Jones — Shaping
Brown — Gluing
Black — Polishing
White — Packaging

Figure 12.1 An edge from worker x to job y indicates that x is suitable for y.

whose vertex set is divided into two sets X and Y, and with an edge set E so that all edges in E are between sets X and Y (see Section 3.3.4).

A job assignment of the kind we are seeking can be represented by replacing an edge from worker x to worker y by a wiggly edge if x is assigned job y. Figure 12.2 shows one such assignment. Note that in this assignment, each vertex is on at most one wiggly edge. This corresponds to each worker being assigned to at most one job and each job being assigned to at most one worker. A set M of edges in a graph G is called a *matching* if each vertex in G is on at most one edge of M. Thus, we seek a matching. In a matching M, a vertex is said to be *saturated* (*M-saturated*) if it is on some edge of M. We seek a matching that saturates every vertex corresponding to a worker. We shall first study matchings for bipartite graphs, and then study them for arbitrary graphs. Figure 12.3 shows a matching in a nonbipartite graph.

Example 12.2 Storing Computer Programs Revisited

In Example 4.10 we also discussed a problem of assigning storage locations to computer programs. Formulating this in general terms, we can think of n programs and m storage locations making up the vertices of a graph. We put an edge between program x and location y if y has sufficient storage capacity for x. We seek an assignment of programs to storage locations such that each storage location gets at most one program and each program is assigned to exactly one location, a location that has sufficient storage capacity for the program. Thus, we seek a matching in the corresponding bipartite graph which saturates all vertices corresponding to programs. In contrast to the situation in Example 4.10, where we were interested in counting the number of such assignments, here we shall be interested in the question of whether or not there is such an assignment.

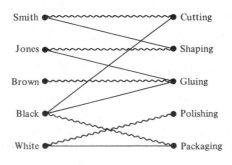

Smith — Cutting
Jones — Shaping
Brown — Gluing
Black — Polishing
White — Packaging

Figure 12.2 A job assignment for the graph of Figure 12.1 shown by wiggly edges.

Figure 12.3 A matching M in a non-bipartite graph.

Example 12.3 Pilots for RAF Planes (Berge [1973], Minieka [1978])

During the Battle of Britain in World War II, the Royal Air Force (RAF) had many pilots from foreign countries. The RAF had to assign two pilots to each plane, and always wanted to assign two pilots to the same plane whose language and training were compatible. The RAF's problem can be translated as follows. Given a collection of pilots available for use on a mission, use these as the vertices of a graph, and join two vertices with an edge if and only if the two pilots can fly together. Then the RAF wanted a matching in this graph. (It is not necessarily bipartite.) Moreover, they were interested in flying as large a number of planes as possible. Thus, they were interested in a *maximum cardinality matching*, a matching of the graph with the largest possible number of edges. In Sections 12.4 and 12.5, we will be interested in solving the combinatorial optimization problem which asks for a maximum cardinality matching.

Example 12.4 Real Estate Transactions (Minieka [1978])

A real estate agent at a given time has a collection X of potential buyers and a collection Y of houses for sale. Let r_{xy} be the revenue to the agent if buyer x purchases house y. From a purely monetary standpoint, the agent wants to match up buyers to houses so as to maximize the sum of the corresponding r_{xy}. We can represent the problem of the agent by letting the vertices X and Y define a *complete bipartite graph* $G = (X, Y, E)$; that is, we take all possible edges between X and Y. We then seek a matching M in G that maximizes the sum of the weights. More generally, we can consider the following *maximum weight matching problem*. Suppose that G is an arbitrary graph (not necessarily bipartite) with a weight (real number) r_{xy} on each edge $\{x, y\}$. If M is a matching of G, we define

$$r(M) = \sum \{r_{xy} : \{x, y\} \in M\}.$$

We seek a *maximum weight matching* of G, a matching M such that for all the matchings M' of G, $r(M) \geq r(M')$. We shall not discuss the general maximum weight matching problem here. For a discussion of this problem, see, for example, Christofides [1975], Lawler [1976], Minieka [1978], or Papadimitriou and Steiglitz [1982]. In Section 13.2.4, we shall see how the maximum weight matching problem arises in connection with the "Chinese postman" problem of Section 11.4.1 and the related computer graph plotting problem of Section 11.4.2.

Example 12.5 The Optimal Assignment Problem

Let us return to the job assignment problem of Example 12.1, but add the simplifying assumption that every worker is suited for every job. Then the "is suited for" graph G is a complete bipartite graph. Let us also assume that worker x is given a rating r_{xy} for his or her potential performance (output) on job y. We seek an assignment of workers to jobs that assigns every worker a job, no more than one per job, and maximizes the sum of the ratings. The problem of finding such an assignment is called the *optimal assignment problem.* If there are at least as many jobs as workers, the problem can be solved. It calls for a matching of the complete bipartite graph G that saturates the set of workers and has at least as large a sum of weights as any other matching that saturates the set of workers. But it is clear that every maximum weight matching saturates the set of workers. Hence, the optimal assignment problem reduces to the maximum weight matching problem for a complete bipartite graph. We return to the optimal assignment problem in Section 13.4.2, where we present an algorithm for solving it. (The algorithm can be understood from the material of Section 12.4, specifically Corollary 12.5.1.)

EXERCISES FOR SECTION 12.1

1. In the graph of Figure 12.4:
   (a) Find a matching that has four edges.
   (b) Find a matching that saturates vertex b.
   (c) Find a matching that is not maximum.
2. Repeat Exercise 1 for the graph of Figure 12.5.
3. In each weighted graph of Figure 12.6, find a maximum weight matching.
4. A company has five positions to fill, clerk (c), typist (t), telephone operator (o), word processor operator (w), and keypuncher (k). There are seven applicants for jobs. The first applicant is qualified for the positions c, t, o; the second for the positions w, k; the third for the positions t, w, k; the fourth for the positions c, o; the fifth for the positions c, t, o, w; the sixth for the position c only; and the seventh for all the positions. Can all the positions be filled? Formulate this

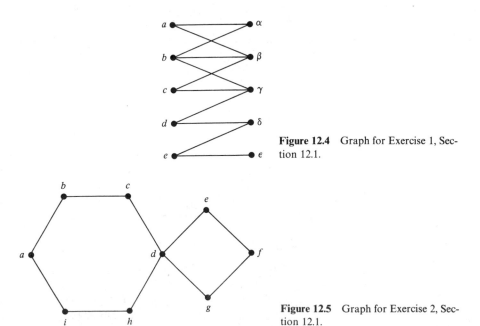

Figure 12.4 Graph for Exercise 1, Section 12.1.

Figure 12.5 Graph for Exercise 2, Section 12.1.

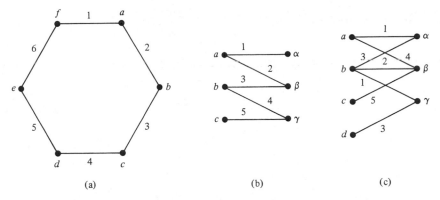

Figure 12.6 Weighted graphs for Exercise 3, Section 12.1.

problem in the language of this section. What are we looking for? You are not asked to solve the problem at this point.

5. A company assigns its workers to temporary jobs. The i, j entry of a matrix gives the distance from worker i's home to the jth job. This matrix is given information. If every worker is suited for every job, find an assignment of workers to jobs that minimizes the sum of the distances the workers have to travel. Formulate this problem in the language of this section. Do not solve. What are we looking for?

6. There are six students looking for rooms in a dormitory, Adams, Boyd, Cohen, Davis, Ellis, and Fisher. Adams likes Boyd, Cohen, Davis, and Fisher; Boyd likes Adams, Davis, Ellis, and Fisher; Cohen likes Adams, Ellis, and Fisher; Davis likes Adams, Boyd, and Ellis; Ellis likes Boyd, Cohen and Davis; and Fisher likes Adams, Boyd, and Cohen. Note that a likes b iff b likes a. We wish to assign roommates, two to a room, such that each person only gets a roommate whom he likes. Can we assign each person to a room? If not, what is the largest number of people we can assign to rooms? Formulate this problem in the language of this section. Do not solve. What are we looking for?

7. Show that to describe a general solution to the maximum weight matching problem for a weighted graph G, we may assume that:
   (a) G has an even number of vertices.
   (b) G is a complete graph.

8. Show that to describe a general solution to the maximum weight matching problem for a bipartite graph G, we may assume that the graph is complete bipartite and both classes have the same number of vertices.

9. Show that the maximum weight matching for a weighted bipartite graph does not necessarily saturate the first class of vertices, even if that class has no more than half the vertices.

12.2 SOME EXISTENCE RESULTS: BIPARTITE MATCHING AND SYSTEMS OF DISTINCT REPRESENTATIVES

12.2.1 Bipartite Matching

In most of the examples of the preceding section, the graph constructed was a bipartite graph $G = (X, Y, E)$. Here we shall ask the question: Given a bipartite graph $G = (X, Y, E)$, under what conditions does there exist a matching that saturates X?

Consider the computer program storage assignment graph of Figure 12.7. Note

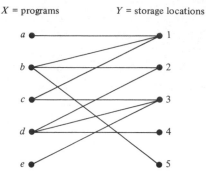

X = programs   Y = storage locations

Figure 12.7 Bipartite graph for computer program storage. The vertices in X are the programs, those in Y are the storage locations, and there is an edge between x in X and y in Y iff y has sufficient capacity to store x.

that the three programs a, c, e have together only two possible locations they could be stored in, locations 1 and 3. Thus, there is no storage assignment that assigns each program to a location of sufficient storage capacity. There is no X-saturating matching. To generalize this example, it is clear that for there to exist an X-saturating matching, if S is any set of vertices in X and $N(S)$ is the *open neighborhood* of S, the set of all vertices y joined to some x in S by an edge, then $N(S)$ must have at least as many elements as S. In our example, S was $\{a, c, e\}$ and $N(S)$ was $\{1, 3\}$. What is startling is that this obvious necessary condition is sufficient as well.

Theorem 12.1 (Philip Hall's Theorem) (Hall [1935]). Let $G = (X, Y, E)$ be a bipartite graph. Then there exists an X-saturating matching if and only if for all subsets S of X, $|N(S)| \geq |S|$.

We will prove Theorem 12.1 in Section 12.4.

To see whether the bipartite graph of Figure 12.1 has an X-saturating matching, where $X = \{$Smith, Jones, Brown, Black, White$\}$, we have to compute $N(S)$ for all subsets S of X. There are $2^{|S|} = 2^5 = 32$ such subsets. To make the computation, we note for instance that

$$N(\{\text{Smith, Jones, Black}\}) = \{\text{Cutting, Shaping, Gluing, Packaging}\},$$

so $N(S)$ has four elements whereas S has three. A similar computation for all 32 cases shows that there is an X-saturating matching. (In this case, finding one directly is faster.)

Example 12.6   A Coding Problem

Each entry in a small file has a three-digit codeword associated with it, using the digits $0, 1, 2, \ldots, n$. Can we associate with each entry in the file just one of the digits of its codeword so that the entry can be uniquely recovered from this single digit? We can formulate this problem graph-theoretically as follows. The vertices of a bipartite graph $G = (X, Y, E)$ consist of X, the set of entries, and Y, the set of digits $0, 1, 2, \ldots, n$. There is an edge from entry x to digit y if y is used in the codeword for x. Figure 12.8 gives an example. We seek an X-saturating matching in the bipartite graph G. In our example, Table 12.1 lists all subsets S of X and the corresponding neighborhood $N(S)$. Note that in each case, $|N(S)| \geq |S|$. Thus, an X-saturating

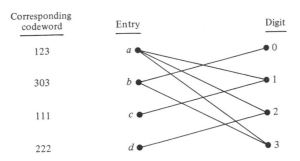

X = set of entries Y = set of digits $0, 1, 2, \ldots, n$

Corresponding codeword   Entry   Digit

123   a 0

303   b 1

111   c 2

222   d 3

Figure 12.8   Bipartite graph for the coding problem.

matching exists. This does not help us to find one. However, it is easy to see that the edges $\{a, 3\}, \{b, 0\}, \{c, 1\}, \{d, 2\}$ form such a matching.

A graph G is called *regular* if every vertex has the same degree, that is, the same number of neighbors.

Corollary 12.1.1.   Suppose that $G = (X, Y, E)$ is a regular bipartite graph with at least one edge. Then G has an X-saturating matching.

Proof. Let S be a subset of X. Let E_1 be the collection of edges leading from vertices of S, and E_2 be the collection of edges leading from vertices of $N(S)$. Since every edge of E_1 leads from a vertex of S to a vertex of $N(S)$, we have $E_1 \subseteq E_2$. Thus, $|E_1| \leq |E_2|$. Moreover, since every vertex has the same number of neighbors, say k, it follows that $|E_1| = k|S|$ and $|E_2| = k|N(S)|$. Thus, $k|S| \leq k|N(S)|$. Since G has an edge, k is positive, so we conclude that $|S| \leq |N(S)|$.   Q.E.D.

As an application of Corollary 12.1.1, suppose in Example 12.6 that every codeword uses exactly k of the digits $0, 1, 2, \ldots, n$ and every digit among $0, 1, 2, \ldots, n$ appears in exactly k codewords. Then there is an association of one digit with each file entry which allows a unique decoding.

Note that the X-saturating matching of a regular bipartite graph $G = (X, Y, E)$ is also Y-saturating. For every edge of G touches both a vertex of X and a vertex of Y. Thus, if k is the number of neighbors of each vertex, $k|X| = |E|$ and $k|Y| = |E|$. Thus, $k|X| = k|Y|$ and $|X| = |Y|$. Thus, every X-saturating matching must also be Y-saturating. A matching that saturates every vertex of a graph is called *perfect*. Perfect

Table 12.1   Subsets S of X and Corresponding Neighborhoods $N(S)$, for Graph of Figure 12.8

S	$\varnothing$	$\{a\}$	$\{b\}$	$\{c\}$	$\{d\}$	$\{a, b\}$	$\{a, c\}$	$\{a, d\}$	$\{b, c\}$
$N(S)$	$\varnothing$	$\{1, 2, 3\}$	$\{0, 3\}$	$\{1\}$	$\{2\}$	$\{0, 1, 2, 3\}$	$\{1, 2, 3\}$	$\{1, 2, 3\}$	$\{0, 1, 3\}$

S	$\{b, d\}$	$\{c, d\}$	$\{a, b, c\}$	$\{a, b, d\}$	$\{a, c, d\}$	$\{b, c, d\}$	$\{a, b, c, d\}$
$N(S)$	$\{0, 2, 3\}$	$\{1, 2\}$	$\{0, 1, 2, 3\}$	$\{0, 1, 2, 3\}$	$\{1, 2, 3\}$	$\{0, 1, 2, 3\}$	$\{0, 1, 2, 3\}$

matchings in job assignments mean that every worker gets a job and every job gets a worker. The question of when in general there exists a perfect matching is discussed further in Section 12.3. (See also Exercises 13, 16, 19, and 21 below.)

12.2.2 Systems of Distinct Representatives

Suppose that $\mathscr{F} = \{S_1, S_2, \ldots, S_p\}$ is a family of sets, not necessarily distinct. Let $T = (a_1, a_2, \ldots, a_p)$ be a p-tuple with $a_1 \in S_1$, $a_2 \in S_2$, $\ldots$, $a_p \in S_p$. Then T is called a *system of representatives* for $\mathscr{F}$. If, in addition, all the a_i are distinct, T is called a *system of distinct representatives* (an SDR) for $\mathscr{F}$. For instance, suppose that

$$\mathscr{F} = \{S_1, S_2, S_3, S_4, S_5\},$$

where

$$S_1 = \{a, b, c\}, \qquad S_2 = \{b, c, d\}, \qquad S_3 = \{c, d, e\},$$
$$S_4 = \{d, e\}, \qquad S_5 = \{e, a, b\}. \tag{12.1}$$

Then $T = (a, b, c, d, e)$ is an SDR for $\mathscr{F}$. Next, suppose that $\mathscr{F} = \{S_1, S_2, S_3, S_4, S_5, S_6\}$, where

$$S_1 = \{a, b\}, \qquad S_2 = \{b, c\}, \qquad S_3 = \{a, b, c\},$$
$$S_4 = \{b, c, d\}, \qquad S_5 = \{a, c\}, \qquad S_6 = \{c, d\}. \tag{12.2}$$

Then $T = (a, b, c, d, a, d)$ is a system of representatives for $\mathscr{F}$. However, there is no SDR, as we shall show below.

Example 12.7   Hospital Internships

In a given year, suppose that p medical school graduates have applied for internships in hospitals. For the ith medical school graduate, let S_i be the set of all hospitals that find i acceptable. Then a system of representatives for the family $\mathscr{F} = \{S_i : i = 1, 2, \ldots, p\}$ would assign each potential intern to a hospital that is willing to take him or her. An SDR would make sure that, in addition, no hospital gets more than one intern. In practice, SDRs could be used in the following way. Modify S_i to include i. Find an SDR. This assigns each i to a hospital or to himself or herself; the latter is interpreted to mean that on the first round, i is not assigned to a hospital. (Exercise 7 asks the reader to show that an SDR exists.) After the initial assignment based on an SDR, the hospitals would be asked to modify their list of acceptable applicants among those not yet assigned. (Hopefully, at least some are assigned in the first round.) Then a new SDR would be found. And so on. A similar procedure could be used to place applicants for admission to graduate school. More complicated procedures would make use of hospitals' and also applicants' ratings or rankings of the alternatives.

The problem of finding an SDR can be formulated graph-theoretically as follows. Suppose that X is the collection of sets S_i in $\mathscr{F}$ and Y is the collection of points in $\cup S_i$. Let $G = (X, Y, E)$, be a bipartite graph, with an edge from x in X to y in Y iff y is in x. Then an SDR is simply an X-saturating matching in G.

For instance, consider the family of sets in (12.2). The corresponding bipartite graph

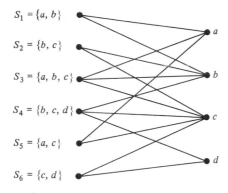

$S_1 = \{a, b\}$

$S_2 = \{b, c\}$

$S_3 = \{a, b, c\}$

$S_4 = \{b, c, d\}$

$S_5 = \{a, c\}$

$S_6 = \{c, d\}$

Figure 12.9   Bipartite graph representing the family of sets of (12.2).

is shown in Figure 12.9. Note that if $S = \{S_1, S_2, S_3, S_5\}$, then $N(S) = \{a, b, c\}$. Thus $|S| > |N(S)|$, so by Philip Hall's Theorem, there is no X-saturating matching and hence no SDR. Using the language of SDRs, we may now reformulate Philip Hall's Theorem as follows.

Corollary 12.1.2.   The family $\mathscr{F} = \{S_1, S_2, \ldots, S_p\}$ possesses an SDR iff for all $k = 1, 2, \ldots, n$, any k S_i's together contain at least k elements of $\cup S_i$.

Example 12.8   Extending Latin Rectangles

A *Latin rectangle* is an $r \times s$ array using the elements $1, 2, \ldots, n$, such that each row and column uses each element from $1, 2, \ldots, n$ at most once. A Latin rectangle is called *complete* if $n = s$. One way to build up an $n \times n$ Latin square is to try building it up one row at a time, that is, by creating a complete $r \times n$ Latin rectangle and then extending it. For instance, consider the 2×6 Latin rectangle of Figure 12.10. Can we extend this to a 6×6 Latin square? More particularly, can we add a third row to obtain a 3×6 Latin rectangle? One approach to this question is to use rook polynomials (see Exercise 13, Section 4.1). Another approach is to ask what numbers could be placed in the new row in the ith column. Let S_i be the set of numbers not yet occurring in the ith column. In our example,

$$S_1 = \{2, 3, 5, 6\}, \qquad S_2 = \{1, 4, 5, 6\}, \qquad S_3 = \{1, 2, 4, 5\},$$
$$S_4 = \{1, 2, 3, 6\}, \qquad S_5 = \{2, 3, 4, 6\}, \qquad S_6 = \{1, 3, 4, 5\}.$$

We want to pick one element from each S_i and we want these elements to be distinct. Thus, we want an SDR. One SDR is $(2, 1, 4, 3, 6, 5)$. Thus, we can use this as a new third row, obtaining the Latin rectangle of Figure 12.11. This idea generalizes. In general, given an $r \times n$ complete Latin rectangle, let S_i be the set of numbers not yet occurring in the ith column. To add an $(r + 1)$st row, we need an SDR for the family of S_i. We shall show that we can always find one.

1	2	3	4	5	6
4	3	6	5	1	2

Figure 12.10   A 2×6 Latin rectangle.

1	2	3	4	5	6
4	3	6	5	1	2
2	1	4	3	6	5

Figure 12.11 A 3 × 6 Latin rectangle obtained by extending the Latin rectangle of Figure 12.10.

Theorem 12.2. If $r < n$, then every $r \times n$ complete Latin rectangle L can be extended to an $(r + 1) \times n$ complete Latin rectangle.

*Proof.** If S_i is defined as in Example 12.8, we shall show that the family of S_i possesses an SDR. Pick k of the sets in this family, $S_{i_1}, S_{i_2}, \ldots, S_{i_k}$. By Corollary 12.1.2, it suffices to show that $A = S_{i_1} \cup S_{i_2} \cup \cdots \cup S_{i_k}$ has at least k elements in it. Each S_{i_j} has $n - r$ elements. Thus, A has $k(n - r)$ elements, including repetitions. Since the $r \times n$ Latin rectangle L is complete, each number in $1, 2, \ldots, n$ appears exactly once in each row, so each number in $1, 2, \ldots, n$ appears exactly r times in L. Thus, each number in $1, 2, \ldots, n$ is in exactly $n - r$ of the sets $S_1, S_2, \ldots, S_n$. Thus, each number in A appears in at most $n - r$ of the sets $S_{i_1}, S_{i_2}, \ldots, S_{i_k}$. Now if we make a list of elements of A including repetitions, we have $k(n - r)$ elements. Each number in A appears in this list at most $n - r$ times. By the pigeonhole principle (Theorem 8.2), there must be at least k distinct numbers in A. Q.E.D.

EXERCISES FOR SECTION 12.2

1. Find $N(S)$ if G and S are as follows.
   (a) G = graph of Figure 12.1 and $S = \{$Smith, White$\}$
   (b) G = graph of Figure 12.8 and $S = \{a, c, d\}$
   (c) G = graph of Figure 12.9 and $S = \{S_1, S_3, S_4\}$.
2. Find $N(S)$ if G and S are as follows.
   (a) G = graph of Figure 12.1 and $S = \{$Cutting, Shaping$\}$
   (b) G = graph of Figure 12.1 and $S = \{$Smith, Black, Polishing$\}$
   (c) G = graph of Figure 12.5 and $S = \{a, b, e\}$.
3. For each bipartite graph $G = (X, Y, E)$ of Figure 12.12, determine if G has an X-saturating matching.
4. (a) Find a bipartite graph corresponding to the family of sets of (12.1).
   (b) *Use Philip Hall's Theorem* to show there is an X-saturating matching in this graph.
5. For each of the following families of sets, find a system of representatives.
   (a) $S_1 = \{a, b, f\}$, $S_2 = \{a\}$, $S_3 = \{a, b, d, f\}$, $S_4 = \{a, b\}$, $S_5 = \{b, f\}$, $S_6 = \{d, e, f\}$, $S_7 = \{a, f\}$;
   (b) $S_1 = \{c\}$, $S_2 = \{a, c\}$, $S_3 = \{a, c\}$, $S_4 = \{a, b, c, d, e\}$;
   (c) $S_1 = \{a_2, a_3\}$, $S_2 = \{a_1, a_4, a_5\}$, $S_3 = \{a_1, a_2, a_4\}$, $S_4 = \{a_2, a_5\}$;
   (d) $S_1 = \{b_1, b_5\}$, $S_2 = \{b_1, b_2, b_3\}$, $S_3 = \{b_3, b_4, b_5\}$, $S_4 = \{b_3, b_4, b_5\}$;
   (e) $S_1 = \{x, y\}$, $S_2 = \{x\}$, $S_3 = \{u, v, w\}$, $S_4 = \{x, y, z\}$, $S_5 = \{u, v\}$, $S_6 = \{y, z\}$;
   (f) $S_1 = \{a, d\}$, $S_2 = \{b, d\}$, $S_3 = \{a, c, d, e\}$, $S_4 = \{a, d\}$, $S_5 = \{a, b\}$.

*The proof can be omitted.

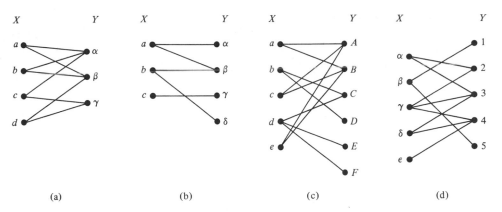

Figure 12.12 Bipartite graphs for exercises of Section 12.2.

6. For each of the families of sets in Exercise 5, determine if there is a system of distinct representatives and if so, find one.

7. In Example 12.7, show that if i is in S_i, an SDR exists.

8. Find the number of SDRs of each of the following families of sets.
   (a) $S_1 = \{1, 2\}, S_2 = \{2, 3\}, S_3 = \{3, 4\}, S_4 = \{4, 5\}, S_5 = \{5, 1\}$;
   (b) $S_1 = \{1, 2\}, S_2 = \{2, 3\}, S_3 = \{3, 4\}, \ldots, S_n = \{n, 1\}$.

9. There are six committees of a state legislature, Finance, Environment, Health, Transportation, Education, and Housing. In Chapter 1 we discussed a meeting assignment problem for these committees. Here, we ask a different question. Suppose that there are 12 legislators who need to be assigned to committees, each to one committee. The following matrix has i, j entry equal to 1 iff the ith legislator would like to serve on the jth committee.

	Finance	Environment	Health	Transportation	Education	Housing
Allen	1	1	1	0	0	0
Barnes	1	1	0	1	1	0
Cash	1	1	1	0	0	0
Dunn	1	0	0	1	1	1
Ecker	0	1	1	0	0	0
Frank	1	1	0	0	0	0
Graham	1	1	1	0	0	0
Hall	1	0	0	0	0	0
Inman	1	1	1	0	0	0
Johnson	1	1	0	0	0	0

Suppose that we want to choose exactly one new member for each committee, choosing only a legislator who would like to serve. Can we do so? (Not every legislator needs to be assigned to a committee and no legislator can be assigned to more than one committee.)

10. Suppose that we are given a collection of codewords and we would like to store the collection on a computer as succinctly as possible, namely by picking one letter from each word to store. Can we do this for the following set of codewords in such a way that the codewords can be uniquely decoded? Words: *abcd, cde, a, b, ce.*

11. Solve Exercise 4, Section 12.1.

12. Consider the following arrays. In each case, can we add one more column to the array, using the numbers 1, 2, 3, 4, 5, 6, 7, so that the entries in this new column are all different and so that the entries in each row of the new array are all different?

(a)
```
123
265
371
436
542
```

(b)
```
67345
12734
73456
34567
```
.

13. **(a)** What circuits Z_n have perfect matchings?
   (b) What chains L_n of n vertices have perfect matchings?
   (c) A wheel graph W_n is obtained from Z_n by adding a vertex adjacent to all vertices of Z_n. What wheel graphs W_n have perfect matchings?

14. There are n men and n women, and each man likes exactly p women and each woman likes exactly p men. Assuming that a likes b iff b likes a, show that the $2n$ people can be paired off into couples so that each man is paired with a woman he likes and who likes him.

15. Is the result of Exercise 14 still true if each man likes at least p women and each woman likes at most p men? Why?

16. (Tucker [1980]) There are n men and n women enrolled in a computer dating service. Suppose that the service has made nm pairings, so that each man dates m different women and each woman dates m different men, and suppose that $m < n$.
   (a) Suppose that we want to schedule all the dates over a period of m nights. Show that it is possible to do so. That is, show that the pairings can be divided into m perfect matchings.
   (b) Show that we can make the division in part (a) one step at a time; that is, show that no matter how the first k perfect matchings are chosen, we can always find a $(k + 1)$st.

17. Let $\mathscr{F} = \{S_1, S_2, \ldots, S_n\}$ be a family of sets that has an SDR.
   (a) If x is in at least one of the S_i, show that there is an SDR that chooses x as the representative of at least one of the S_i.
   (b) For each S_i that contains x, is there necessarily an SDR in which x is picked as the representative of S_i? Why?

18. Let $\mathscr{F} = \{S_1, S_2, \ldots, S_n\}$, where $S_i = \{1, 2, \ldots, n\} - \{i\}$.
   (a) Show that $\mathscr{F}$ has an SDR.
   (b) Show that the number of different SDR's is the number of derangements of n elements, D_n.

19. Is it possible for a tree to have more than one perfect matching? Why?

20. Consider a bipartite graph $G = (X, Y, E)$. Suppose that $m \geq 1$ is the maximum degree of a vertex of G and X_1 is the subset of X consisting of all vertices of degree m. Assume that X_1 is nonempty. Give a proof or a counterexample of the following assertion: There is a matching of G in which all vertices of X_1 are saturated.

21. **(a)** If H is a graph, let $o(H)$ count the number of components of H with an odd number of vertices. If S is a set of vertices in graph G, $G - S$ is the subgraph of G generated by vertices not in S. Show that if G has a perfect matching, then $o(G - S) \leq |S|$ for all $S \subseteq V(G)$. (Tutte [1947] proves this result and also its converse.)
   (b) (Peterson [1891]) Generalizing the ideas of Section 11.2, let an edge $\{a, b\}$ in a graph G be called a *bridge* if removal of the edge, but not its end vertices a and b, increases the number of components. Suppose that G is a graph in which every vertex has degree 3 and suppose that G does not have any bridges. Assuming the converse in part (a), show that G has a perfect matching.

22. Suppose that $A_1, A_2, \ldots, A_m$ and $B_1, B_2, \ldots, B_m$ are two partitions of the same set S with the same number of A_i and B_j and with all $A_i \neq \emptyset$ and all $B_j \neq \emptyset$. (A *partition* of S is a division of all elements of S into disjoint subsets.) Let E be an m-element subset of S such that $A_i \cap E \neq \emptyset$ for all i and $B_j \cap E \neq \emptyset$ for all j. Then, it is clear that $|A_i \cap E| = |B_j \cap E| = 1$ for all i, j. The set E is called a *system of common representatives* (SCR) for the partitions A_i and B_j.

    (a) Find an SCR for the following partitions of $S = \{1, 2, 3, 4, 5\}$: $A_1 = \{1, 2\}$, $A_2 = \{3, 4\}$, $A_3 = \{5\}$; $B_1 = \{1, 3\}$, $B_2 = \{5\}$, $B_3 = \{2, 4\}$.

    (b) Show that an SCR exists for the partitions A_i and B_j if and only if there is a suitable renumbering of the components of the partitions A_i and B_j such that for all i, $A_i \cap B_i \neq \emptyset$.

    (c) Show that the partitions A_i and B_j have an SCR if and only if for all $k = 1, 2, \ldots, m$ and all $i_1, i_2, \ldots, i_k$ and $j_1, j_2, \ldots, j_{k-1}$ from $1, 2, \ldots, m$, the set $A_{i_1} \cup A_{i_2} \cup \cdots \cup A_{i_k}$ is not a subset of the set $B_{j_1} \cup B_{j_2} \cup \cdots \cup B_{j_{k-1}}$. (*Hint*: Form a bipartite graph and construct a matching.)

12.3 THE EXISTENCE OF PERFECT MATCHINGS FOR ARBITRARY GRAPHS*

Example 12.3 (Pilots for RAF Planes) Revisited

Let us consider whether or not, given a group of pilots, it is possible to assign two compatible pilots to each airplane, in such a way that every pilot is assigned to a plane. This is the case if and only if the graph constructed in Example 12.3 has a perfect matching, a matching in which every vertex is saturated. In this section we investigate one condition that guarantees the existence of a perfect matching.

    Theorem 12.3. If a graph G has $2n$ vertices and each vertex has degree $\geq n$, then the graph has a perfect matching.

We shall show why this theorem is true by trying to build up a matching M one step at a time. Here is an algorithm for doing this.

Algorithm 12.1. Finding a Perfect Matching

    Input: A graph G with $2n$ vertices and each vertex having degree $\geq n$.
    Output: A perfect matching.

Step 1. Set $M = \emptyset$.

Step 2. Find any pair of unsaturated vertices a and b which are joined by an edge of G, and place edge $\{a, b\}$ in M.

Step 3. If there is a pair of unsaturated vertices of G joined by an edge of G, return to step 2. Otherwise, go to step 4.

Step 4. If M has n edges, stop and output M. If M does not have n edges, go to step 5.

Step 5. Find a pair of unsaturated vertices a and b in G. These will not be joined by an edge of G. (Why?) Find an edge $\{u, v\}$ in M such that $\{a, u\}$ and $\{b, v\}$ are edges of G. Remove edge $\{u, v\}$ from M and place edges $\{a, u\}$ and $\{b, v\}$ in M. Return to step 4.

*This section can be omitted without loss of continuity.

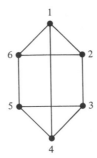

Figure 12.13 Graph used to illustrate Algorithm 12.1.

We illustrate this procedure with the graph of Figure 12.13. We show below why the procedure works. Note that in this graph, there are $2n = 2(3) = 6$ vertices, and each has degree $\geq n = 3$. Start in step 2 by placing edge $\{1, 6\}$ in M. Table 12.2 shows the successive edges of M. Go to step 3. Next, since there is a pair of unsaturated vertices joined by an edge of G, go to step 2. Find such a pair, say 3, 5, and place edge $\{3, 5\}$ in M. Now, there is no pair of unsaturated vertices joined by an edge of G, and since M has fewer than n edges, we proceed from step 3 to step 4 to step 5. Pick vertices 2 and 4, which are unsaturated in G and not joined by an edge of G. Note that 3 and 5 are matched in M, and there are edges $\{2, 3\}$ and $\{4, 5\}$ in G. Thus, remove $\{3, 5\}$ from M, add $\{2, 3\}$ and $\{4, 5\}$, and return to step 4. Since we have $n = 3$ elements in M, we stop, having obtained a perfect matching.

To see why this procedure works, note that at each stage, we have a matching. Moreover, if the algorithm stops, we always end up with n edges in M, and since there are $2n$ vertices in G, we must have a perfect matching. The crucial question is this: How do we know that step 5 works? In particular, how do we know that there always exist u and v as called for in step 5? To see why,* suppose that a and b are unsaturated in M and not joined by an edge of G. There are at present $r < n$ edges in M. If step 5 could not be carried out, then for every edge $\{u, v\}$ in M, at most two of the edges $\{a, u\}$, $\{a, v\}$, $\{b, u\}$, $\{b, v\}$ are in G. The number of edges from a or b to matched edges $\{u, v\}$ in G is thus at most $2r$. Every edge from a or from b goes to some edge in M, for otherwise a or b is joined by an edge of G to an unsaturated vertex of G, and we would not have gotten to step 5 in the algorithm. Thus, degree of a + degree of $b \leq 2r$. Since $r < n$, degree of a + degree of $b < 2n$. But by hypothesis, degree of a + degree of $b \geq n + n = 2n$, which is a contradiction. We conclude that step 5 can always be carried out.

Table 12.2 The Edges in M at each Stage of Algorithm 12.1 Applied to the Graph of Figure 12.13

Stage	1	2	3	4
Edges in M	None	$\{1, 6\}$	$\{1, 6\}, \{3, 5\}$	$\{1, 6\}, \{2, 3\}, \{4, 5\}$

*This argument can be omitted.

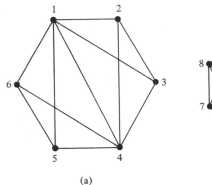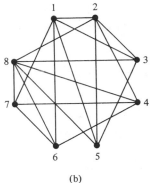

Figure 12.14 Graphs for exercises of Sections 12.3 and 12.4.

(a)   (b)

Let us examine the computational complexity of Algorithm 12.1. To get a crude upper bound, suppose that the ith iteration of the algorithm is the procedure between the time the matching M has $i - 1$ edges and the time it has i edges. Thus, there are n iterations in all. We can keep a record of the unsaturated vertices. Thus, it takes at most e steps to find an edge joining a pair of unsaturated vertices, where $e = |E(G)|$. If there is no such edge, we pick an arbitrary pair of unsaturated vertices a and b. For each edge $\{u, v\}$ in M, we ask if $\{a, u\}$ and $\{b, v\}$ are edges of G or if $\{a, v\}$ and $\{b, u\}$ are edges of G. This requires four questions for each of at most e edges, so at most $4e$ steps. Thus, we have at most $e + 4e = 5e$ steps in each iteration, and at most $n(5e)$ stcps in all. Since $e \le \binom{n}{2} \le n^2$, we have at most $5n^3$ steps, a polynomial bound. In the terminology of Section 2.18, we have an $O(n^3)$ algorithm.

EXERCISES FOR SECTION 12.3

1. Suppose that K_{2p} is the complete graph of $2p$ vertices, $p \ge 2$, and G is obtained from K_{2p} by removing an edge. Does G have a perfect matching?

2. The *complete p-partite graph* $K(n_1, n_2, \ldots, n_p)$ has p classes of vertices, n_i vertices in the ith class, and all vertices in class i joined to all vertices in class j, for $i \ne j$, with no other vertices joined. Which of the following graphs have perfect matchings?
   (a) $K(2, 2)$ **(b)** $K(3, 3)$ **(c)** $K(1, 5)$ **(d)** $K(2, 3)$
   (e) $K(2, 2, 2)$ **(f)** $K(3, 3, 3)$ **(g)** $K(4, 4, 4)$ **(h)** $K(2, 2, 2, 2)$.

3. In Exercise 6, Section 12.1, can all students be assigned roommates?

4. For each graph of Figure 12.14, use Algorithm 12.1 to find a perfect matching.

5. Suppose that G has $2n$ vertices and for all $a \ne b$ such that $\{a, b\}$ is not an edge of G, degree of $a +$ degree of $b \ge 2n$. Show that G has a perfect matching.

12.4 MAXIMUM MATCHINGS AND MINIMUM COVERINGS

12.4.1 Vertex Coverings

A *vertex covering* in a graph, or a *covering* for short, is any set of vertices such that each edge of the graph has at least one of its end vertices in the set. For example, in the graph of Figure 12.13, the set of vertices $\{1, 2, 3, 4, 5\}$ forms a covering. So does the set of vertices $\{1, 3, 5, 6\}$. In this section we shall study *minimum cardinality coverings, minimum coverings* for short. These are coverings that use as few vertices as possible. We shall relate minimum coverings to maximum cardinality matchings, *maximum matchings* for short. Such matchings arose in Example 12.3.

Example 12.9   Police Surveillance

A police officer standing at a street intersection in a city can watch one block in any direction. If we attempt to locate a criminal, we want to be sure that all the blocks in a neighborhood are being watched simultaneously. Assuming that police officers stand only at street intersections, what is the smallest number of police officers we need, and where do we put them? The answer is obtained by letting the street intersections in the neighborhood be vertices of a graph G and joining two vertices by an edge if and only if they are joined by a one-block street. Then we seek a minimum covering of the graph G.

We shall now investigate the relation between matchings and coverings. Let G be a graph, let M be a matching of G, and let K be a covering. Then $|M| \le |K|$. For, given any edge $\alpha = \{x, y\}$ of M, either x or y is in K. Let $f(\alpha)$ be whichever of x and y is in K, picking either if both are in K. Note that since M is a matching, if α and β are two edges of $M, f(\alpha)$ must be different from $f(\beta)$. Why? Thus, we assign each edge of M to a different element of K. Hence, M must have no more elements than K.

It follows that if M^* is a maximum matching and K^* is a minimum covering, then $|M^*| \le |K^*|$. Note that $|M^*|$ can be strictly less than $|K^*|$. In the graph of Figure 12.13, a minimum covering K^* consists of vertices 1, 3, 5, 6, so $|K^*| = 4$. However, a maximum matching consists of three edges. For instance, $M^* = \{\{1, 2\}, \{3, 4\}, \{5, 6\}\}$ is a maximum matching here. Hence, $|M^*| = 3 < |K^*|$. A crucial result is the following.

Theorem 12.4.   If M is a matching and K is a covering and $|M| = |K|$, then K is a minimum covering and M is a maximum matching.

Proof. If M^* is any maximum matching and K^* is any minimum covering, then we have

$$|M| \le |M^*| \le |K^*| \le |K|. \tag{12.3}$$

Since $|M| = |K|$, all terms in (12.3) must be equal, and in particular $|M| = |M^*|$ and $|K| = |K^*|$ hold.   Q.E.D.

For example, consider the graph of Figure 12.9. Here, there is a matching M consisting of the four edges $\{S_1, a\}$, $\{S_3, b\}$, $\{S_4, c\}$, and $\{S_6, d\}$, and a covering K consisting of the four vertices a, b, c, d. Thus, by Theorem 12.4, M is a maximum matching and K is a minimum covering. This example illustrates the following result.

Theorem 12.5   (König [1931]).   Suppose that $G = (X, Y, E)$ is a bipartite graph. Then the number of edges in a maximum matching equals the number of vertices in a minimum covering.

This theorem will be proved in Section 13.3.8.

As a corollary of this theorem, we can now derive Philip Hall's Theorem, Theorem 12.1.

*Proof of Philip Hall's Theorem.** It remains to show that if $|N(S)| \geq |S|$ for all $S \subseteq X$, then there is an X-saturating matching. By Theorem 12.5, since G is bipartite, an X-saturating matching exists if and only if each cover K has $|K| \geq |X|$. Let K be a cover and let $X - S$ be the set of X-vertices in K. Then vertices in S are not in K, so all vertices of $N(S)$ must be in K. It follows that

$$|K| \geq |X - S| + |N(S)| = |X| - |S| + |N(S)| \geq |X|.$$

Hence, $|K| \geq |X|$ for all K, and so an X-saturating matching exists. Q.E.D.

Another corollary of Theorem 12.5 will be important in Section 13.4.2 in our discussion of an algorithm for the optimal assignment problem (Example 12.5).† To state this corollary, let **A** be a matrix of 0's and 1's. A *line* of this matrix is either a row or a column. A set of 0's in this matrix is called *independent* if no two 0's lie in the same line. For instance, suppose that

$$\mathbf{A} = \begin{bmatrix} 1 & 0 & 0 & 1 \\ 0 & 1 & 0 & 0 \\ 1 & 0 & 0 & 1 \\ 1 & 0 & 1 & 1 \end{bmatrix}. \tag{12.4}$$

Then the (2, 1) and (1, 3) entries form an independent set of 0's of **A**, as do the (2, 1), (4, 2), and (3, 3) entries.

Corollary 12.5.1   (König–Egerváry Theorem‡).   If **A** is a matrix of 0's and 1's, a maximum independent set of 0's has the same number of elements as a minimum set of lines covering all the 0's of **A**.

Proof. Build a bipartite graph $G = (X, Y, E)$ by letting X be the rows of **A**, Y the columns of **A**, and joining rows i and j by an edge if and only if the i, j entry of **A** is 0. Then a maximum independent set of 0's of **A** corresponds to a maximum matching of G and a minimum set of lines covering all the 0's of **A** corresponds to a minimum covering of G. Q.E.D.

To illustrate this result, for **A** of (12.4), a maximum independent set of 0's is the three zeros at the (2, 1), (4, 2), and (3, 3) entries. A minimum set of lines covering all the 0's of **A** consists of the second row and the second and third columns.

12.4.2 Edge Coverings

A collection F of edges in a graph is called an *edge covering* if every vertex of the graph is on one of the edges in F. An edge covering is called *minimum* if no other edge covering has

*This proof can be omitted.

†The algorithm can be understood without knowledge of the material in Chapter 13, and can be studied right after Corollary 12.5.1.

‡This theorem is based on work of König [1931] and Egerváry [1931].

fewer edges. Minimum edge coverings have applications to switching functions in computer engineering, to crystal physics, and to other fields (see Deo [1974, p. 184ff.]). We investigate edge coverings and their relations to matchings and vertex coverings in the exercises.

EXERCISES FOR SECTION 12.4

1. In graph (a) of Figure 12.14, find
   (a) a covering of five vertices   (b) a minimum covering.

2. In the graph of Figure 12.1, find
   (a) a covering of six vertices   (b) a minimum covering.

3. In an attempt to improve the security of its computer operations, a company has put in 20 special passwords. Each password is known by exactly two people in the company. Find the smallest set of people who together know all the passwords. Formulate this problem using the terminology of this section.

4. Suppose that $\mathbf{A}$ is a matrix of 0's and 1's. Suppose that we find k lines covering all 0's of $\mathbf{A}$ and we find k independent 0's. What conclusion can we draw?

5. Is the König–Egerváry Theorem still true if the matrix is allowed to have any integers as entries? Why?

6. In each of the following graphs, find an edge covering that is not minimum and an edge covering that is minimum.
   (a) The graph of Figure 12.13
   (b) Graph (a) of Figure 12.14
   (c) Graph (b) of Figure 12.12.

7. (a) Can a graph with isolated vertices have an edge covering?
   (b) What is the smallest conceivable number of edges in an edge covering of a graph of n vertices?

8. A patrol car is parked in the middle of a block and can observe the two street intersections at the end of the block. How would you go about finding the smallest number of patrol cars required to keep all street corners in a neighborhood under surveillance?

9. (Minieka [1978]) A committee is to be chosen with at least one member from each of the 50 states and at least one member from each of the 65 major ethnic groups in the United States. What is the smallest committee that can be constructed from a group of volunteers if the requirements are to be met? To answer this question, let the vertices of a graph be the 50 states and the 65 ethnic groups and let a volunteer correspond to an edge joining his or her state and ethnic group. What are we looking for to answer the question?

10. (a) Can a minimum edge covering of a graph contain a circuit? Why?
    (b) Show that every minimum edge covering of a graph of n vertices has at most $n - 1$ edges.

11. (Gallai [1959]) Suppose that G is a graph without isolated vertices, K^* is a minimum (vertex) covering of G, and I^* is a maximum independent set of vertices. Show that $|K^*| + |I^*| = |V|$.

12. (Norman and Rabin [1959], Gallai [1959]) Suppose that G is a graph without isolated vertices, M^* is a maximum matching, and F^* is a minimum edge covering. Show that $|M^*| + |F^*| = |V|$.

13. Suppose that I is an independent set of vertices in a graph G without isolated vertices and F is an edge covering of G. Show from Exercises 11 and 12 that $|I| \leq |F|$.

14. Suppose that we find an independent set I of vertices in a graph G without isolated vertices and an edge covering F in G such that $|I| = |F|$. What conclusion can we draw? Why?

15. Suppose that $G = (X, Y, E)$ is a bipartite graph without isolated vertices. Let I^* be an independent set with a maximum number of vertices (a *maximum independent set*) and F^* be a minimum edge covering of G. Show that $|I^*| = |F^*|$.

16. (Minieka [1978]) Consider the following algorithms on a graph with no isolated vertices. *Algorithm 1* starts with a matching M and selects any unsaturated vertex x. It adds to M any edge incident to x and repeats the procedure until every vertex is saturated. The resulting set of edges is called C'. *Algorithm 2* starts with an edge covering C and selects any vertex x covered by more than one edge in C. It removes from C an edge that covers x and repeats the procedure until no vertex is covered by more than one edge. The resulting set of edges is called M'.

(a) Show that C' as produced by Algorithm 1 is an edge covering.

(b) Show that M' as produced by Algorithm 2 is a matching.

(c) Show that if M is a maximum matching, then C' is a minimum edge covering.

(d) Show that if C is a minimum edge covering, then M' is a maximum matching.

17. Derive Philip Hall's Theorem (Theorem 12.1) from Theorem 12.4.

12.5 FINDING A MAXIMUM MATCHING

12.5.1 *M*-augmenting Chains

In this section we present a procedure for finding a maximum matching in a graph G. Suppose that M is some matching of G. An *M-alternating chain* in G is a simple chain

$$u_1, e_1, u_2, e_2, \ldots, u_t, e_t, u_{t+1} \tag{12.5}$$

such that e_1 is not in M, e_2 is in M, e_3 is not in M, e_4 is in M, and so on. For instance, consider the matching shown in Figure 12.15 by wiggly edges. Then the chain $a, \{a, b\}, b, \{b, g\}, g, \{g, f\}, f, \{f, e\}, e$ is M-alternating. If an M-alternating chain (12.5) joins two vertices u_1 and u_{t+1} which are not M-saturated (that is, not on any edge of M), then we call the chain an *M-augmenting chain*. Our example is not M-augmenting, since e is M-saturated. However, the chain

$$a, \{a, b\}, b, \{b, g\}, g, \{g, f\}, f, \{f, e\}, e, \{e, d\}, d \tag{12.6}$$

is M-augmenting. Let us now find a new matching M' by deleting from M all edges of M used in the M-augmenting chain (12.6) and adding all edges of (12.6) not in M. Then M' is shown in Figure 12.16. It is indeed a matching. Moreover, M' has one more edge than M. This procedure always gives us a larger matching.

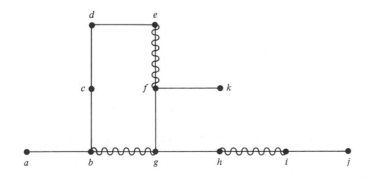

Figure 12.15 A matching M is shown by the wiggly edges.

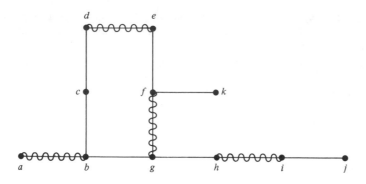

Figure 12.16 The matching M' obtained from the matching M of Figure 12.15 by using the M-augmenting chain (12.6).

Theorem 12.6. Suppose that M is a matching and C is an M-augmenting chain of M. Let M' be M less edges of C in M plus edges of C not in M. Then M' is a matching and $|M'| > |M|$.

Proof. To see that $|M'| > |M|$, note that in an M-augmenting chain (12.5), e_1 and e_t cannot be in M, so t is odd. Hence, we add the edges $e_1, e_3, \ldots, e_t$ and delete the edges $e_2, e_4, \ldots, e_{t-1}$, and we add one more edge than we take away. To see that M' is a matching, note that if j is odd, then edge e_j is added and edge e_{j+1} is subtracted. Thus, if j is odd and $j \neq 1, t+1$, then u_j was previously only on e_{j-1} and is now only on e_j. If j is even, u_j was previously only on e_j and is now only on e_{j-1}. Also, u_1 was previously unmatched and is now only on e_1, and u_{t+1} was previously unmatched and is now only on e_t. Q.E.D.

As a corollary of Theorem 12.6, we observe that if M has an M-augmenting chain, then M could not be a maximum matching. In fact, the converse is also true.

Theorem 12.7 (Berge [1957], Norman and Rabin [1959]). A matching M of G is maximum if and only if G contains no M-augmenting chain.

To apply this theorem, note that the matching M' in Figure 12.16 is not maximum, since there is an M'-augmenting chain

$$j, \{j, i\}, i, \{i, h\}, h, \{h, g\}, g, \{g, f\}, f, \{f, k\}, k. \tag{12.7}$$

If we modify M' by deleting edges $\{i, h\}$ and $\{g, f\}$, the M' edges of the chain (12.7), and adding edges $\{j, i\}$, $\{h, g\}$, and $\{f, k\}$, the non-M' edges of (12.7), we obtain a matching M'' shown in Figure 12.17. There is no M''-augmenting chain, since there is only one unsaturated vertex. Thus, M'' is maximum.

12.5.2 Proof of Theorem 12.7*

To prove Theorem 12.7, it remains to show that if there is a matching M' such that $|M'| > |M|$, then there is an M-augmenting chain. Let H be the subgraph of G consisting of all vertices of G and all edges of G that are in M or in M', but not in both M and M'.

*This subsection can be omitted.

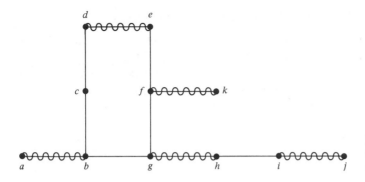

Figure 12.17 The matching M'' obtained from the matching M' of Figure 12.16 by using the M'-augmenting chain (12.7).

Note that in H, there are more M' edges than M edges. Moreover, each vertex of H has at most two neighbors in H, for it can have at most one neighbor in M and at most one neighbor in M'. Using the latter observation, one can show that each connected component of H is either a circuit Z_n or a simple chain L_n of n vertices. Moreover, the edges in Z_n or L_n must alternate between M and M' because M and M' are matchings. It follows that each Z_n has an equal number of edges from M and from M'. Thus, since H has more edges of M' than of M, some component of the form L_n has this property. This L_n must be a chain of the form (12.5) with $e_1, e_3, \ldots, e_t$ in M' and not in M, and $e_2, e_4, \ldots, e_{t-1}$ in M and not in M'. Moreover, u_1 and u_{t+1} cannot be M-saturated, for otherwise L_n would not be a component of H. Hence, the chain L_n is an M-augmenting chain. This completes the proof.

To illustrate this proof, consider the matchings M and M' of Figure 12.18. Then the graph H is also shown in that figure. The chain

$$g, \{g, h\}, h, \{h, i\}, i, \{i, j\}, j$$

is a simple chain of two M' edges and one M edge.

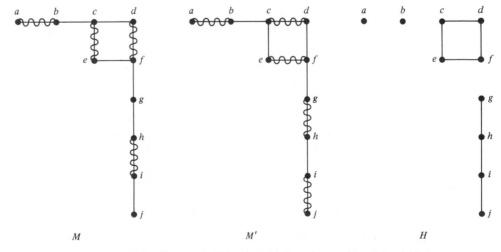

Figure 12.18 The graph H is obtained from the matchings M and M'.

12.5.3 An Algorithm for Finding a Maximum Matching

We next describe an algorithm for finding a maximum matching. This algorithm originates in the work of L. R. Ford and D. R. Fulkerson (see Ford and Fulkerson [1962]). The algorithm has two basic subroutines. Subroutine 1 searches for an M-augmenting chain starting with an unsaturated vertex x and subroutine 2 builds a larger matching M' if subroutine 1 finds an M-augmenting chain. The algorithm chooses an unsaturated vertex x and applies subroutine 1. If an M-augmenting chain starting with x is found, subroutine 2 is called. If no such chain is found, another unsaturated vertex y is used in subroutine 1. The procedure is repeated until either an M-augmenting chain is found or no unsaturated vertices remain. In the latter case, we conclude that there is no M-augmenting chain, so we have a maximum matching.

Subroutine 2 works exactly as described in Theorem 12.6. We now present subroutine 1. It is easiest to describe this subroutine for the case of a bipartite graph G. If G is not bipartite, the procedure is more complicated. It was Edmonds [1965] who first observed that something much more subtle was needed here. (See Exercise 13.) For more details on bipartite and nonbipartite matching, see, for example, Lawler [1976], Minieka [1978] or Papadimitriou and Steiglitz [1982].

Subroutine 1 begins with a matching M and a vertex x unsaturated in M and builds in stages a tree T called an *alternating tree*. The idea is that x is in T and all simple chains in T beginning at x are M-alternating chains. For instance, in Figure 12.15, if $x = a$, one such alternating tree consists of the edges $\{a, b\}$, $\{b, g\}$, $\{g, h\}$, and $\{g, f\}$. The vertices in the alternating tree T are called outer or inner. Vertex y is called *outer* if the unique simple chain* between x and y ends in an edge of M, and *inner* otherwise. The vertex x is called *outer*. If an alternating tree T with an unsaturated vertex $y \neq x$ is found, then the unique simple chain between x and y is an M-augmenting chain. Vertices and edges are added to T until either such a chain is found or no more vertices can be added to T. In the latter case, there is no M-augmenting chain. We now present the subroutine in detail.

> **Algorithm 12.2. Subroutine 1: Searching for an M-augmenting Chain Beginning with Vertex x.**
>
> > *Input:* A bipartite graph G, a matching M of G, and a vertex x unsaturated in M.
> >
> > *Output:* An M-augmenting chain beginning at x or the message that no such chain exists.
>
> *Step 1.* Set $T = \varnothing$ and $T' = \varnothing$. (T is the set of edges in the tree and T' the set of edges definitely not in the tree.) Set $O = \{x\}$ and $I = \varnothing$. (O is the set of outer vertices and I is the set of inner vertices.)
>
> *Step 2.*
> > *Step 2.1.* Among the edges not in T or T', if there is no edge between an outer vertex (a vertex of O) and any other vertex, go to step 3. Otherwise, let $\{u, v\}$ be such an edge with $u \in O$.

---

*We are using the result of Theorem 3.14—that in a tree, there is a unique simple chain between any pair of vertices.

Step 2.2. If vertex v is an inner vertex (is in I), place edge $\{u, v\}$ in T' and repeat step 2.1. If vertex v is neither inner nor outer, place edge $\{u, v\}$ in T, and go to step 2.3. (Since G is bipartite, v cannot be an outer vertex, for otherwise one can show that G has an odd circuit, which is impossible for bipartite graphs by Theorem 3.2. See Exercise 11.)

Step 2.3. If v is unsaturated, stop. The unique chain in T from x to v forms an M-augmenting chain from x to v. If v is saturated, there is a unique edge $\{v, w\}$ in M. Then place $\{v, w\}$ in T, v in I, and w in O. Return to step 2.1.

Step 3. We get to this step only when no further assignment of edges to T or T' is possible. Stop and give the message that there is no M-augmenting chain beginning with x.

Note that the algorithm stops in two ways, having found an M-augmenting chain or having found an alternating tree T where it is impossible to add edges to either T or T'. In the former case, the procedure goes to subroutine 2. In the latter case, one can show that there is no M-augmenting chain starting from the vertex x. We repeat this subroutine for another unsaturated vertex x. If in repeated applications of subroutine 1, we fail to find an M-augmenting chain beginning from an unsaturated vertex x, we conclude that the matching is maximum.

We illustrate the algorithm on the matching M of Figure 12.19. Pick unsaturated vertex x to be a, and call a an outer vertex. This is step 1. Go to step 2.1, and select the edge $\{a, b\}$ that is neither in T nor in T' and joins an outer vertex. Since b is not inner and not outer, in step 2.2 we place this edge in T. Since b is saturated, we consider the unique edge $\{b, g\}$ of M. We place this edge in T, call b inner and g outer, and return to step 2.1. (See Table 12.3 for a summary. For the purposes of this summary, an iteration is considered a return to step 2.1.)

In step 2.1, we consider edges not in T and not in T' and joining an outer vertex. There are two such edges, $\{g, h\}$ and $\{g, f\}$. Let us suppose that we choose edge $\{g, f\}$. Since f is neither inner nor outer, in step 2.2 we place $\{g, f\}$ in T. Now f is saturated, so we consider the unique edge $\{f, e\}$ of M. We place this in T, call f inner and e outer, and return to step 2.1.

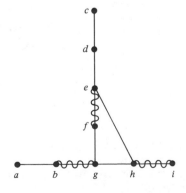

Figure 12.19 A matching M is shown by wiggly edges.

Table 12.3 Steps in Algorithm 12.2 Applied to M of Figure 12.19 and Starting with Vertex $x = a$

Iteration	T	T'	I (inner vertices)	O (outer vertices)
1	$\varnothing$	$\varnothing$	$\varnothing$	a
2	$\{a, b\}$	$\varnothing$	$\varnothing$	a
	$\{a, b\}, \{b, g\}$	$\varnothing$	b	a, g
3	$\{a, b\}, \{b, g\}, \{g, f\}$	$\varnothing$	b	a, g
	$\{a, b\}, \{b, g\}, \{g, f\}, \{f, e\}$	$\varnothing$	b, f	a, g, e
4	$\{a, b\}, \{b, g\}, \{g, f\}, \{f, e\}, \{g, h\}$	$\varnothing$	b, f	a, g, e
	$\{a, b\}, \{b, g\}, \{g, f\}, \{f, e\}, \{g, h\}, \{h, i\}$	$\varnothing$	b, f, h	a, g, e, i
5	$\{a, b\}, \{b, g\}, \{g, f\}, \{f, e\}, \{g, h\}, \{h, i\}$	$\{e, h\}$	b, f, h	a, g, e, i
6	$\{a, b\}, \{b, g\}, \{g, f\}, \{f, e\}, \{g, h\}, \{h, i\}, \{e, d\}$	$\{e, h\}$	b, f, h	a, g, e, i

Next, we again consider edges not in T and not in T' and joining an outer vertex. The possible edges are $\{g, h\}$, $\{e, h\}$, and $\{e, d\}$. Suppose that we choose $\{g, h\}$. Then in step 2.2, since h is neither inner nor outer, we place edge $\{g, h\}$ in T. Then in step 2.3, since h is saturated, we place edge $\{h, i\}$ in T and call h inner and i outer.

We again consider edges not in T and not in T' and joining an outer vertex. The possible edges are $\{e, h\}$ and $\{e, d\}$. Suppose that we pick the former. Then vertex h is inner, so in step 2.2 we place edge $\{e, h\}$ in T', and repeat step 2.1.

In step 2.1 we now choose edge $\{e, d\}$ and in step 2.2 we add edge $\{e, d\}$ to T. Since d is unsaturated, we stop and find the unique chain in T from x to v, that is, from a to d. This is the chain $a, \{a, b\}, b, \{b, g\}, g, \{g, f\}, f, \{f, e\}, e, \{e, d\}, d$. It is an M-augmenting chain.

In closing, we note that it can be shown that the algorithm described can be modified to take on the order of $[\min \{|X|, |Y|\}] \cdot |E|$ steps, given a bipartite graph $G = (X, Y, E)$. Thus, it is a polynomial algorithm in the number of vertices $n = |V|$. For $|E| \le \binom{n}{2} \le n^2$ and $\min \{|X|, |Y|\} \le n$. Thus, the algorithm concludes in a number of steps that is on the order of n^3. In the notation of Section 2.18, we say it is an $O(n^3)$ algorithm. A related algorithm for arbitrary G also takes on the order of n^3 steps. As of this writing, the fastest known algorithms for finding a maximum matching take on the order of $|E| \, |V|^{1/2}$ steps. In terms of n, these algorithms take on the order of $n^{5/2}$ steps. They are due to Hopcroft and Karp [1973] for the bipartite case, and to Micali and Vazirani [1980] for the general case. These algorithms relate matching to network flows. (We discuss this relation in Section 13.3.8.) For a more detailed discussion of the complexity for matching algorithms, see Lawler [1976] or Papadimitriou and Steiglitz [1982].

EXERCISES FOR SECTION 12.5

1. In the matching of Figure 12.20:
   (a) Find an M-alternating chain that is not M-augmenting.
   (b) Find an M-augmenting chain if one exists.
   (c) Use the chain in part (b), if it exists, to find a larger matching.

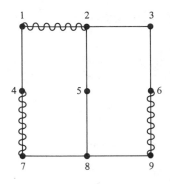

Figure 12.20   Matching for exercises of Section 12.5.

2. Repeat Exercise 1 for the matching of Figure 12.21.
3. Repeat Exercise 1 for the matching of Figure 12.22.
4. Repeat Exercise 1 for the matching of Figure 12.23.
5. Repeat Exercise 1 for the matching of Figure 12.24.
6. Apply subroutine 1 to
   (a) the matching M of Figure 12.20 and the vertex 5;
   (b) the matching M of Figure 12.20 and the vertex 3;
   (c) the matching M of Figure 12.21 and the vertex 12.

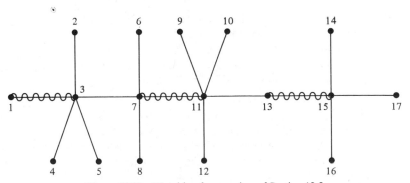

Figure 12.21   Matching for exercises of Section 12.5.

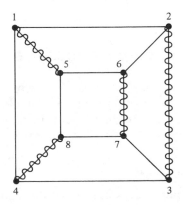

Figure 12.22   Matching for exercises of Section 12.5.

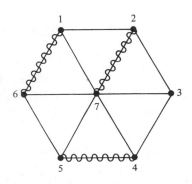

Figure 12.23   Matching for exercises of Section 12.5.

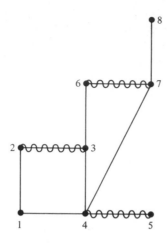

Figure 12.24 Matching for exercises of Section 12.5.

7. Apply subroutine 1 to the matching M of Figure 12.24 by starting with vertex 1 and, in successive iterations, choosing $v = 2$, $v = 4$, $v = 6$, $v = 4$, $v = 8$.

8. Show that if subroutine 1 is applied to the matching of Figure 12.23, starting with the vertex 3, it is possible for the edge $\{u, v\}$ chosen at step 2.1 to join two outer vertices. Why can this happen?

9. Find an alternating tree starting at vertex x in the following situations.
   (a) $x = 6$ and the matching of Figure 12.21;
   (b) $x = 8$ and the matching of Figure 12.20.

10. Show that in the proof of Theorem 12.7, each component of H is either Z_n or L_n.

11. Prove that in step 2.2 of Algorithm 12.2, v cannot be an outer vertex.

12. Prove that Algorithm 12.2 works.

13. Suppose that M is a matching in a graph G. A *blossom* relative to M is an odd-length circuit B of $2k + 1$ vertices having k edges in M. Show that if there are no blossoms relative to M, then subroutine 1 finds an M-augmenting chain beginning at x if there is one. (Thus, the algorithm we have described for finding a maximum matching must be modified only if blossoms are found. The modification due to Edmonds [1965] works by searching for blossoms, shrinking them to single vertices, and searching for M-augmenting chains in the resulting graph.)

12.6 MATCHING AS MANY ELEMENTS OF X AS POSSIBLE

Suppose that $G = (X, Y, E)$ is a bipartite graph. If there is no X-saturating matching in G, we may at least wish to find a matching that matches as large a number of elements of X as possible. Let $m(G)$ be the largest number of elements of X that can be matched in a matching of G. We shall show how to compute $m(G)$. First, we mention an application of this idea.

Example 12.10   Telephone Switching Networks

At a telephone switching station, phone calls are routed through incoming lines to outgoing trunk lines. A *switching network* connects each incoming line to some of the outgoing lines. When a call comes in on an incoming line, it is routed through

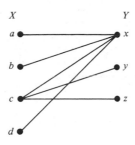

Figure 12.25 A graph G with $\delta(G) = 2$.

the switching network to an outgoing line. The switching network can be represented by a bipartite graph $G = (X, Y, E)$. The vertices of X are the incoming lines, the vertices of Y are the outgoing lines, and there is an edge between an incoming line and an outgoing line if and only if the former is connected to the latter. Suppose that a number of incoming calls come in at once. We want to be able to send all of them out at the same time if possible. If S is the set of incoming lines on which there are calls, we want an assignment of each line of S to an edge of G leading to a different outgoing trunk line. That is, we want an S-saturating matching of G. If this is impossible to find, we want to match a large number of lines in S. In general, we want to design our switching network so that if calls come in on all possible incoming lines, we can match a large number of them with outgoing lines. That is, we want to design a switching network so that it has a maximum matching that has a large number of edges. Put another way, we want to find a bipartite graph $G = (X, Y, E)$ with given X and Y so that $m(G)$ is large enough to be reasonable.

If $S \subseteq X$, let the *deficiency* of S, $\delta(S)$, be defined as $|S| - |N(S)|$ and let the *deficiency* of G, $\delta(G)$, be $\max_{S \subseteq X} \delta(S)$. Note that $\delta(\varnothing) = 0$, so by Philip Hall's Theorem, G has an X-saturating matching if and only if $\delta(G) = 0$.

Theorem 12.8 (König [1931]). If $G = (X, Y, E)$ is a bipartite graph, then $m(G) = |X| - \delta(G)$.

To illustrate this theorem, consider the graph of Figure 12.25. Note that $\delta(\{a, b, d\}) = 2$ and this is maximum, so $\delta(G) = 2$. Thus, $m(G) = |X| - \delta(G) = 4 - 2 = 2$. The largest subset of X that can be matched has two elements. An example of such a set is $\{a, c\}$.

A proof of Theorem 12.8 will be sketched in Exercises 5 and 6.

EXERCISES FOR SECTION 12.6

1. Compute $\delta(G)$ and $m(G)$ for each graph of Figure 12.12.
2. Consider a switching station with 9 incoming lines and 6 outgoing trunk lines. Suppose that by engineering considerations, each incoming line is to be connected in the switching network to exactly two outgoing lines, and each outgoing line can be connected to at most 3 incoming lines.
   (a) Consider any set S of p incoming lines. Show that in the switching network, the sum of the degrees of the vertices in S is at most the sum of the degrees of the vertices in $N(S)$.

(b) Using the result in part (a), show that for any set S of p incoming lines, $N(S)$ has at least $\frac{2}{3}p$ elements.

(c) Conclude that for any set S of p incoming lines, $\delta(S) \leq \frac{1}{3}p$.

(d) Conclude that no matter how the switching network is designed, there is always a matching that matches at least 6 incoming calls.

3. Suppose that the situation of Exercise 2 is modified so that there are 12 incoming lines and 10 outgoing lines. If each incoming line is to be connected to exactly 3 outgoing lines and each outgoing line to at most 4 incoming lines, show that no matter how the switching network is designed, there is always a matching that matches at least 9 incoming calls.

4. Suppose that the situation of Exercise 2 is modified so that there are two kinds of incoming lines, 4 of type I and 4 of type II, and there are 8 outgoing lines, all of the same type. Assume that each type I incoming line is connected to exactly 3 outgoing trunk lines and each type II incoming line to exactly 6 outgoing trunk lines, and assume that each outgoing line is connected to at most 4 incoming lines. Show that no matter how the switching network is designed, there is always a matching that matches at least 7 incoming calls. Do so by considering a set S of p incoming lines of type I and q incoming lines of type II. Show that

$$\delta(S) \leq (p + q) - \frac{3p + 6q}{4}.$$

Conclude that since $p \leq 4$ and $q \geq 0$, $\delta(G) \leq 1$.

5. As a preliminary to the proof of Theorem 12.8, prove the following: A bipartite graph $G = (X, Y, E)$ has a matching of k elements if and only if for all subsets $S \subseteq X$, $|N(S)| \geq |S| + k - |X| = k - |X - S|$. (*Hint:* Modify G by adding $|X| - k$ vertices to Y and joining each new vertex of Y to each vertex of X.)

6. Use the result of Exercise 5 to prove Theorem 12.8.

7. This exercise presents an alternative proof of the König–Egerváry Theorem (Corollary 12.5.1). Let $\mathbf{A}$ be a matrix of 0's and 1's and build a bipartite graph $G = (X, Y, E)$ from $\mathbf{A}$ as in the proof of the König–Egerváry Theorem. Let S be a subset of X, the set of rows, such that $\delta(S) = \delta(G)$.

(a) Show that the rows corresponding to vertices of $X - S$ together with the columns corresponding to vertices of $N(S)$ contain all the 0's in $\mathbf{A}$.

(b) Show that the total number of rows and columns in part (a) is $|X| - \delta(G)$.

(c) Conclude that the minimum number of lines of $\mathbf{A}$ covering all 0's is at most $|X| - \delta(G)$.

(d) Show that the maximum number of independent 0's in G is $|X| - \delta(G)$.

(e) Prove the König–Egerváry Theorem.

REFERENCES FOR CHAPTER 12

BERGE, C., "Two Theorems in Graph Theory," *Proc. Natl. Acad. Sci. USA*, *43* (1957), 842–844.

BERGE, C., *Graphs and Hypergraphs*, American Elsevier, New York, 1973.

CHRISTOFIDES, N., *Graph Theory: An Algorithmic Approach*, Academic Press, New York, 1975.

DEO, N., *Graph Theory with Applications to Engineering and Computer Science*, Prentice-Hall, Englewood Cliffs, N.J., 1974.

EDMONDS, J., "Paths, Trees and Flowers," *Canad. J. Math.*, *17* (1965), 449–467.

EGERVÁRY, E., "Matrixok Kombinatórius Tulajdonságairól," *Mat. Fiz. Lapok*, *38* (1931), 16–28 ("On Combinatorial Properties of Matrices," translated by H. W. Kuhn, Office of Naval Research Logistics Project Report, Dept. of Mathematics, Princeton University, Princeton, N.J., 1953.)

FORD, L. R., and FULKERSON, D. R., *Flows in Networks*, Princeton University Press, Princeton, N.J., 1962.

GALLAI, T., "Über extreme Punkt- und Kantenmengen," *Ann. Univ. Sci. Budap., Eötvös Sect. Math., 2* (1959), 133–138.

HALL, P., "On Representatives of Subsets," *J. Lond. Math. Soc., 10* (1935), 26–30.

HOPCROFT, J. E., and KARP, R. M., "A $n^{5/2}$ Algorithm for Maximum Matching in Bipartite Graphs," *SIAM J. Comput., 2* (1973), 225–231.

KÖNIG, D., "Graphen und Matrizen," *Mat. Fiz. Lapok, 38* (1931), 116–119.

LAWLER, E. L., *Combinatorial Optimization: Networks and Matroids*, Holt, Rinehart and Winston, New York, 1976.

MICALI, S., and VAZIRANI, V. V., "An $O(\sqrt{|V|}\cdot|E|)$ Algorithm for Finding Maximum Matching in General Graphs," *Proceedings of the Twenty-First Annual Symposium on the Foundations of Computer Science*, IEEE, Long Beach, Calif., 1980, pp. 17–27.

MINIEKA, E., *Optimization Algorithms for Networks and Graphs*, Dekker, New York, 1978.

NORMAN, R. Z., and RABIN, M. O., "An Algorithm for a Minimum Cover of a Graph," *Proc. Amer. Math. Soc., 10* (1959), 315–319.

PAPADIMITRIOU, C. H., and STEIGLITZ, K., *Combinatorial Optimization: Algorithms and Complexity*, Prentice-Hall, Englewood Cliffs, N.J., 1982.

PETERSON, J., "Die Theorie der regulären Graphen," *Acta Math., 15* (1891), 193–220.

TUCKER, A. C., *Applied Combinatorics*, Wiley, New York, 1980.

TUTTE, W. T., "The Factorization of Linear Graphs," *J. Lond. Math. Soc., 22* (1947), 107–111.

13 Optimization Problems for Graphs and Networks

In this chapter we present procedures for solving a number of combinatorial optimization problems. We have encountered such problems throughout the book, and discussed some in detail in Sections 12.4–12.6. The problems to be discussed will all be translated into problems involving graphs or digraphs. More generally, they will be translated into problems involving graphs or digraphs in which each edge or arc has one or more *nonnegative* real numbers assigned to it. Such a graph or digraph is called a *network* or a *directed network*, respectively.

It should be mentioned that we have chosen problems to discuss for which good solutions (good algorithms) are known. Not all combinatorial optimization problems have good solutions. The traveling salesman problem, discussed in Sections 2.4 and 11.5, is a case in point. In such a case, as we pointed out in Section 2.18, we seek good algorithms which solve the problem in special cases or which give approximate solutions. Discussion of these approaches is beyond the scope of this book.

13.1 MINIMUM SPANNING TREES

13.1.1 Kruskal's Algorithm

In Examples 3.16–3.18, we described the problem of finding a minimum spanning tree of a graph G with weights, that is, a network. This is a spanning subgraph of G which forms a tree and which has the property that no other spanning tree has a smaller sum of weights on its edges. Here we shall discuss a simple algorithm for finding a minimum spanning tree of a network G. The algorithm will either find a minimum spanning tree or conclude that the network does not have a spanning tree. By Theorem 3.15, the latter implies that the network is disconnected. The problem of finding a minimum spanning tree has a wide variety of applications in the planning of large-scale transportation, communication, or distribution networks. We have mentioned some of these in Section 3.5.4. See Graham and Hell [1982] for a recent survey and references on many applications.

According to Graham and Hell [1982], the earliest algorithms for finding a minimum spanning tree apparently go back to Boruvka [1926a, b], who was interested in the most economical layout for a power-line network. Earlier work on this problem is due to the anthropologist Czekanowski [1909, 1911, 1928] in his work on classification schemes.

Let us recall the problem by continuing with Example 3.17. Suppose that there are five villages in a remote region and the road distances between them are shown in the network of Figure 13.1. We wish to put in telephone lines along the existing roads in such a way that every pair of villages is linked by telephone service and we minimize the total number of miles of telephone line. The problem amounts to finding a minimum spanning tree of the network of Figure 13.1.

In finding a spanning tree of such a network which minimizes the sum of the weights of its edges, we will build up the tree one edge at a time, and we will be *greedy*; that is, we will add edges of smallest weight first. It turns out that being greedy works. Specifically, let us order the edges of the network G in order of increasing weight. In our example, we order them as follows: $\{a, b\}, \{c, d\}, \{d, e\}, \{c, e\}, \{b, d\}, \{b, e\}, \{a, c\}, \{a, e\}$. In case of ties, the ordering is arbitrary. We will examine the edges in this order, for each edge deciding whether or not to include it in the spanning tree T. It is included as long as it does not form a circuit with some already included edges. This algorithm is called Kruskal's algorithm—it is due to Kruskal [1956]—and can be summarized as follows.

Algorithm 13.1.   Kruskal's Minimum Spanning Tree Algorithm

Input: A network G.

Output: A set of edges defining a minimum spanning tree of G or the message that G is disconnected.

Step 1. Order the edges of G in increasing order of weight. (Order arbitrarily in case of ties.) Set $T = \varnothing$. (T is the set of edges of the minimum spanning tree.)

Step 2. Add the first edge to the set T.

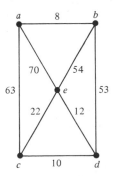

Figure 13.1 A network of road distances between villages.

Step 3. If every edge of G has been examined, stop and output the message that G is disconnected. Otherwise, examine the first unexamined edge in the order and include it in the set T if and only if it does not form a circuit with some edges already in T. If the edge is added to T, go to step 4. Otherwise, repeat step 3.

Step 4. If T has $n - 1$ edges (where n is the number of vertices of G), stop and output T. Otherwise, go to step 3.

Theorem 13.1. If G is a connected network, then Kruskal's algorithm will terminate with a minimum spanning tree T of $n - 1$ edges. If G is a disconnected network, then the algorithm will terminate with the message that G is disconnected after having examined all the edges and not yet having put $n - 1$ edges into T.

Before presenting a proof of Theorem 13.1, we will illustate the algorithm on our example of Figure 13.1. We start by including edge $\{a, b\}$ in T. Then, in step 3, we note that not every edge of G has been examined. We examine edge $\{c, d\}$, adding it to T because it does not form a circuit with edges in T. Now the number of vertices of G is $n = 5$, and T does not yet have $n - 1 = 4$ edges, so we return to step 3. Since the first unexamined edge $\{d, e\}$ does not form a circuit with existing edges in T, we add this edge to T. Now T has three edges, and so we repeat step 3. The next edge examined, $\{c, e\}$, does form a circuit with edges $\{c, d\}$ and $\{d, e\}$ of T, so we do not add it to T. We repeat step 3 for the next edge, $\{b, d\}$. This is added to T. Since T now has four edges in it, we stop and conclude that these edges define a minimum spanning tree of G.

Let us note that Kruskal's algorithm can be applied to a graph G without weights if we simply list the edges in arbitrary order. Then, the algorithm will produce a spanning tree of G or it will conclude that G is disconnected.

Let us consider the computational complexity of Kruskal's algorithm. Disregarding the step of putting the edges in order, the algorithm takes at most e steps, one for each edge.

How many steps does it take to order the edges in terms of increasing weight? This is simply a sorting problem, of the kind discussed in Section 3.6.3. In that section we observed that sorting a set of e items can be performed in $ke \log_2 e$ steps, where k is a constant. Thus, the entire Kruskal algorithm can be performed in at most $e + ke \log_2 e \le e + ke^2$ steps. Note that $e + ke^2 \le n^2 + kn^4$, since $e \le \binom{n}{2} \le n^2$. Thus, the entire Kruskal algorithm can be performed in a number of steps which is a polynomial in either e or n. (In the terminology of Section 2.18, our crude discussion shows that we have an $O(e \log_2 e)$ algorithm. The best implementations of Kruskal's algorithm find a minimum spanning tree in $O(e \log_2 n)$ steps. See Graham and Hell [1982] or Cheriton and Tarjan [1976].)

13.1.2 Proof of Theorem 13.1*

To prove Theorem 13.1, note that at each stage of the algorithm the edges in T define a graph that has no circuits. Suppose that the algorithm terminates with T having $n - 1$ edges. Then by Theorem 3.16, T is a tree, and hence a spanning tree of G. Thus, G is

*This subsection can be omitted.

connected. Hence, if G is not connected, the algorithm will terminate without finding a T of $n - 1$ edges.

If G is connected, we shall show that the algorithm gives rise to a minimum spanning tree. First, we show that the algorithm does give rise to a spanning tree. Then we show that this spanning tree is minimum. The first observation follows trivially from Theorem 3.11 if the algorithm gives us a T of $n - 1$ edges. Suppose that the algorithm terminates without giving us a T of $n - 1$ edges. At the time of termination, let us consider the edges in T. These edges define a spanning subgraph of G, which we shall also refer to as T. This subgraph is not connected, for any connected spanning subgraph of G, having a spanning tree in it, must have at least $n - 1$ edges. Hence, the subgraph T has at least two components. But since G is connected, there is in G an edge $\{x, y\}$ joining vertices in different components of T. Now $\{x, y\}$ cannot form a circuit with edges of T. Hence, when the algorithm came to examine this edge, it should have included it in T. Thus, this situation arises only if the algorithm was applied incorrectly.

We now know that the algorithm gives us a spanning tree T. Let S be a minimum spanning tree of G. If $S = T$, we are done. Suppose that $S \neq T$. Since $S \neq T$ and since both have the same number of edges ($n - 1$), there must be an edge in T that is not in S. In the order of edges by increasing weight, find the first edge $e_1 = \{x, y\}$ that is in T but not in S.

Since S is a spanning tree of G, there is a simple chain $C(x, y)$ from x to y in S. Now adding edge e_1 to S gives us a circuit. Thus, since T has no circuits, there is an edge e_2 on this circuit and hence on the chain $C(x, y)$ which is not in T. Let S' be the set of edges obtained from S by removing e_2 and adding e_1. Then S' defines a graph that is connected (why?) and has n vertices and $n - 1$ edges, so by Theorem 3.11 S' is a spanning tree. We shall now consider two cases.

Case 1: Edge e_2 precedes edge e_1 in the order of edges by increasing weights. In this case, e_2 was not put in T, so it must form a circuit D with edges of T that were examined before e_2. But e_1 is the first edge of T not in S, so all edges in D must be in S since they are in T. Thus, e_2 could not have been put in S, which is a contradiction. Case 1 is impossible.

Case 2: Edge e_1 precedes edge e_2. In this case, e_2 has weight at least as high as the weight of e_1. Thus, S' has its sum of weights at most the sum of weights of S, so S' is a minimum spanning tree since S is. Moreover, S' has one more edge in common with T than does S.

If $T \neq S'$, we repeat the argument for T and S', obtaining a minimum spanning tree S'' with one more edge in common with T than S'. Eventually, we find a minimum spanning tree that is the same as T. This completes the proof.

13.1.3 Prim's Algorithm

There is an alternative algorithm for finding a minimum spanning tree that is also a greedy algorithm. We shall present this algorithm here.

Algorithm 13.2.   Prim's Minimum Spanning Tree Algorithm*

Input: A network G.

Output: A minimum spanning tree of G or the message that G is disconnected.

Step 1. Pick an arbitrary vertex of G and put it in the tree T.

Step 2. Add to T that edge joining a vertex of T to a vertex not in T which has smallest weight among all such edges. Pick arbitrarily in case of ties. If it is impossible to add any edge to T, stop and output the message that G is disconnected.

Step 3. If T has $n - 1$ edges, stop and output T. Otherwise, repeat step 2.

To illustrate this algorithm, let us again consider the graph of Figure 13.1, and start with vertex a. Then we add edge $\{a, b\}$ to T because it has the smallest weight of the edges joining a. Next, we examine edges joining vertices a or b to vertices not in T, namely c, d, or e. Edge $\{b, d\}$ has the smallest weight of all these edges; we add it to T. Next, we examine edges joining vertices a, b, or d to vertices not in T, namely c or e. Edge $\{d, c\}$ has the smallest weight of all of these edges. We add this edge to T. Finally, we examine edges joining a, b, c, or d to the remaining vertex not in T, namely e. Edge $\{d, e\}$ has the smallest weight of all these edges, so we add it to T. Now T has four edges, and we terminate. Note that we found the same T by using Kruskal's algorithm.

Theorem 13.2.   If G is a connected network, then Prim's algorithm will terminate with a minimum spanning tree T of $n - 1$ edges. If G is a disconnected network, then the algorithm will terminate with the message that G is disconnected because it is impossible to add another edge to T.

The proof is left as an exercise (Exercise 11).

[As of this time, the best implementations of Prim's algorithm run in $O(e \log_2 n)$ steps. (See Kershenbaum and Van Slyke [1972], Johnson [1975], or Graham and Hell [1982]).]

EXERCISES FOR SECTION 13.1

1. For each network of Figure 13.2, find a minimum spanning tree using Kruskal's algorithm.
2. Repeat Exercise 1 with Prim's algorithm. Start with vertex a.
3. Apply Kruskal's algorithm to each network of Figure 13.3.
4. Apply Prim's algorithm to each network of Figure 13.3. Start with vertex a.
5. A chemical company has eight storage vats and wants to develop a system of pipelines that makes it possible to move chemicals from any vat to any other vat. The distance in feet between each pair of vats is given in Table 13.1. Determine between which pairs of vats to build pipelines so that chemicals can be moved from any vat to any other vat, and so that the total length of pipe used is minimized.

---

*This algorithm was discovered by Prim [1957] and is usually attributed to him. In fact, it seems to have been previously discovered by Jarník [1930] (Graham and Hell [1982]).

Table 13.1 Distance between Vats

VAT	1	2	3	4	5	6	7	8
1	0	3.9	6.3	2.7	2.1	5.4	6.0	4.5
2	3.9	0	2.7	5.4	3.6	7.8	6.9	3.3
3	6.3	2.7	0	7.8	5.1	7.5	5.7	3.0
4	2.7	5.4	7.8	0	2.1	4.8	4.5	2.7
5	2.1	3.6	5.1	2.1	0	2.7	3.3	2.4
6	5.4	7.8	7.5	4.8	2.7	0	1.8	3.0
7	6.0	6.9	5.7	4.5	3.3	1.8	0	1.5
8	4.5	3.3	3.0	2.7	2.4	3.0	1.5	0

6. In each network of Figure 13.2, find a *maximum spanning tree*, a spanning tree so that no other spanning tree has a larger sum of weights.

7. Modify Kruskal's algorithm so that it finds a maximum spanning tree.

8. If each edge of a network has a different weight, can there be more than one minimum spanning tree? Why?

9. (a) In the network of Figure 13.1, find a minimum spanning tree if it is required that the tree include edge $\{a, e\}$. (Such a problem might arise if we are required to include a particular telephone line, or if one already exists.)

   (b) Repeat for edge $\{c, e\}$.

10. How would you modify Kruskal's algorithm to deal with a situation such as that in Exercise 9, where one or more edges are specified as having to belong to the spanning tree?

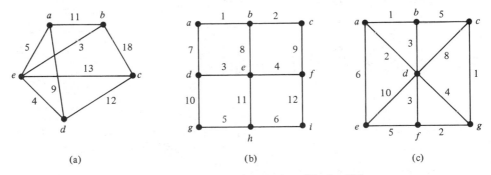

Figure 13.2 Networks for exercises of Section 13.1.

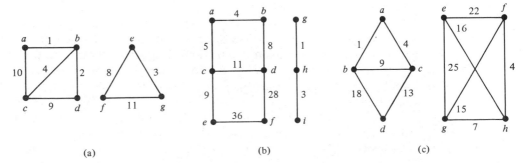

Figure 13.3 Networks for exercises of Section 13.1.

11. Prove that Prim's algorithm works.

12. Show by a crude argument that the computational complexity of Prim's algorithm is bounded by a polynomial in e or in n.

13. Still a third algorithm for finding a minimum spanning tree T provided all weights are distinct works as follows.

> *Step 1.* Set $T = \varnothing$.
>
> *Step 2.* Let G' be the spanning subgraph of G consisting of edges in T.
>
> *Step 3.* For each connected component K of G', find the minimum weight edge of G joining a vertex of K to a vertex of some other component of G'. If there is no such edge, stop and output the message that G is disconnected. Add all the new edges to T. If T has $n - 1$ edges, stop and output T. Otherwise, repeat step 2.

This algorithm has its origin in the work of Boruvka [1926a, b].
(a) Apply this algorithm to each network of Figure 13.2.
(b) Prove that the algorithm works provided all weights are distinct.

14. Show that a minimum spanning tree T is unique if and only if any edge $\{x, y\}$ not in T has larger weight than any edge on the circuit created by adding edge $\{x, y\}$ to T.

15. Let G be a connected graph. A *simple cut set* in G is a set B of edges whose removal disconnects G, but such that no proper subset of B has this property. Let F be a set of edges of G. Show the following.
(a) If F is a simple cut set, then F satisfies the following property C: For every spanning tree H of G, some edge of F is in H.
(b) If F satisfies property C, but no proper subset of F does, then F is a simple cut set.

13.2 THE SHORTEST ROUTE PROBLEM

13.2.1 The Problem

In this section we consider the problem of finding the shortest route between two vertices in a network. We begin with some examples that illustrate the problem.

Example 13.1   Interstate Highways

Suppose that you wish to drive from New York to Los Angeles using only Interstate highways. What is the shortest route to take? This problem can be translated into a network problem as follows. Let the vertices of a graph be junctions of interstate highways and join two such vertices by an edge if there is a single interstate highway joining them and uninterrupted by junctions. Put a real number on the edge $\{x, y\}$ representing the number of miles by interstate highway between vertices x and y. In the resulting network, let the *length* of a chain be defined to be the sum of the numbers (weights) on its edges and take the *distance* $d(x, y)$ between vertices x and y to be the length of the shortest chain between x and y. (If there is no chain between x and y, distance is undefined.) We seek that chain between New York and Los Angeles which is of minimum length, that is, of length equal to

$$d(\text{New York, Los Angeles}).$$

Example 13.2 Planning an Airplane Trip

Suppose that you wish to fly from New York to Bangkok. What is the route that will have you spending as little time in the air as possible? To answer this question, let the vertices of a digraph be cities in the world air transportation network, and draw an arc from x to y if there is a direct flight from city x to city y. Put a real number on arc (x, y) representing the flying time. Then define the *length* of a path from x to y in this directed network as the sum of the numbers (weights) on the arcs, and define the *distance* $d(x, y)$ from x to y as the length of the shortest path from x to y. (Distance is again undefined if there is no path from x to y.) We seek the path of shortest length from New York to Bangkok, that is, the path with length equal to d(New York, Bangkok).

In general, we deal with a network or a directed network, and we seek the shortest chain or path from vertex x to vertex y. We concentrate on directed networks. In Section 13.2.2, we present an algorithm for finding the shortest path from x to y in a directed network. First, let us illustrate the basic ideas. In the directed network of Figure 13.4, the path x, a, b, y has length $1 + 5 + 1 = 7$. The path x, c, d, b, y is shorter; its length is $3 + 1 + 1 + 1 = 6$. Indeed, this is a shortest path from x to y. Hence, $d(x, y) = 6$. Note that we say that x, c, d, b, y is *a* shortest path from x to y. There can be more than one path from x to y of length equal to $d(x, y)$. Here, x, e, d, b, y is another such path. Note also that $d(y, x)$ is undefined; there is no path from y to x.

The problem of finding a shortest path from vertex x to vertex y in a (directed) network is a very common combinatorial problem. According to Goldman [1981], it is perhaps the most widely encountered combinatorial problem in government. Goldman estimates that the shortest path algorithm developed by just one government agency, the Urban Mass Transit Agency in the U.S. Department of Transportation, was regularly applied *billions* of times a year.

13.2.2 Dijkstra's Algorithm

The shortest path algorithm we present is due to Dijkstra [1959]. We present it for a directed network D. Let $w(u, v)$ be the weight on arc (u, v). We recall our standing assumption that weights on arcs in networks are nonnegative. We will need this assumption. It will also be convenient to let $w(u, v)$ be ∞ if there is no arc from u to v. The basic

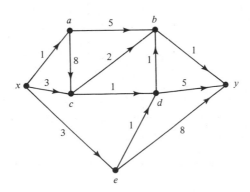

Figure 13.4 A shortest path from x to y is x, c, d, b, y.

idea is that we find at the kth iteration the k vertices $u_1, u_2, \ldots, u_k$ which are the k closest vertices to x in the network, that is, the ones that have (up to ties) the k smallest distances $d(x, u_1), d(x, u_2), \ldots, d(x, u_k)$. We also find for each u_j a shortest path from x to u_j. Having solved this problem for k, we solve it for $k + 1$.

The general idea in going from the kth iteration to the $(k + 1)$st iteration is that for each vertex v not among the k closest vertices, we calculate $d(x, u_j) + w(u_j, v)$ for all j, and find the $(k + 1)$st closest vertex as a v for which this sum is minimized. The rationale for this is that if $x, a_1, a_2, \ldots, a_p, v$ is a shortest path from x to v, then $x, a_1, a_2, \ldots, a_p$ must be a shortest path from x to a_p. If not, we could find a shorter path from x to v by using such a shorter path from x to a_p. Moreover, if weights are all positive, if v is the $(k + 1)$st closest vertex to x and $x, a_1, a_2, \ldots, a_p, v$ is a shortest path from x to v, then a_p must be among the k closest vertices. Even if weights can be zero, the $(k + 1)$st vertex v can be chosen so that there is a shortest path $x, a_1, a_2, \ldots, a_p, v$ from x to v with a_p among the k closest vertices.

At each stage of Dijkstra's algorithm, we keep a list of vertices included in the first k—this defines a class W—and we keep a list of arcs used in the shortest paths from x to u_j—this defines a class B—and we keep a record of $d(x, u_j)$ for all j. We stop once y is added to W. We then use B to construct the path. We are now ready to present the algorithm more formally.

Algorithm 13.3. Dijkstra's Shortest Path Algorithm.*

Input: A directed network D and vertices x and y from D.

Output: A shortest path from x to y or the message that there is no path from x to y.

Step 1. Initially, place vertex x in the class W, let $B = \varnothing$, and let $d(x, x) = 0$.

Step 2.

    Step 2.1. For each vertex u in W and each vertex v not in W, let $\alpha(u, v) = d(x, u) + w(u, v)$. Find u in W and v not in W such that $\alpha(u, v)$ is minimized. (Choose arbitrarily in case of ties.)

    Step 2.2. If the minimum in step 2.1 is ∞, stop and give the message that there is no path from x to y.

    Step 2.3. If the minimum in step 2.1 is not ∞, then place v in W (v is the next vertex chosen), place arc (u, v) in B, and set $d(x, v) = \alpha(u, v)$.

Step 3. If y is not yet in W, return to step 2. If y is in W, stop. A shortest path from x to y can be found by using the unique path of arcs of B which goes from x to y. This can be found by working backward from y.

Let us illustrate this algorithm on the directed network of Figure 13.5. The successive steps in the algorithm are shown in Table 13.2. (The table does not show the $\alpha(u, v)$ which are infinite.) In iteration 2, for instance, arc (x, a) has the smallest α value of those computed, so a is added to W and (x, a) to B and $d(x, a)$ is taken to be $\alpha(x, a)$. In iteration 4, $\alpha(c, e)$ is minimum, so e is added to W and (c, e) to B, and $d(x, e)$ is taken to be $\alpha(c, e)$.

*At this point, we want to present the basic idea of the algorithm. Later, we shall describe how to improve it.

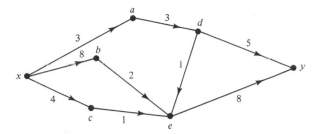

Figure 13.5 A directed network.

And so on. At the seventh iteration, vertex y has been added to W. We now work backward from y, using arcs in B. We find that we got to y from d, to d from a, and to a from x. Thus, we find the path x, (x, a), a, (a, d), d, (d, y), y, which is a shortest path from x to y.

Again we comment on the computational complexity of the algorithm. The algorithm takes at most n iterations, where n is the number of vertices of the directed network. For each iteration adds a vertex. Each iteration involves additions $d(x, u) + w(u, v)$, one for each pair of vertices u in W and v not in W, that is, at most n^2 additions in all. Also, finding the smallest number among the at most n^2 numbers $d(x, u) + w(u, v)$ can be accomplished in at most n^2 comparisons. Hence, at each iteration the algorithm takes at most $2n^2$ steps. Altogether, the algorithm takes at most $2n^3$ steps, a polynomial bound. In the terminology of Section 2.18, we have an $O(n^3)$ algorithm.

Actually, by several simple changes in the procedure, the algorithm can be improved

Table 13.2 Dijkstra's Algorithm Applied to the Directed Network of Figure 13.5

Iteration	Finite numbers $\alpha(u, v)$ computed	Added to W	Added to B	New $d(x, v)$
1	None	x	—	$d(x, x) = 0$
2	$\alpha(x, a) = 0 + 3 = 3$ $\alpha(x, b) = 0 + 8 = 8$ $\alpha(x, c) = 0 + 4 = 4$	a	(x, a)	$d(x, a) = 3$
3	$\alpha(x, b) = 0 + 8 = 8$ $\alpha(x, c) = 0 + 4 = 4$ $\alpha(a, d) = 3 + 3 = 6$	c	(x, c)	$d(x, c) = 4$
4	$\alpha(x, b) = 0 + 8 = 8$ $\alpha(a, d) = 3 + 3 = 6$ $\alpha(c, e) = 4 + 1 = 5$	e	(c, e)	$d(x, e) = 5$
5	$\alpha(x, b) = 0 + 8 = 8$ $\alpha(a, d) = 3 + 3 = 6$ $\alpha(e, y) = 5 + 8 = 13$	d	(a, d)	$d(x, d) = 6$
6	$\alpha(x, b) = 0 + 8 = 8$ $\alpha(e, y) = 5 + 8 = 13$ $\alpha(d, y) = 6 + 5 = 11$	b	(x, b)	$d(x, b) = 8$
7	$\alpha(e, y) = 5 + 8 = 13$ $\alpha(d, y) = 6 + 5 = 11$	y	(d, y)	$d(x, y) = 11$

to be an $O(n^2)$ algorithm.* Suppose that we let u_k be the vertex which is the kth closest. We take $u_1 = x$. Let $\alpha_1(v) = \infty$ for all $v \neq x$, $\alpha_1(x) = 0$, and define

$$\alpha_{k+1}(v) = \min \{\alpha_k(v), \alpha_k(u_k) + w(u_k, v)\}. \tag{13.1}$$

Then it is easy to show (Exercise 16) that

$$\alpha_{k+1}(v) = \min \{d(x, u_j) + w(u_j, v): \quad j = 1, 2, \ldots, k\}. \tag{13.2}$$

At each iteration, we compute the n numbers $\alpha_{k+1}(v)$. For each v, we do this by first doing one addition and then finding one minimum of two numbers. Then we find the minimum of the set of numbers $\alpha_{k+1}(v)$, which can be done in at most n steps. The vertex v which gives us this minimum is u_{k+1}. (Choose arbitrarily in case of ties.) The total number of steps at the iteration which adds u_{k+1} is now at most $n + 2$, so the total number of steps in the entire algorithm is at most $n^2 + 2n$. If we just use the labels $\alpha_k(v)$, we will have to be more careful to compute the set B. Specifically, at each stage that $\alpha_{k+1}(v)$ decreases, we will have to keep track of the vertex u_j such that $\alpha_{k+1}(v)$ is redefined to be $d(x, u_j) + w(u_j, v)$. At the point when u_{k+1} is taken to be v, the corresponding (u_j, v) will be added to B. Details of the improved version of Dijkstra's algorithm are left to the reader.

13.2.3 Applications to Scheduling Problems

Although the shortest route problem has been formulated in the language of distances, the weights do not have to represent distances. Many applications of the shortest route algorithm apply to situations where the arcs correspond to activities of some kind and the weight on an arc corresponds to the cost of the activity. The problem involves finding a sequence of activities that begins at a starting point, accomplishes a desired objective, and minimizes the total cost. Alternatively, the weight is the time involved to carry out the activity and the problem seeks a sequence of activities that accomplishes the desired objective in a minimum amount of time. We have already encountered similar problems in Section 11.6.3. In these situations the network is sometimes called a PERT (for Project Evaluation and Review Technique) network or a CPM (Critical Path Method) network. We illustrate these ideas with the following example.

Example 13.3   A Manufacturing Process

A manufacturing process starts with a piece of raw wood. The wood must be cut into shape, stripped, have holes punched, and be painted. The cutting must precede the hole punching and the stripping must precede the painting. Suppose that cutting takes 1 unit of time; stripping takes 1 unit of time; painting takes 2 units of time for uncut wood and 4 units of time for cut wood; and punching holes takes 3 units of time for unstripped wood, 5 units for stripped but unpainted wood, and 7 units for painted wood. The problem is to find the sequence of activities that will allow completion of the process in as short a period of time as possible. We can let the vertices of a directed network D represent stages in the manufacturing process, for example, raw wood; cut wood; cut and holes punched; cut, stripped, and holes punched; and so on. We take an arc from stage i to stage j if a single activity can

*The rest of this subsection can be omitted.

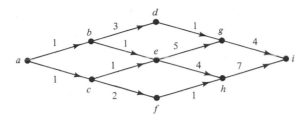

Key:

a, raw wood	*f*, stripped and painted
b, cut	*g*, cut, holes punched, and stripped
c, stripped	*h*, cut, stripped, and painted
d, cut and holes punched	*i*, stripped, cut, holes punched, and painted
e, cut and stripped	

Figure 13.6 Directed network for a manufacturing process. Vertices correspond to stages in the process, arcs to activities taking the process from stage to stage, and weights to times required for the activities.

take the process from stage *i* to stage *j*. Then we put a weight on arc (*i*, *j*) corresponding to the amount of time required to go from *i* to *j*. The directed network in question is shown in Figure 13.6. There are, for example, arcs from *b* to *d* and *e*, because cut wood can next either have holes punched or be stripped. The arc (*b*, *d*) has weight 3 because it corresponds to punching holes in unstripped wood. We seek a shortest path in this network from the raw wood vertex (*a*) to the stripped, cut, holes punched, and painted vertex (*i*).

13.2.4 The "Chinese Postman" Problem Revisited

In Section 11.4.1 we discussed the problem of a mail carrier who wishes to find a smallest number of blocks to walk, and yet cover all the blocks on an assigned route. We formulated the problem as follows. Given a graph *G*, we study *feasible multigraphs H*, multigraphs that have an eulerian closed chain and are obtained from *G* by adding copies of edges of *G*. We seek an *optimal multigraph*, a feasible multigraph with a minimum number of edges. In Section 11.4.2, we observed that the same problem arises in connection with computer graph plotting. In Exercise 8, Section 11.4, the reader was asked to observe that in any optimal multigraph *H*, the newly added edges could be divided up into chains joining vertices of odd degree in *G*, with any such vertex an end vertex of exactly one such chain. Now suppose that for every pair of vertices *u* and *v* of odd degree in *G*, we find between them a shortest chain in *G*. We can use an algorithm like Dijkstra's Algorithm to do this. Let us build a network *G'* by taking as vertices all odd-degree vertices in *G*, joining each pair of vertices by an edge, and putting on this edge a weight equal to the length of the shortest chain between *u* and *v*. To illustrate, suppose that *G* is the graph of Figure 11.26, which is repeated in Figure 13.7. Then there are four vertices of odd degree in *G*, namely *b*, *f*, *h*, and *d*. Between each pair of these, a shortest chain clearly has length 2. Hence, the network *G'* is as shown in the second part of Figure 13.7. Now any optimal multigraph *H* obtained from *G* corresponds to a collection of chains joining pairs of odd-degree vertices in *G*, with each such vertex an end vertex of exactly one such chain. Such a collection of chains defines a perfect matching in the network *G'*. Moreover, an optimum *H* corresponds to a collection of such chains, the sum of whose lengths is as

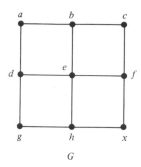

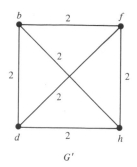

Figure 13.7 The graph G of Figure 11.26 and the corresponding network G'.

small as possible. This collection in turn defines a perfect matching in G' of minimum weight. If G' is changed by taking the negative of each weight, we seek a maximum weight matching in G'. (Such a matching will be perfect. Why?) This is a problem that we have discussed briefly in Examples 12.4 and 12.5. In our example, a minimum weight perfect matching in G' consists of any two edges that do not touch, such as, for example, $\{b, d\}$ and $\{f, h\}$. Then the shortest chains corresponding to the chosen edges will define the optimal multigraph H. Here, a shortest chain between b and d is b, e, d, and one between f and h is f, x, h. If we add to G the edges on these chains, we get the first optimal multigraph H of Figure 11.26. This is now known to be the smallest number of edges that could be added to give an H that has an eulerian closed chain. The eulerian closed chain in H will give us an optimal mail carrier's route in G.

The approach we have described is due to Edmonds (see Edmonds [1965] and Edmonds and Johnson [1973]). It should be pointed out that the procedure described is an efficient one. For Dijkstra's algorithm can be completed in a number of steps of the order of n^2 and the minimum (maximum) weight matching can be found in a number of steps of the order of n^3 (Lawler [1976]).

EXERCISES FOR SECTION 13.2

1. Show that in a digraph, a shortest path from x to y must be a simple path.
2. Show that in a graph, a shortest chain from x to y must be a simple chain.
3. In each directed network of Figure 13.8, use Dijkstra's algorithm (as described in Algorithm 13.3) to find a shortest path from a to z.
4. Find the most efficient manufacturing process in the problem of Example 13.3.
5. A product must be ground, polished, weighed, and inspected. The grinding must precede the polishing and the weighing, and the polishing must precede the inspection. Grinding takes 7 units of time, polishing takes 10 units of time, weighing takes 1 unit of time for an unpolished product and 3 units of time for a polished one, and inspection takes 2 units of time for an unweighed product and 3 units of time for a weighed one. What is the fastest production schedule?
6. A company wants to invest in a fleet of automobiles and is trying to decide on the best strategy for how long to keep a car. After 5 years, it will sell all remaining cars and let an outside firm provide transportation. In planning over the next 5 years, the company estimates that a car bought at the beginning of year i and sold at the beginning of year j will have a net cost

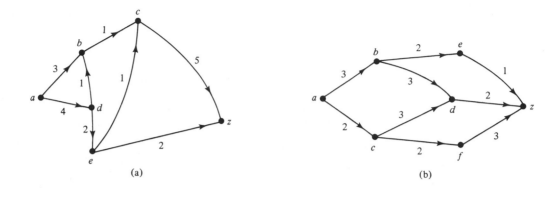

(a)

(b)

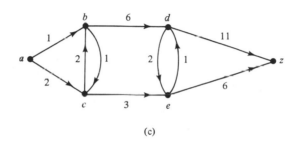

(c)

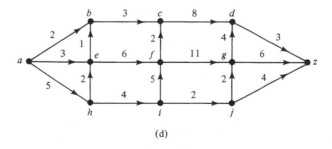

(d)

Figure 13.8 Directed networks for exercises of Sections 13.2 and 13.3.

(purchase price minus trade-in allowance, plus running and maintenance costs) of a_{ij}. The numbers a_{ij} in thousands of dollars are given by the following matrix:

$$
(a_{ij}) = \begin{array}{c} \\ 1 \\ 2 \\ 3 \\ 4 \\ 5 \end{array}
\begin{array}{c} \begin{array}{ccccc} 2 & 3 & 4 & 5 & 6 \end{array} \\
\left[\begin{array}{ccccc}
4 & 6 & 9 & 12 & 16 \\
  & 5 & 7 & 11 & 14 \\
  &   & 6 & 8 & 13 \\
  &   &   & 8 & 11 \\
  &   &   &   & 10
\end{array} \right] \end{array}.
$$

To determine the cheapest strategy for when to buy and sell cars, we let the vertices of a directed network be the numbers 1, 2, 3, 4, 5, 6, include all arcs (i, j) for $i < j$, and we let weight $w(i, j)$ be a_{ij}. The arc (i, j) has the interpretation of buying a car at the beginning of year i and selling it at the beginning of year j. Find the cheapest strategy.

7. Figure 13.9 shows a communication network (arc (i, j) corresponds to a link over which i can communicate directly with j). Suppose that the weight p_{ij} on the arc (i, j) is the probability that the link from i to j is operative. Assuming that defects in links occur independently of each other, the probability that all the links in a path are operative is the product of the link probabilities. Find the most reliable path from a to z. (*Hint:* Consider $-\log p_{ij}$.)

8. It is not efficient to consider all possible paths from i to j in searching for a shortest path. To illustrate the point, suppose that D has n vertices and an arc from every vertex to every other vertex. If x and y are any two vertices of D, find the number of paths from x to y.

9. (Bondy and Murty [1976]) A wolf, a goat, and a cabbage are on one bank of a river and a boatman will take them across, but can only take one at a time. The wolf and the goat cannot be left on one bank of the river together, nor can the goat and the cabbage. How can the boatman get them across the river in the shortest amount of time?

10. (Bondy and Murty [1976]) A man has a full 8-gallon jug of wine and empty jugs that hold 5 and 3 gallons, respectively. What is the fewest number of steps required for the man to divide the wine into two equal amounts?

11. For each graph G of Figure 11.36, find the corresponding network G' as described in the solution to the "Chinese postman" problem in Section 13.2.4. Find a minimum weight perfect matching in G' and translate this into a solution to the "Chinese postman" problem.

12. If all arcs of a directed network have different weights, is the shortest route from x to y necessarily unique?

13. Let D be a digraph. We define distance $d(x, y)$ to be the length of the shortest path from x to y in the directed network obtained from D by putting a weight of 1 on each arc. Discuss how to use the powers of the adjacency matrix of D (Section 3.7) to calculate $d(x, y)$.

14. In Dijkstra's algorithm, show that if direction of arcs is disregarded, the arcs of the set B define a tree rooted at vertex x.

15. Describe a shortest path algorithm for finding, in a directed network D, the shortest paths from a given vertex x to each other vertex.

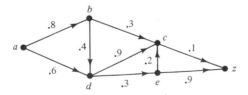

Figure 13.9 Communication network.

16. Show that α_{k+1} defined by (13.1) satisfies (13.2).

17. (a) Write a careful description of the $O(n^2)$ version of Dijkstra's algorithm using the labels $\alpha_k(v)$.

   (b) Apply the algorithm to each directed network of Figure 13.8 to find a shortest path from a to z.

13.3 NETWORK FLOWS

13.3.1 The Maximum Flow Problem

Suppose that D is a directed network and let $c_{ij} = w(i, j)$ be the nonnegative weight on the arc (i, j). In this section we call c_{ij} the *capacity* of the arc (i, j) and interpret it as the maximum amount of some commodity that can "flow" through the arc per unit time in a steady-state situation. The commodity can be finished products, messages, people, oil, trucks, letters, electricity, and so on.

Flows are permitted only in the direction of the arc, that is, from i to j. We fix a *source* vertex s and a *sink* vertex t, and think of a flow starting at s and ending at t. (In all of our examples, s will have no incoming arcs and t no outgoing arcs. But it is not necessary to assume these properties.) Let x_{ij} be the flow through arc (i, j). Then we require that

$$0 \le x_{ij} \le c_{ij}. \tag{13.3}$$

This says that flow is nonnegative and cannot exceed capacity. We also have a *conservation law*, which says that for all vertices $i \ne s, t$, what goes in must go out, that is,

$$\sum_j x_{ij} = \sum_j x_{ji}, \quad i \ne s, t. \tag{13.4}$$

A set of numbers satisfying (13.3) and (13.4) is called an *(s, t) feasible flow*, or an *(s, t) flow*, or just a *flow*. For instance, consider the directed network of Figure 13.10. In part (a) of the figure, the capacities c_{ij} are shown on each arc with the numbers in squares. In part (b), a flow is shown with the numbers in circles. Note that (13.3) and (13.4) hold. For instance, $c_{24} = 2$ and $x_{24} = 0$, so $0 \le x_{24} \le c_{24}$. Also, $\sum_j x_{3j} = x_{35} + x_{36} + x_{37} = 2 + 1 + 0 = 3$, and $\sum_j x_{j3} = x_{23} = 3$. And so on.

Suppose that x_{ij} defines a flow. Let

$$v_t = \sum_j x_{jt} - \sum_j x_{tj}$$

and

$$v_s = \sum_j x_{js} - \sum_j x_{sj}.$$

Note that if we sum the terms

$$\sum_j x_{ji} - \sum_j x_{ij}$$

over all i, Equation (13.4) tells us that all of these terms are 0 except for the cases $i = s$ and $i = t$. Thus,

$$\sum_i \left[\sum_j x_{ji} - \sum_j x_{ij} \right] = v_t + v_s. \tag{13.5}$$

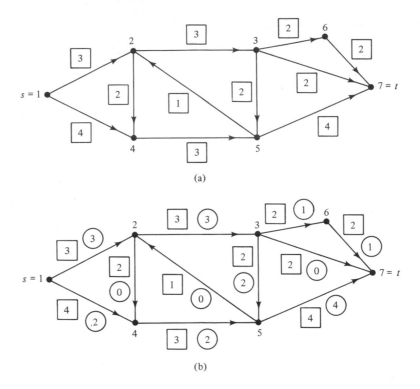

Figure 13.10   Capacities are shown in squares, flows in circles.

However, the left hand side of (13.5) is the same as

$$\alpha = \sum_{i,j} x_{ji} - \sum_{i,j} x_{ij}. \tag{13.6}$$

Since both sums in (13.6) simply sum all flows over all arcs, they are equal and hence $\alpha = 0$. We conclude that $v_s = -v_t$. Thus, there is a number v such that

$$\sum_{j} x_{ji} - \sum_{j} x_{ij} = \begin{cases} -v & \text{if } i = s \\ v & \text{if } i = t \\ 0 & \text{if } i \neq s, t. \end{cases} \tag{13.7}$$

The number v in our example of Figure 13.10(b) is 5. It is called the *value* of the flow. The value represents the total amount of the commodity that can be sent through the network in a given period of time if this flow is used. The problem we will consider is this: Find a flow that has maximum value, a *maximum flow*.

   The standard reference on the theory of flows in networks is the book by Ford and Fulkerson [1962]. Other comprehensive references are Berge and Ghouila-Houri [1965], Frank and Frisch [1972], Hu [1969], Iri [1969], Lawler [1976], Minieka [1978], and Papadimitriou and Steiglitz [1982].

13.3.2 Cuts

Let S and T be two sets that partition the vertex set of the digraph D, that is, $V(D) = S \cup T$ and $S \cap T = \emptyset$. We refer to the partition (S, T) as a *cut*. Equivalently, we think of the cut as the set C of all arcs that go from vertices in S to vertices in T. The set C is called a cut because after the arcs of C are removed, there is no path from any vertex of S to any vertex of T. If x is any vertex of S and y any vertex of T, C is called an (x, y) *cut*. For instance, in Figure 13.10, $S = \{1, 2\}$ and $T = \{3, 4, 5, 6, 7\}$ is a cut, and this is equivalent to the set of arcs $(1, 4)$, $(2, 3)$, $(2, 4)$. (We do not include arc $(5, 2)$ as it goes from a vertex in T to a vertex in S.) Notice that if there are more than two vertices, there are always at least two (x, y) cuts: $S = \{x\}$, $T = V(D) - \{x\}$, and $S = V(D) - \{y\}$, $T = \{y\}$.

In a directed network, if (S, T) is a cut, we can define its *capacity* as $c(S, T) = \sum_{i \in S} \sum_{j \in T} c_{ij}$. In our example above, $c(S, T) = c_{14} + c_{23} + c_{24} = 9$. Notice that in our example, the value of the flow is 5 and the capacity of this cut is 9, which is greater. The next result shows that this is no accident.

Theorem 13.3. In a directed network, the value of any (s, t) flow is $\leq$ the capacity of any (s, t) cut.

Proof. Let x_{ij} be an (s, t) flow and (S, T) be an (s, t) cut. Note that $\sum_j x_{ij} - \sum_j x_{ji}$ is 0 if $i \in S$ and $i \neq s$, and is v if $i = s$. Thus,

$$v = \sum_j x_{sj} - \sum_j x_{js},$$

$$v = \sum_{i \in S} \left[\sum_j x_{ij} - \sum_j x_{ji} \right],$$

$$v = \sum_{i \in S} \sum_{j \in S} [x_{ij} - x_{ji}] + \sum_{i \in S} \sum_{j \in T} [x_{ij} - x_{ji}],$$

$$v = \sum_{i \in S} \sum_{j \in S} x_{ij} - \sum_{i \in S} \sum_{j \in S} x_{ji} + \sum_{i \in S} \sum_{j \in T} [x_{ij} - x_{ji}], \qquad (13.8)$$

$$v = \sum_{i \in S} \sum_{j \in T} (x_{ij} - x_{ji}), \qquad (13.9)$$

because the first two terms of (13.8) are the same. Thus, by (13.9), the value of any flow is the net flow through any cut. Since $x_{ij} \leq c_{ij}$ and $x_{ij} \geq 0$, we have

$$x_{ij} - x_{ji} \leq x_{ij} \leq c_{ij},$$

so (13.9) implies that

$$v \leq \sum_{i \in S} \sum_{j \in T} c_{ij} = c(S, T). \qquad \text{Q.E.D.}$$

Corollary 13.3.1. In a directed network, if (S, T) is an (s, t) cut and x_{ij} is an (s, t) flow, then

$$v = \sum_{i \in S} \sum_{j \in T} [x_{ij} - x_{ji}].$$

Proof. This is a corollary of the proof. Q.E.D.

Figure 13.11 shows another (s, t) flow in the network of Figure 13.10(a). This flow has value 6. Notice that if $S = \{1, 2, 4\}$ and $T = \{3, 5, 6, 7\}$, then $c(S, T) = c_{23} + c_{45} = 6$. Now there can be no (s, t) flow with value more than the capacity of this cut, that is, 6. Hence, the flow shown is a maximum flow. Similarly, this cut must be an (s, t) cut of minimum capacity, a *minimum cut*. Indeed, the same thing must be true any time we find a flow and a cut where the value of the flow is the same as the capacity of the cut.

Theorem 13.4. If x_{ij} is an (s, t) flow with value v and (S, T) is an (s, t) cut with capacity $c(S, T)$, and if $v = c(S, T)$, then x_{ij} is a maximum flow and (S, T) is a minimum cut.

We have used reasoning similar to this in Theorem 12.4, where we argued that since the number of edges in any matching of a graph is less than or equal to the number of vertices in any covering, if we ever find a matching and a covering of the same size, the matching must be maximum and the covering must be minimum.

13.3.3 A Faulty Max Flow Algorithm

Our goal is to describe an algorithm for finding the maximum flow. First we shall present an intuitive, although faulty, technique. If P is a simple path from s to t, we call it an (s, t) *path* and we let the corresponding *unit flow* x_{ij}^P be given by

$$x_{ij}^P = \begin{cases} 1 & \text{if arc } (i, j) \text{ is in } P \\ 0 & \text{otherwise} \end{cases}$$

The idea is to successively add unit flows.

Let us say that an arc (i, j) is *unsaturated* by a flow x_{ij} if $x_{ij} < c_{ij}$, and let us define the *slack* by $s_{ij} = c_{ij} - x_{ij}$. The basic point is that if θ is the minimum slack among arcs of P, then we can add the unit flow x_{ij}^P a total of θ times and obtain a new flow with value increased by θ. Here is the algorithm.

Algorithm 13.4. Max Flow Algorithm—First Attempt

Input: A directed network with a source s and a sink t.

Output: A supposedly maximum (s, t) flow x_{ij}.

Step 1. Set $x_{ij} = 0$, all i, j.

Step 2.

    Step 2.1. Find an (s, t) path P with all arcs unsaturated. If none exists, go to step 3.

    Step 2.2. Compute the slack of each arc of P.

    Step 2.3. Compute θ, the minimum slack of arcs of P.

    Step 2.4. Redefine x_{ij} by adding θ to the flow on arc (i, j) if (i, j) is on P. Return to step 2.1.

Step 3. Stop with the flow x_{ij}.

Let us apply this algorithm to the directed network of Figure 13.10(a). Figure 13.12 shows the successive flows defined by the following iterations. In the first iteration, we

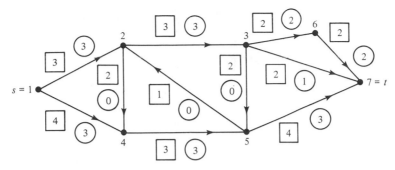

Figure 13.11 Another flow for the directed network of Figure 13.10(a).

take $x_{ij} = 0$, all i, j. We then note that $P = 1, 4, 5, 7$ is an (s, t) path with all arcs unsaturated, and the corresponding θ is 3, for $s_{14} = c_{14} - x_{14} = 4 - 0 = 4$, $s_{45} = c_{45} - x_{45} = 3 - 0 = 3$, and $s_{57} = c_{57} - x_{57} = 4 - 0 = 4$. Hence, we increase x_{14}, x_{45}, and x_{57} by $\theta = 3$, obtaining the second flow in Figure 13.12. In this flow, $P = 1, 2, 3, 7$ is an (s, t) path with all unsaturated arcs. Its θ is 2, for $s_{12} = 3$, $s_{23} = 3$, and $s_{37} = 2$. Thus, we increase x_{12}, x_{23}, and x_{37} by $\theta = 2$, obtaining the third flow in Figure 13.12. In this flow, $P = 1, 2, 3, 6, 7$ has all arcs unsaturated and $\theta = 1$, since $s_{12} = c_{12} - x_{12} = 3 - 2 = 1$, $s_{23} = 1$, $s_{36} = 2$, and $s_{67} = 2$. Thus, we increase x_{12}, x_{23}, x_{36}, and x_{67} by $\theta = 1$, obtaining the fourth flow in Figure 13.12. Since in this flow there are no more (s, t) paths with all arcs unsaturated, we stop. Note that we have obtained a flow of value equal to 6, which we know to be a maximum.

Unfortunately, this algorithm does not necessarily lead to a maximum flow. Let us consider the same directed network. Figure 13.13 shows the successive steps in another use of this algorithm using different (s, t) paths. Notice that after obtaining the fourth flow, we can find no (s, t) path with all arcs unsaturated. Thus, the algorithm stops. However, the flow obtained has value only 5, which is not maximum. What went wrong?

Consider the minimum cut $S = \{1, 2, 4\}$, $T = \{3, 5, 6, 7\}$. One problem is that one of the unit flow paths used, 1, 4, 5, 2, 3, 7, crosses this cut backward, that is, from T to S; equivalently, it crosses it forward from S to T twice. Thus, the 1 unit of flow on P uses up 2 units of capacity—too much capacity is used. We would be better off if we got rid of some of the flow going backward and used more in a forward direction. This suggests a way to improve on our algorithm.

13.3.4 Augmenting Chains

Consider the graph (multigraph) G obtained from a directed network by disregarding directions on arcs. Let C be a chain from s to t in this graph. An arc (i, j) of D is said to be a *forward arc* of C if it is followed from i to j, and a *backward arc* otherwise. For instance, in the directed network D of Figure 13.10, one chain is 1, 4, 5, 3, 6, 7. Here, arc (1, 4) is a forward arc, but arc (3, 5) is backward. If x_{ij} is a flow, C is said to be a *flow-augmenting chain*, or just an *augmenting chain*, *relative to* x_{ij} if $x_{ij} < c_{ij}$ for each forward arc and $x_{ij} > 0$ for each backward arc. The unit flow paths with no saturated arcs, which we discussed in Section 13.3.3, all correspond to flow-augmenting chains with no backward arcs. The chain 1, 4, 5, 3, 6, 7 in Figure 13.10(b) is a flow-augmenting chain relative to the

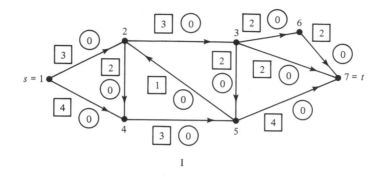

I

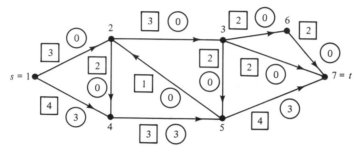

II (from $P = 1, 4, 5, 7$ and $\theta = 3$)

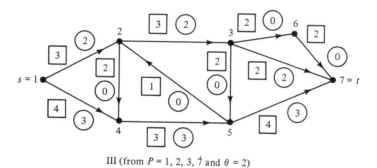

III (from $P = 1, 2, 3, 7$ and $\theta = 2$)

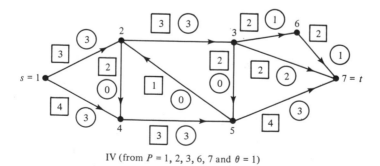

IV (from $P = 1, 2, 3, 6, 7$ and $\theta = 1$)

Figure 13.12 Applying Algorithm 13.4 to the directed network of Figure 13.10(a).

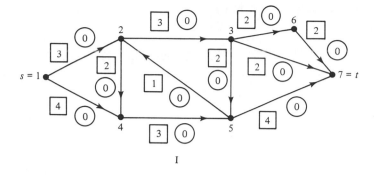

I

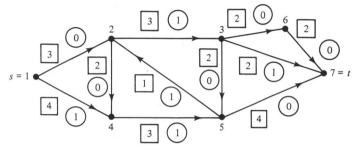

II (from $P = 1, 4, 5, 2, 3, 7$ and $\theta = 1$)

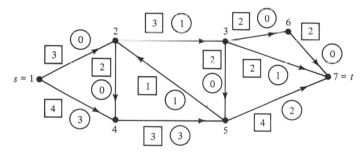

III (from $P = 1, 4, 5, 7$ and $\theta = 2$)

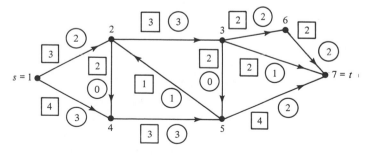

IV (from $P = 1, 2, 3, 6, 7$ and $\theta = 2$)

Figure 13.13 Applying Algorithm 13.4 to the directed network of Figure 13.10(a) using a different choice of (s, t) paths.

flow shown in the figure. For the forward arcs (1, 4), (4, 5), (3, 6), and (6, 7) are all under capacity and the backward arc (3, 5) has positive flow. We can improve on the value of the flow by decreasing the backward flow and increasing the forward flow. Specifically, we record x_{ij} for all backward arcs on the chain and compute and record the slack $s_{ij} = c_{ij} - x_{ij}$ for all forward arcs. If λ is the minimum of these recorded numbers, λ is called the *capacity* of the augmenting chain. We then increase each x_{ij} on a forward arc by λ and decrease each x_{ij} on a backward arc by λ. By choice of λ, the new x_{ij} still is nonnegative and is no higher than the capacity c_{ij}. Moreover, the conservation law (13.4) still holds. Next, we observe that the value of the flow increases by λ, for the chain starts with an edge from x. If this edge is forward, then the flow out of s is increased; if it is backward, then the flow into s is decreased. In any case, the value or net flow out of s is increased. [This argument needs to be expanded in case the flow-augmenting chain is not a simple chain. The expansion is left to the reader (Exercise 24).] However, we will only need simple flow-augmenting chains (Exercise 25). To illustrate with our example of Figure 13.10(b), note that $s_{14} = 2$, $s_{45} = 1$, $s_{36} = 1$, $s_{67} = 1$, and $x_{35} = 2$. Hence, the minimum of these numbers, λ, is 1. We increase x_{14}, x_{45}, x_{36}, and x_{67} by 1, and decrease x_{35} by 1, obtaining a flow of value 6, one more than the value of the flow shown.

We are now ready to state the main results about maximum flows, which will allow us to present our main algorithm. We have already shown that if a flow admits an augmenting chain, then the value of the flow can be increased, so the flow is not maximum. The first result says that if the flow is not maximum, we can find an augmenting chain. (Thus, flow-augmenting chains are analogous to M-augmenting chains for matchings, and the next theorem is analogous to Theorem 12.7. We expand on the relation between network flows and matching in Section 13.3.8.)

Theorem 13.5. An (s, t) flow is maximum if and only if it admits no augmenting chain from s to t.

*Proof.** It remains to suppose that x_{ij} is a flow with no augmenting chain and to show that x_{ij} is a maximum. Let S be the set of vertices j such that there is an augmenting chain from s to j and let T be all other vertices. Note that s is in S because s alone defines an augmenting chain from s to s. Moreover, t is in T, since there is no augmenting chain. By definition of augmenting chain and by definition of S and T, we have for all i in S and all j in T, $x_{ij} = c_{ij}$ and $x_{ji} = 0$. For there is an augmenting chain from s to i, since i is in S. If $x_{ij} < c_{ij}$ or $x_{ji} > 0$, we can add edge $\{i, j\}$ to this chain to find an augmenting chain from s to j, contradicting j in T. Thus, for all i in S and j in T, $x_{ij} - x_{ji} = c_{ij}$.

By Corollary 13.3.1,

$$v = \sum_{i \in S} \sum_{j \in T} [x_{ij} - x_{ji}] = \sum_{i \in S} \sum_{j \in T} c_{ij} = c(S, T).$$

Hence, we have found a cut (S, T) with the same capacity as the value of the flow x_{ij}. By Theorem 13.4, the flow is a maximum. Q.E.D.

A cut (S, T) is called *saturated* relative to a flow x_{ij} if $x_{ij} = c_{ij}$ for all $i \in S, j \in T$, and $x_{ji} = 0$ for all $i \in S, j \in T$.

*The proof can be omitted.

Corollary 13.5.1. If (S, T) is a saturated (s, t) cut relative to flow x_{ij}, then x_{ij} is a maximum (s, t) flow.

Proof. This is a corollary of the proof.   Q.E.D.

The next result is a very famous theorem due to Elias *et al.* [1956] and Ford and Fulkerson [1956].

Theorem 13.6 (The Max Flow Min Cut Theorem). In a directed network, the maximum value of an (s, t) flow equals the minimum capacity of an (s, t) cut.

*Proof.** It follows from Theorem 13.3 that the maximum value of an (s, t) flow is at most the minimum capacity of an (s, t) cut. To show equality, we suppose that x_{ij} is a maximum flow, with value v. Then it can have no flow-augmenting chain, and so, by the proof of Theorem 13.5, we can find an (s, t) cut (S, T) so that $v = c(S, T)$. It follows by Theorem 13.4 that x_{ij} is a maximum flow and (S, T) is a minimum cut.   Q.E.D.

Remark. Our proof of the max flow min cut theorem uses the tacit assumption that there exists a maximum flow. This is easy to prove if all capacities are rational numbers. For then the maximum flow algorithm, to be described in Section 13.3.5, finds a maximum flow. If some capacities are not rational, a maximum flow still exists (even though the maximum flow algorithm, as we describe it, does not necessarily find a maximum flow). See Lawler [1976] or Papadimitriou and Steiglitz [1982] for a proof.

13.3.5 The Max Flow Algorithm

We can now formulate the max flow algorithm.

Algorithm 13.5. The Max Flow Algorithm

> *Input:* A directed network with a source s and a sink t.
> *Output:* A maximum (s, t) flow x_{ij}.

Step 1. Set $x_{ij} = 0$ for all i, j.

Step 2.

> *Step 2.1.* Find a flow-augmenting chain C from s to t. If none exists, go to step 3.
> *Step 2.2.* Compute and record the slack of each forward arc of C and record the flow of each backward arc of C.
> *Step 2.3.* Compute λ, the minimum of the numbers recorded in step 2.2.
> *Step 2.4.* Redefine x_{ij} by adding λ to the flow on all forward arcs of C and subtracting λ from the flow on all backward arcs of C. Return to step 2.1.

Step 3. Stop with the flow x_{ij}.

Suppose that we apply this algorithm to the directed network of Figure 13.10(a). Since every (s, t) path is an augmenting chain with no backward arcs, we can get to the

*The proof can be omitted.

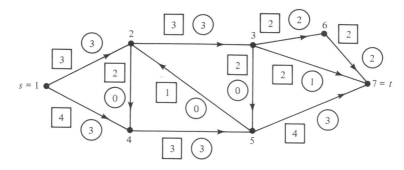

Figure 13.14 The flow obtained from the fourth flow of Figure 13.13 by using the flow-augmenting chain $C = 1, 2, 5, 7$.

fourth flow of Figure 13.13. We then identify the flow-augmenting chain $C = 1, 2, 5, 7$. In step 2.2 we compute $s_{12} = 1$, $s_{57} = 2$, $x_{52} = 1$. Then the minimum, λ, of these numbers is $\lambda = 1$. We increase x_{12} and x_{57} by $\lambda = 1$ and decrease x_{52} by $\lambda = 1$, obtaining the flow of Figure 13.14. In this flow, there is no augmenting chain. We conclude that it is maximum. This agrees with our earlier conclusion, since the value of this flow is 6.

Theorem 13.7. If all capacities in the directed network are rational numbers, the Max Flow Algorithm will attain a maximum flow.

Proof.* Clearly, if all capacities are integers, then at each iteration the number λ is an integer, and the value of the flow increases by λ. Since the value is at most the capacity of the cut defined by $S = \{s\}$, $T = V(D) - \{s\}$, the value cannot keep increasing by an integer amount more than a finite number of times. Hence, there will come a time when there is no augmenting chain from s to t. By Theorem 13.5, the corresponding flow will be maximum. The case of rational capacities can be handled by finding a common denominator δ for all the capacities and considering the directed network obtained from the original one by multiplying all capacities by δ. Q.E.D.

In general, if some capacities are not rational numbers, the algorithm does not necessarily converge in a finite number of steps.† Moreover, it can converge to a flow that is not a maximum, as Ford and Fulkerson [1962] showed. Edmonds and Karp [1972] have shown that if each flow augmentation is made along an augmenting chain with a minimum number of edges, then the algorithm terminates in a finite number of steps and a maximum flow is attained. See Lawler [1976] or Papadimitriou and Steiglitz [1982] for details.

13.3.6 A Labeling Procedure for Finding Augmenting Chains‡

The max flow algorithm as we have described it does not discuss how to find an augmenting chain in step 2.1. In this subsection we describe a labeling procedure for finding such a chain. (The procedure is due to Ford and Fulkerson [1957].) In this procedure, at each

*The proof can be omitted.

†This point is of little practical significance, since computers work with rational numbers.

‡This subsection can be omitted if time is short.

step, a vertex is *scanned* and its neighbors are given *labels*. Vertex j gets label (i^+) or (i^-). The label is determined by finding an augmenting chain from s to j which ends with the edge $\{i, j\}$. The $+$ indicates that (i, j) is a forward arc in this augmenting chain, the $-$ that it is a backward arc. Eventually, if vertex t is labeled, we have an augmenting chain from s to t. If the procedure concludes without labeling vertex t, we will see that no augmenting chain exists, and we conclude that the flow is maximum. The procedure is described in detail as follows.

Algorithm 13.6. **Subroutine: Labeling Algorithm for Finding an Augmenting Chain**

> *Input:* A directed network with a source s and a sink t and an (s, t) flow x_{ij}.
>
> *Output:* An augmenting chain or the message that x_{ij} is a maximum flow.

Step 1. Give vertex s the label $(-)$.

Step 2. Let F be the set of arcs (i, j) such that $s_{ij} > 0$. Let B be the set of arcs (i, j) such that $x_{ij} > 0$. (Note that arcs in F can be used as forward arcs and arcs in B as backward arcs in an augmenting chain.)

Step 3 (*Labeling and Scanning*).

> *Step 3.1.* If all labeled vertices have been scanned, go to step 5.
>
> *Step 3.2.* If not, find a labeled but unscanned vertex i and scan it as follows. For each arc (i, j), if $(i, j) \in F$ and j is unlabeled, give j the label (i^+). For each arc (j, i), if $(j, i) \in B$ and j is unlabeled, give j the label (i^-). Do not label any other neighbors of i. Vertex i has now been scanned.
>
> *Step 3.3.* If vertex t has been labeled, go to step 4. Otherwise, go to step 3.1.

Step 4. Starting at vertex t, use the index labels to construct an augmenting chain. The label on vertex t indicates the next-to-last vertex in this chain, the label on that vertex indicates its predecessor in the chain, and so on. Stop and output this chain.

Step 5. Stop and output the message that the flow x_{ij} is a maximum.

Let us illustrate this algorithm with the flow of Figure 13.10(b). We begin by labeling vertex s by $(-)$. Then we find that F consists of the arcs $(1, 4)$, $(2, 4)$, $(3, 6)$, $(3, 7)$, $(4, 5)$, $(5, 2)$, and $(6, 7)$, and B consists of all arcs except $(2, 4)$, $(5, 2)$, and $(3, 7)$. We now go through the labeling and scanning procedure (step 3). Figure 13.15 shows the labels. Note that vertex $s = 1$ is labeled but unscanned. The only arc $(1, x)$ in F is $(1, 4)$. We therefore label vertex 4 with (1^+). There are no arcs $(x, 1)$ in B. Thus, vertex 1 has been scanned. Since vertex $7 = t$ has not yet been labeled, we find another labeled, but unscanned vertex, namely 4. We now consider the arcs $(4, x)$ in F, namely the arc $(4, 5)$. Thus, vertex 5 gets the label (4^+). We also note that no arc $(x, 4)$ is in B. Thus, vertex 4 has been scanned. Note that t has not yet been labeled, so we find another labeled, but unscanned vertex, namely 5. The arc $(5, 2)$ is in F. Hence, we label vertex 2 with (5^+). The arc $(3, 5)$ is in B, so we label vertex 3 with the label (5^-). Then vertex 5 has been scanned. Now t has not yet been labeled. We find a labeled but unscanned vertex. We have a choice of vertices 2 and 3. Suppose that we pick 3. Scanning 3 leads to the label (3^+) on vertices 6 and 7. Now $t = 7$ has been labeled. (Coincidentally, all vertices have been labeled.) We go to step 4

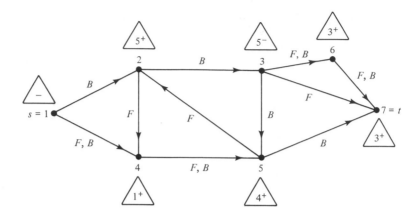

Figure 13.15 The labeling algorithm applied to the flow of Figure 13.10(b). The label is shown in a triangle next to a vertex. The arcs in F and in B are labeled F and B, respectively.

and read backward to find a flow-augmenting chain. In particular, the label (3^+) on vertex 7 sends us back to vertex 3. The label (5^-) here sends us back to vertex 5, the label (4^+) to vertex 4, and the label (1^+) to vertex 1. Thus, we have the flow-augmenting chain 1, 4, 5, 3, 7. In this example we did not need to differentiate a + label from a − label in order to find a flow-augmenting chain. However, had there been two arcs between 3 and 5, (3, 5) and (5, 3), the label 5^- on vertex 3 would have told us to use the backward arc (3, 5). The + and − labels will be useful in modifying the labeling algorithm to compute the number λ needed for the max flow algorithm.

Theorem 13.8. The labeling algorithm finds an augmenting chain if the flow x_{ij} is not a maximum, and ends by concluding that x_{ij} is maximum otherwise.

*Proof.** It is clear that if the algorithm produces a chain from s to t, then the chain is augmenting. We shall show that if t has not yet been labeled and there is no labeled, but unscanned vertex, then the flow is a maximum. Let S consist of all labeled vertices and T of all unlabeled vertices. Then (S, T) is an (s, t) cut. Moreover, every arc from i in S to j in T is saturated and every arc from j in T to i in S has flow 0, for otherwise we could have labeled a vertex of T in scanning the vertices of S. We conclude that (S, T) is a saturated cut. By Corollary 13.5.1, we conclude that x_{ij} is a maximum flow.   Q.E.D.

In closing this subsection, we note that the labeling algorithm can be modified so that at the end, it is easy to compute the number λ needed for the max flow algorithm. At each step where we assign a label to a vertex j, we have just found an augmenting chain from s to j. We then let $\lambda(j)$ be the minimum of the numbers s_{uv} for (u, v) a forward arc of this chain and x_{uv} for (u, v) a backward arc of the chain. We let $\lambda(s)$ be $+\infty$. If we label j by scanning from i, then we can compute $\lambda(j)$ from $\lambda(i)$. In particular, if j gets labeled (i^+), $\lambda(j) = \min\{\lambda(i), s_{ij}\}$, and if j gets labeled (i^-), then $\lambda(j) = \min\{\lambda(i), x_{ji}\}$. Finally, λ is $\lambda(t)$

*The proof can be omitted.

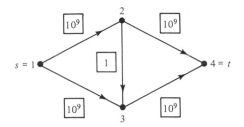

Figure 13.16 A poor choice of augmenting chains leads to 2 billion iterations in the max flow algorithm.

(see Exercise 29). To illustrate, in our example, we would compute the following $\lambda(j)$ in the following order: $\lambda(s) = +\infty$, $\lambda(4) = \min \{\lambda(s), s_{14}\} = 2$, $\lambda(5) = \min \{\lambda(4), s_{45}\} = 1$, $\lambda(2) = \min \{\lambda(5), s_{52}\} = 1$, $\lambda(3) = \min \{\lambda(5), x_{35}\} = 1$, $\lambda(6) = \min \{\lambda(3), s_{36}\} = 1$, $\lambda(7) = \min \{\lambda(3), s_{37}\} = 1$, $\lambda = \lambda(7) = 1$.

13.3.7 Complexity of the Max Flow Algorithm

Let us make a comment on the computational complexity of the max flow algorithm. Let a be the number of arcs in the directed network and v be the maximum flow. The labeling algorithm described in Section 13.3.6 requires at most $2a$ arc inspections in each use to find an augmenting chain—we look at each arc at most once as a forward arc and at most once as a backward arc. If all capacities are integers and we start the maximum flow algorithm with the flow $x_{ij} = 0$ for all i, j, then each flow-augmenting chain increases the value by at least 1, so the number of iterations involving a search for an augmenting chain is at most v. Hence, the number of steps is at most $2av$.* One problem with this computation is that we want to determine complexity solely as a function of the size of the input, not in terms of the solution v. Note that our computation implies that the algorithm might take a long time. Indeed, it can. Consider the directed network of Figure 13.16. Starting with the 0 flow, we could conceivably choose first the augmenting chain 1, 2, 3, 4, then the augmenting chain 1, 3, 2, 4, then 1, 2, 3, 4, then 1, 3, 2, 4, and so on. This would require 2 billion iterations before converging! However, if we happen to choose first the augmenting chain 1, 2, 4, and then the augmenting chain 1, 3, 4, then we finish in two iterations! To avoid these sorts of problems, we change step 3.2 of the labeling algorithm so that vertices are scanned in the same order in which they receive labels. Edmonds and Karp [1972] have shown that in this case, a maximum flow is obtained in at most $an/2$ applications of the labeling algorithm, where n is the number of vertices. Thus, since each use of the labeling algorithm requires at most $2a$ steps, the total number of steps is at most

$$(2a)(an/2) = a^2n \le [n(n - 1)]^2n\dagger.$$

*This disregards the simple computations of computing slacks, and modifying the flow. These are each applied to each arc at most twice in each iteration, so add a constant times av to the number of steps. In each iteration, we also have to compute λ, which is the minimum of a set of no more than a numbers. The total number of computations in computing all the λ's is thus again at most av. Finally, we have to construct the augmenting chain from the labels, which again takes at most a steps in each iteration, or at most av steps in all. In sum, if k is a constant, the total number of steps is thus at most kav, which is $O(av)$ to use the notation of Section 2.18.

$\dagger$As per the previous footnote, a more accurate estimate is at most $ka(an/2)$ steps.

The Edmonds and Karp algorithm has since been improved. See Papadimitriou and Steiglitz [1982, pp. 216–217] for references on improved algorithms.

13.3.8 Matching Revisited*

In this subsection we investigate the relation between network flows and the matchings we studied in Chapter 12. In particular, we show how to prove Theorem 12.5, the result that in a bipartite graph, the number of edges in a maximum matching equals the number of vertices in a minimum covering.

Suppose that $G = (X, Y, E)$ is a bipartite graph. We can make it into a directed network D, called the *associated network*, by adding a source s and a sink t and arcs from s to all vertices of X and from all vertices of Y to t, and by directing all edges between X and Y from X to Y. We put capacity 1 on all new arcs, and capacity ∞ (or a very large number) on all arcs from X to Y. Then it is easy to see that any integer flow (flow all of whose x_{ij} values are integers) from s to t in D corresponds to a matching M in G: We include an edge $\{i, j\}$ in M if and only if there is positive flow along the arc (i, j). Conversely, any matching M in G defines an integer flow x_{ij} in D: For $\{i, j\}$ in E, with i in X, take x_{ij} to be 1 if $\{i, j\}$ is in M, and 0 otherwise; take x_{sk} to be 1 if and only if k is saturated in M, and x_{lt} to be 1 if and only if l is saturated in M. Moreover, the number of edges in the matching M equals the value of the flow x_{ij}. Figure 13.17 illustrates this construction, by showing a bipartite graph G and the associated network. The flow shown in the network corresponds to the matching shown by wiggly edges in G. We summarize these results in the next theorem.

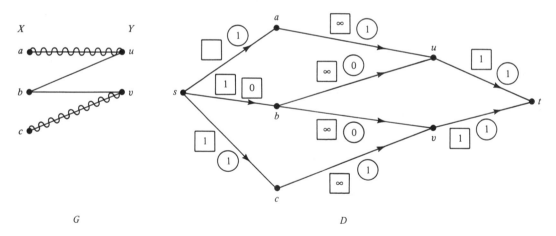

Figure 13.17 A bipartite graph G and its associated network D. The flow in D shown by encircled numbers corresponds to the matching in G shown by wiggly edges.

Theorem 13.9. Suppose that $G = (X, Y, E)$ is a bipartite graph and D is the associated network. Then there is a one-to-one correspondence between integer (s, t) flows in D and matchings in G. Moreover, the value of an integer flow is the same as the number of edges in the corresponding matching.

*This subsection can be omitted.

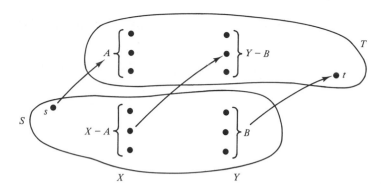

Figure 13.18 The cut (13.10). The only possible arcs in the cut are shown by arrows.

Since matchings in a bipartite graph $G = (X, Y, E)$ correspond to flows in the associated network D, we might ask what coverings in the graph correspond to in the network. The answer is that coverings correspond to (s, t) cuts of finite capacity. To make this precise, suppose $A \subseteq X$ and $B \subseteq Y$. Any covering K in G can be written in the form $A \cup B$ for such A and B. Any (s, t) cut (S, T) of D can be written in the following form:

$$S = \{s\} \cup (X - A) \cup B, \qquad T = \{t\} \cup A \cup (Y - B). \qquad (13.10)$$

For s is in S and t is in T. We simply define A to be all vertices in $T \cap X$ and B to be all vertices in $S \cap Y$ (see Figure 13.18). We now have the following theorem.

Theorem 13.10. Suppose that $G = (X, Y, E)$ is a bipartite graph and D is the associated network. Suppose that $A \subseteq X$ and $B \subseteq Y$. Let $K = A \cup B$ and let

$$S = \{s\} \cup (X - A) \cup B, \qquad T = \{t\} \cup A \cup (Y - B). \qquad (13.10)$$

Then K is a covering of G if and only if (S, T) is an (s, t) cut of D of finite capacity. Moreover, if (S, T) is an (s, t) cut of finite capacity, then $|A \cup B|$ is the capacity of the cut (S, T).

Proof. Suppose that (S, T) defined by (13.10) is an (s, t) cut of finite capacity. Since the cut has finite capacity, there can be no arcs from X to Y in the cut, that is, no arcs from $X - A$ to $Y - B$ (see Figure 13.18). Thus, by (13.10), all arcs from X to Y in D go from A or to B, so all edges of G are joined to A or to B, and so $K = A \cup B$ is a covering. Next, note that $|A \cup B|$ is the capacity of the cut (S, T). For by Figure 13.18, the arcs in the cut are exactly the arcs (s, x) for x in A and (y, t) for y in B. There are $|A \cup B|$ such arcs. Each has unit capacity, so the cut has capacity $|A \cup B|$.

It remains to prove that if $K = A \cup B$ is a covering of G, then (S, T) defined by (13.10) gives an (s, t) cut and it has finite capacity. This is left to the reader (Exercise 26). Q.E.D.

We can now prove Theorem 12.5.

Proof of Theorem 12.5. [Suppose that $G = (X, Y, E)$ is a bipartite graph. Then the number of edges in a maximum matching equals the number of vertices in a minimum covering.]

Consider the associated network. Since all the capacities are integers, it follows from the proof of Theorem 13.7 that there is a maximum flow in which all x_{ij} are integers. The value of this flow will also have to be an integer, and will, by Theorem 13.9, give the number α of edges in the maximum matching. Moreover, by the max flow min cut theorem (Theorem 13.6), this number α is also the minimum capacity of an (s, t) cut (S, T). Suppose that this cut is given by (13.10). Now clearly, an (s, t) cut of minimum capacity has finite capacity, so by Theorem 13.10, (S, T) corresponds via (13.10) to a covering $K = A \cup B$. The number of vertices in K is equal to $|A \cup B|$, which is equal to the capacity of the cut (S, T), which is equal to α. Then we have a covering of α vertices and a matching of α edges, so by Theorem 12.4, the covering must be a minimum. Hence, the number of edges in a maximum matching equals the number of vertices in a minimum covering.   Q.E.D.

EXERCISES FOR SECTION 13.3

1. In each directed network of Figure 13.19, a (potential) (s, t) flow is shown in the circled numbers.
   (a) Which of these flows is feasible? Why?
   (b) If the flow is feasible, find its value.
2. In each directed network D of Figure 13.8, interpret the weights as capacities and find the capacity of the cut (S, T) where $S = \{a, b, e\}$ and $T = V(D) - S$.
3. In each flow x_{ij} of Figure 13.20, compute the slack on each arc.
4. In the flow III of Figure 13.13, which of the following are flow-augmenting chains?
   (a) 1, 2, 3, 7   (b) 1, 2, 5, 7   (c) 1, 4, 5, 7   (d) 1, 4, 2, 3, 7.
5. In the flow III of Figure 13.12, which of the chains of Exercise 4 are flow-augmenting chains?
6. For each flow of Figure 13.20, either show that it is maximum by finding an (s, t) cut (S, T) such that $v = c(S, T)$ or show that it is not maximum by finding an augmenting chain.
7. Give an example of a directed network and a flow x_{ij} for this network which has value 0, but such that x_{ij} is not 0 for all i, j.
8. Apply Algorithm 13.4 to each directed network of Figure 13.8, if weights are interpreted as capacities and $s = a$, $t = z$.
9. Repeat Exercise 8 using Algorithm 13.5.
10. For each flow of Figure 13.20, carefully apply Algorithm 13.6 to search for an augmenting chain.
11. In each application of Algorithm 13.6 in Exercise 10, also calculate the numbers $\lambda(j)$.
12. In directed network (d) of Figure 13.8, let $s = a$ and $t = z$ and define a flow by letting $x_{ab} = x_{bc} = x_{cd} = x_{dz} = 2$ and $x_{ae} = x_{ef} = x_{fg} = x_{gz} = 3$, and $x_{ah} = x_{hi} = x_{ij} = x_{jz} = 2$, and otherwise taking $x_{ij} = 0$. Apply Algorithm 13.6 to search for a flow-augmenting chain.
13. For each bipartite graph G of Figure 12.12:
    (a) Find the associated network D.
    (b) Find an integer (s, t) flow in D and its corresponding matching in G.
    (c) Find an (s, t) cut in D and its corresponding covering in G.
14. A pipeline network sends oil from location A to location B. The oil can go via the northern routh or the southern route. Each route has one junction, with a pipeline going from the junction on the southern route to the junction on the northern route. The first leg of the northern route, from location A to the junction, has a capacity of 400 barrels an hour; the second leg, from the junction to location B, has a capacity of 300 barrels an hour. The first leg

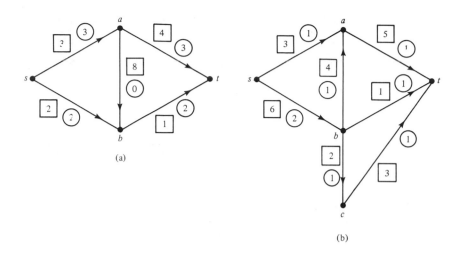

(a)

(b)

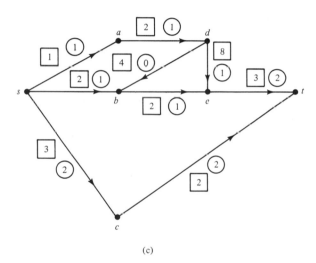

(c)

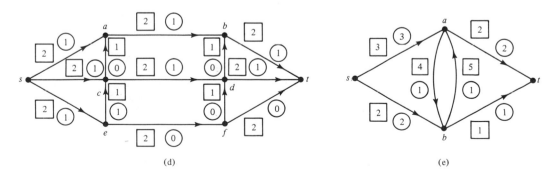

(d)

(e)

Figure 13.19 (Potential) flows for exercises of Section 13.3.

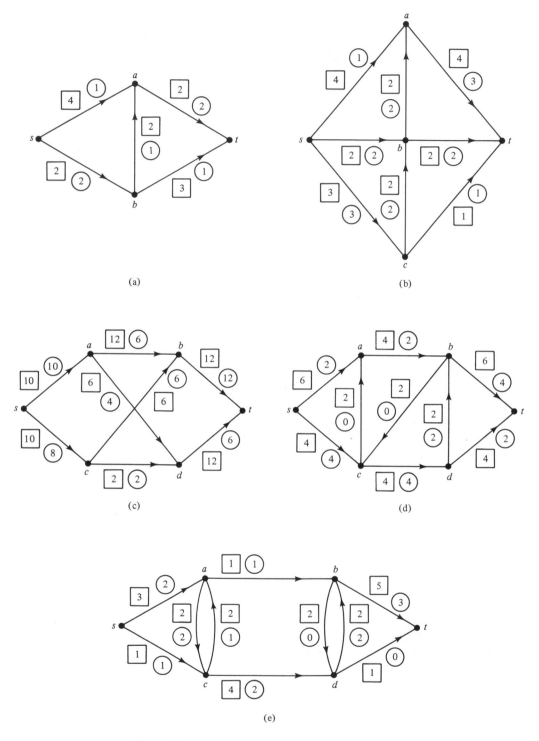

Figure 13.20 Flows for exercises of Sections 13.3 and 13.4.

of the southern route has a 500-barrel/hour capacity, and the second leg a 300-barrel/hour capacity. The pipeline joining the two junctions also has a 300-barrel/hour capacity. What is the largest number of barrels of oil that can be shipped from location A to location B in an hour?

15. In this exercise we build on the notion of reliability of systems discussed in Example 2.13 and in Example 3.8. Suppose that a system is represented by a directed network, with the components corresponding to arcs. Let us say the system works if and only if in the modified network defined by working components, there is an (s, t) flow of value at least v. For each directed network of Figure 13.21, compute $F(x_1 x_2 \cdots x_n)$ as defined in Example 2.13, if v is 3.

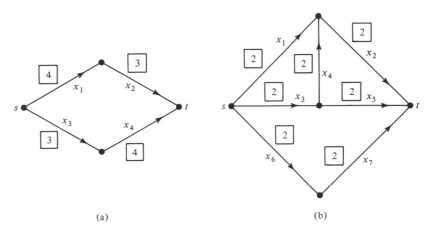

(a)   (b)

Figure 13.21   Directed networks for exercises of Section 13.3.

16. Suppose that D is a directed network and X is a set of sources and Y is a set of sinks. [We assume that $X, Y \subseteq V(D)$ and $X \cap Y = \varnothing$.] An (X, Y) *flow* is a flow where the conservation conditions (13.4) hold only for vertices that are not sources or sinks. The *value* of the flow is defined to be

$$\sum_{\substack{i \in X \\ j \notin X}} x_{ij} - \sum_{\substack{i \in X \\ j \notin X}} x_{ji} .$$

We can find a maximum (X, Y) flow by joining two new vertices, s and t, to D, adding arcs of capacity ∞ from s to all vertices in X and from all vertices in Y to t, and finding a maximum (s, t) flow in the new network. Find a maximum (X, Y) flow in each directed network of Figure 13.22.

17. There are three warehouses, w_1, w_2, and w_3, and three retail outlets, r_1, r_2, and r_3. The warehouses have, respectively, 3000, 4000, and 6000 drums of paint, and the retail outlets have demand for, respectively, 2000, 4000, and 3000 drums. Figure 13.23 shows a freight network, with the capacity on arc (i, j) giving the largest number of drums that can be shipped from location i to location j during a given day. Can all the demands be met if only one day is allowed for shipping from warehouses? (A drum can go along as many arcs as necessary in one day.) If not, what is the largest total demand that can be met? Solve this problem by translating it into a multisource, multisink problem and then into an ordinary network flow problem (see Exercise 16). This is an example of a *transshipment problem*.

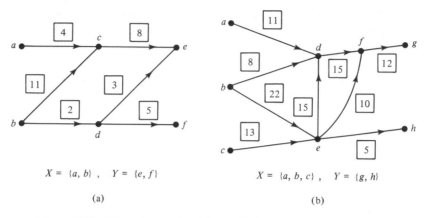

$X = \{a, b\}, \quad Y = \{e, f\}$

$X = \{a, b, c\}, \quad Y = \{g, h\}$

(a)

(b)

Figure 13.22 Directed networks with a set X of sources and a set Y of sinks.

18. Show that if x_{ij} is an (s, t) flow of value v and (S, T) is an (s, t) cut of capacity c, then $v = c$ if and only if for each arc (i, j) from T to S, $x_{ij} = 0$, and for each arc (i, j) from S to T, $x_{ij} = c_{ij}$.

19. Suppose that a flow from s to t is decomposed into unit-flow paths from s to t, and each of these unit-flow paths crosses a given saturated cut exactly once. Show that the flow is maximum.

20. We wish to send messengers from a location s to a location t in a region whose road network is modeled by a digraph. Because some roads may be blocked, we wish each messenger to drive along a route that is totally disjoint from that of all other messengers. How would we find the largest number of messengers who could be sent? (*Hint:* Use unit capacities.)

21. Let D be a strongly connected digraph and let a and b be vertices of D. Show that there are k arc-disjoint paths from a to b if and only if every (a, b) cut has at least k arcs.

22. A flow from a source s to a sink t in an undirected network is defined as a flow from s to t in the directed network obtained by replacing each edge $\{i, j\}$ by the two arcs (i, j) and (j, i) and letting each arc have the same capacity as the corresponding edge. Find a maximum flow from s to t in each network of Figure 13.24.

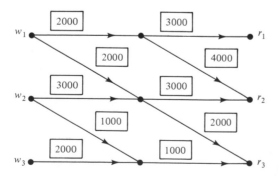

Figure 13.23 Freight network for Exercise 17, Section 13.3.

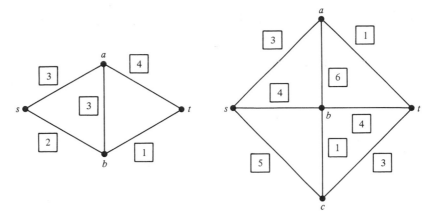

Figure 13.24 Networks for exercises of Section 13.3.

23. Let G be an undirected network whose underlying graph is connected. A *cut* in G is a partition of the vertices into two sets S and T, or equivalently the set of all edges joining a vertex of S to a vertex of T. If s and t are the source and sink of G, respectively, (S, T) is an (s, t) *cut* if $s \in S$ and $t \in T$. Let F be a set of edges of G.
   (a) Show that if F is a simple cut set in the sense of Exercise 15, Section 13.1, then it is a cut.
   (b) Show that if F is an (s, t)-cut with a minimum capacity, then F is a simple cut set.

24. Suppose that C is a flow-augmenting chain and λ is the capacity of C as defined in Section 13.3.4. If we increase each x_{ij} on a forward arc by λ and decrease each x_{ij} on a backward arc by λ, show that even if C is not a simple chain, the value or net flow out of s is increased.

25. Show that in the max flow algorithm, it suffices to find flow-augmenting chains which are simple chains.

26. Complete the proof of Theorem 13.10.

27. Use the results of Section 13.3.8 to prove the König–Egerváry Theorem (Corollary 12.5.1) without first proving Theorem 12.5.

28. Suppose that $G = (X, Y, E)$ is a bipartite graph and M is a matching of G. Use the results of Section 13.3.8 to prove Theorem 12.7, namely, that M is maximum if and only if G contains no M-augmenting chain.

29. Show that in the labeling algorithm, if $\lambda(j)$ is defined as in the discussion following the proof of Theorem 13.8, then $\lambda = \lambda(t)$.

13.4 MINIMUM COST FLOW PROBLEMS

13.4.1 Some Examples

An alternative network flow problem, which has many important special cases, is the following. Suppose that in a directed network, in addition to a nonnegative capacity c_{ij} on each arc, we have a nonnegative cost a_{ij} of shipping one unit of flow from i to j. We want to find a flow sending a fixed nonnegative number v of units from source s to sink t, and

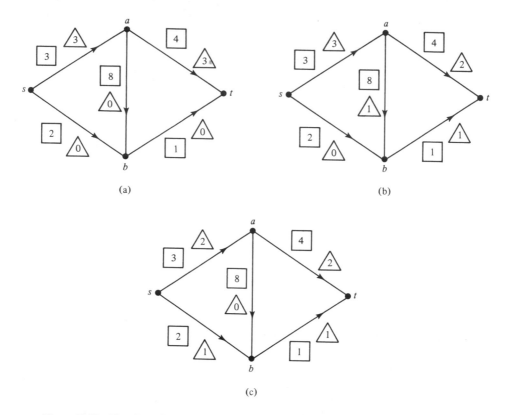

Figure 13.25 The three flows of value 3 for the directed network (a) of Figure 13.19. The capacity is shown in a square, the flow in a triangle.

doing so at minimum cost. That is, we want to find a flow x_{ij} such that its value is v and such that $\sum a_{ij} x_{ij}$ is minimized. We call this a *minimum cost flow problem*.

To illustrate, consider the directed network (a) of Figure 13.19. Consider the number in the square as the capacity and the number in the circle as the cost. There are three flows which attain a value of 3; these are shown in Figure 13.25. Of these flows, flow (c) has the minimum cost, namely

$$2 \cdot 3 + 2 \cdot 3 + 0 \cdot 0 + 1 \cdot 2 + 1 \cdot 2 = 16.$$

Flows (a) and (b) have values 18 and 17, respectively.

Example 13.4   Routing a Traveling Party (Minieka [1978])

A traveling party of 75 people (say a football team) is to go from New York to Honolulu. What is the least cost way to route the party? The solution is obtained by finding a minimum cost flow of 75 from New York to Honolulu in a directed network where the vertices are cities, an arc indicates a direct air link, and there is a capacity constraint (unbooked seats) and a cost constraint (air fare) on each arc. (This assumes that air fares are simple sums along links, which usually would not be true.)

Example 13.5 The Transportation Problem

Imagine that a particular commodity is stored in n warehouses and is to be shipped to m markets. Let a_i be the supply of the commodity at the ith warehouse, let b_j be the demand for the commodity at the jth market, and let a_{ij} be the cost of transporting one unit of the commodity from warehouse i to market j. For simplicity, we assume that $\sum a_i = \sum b_j$, that is, that the total supply equals the total demand. (This assumption can easily be eliminated—see Exercise 7.) The problem is to find a shipping pattern that minimizes the total transportation cost.

This problem can be formulated as follows. Let x_{ij} be the number of units of the commodity shipped from i to j. We seek to minimize

$$\sum_{\substack{i=1 \\ j=1}}^{\substack{m \\ n}} a_{ij} x_{ij}$$

subject to the following constraints: For every i,

$$\sum_{j=1}^{m} x_{ij} \le a_i, \qquad (13.11)$$

and for every j,

$$\sum_{i=1}^{n} x_{ij} \ge b_j. \qquad (13.12)$$

Constraint (13.11) says that the total amount of commodity shipped from the ith warehouse is at most the amount there, and constraint (13.12) says that the total amount of commodity shipped to the jth market is at least the amount demanded. Note that since $\sum a_i = \sum b_j$, any solution satisfying (13.11) and (13.12) for all i and j will also satisfy

$$\sum_{j=1}^{m} x_{ij} = a_i \qquad (13.13)$$

and

$$\sum_{i=1}^{n} x_{ij} = b_j. \qquad (13.14)$$

We can look at the transportation problem as a minimum cost flow problem. Draw a digraph with vertices the n warehouses $w_1, w_2, \ldots, w_n$ and the m markets k_1, $k_2, \ldots, k_m$. Add a source vertex s and a sink vertex t, and include arcs from each warehouse to each market, from the source to all warehouses, and from each market to the sink (see Figure 13.26). On arc (w_i, k_j), place a cost a_{ij} and a capacity $c_{ij} = \infty$ (or a very large number). On arc (s, w_i), place a cost of 0 and a capacity of a_i, and on arc (k_j, t), place a cost of 0 and a capacity of b_j. Because we have constraints (13.11) and (13.12), it is easy to see that we have a minimum cost flow problem for this directed network: We seek a flow of value equal to $\sum a_i = \sum b_j$, at minimum cost.

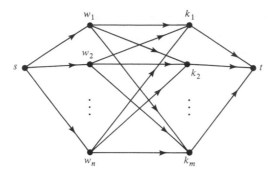

Figure 13.26 A directed network for the transportation problem.

Example 13.6 The Optimal Assignment Problem

In Example 12.5 we discussed the following job assignment problem. There are n workers and m jobs, every worker is suited for every job, and worker i's potential performance on job j is given a rating r_{ij}. We wish to assign workers to jobs so as to maximize the sum of the performance ratings. Let us assume that $m = n$ and let x_{ij} be a variable that is 1 if worker i is assigned to job j and 0 otherwise. Then, since $m = n$, we want to maximize

$$\sum_{\substack{i=1 \\ j=1}}^{\substack{m \\ n}} r_{ij} x_{ij}.$$

We require that no two workers get the same job, that is, that

$$\sum_{i=1}^{n} x_{ij} \le 1. \tag{13.15}$$

We also require that every worker get a job, that is, that

$$\sum_{j=1}^{n} x_{ij} \ge 1. \tag{13.16}$$

Then since the number of workers equals the number of jobs, (13.15) and (13.16) give us the constraints

$$\sum_{i=1}^{n} x_{ij} = 1 \quad \text{and} \quad \sum_{j=1}^{n} x_{ij} = 1. \tag{13.17}$$

We may think of this as a transportation problem by letting the workers correspond to the warehouses and the jobs to the markets. We then take all a_i and b_j equal to 1, and let the cost be $c_{ij} = -r_{ij}$. We seek to maximize $\sum_{i,j} r_{ij} x_{ij}$ or minimize $\sum_{i,j} c_{ij} x_{ij}$. Hence, this optimal assignment problem is also a minimum cost flow problem. [One difference is that we have the additional requirement that $x_{ij} = 0$ or 1. However, it is possible to show that if a minimum cost flow problem has integer capacities, costs, and value v, then some optimal solution x_{ij} is in integers, and standard algorithms produce integer solutions (see below). Then the requirement (13.17) and the added requirement $x_{ij} \ge 0$ are sufficient to give us $x_{ij} = 0$ or 1.]

The max flow algorithm we have described in Section 13.3.5 is readily modified to give an efficient algorithm for solving the minimum cost flow problem. (This algorithm will produce an integer solution if capacities, costs, and value v are integers.) For details, see, for example, Lawler [1976], or Minieka [1978], or Papadimitriou and Steiglitz [1982]. In special cases such as the transportation problem or the optimal assignment problem, there are more efficient algorithms. We close by presenting an algorithm for solving the optimal assignment problem.

13.4.2 An Algorithm for the Optimal Assignment Problem

Suppose that we are given an $n \times n$ matrix (c_{ij}) and we wish to find numbers $x_{ii} = 0$ or 1 such that we minimize $\sum_{i,j} c_{ij} x_{ij}$ and such that (13.17) holds. Note that if a constant p_k is subtracted from all entries in the kth row of (c_{ij}), giving rise to a matrix (c'_{ij}), then

$$\sum_{i,j} c'_{ij} x_{ij} = \sum_{i,j} c_{ij} x_{ij} - p_k \sum_j x_{kj} = \sum_{i,j} c_{ij} x_{ij} - p_k,$$

by (13.17). Thus the assignment x_{ij} that minimizes $\sum_{i,j} c_{ij} x_{ij}$ will also minimize $\sum_{i,j} c'_{ij} x_{ij}$. The same is true if a constant q_l is subtracted from each entry in the lth column of (c_{ij}). Indeed, the same is true if we subtract constants p_i from the ith row of (c_{ij}) for all i and q_j from the jth column of (c_{ij}) for all j.

Using this observation, we let p_i be the minimum element in the ith row of (c_{ij}). For each i, we subtract p_i from each element in the ith row of (c_{ij}), obtaining a matrix (c'_{ij}). We now let q_j be the minimum element in the jth column of (c'_{ij}) and, for each j, subtract q_j from each element in the jth column of (c'_{ij}), obtaining a matrix $(\bar{c}_{ij})$. This is called the *reduced matrix*. By what we have observed, we may as well solve the optimal assignment problem using the reduced matrix.

To illustrate this procedure, suppose that $n = 4$ and

$$(c_{ij}) = \begin{bmatrix} 12 & 14 & 15 & 14 \\ 9 & 6 & 11 & 8 \\ 10 & 9 & 16 & 14 \\ 12 & 13 & 13 & 10 \end{bmatrix}. \tag{13.18}$$

Then $p_1 = 12, p_2 = 6, p_3 = 9, p_4 = 10$, and

$$(c'_{ij}) = \begin{bmatrix} 0 & 2 & 3 & 2 \\ 3 & 0 & 5 & 2 \\ 1 & 0 & 7 & 5 \\ 2 & 3 & 3 & 0 \end{bmatrix}.$$

Now $q_1 = 0, q_2 = 0, q_3 = 3$, and $q_4 = 0$, so

$$(\bar{c}_{ij}) = \begin{bmatrix} 0 & 2 & 0 & 2 \\ 3 & 0 & 2 & 2 \\ 1 & 0 & 4 & 5 \\ 2 & 3 & 0 & 0 \end{bmatrix}. \tag{13.19}$$

Note that the reduced matrix $(\bar{c}_{ij})$ always has nonnegative entries. (Why?) A job assignment x_{ij} giving each worker one job and each job one worker corresponds to a choice of n entries of this matrix, one in each row and one in each column. We have $x_{ij} = 1$ if and only if the i, j entry is picked. Now suppose that we can find a choice of n entries, one in each row and one in each column, so that all the entries are 0. Let us take $x_{ij} = 1$ for precisely these n i, j entries and $x_{ij} = 0$ otherwise. Then the corresponding $\sum \bar{c}_{ij} x_{ij}$ will be 0, because when x_{ij} is 1, $\bar{c}_{ij} = 0$. For instance, if

$$(\bar{c}_{ij}) = \begin{bmatrix} 3 & 0 & 1 \\ 0 & 5 & 4 \\ 0 & 3 & 0 \end{bmatrix},$$

we can pick the 1, 2, the 2, 1, and the 3, 3 entries. If we take $x_{12} = 1$, $x_{21} = 1$, $x_{33} = 1$, and $x_{ij} = 0$ otherwise, then $\sum \bar{c}_{ij} x_{ij} = 0$. This is clearly a minimum, since $\sum \bar{c}_{ij} x_{ij} \geq 0$ because $\bar{c}_{ij} \geq 0$.

Recall from Section 12.4.1 that an *independent set of 0's* in a matrix is a collection of 0's, no two of which are in the same row and no two of which are in the same column. What we have just observed is that if we can find an independent set of n 0's in $(\bar{c}_{ij})$, then we can find an optimal job assignment by taking the corresponding x_{ij} to be 1. Since there can be no independent set of more than n 0's in $(\bar{c}_{ij})$, we look for a maximum independent set of 0's. How do we find such a set? Let us change all positive entries in $(\bar{c}_{ij})$ to 1. Then since $\bar{c}_{ij}$ is nonnegative, the resulting matrix is a matrix of 0's and 1's. Recall from Section 12.4.1 that a *line* of a matrix is either a row or a column. Then by the König–Egerváry Theorem (Corollary 12.5.1), the maximum number of independent 0's in $(\bar{c}_{ij})$, equivalently in its modified matrix of 0's and 1's, is equal to the minimum number of lines which cover all the 0's. Thus we can check to see if there is a set of n independent 0's by checking to see if there is a set of n lines which cover all the 0's. Alternatively, from the proof of the König–Egerváry Theorem, we recall that a maximum independent set of 0's in the matrix $(\bar{c}_{ij})$, equivalently in the modified matrix of 0's and 1's, corresponds to a maximum matching in the bipartite graph $G = (X, Y, E)$, where $X = Y = \{1, 2, \ldots, n\}$ and where there is an edge between i in X and j in Y iff $\bar{c}_{ij} = 0$. Thus, we can apply the maximum matching algorithm of Section 12.5.3 to see if there is a set of n independent 0's.

In our example of Equation (13.18), we have derived $(\bar{c}_{ij})$ in Equation (13.19). Note that the minimum number of lines covering all the 0's in $(\bar{c}_{ij})$ is 3: Use the first and fourth rows and the second column (see Figure 13.27). Thus, the maximum independent set of 0's has three elements, not enough to give us a job assignment, as we need four independent 0's.

The algorithm for solving the optimal assignment problem proceeds by successively modifying the reduced matrix $(\bar{c}_{ij})$ so that eventually we obtain one where we can find n independent 0's. The modification step is to use the minimum covering of 0's by lines, such as shown in Figure 13.27, and find the smallest uncovered element. Then subtract this from each uncovered element and add it to each twice-covered element and find a new matrix in which to search for independent 0's. In our example, the smallest uncovered element is 1, the 3, 1 element. We subtract 1 from the 2, 1, the 2, 3, the 2, 4, the 3, 1, the 3, 3, and the 3, 4 entries of $(\bar{c}_{ij})$, and add it to the 1, 2 and the 4, 2 entries, obtaining the new reduced matrix

$$\begin{bmatrix} 0 & 3 & 0 & 2 \\ 2 & 0 & 1 & 1 \\ 0 & 0 & 3 & 4 \\ 2 & 4 & 0 & 0 \end{bmatrix}. \tag{13.20}$$

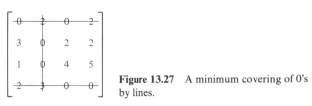

Figure 13.27 A minimum covering of 0's by lines.

It is not hard to show that the new reduced matrix has been obtained from the preceding one by adding or subtracting a constant from different rows or columns (Exercise 8). Thus, solving the optimal assignment problem with this matrix is the same as solving it with the previous reduced matrix. Also, the new reduced matrix has all entries nonnegative. Thus, once again with this new matrix, we can seek an independent set of n 0's. We seek this by finding a minimum covering by lines. In (13.20), a minimum covering uses four lines, and hence there must be an independent set of four 0's. One such set consists of the 1, 3 entry, the 2, 2 entry, the 3, 1 entry, and the 4, 4 entry. Hence, we can find an optimal assignment by letting $x_{13} = 1$, $x_{22} = 1$, $x_{31} = 1$, $x_{44} = 1$, and all other $x_{ij} = 0$. If we had not found a set of n independent 0's, we would have repeated the modification of the reduced matrix again.

The algorithm we have described is called the *Hungarian algorithm*, and is due to Kuhn [1955]. We summarize it as follows.

Algorithm 13.7.   The Hungarian Algorithm for the Optimal Assignment Problem

> *Input:* An $n \times n$ matrix (c_{ij}).
>
> *Output:* An optimal assignment x_{ij}.

Step 1 (Initialization).
> *Step 1.1.* For each i, let p_i be the minimum element in the ith row of (c_{ij}).
> *Step 1.2.* Let (c'_{ij}) be computed from (c_{ij}) by subtracting p_i from each element of the ith row, for all i.
> *Step 1.3.* For each j, let q_j be the minimum element in the jth column of (c'_{ij}).
> *Step 1.4.* Let $(\bar{c}_{ij})$ be computed from (c'_{ij}) by subtracting q_j from each element of the jth column, for all j.

Step 2.
> *Step 2.1.* Find a minimum collection of lines covering the 0's of $(\bar{c}_{ij})$.
> *Step 2.2.* If this collection of lines has fewer than n elements, go to step 3. Otherwise, go to step 4.

Step 3 (Modification of Reduced Matrix).
> *Step 3.1.* Using the covering obtained in step 2.1, let p be the smallest uncovered element in $(\bar{c}_{ij})$.
> *Step 3.2.* Change the reduced matrix $(\bar{c}_{ij})$ by subtracting p from each uncovered element and adding p to each twice-covered element. Return to step 2.

Step 4.
> *Step 4.1.* Find a set of n independent 0's in $(\bar{c}_{ij})$.
> *Step 4.2.* Let x_{ij} be 1 if the i, j entry of $(\bar{c}_{ij})$ is one of the independent 0's and let x_{ij} be 0 otherwise. Output this solution x_{ij} and stop.

Theorem 13.11. If all of the c_{ij} are integers, the Hungarian algorithm gives an optimal assignment.

*Proof.** We have already observed that if the algorithm gives an assignment x_{ij}, this must be optimal. But how do we know the algorithm will ever give an assignment? The reason it does is because the reduced matrix always has nonnegative integer entries and because at each modification, the sum of the entries in this matrix decreases by an integer at least 1, as we shall show below. Thus, in a finite number of steps, if we have not yet reached an optimal solution, all the entries of the reduced matrix will be 0. In this case, there is of course a collection of n independent 0's and so we find an optimal solution.

We now show that if $(\bar{d}_{ij})$ is the modified reduced matrix obtained in step 3 from the reduced matrix $(\bar{c}_{ij})$, then

$$\sum_{i,j} \bar{c}_{ij} - \sum_{i,j} \bar{d}_{ij} = \text{an integer} \geq 1.$$

Recall that we have a covering of $(\bar{c}_{ij})$ by $k < n$ lines. Let S_r be the set of uncovered rows of $(\bar{c}_{ij})$, S_c be the set of uncovered columns, $\bar{S}_r$ be the set of covered rows, and $\bar{S}_c$ the set of covered columns. Let $\alpha = |S_r|$ and $\beta = |S_c|$. Then $k = (n-\alpha) + (n-\beta)$. Also, recall that p is the smallest uncovered entry of $(\bar{c}_{ij})$. Then, remembering how we get $(\bar{d}_{ij})$ from $(\bar{c}_{ij})$, we have

$$\sum_{i,j} \bar{c}_{ij} - \sum_{i,j} \bar{d}_{ij} = \sum_{\substack{i \in S_r \\ j \in S_c}} [\bar{c}_{ij} - \bar{d}_{ij}] + \sum_{\substack{i \in S_r \\ j \in \bar{S}_c}} [\bar{c}_{ij} - \bar{d}_{ij}]$$

$$+ \sum_{\substack{i \in \bar{S}_r \\ j \in S_c}} [\bar{c}_{ij} - \bar{d}_{ij}] + \sum_{\substack{i \in \bar{S}_r \\ j \in \bar{S}_c}} [\bar{c}_{ij} - \bar{d}_{ij}]$$

$$= \sum_{\substack{i \in S_r \\ j \in S_c}} p + \sum_{\substack{i \in S_r \\ j \in \bar{S}_c}} 0 + \sum_{\substack{i \in \bar{S}_r \\ j \in S_c}} 0 + \sum_{\substack{i \in \bar{S}_r \\ j \in \bar{S}_c}} (-p)$$

$$= \alpha\beta p - (n-\alpha)(n-\beta)p$$

$$= n(\alpha + \beta - n)p.$$

But $\alpha + \beta$ is the number of uncovered rows and columns, so

$$\alpha + \beta - n = (2n - k) - n = n - k > 0,$$

since $k = (n-\alpha) + (n-\beta)$ and $k < n$. Thus,

$$\sum_{i,j} \bar{c}_{ij} - \sum_{i,j} \bar{d}_{ij} = n(\alpha + \beta - n)p = \text{an integer} \geq 1,$$

since n, $\alpha + \beta - n$, and p are all positive integers. Q.E.D.

We close by observing that if there are n workers and n jobs, the Hungarian algorithm can be implemented in $O(n^3)$ time, to use the terminology of Section 2.18. For a discussion of this point, see Papadimitriou and Steiglitz [1982].

---

*The proof can be omitted.

1. In each directed network of Figure 13.20, consider the number in the circle as the cost rather than the flow, and consider the number in the square as the capacity as usual. In each case, find a minimum cost flow of value 3 from s to t.

2. (The Caterer Problem)   A caterer knows in advance that for the next n days, a_j napkins will be needed on the jth day, $j = 1, 2, \ldots, n$. For any given day, the caterer can buy new napkins or use laundered napkins. Laundering of napkins can be done either by quick service, which takes q days, or by slow service, which takes r days. The cost of a new napkin is c cents, the cost of laundering a napkin quickly is d cents, and the cost of laundering one slowly is e cents. Starting with no napkins, how would the caterer meet the napkin requirement with minimum cost? Set this up as a minimum cost flow problem. (*Note:* An analogous problem, which predates this problem historically, involves aircraft maintenance, with either quick or slow overhaul of engines.)

3. Suppose that we have two warehouses and two markets. There are 10 spools of wire at the first warehouse and 14 at the second. Moreover, 13 are required at the first market and 11 at the second. If the following matrix gives the transportation costs, find the minimum cost transportation schedule.

Factory

		1	2
Warehouse	1	100	84
	2	69	75

4. Each of the following matrices give the cost c_{ij} of using worker i on job j. Use the Hungarian algorithm to find a minimum cost job assignment.

(a)

	1	2	3	4
1	8	3	2	4
2	10	9	3	6
3	2	1	1	5
4	3	8	2	1

(b)

	1	2	3	4
1	17	5	8	11
2	3	9	2	10
3	4	2	8	6
4	7	6	4	5

(c)

	1	2	3	4	5
1	8	7	5	11	4
2	9	7	6	11	3
3	12	9	4	8	2
4	1	2	3	5	6
5	11	4	2	8	2

5. The following matrix gives the rating r_{ij} of worker i on job j. Use the Hungarian algorithm to find an optimum (that is, highest rated) job assignment.

	1	2	3	4	5
1	8	7	5	9	6
2	11	9	7	4	8
3	12	6	7	5	10
4	9	7	6	9	6
5	3	9	8	9	8

6. A company has purchased five new machines of different types. In its factory, there are five locations where the machines can be located. For a given machine, its location in a particular spot would have an effect on the ease of handling materials. For instance, if the machine is near the work center that produces materials for it, this would be efficient. For machine i at location

j, the hourly cost of handling materials to be brought to the machine can be estimated. The following matrix gives this information.

Location

	1	2	3	4	5
Machine 1	4	6	8	5	7
2	2	4	6	9	5
3	1	7	6	8	3
4	4	6	7	5	7
5	10	4	5	4	5

How would we assign machines to locations so as to minimize the resulting materials handling costs?

7. Consider the transportation problem (Example 13.5).
   (a) Show that if $\sum a_i < \sum b_j$, there is no shipping pattern that meets requirements (13.11) and (13.12).
   (b) If $\sum a_i > \sum b_j$, show that we may as well assume $\sum a_i = \sum b_j$ by creating a new $(m + 1)$st market, setting $b_{m+1} = \sum_{i=1}^{n} a_i - \sum_{j=1}^{m} b_j$, adding arcs from each warehouse to the new market, and letting the capacities of all new arcs be 0.

8. Show that in the Hungarian algorithm, the new reduced matrix is obtained from the old by adding or subtracting a constant from different rows or columns.

9. In a directed network with costs, modify the definition of an augmenting chain to allow it to start at a vertex other than s and end at a vertex other than t, define the cost of an augmenting chain to be the sum of the costs of forward arcs minus the sum of the costs of backward arcs. An *augmenting circuit* is an augmenting chain that forms a circuit.
   (a) In each example for Exercise 1, find a flow of value 3 which *does not* have minimum cost and find a flow augmenting circuit with negative cost.
   (b) Show that if a flow of value v is of minimum cost, it admits no flow augmenting circuit with negative cost.
   (c) Show that if a flow of value v admits no flow-augmenting circuit of negative cost, the flow has minimum cost.

10. We can reduce a transportation problem (Example 13.5) to a maximum weight matching problem (Section 12.1) as follows. Build a bipartite graph $G = (X, Y, E)$ of $2 \sum a_i$ vertices, as follows. Let X consist of a_i copies of the ith warehouse, $i = 1, 2, \ldots, n$, and let Y consist of b_j copies of the jth market, $j = 1, 2, \ldots, m$. G has all possible edges between vertices in X and in Y. On the edge joining a copy of the ith warehouse to a copy of the jth market, place a weight equal to $K - a_{ij}$, where K is sufficiently large. Show that an optimal solution to the transportation problem corresponds to a maximum weight matching in G.

REFERENCES FOR CHAPTER 13

BERGE, C., and GHOUILA-HOURI, A., *Programming, Games and Transportation Networks*, Wiley, New York, 1965.

BONDY, J. A., and MURTY, U. S. R., *Graph Theory with Applications*, American Elsevier, New York, 1976.

BORUVKA, O., "O Jistém Problému Minimálním," *Práce Mor. Prírodoved. Spol. v Brne (Acta Soc. Sci. Nat. Moravicae)*, 3 (1926a), 37–58.

BORUVKA, O., "Príspevek k. Resení Otázky Ekonomické Stavby Elektrovodních Sítí," *Elektrotech. Obzor*, 15 (1926b), 153–154.

CHERITON, D., and TARJAN, R. E., "Finding Minimum Spanning Trees," *SIAM J. Comput.*, *5* (1976), 724–742.

CZEKANOWSKI, J., "Zur Differentialdiagnose der Neandertalgruppe," *Korrespondenzbl. Dtsch. Ges. Anthrop., Ethn. Urg., 40* (1909), 44–47.

CZEKANOWSKI, J., "Objektive Kriterien in der Ethnologie," *Ebenda, 43* (1911), 71–75.

CZEKANOWSKI, J., "Das Typenfrequenzgesetz," *Anthropol. Anz., 5* (1928), 15–20.

DIJKSTRA, E. W., "A Note on Two Problems in Connexion with Graphs," *Numer. Math., 1* (1959), 269–271.

EDMONDS, J., "The Chinese Postman Problem," *Oper. Res., 13*, Suppl. 1 (1965), 373.

EDMONDS, J., and JOHNSON, E. L., "Matching, Euler Tours and the Chinese Postman," *Math. Program., 5* (1973), 88–124.

EDMONDS, J., and KARP, R. M., "Theoretical Improvements in Algorithmic Efficiency for Network Flow Problems," *J. ACM, 19* (1972), 248–264.

ELIAS, P., FEINSTEIN, A., and SHANNON, C. E., "Note on Maximum Flow through a Network," *IRE Trans. Inf. Theory, IT-2* (1956), 117–119.

FORD, L. R., and FULKERSON, D. R., "Maximal Flow through a Network," *Canad. J. Math., 8* (1956), 399–404.

FORD, L. R., and FULKERSON, D. R., "A Simple Algorithm for Finding Maximal Network Flows and an Application to the Hitchcock Problem," *Canad. J. Math., 9* (1957), 210–218.

FORD, L. R., and FULKERSON, D. R., *Flows in Networks*, Princeton University Press, Princeton, N.J., 1962.

FRANK, H., and FRISCH, I., *Communication, Transportation and Flow Networks*, Addison-Wesley, Reading, Mass., 1972.

GOLDMAN, A. J., "Discrete Mathematics in Government," lecture presented at SIAM Symposium on Applications of Discrete Mathematics, Troy, N.Y., June 1981.

GRAHAM, R. L., and HELL, P., "On the History of the Minimum Spanning Tree Problem," mimeographed, Bell Laboratories, Murray Hill, N.J., 1982. (To appear in *Annals of the History of Computing*.)

HU, T. C., *Integer Programming and Network Flows*, Addison-Wesley, Reading, Mass., 1969.

IRI, M., *Network Flows, Transportation and Scheduling*, Academic Press, New York, 1969.

JARNÍK, V., "O Jistém Problému Minimálním," *Práce Mor. Prírodoved Spol. v Brne (Acta Soc. Sci. Nat. Moravicae), 6* (1930), 57–63.

JOHNSON, D. B., "Priority Queues with Update and Minimum Spanning Trees," *Inf. Process. Lett., 4* (1975), 53–57.

KERSHENBAUM, A., and VAN SLYKE, R., "Computing Minimum Spanning Trees Efficiently," ACM 72, *Proceedings of the Annual ACM Conference*, 1972, pp. 518–527.

KRUSKAL, J. B., "On the Shortest Spanning Tree of a Graph and the Traveling Salesman Problem," *Proc. Amer. Math. Soc., 7* (1956), 48–50.

KUHN, H. W., "The Hungarian Method for the Assignment Problem," *Naval Res. Logist. Quart., 2* (1955), 83–97.

LAWLER, E. L., *Combinatorial Optimization: Networks and Matroids*, Holt, Rinehart and Winston, New York, 1976.

MINIEKA, E., *Optimization Algorithms for Networks and Graphs*, Dekker, New York, 1978.

PAPADIMITRIOU, C. H., and STEIGLITZ, K., *Combinatorial Optimization: Algorithms and Complexity*, Prentice-Hall, Englewood Cliffs, N.J., 1982.

PRIM, R. C., "Shortest Connection Networks and Some Generalizations," *Bell Syst. Tech. J., 36* (1957), 1389–1401.

Answers
to Selected Exercises

Chapter 1

1. $\begin{bmatrix} 1 & 2 & 3 & 4 \\ 2 & 3 & 4 & 1 \\ 3 & 4 & 1 & 2 \\ 4 & 1 & 2 & 3 \end{bmatrix};$ **4(b).** $\begin{bmatrix} 1 & 2 & 3 & 4 \\ 2 & 1 & 4 & 3 \\ 3 & 4 & 1 & 2 \\ 4 & 3 & 2 & 1 \end{bmatrix} \begin{bmatrix} d & a & b & c \\ c & b & a & d \\ a & d & c & b \\ b & c & d & a \end{bmatrix};$ **7(b).** yes;

12(b). time 1: English and physics; time 2: calculus; time 3: history.

Chapter 2

Section 2.1. **1.** Yes: $17,576 < 35,000$; **4.** $8^3 \times 10^5$; $8 \times 2 \times 10 \times 8^3 \times 10^5$; **8.** $10^6 - 9^6$; **11.** $2^{2^{n-1}}$; **13(b).** $2^{m-1}2^{p-m}$.

Section 2.2. **2.** 230; **5.** $26^4 + 25^5$.

Section 2.3. **2.** 24; **4(b).** 710; **6(a).** 36.

Section 2.4. **2.** 4.9×10^6; **6.** best order is 2, 3, 1; **8(b).** $\dfrac{n+1}{2} \times 3 \times 10^{-11}$.

Section 2.5. **2(b).** 210; **2(c).** 49; **5(a).** $9 \times 9 \times 8 \times 7$.

Section 2.6. **3.** $2^{10} - 1$; **5(a).** 2^{2^3}.

Section 2.7. **2.** $C(50, 7)$; **3(b).** 35; **8(a).** $C(8, 4)$; **8(b).** $2^8 - 2$; **11.** $C(21, 5)C(5, 3)8! + C(21, 4)C(5, 4)8! + C(21, 3)C(5, 5)8!$; **15(a).** $[C(3, 1)C(27, 2) + C(3, 2)C(27, 1) + C(3, 3)C(27, 0)]$ $\times [C(12, 1)C(138, 2) + C(12, 2)C(138, 1) + C(12, 3)C(138, 0)].$

Section 2.8. **1.** 1 7 21 35 35 21 7 1;
5. use repeated applications of Theorem 2.2;   **8(a).** in 1, 1, 1, t could be 0, 1, or 2.

Section 2.9. **1(b).** No;   **3(a).** $\frac{3}{8}$;   **6.** $\frac{1}{13}$;   **9.** $\frac{5}{8}$;   **11.** $\frac{1}{4}$;   **14(b).** $C(6, 2)/2^6 + C(6, 4)/2^6$.

Section 2.10. **2(a).** $P^R(2, 4) = 2^4$;   **2(c).** $C^R(2, 4) = 5$;   **5.** $C^R(5, 12) = 1820$.

Section 2.11. **3(a).** $2^3 = 8$;   **3(d).** $C(4, 3) = 4$;   **4(a).** $S(3, 1) + S(3, 2) = 4$;
4(d). Number of partitions of 3 into two or fewer parts $= 2$;   **5(a).** $2!S(3, 2) = 6$;
5(d). $C(2, 1) = 2$;   **6(a).** $S(3, 2) = 3$;   **6(d).** Number of partitions of 3 into exactly two parts $= 1$;
9(b). 1;   **13.** $C(35, 30)$;
16. $4!S(6, 4)$ assuming both jobs and workers are distinguishable;   **18.** $1/C(11, 8)$.

Section 2.12. **1(a).** 630;   **1(d).** 90/729;   **5(a).** $C(6; 3, 1, 2)$;
8. $C(11; 3, 2, 1, 1, 1, 1, 1, 1)$;   **9.** $4C(10; 4, 2, 2, 2) + C(4, 2)C(10; 3, 3, 2, 2)$;
12(a). $C(4; 2, 1, 1)$;   **15(b).** 4.

Section 2.13. **1(a).** 6;   **1(b).** 60;   **1(c).** CAAGCUGGUC;   **2.** $C(10; 2, 3, 3, 2)$;
10. yes: 00101010 and 01001010 have same break-up.

Section 2.14. **1.** $8/C(12, 5)$;   **6(a).** $7/4^4$;   **7.** $C(17; 13, 4)$.

Section 2.15. **1(b).** $a^3 + 6a^2b + 12ab^2 + 8b^3$;
3. $C(12, 10)C(4, 0) + C(12, 9)C(4, 1) + \cdots + C(12, 6)C(4, 4)$;   **7.** 20;   **14.** 5^n;
18. *Hint:* Start with $(x + 2)^n$ and differentiate.

Section 2.16. **1(a).** $\{1, 2\}, \{1, 3\}, \{1, 2, 3\}$;
2(a). $\{1, 2, 5\}, \{1, 3\}, \{2, 3\}, \{1, 4\}, \{2, 4\}, \{3, 4\}, \{4, 5\}$;   **3(b).** $\frac{1}{5}$ for each;   **6(a).** yes;
10. nonpermanent: $\dfrac{9!9!}{3!16!}$.

Section 2.17. **1(a).** 2341 precedes 3412;   **2(c).** 124635;   **5(b).** 3.

Section 2.18. **1(b).** yes;   **1(d).** yes;   **3.** $f \leq k_1 g \leq k_1 k_2 h$;
7(a). yes;   **8(a).** yes;   **8(f).** yes;   **8(i).** yes;
11. The third is "big oh" of the other two; the second is "big oh" of the first.

Additional Exercises for Chapter 2.   No answers provided.

Chapter 3

Section 3.1. **1.** $V = \{$Chicago (C), Springfield (S), Albany (A), New York (N), Miami (M)$\}$;
$A = \{$(C, S), (S, C), (C, N), (N, C), (A, N), (N, A), (C, M), (M, C), (N, M), (M, N)$\}$;
4(b). In G_3, $E = \{\{u, v\}, \{v, w\}, \{u, w\}, \{x, y\}, \{x, z\}, \{y, z\}\}$;
10(a). yes;   **10(b).** no;   **12.** 15;   **15.** yes;   **17.** yes.

Section 3.2. **1(d).** yes;   **5.** D_4: no, D_8: yes;   **6(a).** no;   **6(e).** yes;
9(b). D_4: yes, D_8: yes;   **10(c).** D_4: yes, D_8: yes;
13(b). $\{1, 2, 3, 4\}, \{5\}, \{6\}, \{7\}, \{8, 9, 10\}, \{11\}, \{12\}$;
18. *Hint:* Use induction on the number of vertices;   **23(a).** yes;   **28(a).** 9;   **28(b)** $(n - 1)^2$.

Section 3.3. **3(b).** yes;   **4(b).** 3;   **6.** no;   **9.** 4;   **14.** no;   **18(b).** yes;
21. $\chi \geq \omega$;   **28(a).** 4;   **32(b).** Z_n, n odd, $n \geq 3$;   **33(a).** $\omega(G) = \alpha(G^c)$;
34(c). Z_5 plus a vertex adjacent to two consecutive vertices of Z_5 .

Section 3.4. **1(a).** $x(x-1)^3$; **2(a).** 24; **6.** $[P(I_2, x) - P(I_1, x)][P(I_1, x) - 2]^2$;
9(a). $5 \cdot 4!$; **11.** 2; **13(a).** 0; **15(d).** $P(x) \neq x^n$ and the sum of the coefficients is not zero;
20(b). yes; **25(c).** $(-1)^{n-1}(n-1)!$

Section 3.5. **3(a).** 11; **7.** $n - k$; **10.** 16;
15. There are too few edges to have a spanning tree; alternatively, the deleted edge was the only simple chain between its end vertices;
18. 2 if $n \geq 2$, since $2(2-1)^{n-1} > 0$, $1(1-1)^{n-1} = 0$; 1 if $n = 1$;
23. *Hint:* The sum of the degrees is $4k + m$ and we have a tree;
26(b). yes; **28(a).** 6; **29(b).** $\binom{8}{2}6! = 20,160$.

Section 3.6. **2(a).** $a{:}0$, b, $c{:}1$, d, e, f, $g{:}2$, h, i, j, $k{:}3$; **3(a).** 3; **4(a).** $\{d, e, h, i\}$;
6(a). The sons of vertex 1 are 2 and 3 and the sons of vertex 2 are 4 and 5; **10(a).** 6;
13. $[1 \cdot 2^0 + 2 \cdot 2^1 + 3 \cdot 2^2 + \cdots + (h+1)2^h]/n$, where h is the height;
17. 3 4 1 2, 3 1 4 2, 3 1 2 4, 1 3 2 4, 1 2 3 4, 1 2 3 4; **22.** 10; **28(b).** 240.

Section 3.7. **1(a).** For D_1:

$$
\begin{array}{c c}
 & \begin{array}{cccc} u & v & w & x \end{array} \\
\begin{array}{c} u \\ v \\ w \\ x \end{array} &
\left[\begin{array}{cccc}
0 & 1 & 1 & 0 \\
0 & 0 & 1 & 0 \\
0 & 0 & 0 & 1 \\
1 & 0 & 0 & 0
\end{array}\right];
\end{array}
$$

2(h).

$$
\begin{array}{c c}
 & \begin{array}{ccccc} \{a, b\} & \{b, c\} & \{a, d\} & \{b, e\} & \{d, e\} \end{array} \\
\begin{array}{c} a \\ b \\ c \\ d \\ e \end{array} &
\left[\begin{array}{ccccc}
1 & 0 & 1 & 0 & 0 \\
1 & 1 & 0 & 1 & 0 \\
0 & 1 & 0 & 0 & 0 \\
0 & 0 & 1 & 0 & 1 \\
0 & 0 & 0 & 1 & 1
\end{array}\right];
\end{array}
$$

7. for G_1:

$$
\begin{array}{c c}
 & \begin{array}{ccc} u & v & w \end{array} \\
\begin{array}{c} u \\ v \\ w \end{array} &
\left[\begin{array}{ccc}
0 & 1 & 0 \\
1 & 0 & 1 \\
0 & 1 & 0
\end{array}\right];
\end{array}
$$

11. for D_4:

$$
\begin{array}{c c}
 & \begin{array}{cccccc} u & v & w & x & y & z \end{array} \\
\begin{array}{c} u \\ v \\ w \\ x \\ y \\ z \end{array} &
\left[\begin{array}{cccccc}
1 & 1 & 1 & 1 & 1 & 1 \\
1 & 1 & 1 & 1 & 1 & 1 \\
1 & 1 & 1 & 1 & 1 & 1 \\
0 & 0 & 0 & 1 & 1 & 1 \\
0 & 0 & 0 & 1 & 1 & 1 \\
0 & 0 & 0 & 1 & 1 & 1
\end{array}\right];
\end{array}
$$

17(a). There is a path from i to j if and only if there is a path of length at most $n - 1$;
20(a). j is in the strong component containing i if and only if $r_{ij} = 1$ and $r_{ji} = 1$;
26. it gives the number of vertices i and j have in common;
29. yes: take Z_4 as in Figure 3.24 and append x adjacent to a and b and y adjacent to b and c; repeat with Z_4 and x as above, but take y adjacent to c and d; relabel edges.

Chapter 4

Section 4.1. **1(b).** $\sum\limits_{k=0}^{\infty} \dfrac{2^k x^k}{k!}$; **1(d).** $1 + 2x + \sum\limits_{k=2}^{\infty} \dfrac{x^k}{k!}$; **2(b).** $\sum\limits_{k=0}^{\infty} x^{k+3}$;

2(c). $x^3 - \dfrac{x^9}{3!} + \dfrac{x^{15}}{5!} - \cdots$; **3(a).** $1 + x + x^2$; **3(d).** $-x + \dfrac{1}{1-x}$;

3(j). $-1 - x + e^x$; **4(c).** $(0, 0, 0, 0, 0, 1, 1, 1, \ldots)$; **4(h).** $\left(2, 1, \dfrac{1}{2!}, \dfrac{1}{3!}, \ldots, \dfrac{1}{k!}, \ldots\right)$;

4(k). $(1, 0, 1, 0, 1, 0, \ldots)$; **5(a).** 1; **6(a).** 0; **9.** 3^n;
11(b). $1 + 8x + 12x^2$; **12(b).** $1 + 16x + 72x^2 + 96x^3 + 24x^4$.

Section 4.2. **1(a).** $\left(0, 0, 1, 0, -\dfrac{1}{3!}, 0, \dfrac{1}{5!}, 0, \cdots\right)$;

1(b). $(1, -\frac{1}{2}, \frac{1}{3}, -\frac{1}{4}, \ldots)$; **1(e).** $(3, 10, 6, 2, 2, 2, \ldots)$;

2(a). $(0, 1, 2, 3, \ldots)$; **3(b).** $a_k = \sum\limits_{i=0}^{k} i$; **4(a).** $a_k = 6(k+1)$;

7(a). $(1, 3, 4, 8, 16, \ldots)$; **8(a).** $A(x) + (7 - a_2)x^2$;

10(a). $\dfrac{1}{(1-x)^2} + \dfrac{1}{1-x} = \dfrac{2-x}{(1-x)^2}$; **11(b).** $\dfrac{2x}{(1-x)^3}$;

17. $R(x, B) = 1 + 6x + 11x^2 + 8x^3 + 2x^4$.

Section 4.3. **1(a).** $(1 + x + x^2)^2(1 + x + x^2 + x^3)^2$, coefficient of x^5;
1(e). $(1 + x + x^2 + x^3 + x^4 + x^5)^2(x^3 + x^4 + x^5 + x^6 + x^7)$, coefficient of x^{10};
2. 8; **8.** $(1 + x + x^2 + x^3 + x^4)^3(1 + x + x^2 + x^3)(1 + x + \cdots + x^8)(1 + x + \cdots + x^{13})$;
12. $(1 + x)(1 + x^2)(1 + x^3) \cdots (1 + x^k)$.

Section 4.4. **1(b).** -4; **2(a).** 56; **3.** $-\dfrac{1}{9}$; **6.** $\dbinom{16}{12}$; **11.** 5^7;

13(b). $\dbinom{p + k - 1}{k}$; **16(b).** $\dbinom{n-1}{r-1}$.

Section 4.5. **1(b).** e^{5x}; **1(c).** $e^x - x$; **2(a).** $a_k = 3k!$; **2(f).** $a_k = 1 + 4^k$;

6(a). $\left(1 + x + \dfrac{x^2}{2!} + \dfrac{x^3}{3!}\right)(1 + x)^3$, coefficient of $\dfrac{x^3}{3!}$;

6(d). $(1 + x + x^2 + \cdots + x^{2n})^3$, coefficient of x^{3n};
10. $(e^x - 1)^p$; **12.** $\frac{1}{2}[5^k + 3^k]$;

17(b). $\left[\dfrac{3^0}{0!} + \dfrac{3^1}{1!} + \dfrac{3^2}{2!} + \dfrac{3^3}{3!}\right]3!$.

Section 4.6. **1(a).** $G(x) = \frac{1}{3} + \frac{1}{3}x + \frac{1}{3}x^2$, $E = 1$, $V = \frac{2}{3}$;

3(a). $\dfrac{px}{1 + px - x}$; **4(b).** $E = \dfrac{qm}{p}$, $V = \dfrac{q^2 m}{p^2} + \dfrac{qm}{p}$.

Section 4.7. **1(d).** Coleman: $0, \frac{4}{8}, \frac{4}{8}, \frac{4}{8}$; Banzhaf: $0, \frac{4}{12}, \frac{4}{12}, \frac{4}{12}$;
2(b). $[5; 4, 2, 1, 1]$; **4(c).** $\frac{5}{12}, \frac{3}{12}, \frac{3}{12}, \frac{1}{12}$.

Chapter 5

Section 5.1. **1.** $a_5 = 485$; **2.** $s_4 = 1400$; **3.** $s_4 = 1464.10$; **4(b).** 528;
12(a). $7! - D_7$; **12(b).** 1; **17(a).** $(D_5)^2$; **17(b).** $(5!)^2$;
22. $f(n + 1) = (2n + 1)f(n)$; **25.** $f(n + 1) = f(n) + 2n$, $n \geq 1, f(1) = 2$;
29(a). $n!$; **31.** $C_3 = 4$.

1(a). linear;   **1(e).** not linear;

2(a). not homogeneous;   **2(e).** homogeneous;

3(a). has constant coefficients;   **3(f).** does not have constant coefficients;

4(a). $x^2 - 2x + 1 = 0$;   **4(f).** $x^2 + x - 2 = 0$;   **5(a).** 1, 1;   **5(i).** 1, 2, 3;   **9(a).** a solution;

10(c). not a solution;   **11(a).** $a_n = 1 + n$;   **11(e).** $h_n = \frac{11}{2} \cdot 2^n + \frac{9}{2} \cdot (-2)^n$;

15(a). $a_n = \lambda_1(2i)^n + \lambda_2(-2i)^n$;   **15(b).** $a_n = (-\frac{5}{2}i)(2i)^n + (\frac{5}{2}i)(-2i)^n$;

19. 12 is a characteristic root of multiplicity 3;   **23(a).** $a_n = \frac{7}{8}n(2^n) + (-\frac{3}{8})n^2(2^n)$;

24(a). $a_n = 5^n - \frac{3}{5}n5^n$.

Section 5.3.   **2(a).** $a_k = 2k + 1$;   **3(a).** $a_n = n + 1$;

3(d). $G(x) = \dfrac{2x + 1}{1 + x - 2x^2}$;   **4(a).** $2 \cdot 4^k - 2^k$;   **4(e).** $1 - \left(\dfrac{1}{2}\right)^k$;

9. generating function is $G(x) = \dfrac{x + 3}{-x^2 - x + 1}$,

$$a_k = \left(\frac{3 + \sqrt{5}}{2}\right)\left(\frac{1 + \sqrt{5}}{2}\right)^k + \left(\frac{3 - \sqrt{5}}{2}\right)\left(\frac{1 - \sqrt{5}}{2}\right)^k;$$

15. $G(x) = \dfrac{2(x - 2)}{(x - 1)(x^2 - 2x + 2)}$;   **19.** $G(x) = \dfrac{x^3}{(1 - 2x)(x^2 + 1)}$;

27. $1 + 7x + 13x^2 + 6x^3$.

Section 5.4.   **2.** $u_5 = 42$;   **4.** $R_4 = 4$;   **5.** $h_4 = 36$;

7(a). $C(x) - x^2 = A(x) + B(x) - x$;   **7(b).** $A(x) = 4xC(x)$;

7(c). $B(x) = x\{[C(x)]^2 + 1\}$;   **7(d).** $C(x) - x^2 = 4xC(x) + x[C(x)]^2$;

12(a). $q_3 = 5$;   **15(i).** $B(x) = x[U(x)]^2$;   **18(c).** $A(x)B(x) = \dfrac{x^3}{1 - 2x}$.

Section 5.5.   **1(a).** $a_n = \frac{15}{2}n^{\log_4 5} - \frac{3}{2}$;

6(b). use $b_k < n \le b^{k+1}$ and $f(n) \le f(b^{k+1})$ to show that $f(n) \le c[\log_b n + 2]$;

8. $f(n) \le f(\frac{n}{2}) + 1$, $n = 2^k$, so $f(n) \le \log_2 n + 2$;

11(a). $f(n) \le (c + 1)n \log_2 n + (c + 1)n$.

Chapter 6

Section 6.1.   **1.** 65;   **7.** 15;   **10.** 6233;

13. $7^8 - \binom{7}{1}(7 - 1)^8 + \binom{7}{2}(7 - 2)^8 \mp \cdots + \binom{7}{6}(7 - 6)^8 - 0$;

16(d). $P(G, x) = x^4 - 4x^3 + 5x^2 - 2x$;   **19(a).** 6;   **24(b).** 3 times;   **26.** 21;

32. $b_n = n! - \binom{n-1}{1}(n - 1)! + \binom{n-1}{2}(n - 2)! - \binom{n-1}{3}(n - 3)! \pm \cdots + (-1)^{n-1}\binom{n-1}{n-1}1!$

Section 6.2.   **1.** 22%;   **4.** 7;   **8.** $\frac{1}{2}$;   **13(a).** 315;   **17.** 7/24;

21. 128;   **31.** 9.

Chapter 7

Section 7.1.   **1(c).** transitivity;   **1(h).** symmetry;   **3(a).** $\{1, 2\}, \{3, 4\}$;

7. $\{bb\}, \{rr\}, \{pp\}, \{bp, pb\}, \{br, rb\}, \{pr, rp\}$;   **11(a).** there are 3 others;   **14.** 3.

Section 7.2.   **1(a).** $\begin{pmatrix} 1 & 2 & 3 & 4 & 5 & 6 & 7 \\ 1 & 2 & 4 & 6 & 7 & 5 & 3 \end{pmatrix}$;   **2(a).** $\begin{pmatrix} 1 & 2 & 3 & 4 \\ 3 & 2 & 4 & 1 \end{pmatrix}$;

5(a). all;   **5(f).** G1, G2;   **8(a).** no;   **10(a).** $\{1, 5\}, \{2, 4\}, \{3\}$;

11(a). reflection in a diagonal from upper right to lower left;

12(a). $\begin{pmatrix} 1 & 2 & 3 & 4 \\ 4 & 3 & 2 & 1 \end{pmatrix}$; **18(a).** $\left\{ \begin{pmatrix} 1 & 2 & 3 & 4 \\ 1 & 2 & 3 & 4 \end{pmatrix}, \begin{pmatrix} 1 & 2 & 3 & 4 \\ 4 & 3 & 2 & 1 \end{pmatrix} \right\}$.

Section 7.3. **3(a).** 3: $\{1, 5\}, \{2, 4\}, \{3\}$;

4(a). (i) St $(1) = \left\{ \begin{pmatrix} 1 & 2 & 3 & 4 & 5 \\ 1 & 2 & 3 & 4 & 5 \end{pmatrix} \right\}$, **(ii)** $C(1) = \{1, 5\}$; **8(b).** 12.

Section 7.4. **2.** 3^8; **4(a).** black; **5(a).** r;
6(a). no; **6(g).** 8; **7(a).** no; **7(f).** 4; **10(e).** 6;

16(c). if π_i is $\begin{pmatrix} \{1, 2\} & \{1, 3\} & \{2, 3\} \\ \{1, 2\} & \{2, 3\} & \{1, 3\} \end{pmatrix}$, then Inv (π_i^*) is 4; **20.** 24.

Section 7.5. **1(a).** $(1)(2)(34657)$; **3(a).** cyc (π) is 5 and 3, respectively;
4(a). x_1^5 and $x_1 x_2^2$, respectively. **5(a).** $\frac{1}{2}(x_1^5 + x_1 x_2^2)$; **9(a).** 10; **12(a).** 6;
20(b). x_2^4; **24.** $(12)(13)(14)(15)(16)$.

Section 7.6. **1.** The second has weight 54; **2(a).** g of part (a) has weight $v^2 u^2$;
4. $3a^3b + 2ab^3$; **8.** 3280; **13.** 11; **16(e).** $1 + x$;
17(c). if the entries of (x_1, x_2, x_3) give the colors of a, b, c respectively, then equivalence classes are
$\{(0, 0, 0)\}, \{(0, 0, 1)\}, \{(0, 1, 0), (1, 0, 0)\}, \{(0, 1, 1), (1, 0, 1)\}, \{(1, 1, 0)\}, \{(1, 1, 1)\}$.

Chapter 8

Section 8.1. **1(a).** 27; **2(a).** 53; **9.** largest has at least 15, smallest at most 14;
11. for graph (a), $\alpha = 2$, $\chi = 4$;
14(a). longest increasing is 5, 8 or 3, 8 and longest decreasing is 5, 3, 2, 1;
17. let the pigeons be the 81 hours and the five holes be days 1 and 2, 3 and 4, 5 and 6, 7 and 8, and
9 and 10;
25. if there are n people, each has at most $n - 1$ acquaintances.

Section 8.2. **1(a).** $X = \{\{a, b\}, \{b, c\}, \{c, d\}\}$, $Y = \{\{a, c\}, \{b, d\}, \{a, d\}\}$;
2(a). 2; **4(a).** $\{a, e, c\}$ is independent; **8.** $m = 6$; **13.** p;
18. take $|V(G)| \geq R(3, 3, 3, 12; 2)$; **23.** 19; **26(a).** 4.

Section 8.3. **2(a).** $\{b, c, d, e\}$; **3.** Take $n \geq R(5, 5)$; **8(a).** 5; **9(a).** 5;
13(a). use each color twice; **16(a).** $\sigma(2, 1)$ consists of the sets $\{1, 3\}, \{1, 4\}, \{2, 4\}$;
18(a). $\sigma(2, 1, 3)$ consists of the set $\{1, 3, 5\}$; **21(a).** no; **21(c).** yes;
25(a). $f_1(a) = 6, f_1(b) = 5, f_1(c) = 4, f_1(d) = 3, f_1(e) = 2, f_1(f) = 1, f_2(a) = 2, f_2(b) = 1, f_2(c) = 4,$
$f_2(d) = 3, f_2(e) = 6, f_2(f) = 5$;
26(b). 2; **31(b).** digraph (c) of Figure 8.24: yes; digraph (c) of Figure 8.25: no.

Chapter 9

Section 9.1. No exercises.

Section 9.2. **1(a).** yes; **3(a).** yes; **4(a).** yes; **5(a).** cannot be sure;
5(c). cannot be sure; **8.** yes; **11(a).** no; **11(b).** it is at most 6.

Section 9.3. **1(a).** 2; **2(a).** $a + b = 1, a \times b = 0$; **5(c).** no;

6(a).

+	0	1	2	3	4
0	0	1	2	3	4
1	1	2	3	4	0
2	2	3	4	0	1
3	3	4	0	1	2
4	4	0	1	2	3

×	0	1	2	3	4
0	0	0	0	0	0
1	0	1	2	3	4
2	0	2	4	1	3
3	0	3	1	4	2
4	0	4	3	2	1

;

7(b). 2; **9(a).** 3; 7; **11(a).** no.

Section 9.4. **1(a).** not a BIBD; **2(a).** $b = 21, r = 14$; **3(a).** $bk \neq vr$; **7.** no: $b \geq v$ fails; **9(a).**

$$
\begin{array}{c c}
 & \begin{array}{cccc} \{1, 2\} & \{1, 3\} & \{2, 3\} & \{1, 2, 3\} \end{array} \\
\begin{array}{c} 1 \\ 2 \\ 3 \end{array} &
\left[\begin{array}{cccc}
1 & 1 & 0 & 1 \\
1 & 0 & 1 & 1 \\
0 & 1 & 1 & 1
\end{array} \right] ;
\end{array}
$$

10(a).

$$
\left[\begin{array}{cccc}
3 & 2 & 2 & 2 \\
2 & 3 & 2 & 2 \\
2 & 2 & 3 & 2 \\
2 & 2 & 2 & 3
\end{array} \right] ;
$$

13. 651; **16.** $b = 26, v = 13, r = 20, k = 10, \lambda = 15$; **21(a).** this is a $(4m - 1, 2m - 1, m - 1)$-design, $m = 2^3$; **25(a).** no: $k - \lambda$ is not a square; **31.** take two copies of each block of a $(31, 15, 7)$-design.

Section 9.5. **1(d).** (P3); **2(a).** There are 7 distinct points, no 3 of which lie on the same line; **4.** 21; **8(a).** no; **9(a).** $v = 31, k = 6, \lambda = 1$; **10(a).** yes (Corollary 9.24.1); **13(a).** yes (but cannot be sure); **15(a).** 1; **16(a).** 11; **19(b).** if we take $U_3 = \{1, 3, 5, 7\}$, $V_2 = \{2, 3, 4, 13\}$, $W_{11} = \{3, 6, 8, 11\}$, $W_{21} = \{3, 9, 10, 12\}$, then the point 3 is associated with $(3, 2)$ and $(3, 2, 1, 1)$; **19(c).** $a_{32}^{(1)} = 1$, $a_{32}^{(2)} = 1$; **20(a).** $(2, 3)$ is associated with $(2, 3, 1, 2)$; **20(b).** $W_{12} = \{(1, 2), (2, 1), (3, 3)\}$; **20(e).** W_{12} is now $\{(1, 2), (2, 1), (3, 3), w_1\}$, the finite points are all (i, j) with $1 \leq i, j \leq 3$, and the infinite points are u, v, w_1, w_2; **20(f).** $m^2 + m + 1$ lines, including the line at infinity.

Chapter 10

Section 10.1. No exercises.

Section 10.2. **1(a).** (49, 73); **1(d).** 39, 58, 29, 34, 17, 19, 22, 27, 41, 61, 44, 59, 43, 57; **2(a).** (8, 14, 14); **5(a).** AB; **5(c).** error; **6(a).** AZZ; **6(d).** DELETE; **9(a). (i)** 101; **9(b). (ii)** 0110.

Section 10.3. **1(a).** 3; **2(a).** detect 2, correct 1; **3(a). (i)** 4; **4.** $2^8/37$; **7(b).** $100(.1)(.9)^{99}$; **10.** probability of 0 errors is .73, of 0 or 1 errors is .97, so $d = 3$; **14(b).** 1 error; **17(b).** 2 errors (use horizontal completeness).

Section 10.4. **1(a).** 11110; **2(b).** 1011; **4(a).** 1111111;

5(b). the $3 \rightarrow 6$ double repetition code; **7(b)** $.972$; **9(a).** 2;

10(a). $\begin{pmatrix} 1 & 1 & 1 & 0 \\ 1 & 0 & 0 & 1 \end{pmatrix}$; **11(a).** $\begin{bmatrix} 1 & 0 & 0 & 1 & 1 & 0 \\ 0 & 1 & 0 & 1 & 0 & 1 \\ 0 & 0 & 1 & 0 & 1 & 1 \end{bmatrix}$;

12(a). $x_3 = x_1 + x_2, x_4 = x_1$; **14(a).** no error; **15(c).** yes—an error;
17(b). 111000; **18(c).** not possible; **20(a).** 1001100;
24. 111 if $p = 2$; **32.** use $|C| = 2^{2^p - 1 - p}, t = 1$; **34(a).** (i) 1401;
34(b). (ii) 102102; **35(b).** $d(\mathbf{x}, \mathbf{y}) = \text{wt } (\mathbf{x} - \mathbf{y})$.

Section 10.5. **2(e).** $2(r - \lambda) - 1$; **4(a).** no; **5.** yes; **9(b).** 7; **14(a).** $2m$;
14(b). $4m - 1$ if $i = j, 2m - 1$ otherwise; **20(b).** no;

24. $\mathbf{M} = \begin{pmatrix} 1 & 0 & 1 \\ 0 & 1 & 1 \end{pmatrix}$; **27(a).** the distance between codewords of equal weight is even.

Chapter 11

Section 11.1. **1(a).** One example: assign labels $a = 1, e = 2, j = 3, k = 4, f = 5, g = 6, h = 7$,
$l = 8, m = 9, i = 10, d = 11, c = 12, b = 13$ and mark the edges $\{a, e\}, \{e, j\}, \{j, k\}, \{k, f\}, \{f, g\}$,
$\{g, h\}, \{h, l\}, \{l, m\}, \{m, i\}, \{i, d\}, \{d, c\}, \{c, b\}$;
3(a). connected; **4(a).** one example: use the marked edges in answer to 1(a);
6. it is a spanning forest; **8(a).** no; **10.** yes.

Section 11.2. **1(a).** none; **1(f).** $\{c, e\}$;
2(a). orientation based on answer to Exercise 1(a), Section 11.1 orients 1 to 2 to 3 ... to 13 and all
other edges from higher number to lower number;
7. $V = \{x, a, b, c\}, E = \{\{x, a\}, \{x, b\}, \{x, c\}\}$; **8(a).** none; **8(f).** c, e;
14(a). for digraph (a): essentially the only orientations are: (1) which uses arcs $(a, b), (b, c), (c, f)$,
$(f, e), (e, d), (d, a)$, and (b, e), or (2) which uses arcs $(a, b), (b, e), (e, d), (d, a), (e, f), (f, c)$, and (c, b);
both are equally efficient;
15(a). for digraph (a): orientation (2) above is best;
16(a). for digraph (a): orientation (2) above is best;
17(a). for digraph (a): orientation (2) above is best;
18. D_1: category 3, D_4: category 2;
19. if $V = \{a, b, c, d\}$ and $A = \{(a, b), (b, c), (c, d), (d, a)\}$, any arc is (3, 2);
20. in previous example, any vertex is (3, 2); **21(a).** 1.

Section 11.3. **2.** G_1: $a, b, e, f, b, c, d, c, d, f, g, a$; G_3: none;
3. G_3: none; **4.** D_1: $a, b, d, c, d, f, e, c, a$; D_2: none;
5. D_2: c, a, b, d, f, e, c, d; **10(a).** yes; **14.** no; consider D_2 of Figure 11.23.

Section 11.4. **2(a).** add edge $\{c, f\}$; **5(a).** CAAGCUGGUC;
6(a). yes; $A_1 A_1 A_2 A_2 A_2 A_3 A_3 A_3 A_3 A_3 A_2 A_1$;
11(a). say B is an interior extended base of a U, C (G) fragment; then both B and the preceding
extended base end in G (U, C), so B is on the second list;
12. ends in A; **14(b).** if there is a second abnormal fragment, it is B alone.

Section 11.5. **1(a).** a, b, d, f, e, c, a; **2(a).** e, c, d, a, b;
3(a). a, b, c, d, e, f, a; **4(a).** a, d, b, c; **9(a)** yes;
10. for (a) of Figure 11.42: complete graph;
11. for (a) of Figure 11.42: yes; **14.** for (a) of Figure 11.44: no;
15. for (a) of Figure 11.44: no; **16.** Use Theorem 11.7.

Section 11.6. **1(a).** One is 1, 2, 3, 4, 5; **4.** i beats j iff $i < j$;
5(a). $a = 1, b = 2, c = 3, d = 4, e = 5, f = 6$; **6.** 3, 1, 2;
9. if C is a, b, f, d, c, e, a, then H has edges $\{b, d\}$ to $\{e, f\}$ and $\{a, c\}$ to $\{e, f\}$ and is 2-colorable;
12(b). (0, 1, 2, 3, 4); **14(a).** no; **20(a).** $(0, 1, 2, \cdots, n - 1)$;
22. start with any vertex x and find the longest simple path heading into x; this must start at a vertex with no incoming arcs.

Chapter 12

Section 12.1. **1.** $\{a, \alpha\}, \{b, \beta\}, \{c, \gamma\}, \{d, \delta\}$ does all; **3(a).** $\{a, b\}, \{c, d\}, \{e, f\}$; **5.** find a minimum weight matching; **7(b).** put a very small weight on all edges not in the graph.

Section 12.2. **1(a).** {Cutting, Shaping, Polishing, Packaging};
2(a). {Smith, Jones, Black}; **3(a).** no; **5(a).** (a, a, a, a, b, d, a); **6(a).** no SDR;
12(a). yes; **15.** yes: show each likes exactly p; **19.** no.

Section 12.3. **2(a).** yes; **2(f).** no.

Section 12.4. **1(a).** $\{1, 2, 3, 4, 5\}$; **1(b).** $\{1, 2, 4, 5\}$;
6(a). minimum $\{2, 6\}, \{3, 5\}, \{1, 4\}$; **9.** minimum edge covering;
11. If I is independent, $V - I$ is a vertex cover and if K is a vertex cover, $V - K$ is independent;
13. $|M^*| \le |K^*|$ implies $|I| \le |I^*| \le |F^*| \le |F|$.

Section 12.5. **1(a).** 8, $\{8, 9\}$, 9, $\{9, 6\}$, 6; **1(b).** 8, $\{8, 9\}$, 9, $\{9, 6\}$, 6, $\{6, 3\}$, 3;
1(c). add edges $\{8, 9\}$ and $\{6, 3\}$, delete edge $\{9, 6\}$;
6(a). pick vertex 8, place edge $\{5, 8\}$ in T, note 8 is unsaturated, and note 5, $\{5, 8\}$, 8 is an M-augmenting chain;
9(a). use edges $\{6, 7\}, \{7, 11\}, \{11, 9\}$.

Section 12.6. **1(a).** $\delta(G) = 1, m(G) = 3$; **2(b).** $2p \le 3 | N(S)|$;
2(c). $|S| - |N(S)| \le p - \frac{2}{3}p$; **2(d).** $m(G) = |X| - \delta(G) \ge 9 - \frac{1}{3}(9) = 6$.

Chapter 13

Section 13.1. **1(a).** Add edges in the order $\{b, e\}, \{d, e\}, \{a, e\}, \{c, d\}$;
2(a). add edges $\{a, e\}, \{e, b\}, \{e, d\}, \{d, c\}$;
3(a). terminate with message disconnected; T ends up with 5 edges;
6(a). edges $\{b, c\}, \{c, e\}, \{c, d\}, \{a, b\}$; **9(a).** edges $\{a, e\}, \{a, b\}, \{c, d\}, \{d, e\}$;
13(a). for network (c): G' has edges $\{a, b\}, \{c, g\}, \{d, a\}, \{e, f\}, \{f, g\}$; in the next iteration we add $\{d, f\}$ and obtain a minimum spanning tree.

Section 13.2. **3(a).** We successively add to W: a, b, d, c, e, z, obtaining the path a, d, e, z;
5. grind, weigh, polish, inspect; **8.** $(n - 2)!$;
10. draw a network whose vertices are vectors (x_1, x_2, x_3) with $x_i =$ number of gallons in jug i, and find a shortest path from (8, 0, 0) to (4, 4, 0);
13. let d_{ij} be the distance from i to j and let $\mathbf{A} = $ adjacency matrix; then d_{ij} is smallest k such that the i, j entry of $\mathbf{A}^k$ is nonzero.

Section 13.3. **1.** for network (c): **(a)** feasible, **(b)** value 4; **2(a).** 8;
3(a). $s_{sa} = 3, s_{sb} = 0, s_{ba} = 1, s_{at} = 0, s_{bt} = 2$; **4(a).** yes;
5(b). no; **6(a).** augmenting chain s, a, b, t;
13(b). for graph (a), let the flow be 1 on arcs $(s, a), (s, b), (s, d), (a, \alpha), (b, \beta), (d, \gamma), (\alpha, t), (\beta, t), (\gamma, t)$, and 0 otherwise; the matching is $\{a, \alpha\}, \{b, \beta\}, \{d, \gamma\}$;
13(c). the (s, t) cut $S = \{s, c, d, \alpha, \beta, \gamma\}, T = \{t, a, b\}$ has corresponding covering $\{a, b, \alpha, \beta, \gamma\}$;

15(a). $F(0011) = 1$, $F(0010) = 0$, and so on;

16(a). $x_{ac} = 4$, $x_{bc} = 4$, $x_{bd} = 2$, $x_{ce} = 8$, $x_{df} = 2$, rest $= 0$;

22(a). $x_{sa} = 3$, $x_{sb} = 2$, $x_{at} = 4$, $x_{bt} = 1$, $x_{ba} = 1$, rest $= 0$;

25. argue by induction on the length of C that if C is flow-augmenting, it contains a simple flow-augmenting chain.

Section 13.4. **1(a).** $x_{sa} = x_{at} = 2$, $x_{sb} = x_{bt} = 1$, $x_{ba} = 0$, or $x_{sa} = x_{at} = 1$, $x_{sb} = x_{bt} = 2$, $x_{ba} = 0$; **4(a).** worker 1 to job 2, 2 to 3, 3 to 1, and 4 to 4;

7(a). (13.11) and (13.12) give us $\sum_{i=1}^{n} \sum_{j=1}^{m} x_{ij} \leq \sum_{i=1}^{n} a_i < \sum_{j=1}^{m} b_j \leq \sum_{j=1}^{m} \sum_{i=1}^{n} x_{ij}$;

9(a). In (a), such a flow has $x_{sa} = 1$, $x_{at} = 2$, $x_{sb} = 2$, $x_{ba} = 1$, $x_{bt} = 1$ and a negative cost augmenting circuit is t, a, b, t.

Author Index

Subject Index

Algorithm (*cont.*)
nondeterministic, 70
for one-way streets, 452
for optimal assignment problem, 565–568
for perfect matching, 509–511
for planarity, 103, 448
polynomial, 23, 68
Prim's, 529–530, 532
quadratic, 68, 445
for shortest route, 533–536
for sorting, 136–138, 208, 248, 528
for strong connectedness, 91, 445, 448
for summing entries of an array, 196
for topological sorting, 489
for 2-coloring, 105–107
Algorithms, treatment of, xiv–xv
$\alpha(G)$, 322
Alternating chain, 515
M-alternating chain, 515
Alternating tree, 518
Amino acids, 13, 14, 15, 38–39, 437
Arc, 80
backward, 545
forward, 545
(i, j)-, 456
multiple, 82
saturated, 544
Archaeology, 125
Arc list, 142
Arithmetic, binary, 172, 376
Arithmetic used in computers, 376
Arranging numbers into increasing order, 247 (*see also* Sorting)
Art, 202, 397
Associativity, 288, 374
Asymptotically dominates, 68
Asymptotic analysis, 67–70
Augmenting chain (matching), 515–520
M-augmenting chain, 515–520
Augmenting chain (network flows), 545–549
capacity of, 546
defined, 545
flow augmenting, 545
labeling algorithm for, 550–553
M-augmenting chain, 548
Augmenting circuit, 570
Australian government, 59–60, 61–62, 191
Automorphism group, 293, 297
Automorphism of a graph, 293, 297
Average case complexity, 24
Average value, 321

B

Balanced incomplete block designs, 381–396
building new ones from existing ones, 392–393
conditions for existence, 385
definition, 382, 386
symmetric, 390–392
and Latin squares, 400–401
and projective planes, 400–401
Balls in cells (*see* Distinguishable balls and cells; Occupancy problems)
Bank courier, 23, 479
Banzhaf power index, 60, 188–192
defined, 189
Base:
in DNA chain, 13
extended, 469
interior extended, 469
in RNA chain, 49
BASIC, 20, 31–32
Benzene rings, 239–242
Berge Conjecture (strong), 104
Bernoulli, 9
Bernoulli trials, 185–186, 187, 188, 414
BIBD, 382
symmetric, 390
Binary arithmetic, 172, 376
Binary block code, 3, 411 (*see also* Code)
Binary code, 3, 13, 411 (*see also* Code)
group, 419
Binary digit, 3
Binary search algorithm, 140, 248 (*see also* Searching)
Binary search tree, 133, 134, 249, 343
Binary tree (*see* Tree, binary)
Binomial coefficient, 9
defined, 29
generalized, 172
Binomial expansion, 9, 56–58
Binomial theorem, 172–177
Bit, 3
Bit string, 3, 13, 15
and binary trees, 139
complement of, 287
defined, 3
of even parity, 58
Block, 356, 357, 383, 384
defined, 359, 382
message, 408, 409

Circuit, electric (*cont.*)
    design of, 17, 101, 131, 283, 301, 302
    integrated, 304, 310
    printed, 101
Circuit (in a graph or network), 90
    augmenting, 570
    hamiltonian, 478, 480–483, 486, 491
    length of, 90
Circuit matrix, 145
Classification schemes, 527
CL CHAIN (*x*, *x*), 459–460
Clique, 110, 320, 327
Clique number, 110, 320
Closure, Axiom of, 288, 374
Closure of a graph, 482,
$C(n; n_1, n_2, \ldots, n_k)$, 47
$C(n, r)$, 28
Coalition, 59
    losing, 59
    winning, 59.
        minimal, 64
Code:
    binary, 3, 13, 411
    binary block, 411
    binary group, 419
    binary *n*-, 411
    block, 3, 409
    equivalent, 423, 426
    error-correcting, 410–416, 427–439
        perfect, 423, 426
        t-error-correcting, 412
    error-detecting, 411, 426
    formal definition in coding theory, 409
    4*m*-Hadamard, 436
    (4*m* − 1)-Hadamard, 435
    genetic, 406, 470
    group, 419
    Hadamard, 427-428
    Hamming, 423, 425
    linear, 416–427
    matrix, 408–409
    Morse, 3–4, 15
    (*n*, *d*)-code, 413, 427, 433–437
    parity check, 411, 417, 418, 419–420, 421
    permutation, 409
    q-ary, 411, 415, 415–416, 426–427
    q-ary block *n*-code, 415
    Reed-Muller, 437
    repetition, 416–417, 417–418, 419, 420, 421,
        422
    rich, 427, 433–437

Code alphabet, 339, 407
    unambiguous, 339
Codeword, 179, 420, 421
    distance between, 411
    and duration of messages, 198
    legitimate, 197–198, 206–207, 220–221,
        274–275
    representing by a digit, 502–503, 507
    with specified distribution of letters and
        following, 470–472
Codewords, set of:
    formal definition in coding theory, 409
    rich, 427
Coding theory, 406–441
Coin tossing, 35, 185, 187, 278
Coleman power index, 60, 188–192
    defined, 189
Coloring of graphs, 97–111, 111–122
    improper, 260
    k-coloring, 97
    m-tuple coloring, 110
    proper, 260
    2-coloring, 104–107
Colorings of a collection of objects, 297
    distinct colorings, 297–304
    equivalent colorings, 299–302
    as graphs, 297, 303–304
    inventory of, 311–312
    weight of, 311
Colorings of a 2 × 2 array, 282, 286, 289,
    290–291, 292–293, 297, 299, 300, 306,
    311, 312, 313
Coloring trees, 284, 297, 298, 300, 307, 308
Column factor, 360
Combinations, 28–32
    algorithm for generating, 67
    counting, 165–172, 177, 263–264
Combinatorial optimization, 497–571, 526
Combinatorial proof, 29
Combinatorics:
    applications of, xiii
    basic tools of, xiv, 13–148
    and computing, xiii
Commodity, 563
    flow of, 541, 542
Communications (*see* Code; Duration of
    messages; Information transmission;
    Network, communication; Telecom-
    munications; Telephones)
Communications with computers, 406
Commutativity, 293, 374

D

Data structures, 79, 141–145
da Vinci, Leonardo, 202
D^c, 97
Deadheading time, 467
de Bruijn diagram, 473
de Bruijn sequence, 473–475
  (p, n) de Bruijn sequence, 473
Decisionmaking, 346–348, 486, 488
Decoding, 406, 407–410, 419
  efficient, 416
Defective products, 185
Deficiency, 523
Degree, 110 (*see also* Indegree; Outdegree)
  sum of, 124, 330
  total, 457
Deliveries, 465
  mail, 122, 465–466
della Francesca, Piero, 202
De Moivre, 149, 254
Density of edges (lines), 79, 90
Dependencies among the chapters and sections of the book, xvi
Depth-first search, 104, 107, 443–448, 452, 460, 489
  computational complexity of, 445
  directed, 445
Depth-first search spanning tree, 444, 447
Derangements, 203–206, 223–226, 229, 262–263, 271–272
  defined, 204
Descendant, 133
Design (*see* Balanced incomplete block designs; Experimental design)
DFSEARCH (v, u), 445
Die tossing, 34, 36–37
Difference equations, 210
Differential equations, 210, 224
Digraph:
  acyclic, 489
  asymmetric, 346
  complementary, 97
  complete symmetric, 89
  defined, 80
  equipathic, 141
  isomorphic, 81, 85
  labeled, 83–86
    number of, 158, 159
  program, 79, 88, 90

representation in the computer, 141–145
strongly connected, 90–91, 144, 445, 448, 560
transitive, 346, 487
unilaterally connected, 94, 145
unipathic, 96
weakly connected, 94, 145, 461
Dijkstra's algorithm, 533–536, 537, 538, 540, 541
Dirichlet, Peter, 319
Dirichlet drawer principle, 319
Discrete mathematics, xv
Dishwashing detergents, 384
Distance:
  between bit strings, 411 (*see* Hamming distance)
  in digraph or graph, 107, 455, 540
    average, 456
    maximum, 455
  in directed network, 533
  home to job, 501
  in network, 532
Distinguishable balls and cells, 40–47
  distinguishable balls and distinguishable cells, 41–42, 182–183, 258–259, 279
  distinguishable balls and indistinguishable cells, 43, 182–183
  indistinguishable balls and distinguishable cells, 42–43, 169, 174
  indistinguishable balls and indistinguishable cells, 43–44
Distributivity, 375
Divide-and-conquer algorithms, 245–250
Divine proportion, 202
D_n, 204
DNA (chain), 13, 14, 15, 16, 38–39, 437, 470
  [*see also* RNA (chain)]
Dodecahedron, following the edges of, 478–479
Drug testing experiments, 2–3, 9–10, 356, 360–361
Duality principle, 398
Dual statement, 398
Dual theorem, 398
Duration of messages, 198, 214, 216

E

Ecosystems, 79
Edge, 82
  multiple, 82

Edge chromatic number, 110
Edge-colorable, 110
Einstein, 45, 46, 397
Electoral votes, 63
Electrical distribution system, 132 (*see also* Network, electrical; Network, power line)
Electrical outlets, 321
Electronic banking, 406
Elevators, 44
Emissions testing, 255, 271, 278, 279 (*see also* Pollution)
Emperor Yu, 8
Encoding, 406, 407–410
    efficient, 416
Energy, 95–96 (*see also* Network, electrical; Network, pipeline; Network, power line; Pipeline, gas)
    demand for, 95–96
    fuel economy, 363–364
Enumeration, 4, 6, 8, 21
Enzyme, complete digest by, 51–54, 469–470, 477–478
    ambiguous, 53, 477
Equivalence class, 285–286
    of colorings (*see* Pattern, of colorings)
    containing an element, 285
    counting (*see* Polya theory)
Equivalence relation, 281–287
    defined, 281
    induced by a permutation group, 290–291
Erastothenes, Sieve of, 256
Erdös-Szekeres Theorem, 323, 325
Error-correcting codes, 410–416, 427–439
    perfect, 423, 426
    t-error-correcting codes, 412
Error correction, 406, 411, 412, 417, 419–423 (*see also* Error-correcting codes)
Error-detecting codes, 411, 416
Error detection, 406, 409, 411, 417, 422 (*see also* Error-detecting codes)
Errors in transmission, 407
    probability of, 411, 414, 415, 417–418, 422–423
Euclid, 9
Euclid's parallel postulate, 397, 403
Euler, L., 9, 149, 169, 268, 368, 369, 457, 458, 461
Eulerian chain, 442, 443, 457–478
    defined, 458
    eulerian closed chain, 458, 538

algorithm for finding, 458–460
Eulerian graph, 178
Eulerian path, 457–478
    defined, 458
    eulerian closed path, 461
Euler ϕ function, 268–269
Event, 34
Exam scheduling, 6, 98
Exercises, role of in book, xvi
Existence problem, the, 319–496, 1, 2, 5, 357, 363, 383, 397, 407, 497
    in graph theory, 442–496
Expectation (*see* Expected value)
Expected value, 186, 187
    existence of, 186
Experimental design, 2–3, 266, 267, 278, 356–405, 396 (*see also* Balanced incomplete block designs; Drug testing experiments; names of specific experimental designs)
Experimental unit, 356, 357, 384

F

Facilities design, 101, 490–491
Fano plane, 396, 397
Father of (in depth-first search), 443
Fault diagnosis, 79
Fault-tolerant computer systems, 79, 411
Fermat, 9, 33
Fermi-Dirac statistics, 45, 46
Fibonacci numbers, 200–203, 209, 213–214, 228
Fibonacci search, 202
Fibonacci sequence, 201
    growth rate of, 201
Field, 374
    finite, 373, 374–377
        existence of, 377
    Galois, 377
    isomorphic, 377
File search, 24, 133–135, 140, 248, 343–346
Fisher, R. A., 356
Fisher's inequality, 386, 387–388
Fleet of automobiles, 538–540
Floating point representation, 19
Flowchart, 79
Flows in networks, 125, 541–570
    algorithm for max flow, 549–550, 550–554
    conservation of flow, 541
    and covering, 555

number of, 155, 159
mixed, 454
optimization problems for, 526–571
planar, 101, 448, 490–491
regular, 485, 503
relationship, 490
representation in the computer, 141–145, 443
and scheduling, 5–8, 97–98, 113–114
similarity, 80
tour, 99, 103–104, 108
2-colorable, 104–107
water-light-gas, 101
weakly α-perfect, 111
weakly γ-perfect, 110
Graph plotting, computerized, 467, 499, 537
Graph Ramsey numbers, 333
Gravitation, theory of, 397
Group, 289, 375, 419
automorphism, 293
group code, 419
permutation, 287–293
symmetric, 289
Guessing abilities (see Psychic experiments)

H

Hadamard code, 427–428
4m-Hadamard code, 436
(4m-1)-Hadamard code, 435
Hadamard design, 391–392, 427
construction of, 428–433
of dimension m, 391, 427
Hadamard matrix, 428–433, 434
normalized, 429
order of, 428
Hamilton, Sir William Rowan, 9, 76, 478
Hamiltonian chain, 442, 478–494
Hamiltonian circuit, 478, 480–483, 486, 491
Hamiltonian cycle, 478, 483
Hamiltonian path, 478–494
Hamming bound, 413–414, 427
Hamming code, 423, 425
Hamming distance, 411, 412
for linear codes, 419
for q-ary codes, 415
Hamming weight, 419
Hatcheck Problem, 203–204, 205, 271–272, 278
Heap sort, 136

Height of a rooted tree, 133, 134, 135, 136
Hexagons, configuration of, 239–242
Highway construction, 126
History of combinatorics, 8–9
Homeomorphic graphs, 102
Homogeneity of bimetallic objects, 127
Homogeneous polynomial, 177
Hospital deliveries, 44
Hospital internships, 504
How to use the book, xiv–xvi
Hungarian algorithm, 567, 568, 570

I

IBM 650, 411, 417
Identity, 288, 375
additive, 375
multiplicative, 375
(i, j)-arc, 456
(i, j)-vertex, 456
Immunity to diseases, 266, 278
I_n, 114
Incidence matrix:
for block designs, 386, 387, 427
of the complementary design, 394
of a Hadamard design, 428
for digraphs, 144
for graphs, 142
point-set, 142
Inclusion and exclusion, principle of, 252–280
stated, 254
Indegree, 456, 457, 461
Independence number (vertex), 110, 322
Independent set of 0's in a matrix, 513, 566
Independent set of vertices, 110, 322, 327, 339–341, 514
maximum, 322, 514, 515
Index registers, 98, 109
Indistinguishable men and indistinguishable women, 167
Induction, mathematical, xiv
Information exchange, 343
Information retrieval, 80, 93, 343–346
Information transmission, 198, 406–407 (see also Code)
errors in, 407, 411, 414, 415, 417–418, 422–423
Initial condition, 195, 210
Inner product, 387, 428
Inner vertex, 518

Network (*cont.*)
  communication, 77–78, 88, 90, 95, 96, 125, 527, 540
  computer, 125
  CPM, 536
  defined, 526
  directed, 526
  distribution, 527
  electrical, 9, 17, 76, 78, 125, 126, 131, 132, 142, 467, 527
  flows in (*see* Flows in networks)
  optimization problems for, 526–571
  packet-switched, 341–343, 350
  PERT, 536
  physical, 78
  pipeline, 4–5, 15, 78, 125, 126, 530, 556–559
  power line, 527
  reliability of, 126, 540, 559
  switching, 142, 324, 341–343, 350, 522–523, 523–524
  telephone, 78, 125, 126, 324, 333, 341–343, 350, 522–523, 523–524, 527
  transportation, 77, 88, 90, 125, 449, 527, 532–541, 560, 562, 563, 564, 565, 570
  vulnerability of, 456
Niche overlap graph (*see* Competition graph)
Normal product, 339
NP, 70
NP-complete problem, 23, 70
NP-hard problem, 23, 70
n-set, 20
Number theory, 44, 169, 268–269
Numerical analysis, 202

O

Occupancy problems, 40–47, 169, 174, 182–183, 258–259, 275, 279 (*see also* distinguishable balls and cells)
  with a specified distribution, 47–49
Octapeptide, 71
$O(g)$, 68
Oil discovery, 208, 216, 228
$\omega(G)$, 320
$O(n)$, 68
$O(n^2)$, 68
One-person, one-vote, 190
One-way street assignment, 448–457
  efficient, 453–454, 455–456
  inefficient, 454

One-way street problem, 448–457
  with some streets two-way, 454–455
Open neighborhood, 502
Operation, 374
  binary, 374
Operations research, xv, 24, 102, 125, 490
    (*see also* Network and Scheduling)
Optimal assignment problem, 499–500, 513, 564–568
Optimization, combinatorial, 497–571
Optimization problem, the, 1, 4, 6, 327, 407, 497, 497–571
  for graphs and networks, 526–571
Orbit, 291
Organic molecules, 284, 297, 298, 308, 311, 314 (*see also* Chemistry, organic)
Orientation, 88, 449
  strongly connected, 442, 449–454, 455–456
  unilaterally connected, 454
  weakly connected, 454
Orthogonal Latin squares, 362–371
  complete family of, 367, 368, 400–401
    construction of, 377–380
    and projective planes, 400–401, 403–404
    and symmetric balanced incomplete block designs, 400–401
  family of, 363, 366, 368, 370
    and q-ary codes, 416
  pair of, 363, 369
Outdegree, 457, 461
Outer vertex, 518

P

P, 70
Pair, comparison experiment, 79, 486, 488
Pappas of Alexandria, 397
Parity check code, 411, 417, 418, 419–420, 421
Parity check digit, 417, 418
Parity check equation, 421, 422
Parity check matrix, 419, 422
Partial fractions, method of, 221, 223, 227
Partial order, 346–348
  dimension of, 346–348
  strict, 346
Partition of integers, 42, 43, 44, 169, 171–172, 270

"Postman" problem (*see* "Chinese postman" problem)

Power in simple games, 58–65, 188–192 [*see also* names of individual power indices (Banzhaf, Coleman, Shapley-Shubik)]

Power series, 150–152
  formal, 150, 178

Preference, 79, 346, 347, 486, 488
  among cities, 493
  among geographical areas, 488
  for vanilla puddings, 352

Prerequisites for the book, xiv

Presidential primary, 171

Prime number, 256, 368

Prime power decomposition, 368

Prim's algorithm, 529–530, 532

Printers and minicomputers, 322

$P^R(m, r)$, 38

$\Pr(n; n_1, n_2, \ldots, n_k)$, 47

Probability, 9, 32, 33–38
  defined, 33, 34
  equally likely outcomes, 33, 34
  that exactly m events occur, 275
  generating function for, 185–188
  that none of a collection of events occurs, 258

Product, 288

Product rule, 15

Program blocks, 78

Program digraph, 79, 88, 90

Project Evaluation and Review Technique (PERT), 536

Projection, 399

Projective plane, 396, 397
  finite, 396–404
    nondegenerate, 397
  and Latin squares, 400–401, 403–404
  order of, 400
  and symmetric balanced incomplete block designs, 400–401

Properties, 253
  objects with an even (odd) number, 275
  objects with exactly m, 270–280
  objects with at least m, 279
  objects with none (*see* Inclusion and exclusion, principle of)

Prosthodontics, 360

Proteins, 13, 437 (*see also* Amino acids)

Psychic experiments, 204, 272–273, 279

Psychology, 79, 272–273 (*see also* Preference)

Q

q-ary code, 411, 415, 415–416, 426–427
  and Latin squares, 416

Quadratic algorithm, 68, 445

Queue, 106, 448

Quik sort, 136, 140–141, 245, 249

$[q; v_1, v_2, \ldots, v_n]$, 59

R

Rabbi Ibn Ezra, 9

Rabbit breeding, 200–201

Radiation, 40, 46, 50, 170, 275

Radio frequency assignment, 99, 109, 110

Radius of convergence, 152

RAF planes, pilots for, 499, 509

Ramsey numbers, 327–331
  bounds on, 328, 329–331
  generalized, 332, 333
    existence of, 336
  graph, 333
  known, 328

Ramsey property:
  graph, 333
  $(p_1, p_2, \ldots, p_t; r)$, 332
  (p, q), 326
  $(p, q; r)$, 332

Ramsey's Theorem, 326
  generalizations, 331–333
  proof of, 330

Ramsey theory, 325–353
  applications of, 336–353

Random coil, 236

Randomizing, 357, 358, 359

r-combination, 28

Reachability matrix:
  of digraph, 144
  of graph, 145

Reachable, 89

Reaching, 88–89

Real estate transactions, 499

Recurrence relation, 33, 194–251
  basic solution of, 215
  characteristic equation of, 210
  defined, 195
  general solution of, 212
  homogeneous, 210
  initial condition for, 195

involving convolutions, 230–245

involving more than one sequence, 206–207, 226–227

linear, 210

linear homogeneous with constant coefficients, 210–218, 222

relation between solution methods, 230

solution by:

    generating functions, 218–230

    iteration, 195

    method of characteristic roots, 210–218

solution of, 195

systems of, 206–207, 226–227

weighted sum of solutions to, 211

Recursive programming, 445

Reduced matrix, 565

Reduction theorems, 114–117, 121, 162–163, 221–222, 245 (*see also* Divide-and-conquer algorithms)

Reed-Muller codes, 437

Reflexivity, 281

Regions in the plane, 198–200

Regular (of degree *k*), 485, 503

Relation:

    binary, 281

    equivalence, 281–287

Relatively prime integers, 268

Reliability of systems (or networks), 35–36, 92, 126, 540, 559

Rencontres, 204

    le problème des, 204, 278

Repetition code, 416–417, 417–418, 419, 420, 421, 422

Retina (*see* Vision)

$R(G_1, G_2, \ldots, G_t)$, 333

RNA (chain), 49, 51–54, 181–182, 273–274, 469–470, 470, 477–478 [*see also* DNA (chain)]

    ambiguous, 53, 477

    primary structure in, 236

    as random coil, 236

    secondary structure in, 236–239

Road system, 92, 126, 532 (*see also* Network, transportation; One-way street problem; Shortest route problem)

Robbins' Theorem, 448–451, 452

Robot, 23, 479

Rook, 156

    assignment of nontaking, 265

    taking, 156

Rook polynomial, 156–157, 159, 162–163, 209, 221–222, 245, 265–266, 497, 505

Room assignments, 47–48, 501, 511

Root, 133

Rotating drum problem, 474–475

Routing problems, 99 (*see also* Shortest route)

Row factor, 360

r-permutation, 26–27

$R(p, q)$, 327

$R(p, q; r)$, 332

S

Sample space, 34

Sampling:

    with replacement, 38–40

    without replacement, 165–169, 173

    with and without replacement, 39

Sampling survey, 167–168, 170

Sanitation, 99, 103–104, 108, 467–469

Saturated arc, 544

Saturated hydrocarbons, 128–129, 230

Saturated vertex, 498

    M-saturated, 498

Saturated vertex set:

    X-saturated matching, 501, 504

Scanned vertex, 551

Scheduling, 536–537

    computer system, 23–24, 479

    exams, 5–8, 97–98, 113–114

    industrial processes, 479–480, 536–537

    meetings, 5–8, 97–98, 113–114

    in operations research, 24, 102, 490

    periodic, 8

    production, 479–480, 538

    workforce, 8

Score, 493

Score sequence, 487, 493, 494

SDR (*see* System of distinct representatives)

Searching, 24, 133–135, 140, 202, 248, 343–346

Search strategy, 343

Secretarial help, 325

Security Council (*see* U.N. Security Council)

Seriation, 125

Sex distribution, 44

Sex of children, 34, 35, 279

Shannon capacity (*see* Capacity, of channel)

Shapley-Shubik power index, 60–62, 188–192